# Heat Transfer Analysis: Recent Challenges and Applications—2nd Edition

# Heat Transfer Analysis: Recent Challenges and Applications—2nd Edition

Guest Editors

**Andrzej Frąckowiak**
**Bartosz Ciupek**
**Łukasz Brodzik**
**Wojciech Prokopowicz**

MDPI

Basel • Beijing • Wuhan • Barcelona • Belgrade • Novi Sad • Cluj • Manchester

*Guest Editors*
Andrzej Frąckowiak
Institute of Thermal Engineering
Poznan University of Technology
Poznan
Poland

Bartosz Ciupek
Department of Fuels and Renewable Energy
Poznan University of Technology
Poznan
Poland

Łukasz Brodzik
Institute of Thermal Engineering
Poznan University of Technology
Poznan
Poland

Wojciech Prokopowicz
Ministry of National Defense
Warsaw
Poland

*Editorial Office*
MDPI AG
Grosspeteranlage 5
4052 Basel, Switzerland

This is a reprint of the Special Issue, published open access by the journal *Energies* (ISSN 1996-1073), freely accessible at: https://www.mdpi.com/journal/energies/special_issues/434940DQJ8.

For citation purposes, cite each article independently as indicated on the article page online and as indicated below:

Lastname, A.A.; Lastname, B.B. Article Title. *Journal Name* **Year**, *Volume Number*, Page Range.

**ISBN 978-3-7258-8403-2 (Hbk)**
**ISBN 978-3-7258-8404-9 (PDF)**
**https://doi.org/10.3390/books978-3-7258-8404-9**

# Contents

# About the Editors

**Andrzej Frąckowiak**

Andrzej Frąckowiak is employed as a full professor at the Institute of Thermal Energy, Poznań University of Technology. His research focuses on heat transfer and computational fluid mechanics, with particular attention paid to the modeling of heat transfer processes through nanotube-coated surfaces and the inverse cooling of gas turbine blades using porous materials. He has conducted his research in collaboration with, and as part of research fellowships at, the University of Stuttgart, the Institute for Thermodynamics of Air and Space, the Technical University of Munich, Forschungseinheit M6, and the Free University of Berlin, Institute of Experimental Physics.

**Bartosz Ciupek**

Bartosz Ciupek is employed as an assistant professor at the Institute of Thermal Energy, Poznań University of Technology. His research focuses on the design and operation of heating systems and equipment, particularly boilers. He is currently conducting research on improving air quality and reducing pollutant emissions from solid fuel boilers through chemical modification of the combustion process, the use of exhaust gas cleaning systems, and alternative environmentally friendly solid fuels. He also works on improving the thermal and environmental performance of aircraft engines. He conducts his research in collaboration with, and through research fellowships with, the Cracow University of Technology, the Military University of Technology in Warsaw, the Kielce University of Technology, and the University of Kalisz. He holds a scholarship from the City of Poznań and has authored 18 patents.

**Łukasz Brodzik**

Łukasz Brodzik is employed as an assistant professor at the Institute of Thermal Energy, Poznań University of Technology. His research activities are focused on two fields, the first of which is space technology. In this field, he has experience solving problems with heating orbiter heat shields. His research focuses on heat transfer via thermal radiation for porous coatings in high-temperature environments, as well as their impact on structures in the event of damage. The second field is aviation. In this case, the analyses conducted concern two aspects. The first is research on miniature turbojet engines for co-combustion of aviation kerosene and hydrogen. The analyses focus on the impact of hydrogen on engine operation, primarily in terms of temperature changes in individual sections. The second aspect of our aviation sector activities is aircraft performance analysis, which involves calculating flight parameters, primarily for smaller aircraft.

**Wojciech Prokopowicz**

Wojciech Prokopowicz graduated from the Faculty of Electronics at the Military University of Technology in Warsaw, earning a Master of Science. He obtained his Ph.D. in technical sciences in 2016 at the Faculty of Machines and Transport at Poznań University of Technology. He specializes in non-destructive testing of composite structures and vibroacoustics of machines and devices. Currently, his research focuses on the use of space segment in military operations and the alternative applications of unmanned aerial vehicles. His work also covers energy consumption prediction and decision-making support using artificial intelligence algorithms. He has authored numerous publications on these topics. He participated in the introduction of the F-16 aircraft into the Polish Air Force. He graduated from the Air Force Safety University in the United States and served as Flight

Safety Inspector at the 33rd Air Transport Base in Powidz. As an Assistant Professor at the Faculty of Military Studies in the Department of Air Force and Air Defense at the War Studies University in Warsaw, he lectured on the use of space assets in Air Force operations. He held the position of the Head of the Department in the Aviation Technology Directorate of the Inspectorate for Armed Forces Support of the Polish Armed Forces. Currently, he serves as a Chief Specialist in the Team of the Plenipotentiary of the Minister of National Defense for Aviation at the Ministry of National Defense, where he deals with issues related to the support for aircraft operations.

# Preface

This Reprint brings together selected contributions addressing contemporary issues in applied heat transfer. Its scope covers theoretical investigations, modeling, and optimization of heat transfer processes, as well as applied solutions implemented in energy devices and installations. The main purpose is to provide a comprehensive overview of innovative approaches that not only improve thermal efficiency but also support environmental protection by reducing the consumption of conventional energy sources.

The motivation for preparing this Reprint arises from the dynamic development of modern energy technologies and the growing need for sustainable practices. Heat transfer remains a central area of research with direct impact on efficiency, performance, and ecological responsibility in power engineering and related fields.

This Reprint is intended for researchers, engineers, and professionals who deal with the challenges of designing, optimizing, and implementing energy systems. It may also be of interest to graduate students and decision makers seeking insight into current research trends and technological advances in heat transfer applications.

**Andrzej Frąckowiak, Bartosz Ciupek, Łukasz Brodzik, and Wojciech Prokopowicz**
*Guest Editors*

*energies*

MDPI

*Article*

# Correlation-Based Single-Phase Heat Transfer Assessment of Binary HFE/Ethyl Acetate Mixtures in Minichannels

**Artur Piasecki [1] and Magdalena Piasecka [2,*]**

[1] Faculty of Environmental Engineering, Geomatics and Renewable Energy, Kielce University of Technology, 25-314 Kielce, Poland; apiasecki@tu.kielce.pl
[2] Faculty of Mechatronics and Mechanical Engineering, Kielce University of Technology, 25-314 Kielce, Poland
* Correspondence: tmpmj@tu.kielce.pl; Tel.: +48-41-34-24-320

**Abstract**

This work presents a correlation-based framework for comparative assessment of single-phase forced convection of binary hydrofluoroether/ethyl acetate (HFE/EA) mixtures in rectangular minichannels. Density, kinematic viscosity, and thermal conductivity were measured at 293.1, 313.1, and 328.1 K for selected compositions of HFE-7100/EA, HFE-7300/EA, and HFE-73DE/EA. Because several DSC-derived mixture-specific heat values were not sufficiently reliable for direct use, the mixture-specific heat capacity was estimated from literature-supported pure-component values using an ideal-mixture, mass-fraction-weighted approximation and used only for evaluation of the Prandtl number. Heat transfer was then assessed using the Sieder–Tate correlation for laminar thermally developing flow in two representative minichannel geometries. The highest predicted values for the HFE-7100/EA and HFE-7300/EA families were obtained for the most EA-rich retained compositions, whereas in the retained HFE-73DE/EA subset, the 50/50 mixture performed best because of its higher thermal conductivity. Validation against an experimental dataset for pure HFE-7100 in the short module showed systematic overprediction, with a mean relative difference of −13.44% and a MAPE of 15.6%. The calculated values should, therefore, be used for relative comparison rather than treated as unbiased absolute predictions.

**Keywords:** binary mixtures; thermophysical properties; dielectric fluids; minichannel; heat transfer coefficient; Nusselt number; single-phase forced convection

## 1. Introduction

Heat transfer devices are used in a wide range of technologies, including refrigeration, HVAC systems, thermal management of high-power electronics, cryogenic equipment, photovoltaic systems, and industrial processes. Their operating efficiency is largely governed by the properties of the selected coolant or process fluid. In practice, the fluid must be compatible with the required temperature and pressure levels, expected flow conditions, construction materials, and safety requirements. Fluorinated liquids, including hydrofluoroethers (HFEs), perfluoropolyethers (PFPEs), and Fluorinert-type fluids (FCs), are often considered for such applications because they combine dielectric behaviour with good chemical inertness and thermal resistance. Their normal boiling points cover a broad temperature interval, approximately from 300 K to more than 500 K, which makes them suitable for the thermal control of electronic and optoelectronic devices. A drawback of these fluids is their relatively poor heat conduction and, in many cases, high price. For this reason, increasing attention is being paid to liquid blends whose properties can be adjusted

 https://doi.org/10.3390/en19102291

for a given thermal process. In particular, binary systems composed of fluorinated liquids and organic solvents, or two fluorinated components, offer a practical route to modify boiling point, density, viscosity, and specific heat while preserving acceptable material compatibility and ease of circulation in typical pumping systems.

The work presented in [1] proposes the use of differential scanning calorimetry (DSC) for acquiring atmospheric-pressure boiling-point and vapor–liquid equilibrium data for binary liquid mixtures. The approach is especially relevant to mixtures whose components exhibit large differences in volatility, since such systems are susceptible to pre-evaporation effects during measurement. These errors were limited by optimising the sample size and heating programme. The methodology was first assessed for several model mixtures and was then used to investigate the previously undocumented carvacrol–DMSO binary system. The experimental boiling line was correlated with the NRTL thermodynamic model, which confirmed significant non-ideal behaviour and suggested the presence of an azeotropic point.

Recent studies show a clear shift towards computational and machine-learning tools for predicting properties of HFE-based and related fluid mixtures. For example, Aminian et al. [2] combined ab initio calculations with the Peng–Robinson equation of state to evaluate ideal-gas heat capacities and critical properties of HFE-7000–7500 compounds. In another contribution, Urata et al. [3] applied artificial neural networks to estimate vapour–liquid equilibrium behaviour in HFE-containing mixtures. Similar data-oriented strategies have been extended to other refrigerant and hydrocarbon systems. Wu et al. [4] proposed a machine-learning-assisted group-contribution approach for predicting critical temperatures of binary refrigerant mixtures from molecular information, while Soria-Lopez et al. [5] reported improved viscosity predictions for binary alkane mixtures compared with traditional mixing rules, owing to the ability of ML models to capture nonlinear dependence on composition.

Reliable high-pressure property data are available for several HFE-containing binary systems. For the HFE-7200 + 2-propanol mixture, Muñoz-Rujas et al. [6] measured thermophysical properties between 293.15 and 393.15 K and at pressures from 0.1 to 140 MPa. The authors represented the density data using a Tait-type equation, calculated related derivative properties, and supplemented the analysis with speed-of-sound results at 0.1 MPa. In another study [7], analogous density and sound-speed measurements were reported for HFE-7500 + diisopropyl ether mixtures up to 100 MPa. Thermodynamic properties of hydrofluoroether systems with selected organic solvents were further investigated by Ogawa et al. [8], who observed positive excess molar volumes and excess enthalpies dependent on composition.

From an engineering viewpoint, zeotropic mixtures offer promising opportunities for improving heat transfer performance. In pulsating heat pipes, Xu et al. [9] observed that partially immiscible zeotropic fluids can form dispersed emulsion structures, leading to enhanced heat transfer. For condensation processes, Eissa et al. [10] formulated a generalised non-equilibrium heat transfer model for binary zeotropic mixtures. Their approach includes vapour-side mass diffusion and interfacial temperature effects, and its validation against 871 experimental data points showed strong predictive capability, with 92% of the results lying within a $\pm 30\%$ deviation band. This makes the model relevant for compact heat-exchanger analysis and design.

Tagliaferri et al. [11] showed that, for CFD of bubbling fluidized bidisperse mixtures, a first-order upwind discretization can introduce strong numerical diffusion that suppresses bubble holdup and mixing, whereas a second-order upwind scheme yields more realistic bubble dynamics and correctly predicts the transition to a fully mixed state; the restitution coefficient has a minor influence except near the quasi-ideal value $e \approx 0.99$, and explicit

https://doi.org/10.3390/en19102291

vs implicit time integration mainly affects the transient duration rather than the pseudo-steady behaviour.

Additional studies have addressed related features of binary liquid systems, such as intermolecular effects, equilibrium characteristics, and behaviour at phase boundaries. Verma et al. [12] used density, ultrasonic velocity, and refractive-index measurements to show pronounced donor–acceptor interactions in ether/hydrocarbon mixtures. Tsai et al. [13], in turn, evaluated the ability of molecular dynamics simulations to predict transport properties in alcohol/water systems and indicated which force-field combinations most accurately describe diffusion coefficients and density variations.

In previous work [14], the authors combined thermophysical property measurements with validated conjugate CFD to assess HFE-73DE/ethyl acetate mixtures for laminar minichannel cooling and to identify compositions offering a heat transfer/pressure-drop compromise. In that study, the thermophysical properties of the working fluids were determined experimentally at the Faculty of Civil Engineering, Mechanics and Petrochemistry, Warsaw University of Technology in Płock, Poland. Liquid density was measured with a Mettler Toledo Densito densimeter in accordance with ASTM D1250 [15], while kinematic viscosity over the investigated temperature range was determined using a Ubbelohde capillary viscometer thermostated in water baths according to EN ISO 3104 [16]. The initial boiling point was determined by atmospheric distillation following ASTM D86 [17]. Thermal measurements were performed by differential scanning calorimetry (DSC) using a Netzsch Maia 200 F3 instrument, with an empty crucible as the reference and liquid sample masses of approximately 60–70 mg [18,19].

These measurements provided the experimental property basis for the present comparative assessment. The mixtures were prepared in the laboratory from commercially available components rather than purchased as ready-made formulations. Because this study aims at a correlation-based comparison under identical hydraulic assumptions, directly measured temperature-dependent properties were treated as the primary input data wherever their consistency was confirmed. This is particularly important for dielectric multicomponent systems, for which simple mixing rules may be insufficient to represent the actual behaviour over the full temperature and composition range. The present work uses that earlier experimental property base as an input layer for a simplified correlation-based comparison, rather than repeating the full conjugate CFD framework reported previously [14]. However, because several DSC-derived mixture-specific heat values were found to be insufficiently reliable for direct use in the final calculations, the specific heat capacity of the mixtures was estimated from pure-component data using an ideal-mixture, mass-fraction-weighted mixing rule.

For each mixture series (HFE-7100/EA, HFE-7300/EA, and HFE-73DE/EA), mixtures were considered at selected mass fractions (first component/second component) and characterised at 293.1, 313.1, and 328.1 K under near-atmospheric pressure. The dataset used in the final comparative analysis includes experimentally determined density, kinematic viscosity, and thermal conductivity, while specific heat was estimated from pure-component mass fractions for calculation of the Prandtl number. This article presents a transparent, correlation-based workflow for converting the measured thermophysical properties into comparable single-phase forced-convection heat transfer characteristics for rectangular minichannels. The novelty lies in combining a dedicated property dataset for HFE/EA mixtures with a consistent minichannel assessment framework applied to two representative module geometries and in anchoring the assessment with an independent validation dataset for pure HFE-7100 obtained in the short multi-minichannel module using IR thermography and an energy-balance data reduction.

https://doi.org/10.3390/en19102291

## 2. Characteristics of Pure Liquids and Selected Binary Mixtures

The present study considers three binary mixture series based on hydrofluoroethers (HFEs) and an organic ester: 3M™ Novec™ HFE-7100/Ethyl acetate EA, 3M™ Novec™ HFE-7300/Ethyl acetate EA, and 3M™ Novec™ HFE-73DE/Ethyl acetate EA. These liquids were selected because they are electrically insulating, chemically stable, and relevant to thermal-management and compact heat-exchanger applications, while the ester co-solvent enables tunability of density, viscosity, and boiling behaviour.

Table 1 provides selected reference properties of the base fluids (reported at 298 K unless stated otherwise). For HFE-7100 [20], HFE-7300 [21], and HFE-73DE [22], the listed densities, viscosities, and boiling temperatures illustrate the range of base-fluid properties relevant to minichannel applications. Ethyl acetate [23] is included as the organic co-solvent used to modify the density, viscosity, and boiling behaviour of the analysed HFE/EA mixtures.

**Table 1.** Selected reference properties of the base fluids [20–23].

| Physical Properties | HFE-7100 | HFE-7300 | HFE-73DE | EA |
|---|---|---|---|---|
| Density, $kg/m^3$ | 1520 | 1660 | 1280 | 902 |
| Viscosity, $mm^2/s$ | 0.401 | 0.71 | 0.403 | 0.463 |
| Boiling temperature, K | 334.1 | 371.1 | 321.1 | 350.2 |

The HFE-7100/EA, HFE-7300/EA, and HFE-73DE/EA systems were considered at mass fractions of 10/90, 25/75, 50/50, and 75/25 and at three temperature levels. The 10/90 composition was available only for HFE-7100/EA and HFE-73DE/EA. The 75/25 HFE-73DE/EA case was excluded from the final dataset, and all remaining compositions were used as input to the correlation-based calculations reported in the following tables.

## 3. Experimental Rig and Minichannel Modules

### 3.1. General Information

The present study comprises (i) a correlation-based assessment of selected binary mixtures using experimentally measured mixture properties, performed for two minichannel modules—a long reference module (Setup #1) and a short multi-minichannel module (Setup #2), and (ii) an experimental validation dataset acquired in the short module (Setup #2) using a pure reference fluid (HFE-7100). For the flow of either mixtures or pure liquids in the single minichannel (Setup #1) or in the central minichannel of the parallel-channel group (Setup #2), laminar single-phase forced convection was assumed.

Setup #1, incorporating the long minichannel, had been investigated in the authors' previous experiments using liquid crystal thermography to obtain the outer-wall temperature distribution of the heated foil. The corresponding data were not considered in the present study [24].

The configuration with a short minichannel module (Setup #2) was used in recent experimental studies by the authors, in which infrared thermography was employed to measure the temperature of the outer surface of the heated foil [25]. These IR-based data were selected for the present analysis because they provide a more complete and reliable wall-temperature field. In the liquid-crystal thermography measurements, portions of the foil surface could not be evaluated when the local temperature fell outside the active temperature range of the liquid-crystal coating; therefore, this technique was not adopted for the validation dataset used here.

Uncertainty estimates for the principal measured variables were reported in [24,25]. Since the heat transfer coefficient is calculated from temperature-dependent quantities, its uncertainty is largely governed by the precision of the temperature measurements.

https://doi.org/10.3390/en19102291

It should also be noted that the mixtures addressed in this article were not tested experimentally in the heat transfer rig. Therefore, the heat transfer coefficients and Nusselt numbers reported for the mixtures are calculated estimates obtained from the measured thermophysical properties and a selected literature correlation, applied to the geometries and operating assumptions defined in this work.

*3.2. Experimental Rig with a Minichannel Module*

The validation experiments were performed in single-phase forced convection. The working fluid was circulated through the test section at controlled inlet temperature and mass flow rate, while the heat input was supplied electrically at the heated wall of the minichannel module (Setup #2). Inlet and outlet temperatures and pressures were monitored continuously, while infrared (IR) thermography was used to capture the temperature distribution over the external side of the heater foil, yielding spatially resolved thermal data. The validation dataset for the reference fluid (HFE-7100) was collected using Setup #2.

A schematic of the experimental loop and the interchangeable test sections is shown in Figure 1.

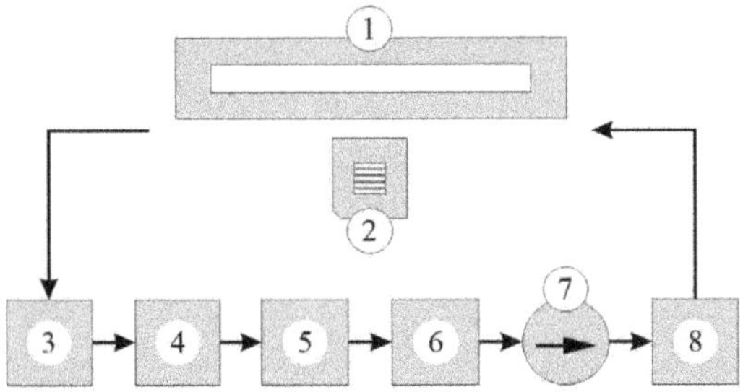


**Figure 1.** Schematic diagram of the experimental facility: 1—Setup #1 (long module with the minichannel); 2—Setup #2 (short module with the 5 minichannels); #3—condenser; #4—pressure-control unit; #5—deaerator; #6—filter; #7—pump; #8—mass flow meter.

In the experimental loop, the working fluid was pumped to a pressure-control vessel, which also acted as an expansion chamber. The vessel was divided into two zones: one filled with liquid and the other with compressed air, allowing the required system pressure to be maintained. After leaving this unit, the fluid flowed through a preheater, where it was brought to the prescribed inlet temperature, usually slightly below the saturation point, before entering the test section. Further details of the experimental facility are given in [24,25]. All tests were performed under steady-state conditions for several values of electrical power delivered to the heater, which produced different imposed heat-flux levels.

*3.3. Reference Minichannel Modules*

Both modules comprise minichannels with a depth of 1 mm. In each minichannel, the heated wall was made of Haynes-230 alloy foil, approximately 0.1 mm in thickness, the opposite wall was a glass window, and the side walls were made of PTFE. This work focuses on the short module with five parallel minichannels (Setup #2), which is schematically shown in Figure 2.

The principal geometrical dimensions of both modules are summarised in Table 2. Setup #1 employs a single minichannel with a length of 0.36 m, whereas Setup #2 uses five parallel minichannels, each 0.043 m long; in both modules, the minichannel depth is 1 mm.

https://doi.org/10.3390/en19102291

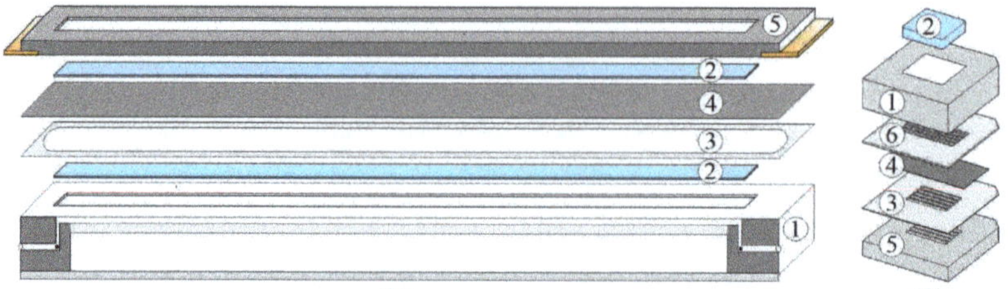


**Figure 2.** Schematic diagrams of (**a**) the long module with a single minichannel (Setup #1), (**b**) the short module with five parallel minichannels (Setup #2), 1—body, 2—glass plate, 3—PTFE minichannel insert, 4—heated foil, 5—cover plate, and 6—PTFE electrically insulating insert.

**Table 2.** Geometrical characteristics of the test modules.

| | **Setup #1** | **Setup #2** |
|---|---|---|
| Number of minichannels (-) | 1 | 5 |
| Minichannel length L (m) | 0.36 | 0.043 |
| Minichannel width w (m) | 0.004 | 0.006 |
| Minichannel depth s (m) | 0.001 | 0.001 |

## 4. Calculations and Correlation-Based Assessment Method

To compare selected mixtures of HFE-7100, HFE-7300, HFE-73DE, and ethyl acetate (EA) under identical hydraulic conditions and a fixed operating point, a correlation-based assessment approach was adopted for single-phase forced convection in a reference rectangular minichannel. The assessment geometry and operating point were fixed as follows: channel width $w = 0.004$ m, channel depth $s = 0.001$ m, length $L = 0.36$ m, hydraulic diameter $D_h = 1.6$ mm, and mass flux $G = 200$ kg/(m$^2$·s).

All thermophysical property measurements for the mixtures were carried out at the Faculty of Civil Engineering, Mechanics and Petrochemistry, Warsaw University of Technology in Płock (Poland). The dataset was collected at three temperature levels (293.1, 313.1, and 328.1 K). In the final comparative calculations, experimentally determined density, kinematic viscosity, and thermal conductivity were used directly. However, several DSC-derived mixture-specific heat values were found to be insufficiently reliable for direct use in the correlation-based calculations, mainly due to volatility-related measurement artefacts and the lack of a repeat verification campaign at the revision stage. Therefore, in the present manuscript, specific heat was not taken directly from the disputed mixture DSC results. Instead, $c_p$ values for the mixtures were estimated from pure-component mass fractions and used only for evaluation of the Prandtl number within the comparative assessment framework. Details of the original property measurement procedures are given in [14].

For each mixture and temperature, the dynamic viscosity was computed from density and kinematic viscosity. The Reynolds and Prandtl numbers were then calculated according to the following relations:

$$Re = \frac{G \cdot D_h}{\mu} \tag{1}$$

$$Pr = \frac{\mu c_p}{k} \tag{2}$$

where $D_h$ is the hydraulic diameter of the minichannel, $G$ is mass flux, $\mu$ is the dynamic viscosity, $c_p$ is the specific heat, and $k$ is the thermal conductivity of the fluid.

https://doi.org/10.3390/en19102291

The average Nusselt number $Nu_{corr}$ was estimated using the classical Sieder–Tate-type correlation [26] for laminar internal flow with thermal development effects:

$$Nu_{corr} = 1.86\left(Re \cdot Pr \cdot \frac{D_h}{L}\right)^{\frac{1}{3}} \left(\frac{\mu}{\mu_w}\right)^{0.14} \tag{3}$$

where $L$ is the heated length of the minichannel.

The viscosity ratio is the Sieder–Tate correction accounting for property variations between the bulk fluid and the wall: $\mu$ is the dynamic viscosity evaluated at the bulk (mean) fluid temperature, while $\mu_w$ is the dynamic viscosity evaluated at the wall temperature. In the present calculations, this viscosity-ratio correction was not applied, and $(\mu/\mu_w)^{0.14}$ was set to unity. This choice keeps the workflow uniform across all mixtures and temperature levels and avoids introducing an additional wall-property model not supported by direct mixture data. Its effect is modest unless the wall-to-bulk viscosity contrast is strong: for $\mu/\mu_w$ = 0.8, 1.2, 1.5, and 2.0, the correction factor equals 0.969, 1.026, 1.058, and 1.102, respectively. Therefore, omission of the term should be interpreted as acceptable for first-stage comparative assessment, but not as a substitute for a detailed local-property treatment. The use of the hydraulic diameter to extend tube correlations to non-circular ducts is standard practice in preliminary internal-flow heat transfer estimates, while detailed aspect-ratio-dependent solutions for fully developed laminar duct flow can be found in dedicated duct-flow sources [27]. Accordingly, the Sieder–Tate correlation is used here as a uniform comparative metric across all cases, not as a geometry-specific predictive model for rectangular minichannels.

Finally, the heat transfer coefficient was obtained from the definition of the Nusselt number $Nu$ ($Nu = h \cdot D_h/k$, where $h$ is the heat transfer coefficient). Using the Sieder–Tate correlation to compute $Nu_{corr}$, the corresponding coefficient $h_{corr}$ is

$$h_{corr} = \frac{Nu_{corr} \cdot k}{D_h} \tag{4}$$

where $k$ is the thermal conductivity evaluated at the bulk (reference) fluid temperature.

The assessment is restricted to single-phase forced convection with no phase change, no conjugate heat conduction in the wall, and no pressure-drop constraint included in the ranking criterion. Properties are evaluated at discrete temperatures (293.1/313.1/328.1 K) and assumed uniform for each case. The adopted correlation targets laminar flow with thermal development along the heated length; therefore, the computed $Nu$ and $h$ should be interpreted as relative heat transfer metrics under the specified geometry and operating point, suitable for experimental validation.

Because the present framework is intended for comparative assessment rather than high-fidelity local prediction, the adopted property treatment was kept deliberately simple and internally consistent across all cases. Density, kinematic viscosity, and thermal conductivity were taken directly from the experimentally determined mixture dataset at each temperature level, whereas mixture-specific heat capacity was estimated from pure-component data only for evaluation of the Prandtl number. This approach preserves the measured transport-property trends while avoiding direct use of disputed DSC-derived mixture $c_p$ values in the final calculations.

## 5. Results and Analysis

### 5.1. *Heat Transfer Assessment Outputs for Investigated Mixtures*

This section presents the measured thermophysical property trends and the resulting heat transfer assessment outputs for the investigated mixtures.

https://doi.org/10.3390/en19102291

For each mixture series (HFE-7100/EA, HFE-7300/EA, and HFE-73DE/EA), mixtures were considered at selected mass fractions (first component/second component) and characterised at 293.1, 313.1, and 328.1 K under near-atmospheric pressure. The dataset used in the final comparative analysis includes experimentally determined density, kinematic viscosity, and thermal conductivity, while specific heat was estimated from pure-component mass fractions. These properties were used to evaluate the Reynolds and Prandtl numbers. For each mixture and temperature, dynamic viscosity was calculated from the experimentally determined density and kinematic viscosity and then used in the Reynolds- and Prandtl-number calculations.

The specific heat capacity of each mixture was estimated using an ideal-mixture, mass-fraction-weighted mixing approximation, according to the following relation [28]:

$$c_{p,mix}(T) = \sum w_i \cdot c_{p,i}(T) \tag{5}$$

where $c_{p,mix}(T)$ is the estimated isobaric specific heat capacity of the mixture at temperature $T$, $w_i$ is the mass fraction of component $i$, and $c_{p,i}(T)$ is the isobaric specific heat capacity of pure component $i$ at the same temperature. This approximation corresponds to neglecting the excess heat-capacity contribution of the mixture.

Pure-component $c_p$ values were assembled as follows. For ethyl acetate, liquid $c_p$ data from the NIST Chemistry WebBook [23] were used to assign near-ambient values at 293.1, 313.1, and 328.1 K. For HFE-7300, ambient-pressure liquid $c_p$ data from the thermodynamic study of Kocian et al. [29] were interpolated to the three target temperatures. For HFE-7100, a literature-backed reference value of 1183 J/(kg·K) at 25 °C [30] was retained as a narrow-range approximation across the analysed temperatures. For HFE-73DE, no reliable public temperature-dependent $c_p(T)$ dataset was identified; therefore, the reference value of 1201 J/(kg·K), reported in [31], was used as a constant approximation. Because $c_p$ enters the adopted screening only through $Pr$ and then through the one-third power in Equation (3), the resulting sensitivity of $Nu_{corr}$ and $h_{corr}$ to residual $c_p$ uncertainty is limited. These assumptions should nevertheless be treated as a limitation of the present comparative screening framework.

Table 3 summarises the physical properties of the mixtures at 293.1 K, Table 4 shows the data at 313.1 K, whereas Table 5 concerns data at 328.1 K. Within the analysed composition ranges, density, viscosity, thermal conductivity, and specific heat vary systematically with both temperature and composition, providing a consistent basis for comparative thermal-hydraulic assessment.

**Table 3.** Summary of experimentally determined properties ($\rho$, $\nu$, $k$), estimated specific heat capacity $c_p$ obtained from the mass-fraction-weighted mixing rule, and assessment outputs for all selected mixtures and compositions retained for analysis at $T$ = 293.1 K (reference Setup #1: $w$ = 4 mm, $s$ = 1 mm, $L$ = 0.36 m, $D_h$ = 1.6 mm, $G$ = 200 kg/(m$^2$·s)).

| Mixture | Mass Fraction (%) | $\rho$ (kg/m$^3$) | $\mu$ (mPa·s) | $k$ (W/(m·K)) | $c_p$ (J/(kg·K)) | $Re$ (–) | $Nu_{corr}$ (–) | $h_{corr}$ (W/(m$^2$·K)) |
|---|---|---|---|---|---|---|---|---|
| HFE-73DE/EA | 10/90 | 931 | 0.4731 | 0.165 | 1848.5 | 676.4 | 4.68 | 482.7 |
| HFE-73DE/EA | 25/75 | 969 | 0.4877 | 0.189 | 1740.6 | 656.1 | 4.38 | 517.9 |
| HFE-73DE/EA | 50/50 | 1035 | 0.4839 | 0.207 | 1560.8 | 661.3 | 4.10 | 530.6 |
| HFE-7100/EA | 10/90 | 940 | 0.5194 | 0.125 | 1846.8 | 616.1 | 5.13 | 401.0 |
| HFE-7100/EA | 25/75 | 1017 | 0.5039 | 0.116 | 1736.1 | 635.0 | 5.15 | 373.7 |
| HFE-7100/EA | 50/50 | 1111 | 0.5341 | 0.105 | 1551.8 | 599.1 | 5.13 | 336.9 |
| HFE-7100/EA | 75/25 | 1216 | 0.5787 | 0.089 | 1367.4 | 553.0 | 5.20 | 289.2 |
| HFE-7300/EA | 25/75 | 1006 | 0.5005 | 0.133 | 1719.3 | 639.4 | 4.91 | 408.1 |
| HFE-7300/EA | 50/50 | 1154 | 0.578 | 0.101 | 1518.0 | 553.6 | 5.16 | 325.9 |
| HFE-7300/EA | 75/25 | 1348 | 0.741 | 0.087 | 1316.8 | 431.8 | 5.17 | 281.3 |

https://doi.org/10.3390/en19102291

**Table 4.** Summary of experimentally determined properties ($\rho$, $\nu$, $k$), estimated specific heat capacity obtained from the mass-fraction-weighted mixing rule, and assessment outputs for all selected mixtures and compositions retained for analysis at $T$ = 313.1 K (reference Setup #1: $w$ = 4 mm, s = 1 mm, $L$ = 0.36 m, $D_h$ = 1.6 mm, $G$ = 200 kg/(m$^2$·s)).

| Mixture | Mass Fraction (%) | $\rho$ (kg/m$^3$) | $\mu$ (mPa·s) | $k$ (W/(m·K)) | $c_p$ (J/(kg·K)) | $Re$ (–) | $Nu_{corr}$ (–) | $h_{corr}$ (W/(m$^2$·K)) |
|---|---|---|---|---|---|---|---|---|
| HFE-73DE/EA | 10/90 | 902.1 | 0.3804 | 0.155 | 1852.7 | 841.2 | 4.78 | 463.3 |
| HFE-73DE/EA | 25/75 | 951.7 | 0.399 | 0.177 | 1744.1 | 802.0 | 4.48 | 496.1 |
| HFE-73DE/EA | 50/50 | 933.0 | 0.3613 | 0.194 | 1563.0 | 885.7 | 4.19 | 508.4 |
| HFE-7100/EA | 10/90 | 888.0 | 0.3719 | 0.123 | 1850.9 | 860.4 | 5.16 | 397.0 |
| HFE-7100/EA | 25/75 | 785.2 | 0.3239 | 0.111 | 1739.6 | 988.0 | 5.23 | 363.1 |
| HFE-7100/EA | 50/50 | 1044.0 | 0.4143 | 0.1 | 1554.0 | 772.4 | 5.22 | 326.2 |
| HFE-7100/EA | 75/25 | 1042.3 | 0.4158 | 0.082 | 1368.5 | 769.6 | 5.35 | 274.0 |
| HFE-7300/EA | 25/75 | 1067.4 | 0.4475 | 0.107 | 1730.2 | 715.1 | 5.29 | 353.7 |
| HFE-7300/EA | 50/50 | 1138.0 | 0.4603 | 0.095 | 1535.3 | 695.2 | 5.29 | 314.0 |
| HFE-7300/EA | 75/25 | 1470.8 | 0.6413 | 0.08 | 1340.4 | 499.0 | 5.35 | 267.6 |

**Table 5.** Summary of experimentally determined properties ($\rho$, $\nu$, $k$), estimated specific heat capacity obtained from the mass-fraction-weighted mixing rule, and assessment outputs for all selected mixtures and compositions retained for analysis at $T$ = 328.1 K. (reference Setup #1: $w$ = 4 mm, s = 1 mm, $L$ = 0.36 m, $D_h$ = 1.6 mm, $G$ = 200 kg/(m$^2$·s)).

| Mixture | Mass Fraction (%) | $\rho$ (kg/m$^3$) | $\mu$ (mPa·s) | $k$ (W/(m·K)) | $c_p$ (J/(kg·K)) | $Re$ (–) | $Nu_{corr}$ (–) | $h_{corr}$ (W/(m$^2$·K)) |
|---|---|---|---|---|---|---|---|---|
| HFE-73DE/EA | 10/90 | 894.4 | 0.3292 | 0.148 | 1858.8 | 972.1 | 4.86 | 449.7 |
| HFE-73DE/EA | 25/75 | 951.6 | 0.3543 | 0.169 | 1749.2 | 903.2 | 4.56 | 481.5 |
| HFE-73DE/EA | 50/50 | 916.3 | 0.3478 | 0.185 | 1566.5 | 920.1 | 4.26 | 492.9 |
| HFE-7100/EA | 10/90 | 888.0 | 0.327 | 0.12 | 1857.0 | 978.6 | 5.21 | 390.9 |
| HFE-7100/EA | 25/75 | 766.3 | 0.2802 | 0.11 | 1744.7 | 1142.0 | 5.26 | 361.3 |
| HFE-7100/EA | 50/50 | 1030.0 | 0.3594 | 0.097 | 1557.5 | 890.4 | 5.28 | 319.9 |
| HFE-7100/EA | 75/25 | 1022.0 | 0.3431 | 0.081 | 1370.2 | 932.7 | 5.37 | 271.8 |
| HFE-7300/EA | 25/75 | 1084.0 | 0.4036 | 0.097 | 1740.9 | 792.9 | 5.48 | 332.0 |
| HFE-7300/EA | 50/50 | 1107.2 | 0.3906 | 0.093 | 1550.0 | 819.3 | 5.34 | 310.6 |
| HFE-7300/EA | 75/25 | 1458.6 | 0.5479 | 0.073 | 1359.0 | 584.0 | 5.54 | 252.9 |

### 5.2. Thermophysical Property Trends

Across all mixture series, density decreases with temperature and changes systematically with composition. Kinematic viscosity also decreases with temperature, which increases the Reynolds number under the fixed mass-flux conditions adopted here. However, because the viscosity-ratio correction term was set to unity, the adopted correlation yields $Re \cdot Pr$ independent of viscosity for fixed geometry and mass flux. As a result, $Nu_{corr} \propto (c_p/k)^{1/3}$, while $h_{corr} \propto c_p{}^{1/3}\, k^{2/3}$. Under these assumptions, thermal conductivity is the dominant property controlling the predicted $h_{corr}$ levels, while viscosity primarily affects the hydraulic interpretation and verification of the laminar-flow range.

### 5.3. Correlation-Based Assessment Results for Setup #1 (Long Test Section)

The long-channel assessment (Tables 3–5), together with Figure 3, shows clear differences between mixture families within the compositions retained for analysis. Under the fixed reference operating point, EA-rich compositions generally yield higher correlation-based heat transfer coefficients.

https://doi.org/10.3390/en19102291

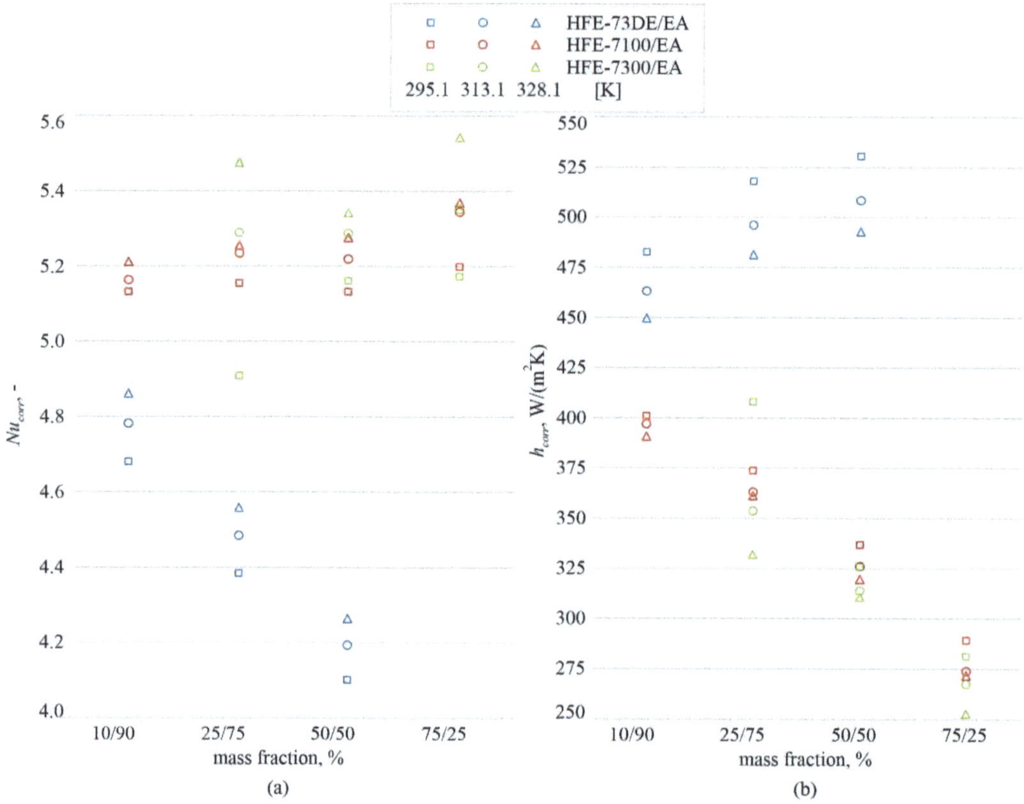


**Figure 3.** Correlation-based estimates under the Setup #1 reference conditions: (**a**) Nusselt number versus mass fraction; (**b**) heat transfer coefficient versus mass fraction for the binary mixtures.

Under the present assumptions, this trend reflects primarily the effect of thermal conductivity and, to a lesser extent, the contribution of $c_p$ through the corrected Prandtl number treatment. Reynolds number is still reported for hydraulic interpretation, but it is not the dominant factor in the ranking of $h_{corr}$ when the viscosity-ratio correction is set to unity.

For the HFE-7100/EA family, the highest $h_{corr}$ values occur at the most EA-rich composition available (10/90) across the investigated temperature range. For HFE-7300/EA, the highest $h_{corr}$ values occur at 25/75 throughout the analysed range. For HFE-73DE/EA, the comparison is limited to 10/90–50/50 because the 75/25 composition was excluded; within the retained set, the 50/50 mixture yields the highest $h_{corr}$ at all three temperature levels, followed by 25/75 and 10/90. This difference relative to the HFE-7100/EA and HFE-7300/EA families follows from the stronger contribution of thermal conductivity in the adopted correlation-based ranking.

Using the reference geometry and operating point defined in Section 4, Reynolds and Prandtl numbers were computed for each mixture at 293.1, 313.1, and 328.1 K. In the corrected $c_p$ treatment, an increase in temperature does not automatically increase $h_{corr}$. In most of the retained cases, $h_{corr}$ decreases with temperature because the reduction in thermal conductivity outweighs the relatively weak one-third-power influence of $c_p$. Figure 3 summarises the correlation-based Nusselt number and heat transfer coefficient as functions of mass fraction under the Setup #1 reference conditions.

https://doi.org/10.3390/en19102291

### 5.4. Correlation-Based Assessment Results for Setup #2 (Short Test Section)

For the short-module geometry, the same correlation framework was applied using the module-specific hydraulic diameter and heated length. In contrast to the long-module assessment, the mass flux was set to the value corresponding to the selected convection-only operating point used for validation (Section 6). This ensures that the correlation-based estimates for Setup #2 are evaluated under the same hydraulic conditions as the experimental dataset, enabling a consistent comparison.

For the Setup #2 assessment (Tables 6–8 and Figure 4), the reference conditions were defined by the Setup #2 geometry (five parallel minichannels with $w = 0.006$, $s = 0.001$ m, and $L = 0.043$ m; Table 2). The hydraulic diameter was evaluated accordingly, and mixture properties at each temperature level were treated as constant reference values, in analogy to the long-module calculations. Based on the experimental operating conditions, a fixed mass flux of $G = 372.5$ kg/(m$^2$·s) was adopted.

**Table 6.** Summary of correlation-based assessment outputs for the selected mixtures and compositions retained for analysis at $T = 293.1$ K under the Setup #2 reference conditions.

| Mixture | Mass Fraction (%) | $Re$ (–) | $Nu_{corr}$ (–) | $h_{corr}$ (W/(m$^2$·K)) |
|---|---|---|---|---|
| HFE-73DE/EA | 10/90 | 1939.7 | 12.24 | 1178.4 |
| HFE-73DE/EA | 25/75 | 1802.4 | 11.47 | 1264.5 |
| HFE-73DE/EA | 50/50 | 1836.0 | 10.73 | 1295.6 |
| HFE-7100/EA | 10/90 | 1952.9 | 13.43 | 979.0 |
| HFE-7100/EA | 25/75 | 2279.3 | 13.48 | 912.4 |
| HFE-7100/EA | 50/50 | 1776.9 | 13.43 | 822.4 |
| HFE-7100/EA | 75/25 | 1861.0 | 13.60 | 706.2 |
| HFE-7300/EA | 25/75 | 1582.3 | 12.84 | 996.3 |
| HFE-7300/EA | 50/50 | 1635.0 | 13.50 | 795.6 |
| HFE-7300/EA | 75/25 | 1165.5 | 13.53 | 686.9 |

**Table 7.** Summary of correlation-based assessment outputs for the selected mixtures and compositions retained for analysis at $T = 313.1$ K under the Setup #2 reference conditions.

| Mixture | Mass Fraction (%) | $Re$ (–) | $Nu_{corr}$ (–) | $h_{corr}$ (W/(m$^2$·K)) |
|---|---|---|---|---|
| HFE-73DE/EA | 10/90 | 1678.5 | 12.51 | 1131.2 |
| HFE-73DE/EA | 25/75 | 1600.4 | 11.73 | 1211.2 |
| HFE-73DE/EA | 50/50 | 1767.4 | 10.97 | 1241.4 |
| HFE-7100/EA | 10/90 | 1717.2 | 13.51 | 969.2 |
| HFE-7100/EA | 25/75 | 1971.2 | 13.70 | 886.6 |
| HFE-7100/EA | 50/50 | 1541.5 | 13.65 | 796.5 |
| HFE-7100/EA | 75/25 | 1535.8 | 13.98 | 668.9 |
| HFE-7300/EA | 25/75 | 1426.9 | 13.84 | 863.6 |
| HFE-7300/EA | 50/50 | 1387.3 | 13.83 | 766.6 |
| HFE-7300/EA | 75/25 | 995.8 | 14.00 | 653.4 |

Tables 6–8 compile the correlation-based assessment outputs for the selected mixtures and compositions retained for analysis under the Setup #2 reference conditions at $T = 293.1$, 313.1, and 328.1 K. Figure 4 illustrates the corresponding Nusselt number and heat transfer coefficient as functions of mass fraction for the binary mixtures under the same Setup #2 reference conditions. The same property interpretation applies here as in Setup #1: the shorter heated length increases the absolute $Nu_{corr}$ and $h_{corr}$ levels, whereas the relative ranking remains governed primarily by the $c_p - k$ combination embedded in the adopted correlation.

https://doi.org/10.3390/en19102291

**Table 8.** Summary of correlation-based assessment outputs for the selected mixtures and compositions retained for analysis at $T$ = 328.1 K under the Setup #2 reference conditions.

| Mixture | Mass Fraction (%) | $Re$ (–) | $Nu_{corr}$ (–) | $h_{corr}$ (W/(m$^2$·K)) |
|---|---|---|---|---|
| HFE-73DE/EA | 10/90 | 1349.9 | 12.72 | 1098.1 |
| HFE-73DE/EA | 25/75 | 1309.4 | 11.92 | 1175.5 |
| HFE-73DE/EA | 50/50 | 1319.5 | 11.15 | 1203.5 |
| HFE-7100/EA | 10/90 | 1229.5 | 13.64 | 954.5 |
| HFE-7100/EA | 25/75 | 1267.1 | 13.75 | 882.1 |
| HFE-7100/EA | 50/50 | 1195.5 | 13.80 | 781.1 |
| HFE-7100/EA | 75/25 | 1103.5 | 14.05 | 663.7 |
| HFE-7300/EA | 25/75 | 1275.8 | 14.33 | 810.6 |
| HFE-7300/EA | 50/50 | 1104.9 | 13.98 | 758.2 |
| HFE-7300/EA | 75/25 | 861.7 | 14.50 | 617.5 |

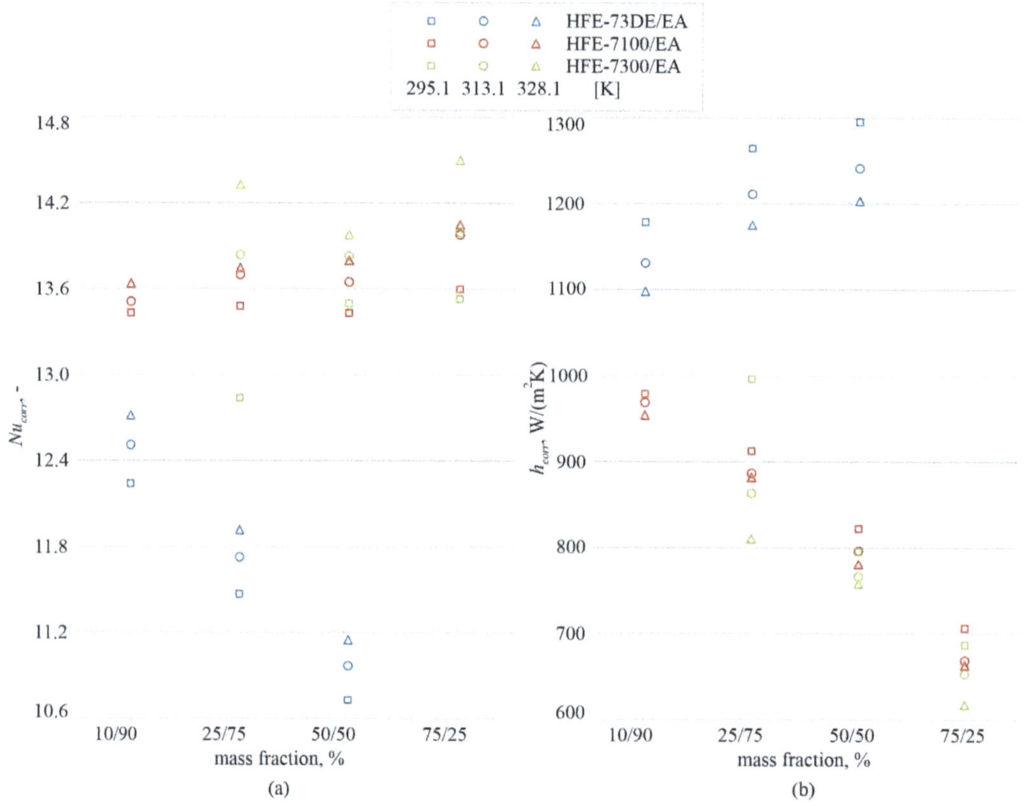


**Figure 4.** Correlation-based estimates under the Setup #2 reference conditions: (**a**) Nusselt number versus mass fraction; (**b**) heat transfer coefficient versus mass fraction for the binary mixtures.

The results in Tables 6–8 and Figure 4 show that the mixture ranking is not specific to the long-channel geometry. The same ranking pattern is retained in the short-module geometry, although the absolute $Nu_{corr}$ and $h_{corr}$ levels are higher because of the shorter heated length. Under the present corrected $c_p$ treatment and with the viscosity-ratio term set to unity, the relative ranking is governed primarily by thermal conductivity, with a weaker contribution from $c_p$ through its one-third-power effect. The validation presented in Section 6 shows that the correlation systematically overpredicts $h_{corr}$ for HFE-7100 under the tested conditions; therefore, the reported absolute values should be treated cautiously, whereas the comparative ranking remains the main intended output of the framework.

## 6. Validation of the Correlation-Based Estimates Against the Experimental Dataset

The correlation-based estimates were validated against an experimental dataset obtained for pure HFE-7100 in the short multi-minichannel module (Setup #2, five parallel minichannels) under single-phase forced convection. The selected convection-only operating point corresponded to a mean mass flux of $G_{mean}$ = 372.5 (kg/(m$^2$·s)), inlet and outlet bulk-fluid temperatures of $T_{f,in}$ = 287.3 K and $T_{f,out}$ = 288.5 K, inlet and outlet gauge pressures of $p_{in}$ = 25,559.9 Pa and $p_{out}$ = 24,399.2 Pa, and an imposed heat flux of $q''$ = 6473.2 W/m$^2$ calculated from the measured electrical current and voltage. The IR-measured outer-foil temperature ranged from 294.7 K to 303.7 K; the ambient-air temperature was 294.9 K, and the atmospheric pressure was 0.96 bar.

The experimental heat transfer coefficient was evaluated from the spatial wall-temperature field measured by IR thermography. For each frame, the local coefficient was computed as

$$h_{exp}(x) = \frac{q''}{\left[T_F(x) - T_f(x) - \Delta T_{cond}\right]} \tag{6}$$

where $T_F(x)$ is the local outer-surface temperature of the heated foil measured by infrared thermography, $T_f(x)$ is the linear HFE-7100 bulk-temperature profile between the measured inlet/outlet temperatures (measured by K-type thermocouples placed in the inlet and outlet of the minichannel collector), and $q''$ is determined from the measured current and voltage, the heated area, and the assumed heat-loss correction.

Furthermore, $\Delta T_{cond}$ accounts for conduction through the heater foil and is calculated as follows:

$$\Delta T_{cond} = \frac{q'' \cdot \delta_F}{k_F} \tag{7}$$

where $\delta_F$ is the heater foil thickness and $k_F$ is the foil conductivity.

A minimal quality criterion was applied (temperature difference in Equation (6) > 0.5 K and $h_{exp} > 0$), and $h_{exp,mean}$ was obtained by averaging over the retained axial positions. The correlation-based heat transfer coefficient $h_{corr}$ was computed by first evaluating $Re$ and $Pr$ from Equations (1) and (2), then calculating $Nu_{corr}$ using the laminar thermally developing correlation (3), and finally, converting $Nu_{corr}$ to $h_{corr}$. For this first-stage validation, $\mu$, $c_p$, and $k$ were treated as constant reference values for HFE-7100 (near 298.1 K). For the selected 30 low-heat-flux frames (convection-only subset), the mean $h_{exp,mean}$ was 480.6 $\pm$ 11.7 W/(m$^2$ K), while the mean $h_{corr}$ was 555.2 $\pm$ 0.1 W/(m$^2$ K). The mean relative difference was defined as

$$\Delta h_{rel} = \frac{(h_{exp} - h_{corr})}{h_{corr}} \cdot 100\%, \tag{8}$$

and equalled $-$13.44% (median $-$14.15%; 10th–90th percentiles $-$15.23% to $-$10.72%). The RMSE was 75.5 W/(m$^2$ K), and the MAPE (relative to $h_{exp}$) was 15.6%. This level of discrepancy is not negligible and confirms that the adopted correlation systematically overpredicts the absolute heat transfer coefficient for the validated case. The most likely contributors are the following: use of a circular-tube laminar entrance correlation with hydraulic diameter for a rectangular five-minichannel module, neglect of the Sieder–Tate viscosity correction, treatment of properties as spatially uniform, and omission of conjugate wall-conduction, and other module-specific effects from the reduced comparison model. For this reason, the

https://doi.org/10.3390/en19102291

framework is retained as a comparative screening tool under fixed assumptions, not as a bias-free predictive correlation for absolute heat transfer coefficient values.

Table 9 lists the 30 low-heat-flux (convection-only) validation cases for Setup #2, including the mass flux $G$, imposed heat flux $q''$, the experimentally obtained mean heat transfer coefficient $h_{exp,mean}$, the corresponding correlation-based value $h_{corr}$, and the resulting relative difference $\Delta h_{rel}$.

**Table 9.** Validation dataset for the short module (Setup #2, HFE-7100): experimental mean heat transfer coefficient $h_{exp,mean}$, compared with the correlation-based value $h_{corr}$ calculated using Equation (3), and the resulting relative difference $\Delta h_{rel}$ calculated from Equation (8).

| $G$ (kg/(m$^2$·s)) | $q''$ (W/m$^2$) | $h_{exp,mean}$ (W/(m$^2$·K)) | $h_{corr}$ (W/(m$^2$·K)) | $\Delta h_{rel}$ (%) |
|---|---|---|---|---|
| 372.2 | 4291.5 | 470.8 | 555.1 | −15.2 |
| 372.1 | 4859.2 | 503.8 | 555.1 | −9.2 |
| 372.2 | 5773.7 | 520.7 | 555.1 | −6.2 |
| 372.7 | 6144.2 | 470.8 | 555.4 | −15.2 |
| 372.4 | 6163.8 | 473.2 | 555.2 | −14.8 |
| 372.4 | 6169.3 | 476.4 | 555.2 | −14.2 |
| 372.1 | 6176.0 | 468.1 | 555.1 | −15.7 |
| 372.3 | 6181.7 | 478.1 | 555.2 | −13.9 |
| 372.1 | 6181.8 | 478.4 | 555.1 | −13.8 |
| 372.4 | 6184.6 | 475.9 | 555.2 | −14.3 |
| 372.5 | 6186.8 | 473.3 | 555.3 | −14.8 |
| 372.3 | 6187.8 | 466.8 | 555.2 | −15.9 |
| 372.6 | 6198.7 | 475.1 | 555.3 | −14.4 |
| 372.5 | 6199.8 | 473.5 | 555.3 | −14.7 |
| 372.5 | 6208.4 | 476.0 | 555.3 | −14.3 |
| 372.5 | 6209.7 | 482.0 | 555.3 | −13.2 |
| 372.6 | 6210.7 | 480.9 | 555.3 | −13.4 |
| 372.3 | 6214.7 | 476.8 | 555.2 | −14.1 |
| 372.4 | 6217.4 | 470.6 | 555.2 | −15.2 |
| 372.4 | 6220.7 | 476.3 | 555.2 | −14.2 |
| 372.3 | 6223.7 | 489.1 | 555.2 | −11.9 |
| 372.6 | 6226.7 | 502.6 | 555.3 | −9.5 |
| 372.4 | 6227.8 | 479.5 | 555.2 | −13.6 |
| 372.0 | 6229.8 | 487.1 | 555.0 | −12.2 |
| 372.4 | 6230.8 | 475.2 | 555.2 | −14.4 |
| 372.4 | 6232.8 | 474.1 | 555.2 | −14.6 |
| 372.7 | 6232.8 | 480.5 | 555.4 | −13.5 |
| 372.6 | 6241.8 | 484.5 | 555.3 | −12.8 |
| 372.6 | 6242.5 | 482.6 | 555.3 | −13.1 |
| 372.5 | 6247.8 | 495.0 | 555.3 | −10.9 |

## 7. Conclusions

Thermophysical property data for three binary HFE/ethyl acetate systems (HFE-7100/EA, HFE-7300/EA, and HFE-73DE/EA) were compiled at 293.1, 313.1, and 328.1 K to support comparative single-phase forced-convection analysis in minichannel geometries. In the final assessment workflow, experimentally determined density, viscosity, and thermal conductivity were used directly, whereas mixture-specific heat capacity was estimated from pure-component mass fractions using a literature-supported ideal-mixture approximation solely for evaluation of the Prandtl number.

Under the adopted assumptions, the calculated correlation-based heat transfer coefficients span about 253–531 W/(m$^2$·K) for the long reference module (Setup #1) and about 618–1296 W/(m$^2$·K) for the short multi-minichannel module (Setup #2). For the

https://doi.org/10.3390/en19102291

HFE-7100/EA and HFE-7300/EA families, the highest $h_{corr}$ values were obtained for the most EA-rich retained compositions, whereas within the retained HFE-73DE/EA subset, the 50/50 composition gave the highest $h_{corr}$ because of its more favourable thermal conductivity level. In the corrected $c_p$ treatment, the ranking is governed primarily by thermal conductivity, while $c_p$ contributes more weakly through the one-third-power dependence embedded in the adopted correlation.

Validation against the short-module dataset for pure HFE-7100 showed a mean relative difference of −13.44% and a MAPE of 15.6%, confirming that the simplified framework systematically overpredicts the absolute heat transfer coefficient under the tested conditions. The most likely reasons are the use of a tube-derived entrance correlation with hydraulic diameter for a rectangular five-minichannel module, neglect of the viscosity-ratio correction, and omission of conjugate and module-specific effects. Accordingly, the proposed framework should be interpreted as a comparative screening tool under fixed assumptions rather than as a geometry-exact predictive model for absolute heat transfer coefficient values.

The dataset and workflow support early-stage selection of binary (and potentially more complex) dielectric working fluids for compact heat exchangers and mini-/microchannel thermal management, where limited space, low fluid charge, and electrical safety are key constraints. Application areas include electronics and power electronics cooling, HVAC&R components with confined internal passages, and compact laboratory- or process-scale heat-exchanger modules requiring a balance between heat transfer performance, pressure-drop (pumping-power) penalties, and the target operating temperature range.

Limitations include the restriction to single-phase forced convection, the use of a tube-derived correlation applied via the hydraulic diameter, and the simplified treatment of thermophysical properties at each temperature level. Further work should extend validation to additional operating points and consider rectangular-duct-specific laminar formulations, temperature-dependent properties, and joint heat transfer/pressure-drop criteria for mixture down-selection.

**Author Contributions:** Conceptualization, A.P.; Methodology, A.P.; Software, M.P.; Validation, M.P.; Formal analysis, A.P. and M.P.; Investigation, A.P.; Resources, A.P.; Data curation, A.P. and M.P.; Writing—original draft, A.P. and M.P.; Writing—review & editing, A.P. and M.P.; Visualization, A.P.; Supervision, M.P.; Project administration, M.P.; Funding acquisition, M.P. All authors have read and agreed to the published version of the manuscript.

**Funding:** This research was funded by the National Science Centre, Poland, grant no. UMO-2025/57/B/ST8/00907. For the purpose of Open Access, the authors have applied a CC BY public copyright licence to the Author Accepted Manuscript (AAM) version arising from this submission.

**Data Availability Statement:** The original contributions presented in this study are included in the article. Further inquiries can be directed to the corresponding author.

**Conflicts of Interest:** The authors declare no conflicts of interest.

## Nomenclature

Symbols

| | |
|---|---|
| $c_p$ | isobaric specific heat capacity (J·kg$^{-1}$·K$^{-1}$) |
| $c_{p,i}$ | isobaric specific heat capacity of pure component $i$, (J·kg$^{-1}$·K$^{-1}$) |
| $c_{p,mix}$ | estimated isobaric specific heat capacity of the mixture (J·kg$^{-1}$·K$^{-1}$) |
| $D_h$ | hydraulic diameter, $D_h = 2w \cdot s/(w + s)$ (m) |
| $G$ | mass flux (kg·m$^{-2}$·s$^{-1}$) |

https://doi.org/10.3390/en19102291

| | |
|---|---|
| $h_{exp}$ | experimental heat transfer coefficient, calculated according to Equation (6) ($W \cdot m^{-2} \cdot K^{-1}$) |
| $h_{corr}$ | correlation-based heat transfer coefficient ($W \cdot m^{-2} \cdot K^{-1}$) |
| $h_{exp,mean}$ | experimentally obtained mean heat transfer coefficient ($W \cdot m^{-2} \cdot K^{-1}$) |
| $I$ | electric current supplied to the heater (A) |
| $k$ | thermal conductivity ($W \cdot m^{-1} \cdot K^{-1}$) |
| $L$ | minichannel length (m) |
| $Nu_{exp}$ | experimental Nusselt number, $Nu_{exp} = h_{exp} \cdot D_h / k$ |
| $Nu_{corr}$ | correlation-based Nusselt number (–) |
| $p$ | pressure (Pa) |
| $Pr$ | Prandtl number, $Pr = \mu \cdot c_p / k$ (–) |
| $Q''$ | heat flux at the heated wall ($W \cdot m^{-2}$) |
| $Re$ | Reynolds number, $Re = G \cdot D_h / \mu$ (–) |
| $s$ | minichannel depth (m) |
| $T$ | temperature (K) |
| $w$ | minichannel width (m) |
| $w_i$ | mass fraction of component $i$ in the mixture (-) |
| $x$ | axial coordinate along the minichannel (m) |
| Greek letters | |
| $\Delta T_{cond}$ | temperature drop across the heater foil due to one-dimensional conduction $\Delta T_{cond} = q'' \cdot \delta / k_F$ (K) |
| $\Delta h_{rel}$ | relative difference between $h_{exp,mean}$ and $h_{corr}$ (%) |
| $\mu$ | dynamic viscosity (Pa·s) |
| $\mu_w$ | dynamic viscosity evaluated at the wall temperature (Pa·s) |
| $\nu$ | kinematic viscosity ($m^2 \cdot s^{-1}$) |
| $\rho$ | density ($kg \cdot m^{-3}$) |
| $\delta$ | heater foil thickness (m) |
| Subscripts | |
| *corr* | value calculated from a correlation |
| *exp* | value obtained from experiment |
| *F* | foil (heater wall) |
| *f* | fluid (bulk) |
| *i* | component index |
| *in/out* | inlet/outlet |
| *max* | maximum value |
| *mean* | mean (averaged) value |
| *min* | minimum value |
| *mix* | mixture |
| Abbreviations | |
| EA | ethyl acetate |
| HFE | hydrofluoroether |
| IR | infrared thermography |
| LCT | liquid crystal thermography |

## References

1. von Westarp, W.G.; Jupke, A. Improved methodology for the measurement of vapor–liquid equilibria of binary mixtures with large boiling point differences via differential scanning calorimetry. *Thermochim. Acta* **2026**, *757*, 180222. [CrossRef]
2. Aminian, A.; Celný, D.; Mickoleit, E.; Jäger, A.; Vinš, V. Ideal Gas Heat Capacity and Critical Properties of HFE-Type Engineering Fluids: Ab Initio Predictions of Cp,ig, Modelling of Phase Behaviour and Thermodynamic Properties Using Peng–Robinson and Volume-Translated Peng–Robinson Equations of State. *Int. J. Thermophys.* **2022**, *43*, 87. [CrossRef]
3. Urata, S.; Takada, A.; Murata, J.; Hiaki, T.; Sekiya, A. Prediction of vapour–liquid equilibrium for binary systems containing HFEs by using artificial neural network. *Fluid Phase Equilibria* **2002**, *199*, 63–78. [CrossRef]
4. Wu, J.; Pan, Y.; Ren, J.; Li, Q. Exploring structure–property relationships of critical temperatures for binary refrigerant mixtures via group contribution and machine learning. *DeCarbon* **2025**, *9*, 100123. [CrossRef]

https://doi.org/10.3390/en19102291

5. Soria-Lopez, A.; Simal-Gandara, J.; Mejuto, J.C. Viscosity mixing rules and machine learning-based models for predicting the viscosity of liquid binary mixtures of aliphatic alkanes. *J. Mol. Liq.* **2025**, *437*, 128401. [CrossRef]
6. Muñoz-Rujas, N.; Aguilar, F.; Garcia-Alonso, J.M.; Montero, E.A. Thermodynamics of binary mixtures 1-ethoxy-1,1,2,2,3,3,4,4,4-nonafluorobutane (HFE-7200) + 2-propanol: High-pressure density, speed of sound and derivative properties. *J. Chem. Thermodyn.* **2019**, *131*, 630–647. [CrossRef]
7. Muñoz-Rujas, N.; Bazile, J.P.; Aguilar, F.; Galliero, G.; Montero, E.; Daridon, J.-L. Speed of sound, density and derivative properties of binary mixtures HFE-7500 + diisopropyl ether under high pressure. *J. Chem. Thermodyn.* **2019**, *128*, 19–33. [CrossRef]
8. Ogawa, H.; Karashima, S.; Takigawa, T.; Murakami, S. Excess molar enthalpies and volumes of binary mixtures of two hydrofluoroethers with hexane, or benzene, or ethanol, or 1-propanol, or 2-butanone at T=298.15 K. *J. Chem. Thermodyn.* **2003**, *35*, 763–774. [CrossRef]
9. Xu, R.; Zhang, C.; Chen, H.; Wu, Q.; Wang, R. Heat transfer performance of pulsating heat pipe with zeotropic immiscible binary mixtures. *Int. J. Heat Mass Transf.* **2019**, *137*, 31–41. [CrossRef]
10. Eissa, M.S.; Kotb, A.; Liu, L.; Wang, S. The prediction of binary zeotropic mixtures in-tube flow condensation—A generalized non-equilibrium heat transfer model. *Energy Convers. Manag.* **2026**, *347*, 120562. [CrossRef]
11. Tagliaferri, C.; Mazzei, L.; Lettieri, P.; Marzocchella, A.; Olivieri, G.; Salatino, P. CFD simulation of bubbling fluidized bidisperse mixtures: Effect of integration methods and restitution coefficient. *Chem. Eng. Sci.* **2013**, *102*, 324–334. [CrossRef]
12. Verma, S.; Rani, M.; Lee, Y.; Maken, S. Thermophysical properties of binary mixtures of diethyl ether as oxygenate with cyclohexane and aromatic hydrocarbons. *J. Mol. Liq.* **2023**, *387*, 122663. [CrossRef]
13. Tsai, M.-Y.; Wu, Y.-Y.; Lin, L.-C. Transport properties of alcohol/water mixtures: Evaluation of molecular potentials. *J. Mol. Liq.* **2025**, *433*, 127870. [CrossRef]
14. Piasecki, A.; Maciejewska, B.; Piasecka, M.; Grabowski, M.; Grabowski, P. Evaluation of HFE-73DE/ethyl acetate mixtures for use in minichannel heat exchangers. *Energies* **2026**, *19*, 110. [CrossRef]
15. *ASTM D1250*; Standard Guide for Use of the Petroleum Measurement Tables. ASTM International: West Conshohocken, PA, USA, 2013.
16. *EN ISO 3104:2020*; Petroleum Products—Transparent and Opaque Liquids—Determination of Kinematic Viscosity and Calculation of Dynamic Viscosity. ISO: Geneva, Switzerland, 2020.
17. *ASTM D86*; Standard Test Method for Distillation of Petroleum Products at Atmospheric Pressure. ASTM International: West Conshohocken, PA, USA, 2012.
18. Makomaski, G.; Ciesińska, W.; Zieliński, J. Thermal properties of pitch-polymer compositions and derived activated carbons. *J. Therm. Anal. Calorim.* **2012**, *109*, 767–772. [CrossRef]
19. Pecchi, M.; Goldfarb, J.L.; Baratieri, M. Hydrothermal carbonization enthalpy using differential scanning calorimetry: Assessing the accuracy of the exhaust sample method. *Thermochim. Acta* **2022**, *718*, 179388. [CrossRef]
20. 3M™ Novec™ 7100 Engineered Fluid, Product Information, 3M. Available online: https://multimedia.3m.com/mws/mediawebserver?mwsId=SSSSSu9n_zu8l00xl8mBm8mePv70k17zHvu9lxtD7xt1evSSSSSS-(accessed on 1 August 2025).
21. 3M™ Novec™ 7300 Engineered Fluid, Product Information, 3M. Available online: https://multimedia.3m.com/mws/mediawebserver?mwsId=SSSSSu9n_zu8l00x4Y_9lx2U4v70k17zHvu9lxtD7xt1evSSSSSS- (accessed on 1 August 2025).
22. 3M™ Novec™ 73DE Engineered Fluid, Product Information, 3M. Technical Data, May 2016. Available online: https://multimedia.3m.com/mws/mediawebserver?mwsId=SSSSSu9n_zu8l00xm8_el8m9lv70k17zHvu9lxtD7xt1evSSSSSS- (accessed on 1 August 2025).
23. NIST Chemistry WebBook. *Ethyl Acetate (CAS 141-78-6)*; NIST: Gaithersburg, MD, USA, 2026. Available online: https://webbook.nist.gov/cgi/cbook.cgi?ID=C141786 (accessed on 1 October 2025).
24. Piasecka, M.; Piasecki, A.; Maciejewska, B. Liquid Crystal Thermography and Infrared Thermography Application in Heat Transfer Research on Flow Boiling in Minichannels. *Energies* **2025**, *18*, 940. [CrossRef]
25. Piasecka, M.; Maciejewska, B.; Michalski, D.; Dadas, N.; Piasecki, A. Investigations of Flow Boiling in Mini-Channels: Heat Transfer Calculations with Temperature Uncertainty Analyses. *Energies* **2024**, *17*, 791. [CrossRef]
26. Sieder, E.N.; Tate, G.E. Heat Transfer and Pressure Drop of Liquids in Tubes. *Ind. Eng. Chem.* **1936**, *28*, 1429–1435. [CrossRef]
27. Shah, R.K.; London, A.L. *Laminar Flow Forced Convection in Ducts: A Source Book for Compact Heat Exchanger Analytical Data*; Academic Press: New York, NY, USA, 1978.
28. Kleiber, M.; Joh, R. D1 Calculation Methods for Thermophysical Properties. In *VDI Heat Atlas*; Springer: Berlin/Heidelberg, Germany, 2010.
29. Kocian, Š.; Knöbel, E.; Klink, S.; Prokopová, O.; Vrbka, P.; Čenský, M.; Fulem, M.; Růžička, K.; Richter, M.; Jäger, A.; et al. Thermodynamic Properties of HFE-7300. *Int. J. Thermophys.* **2026**, *47*, 45. [CrossRef]

https://doi.org/10.3390/en19102291

30. Zheng, Y.; Wei, Z.; Song, X. Measurements of Isobaric Heat Capacities for HFE-7000 and HFE-7100 at Different Temperatures and Pressures. *Fluid Phase Equilibria* **2016**, *425*, 335–341. [CrossRef]
31. Best Technology. Solvent Cleaning with 3M™ Novec™ Engineered Fluids & Solvents. n.d. Available online: https://www.besttechnologyinc.com/3m-novec-engineered-fluids-solvents/cleaning-vapor-degreasing/ (accessed on 17 April 2026).

*energies*

MDPI

*Article*

# Comparison of Analytical and Numerical Methods for Predicting the Shell-Side Heat Transfer Coefficient in Heat Exchanger with Segmental Baffles

**Janusz T. Cieśliński *, Jacek Barański, Kamil Stasiak, Krzysztof Tesch and Paweł Dąbrowski**

Institute of Energy, Gdańsk University of Technology, Narutowicza 11/12, 80-233 Gdańsk, Poland; jacek.baranski@pg.edu.pl (J.B.); kamil.stasiak@pg.edu.pl (K.S.); krzysztof.tesch@pg.edu.pl (K.T.); pawel.dabrowski@pg.edu.pl (P.D.)

* Correspondence: jcieslin@pg.edu.pl

**Abstract**

This study reports the calculated values results of the shell-side heat transfer coefficient for a shell-and-tube heat exchanger with an inner shell diameter of 200 mm and a length of 518 mm, containing 85 tubes arranged in a staggered layout. Shell-side cross-flow was generated by nine standard segmental baffles with a 25% baffle cut and a baffle pitch of 48 mm. In particular, the effect of 13 combinations of shell-to-baffle and baffle-to-tube gaps on the heat transfer coefficient was investigated. Moreover, the influence of sealing strips and tube bundle diameter on the heat transfer coefficient was also examined. The calculations were carried out using three different approaches, namely the Gaddis-Gnielinski method, the extended Bell-Delaware method, and Aspen EDR code. Numerical simulations for an idealized heat exchanger were also conducted using Ansys Fluent and OpenFOAM. As far as the authors are aware, this is the first study to compare two computational methods widely regarded as reference approaches for shell-and-tube heat exchangers, namely the Bell-Delaware and the Gaddis-Gnielinski approaches. The results obtained using the Aspen EDR code, a widely recognized software tool for modeling and design of heat exchangers, were evaluated against the forecast of the Bell-Delaware and Gaddis-Gnielinski approaches.

**Keywords:** heat transfer coefficient; segmental baffles; leakage; shell-and-tube heat exchanger

## 1. Introduction

In most thermal systems, heat exchangers serve as essential components. Some, such as car radiators, are produced in tens of millions of units each year. Due to dwindling resources of raw materials and energy, computational methods for heat exchangers have been continuously developed for over a hundred years.

Several parameters affect the heat transfer coefficient (HTC) inside shell. Apart from baffle spacing and baffle cut, these parameters include tube layout and tube pitch, the tube bundle diameter, the gap between the baffle and the tube, the gap between the shell and the baffle, and the number of sealing strips [1–4]. Each of these parameters influences the shell-side flow pattern and, consequently, the obtained HTC.

### *1.1. Analytical Approaches*

A considerable number of models for analytical prediction of shell-side HTCs have been published in the literature [5]. Most modern analytical models of shell-and-tube heat exchangers (STHEx) are founded on Tinker's approach, which assumes that the fluid stream entering the shell is divided into four main streams, namely the cross-flow, the baffle-to-tube leakage flow, the shell-to-baffle flow and the bypass flow [6,7]. The most important parameters affecting the formation of these streams are baffle spacing and baffle cut [8]. Consequently, most studies concentrate on baffling [9]. Among the analytical models, the Kern model [10], the Donohue model [11], the Žhukauskas model [12], the extended Bell-Delaware (Bell-D) approach [13], and the Gaddis-Gnielinski approach, known as the Verein Deutscher Ingenieure (VDI) method [14], are most frequently chosen as reference approaches. Roetzel and Lee [15] showed that the Žhukauskas model predicts the shell-side HTC, as determined from their empirical dispersion model, to within ±5%. Kücük et al. [16] experimentally demonstrated that the Kern and VDI methods predict the HTC within a range −15% to +10% for a mini-channel heat exchanger. Abdelkader and Zubair [17] calculated the HTC using the Kern, Bell-D, and Wills et al. methods [18] across a range of baffle numbers, mass flow rates, tube bundle layouts, fluid thermal properties, as well as baffle cuts. They established that the Bell-D approach most accurately predicts the published experimental findings. Abbas et al. [19] experimentally established that the Donohue and Kern approaches underpredicted their experimental findings by approximately 40–48%. Recently, Di Bono et al. [20] determined the HTC using the Kern and Bell-D approaches, utilizing the technical specifications of the STHEx experimentally examined by Zhang et al. [21]. Moreover, Di Bono et al. compared the calculation results obtained using the Kern and Bell-D approaches with those generated by Alperen et al. [22] using the standard HTRI Xchanger code, for the technical specifications of the STHEx examined by Zhang et al. [21]. The study found that the Bell-D approach predicts the experimental findings reported by Zhang et al. exceptionally well, whereas the Kern method and the HTRI Xchanger code overestimate the experimental HTC by about 10% and 21% for the same mass flow rate. Moreover, it was established that, under identical baffle pitch and the count of sealing strips, the HTC estimated by the Bell-D approach may differ by approximately 50% to 180% compared with that calculated by the Kern approach. On the basis of the literature review, the Bell-D approach emerges as the most commonly cited reference model.

### *1.2. Numerical Approaches*

Increasingly, with advances in computing power, numerical methods are being used more often, including those based on commercial software such as Ansys Fluent, as well as tools that offer greater freedom in defining the mathematical description, such as the OpenFOAM software. Mohammadi et al. [23] used Fluent 6.1 code to investigate the impact of baffle cut position on the HTC, defined in a manner similar to the VDI method. It was found that the discrepancy between the Computational Fluid Dynamics (CFD) results and the predictions of the VDI method ranged from −50% to +50% for fluids with Prandtl number $0.7 \leq Pr \leq 206$. It was emphasized that the baffle-to-tube and bypass streams can explain the differences between performance of vertical and horizontal configurations of segmental baffles. Ozden and Tari [24] used Ansys Fluent 6.3 package to examine the influence of baffle cut on the shell-side HTC for a small (seven tubes) STHEx. It was observed that both methods, namely the Bell-D and Kern approaches, underestimate the HTC relative to the numerically calculated values, with maximum deviations of 52% and 35%, respectively. Mellal et al. [25] used COMSOL Multiphysics 5.1 software to examine the influence of baffle spacing and orientation, as well as the number of baffles for an unconventional 38% baffle cut on the HTC of a small—only nine tubes—STHEx. By contrast to the findings of Ozden and Tari [24], the Kern method overpredicts the CFD results with a maximum deviation of about 20.8%. Alperen et al. [22] used the HTRI

https://doi.org/10.3390/en19092114

Xchanger software [13] to calculate the HTC related to the heat exchanger examined experimentally by Zhang et al. [21]. Two additional calculation methods employed were the Kern and Bell-D approaches. The empirical results reported by Zhang et al. served as the benchmark for evaluating the results. The predictions indicate that the Kern approach and HTRI Xchanger code overestimate the HTC across the entire mass flow rate range considered, from about 10% at lower mass flow rates to a maximum of approximately 35% at higher mass flow rates. Conversely, the Bell-D method yields HTC values that are, on average, approximately 11% lower, and its agreement with the experimental results improves with increasing mass flow rate. Gómez et al. [26] proposed an OpenFOAM approach to support the design of STHEx. The calculation findings were compared with empirical results from an existing heat exchanger. The difference between measured and calculated heat transfer flow rate was less than 4.5%. He et al. [27] proposed an OpenFOAM solver for 3D numerical calculations of STHEx using the concept of the porous media model. The CFD predictions were experimentally validated with a maximum deviation of ±13.3%. Arumsari and Ginting [28] also used the HTRI Xchanger software to estimate the shell-side HTC. The HTC calculated using HTRI Xchanger software was approximately 83% higher than that obtained from the Gnielinski correlation [29]. Wang et al. [30] used Ansys Fluent to study the impact of leakage between the shell and the baffle, as well as between the baffle and the tubes, on the flow pattern inside the shell and the thermal performance of the STHEx. The deviation between the simulation results and the predictions from the Bell-D approach was below 7.1%. Adebayo et al. [31] employed Ansys Fluent to examine the performance of a small seven-tube STHEx equipped with segmental baffles and an unconventional 36% baffle cut. It was established that the Kern method underpredicts the HTC, with a maximum deviation of about 25% at the highest mass flow rate. This result may appear questionable, since the Kern method is generally considered applicable only to a 25% baffle cut [32]. Hoang et al. [33] used Ansys Fluent to study the performance of STHEx with new types of baffles. For the validation test of the STHEx with segmental baffles (23% baffle cut, 70 mm spacing), it was revealed that the mean deviation between the numerically predicted HTC and the value estimated by the Bell-D approach, used as the reference, did not exceed 6.7%. Ullah et al. [34] used Aspen EDR to design a reference model of the STHE using industrial data. The literature review indicates that the most frequently used approach for modeling STHEx is based on Ansys Fluent, while studies employing OpenFOAM are comparatively rare.

The above review reveals that no published study has been identified in which the two methods regarded as references approaches, namely the VDI and Bell-D methods, were directly compared. Moreover, no study has been identified in which the HTC was determined using not only the Bell-D and VDI methods, but also the Aspen EDR code, which is founded on the same underlying principles as the other two approaches. The only study in which Aspen EDR was employed as a reference benchmark is that reported in [19]. Another novel contribution is the comparison of the shell-side HTC predictions obtained numerically using two CFD codes, namely OpenFOAM and Ansys Fluent, with those produced by three reference methods for a quasi-ideal STHEx, defined as a configuration without any clearances, but with a bypass stream around the tube bundle. The calculations placed particular emphasis on the influence of the clearances on the resulting HTC.

## 2. Model Heat Exchanger

The geometry and dimensions of the analyzed STHEx were adopted from [35]. Figure 1 illustrates the flow arrangement and the principal dimensions of the tested STHEx, whereas Table 1 summarizes the dimensions and the operating conditions.

https://doi.org/10.3390/en19092114

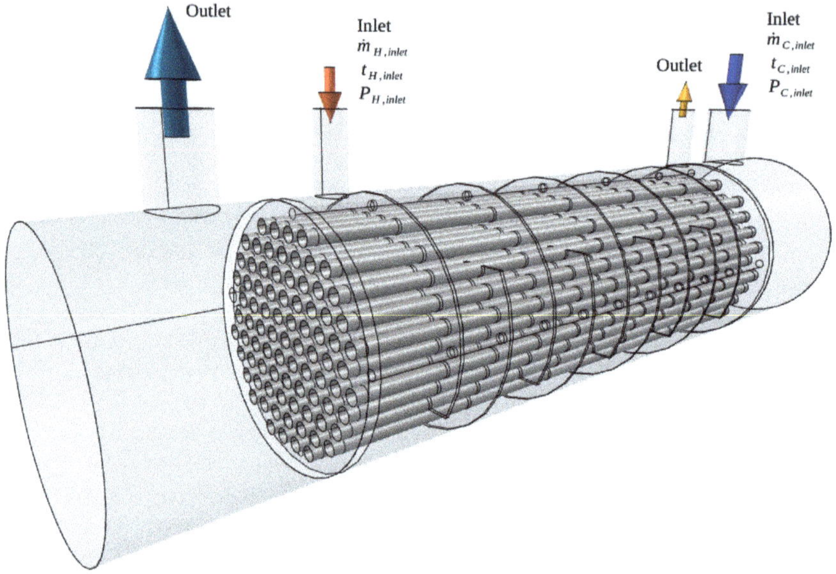


(a)

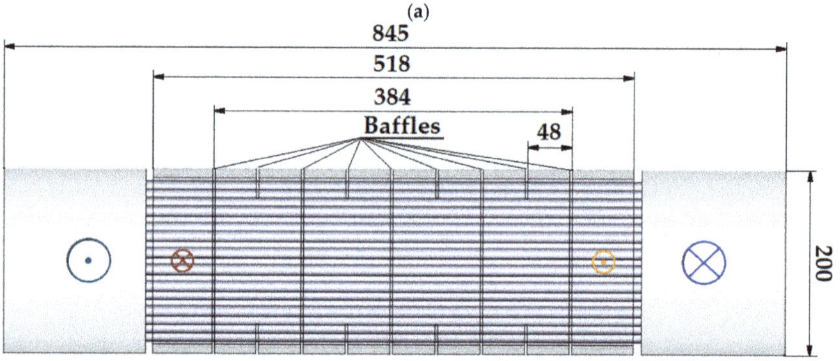


(b)

**Figure 1.** The model STHEx: (**a**) flow arrangement, (**b**) principal dimensions.

**Table 1.** Dimensions and flow parameters of the tested STHEx [35].

| Item | Value/Type |
|---|---|
| Shell inner diameter | 200 mm |
| Length of shell | 518 mm |
| Total tube number | 85 |
| OD/ID tube diameter | 12/10 mm |
| Tube layout | Triangular (30°) |
| Tube pitch | 1.5 |
| Total number of baffles | 9 |
| Baffle cut | 25% |
| Central baffle spacing | 48 mm |
| Inlet and outlet sections | 67 mm |
| Hot water mass flow rate | 1–6 kg/s |
| Cold water mass flow rate | 3 kg/s |
| Mean hot water temperature | 69 °C |

Table 2 illustrates the gap combinations used in the analysis.

https://doi.org/10.3390/en19092114

**Table 2.** The set of the analyzed clearances [35].

| Case | $\delta_d = (d_B - d_o/2)$ | $d_B$ | $\delta_D = (D_s - D_1)/2$ | $D_1$ |
|---|---|---|---|---|
| A | 0.0 | 0.01200 | 0.0 | 0.2000 |
| B1 | 0.00012 | 0.01224 | 0.0 | 0.2000 |
| B2 | 0.00025 | 0.01250 | 0.0 | 0.2000 |
| B3 | 0.00050 | 0.01300 | 0.0 | 0.2000 |
| B4 | 0.00100 | 0.01400 | 0.0 | 0.2000 |
| C1 | 0.0 | 0.00012 | 0.0001 | 0.1998 |
| C2 | 0.0 | 0.00012 | 0.0007 | 0.1986 |
| C3 | 0.0 | 0.00012 | 0.0012 | 0.1976 |
| C4 | 0.0 | 0.00012 | 0.0025 | 0.1950 |
| C5 | 0.0 | 0.00012 | 0.0050 | 0.1900 |
| D1 | 0.00012 | 0.01224 | 0.0001 | 0.1998 |
| D2 | 0.00025 | 0.01250 | 0.0007 | 0.1986 |
| D3 | 0.00050 | 0.0130 | 0.0012 | 0.1976 |
| D4 | 0.00100 | 0.014 | 0.0025 | 0.1950 |

## 3. Materials and Methods

### *3.1. Hand Calculation Methods*

This study employed two widely accepted and commonly used manual calculation methods: the Bell-D and VDI approaches. Both methods allow for input of thermal-flow parameters, such as mass flow rate and temperature, as well as very detailed data concerning the geometry of the STHEx.

#### 3.1.1. The Bell-Delaware Method

Essentially, the Bell-D method is based on determining the HTC for an ideal tube bundle and applying numerous correction factors that account for, among other things, the effects of clearances between the baffle and the shell, as well as between the baffle and the tube bundle. As the procedures used to determine the correction factors are highly complex, their detailed presentation is beyond the scope of this article. The computational milestones of the Bell-D method are presented in [13,36], while the calculations in this study were based on [37], which provides a step-by-step algorithm for calculating the HTC.

#### 3.1.2. The VDI Method

In essence, the VDI method involves determining the HTC for a single tube and then using several correction factors that account for the complex geometry of the STHEx. Similarly to the Bell-D method, the procedures used to predict the HTC are highly intricate; therefore, their discussion is beyond the scope of this paper. The key computational steps of the VDI method are presented in [14,36], while calculations in this study were based on [38], which provides a step-by-step algorithm for predicting the HTC.

### *3.2. Numerical Methods*

#### 3.2.1. OpenFOAM Approach

The system of equations governing flow with heat transfer can be expressed as

$$\frac{\partial \rho}{\partial \tau} + \nabla \cdot (\rho \boldsymbol{u}) = 0 \tag{1}$$

https://doi.org/10.3390/en19092114

$$\frac{\partial}{\partial \tau}(\rho \boldsymbol{u}) + \nabla \cdot (\rho \boldsymbol{u}\boldsymbol{u}) = \rho \boldsymbol{g} - \nabla P + \nabla \cdot (\mu \nabla \boldsymbol{u}) + \nabla \cdot \left(\mu \mathrm{dev}_2 (\nabla \boldsymbol{u})^T\right) \tag{2}$$

$$\frac{\partial}{\partial \tau}(\rho (K+h)) + \nabla \cdot (\rho \boldsymbol{u}(K+h)) = \frac{\partial P}{\partial \tau} + \rho \boldsymbol{g} \cdot \boldsymbol{u} + \nabla \cdot (\lambda \nabla h) \tag{3}$$

In the system in Equations (1)–(3), $\boldsymbol{u}$ is the velocity vector, $\mathrm{dev}_2\ \mathbf{T} = \mathbf{T} - 2/3\ \delta \mathrm{tr}\mathbf{T}$, and $\delta$ is the unity tensor. The kinetic energy is of the form $K = 1/2\ ||\mathbf{u}||^2$.

The boundary conditions, such as inlets and outlets to individual domains, are shown in Figure 1. All other surfaces are defined as no-slip walls. The averaged transport Equations (1)–(3) were solved using the Reynolds-Averaged Simulation approach to turbulence, together with the *k*-$\omega$ SST turbulence model [39]. The open-source software OpenFOAM 2312 [40] was used to solve the discretized equations by means of the Finite Volume Method (FVM). The computational Cartesian mesh consisted of 9.92 million nodes—Figure 2. Mesh details near the walls are presented in Figure 3. The SIMPLE algorithm and the GAMG solver were used to solve the system Equations (1)–(3). More details about the OpenFOAM procedure are discussed in [35,36].

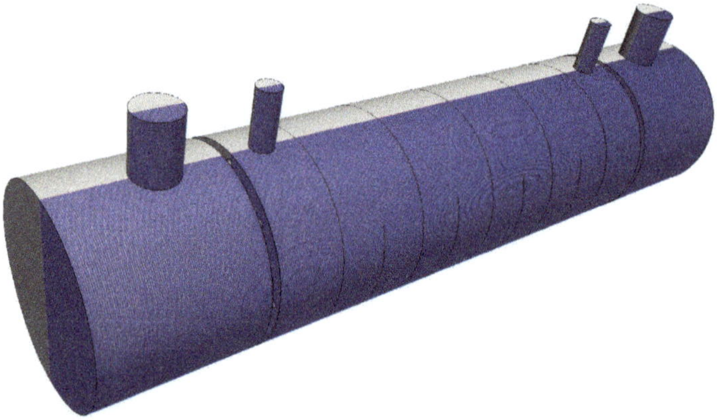

**Figure 2.** Computation domain for OpenFOAM.

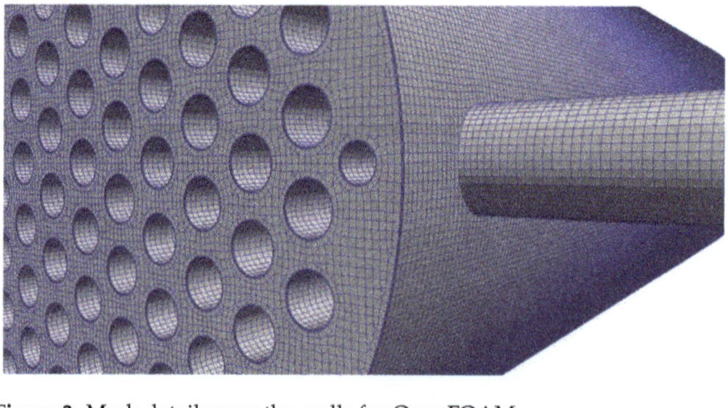

**Figure 3.** Mesh details near the walls for OpenFOAM.

Figure 4 presents the effect of the computational mesh on the outlet temperature of both hot, $t_H$, and cold, $t_C$, fluids. Six cases were examined, spanning from 2.2 million to almost 11 million nodes.

Figure 4 shows that the exit temperature of both hot and cold fluids vary with mesh refinement up to 9.9 million nodes. Further refinement of the mesh does not lead to appreciable changes. A mesh of 9.92 million nodes was therefore selected for subsequent

https://doi.org/10.3390/en19092114

analysis. The mesh sensitivity was investigated for $\dot{m}_H = 2.707$ kg/s, $t_{H,in}$ = 77.82 °C, $\dot{m}_C = 3.173$ kg/s, and $t_{C,in}$ = 9.81 °C.

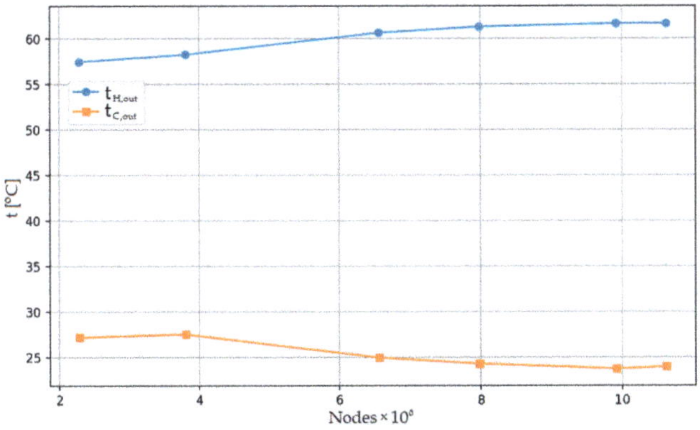


**Figure 4.** Mesh sensitivity for OpenFOAM.

3.2.2. Ansys Fluent Approach

Ansys Fluent 21.1 was utilized for conducting the numerical simulations [41]. For the case considered, the system of conservation equations was resolved in the following form

$$\nabla \cdot \boldsymbol{u} = 0 \tag{4}$$

$$\rho(\boldsymbol{u} \cdot \nabla \boldsymbol{u}) = -\nabla P + \nabla \cdot \boldsymbol{\tau} \tag{5}$$

$$\rho c_p(\boldsymbol{u} \cdot \nabla T) = \nabla \cdot \left(\lambda_{eff} \nabla T\right) \tag{6}$$

where $\boldsymbol{\tau}$ represents the stress tensor including the contributions of molecular and turbulent viscosity of the fluid.

Similarly to the OpenFOAM approach, the boundary conditions, including the inlets and outlets to individual domains, are shown in Figure 1. All other surfaces are defined as no-slip walls. The system of Equations (4)–(6) was solved by means the FVM together with the $k$-$\varepsilon$ turbulence model. The SIMPLE algorithm was applied to solve the pressure–velocity coupling. The computational domain—shown in Figure 5—was generated using the ANSYS Gambit 2.4.6 program (Fluent Inc., Lebanon, NH, USA). Due to geometric symmetry, only half of the domain was simulated. More details about the Ansys Fluent procedure are discussed in [35,36].

Figure 6 shows the effect of mesh resolution on the exit temperature of both hot, $t_{H,out}$ and cold, $t_{C,out}$ fluids. Four cases were considered, with 1.538 million–5.422 million finite volumes.

Figure 6 indicates that increasing the number of finite volumes from 3.071 million to 5.422 million resulted in an insignificant change in the predicted exit temperatures of both fluids. The mesh sensitivity was investigated for $\dot{m}_H = 3.0$ kg/s, $t_{H,in}$ = 75.80 °C, $\dot{m}_C = 3.0$ kg/s, and $t_{C,in}$ = 10.0 °C. Since the computation time for the mesh with 5.422 million finite volumes did not differ substantially from that for the mesh with 3.071 million finite volumes, the mesh with the larger number of finite volumes was adopted for further calculations.

Because the $k$-$\omega$ SST and $k$-$\varepsilon$ models are widely used in numerical heat transfer simulations, it was decided to use both in this study. In the modeling of the STHEx, the $k$-$\varepsilon$ model proves effective as a tool for rapid, stable calculations of the system's overall parameters, whereas the $k$-$\omega$ SST model is more suitable in cases where high local accuracy is crucial—

https://doi.org/10.3390/en19092114

particularly in wall-adjacent regions and in intricate flow inside the shell. Furthermore, the ability of the $k$-$\omega$ SST model to accurately capture near-wall behavior and flow separation phenomena, which predominate inside the shell, has been well documented. This is because the model combines the advantages of the $k$-$\omega$ formulation in near-wall regions and those of the $k$-$\varepsilon$ formulation in the bulk flow. However, the $k$-$\varepsilon$ model is a well-established and reliable approach commonly employed in industrial simulations involving fully developed turbulence, and its computational efficiency and extensive validation studies in the field of heat exchangers make it a suitable reference model for comparative purposes.

**Figure 5.** Computational domain and unstructured tetrahedral mesh developed in Ansys Gambit pre-processor.

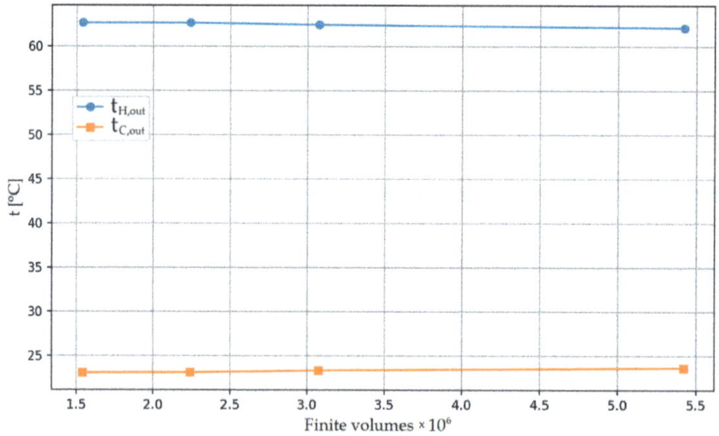


**Figure 6.** Mesh sensitivity for Ansys Fluent.

### 3.3. Industrial Software

The academic license under which the calculations were performed does not permit disclosure of the specific correlations used in Aspen Exchanger Design and Rating (Aspen EDR) V15.0 [42]. The shell-side thermohydraulic analysis is performed based on the concept of flow paths in the STHEx originally proposed by Tinker [5,6] and subsequently refined by Palen and Taborek [43]. Aspen EDR is intended primarily for heat exchanger design. However, as demonstrated by the calculations carried out in this study, it also proves to be a highly effective tool for thermohydraulic analysis.

*3.4. Water Properties*

The properties of hot and cold water were determined using correlations developed on the basis of the database reported in [44]:

- Dynamic viscosity.

$$\mu = 2.2551418931 - 3.3948447154 \times 10^{-2}T + 2.0532435429 \times 10^{-4}T^2 - 6.2294811154 \times 10^{-7}T^3 + 9.4740755112 \times 10^{-10}T^4 - 5.7750953816 \times 10^{-13}T^5 \quad (7)$$

- Thermal conductivity.

$$\lambda = 2.7689426052 \times 10 - 4.1570211623 \times 10^{-1}T + 2.4968651122 \times 10^{-3}T^2 - 7.3889761073 \times 10^{-6}T^3 + 1.0839020278 \times 10^{-8}T^4 - 6.3292341999 \times 10^{-12}T^5 \quad (8)$$

- Density.

$$\rho = -5.8593636805 \times 10^3 + 9.8968557714 \times 10T - 5.7474749375 \times 10^{-1}T^2 + 1.6856053180 \times 10^{-3}T^3 - 2.4989340377 \times 10^{-6}T^4 + 1.4908022754 \times 10^{-9}T^5 \quad (9)$$

- Specific heat.

$$c_p = 1.8561445426 \times 10^5 - 2.7374285462 \times 10^3T + 1.6544683581 \times 10T^2 - 5.0060497615 \times 10^{-2}T^3 + 7.5807970466 \times 10^{-5}T^4 - 4.5941525532 \times 10^{-8}T^5 \quad (10)$$

- Prandtl number.

$$Pr = 1130.643576 - 10.068594T + 0.030127T^2 - 3 \times 10^{-3}T^3 \quad (11)$$

The applicable range of Equations (7)–(11) is 10 °C to 80 °C. Within the considered temperature range, the coefficient of determination ($R^2$) equals 1.00000 for dynamic viscosity, thermal conductivity, and density, whereas for specific heat is 0.99987.

## 4. Results

*4.1. An Idealized Heat Exchanger*

An idealized STHEx is characterized by the absence of leakage between the baffle and the tubes, as well as between the baffle and the shell. However, a bypass stream occurs between the tube bundle and the shell. This fluid stream flows around the outer tubes of the bundle and is therefore not fully utilized for heat transfer. Figure 7 illustrates a comparison of the HTC predicted by all five methods for case A (Table 2).

Figure 7 exhibits that the VDI method predicts the highest HTC, whereas Ansys Fluent yields the lowest value. The discrepancy between the VDI method predictions and the Ansys Fluent results increases as mass flow rate rises, reaching 45% at the highest mass flow rate. The solid black line represents the arithmetic average of the five methods considered, hereafter referred to as the mean. With the exception of the VDI method, the HTCs predicted by the Bell-D method, CFD-OpenFOAM, Ansys Fluent, and Aspen EDR deviate from the mean by no more than ±20%, which indicates close agreement among these methods. The maximum deviation between the values predicted by the VDI method and the mean value was about 34%.

Figure 8 illustrates the HTC predicted by the $k$-$\varepsilon$ and $k$-$\omega$ SST turbulence models for the idealized STHEx corresponding to case A in Table 2.

https://doi.org/10.3390/en19092114

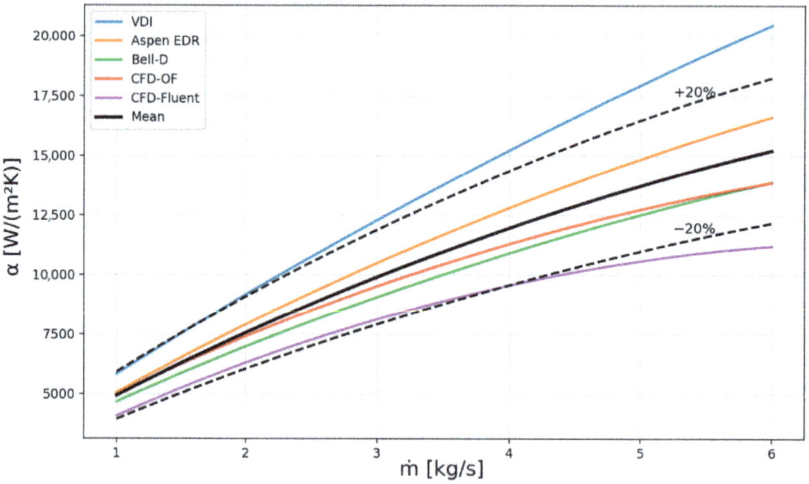


**Figure 7.** Shell-side HTC from considered five methods.

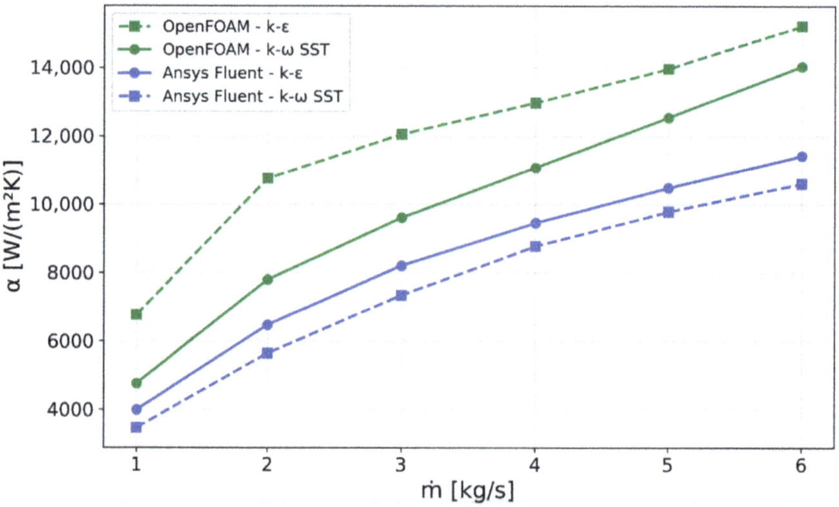


**Figure 8.** Comparison of the HTC for different turbulence models.

Figure 8 exhibits that the $k$-$\varepsilon$ formulation implemented in OpenFOAM predicts the highest HTC, while the $k$-$\omega$ SST formulation incorporated in Ansys Fluent yields the lowest. In both CFD approaches, the $k$-$\varepsilon$ turbulence model predicts a higher HTC than the $k$-$\omega$ SST. In the case of OpenFOAM, the discrepancy between the HTC predictions of the $k$-$\varepsilon$ and the $k$-$\omega$ SST models decreases markedly as the mass flow rate increases, amounting to 28% for the lower mass flow rate and 15% for higher mass flow rate. In Ansys Fluent, the discrepancy between the results using the two turbulence models also decreases as the mass flow rate increases, declining from approximately 13% to 7%. The discrepancy between the OpenFOAM predictions and the Ansys Fluent results for the $k$-$\omega$ SST turbulence model decreases slightly as the mass flow rate increases, dropping from approximately 28% to 21%. Similarly, the discrepancy between the OpenFOAM predictions and the Ansys Fluent results for the $k$-$\varepsilon$ model decreases as the mass flow rate increases, falling from approximately 39% to 27%.

https://doi.org/10.3390/en19092114

Generally, it can be stated that, in Ansys Fluent, the use of either the *k*-ε or *k*-ω SST turbulence model does not significantly affect the predicted HTC, whereas for OpenFOAM, the differences in the resulting HTCs are much greater. Importantly, in the case considered, the results obtained with the *k*-ω SST formulation in OpenFOAM and the *k*-ε formulation in Ansys Fluent were more consistent than those obtained using the same turbulence model across different computational codes. Therefore, different turbulence models were adopted for the subsequent analysis in the two codes: the *k*-ω SST in OpenFOAM and the *k*-ε in Ansys Fluent. It should also be borne in mind that the *k*-ω SST model provides better handling of near-wall flows, adverse pressure gradients, and flow separation. However, the enhanced predictive capabilities of the *k*-ω SST model are achieved at the expense of increased computational time. This is because the model is more sensitive to the quality of the computational mesh and requires a correspondingly denser mesh near the wall.

Figure 9 exhibits a cross-sectional view of the idealized STHEx, whereas Figure 10 illustrates the HTC for the cases without and with sealing strips.

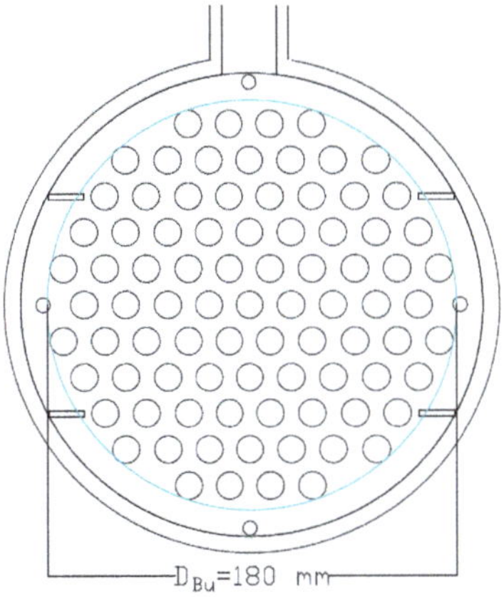


**Figure 9.** The idealized STHEx with two pairs of sealing strips.

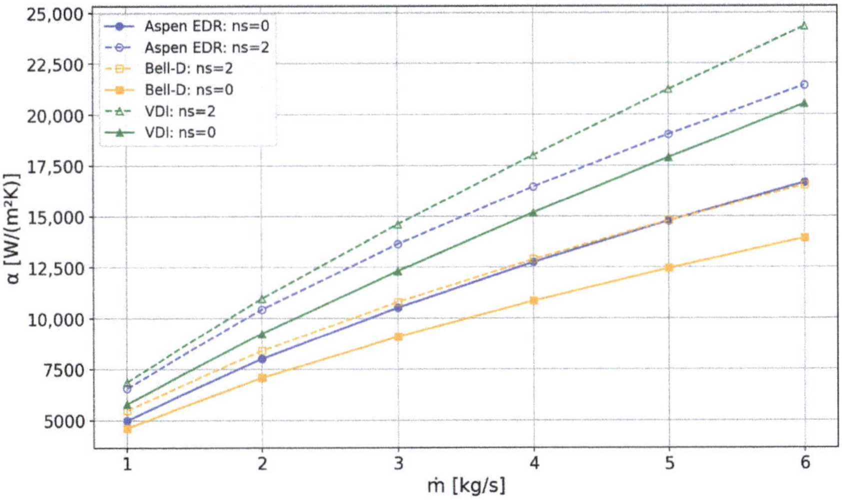


**Figure 10.** Influence of sealing strips on the HTC.

https://doi.org/10.3390/en19092114

Figure 10 exhibits that all three considered methods predict a rise in HTC for the STHEx equipped with sealing strips, as expected, since the sealing strips redirect the bypass stream into cross-flow. While the VDI and Bell-D methods predict virtually the same increase in HTC—18.6% and 18.7%, respectively—regardless of mass flow rate, Aspen EDR yields a rise in HTC of about 30%, which is slightly higher at lower mass flow rates.

Figure 11 illustrates the temperature distribution obtained from numerical simulations carried out in Ansys Fluent for $\dot{m}_H = 3$ kg/s and $\dot{m}_C = 3$ kg/s, respectively, with inlet temperatures of $t_{H,in}$ = 75.8 °C and $t_{C,in}$ = 10 °C, respectively.

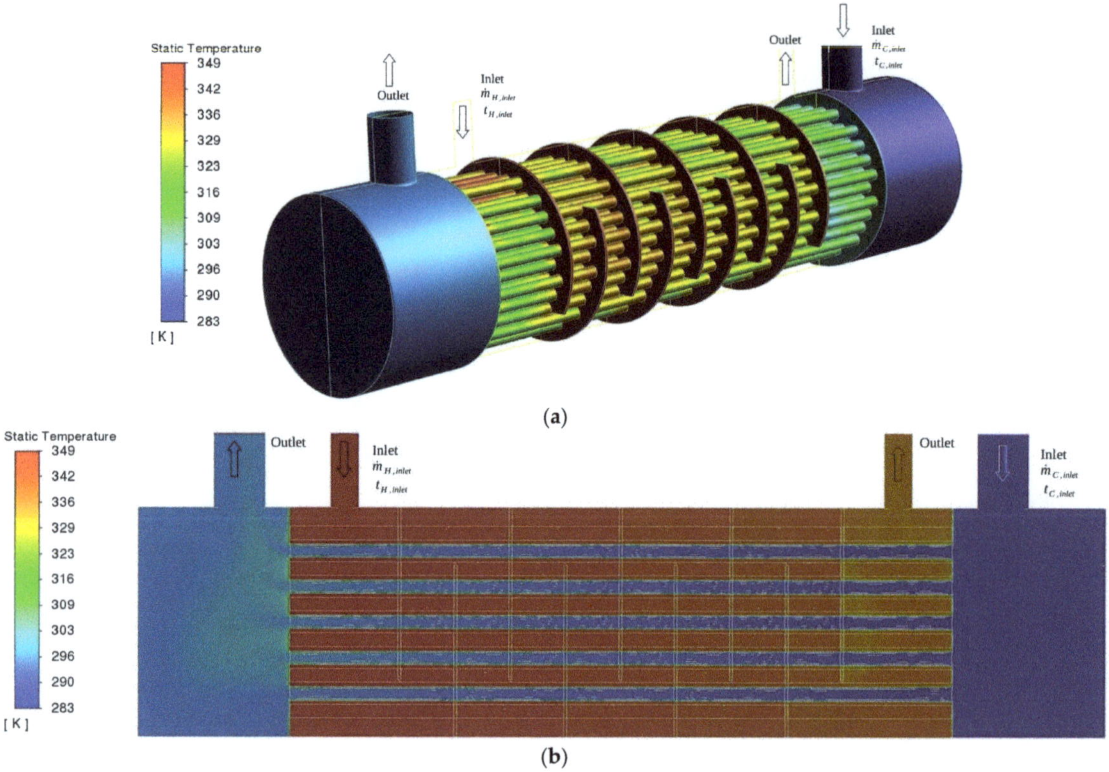


**Figure 11.** Temperature distribution: (**a**) axonometric view; (**b**) longitudinal cross section.

In Figure 11a, the temperature distribution of the baffles is particularly noteworthy. The first three baffles on the hot water inlet side have nearly identical high temperatures, whereas from the fourth baffle onward a clear drop in the baffle temperature can be observed. In Figure 11b, it is characteristic that the hot water temperature drops very sharply already in the first section of the STHEx, which is expected since the hot water comes into contact there with tube surfaces at the lowest temperature, resulting from cold water inlet temperature.

Figure 12 shows the flow map inside the shell of the idealized STHEx for $\dot{m}_H = 3.0$ kg/s.

Figure 12 exhibits that the hot water flow is relatively uniform throughout most of the exchanger, with the exception of the region near the outlet port. The velocity distribution at the outlet port is more uniform than at the inlet port, where a clearly higher velocity is observed along the port axis. Since the case considered involves an idealized STHEx, Figure 12 shows no leakage flows associated with the gaps between the baffle and the tubes or between the baffle and the shell.

Figure 13 shows the velocity distribution in the idealized STHEx.

**Figure 12.** Streamlines of hot water in the idealized STHEx.

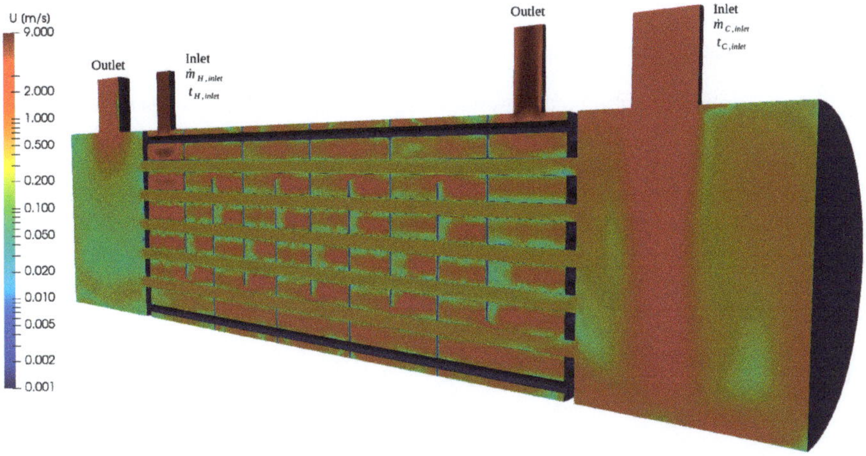


**Figure 13.** Velocity distribution in the idealized STHEx.

Figure 13 shows the velocity distributions in the form of vector magnitude in a longitudinal view of the entire unit. The distribution is presented on a logarithmic scale, which permits better visualization of the full range of velocities compared to a linear scale. Figure 13 shows that there are no stagnation zones on either side of the exchanger, which would be represented by dark blue color. In both computational domains, the peak velocities are located in the entrance and exit ports. Localized velocity drops are also visible where the water flows around the baffles. Figure 13 shows that, although the baffles cause the flow to decelerate, these regions are not true stagnation zones, as the velocity remains approximately 0.2 m/s. This local reduction in velocity is accompanied by a slight change in temperature, as shown in Figure 11b. These velocity variations are only visible on a logarithmic scale; otherwise, the distribution would appear uniform.

*4.2. Heat Exchanger with Leakage*

Figure 14 compares the shell-side HTC from the VDI, Bell-D, and Aspen EDR approaches for cases B1 to B4 (Table 2), i.e., for the instances without leakage between the baffle and the shell and with leakage between the baffle and the tubes.

https://doi.org/10.3390/en19092114

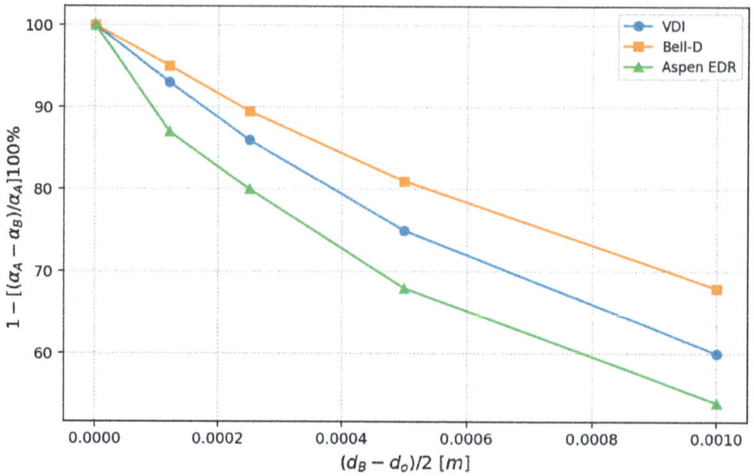


**Figure 14.** Effect of baffle-to-tube leakage on the HTC.

As illustrated in Figure 14, the results show that the HTC decreases as the baffle-to-tube gap increases. This is expected, since baffle-to-tube leakage reduces cross-flow. It should be noted that the decrease in the HTC becomes more pronounced as the clearances increase. At the maximum gap ($\delta_d = 1$ mm), the Bell-D, VDI, and Aspen EDR approaches predict an HTC reduction of about 32%, 40%, and 46%, respectively.

Figure 15 exhibits the variation in the HTC obtained using the VDI, Bell-D, and Aspen EDR approaches with the shell-to-baffle gap for the cases C1–C5 relative to the reference case A (Table 2).

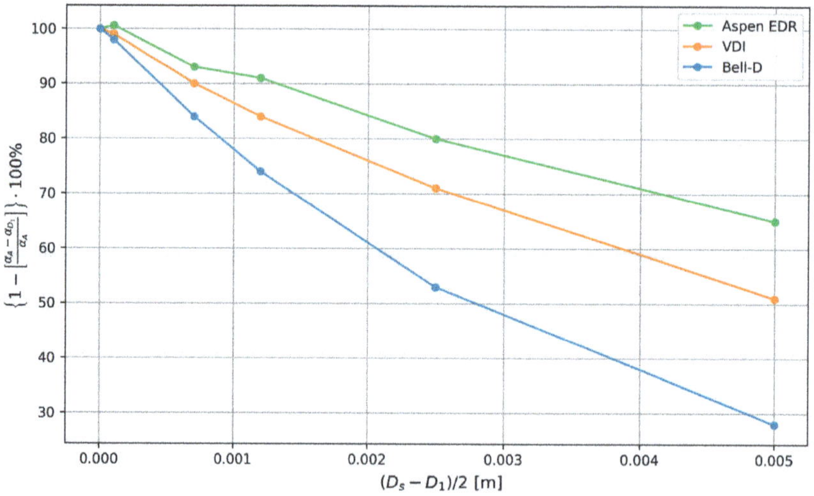


**Figure 15.** Influence of shell-to-baffle leakage on the HTC.

As can be seen from Figure 15, the HTC decreases as the shell-to-baffle gap increases, irrespective of the calculation method applied. This behavior is expected, since a larger shell-to-baffle gap promotes a greater bypass stream, which does not contribute effectively to heat transfer from the tube bundle. The individual methods exhibit different sensitivities to increasing clearances. The most conservative results are obtained with the Bell-D method,

https://doi.org/10.3390/en19092114

which indicates a 72% decrease in HTC for the largest shell-to-baffle gap ($\delta_D = 5$ mm). Aspen EDR is the least sensitive to increasing shell-to-baffle clearances and predicts an HTC decrease of approximately 35% for the largest clearance. The VDI method predicts an HTC reduction of about 50% for the largest clearance.

Figure 16 shows the HTC calculated according to the VDI, Bell-D, and Aspen EDR approaches for cases A, D1, and D4 listed in Table 2, which correspond to conditions involving leakage between the baffle and the tubes, as well as between the baffle and the shell.

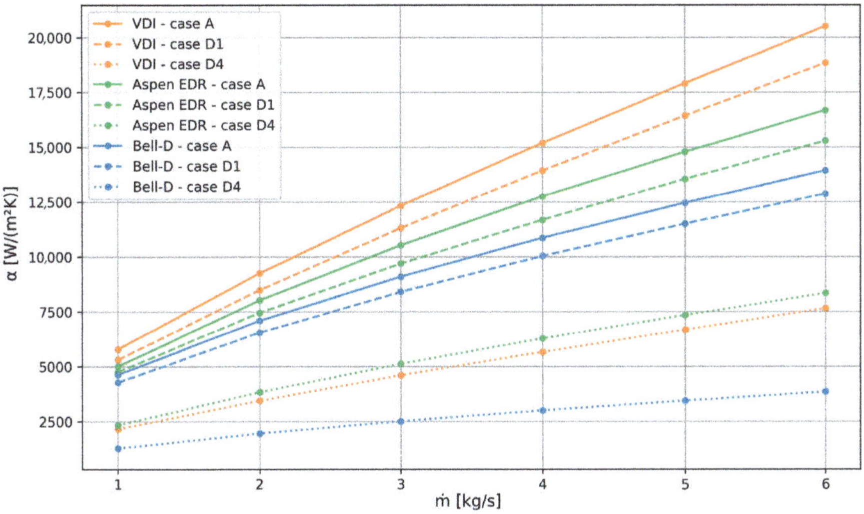


**Figure 16.** Influence of two leakage types on the HTC.

As can be seen from Figure 16, all three considered methods show a rise in HTC with higher mass flow rate, irrespective of the gap size, from case A with no leakage (solid lines) to case D4 with the maximum values of both types of leakage (dotted lines). Importantly, for case D1 (i.e., for the combination of the smallest gaps, represented by the dashed lines), the considered methods exhibit the same HTC tanking as in case A: the VDI approach yields the highest values, Aspen EDR intermediate, while the Bell-D approach presents the lowest. For case D4, the HTC ranking differs: Aspen EDR predicts the highest HTC values, the VDI method yields intermediate values, while the Bell-D approach gives the lowest.

Figure 17 shows the influence of both types of leakage on the HTC, as calculated using the VDI, Bell-D, and Aspen EDR approaches for cases D1 and D4 relative to the reference case A for each method (Table 2).

Figure 17 shows clearly that a reduction in the HTC is predicted by all methods of about 8% for case D1 relative to case A within each respective method. For case D4, however, the situation differs: Aspen EDR, the VDI and Bell-D approaches predict HTC reductions of 55%, 60%, and 65%, respectively, relative to case A.

https://doi.org/10.3390/en19092114

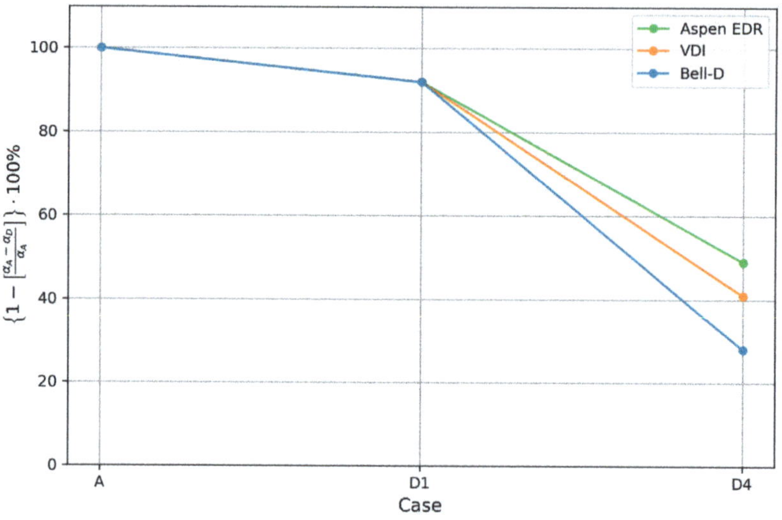


**Figure 17.** Relative impact of both clearance types on shell-side HTC.

## 5. Conclusions

- Two analytical approaches, namely the VDI method and the extended Bell-D method, two software tools, namely OpenFOAM and Ansys Fluent, as well as Aspen EDR were applied to estimate the shell-side HTC in the STHEx with segmental baffles, considering a wide range of gaps between the baffle and the shell and between the baffle and the tubes as well as varying fluid mass flow rates. This comparative analysis, using analytical methods as the reference benchmark, represents a preliminary stage of a comprehensive study that includes an extensive experimental component for validation.
- For the idealized STHEx, with the exception of the VDI method, the remaining four methods predict the HTC within ±20% of the mean, defined as the arithmetic average of the results obtained using the five methods considered. The VDI approach overestimates the HTC by approximately 34%.
- The predictions showed that using two pairs of sealing strips in the idealized STHEx increased the HTC by more than 18% in the VDI and Bell-D approaches, and approximately 30% in Aspen EDR.
- The calculations showed a strong negative effect of leakage between the baffle and the tubes on the HTC. Specifically, when no shell-to-baffle gap occurs, Aspen EDR predicts a 46% reduction in HTC relative to the idealized STHEx.
- Shell-to-baffle clearances have a similarly negative impact on the HTC. For the largest clearance-to-shell radius ratio of 5%, and with no leakage between the baffle and the tubes, the Bell-D method predicts an HTC reduction of up to 72% relative to the idealized STHEx.
- In real STHEx units, both types of leakage occur due to assembly reasons. For the configuration of maximum clearances considered in the calculations, the HTC decreased by at least 50%. The most conservative prediction was obtained with the Bell-Delaware method, which indicated an HTC reduction of about 65% compared with the quasi-ideal STHEx.
- Both CFD approaches used—OpenFOAM and Ansys Fluent—were found to be highly sensitive to the method used to average the fluid temperature from the computed temperature field (Figure 11), which is a necessary step in determining the HTC.

https://doi.org/10.3390/en19092114

Future work will include systematic experimental investigations of STHEx units with different baffle configurations.

**Author Contributions:** Conceptualization, J.T.C.; methodology, J.T.C.; software, J.B., P.D., K.S. and K.T.; validation, J.T.C., J.B., P.D., K.S. and K.T.; formal analysis, J.T.C.; investigation, J.T.C., J.B., P.D., K.S. and K.T.; data curation, J.T.C., J.B., P.D., K.S. and K.T.; writing—original draft preparation, J.T.C.; writing—review and editing, J.T.C., P.D., J.B. and K.T. All authors have read and agreed to the published version of the manuscript.

**Funding:** This research received no external funding.

**Data Availability Statement:** The original contributions presented in this study are included in the article. Further inquiries can be directed to the corresponding author.

**Conflicts of Interest:** The authors declare no conflicts of interest.

## Nomenclature

| | | |
|---|---|---|
| $d_o$ | Tube diameter | [m] |
| $d_B$ | Baffle tube hole diameter | [m] |
| $D_1$ | Baffle diameter | [m] |
| $D_{Bu}$ | Tube bundle diameter | [m] |
| $D_s$ | Shell diameter | [m] |
| g | Acceleration due to gravity | [m/s$^2$] |
| h | Specific enthalpy | [J/kg] |
| $\dot{m}$ | Mass flow rate | [kg/s] |
| P | Pressure | [Pa] |
| T | Temperature | [K] |
| u | Velocity | [m/s] |
| Greek letters | | |
| $\alpha$ | Heat transfer coefficient | [W/(m$^2$K)] |
| $\delta$ | Clearance | [m] |
| $\lambda$ | Thermal conductivity | [W/(mK)] |
| $\rho$ | Density | [kg/m$^3$] |
| $\mu$ | Dynamic viscosity | [Pas] |
| Subscripts | | |
| A | Case A | |
| C | Cold | |
| eff | Effective | |
| H | Hot | |
| in | Inlet | |
| out | Outlet | |
| s | Shell | |

## Abbreviations

| | |
|---|---|
| Bell-D | Bell-Delaware |
| CFD | Computational Fluid Dynamics |
| FVM | Finite Volume Method |
| HTC | Heat Transfer Coefficient |
| ID | Inside Diameter |
| OD | Outside Diameter |
| OF | OpenFOAM |
| STHEx | Shell-and-Tube Heat Exchanger |
| VDI | Verein Deutscher Ingenieure |

https://doi.org/10.3390/en19092114

## References


1. Gupta, J.P. *Heat Exchanger and Pressure Vessel Technology*; Hemisphere Publishing Corporation: Washington, DC, USA, 1986.
2. Shah, R.K.; Sekulić, D.P. *Fundamentals of Heat Exchanger Design*; Wiley & Sons, Inc.: Hoboken, NJ, USA, 2003.
3. Thulukkanam, K. *Heat Exchanger Design Handbook*, 2nd ed.; CRC Press: Boca Raton, FL, USA, 2013.
4. Kakaç, S.; Liu, H.; Pramuanjaroenkij, A. *Heat Exchangers: Selection, Rating, and Thermal Design*, 4th ed.; CRC Press: Boca Raton, FL, USA; London, UK; New York, NY, USA, 2020.
5. Taborek, J. Survey of Shell-Side Flow Correlations. In *Heat Exchanger Design Handbook*; Begell House, Inc.: Danbury, CT, USA, 2025.
6. Tinker, T. *Shell Side Characteristics of Shell and Tube Heat Exchanger, Parts I–III, General Discussion of Heat Transfer*; Institution of Mechanical Engineers: London, UK; ASME: New York, NY, USA, 1951; pp. 84–110.
7. Tinker, T. Shell-Side Characteristics of Shell and Tube Heat Exchanger. *J. Heat Transf.* **1958**, *80*, 36–52.
8. Mukherjee, R. Effectively design shell-and-tube heat exchangers. *Chem. Eng. Prog.* **1998**, *94*, 21–37.
9. Marzouk, S.A.; Abou Al-Sood, M.M.; El-Said, E.M.S.; Younes, M.M.; El-Fakharany, M.K. A comprehensive review of methods of heat transfer enhancement in shell and tube heat exchangers. *J. Therm. Anal. Calorim.* **2023**, *148*, 7539–7578. [CrossRef]
10. Kern, D.Q. *Kern's Process Heat Transfer*, 2nd ed.; Wiley & Sons, Inc.: Hoboken, NJ, USA, 2019.
11. Donohue, D.A. Heat Transfer and Pressure Drop in Heat Exchangers. *Ind. Eng. Chem.* **1949**, *41*, 2499–2511. [CrossRef]
12. Žukauskas, A. Heat transfer from tubes in crossflow. *Adv. Heat Transf.* **1972**, *8*, 93–160.
13. Serth, R.W.; Lestina, T.G. *Process Heat Transfer*; Elsevier: Amsterdam, The Netherlands, 2014.
14. Gaddis, E.S.; Gnielinski, V. G8 Shell-Side Heat Transfer in Baffled Shell-and-Tube Heat Exchangers. In *VDI Heat Atlas*; Springer: Berlin/Heidelberg, Germany, 2010; pp. 731–741.
15. Roetzel, W.; Lee, D. Experimental investigation of leakage in shell-and-tube heat exchangers with segmental baffles. *Int. J. Heat Mass Transf.* **1993**, *36*, 3765–3771.
16. Kücük, H.; Ünverdi, M.; Yılmaz, M.S. Experimental investigation of shell side heat transfer and pressure drop in a mini-channel shell and tube heat exchanger. *Int. J. Heat Mass Transf.* **2019**, *143*, 118493. [CrossRef]
17. Abdelkader, B.A.; Zubair, S.M. The Effect of a Number of Baffles on the Performance of Shell-and-Tube Heat Exchangers. *Heat Transf. Eng.* **2019**, *40*, 39–52. [CrossRef]
18. Wills, M.J.N.; Johnston, D.; Harwell, A.E.R.E. A new and accurate hand calculation method for shell-side pressure drop and flow distribution. In *Proceedings of the 22nd National Heat Transfer Conference, HTD*; ASME: New York, NY, USA, 1984; Volume 36.
19. Abbas, A.; Ismail, T.; Ayub, Z.; Ayub, A.; Cheema, T.A.; Li, W.; Ye, J. Shell Side Single-Phase Experimental Heat Transfer Analysis of a Vertically Oriented Single Segmental Baffle Bundle with Dimpled Tubes. *J. Therm. Sci. Eng. Appl.* **2020**, *12*, 011017. [CrossRef]
20. Di Bono, G.; Corcione, M.; Quintino, A. Systematic comparative analysis of Kern and Bell-Delaware methods for the design of shell-and-tube heat exchangers. *Appl. Therm. Eng.* **2025**, *278*, 127327. [CrossRef]
21. Zhang, J.F.; Li, B.; Huang, W.J.; Lei, Y.G.; He, Y.L.; Tao, W.Q. Experimental performance comparison of shell-side heat transfer for shell-and-tube heat exchangers with middle-overlapped helical baffles and segmental baffles. *Chem. Eng. Sci.* **2009**, *64*, 1643–1653. [CrossRef]
22. Alperen, M.A.; Kayabaşi, E.; Hüseyin Kurt, H. Detailed comparison of the methods used in the heat transfer coefficient and pressure loss calculation of shell side of shell and tube heat exchangers with the experimental results. *Energy Sources Part A Recovery Util. Environ. Eff.* **2023**, *45*, 5661–5680. [CrossRef]
23. Mohammadi, K.; Heidemann, W.; Müller-Steinhagen, H. Numerical Investigation of the Effect of Baffle Orientation on Heat Transfer and Pressure Drop in a Shell and Tube Heat Exchanger with Leakage Flows. *Heat Transf. Eng.* **2009**, *30*, 1123–1135. [CrossRef]
24. Ozden, E.; Tari, I. Shell side CFD analysis of a small shell-and-tube heat exchanger. *Energy Convers. Manag.* **2010**, *51*, 1004–1014. [CrossRef]
25. Mellal, M.; Benzeguir, R.; Sahel, D.; Ameur, H. Hydro-thermal shell-side performance evaluation of a shell and tube heat exchanger under different baffle arrangement and orientation. *Int. J. Therm. Sci.* **2017**, *121*, 138–149. [CrossRef]
26. Gómez, A.; Montañés, C.; Cámara, M.; Cubero, A.; Fueyo, N.; Muñoz, J.M. An OpenFOAM-based model for heat-exchanger design in the Cloud. *Appl. Therm. Eng.* **2018**, *139*, 239–255. [CrossRef]
27. He, S.; Wang, M.; Tian, W.; Qiu, S.; Su, S.G. Development of an OpenFOAM solver for numerical simulations of shell-and-tube heat exchangers based on porous media model. *Appl. Therm. Eng.* **2022**, *210*, 118389. [CrossRef]
28. Arumsari, A.G.; Ginting, P.J. Analysis of Heat Transfer Coefficient of Shell and Tube on Heat Exchanger Using Heat Transfer Research Inch (HTRI) Software. *Formosa J. Sustain. Res.* **2023**, *2*, 1175–1184. [CrossRef]
29. Gnielinski, V. New Equations for Heat and Mass Transfer in Turbulent Pipe and Channel Flow. *Int. Chem. Eng.* **1976**, *16*, 359–368.
30. Wang, Y.; Liu, K.; Chen, W.; Chen, J.; Wang, K.; Zhou, G. Analysis of stream flow and its impact on thermal performance in shell-side of heat exchangers based on flow dead zone. *Appl. Therm. Eng.* **2024**, *249*, 123414. [CrossRef]
31. Adebayo, J.K.; Dare, A.A.; Petinrin, M.O. Performance characteristics of shell and tube heat exchanger using sectoral baffle. *J. Eng. Appl. Sci.* **2025**, *72*, 13. [CrossRef]


https://doi.org/10.3390/en19092114

32. Prithiviraj, M.; Andrews, M.J. Comparison of a Three-Dimensional Numerical Model with Existing Methods for Prediction of Flow in Shell-and-Tube Heat Exchangers. *Heat Transf. Eng.* **1999**, *20*, 15–19.
33. Hoang, H.M.K.; Cao, H.-L.; Pham, P.M.Q.; Hajjar, A.; Nguyen, V.L. A novel shell-and-tube heat exchanger design with alternative inclined baffles. *Case Stud. Therm. Eng.* **2025**, *65*, 105542. [CrossRef]
34. Ullah, Z.; Ahmad, I.; Samad, A.; Saghir, H.; Ahmad, F.; Kano, M.; Caliskan, H.; Caliskan, N.; Hong, H. Artificial intelligence assisted prediction of optimum operating conditions of shell and tube heat exchangers: A grey-box approach. *CAAI Trans. Intell. Technol.* **2025**, *10*, 349–358. [CrossRef]
35. Cieśliński, J.T.; Stasiak, K.; Tesch, K.; Barański, J.; Dąbrowski, P. A Comparative Analysis of Calculation Methods for Flow Resistance in Shell-and-Tube Heat Exchanger with Segmental Baffles. *Energies* **2026**, *19*, 1852. [CrossRef]
36. Cieśliński, J.T.; Barański, J.; Dąbrowski, P.; Fabrykiewicz, M.; Stasiak, K.; Tesch, K. Heat Transfer and Pressure Drop in a Shell-and-Tube Heat Exchanger with Segmental Baffles. *Energies* **2026**, *19*, 1760. [CrossRef]
37. Thome, J. *Wolverine Heat Transfer Engineering Data Book III*; Wolverine Tube, Inc.: Shawnee, OK, USA, 2010.
38. Gnielinski, V.; Gaddis, E.S. Berechnung des mittleren Wärmeübergangskoeffizienten im Aussenraum von Rohrbündelwärmeaustauschern mit Segment-Umlenkblechen. *Verfahr.-Tech.* **1978**, *12*, 211–217.
39. Menter, F. Two-equations eddy-viscosity turbulence models for engineering applications. *AIAA-J.* **1994**, *32*, 1598–1605. [CrossRef] [PubMed]
40. OpenFOAM. Available online: https://www.openfoam.com (accessed on 11 June 2024).
41. *ANSYS Fluent*, version 21.1. Fluent 21.1 user's guide. ANSYS Inc.: Canonsburg, PA, USA, 2018.
42. AspenTech. *Aspen Exchanger Design and Rating, V15.0*; AspenTech Inc.: Bedford, MA, USA, 2023.
43. Palen, J.W.; Taborek, J. Solution of Shellside Flow Pressure Drop and Heat Transfer by Stream Analysis Method. *Chem. Eng. Prog. Symp. Ser.* **1969**, *65*, 53–63.
44. Lemmon, E.W.; Huber, M.L.; McLinden, M.O. *NIST Standard Reference Database 23, Reference Fluid Thermodynamic and Transport Properties (REFPROP), Version 9.0*; National Institute of Standards and Technology: Gaithersburg, MD, USA, 2010.




https://doi.org/10.3390/en19092114

*energies*

MDPI

*Article*

# SHAP-Based Insights into Environmental and Economic Performance of a Shower Heat Exchanger Under Unbalanced Flow Conditions: A Feasibility Study

**Sabina Kordana-Obuch * and Mariusz Starzec**

Department of Infrastructure and Water Management, Rzeszow University of Technology, al. Powstańców Warszawy 6, 35-029 Rzeszow, Poland; mstarzec1990@prz.edu.pl
* Correspondence: sk@prz.edu.pl

**Abstract**

Heat recovery from greywater is one solution for improving the energy efficiency of buildings and reducing greenhouse gas emissions. Particular attention is paid to systems utilizing heat from shower water, which, due to its high temperature and regularity, represents a promising energy source. However, the interplay of parameters determining the financial and environmental effectiveness of such a solution has not yet been fully explored. Therefore, the aim of this paper was to identify key variables influencing the feasibility of using a shower heat exchanger operating under unbalanced flow conditions and to assess the consistency between financial and environmental effects. The analyzed net present values ranged from −€1381 to €52,168. Greenhouse gas emission reduction values ranged between 61 $kgCO_2e$ and 37,207 $kgCO_2e$. The analysis was conducted using predictive modeling and the SHAP (SHapley Additive exPlanations) method, which allows for the interpretation of the impact of individual variables on the forecasted net present value and potential greenhouse gas emission reduction. A global analysis was carried out to determine the relative importance of variables, as well as a local analysis for selected cases. The results showed that operational variables related to shower use, particularly shower length and mixed water flow rate, significantly influenced the prediction results of both models. In the case of emission reduction, greenhouse gas emission intensity and its change over time also had a significant impact, whilst the financial effects were determined by the energy price from the perspective of the subsequent years of the system's operation.

**Keywords:** European Union; greenhouse gas emission reduction; heat recovery; net present value; residential building; SHAP; vertical heat exchanger; XGBoost

## 1. Introduction

Buildings, including residential buildings, are responsible for a significant portion of final energy consumption and greenhouse gas emissions. It is worth noting that one of the most energy-intensive processes in the construction sector is the preparation of domestic hot water (DHW) [1], the heating of which generates additional high costs. Unlike space heating, where energy demand systematically decreases with improving energy standards in buildings [2], the share of energy allocated to DHW is becoming an increasingly dominant component of buildings' energy balances. In highly energy-efficient buildings, it can even significantly exceed the demand for other purposes. For example, research by Hall et al. [3] has shown that water-related energy use can in some cases reach as much as 7 kWh/person/day, accounting for up to 50% of energy consumption in households. As a result, technologies based on renewable energy sources [4] and enabling energy consumption reduction and recovery at the end-user level [5–7] are becoming increasingly desirable. Furthermore, growing energy consumption in cities increases the need to implement systems that balance energy efficiency with thermal comfort [8]. This, in turn, implies an increased demand for detailed analyses enabling a reliable assessment of the energy, environmental, and economic effects of the proposed solutions.

One solution for reducing energy consumption for DHW is a shower heat exchanger (SHE) [9]. During normal operation of the installation, the thermal energy contained in greywater is permanently lost, and its energy potential remains unused. By using SHEs, the thermal energy contained in shower water is transferred to the water, which is then used during the same shower [10]. This is one of the few examples of measures that do not require changing user habits and do not affect the comfort of using the shower system. Previous studies on assessing the effectiveness of SHEs, determining their financial efficiency, or reducing $CO_2$ emission [11–13] have focused mainly on the use of aggregated, simplified indicators and lacked a deeper analysis of the impact of individual operating and system parameters on the analysis results. At the same time, most studies only covered balanced flow conditions [14]. This means that even with advanced computational tools [15], the scope of analytical conclusions and their practical applicability is clearly limited. In installation practice, systems operating under unbalanced flow conditions are present [16]. In such a situation, the water flow rate through the SHE is significantly lower than the drain water flow rate, which is a consequence of supplying preheated water to only one of the devices (the shower mixing valve (SMV) or the DHW heater), while the other is supplied with cold water. Despite the limited possibilities of thermal energy recovery compared to the balanced flow variant, such solutions are used to reduce the complexity of the hydraulic system or to adapt the installation to existing DHW preparation systems in which parallel supply of both devices is impossible or technically unjustified.

Despite the growing number of publications on SHEs, their operation under unbalanced flow conditions remains relatively poorly understood, and their impact on key environmental and financial indicators is often marginalized, simplified, and sometimes even overlooked. Relying on balanced flow conditions, which are often impossible to apply in installation practice, leads to systematic overestimation of the achieved energy and economic effects. Furthermore, commonly used analysis methods rarely enable clear identification of the parameters determining the environmental and financial benefits achieved, limiting their usefulness both in the design phase of shower heat recovery systems and in implementing measures supporting the realization of this type of technology.

In response to these limitations, this paper adopts an approach aimed at systematic and quantitative decomposition of the impact of input variables on environmental and financial effects in conditions close to real-world conditions, with the main goal being to analyze the impact of variables determining SHE operation under unbalanced flow conditions on the projected greenhouse gas emission reduction ($R_{GHG}$) and the financial efficiency of the investment expressed as net present value ($NPV$). The novelty of the study lies in the parallel, coherent approach to both dimensions based on a partially common set of input variables and in the identification of potential discrepancies between financial and environmental optimality, which is poorly recognized in the existing literature. The analysis was conducted for a system configuration that assumed supplying preheated water exclusively to the shower mixing valve. This choice was dictated by its widespread use in installation practice and limited interference with the existing DHW heating system. The vertical heat exchanger was chosen as the research object due to its dominant position among solutions used in installation practice and its more favorable heat transfer

https://doi.org/10.3390/en19081845

parameters compared to horizontal SHEs [17]. Furthermore, analyses of this shower heat recovery system configuration available in the literature focus primarily on horizontal heat exchangers [18], while vertical units, despite their widespread use in practice, remain relatively underexplored.

Interpretable machine-learning methods, along with SHAP analysis, were used in the analysis, which enabled transparent identification and quantification of the impact of individual input variables on the analysis results. The use of separate models for both dependent variables ($R_{GHG}$, $NPV$) allowed for the identification of key parameters determining their variability, which made it possible to assess the consistency between environmental and financial effects and to identify potential differences in the mechanisms shaping these effects. This approach represents a significant extension of the current state of knowledge, providing tools to support both design optimization and the development of policies supporting the implementation of heat recovery technologies in real-world conditions.

The research described here is a continuation of analyses conducted to assess the energy efficiency of vertical shower heat exchangers under various flow conditions, the results of which were described in [19]. The scope of this paper includes:

- Evaluation of the financial efficiency of using a vertical shower heat exchanger in a configuration where preheated water is supplied exclusively to the SMV.
- Estimation of the potential greenhouse gas emission reduction under similar operating conditions of the SHE.
- Development of eXtreme Gradient Boosting (XGBoost) models to predict $NPV$ and $R_{GHG}$.
- Evaluation of the impact of individual input variables on the expected $NPV$ and projected greenhouse gas emission reduction.

## 2. Materials and Methods

### *2.1. Environmental and Financial Analysis*

#### 2.1.1. Evaluation of the Performance of a Shower Heat Exchanger Under Unbalanced Flow Conditions

The analysis included an evaluation of the performance of a vertical "pipe-in-pipe" SHE, 1.68 m long and 50 mm internal diameter. A detailed description of the unit, along with a description of the measuring devices, can be found in [19]. Vertical shower heat exchangers are commercially available units used in residential buildings. Due to the need to meet specific installation requirements, primarily the need to install them in a vertical section of the sewage system, their application is easier in newly designed buildings, where adequate installation space can be provided for this type of unit. Failure to incorporate a vertical SHE into the building design may require a more complex retrofit of the existing facility, particularly in systems supplying preheated water to a DHW heater. In some cases, installing a vertical heat exchanger may even prove impossible.

Based on the input dataset, marked in green in Figure 1, a series of experimental studies were conducted to determine the preheated water temperature ($T_{pw}$) (Figure 2). The studies were conducted for a system configuration that supplied preheated water exclusively to the shower mixing valve (Figure 3). In practice, this type of configuration is often used in facilities where interfering with the existing domestic hot water system is difficult or would require significant capital expenditure. It should be noted that the heat recovery system configuration under consideration involves minimal interference with the existing internal water and sewage system and is less complex hydraulically than systems supplying a DHW heater. This makes the solution easier to implement in retrofit projects for existing buildings. In existing DHW systems, the possibility of interfering with the hydraulic system is usually limited, as a result of which preheated water can

https://doi.org/10.3390/en19081845

only be supplied to selected points of the internal system. Due to the location of the SHE below the shower, this is most often an SMV. This arrangement of preheated water pipes results in unbalanced flow conditions in the heat exchanger. It should also be noted that a configuration ensuring balanced flow requires more complex modifications to the water and sewage system, which involves the purchase of additional pipes, fittings, and often interventions in the building structure. This, in turn, significantly increases investment outlays. Therefore, solutions characterized by unbalanced flow conditions represent a practical compromise between energy efficiency and the feasibility of a drain water heat recovery system. This justifies their analysis as a scenario representative of a wide range of real-world applications. Under such conditions, the heated water flow rate is significantly lower than the drain water flow rate, which results from the fact that only a portion of the water used for showering flows through the heat exchanger. As a result, the achieved temperature effectiveness of the unit is higher than when the SHE operates under balanced flow conditions. However, the heat transfer rate is reduced in such conditions [19], which, in turn, affects the achieved financial and environmental effects.

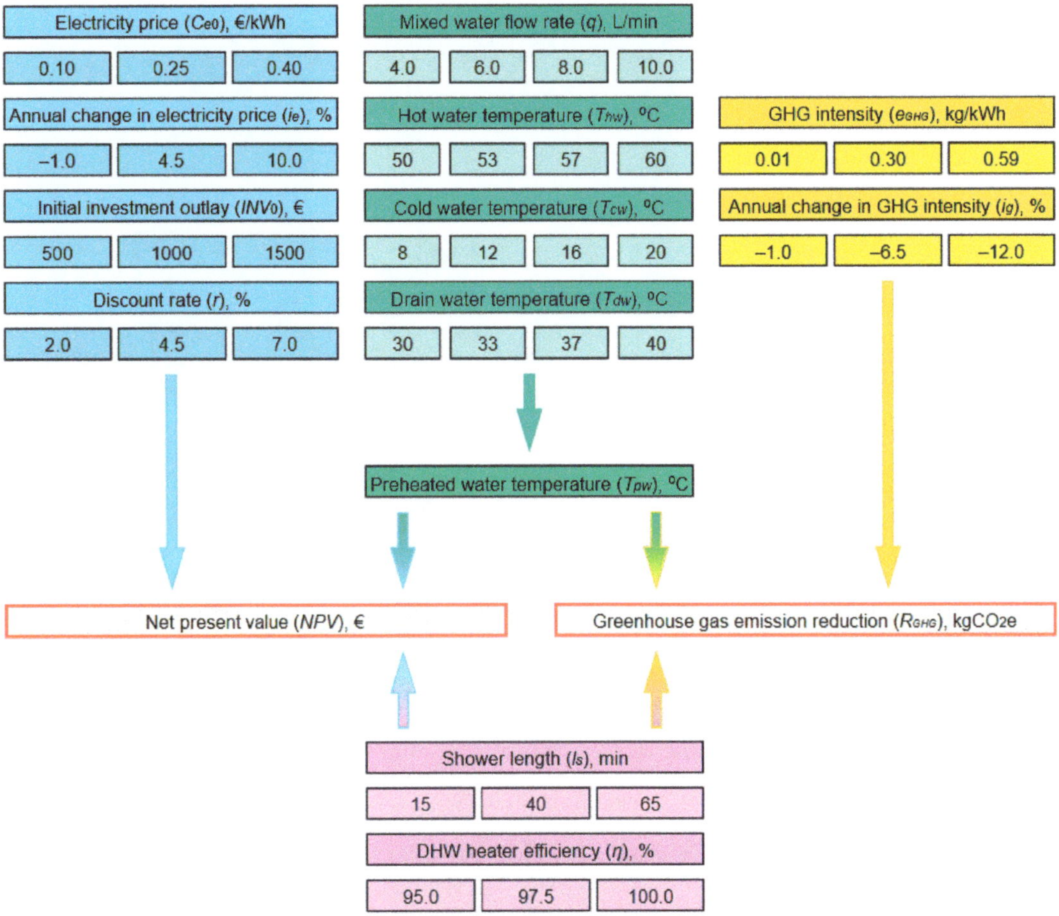


**Figure 1.** Input parameters with their assumed values and their corresponding output parameters (GHG—greenhouse gas; DHW—domestic hot water).

https://doi.org/10.3390/en19081845

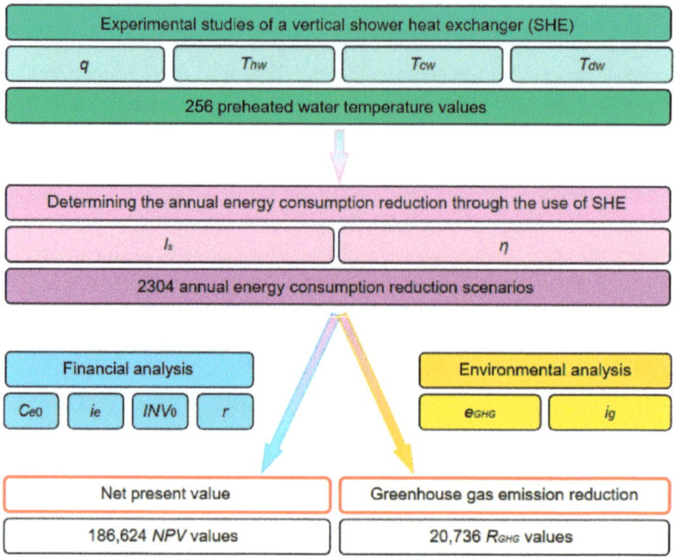


**Figure 2.** Research stages. Colors and designations are consistent with Figure 1.

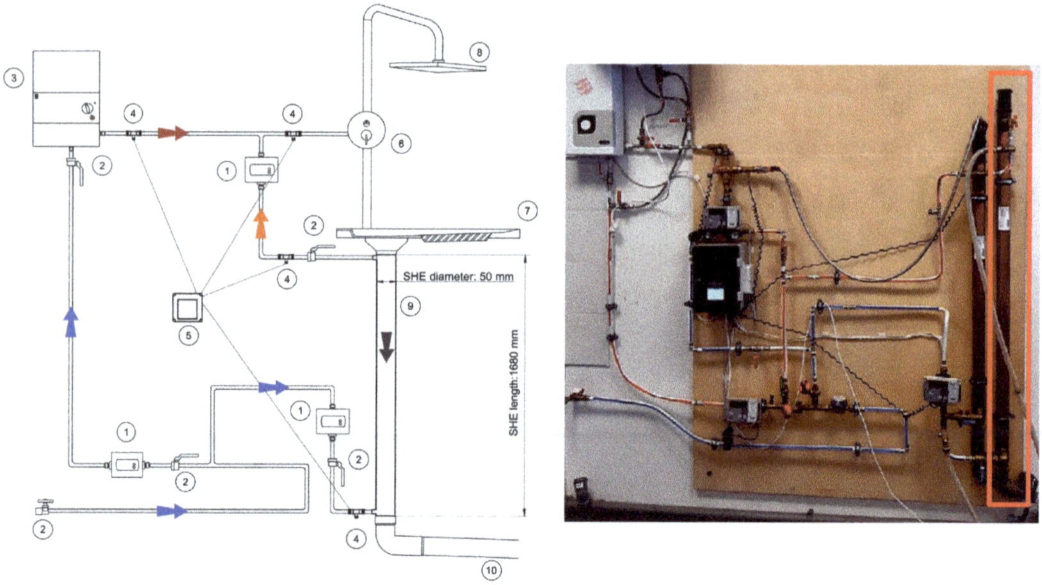


(**a**) (**b**)

**Figure 3.** Shower drain water heat recovery system: (**a**) System configuration diagram; (**b**) Research stand with a heat exchanger installed; the red box indicates the investigated unit (1—flowmeter; 2—cut-off valve; 3—domestic water heater; 4—temperature sensors; 5—data recorder; 6—shower mixing valve; 7—shower tray; 8—shower head; 9—shower heat exchanger; 10—drain pipe; SHE—shower heat exchanger).

For the assumed range of input variables, the water flow rate through the vertical SHE constituted several tens of percent of the drain water flow rate. Its value resulted from the assumed media temperatures in the individual sections of the system and the

https://doi.org/10.3390/en19081845

mixed water flow rate ($q$). The tests were conducted over a 12-month period to reflect the annual variability of cold water temperature. The remaining parameters were adjusted accordingly to represent different combinations of input variables. Therefore, while the mixed water flow rate from the shower head was treated as an independent variable, the ratio of water flow rate to drain water flow rate in the SHE was not analyzed as a separate independent parameter. The unbalanced flow conditions resulted from the adopted thermal and hydraulic assumptions.

A total of 256 preheated water temperature values ($T_{pw}$) were obtained from experiments conducted across different combinations of input parameters. This, along with the cold water ($T_{cw}$) and hot water ($T_{hw}$) temperatures, as well as the mixed water flow rate ($q$), formed the basis for further calculations, which were performed over a fifteen-year time horizon. Annual energy savings ($P_y$), resulting from heat recovery from shower water, were calculated as the difference between the energy consumption in a building without a heat exchanger and one with the SHE installed. For the purposes of the analysis, it was assumed that water was heated using an electric DHW heater, as this is the most advantageous option, as noted by Ali et al. [20]. The domestic hot water flow rate to the shower mixing valve was determined based on the heat balance equation. For both the DHW heater efficiency ($\eta$) and the total daily shower length ($l_s$), three representative values of these parameters were assumed (marked in pink in Figure 1), which allowed for the consideration of various building operating scenarios. This approach enabled the mapping of diverse household shower usage conditions and the resulting annual energy consumption for DHW heating. As a result, the analysis included 2304 scenarios for reducing energy consumption for shower water heating. For example, Figure 4 shows the distribution of 256 annual energy consumption reduction values obtained assuming the most unfavorable scenario in terms of shower length, i.e., for $l_s$ = 15 min. The results correspond to a DHW heater efficiency of 95%.

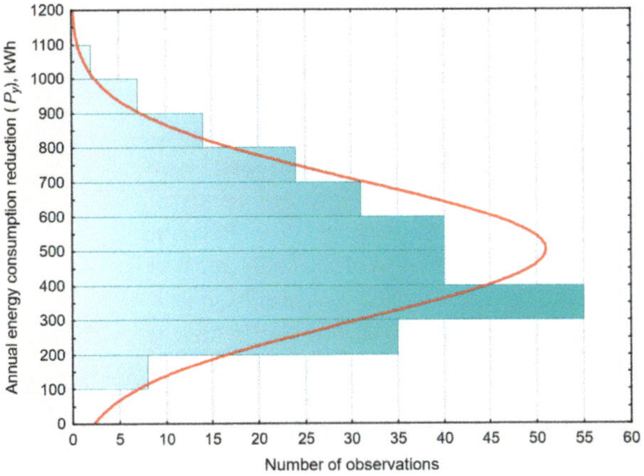


(a)

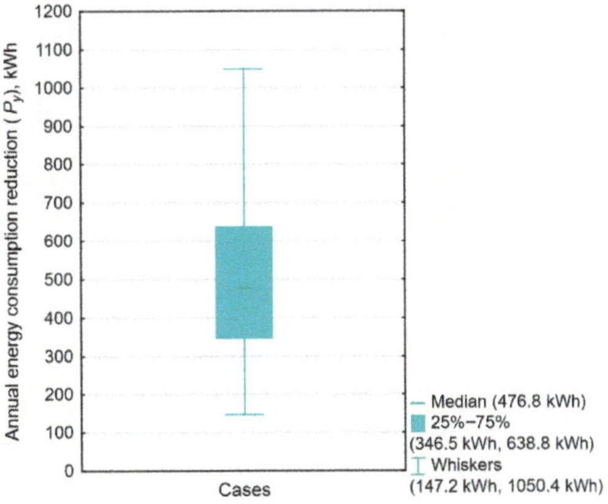


(b)

**Figure 4.** Distribution of annual energy savings resulting from the installation of the SHE ($l_s$ = 15 min, $\eta$ = 95%): (**a**) histogram; (**b**) box-plot.

### 2.1.2. Net Present Value

The financial efficiency of the project involving the installation of the SHE in an existing residential building was determined based on net present value ($NPV$). This indicator can

https://doi.org/10.3390/en19081845

be determined as the difference between the discounted cash flows generated by the project and the investment outlay incurred, as described by Equation (1).

$$NPV = \sum_{t=1}^{n} \frac{P_y \cdot C_{e0} \cdot (1 + i_e)^t}{(1 + r)^t} - INV_0, \tag{1}$$

where $NPV$ is the net present value, €; $P_y$ is the annual energy saving resulting from the installation of the SHE, kWh; $C_{e0}$ is the initial electricity price, € per kWh; $i_e$ is the annual change in electricity price; $r$ is the discount rate; $t$ is the year of operation of the SHE; $n$ is the service life of the SHE; $INV_0$ is the initial investment outlay, €.

In the case under consideration, positive cash flows represented savings resulting from reduced energy consumption for DHW heating. This indicator is a commonly used metric in investment profitability assessments, as it accounts for the time value of money and enables a clear determination of whether an investment will be financially beneficial. Therefore, $NPV$ is widely used in analyses of the implementation of pro-ecological technologies in residential buildings [21], including those aimed at improving energy efficiency [22].

In addition to the system's technical parameters, such as flow rate and media temperatures, system operating time, and DHW heater efficiency, the analysis also considered additional variables that significantly impact the financial benefits achieved. The assumed values for these variables are highlighted in blue in Figure 1. In each case, three values of the initial electricity price ($C_{e0}$), the annual change in electricity price ($i_e$), the initial investment outlay ($INV_0$), and the discount rate ($r$) were considered. As a result, the financial analysis was based on 186,624 cases characterized by varying energy demand for DHW heating and diverse economic assumptions.

The initial electricity price ($C_{e0}$) was assumed based on Eurostat data for European Union (EU) countries [23]. These data covered households with annual energy consumption between 2500 and 5000 kWh. According to these data [23], the highest electricity price in the first half of 2025 was recorded in Germany ($C_e$ = €0.3835 per kWh), and the lowest in Hungary ($C_e$ = €0.1040 per kWh). Based on data from recent years [24], the dynamics of annual changes in electricity prices in individual countries were additionally analyzed, and annual percentage changes in electricity prices ($i_e$) were assumed. Changes in the shorter term were not analyzed to avoid seasonal effects. Short-term, sharp price fluctuations occurring in some countries in individual years were also omitted. Such jumps are incidental and, therefore, do not provide a reliable basis for multi-year forecasting. Initial investment outlays ($INV_0$), including the purchase of the SHE and its installation, were estimated considering a wide range of required installation work. Notably, the heat recovery system configuration under consideration is characterized by lower hydraulic complexity than configurations involving preheated water supply to a DHW heater, thereby potentially leading to lower investment outlays than more complex systems. The adopted range of $INV_0$ values allowed for cost variability resulting from differences in installation conditions, location, and market prices of equipment and labor. The adopted discount rate ($r$) values reflect the variation in this parameter across EU countries [25]. The analysis adopted an approach based on discount rates interpreted as real rates. Values close to the lowest $r$ levels observed in some EU countries, reflecting the overall downward trend, were considered, as well as higher discount rates corresponding to more conservative investor assumptions. This approach allowed for the incorporation of economic uncertainty into the analysis and increased the universality of the obtained results, regardless of the adopted scenario of changes in energy prices.

https://doi.org/10.3390/en19081845

2.1.3. Greenhouse Gas Emission Reduction

The reduction in greenhouse gas emission was estimated based on the projected reduction in energy consumption resulting from the use of SHE ($P_y$) and GHG emission intensity of electricity generation ($e_{GHG}$), according to Equation (2).

$$R_{GHG} = \sum_{t=1}^{n} P_y \cdot e_{GHG} \cdot \left(1 + i_g\right)^t, \quad (2)$$

where $R_{GHG}$ is the greenhouse gas emission reduction, $kgCO_2e$; $e_{GHG}$ is the GHG emission intensity of electricity generation, $kgCO_2e/kWh$, and $i_g$ is the annual change in the value of $e_{GHG}$ (the study included only negative values).

For both $e_{GHG}$ and $i_g$, three values of the input variables were considered, resulting in the analysis of greenhouse gas emission reduction based on 20,736 model cases. As with the initial electricity prices ($C_{e0}$), the initial $e_{GHG}$ value was assumed based on current data for individual EU countries [26]. According to the European Environment Agency [26], in 2024, the lowest $e_{GHG}$ value was in Sweden ($e_{GHG}$ = 0.007 $kgCO_2e/kWh$), while the highest was in Estonia ($e_{GHG}$ = 0.585 $kgCO_2e/kWh$). Additionally, the dynamics of changes in this parameter in recent years were analyzed [26], and based on this, annual changes in greenhouse gas emission intensity of electricity generation ($i_g$) were assumed. Although increases in $e_{GHG}$ values have been observed in some EU countries in recent years, a long-term downward trend was assumed for this variable. For this reason, all assumed $i_g$ values were negative (Figure 1). This is due to the ongoing decarbonization of the energy sector, the increasing share of renewable sources in electricity production, and tightening climate policy. Short-term increases in emissions are usually temporary and result from extraordinary circumstances. Assuming a long-term decreasing trend enabled a more realistic assessment of the future environmental impact of vertical SHEs.

## 2.2. *Prediction Models and Their Interpretation*

2.2.1. eXtreme Gradient Boosting

A variety of machine-learning models are used in contemporary research [27,28]. This also applies to environmental and economic analyses. Popular models include eXtreme Gradient Boosting (XGBoost), which is successfully used both to predict $CO_2$ emissions [29] and in financial risk analyses [30]. It is also used to predict the thermal-hydraulic characteristics of heat exchangers [31], including SHEs [32]. Previous studies by the authors [19] involving the use of various predictive models to assess SHE performance also showed that the use of XGBoost models allows for obtaining more favorable model fitting metric values than other machine-learning models. This analysis was conducted using XGBoost version 2.1.2.

XGBoost is an ensemble learning algorithm based on gradient boosting of decision trees, in which successive trees are constructed sequentially to minimize a predefined loss function. Each subsequent tree is trained to reduce the residual errors of the previous model, allowing the algorithm to capture complex, nonlinear relationships between variables. In addition to the loss function, XGBoost incorporates a regularization term that penalizes model complexity, improving generalization and reducing the risk of overfitting. As a result, the final prediction is obtained as a sum of contributions from all individual trees. The XGBoost method allows for modeling complex, nonlinear relationships and interactions between variables, making it an appropriate tool for analyzing economic and environmental processes. Its additional advantage is high predictive accuracy and robustness to varying input data scales, which is crucial when considering both hydraulic and technical parameters, as well as economic and environmental ones [33]. Two predictive models were developed as part of the study, enabling the parallel assessment of financial

https://doi.org/10.3390/en19081845

and environmental effects. The first, based on ten input variables, predicted the net present value of a project involving the installation of an SHE operating under unbalanced flow conditions. This enabled the identification of key factors influencing its financial viability. The second model was used to forecast the cumulative reduction in greenhouse gas emissions over a 15-year time horizon, allowing for the assessment of long-term environmental effects resulting from the SHE installation. In this case, the model was based on eight input variables.

The SHAP (SHapley Additive exPlanations) method was used to interpret the models and determine the impact of individual independent variables on their predictions [34]. This method enabled a quantitative assessment of the significance of individual variables and the identification of the direction and strength of their impact on model results. This led to a better understanding of the mechanisms shaping both the project's financial efficiency and its environmental impact.

#### 2.2.2. SHAP Analysis

The SHAP (SHapley Additive exPlanations) method is based on Shapley values derived from cooperative game theory. For a given observation, the model prediction can be expressed as:

$$f(x) = \varphi_0 + \sum_i \varphi_i, \tag{3}$$

where $\varphi_0$ is the base value, and $\varphi_i$ represents the contribution of the $i$-th feature.

This formulation enables an additive decomposition of the prediction into the effects of individual input variables and ensures a comparable interpretation of variable effects regardless of their scale [35]. SHAP analysis assigns each variable a contribution to the prediction result for individual observations and globally for the entire model [35]. In the local analysis, the distribution of SHAP values for individual predictions is examined to identify the direction and strength of the influence of individual features. This paper presents the results of such an analysis for three observations representative of each model. In each case, a positive SHAP value indicates an increase in the model's prediction value, while a negative value results in a decrease. The global analysis, on the other hand, is based on the mean absolute SHAP values for each variable, which allows for determining its overall importance. The results of these analyses for both models are presented in the paper as plots.

## 3. Results

### 3.1. Financial Analysis

#### 3.1.1. Results of the Net Present Value Calculation

Analysis of 186,624 combinations of input variables showed that for the studied range, their *NPVs* remained between −€1381 and €52,168 (Figure 5). Negative values were obtained for approximately 24,000 cases, which constituted less than 13% of all combinations considered. The mean *NPV* reached €4317, and the median value was €2445, which indicates a right-sided asymmetry of the results distribution. The skewness value in this case is 2.30. Most of the results are relatively low compared to the maximum value (almost 75% of the results fall within the €0–€10,000 range). However, there is a group of combinations of input variables for which the *NPV* reached very high values, on the order of tens of thousands of euros. When analyzing the distribution of *NPVs*, it can also be seen that 25% of the obtained results exceed €5995. The same proportion of results is below €672, implying an interquartile range of nearly €5500. This indicates significant variability in results and high economic uncertainty of the project, resulting from the high sensitivity of *NPV* to the adopted model assumptions. The relatively low value of the lower quartile suggests that in a significant portion of the scenarios, the project generates only a small

https://doi.org/10.3390/en19081845

surplus of positive cash flows. A high upper quartile, on the other hand, indicates the possibility of achieving very favorable results under favorable conditions. This confirms that the project's profitability is asymmetric, with high-risk scenarios dominating, while fewer but very favorable observations are present.

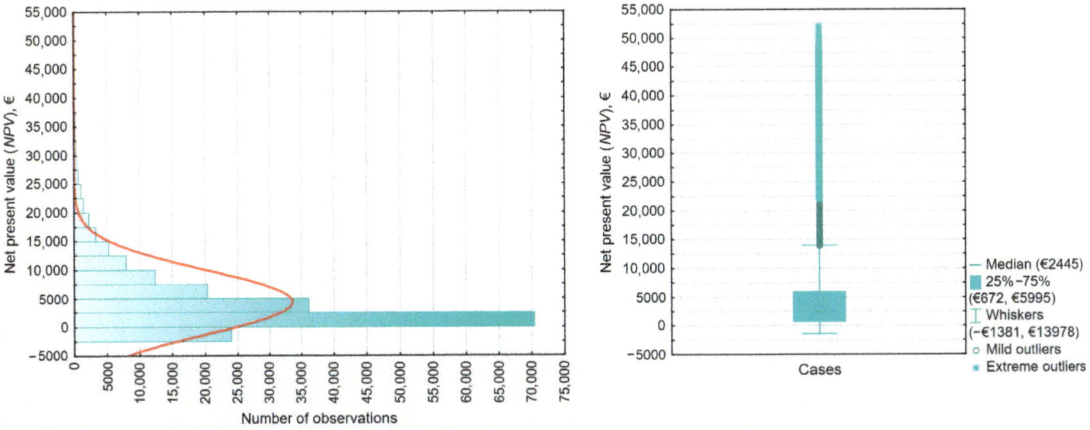


(**a**) (**b**)

**Figure 5.** Net present value distribution: (**a**) histogram; (**b**) box-plot.

The above observations are also confirmed by the obtained kurtosis value of 7.4. Its high value indicates a clearly leptokurtic nature of the distribution, i.e., a strong concentration of observations around central values combined with a large number of outliers. High kurtosis combined with right-sided skewness indicates a low-probability, high-payoff profile, where most scenarios generate moderate results, but there is a certain probability of very large, positive deviations from the median, which should be considered in decision analysis.

As mentioned earlier, the observed high variability, right-sided asymmetry, and leptokurtic nature of the *NPV* distribution indicate that the project's profitability is largely determined by the presence of specific combinations of input variables. This distribution structure suggests nonlinear and potentially heterogeneous model sensitivity to changes in input variables, which complicates the identification of key risk factors. Nonlinearity and extreme values justify explainability methods like SHAP. For this purpose, the XGBoost model was developed. This model formed the basis for the SHAP analysis performed in the next stage of the research, enabling the identification of variables that most significantly determine the *NPV* and are responsible for the observed extreme scenarios.

### 3.1.2. Results of the SHAP Analysis

The XGBoost model was developed to simulate the relationship between input variables and *NPV*. The input data was divided into three datasets (training, validation, and test) in a 70/15/15 ratio, with each sample appearing in only one set, which excluded the risk of data leakage. The model was assessed using standard model fitting metrics, i.e., coefficient of determination ($R^2$), root mean square error (*RMSE*), and mean absolute error (*MAE*) [36]. The values of the metrics for the selected model are summarized in Table 1.

https://doi.org/10.3390/en19081845

**Table 1.** Model fitting metrics for *NPV* prediction. *RMSE* and *MAE* are reported in euros.

| | $R^2$ | *RMSE* | *MAE* |
|---|---|---|---|
| Training dataset | 0.999 | 19.99 | 13.24 |
| Validation dataset | 0.999 | 37.53 | 23.00 |
| Test dataset | 0.999 | 38.12 | 23.24 |

The results indicate a very good model fit to the data for the *NPV* prediction task, which in the analyzed case ranged from −€1381 to €52,168 (Figure 6). The coefficient of determination ($R^2$) reached almost identical, very high values (0.999) for all datasets, indicating good model generalization. Such high values of the coefficient of determination may suggest the risk of model overfitting or data leakage. In the present study, however, this is primarily a consequence of the nature of the input data. Net present values were determined based on deterministic relationships resulting from the adopted equations. The XGBoost model, therefore, reproduces existing mathematical relationships, leading to very high prediction accuracy. Achieving $R^2$ results close to 1 may also be due to the wide range of output variable values compared to the relatively small prediction errors.

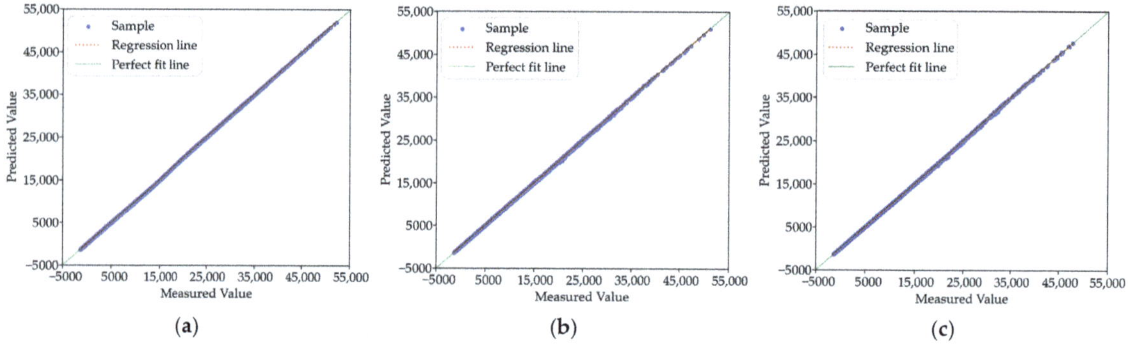


**Figure 6.** Comparison of predicted and observed *NPVs*: (**a**) training dataset; (**b**) validation dataset; (**c**) test dataset.

The high quality of the model, while maintaining realistic error estimation, was also confirmed by the values of the mean absolute and root mean square errors, which remained relatively low compared to the range of the predicted net present value. This demonstrates that the model is characterized by high predictive accuracy across the entire range of *NPVs* considered, as slight increases in *RMSE* and *MAE* on the validation and test datasets relative to the training dataset are expected. These results indicate that the model generalizes effectively, and the high $R^2$ is not the result of data leakage or overfitting.

To identify the key variables determining the financial profitability of the investment expressed as *NPV*, an analysis of the model's explainability was conducted using SHAP analysis. The prediction value can be interpreted as the sum of the baseline value and the contributions of all independent variables, expressed by their SHAP values. The model's base value, as shown in Figure 7, was €4322, meaning that individual input variables shifted the *NPV* forecast relative to this reference level. The direction of this shift was determined by the values of individual parameters.

The global analysis showed that the shower length ($l_s$) had the greatest impact on the *NPV*, with the mean absolute SHAP value exceeding €2200. This result confirms that the duration of installation use directly impacts the amount of energy saved and, consequently, the amount of positive cash flows generated by the project. The second most significant variable was the initial energy price ($C_{e0}$). However, it should be noted that the mean

https://doi.org/10.3390/en19081845

|SHAP value| for this variable is only €87 lower than for $l_s$, suggesting that both variables are crucial to the analysis results. Considering that the percentage change in electricity prices ($i_e$) is ranked next in importance, it can be assumed that long-term market conditions, reflected in the forecasted dynamics of energy prices, may significantly determine the economic impact of the investment. The mixed water flow rate ($q$) has only a slightly lower impact on the net present value, with its mean absolute SHAP value, as with $i_e$, being approximately €1500. The hierarchy of features indicates that the financial efficiency of the project is determined mainly by operational variables related to shower use intensity and by economic conditions.

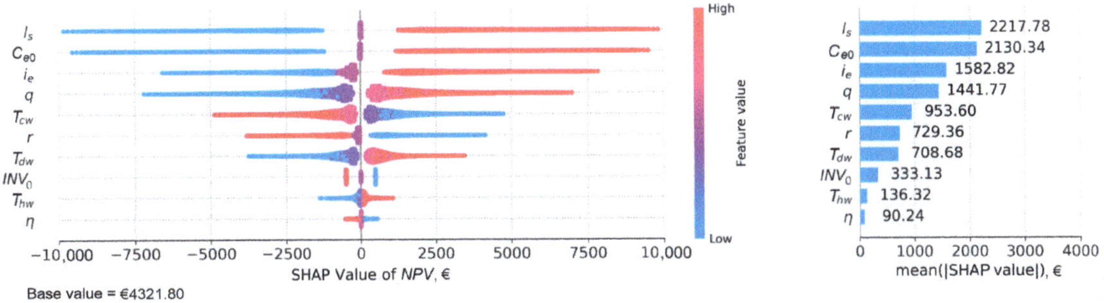


**Figure 7.** SHAP analysis of projected net present value (designations as in Figure 1).

Further bar plot analysis showed that the significance of the remaining input variables is significantly lower, and the mean |SHAP value| does not exceed €1000 in any case. For cold water temperature ($T_{cw}$), discount rate ($r$), and drain water temperature ($T_{dw}$), the mean absolute SHAP value was above €700. On the other hand, for the remaining variables, i.e., initial investment outlays ($INV_0$), domestic hot water temperature ($T_{hw}$), and DHW heater efficiency ($\eta$), it did not exceed €350. The hierarchy of significance, therefore, proves that in the case of a vertical SHE operating under unbalanced flow conditions, operational and market factors dominate over technical parameters, such as the efficiency of the primary heat source in the installation.

Beeswarm plot analysis allowed us to interpret the direction of the influence of individual variables relative to the base value (€4322). For shower length, initial electricity price, and its rate of increase, the *NPV* increased with increasing values of these variables, as evidenced by the location of the red points in Figure 7 to the right of the baseline value. A similar influence was observed for mixed water flow rate, drain water temperature, and DHW temperature. Increasing their values resulted in an increase in the projected investment profitability, resulting from greater energy recovery potential and higher financial savings.

For cold water temperature, discount rate, and DHW heater efficiency, lower values (blue points) are associated with higher *NPV*, indicating that increasing these variables negatively affects the outcome. Higher cold water temperature reduces the energy demand for heating, resulting in reduced potential savings. A higher $r$ value reduces future cash flows, which worsens the *NPV*. In turn, the higher efficiency of the DHW heater reduces the difference in energy consumption between the reference variant and the variant with the vertical SHE, limiting the benefits of using a heat exchanger (Q-Blue B.V., Emmen, The Netherlands).

https://doi.org/10.3390/en19081845

### 3.2. Environmental Analysis

#### 3.2.1. Results of the Greenhouse Gas Emission Reduction Calculation

An analysis of 20,736 scenarios for the operation of a shower system in a building showed that installing a vertical shower heat exchanger operating under unbalanced flow conditions could reduce 15-year greenhouse gas emissions by up to 37,207 kgCO2e (Figure 8). However, it should be noted that the emission reduction potential varies depending on the national energy mix and the adopted decarbonization scenario for the coming years. The average $R_{GHG}$ value obtained for European Union countries was 4185 $kgCO_2e$, while the median result was 2405 $kgCO_2e$. On the other hand, the minimum $R_{GHG}$ value, obtained assuming the lowest input GHG emission intensity ($e_{GHG}$) and the highest assumed reduction in this variable in subsequent years ($i_g$), was only 61 $kgCO_2e$. The mean-to-median ratio itself suggests a strongly asymmetric distribution of results, as confirmed by a positive skewness of 2.1 and an increased kurtosis of 5.9. This suggests that in most of the scenarios considered, projected emission reduction is moderate, while in a limited number of cases, primarily related to high-emission countries, very high outliers appear, significantly inflating the mean. Therefore, also in this case, nonlinearity and extreme values justify the use of model interpretability methods. Importantly, however, despite the clear asymmetry of the $R_{GHG}$ distribution, the *NPV* distribution is characterized by even greater skewness and kurtosis, which is most likely a consequence of the significant increase in energy prices over time.

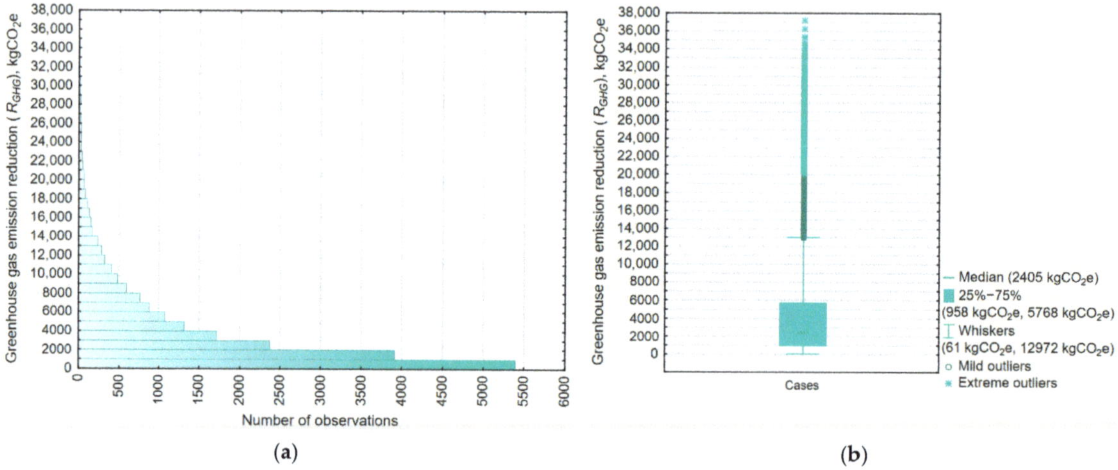


**Figure 8.** Distribution of projected greenhouse gas emission reduction: (**a**) histogram; (**b**) box-plot.

Further statistical analysis confirms significant uncertainty regarding the actual scale of the environmental effect. The interquartile range exceeds 4800 $kgCO_2e$. One-quarter of the results are lower than 958 $kgCO_2e$, and 75% of the obtained $R_{GHG}$ values do not reach 5768 $kgCO_2e$, a value representing less than 18% of the potential maximum greenhouse gas emission reduction. The results, therefore, have a significant impact on the assessment of the environmental efficiency of the shower heat recovery technology. Relying solely on average $R_{GHG}$ values may lead to an overestimation of the expected environmental benefits in countries with low-emission energy systems.

#### 3.2.2. Results of the SHAP Analysis

To extend the analysis of greenhouse gas emission reduction by using a vertical SHE operating under unbalanced flow conditions, a predictive model based on the XGBoost

https://doi.org/10.3390/en19081845

algorithm was developed and then used to conduct an explainability analysis using SHAP. The main goal of the model, similar to the financial analysis, was to identify the relative importance of individual input variables and the direction of their influence on the $R_{GHG}$ value. This approach allowed the identification of which parameters are most decisive for the environmental performance of SHEs in different countries and energy transition scenarios. As a result, the model provides a tool supporting both the interpretation of the obtained greenhouse gas emission reduction results and the formulation of recommendations for climate and energy policy.

As with the net present value, the input data was split 70/15/15 to prevent data leakage. The model fitting metrics, summarized in Table 2, indicate the model's high ability to reproduce the analyzed relationships. This is also confirmed by Figure 9. The coefficient of determination of 0.999 indicates the model's high explanatory power of data variability and the absence of significant overfitting, as high predictive accuracy characterizes all datasets, including those not used in the training process. It should be noted, however, that also in the case of $R_{GHG}$ predictions, the XGBoost model reproduced the existing mathematical relationships. The model's high quality is also confirmed by the *RMSE* and *MAE* values, which are relatively low compared to the range of analyzed $R_{GHG}$ values, which reach tens of thousands of kg$CO_2$e. Therefore, it can be concluded that using the model as a reliable basis for performing SHAP analysis is fully justified.

**Table 2.** Model fitting metrics for $R_{GHG}$ prediction. *RMSE* and *MAE* are reported in kilograms of $CO_2$ equivalent (kg$CO_2$e).

| | $R^2$ | *RMSE* | *MAE* |
|---|---|---|---|
| Training dataset | 0.999 | 17.69 | 11.65 |
| Validation dataset | 0.999 | 33.89 | 20.74 |
| Test dataset | 0.999 | 33.26 | 21.26 |

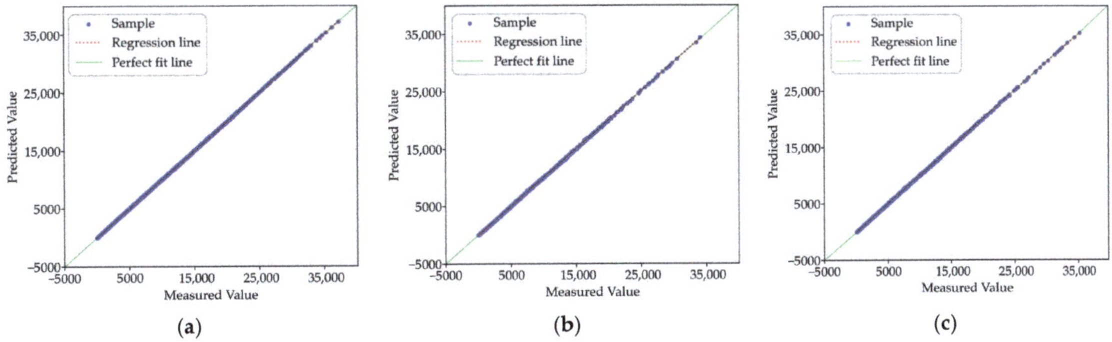


**Figure 9.** Comparison of predicted and observed $R_{GHG}$ values: (**a**) training dataset; (**b**) validation dataset; (**c**) test dataset.

The global SHAP analysis (Figure 10) clearly demonstrated that the key factor determining greenhouse gas emission reduction through the use of the vertical SHE is $e_{GHG}$, with an average absolute SHAP value of nearly 2500 kg$CO_2$e. This variable, therefore, contributes the most to explaining the variability of the developed XGBoost model, confirming previous conclusions about the significant importance of the national energy mix for the environmental performance of SHEs. Reducing the energy source's carbon intensity is equivalent to limiting the potential for avoided emissions per unit of recovered thermal energy, resulting in significant variation in effects between countries with different energy generation structures. This is evidenced by the beeswarm plot analysis, which allows for a detailed interpretation of the direction and scale of the impact of individual input variables on deviations from the base value (3804 $kgCO_2e$). In the case of greenhouse gas emission intensity, this range is the widest. Absolute SHAP values for individual observations reach almost 10,000 $kgCO_2e$, for both negative and positive values of this parameter. This wide range confirms that the variable can significantly increase or decrease the projected emission reduction ($R_{GHG}$) relative to the base value, depending on local conditions. The presence of high SHAP points (marked in red to the right of the base value) indicates that the high emission intensity of the energy system significantly increases the reduction effect of a vertical SHE operating under unbalanced flow conditions.

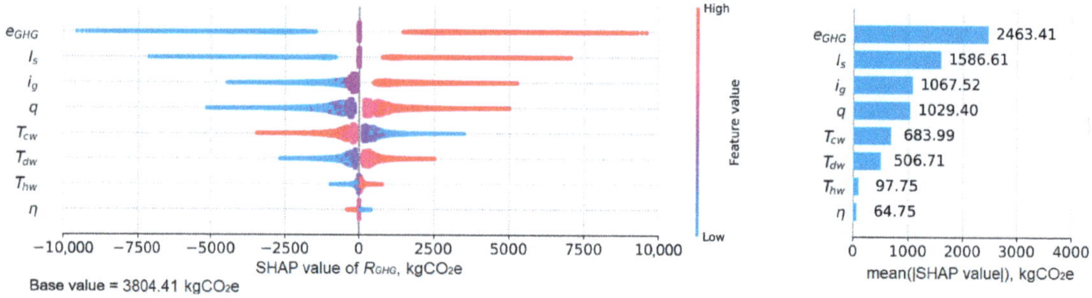


**Figure 10.** SHAP analysis of projected greenhouse gas emission reduction (designations as in Figure 1).

The second most significant factor was shower length ($l_s$), which directly impacts the amount of water used and, consequently, the amount of recoverable energy. In this case, however, the mean |SHAP value| is approximately 35% lower. This result indicates that user habits may be nearly as important as technological factors, highlighting the importance of consumption patterns in environmental efficiency analyses of projects that reduce energy consumption for water heating. Next in line were the annual change in GHG emission intensity ($i_g$), essentially a reduction in this indicator, and the mixed water flow rate ($q$). In both cases, the mean |SHAP value| exceeded 1000 $kgCO_2e$. This indicates that both energy transition dynamics and the hydraulic parameters of the building installation significantly influence long-term emission reduction potential. It should also be noted that all three factors demonstrate a similar direction of influence on the $R_{GHG}$ prediction result as greenhouse gas emission intensity, although the range of SHAP values for individual observations is significantly smaller. An increase in water consumption per shower, resulting from extending the shower length or increasing the mixed water flow rate from the shower head, translates into an increase in the amount of recoverable thermal energy and, consequently, a higher emission reduction. The same applies to the variable $i_g$. The slow rate of decarbonization of the energy sector (lower emission reduction in subsequent years) may significantly enhance the environmental performance of a vertical SHE-equipped installation, especially since some observations indicate SHAP values exceeding 5000 $kgCO_2e$.

Significantly lower results were obtained for media temperatures in individual sections of the system. For cold water and drain water, they were in the range of several hundred kilograms, while the impact of DHW temperature was negligible, as the mean |SHAP value| did not reach 100 $kgCO_2e$. The significantly lower significance of temperature variables suggests that, within the considered range of their variability, they do not constitute a major limitation to the efficiency of the heat recovery system. The variable with the least impact on the projected greenhouse gas emission reduction was the efficiency of

https://doi.org/10.3390/en19081845

the DHW heater, which is likely due to the high efficiency of electric DHW heaters and the narrow range of analyzed values of this parameter.

### 3.3. Evaluation of Extreme Cases and Medians

To broaden the interpretation of the results, an analysis of local SHAP values was conducted for three characteristic observations, corresponding to the maximum, minimum, and median values of the distribution. This analysis was performed for both models. This approach enabled an assessment of the impact of combinations of input variables on the $NPV$ and $R_{GHG}$ calculation results.

For observations corresponding to the maximum $NPV$ and $R_{GHG}$ forecasts (Figure 11), all input variables contributed to increasing the forecast compared to the baseline value. This suggests that the greatest benefits from installing a vertical SHE on the shower's greywater outlet are achieved under conditions of cumulative favorable factors, including operational factors (a large amount of water consumption), economic factors (high energy prices and their rapid growth), and environmental factors (high GHG emission intensity throughout the heat exchanger's lifetime). For the maximum $NPV$ forecast, only a slight change in the hierarchy of importance of individual input variables was observed. The importance of investment outlays decreased relative to the average, resulting in this variable being moved to the last position. This means that under conditions of high positive financial flows resulting from significant energy savings for heating DHW, the $INV_0$ value loses its relative importance compared to the scale of generated financial savings. This phenomenon is further compounded by the fact that when preheated water is supplied exclusively to the SMV, the scope of installation work is small, significantly reducing the initial investment costs. In the case of $R_{GHG}$, on the other hand, the importance structure of input variables remained consistent with the global analysis.

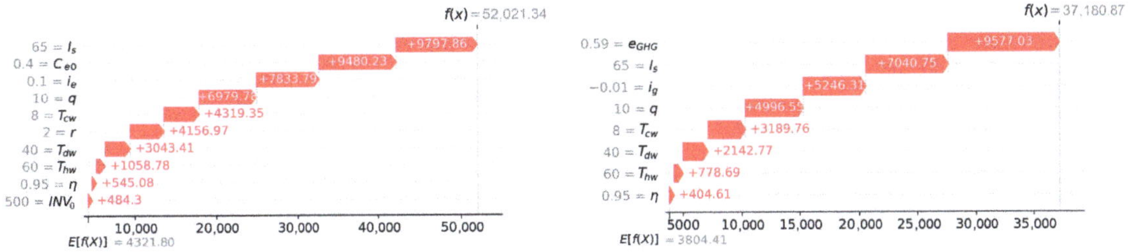


(**a**) (**b**)

**Figure 11.** SHAP analysis results for maximum cases: (**a**) $NPV$; (**b**) $R_{GHG}$ (designations as in Figure 1).

For observations corresponding to the minimum $NPV$ and $R_{GHG}$ values (Figure 12), all variables contributed to lower forecasted results. Therefore, unlike the maximum forecasts, the lowest results are a consequence of the coexistence of unfavorable conditions related to low shower system use intensity, low energy prices throughout the unit's life cycle, and parameters limiting heat recovery potential. With respect to the minimum $NPV$ forecast, a slight change in the structure of variable significance was noted, consisting of an increased influence of mixed water flow rate ($q$) relative to the annual increase in electricity prices ($i_e$). A similar change occurred in the second model, where water flow rate proved more significant than changes in greenhouse gas emission intensity ($i_g$). This suggests that under conditions of extremely low financial savings, current operating parameters may be more important than long-term factors such as price dynamics or decarbonization.

https://doi.org/10.3390/en19081845

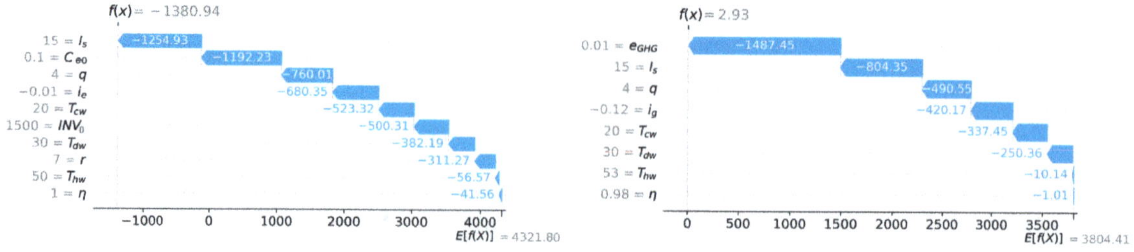


(**a**) (**b**)

**Figure 12.** SHAP analysis results for minimal cases: (**a**) *NPV*; (**b**) $R_{GHG}$ (designations as in Figure 1).

The most complex relationships were observed for observations corresponding to the median forecasts (Figure 13). In the case of the median *NPV*, the initial electricity price had the greatest impact, with a low value resulting in a lower forecast compared to the baseline. A similar trend was observed for high cold water temperature. High initial investment outlays, a moderate increase in electricity prices, high DHW heater efficiency, and low DHW temperature also contributed to the lower *NPV* forecast. On the other hand, a significant shower length increased *NPV* compared to the baseline by almost €2500. High drain water temperature, a low discount rate, and a relatively high mixed water flow rate also contributed to the increase in *NPV* forecast. This pattern indicates that, under average conditions, the effects of individual variables partially offset each other. Some of them enhance the project's financial efficiency, while others reduce it.

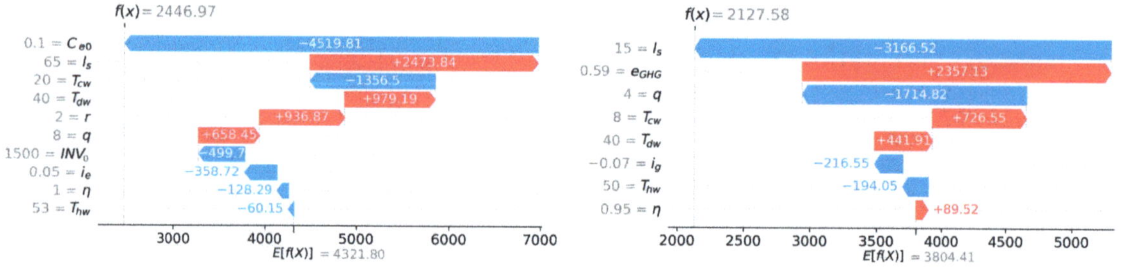


(**a**) (**b**)

**Figure 13.** SHAP analysis results for median cases: (**a**) *NPV*; (**b**) $R_{GHG}$ (designations as in Figure 1).

An equally complex hierarchy of significance was observed for the median $R_{GHG}$. In the considered scenario, however, the minimum shower length and low water flow rate resulted in a reduction in the environmental impact compared to the baseline value by almost 3167 kg$CO_2$e and 1715 kg$CO_2$e, respectively. The projected level of greenhouse gas emission reduction was significantly increased by a high GHG emission intensity of electricity generation. Low cold water temperature and high drain water temperature also contributed to the improvement in the environmental impact, while the efficiency of the DHW heater had only a minimal positive effect. On the other hand, the significant annual reduction in $e_{GHG}$ and low domestic hot water temperature reduced the projected $R_{GHG}$ value by approximately 200 kg$CO_2$e. The different directions of influence in the above observations confirm that the individual scenarios are the result of overlapping variables with opposing impacts. Unlike the extreme cases, where a unidirectional cumulative effect dominates, other cases reflect the true complexity of interactions between individual input variables. These results highlight the importance of local analysis in interpreting predictive models. Only such an approach allows us to understand how observations with

https://doi.org/10.3390/en19081845

similar environmental or economic outcomes can result from different configurations of input variables.

### 3.4. Assessment of Consistency Between Financial and Environmental Effects

The assessment of consistency between financial and environmental effects was based on a comparison of the SHAP analysis results for two independent predictive models: the *NPV* model and the $R_{GHG}$ model. Due to the different number of input variables included in both models, the comparison did not apply to full observations, but only to the part of the feature space common to both analyses. This included six input variables related to the operating and technical conditions of the shower installation ($l_s$, $q$, $T_{cw}$, $T_{dw}$, $T_{hw}$, $\eta$). For each combination of values of these six common input variables in the *NPV* model, there were 81 scenarios resulting from the combination of four additional economic variables, each of which assumed three values. Similarly, in the $R_{GHG}$ model, for the same combination of common variables, there were nine scenarios resulting from the combination of two additional environmental variables, also analyzed for three different values. Therefore, representative cases were determined not for a single identical observation in both models, but within sets of observations with the same values of the common variables. For each of the analyzed common variables, SHAP values corresponding to three representative points of the prediction distribution were compared: the minimum value, the median value, and the maximum value. Detailed results for shower length ($l_s$), which turned out to be the most significant common variable in both models, are presented in Figure 14. Results for the remaining common variables are provided in Appendix A.

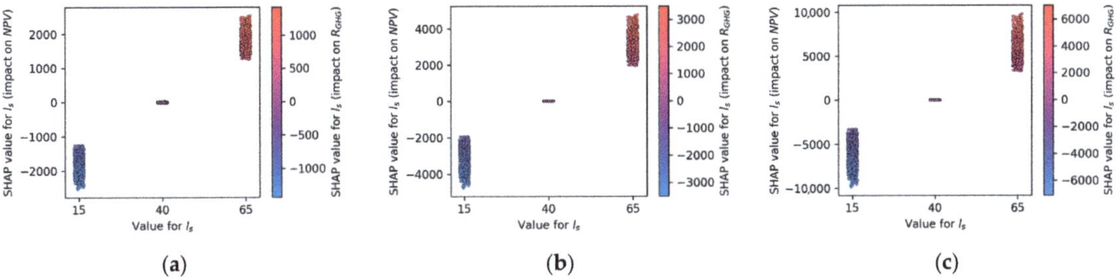

**Figure 14.** SHAP-based comparison of the impact of shower length ($l_s$) on *NPV* and $R_{GHG}$: (**a**) minimum; (**b**) median; (**c**) maximum.

The results indicated high consistency between the financial and environmental effects. In the minimum, median, and maximum cases considered, common input variables exhibited the same direction of influence on the prediction values of both models. This means that factors increasing the predicted *NPV* simultaneously increased the predicted greenhouse gas emission reduction, while factors reducing the financial effect had a similar effect on the environmental effect.

Furthermore, the relative importance of common variables was found to be consistent. In the global analysis, shower length ($l_s$) had the highest significance in both models, with subsequent common variables maintaining a similar position in terms of influence. This confirms that both financial and environmental benefits are largely determined by the same operational mechanisms, primarily related to the intensity of shower system use.

It should be emphasized, however, that in real-world conditions, the profitability of the project will also be determined by other input variables. In this context, the results obtained for the median scenarios (Figure 13) are particularly important, as they can be considered the most representative of actual operating conditions. The SHAP analysis showed that the relationships between input variables and *NPV* and $R_{GHG}$ are more complex and less clear-

https://doi.org/10.3390/en19081845

cut than in extreme cases. For key common variables, the general trend is maintained. For example, a longer shower increases the result, while a lower $l_s$ value results in a reduction in the expected results. However, these variables do not fully determine the final values, as parameters specific to both models also play a significant role. Consequently, the observed variability in results under median conditions results from the overlapping influence of common variables and model-specific variables, whose impacts are not mutually offset. This leads to a situation in which the general trends remain unchanged, but their final effect is modified depending on the configuration of external parameters.

This indicates the necessity of apparent trade-offs between financial and environmental efficiency. However, it should be noted that not all input variables are controllable from the perspective of the user or system designer. Variables such as electricity prices and their variability over time, or greenhouse gas emission intensity, are external factors independent of the installation's operation. In the case of operational variables such as $l_s$ or $q$, which the user can actually influence, a high degree of consistency is observed in the impact on $NPV$ and $R_{GHG}$. Therefore, the observed discrepancies between both effects result largely from the system's location, not from direct user decisions. The lack of full consistency between $NPV$ and emission reduction is, therefore, not due to conflicting operational objectives, but rather to the overlapping influence of controllable and uncontrollable parameters. The final effect is a result of user decisions and external conditions.

## 4. Discussion

The results of the analysis confirmed that the use of XGBoost models combined with SHAP analysis is an effective tool for identifying key variables determining the financial and environmental performance of a shower heat recovery system operating under unbalanced flow conditions. By quantifying the impact of individual input variables on $NPV$ and $R_{GHG}$, this approach can provide a basis for transparent system optimization [8]. It allows not only for identifying the most important variables but also for understanding their impact in various shower system operation scenarios.

The global SHAP analysis showed that for both dependent variables considered, variables related to the intensity of shower use, i.e., shower length ($l_s$) and mixed water flow rate ($q$), significantly impact the forecast result. These parameters directly influence the amount of tap water used and, therefore, the amount of recoverable thermal energy. Local analysis of individual shower system scenarios confirmed that increasing water consumption in the shower system not only increases the potential for GHG emission reduction but also translates into increased expected financial savings. This convergence suggests that, in many cases, improving the environmental performance is correlated with economic attractiveness.

However, it is important to remember the variables that were not included in both models. In the case of GHG emission reduction, the greenhouse gas emission intensity of electricity generation ($e_{GHG}$) and the rate of its decline over the subsequent years of vertical SHE use play a significant role. These variables determine the "environmental value" of each unit of saved energy. On the other hand, the level and dynamics of energy price changes, and to a lesser extent, the discount rate, are crucial for financial profitability. This means that the environmental effects of drain water heat recovery strongly depend on the structure of the energy system, while the financial effects depend on market conditions. The observed differences in the mechanisms shaping the net present value and potential GHG emission reduction demonstrate that maximizing environmental benefits will not, in many cases, equate to maximizing financial benefits, particularly in the case of electricity production in low-emission systems and at relatively low energy prices.

https://doi.org/10.3390/en19081845

Interpreting the SHAP analysis results in a practical context, it can be concluded that the application of a vertical SHE is most economically and environmentally beneficial under conditions of intensive shower use. The consistency analysis results indicate that, despite differences in the model structures, common operational and technical variables exhibit consistent effects on both predicted variables. In particular, parameters related to the intensity of shower system use, such as shower length ($l_s$) and mixed water flow rate ($q$), simultaneously increase both the *NPV* and the potential reduction in greenhouse gas emission. This means that within the common part of the feature space, the financial and environmental effects are determined by converging physical mechanisms related to the amount of recovered energy. Electricity prices and their rate of increase over time also play a significant role in the investment profitability. Under conditions of high energy prices, the financial savings achieved will be the greatest, resulting in an increase in the *NPV*. The environmental effect, however, will be most visible in countries with the highest emission intensity of energy used for water heating, where the potential for GHG emission reduction will be the greatest. On the other hand, in countries that, thanks to conscious energy policies, have eliminated fossil fuels and replaced them with renewable sources and nuclear energy, the environmental impact of using vertical SHEs will be limited. Furthermore, reducing the amount of water used limits the amount of recoverable thermal energy, which impacts both the environmental and financial benefits. This means that the greatest benefits from these systems can be achieved in buildings with high and regular water consumption per shower, and in conditions of relatively high electricity prices and high emissions from their generation.

However, when considering the results of the analysis, it is important to remember the limitations of the adopted research methodology. The dataset used was synthetic in nature. In real-world operating conditions, greater data variability and the presence of random factors can lead to reduced accuracy of predictive models. Therefore, the results should be interpreted as an assessment of potential relationships between variables rather than as direct forecasts for actual installations.

Another significant limitation was that the analysis was based on an electric DHW heater, which significantly determined the results. For natural gas-powered devices, different results can be expected for both the financial and environmental performance of the tested solution. On the one hand, a lower natural gas price may lead to lower projected financial savings, and thus a reduction in *NPV*. Eurostat statistics [37] indicate that in the first half of 2025, the price of natural gas for households in the EU ranged from €0.0307 per kWh (Hungary) to €0.2128 per kWh (Sweden). The maximum price of natural gas was, therefore, almost half that of electricity. On the other hand, greenhouse gas emissions from natural gas combustion are typically lower than those from electricity generation in systems based on a high-emission energy mix. This may result in limited emission reduction potential. It should be noted, however, that in systems with very low emissions, using electricity to heat DHW may be more environmentally beneficial than burning natural gas. Data from the European Commission [38] indicate that the GHG emission factor for natural gas is approximately 0.20 kgCO2e/kWh. In countries with lower $e_{GHG}$ values, the use of SHE could contribute to greater reductions in greenhouse gas emissions when used in conjunction with a gas DHW heater.

It should also be emphasized that the results of SHAP analysis are dependent on the assumed ranges of input variables. Feature importance reflects the influence of a variable within the analyzed dataset, not its universal significance. This means that changing the ranges of input variable values can lead to a change in their relative importance in the model. This is particularly true for variables with a wide range of values, such as shower length. Their high position in the importance hierarchy does not always result from their

https://doi.org/10.3390/en19081845

actual importance. It may also be a consequence of a greater influence on the variability of the results. Therefore, the obtained variable rankings should be interpreted in the context of the assumed ranges of input variables. With different shower usage scenarios or with changed environmental or financial parameters, the hierarchy of importance of individual variables may change.

It is also important to note that the analysis focused on one specific type of shower heat exchanger and a specific system configuration, which limits the possibility of directly generalizing the obtained results to all solutions used in practice. For other SHE geometries, the absolute values of the achieved environmental and financial effects may be slightly different. However, the identified relationships, particularly those concerning the direction of the influence of individual input variables, are more universal and can be applied to other types of SHEs heating water supplied to the shower mixing valve.

The identified relationships should be considered in the context of the growing demand for energy in various sectors, as societal development and the functioning of a modern economy are largely dependent on access to energy, which should be both reliable, safe, and economically affordable [39]. It should be noted that one of the most energy-intensive areas of municipal infrastructure is the water and sewage sector. This is due, among other things, to growing requirements for water quality, climate change, and the occurrence of extreme events [40]. The need to ensure high system reliability is also important. In this context, water losses are a significant problem [41], which necessitates the collection, treatment, and transportation of additional volumes of water, increasing total energy consumption. In such conditions, the utilization of thermal energy carried by wastewater becomes particularly important, especially since, as Arsenyeva et al. [42] noted, even a few percent of the heat demand can be met through heat recovery from wastewater. The possibilities of heat recovery from wastewater are not limited to single-family homes. This technology demonstrates potential applications in many other facilities, such as multi-family buildings, hotels, and industrial facilities, where the amount of available waste energy can be significantly higher. The literature provides recommendations for the application of wastewater heat recovery systems in facilities with high hot water consumption. These may include hotels, for example, where decisions regarding the location of the heat recovery system should, in addition to energy demand and heat recovery potential, also consider the region's tourism potential [43]. Heat recovery from sanitary appliances other than showers is also justified. For example, Zanganeh et al. [44] emphasized the importance of heat recovery in commercial dishwashers. It is, therefore, clear that the range of possible applications of heat recovery technology from wastewater, primarily greywater, is much broader than that described in this paper.

This wide variety of heat recovery system configurations and operating conditions demonstrates that the use of predictive models combined with SHAP analysis can be a valuable tool in future research, enabling the identification of key parameters determining both the energy efficiency, environmental performance, and economic efficiency of various wastewater heat recovery systems. These studies should consider not only small local installations but also systems based on heat pumps [45]. Particular emphasis should be placed on hybrid systems combining various renewable and waste energy sources, such as wastewater and air heat pumps, as well as solar collectors [46]. Considering heat storage systems [47] would also be beneficial, as such a solution could further increase the flexibility of wastewater heat recovery systems and enable better adjustment to fluctuating demand. Integrating wastewater heat recovery systems with other technologies can increase both the energy and environmental efficiency as well as financial viability. For this reason, it is also worth examining available forms of support. As noted by Ogarek et al. [48], taking

https://doi.org/10.3390/en19081845

Poland as an example, the fastest-growing renewable energy sector in this region is the one receiving the highest subsidies.

Future research should also consider the development of heat recovery technology from wastewater, including greywater. Technologies used in other energy sectors could provide inspiration for new SHE solutions. For example, Ahmad and Alsalmah [49] proposed a two-pipe solar-assisted heat exchanger. The development of such technologies defines the direction in which modern energy production is heading. These are hybrid solutions, combining various energy-harvesting mechanisms. The research described in this paper suggests that even for more complex systems for heat recovery from drain water, an approach based on SHAP analysis could be a useful tool supporting their design and optimization.

Integrating the heat recovery system under consideration with broader building energy management strategies may also be an interesting direction for further research, as Li et al. [50] noted that demand-side flexibility depends on the interaction between a number of parameters that determine the building's energy profile. Although the demonstrated vertical SHE application affects only the energy demand for DHW heating, its significance can be considered in the broader context of building energy performance.

The results can also be considered in the broader context of achieving sustainable development objectives and global climate goals. Previous research indicates that modernizing and adapting existing systems to new environmental requirements is a crucial element of the transition to more sustainable energy systems [51]. Undoubtedly, SHEs also meet this requirement, contributing to rational resource management, increasing the energy efficiency of buildings, and reducing greenhouse gas emissions. Such solutions are also consistent with the concept of a circular economy, which strives to maximize the use of energy already present in the system [52]. The development and optimization of wastewater heat recovery technologies can also support the implementation of selected sustainable development goals (SDGs). Therefore, it is hoped that the integration of sustainable energy saving techniques with modern analytical methods and predictive modeling will be an important element of the transformation towards a more environmentally friendly and resilient energy future, which is particularly important given the growing exposure to the effects of climate change [53].

## 5. Conclusions

The analysis demonstrated that heating the water supplied to the shower mixing valve using a vertical heat exchanger enables efficient heat recovery from drain water. For the considered range of input parameter variability, *NPVs* ranged from −€1381 to €52,168. Meanwhile, the greenhouse gas emission reduction over the 15-year operation period ranged from 61 kg$CO_2$e to 37,207 kg$CO_2$e. Using predictive modeling in conjunction with SHAP analysis, it was possible to identify key parameters influencing the financial and environmental impacts of this solution, as well as assess the consistency between these effects. It was also possible to determine the relative importance of individual input variables influencing the project's net present value and potential greenhouse gas emission reduction. The global SHAP analysis was supplemented with local analysis for selected cases, allowing for a better understanding of how individual input variables influence the predictions of both models. Key findings from these analyses include:

1. Greenhouse gas emission intensity of electricity generation and shower length are among the most important variables shaping the potential GHG emission reduction.
2. Annual change in GHG emission intensity and mixed water flow rate also have a significant impact on environmental effects.

https://doi.org/10.3390/en19081845

3. In the case of financial effects, in addition to water consumption per shower, the price of electricity and its dynamics over the entire operational life of the heat recovery installation are also crucial.
4. The direction of the impact of individual operating parameters on the projected greenhouse gas emission reduction and net present value is consistent. However, differences in the relative importance of individual variables can be noted, as confirmed by local analysis, indicating certain changes in the impact hierarchy depending on the scenario analyzed.

It should be noted, however, that the results are subject to certain limitations resulting from the model assumptions and the scope of the analyzed data. The analysis was conducted for specific technical and operational conditions of the drain water heat recovery system. Other methods of connecting the shower heat exchanger, as well as other SHE models, were not considered. Furthermore, the predictive models considered a limited number of parameters, meaning that in real-world conditions, other factors, such as the variability of water consumption profiles in the building, may also influence the system's efficiency. Another limitation may be that the local analysis was based on selected characteristic cases, which, while providing a better understanding of the model's operation, do not reflect the full variability of all possible operating scenarios.

It is also important to note the technical limitations and potential implementation risks of vertical SHEs, particularly in existing buildings. These limitations may result from local conditions, variability in operating parameters, and the need for integration with existing building systems. In each case, they may impact the actual financial and environmental performance of the solution.

Future research would benefit from conducting analyses combining predictive modeling with interpretation tools. It is also worthwhile to expand the scope of analyses to include other configurations of drain water heat recovery systems, as well as solutions used in other areas of water and sewage infrastructure. Future research could also include a larger number of scenarios, allowing for a more comprehensive assessment of the impact of variability in individual parameters on the system's financial efficiency and its potential to reduce greenhouse gas emissions.

**Author Contributions:** Conceptualization, S.K.-O. and M.S.; methodology, S.K.-O. and M.S.; software, S.K.-O. and M.S.; validation, S.K.-O. and M.S.; formal analysis, S.K.-O. and M.S.; investigation, S.K.-O. and M.S.; resources, S.K.-O. and M.S.; data curation, S.K.-O. and M.S.; writing—original draft preparation, S.K.-O. and M.S.; writing—review and editing, S.K.-O. and M.S.; visualization, S.K.-O. and M.S. All authors have read and agreed to the published version of the manuscript.

**Funding:** This research was financed by the Minister of Science and Higher Education of the Republic of Poland within the "Regional Excellence Initiative" program for the years 2024–2027 (RID/SP/0032/2024/01).

**Data Availability Statement:** Data are contained within this paper.

**Conflicts of Interest:** The authors declare no conflicts of interest. The funders had no role in the design of the study; in the collection, analyses, or interpretation of data; in the writing of the manuscript; or in the decision to publish the results.

## Abbreviations

The following abbreviations are used in this manuscript:

| | |
|---|---|
| $CO_2$ | carbon dioxide |
| DHW | domestic hot water |
| EU | European Union |

https://doi.org/10.3390/en19081845

| | |
|---|---|
| GHG | greenhouse gas |
| *MAE* | mean absolute error |
| *NPV* | net present value |
| *RMSE* | root mean square error |
| SDG | Sustainable Development Goal |
| SHAP | SHapley Additive Explanations |
| SHE | shower heat exchanger |
| SMV | shower mixing valve |

## Appendix A

Figures A1–A5 present SHAP-based comparisons of the impact of selected variables on *NPV* and $R_{GHG}$.

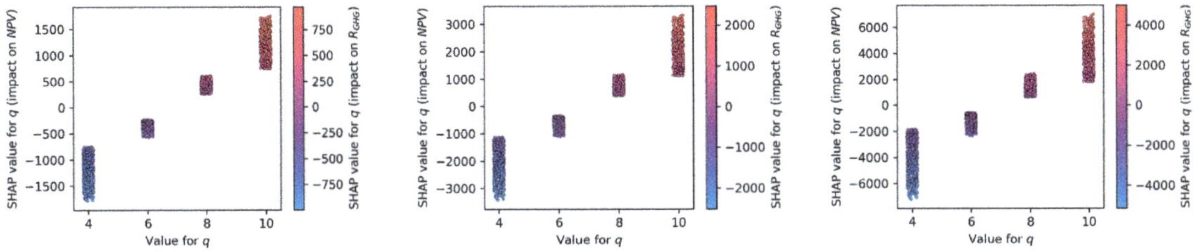


(a) (b) (c)

**Figure A1.** SHAP-based comparison of the impact of mixed water flow rate ($q$) on *NPV* and $R_{GHG}$: (**a**) minimum; (**b**) median; (**c**) maximum.

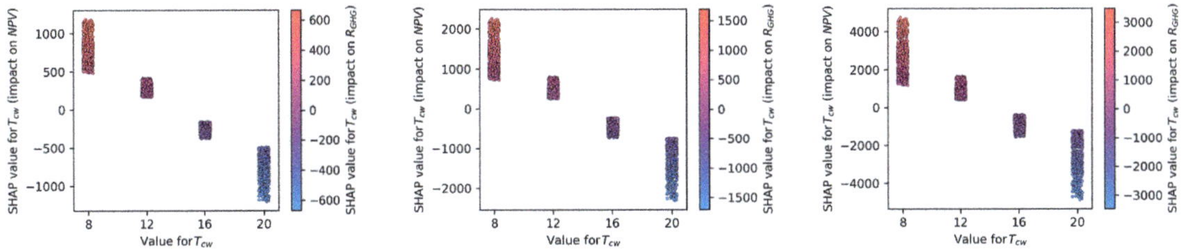


(a) (b) (c)

**Figure A2.** SHAP-based comparison of the impact of cold water temperature ($T_{cw}$) on *NPV* and $R_{GHG}$: (**a**) minimum; (**b**) median; (**c**) maximum.

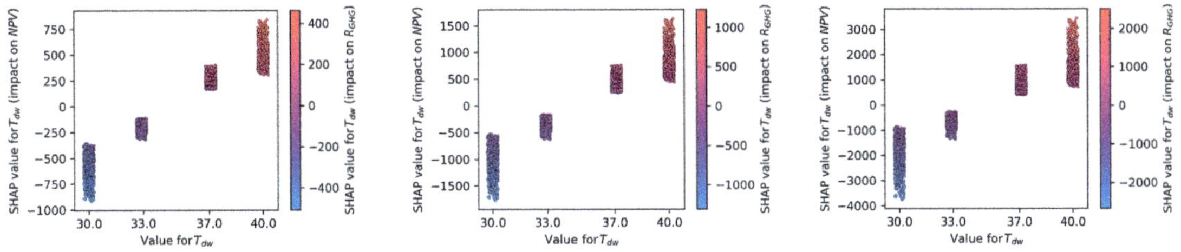


(a) (b) (c)

**Figure A3.** SHAP-based comparison of the impact of drain water temperature ($T_{dw}$) on *NPV* and $R_{GHG}$: (**a**) minimum; (**b**) median; (**c**) maximum.

https://doi.org/10.3390/en19081845

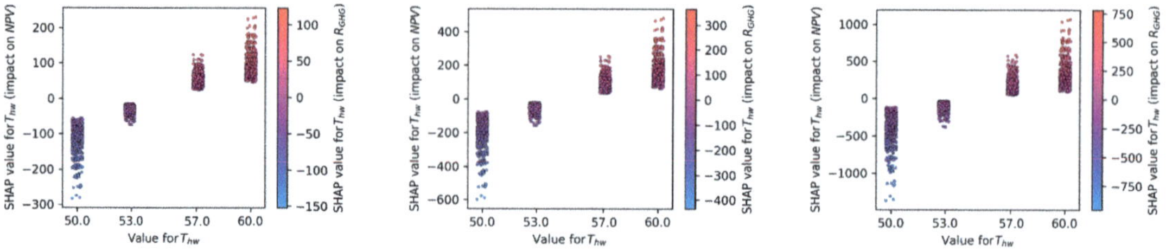


(a) (b) (c)

**Figure A4.** SHAP-based comparison of the impact of DHW temperature ($T_{hw}$) on *NPV* and $R_{GHG}$: (**a**) minimum; (**b**) median; (**c**) maximum.

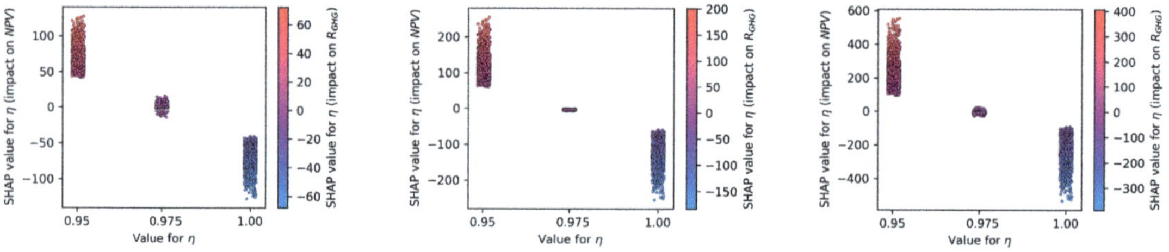


(a) (b) (c)

**Figure A5.** SHAP-based comparison of the impact of DHW heater efficiency ($\eta$) on *NPV* and $R_{GHG}$: (**a**) minimum; (**b**) median; (**c**) maximum.

## References

1. Hall, R.J.; Kenway, S.; Memon, F.A.; O'Brien, K.R. Trade-offs between total and peak energy consumption due to plumbing layouts in Australian and UK domestic hot water systems. *J. Clean. Prod.* **2025**, *518*, 145941. [CrossRef]
2. Voznyak, O.; Dudkiewicz, E.; Laska, M.; Antypov, I.; Spodyniuk, N.; Sukholova, I.; Savchenko, O. A New Approach to the Economic Evaluation of Thermomodernization: Annual Assessment Based on the Example of Production Space. *Energies* **2024**, *17*, 2105. [CrossRef]
3. Hall, R.J.; Kenway, S.; O'Brien, K.R.; Memon, F.A. Quantification of residential water-related energy needs cohesion, validation and global representation to unlock efficiency gains. *Renew. Sustain. Energy Rev.* **2025**, *207*, 114906. [CrossRef]
4. Bednarz, K.; Stec, A.; Słyś, D. Research Trends in Renewable Energy Sources in Academic Infrastructure: A Bibliometric Analysis. *Energies* **2025**, *18*, 5812. [CrossRef]
5. Enayatollahi, R.; Appleby, A. Development of a wastewater heat recovery unit for residential buildings in New Zealand. *IOP Conf. Ser. Earth Environ. Sci.* **2025**, *1500*, 012059. [CrossRef]
6. Kaewluan, S.; Unsomsri, N.; Wiriyasart, S. Thermal Performance and Energy Consumption of Air Conditioning Systems with Heat Exchangers for Water Heater Efficiency Improvement. *ES Energy Environ.* **2025**, *29*, 1749. [CrossRef]
7. Tauseef, M.; Sharqawy, M.F. Thermal and Hydraulic Performance Analysis of a Waste Heat Recovery Device for Domestic Furnace Exhaust Gases. *Green Energy Environ. Technol.* **2026**, *5*, 1–20. [CrossRef]
8. Li, Y.; Zhang, H.; Shen, X.; Qu, K. Interpretable machine learning for predicting and optimizing residential building performance in cold regions. *Energy Build.* **2025**, *347*, 116321. [CrossRef]
9. Piotrowska, B.; Słyś, D. Variant analysis of financial and energy efficiency of the heat recovery system and domestic hot water preparation for a single-family building: The case of Poland. *J. Build. Eng.* **2023**, *65*, 105769. [CrossRef]
10. Silva, E.; Ferreira, V.; Cunha, J.; Monteiro, E. CFD Analysis on the Performance of Twisted Tapes in Horizontal Heat Exchangers for Shower Water Heat Recovery. *J. Heat Mass Transf. Res.* **2025**, *12*, 137–150. [CrossRef]
11. Forero, J.D.; Vigoya, M.F.; Casseres, D.M. Analysis of a Waste Heat Recovery System from Hot Drainage Water Using a Heat Recovery Concentric Tube Heat Exchanger Through CFD Simulation. *Int. Rev. Mech. Eng.* **2025**, *19*, 318–326. [CrossRef]
12. Kowalik, R.; Sidło, W. Drain water heat recovery for domestic hot water in polish residential buildings: Energy, economic and environmental assessment. *Desalin. Water Treat.* **2025**, *324*, 101475. [CrossRef]
13. Piotrowska, B.; Słyś, D. Analysis of the Life Cycle Cost of a Heat Recovery System from Greywater Using a Vertical "Tube-in-Tube" Heat Exchanger: Case Study of Poland. *Resources* **2023**, *12*, 100. [CrossRef]

https://doi.org/10.3390/en19081845

14. Stec, A.; Piotrowska, B.; Słyś, D. Greywater as an undervalued tool for energy efficiency enhancement and decarbonizing the construction sector: A case study of a dormitory in Poland. *Energy Build.* **2025**, *330*, 115337. [CrossRef]
15. Stec, A.; Pochwat, K. Assessment of the efficiency of shower wastewater heat exchangers using machine learning-based methods. *J. Build. Eng.* **2025**, *103*, 112062. [CrossRef]
16. Purghel, D.M.; Teodosiu, C.I. Numerical Study on Shower Water Heat Recovery. *IOP Conf. Ser. Earth Environ. Sci.* **2025**, *1546*, 012025. [CrossRef]
17. Krishnan, E.N.; Muneeshwaran, M.; Rendall, J.; Brechtl, J.; Nawaz, K. Experimental investigation on heat transfer performance of drain water heat recovery heat exchangers. *Energy Build.* **2025**, *346*, 116213. [CrossRef]
18. Zhang, X.; Rhee, K.N.; Jung, G.J.; Kim, C. Exploring energy efficiency and savings potential of a horizontal domestic drain water heat recovery system in high-rise apartment buildings. *Energy Build.* **2024**, *325*, 115038. [CrossRef]
19. Kordana-Obuch, S.; Piotrowska, B.; Starzec, M. Management of waste heat in residential buildings: Predictive modelling and sensitivity analysis of variables characterising shower heat exchanger conditions. *Energy* **2025**, *318*, 134743. [CrossRef]
20. Ali, S.; Faraj, J.; Khaled, M. Heat recovery from hot water drainage: CFD simulation, energy efficiency and exergoeconomic analysis of concentric tube heat exchangers. *J. Build. Eng.* **2024**, *98*, 111034. [CrossRef]
21. Dudkiewicz, E.; Ludwińska, A. Family Dwelling House Localization in Poland as a Factor Influencing the Economic Effect of Rainwater Harvesting System with Underground Tank. *Sustainability* **2023**, *15*, 10687. [CrossRef]
22. Karimi, M.; Heidarigoujani, H.; Jahangiri, M.; Anaraki, M.T.; Janaki, D.M. Heating the Future: Solar Hot Water Collectors for Energy-Efficient Homes in Sweden. *Energy Eng.* **2026**, *123*, 6. [CrossRef]
23. Electricity Price Statistics. Available online: https://ec.europa.eu/eurostat/statistics-explained/index.php?title=Electricity_price_statistics (accessed on 17 December 2025).
24. Electricity Prices for Household Consumers—Bi-Annual Data. Available online: https://ec.europa.eu/eurostat/databrowser/view/nrg_pc_204/default/table?lang=en (accessed on 17 December 2025).
25. Commission Notice on Current State Aid Recovery Interest Rates and Reference/Discount Rates Applicable as from 1 January 2026 (Published in Accordance with Article 10 of Commission Regulation (EC) No 794/2004). Available online: https://eur-lex.europa.eu/eli/C/2025/6782/oj/eng (accessed on 7 January 2026).
26. Greenhouse Gas Emission Intensity of Electricity Generation in Europe. Available online: https://www.eea.europa.eu/en/analysis/indicators/greenhouse-gas-emission-intensity-of-1 (accessed on 17 December 2025).
27. Antosz, K.; Kozłowski, E.; Sęp, J.; Prucnal, S. Application of Machine Learning to the Prediction of Surface Roughness in the Milling Process on the Basis of Sensor Signals. *Materials* **2025**, *18*, 148. [CrossRef]
28. Mądziel, M. Investigating Real-world Emissions from Liquefied Petroleum Gas-fueled Vehicles: A Modeling Approach that Utilizes Portable Emissions Measurement Systems. *Open Transp. J.* **2025**, *19*, e26671212367963. [CrossRef]
29. Xu, Z.; Zhou, X.; Wang, Y. The implementation of machine learning models for predicting $CO_2$ emission in carbon capture and storage systems. *Sustain. Comput. Inform. Syst.* **2026**, *49*, 101264. [CrossRef]
30. Zhang, T.; Zhu, W.; Wu, Y.; Wu, Z.; Zhang, C.; Hu, X. An explainable financial risk early warning model based on the DS-XGBoost model. *Financ. Res. Lett.* **2023**, *56*, 104045. [CrossRef]
31. Liu, J.; Uglanov, D.A.; Kedam, N.; Zheng, G.; Shimanov, A.A.; Shimanova, A.B.; Lian, Z. Machine learning prediction of thermal–hydraulic (j–f) characteristics in compact heat exchangers with six fin configurations. *Appl. Therm. Eng.* **2025**, *280*, 128491. [CrossRef]
32. Piotrowska, B.; Kordana-Obuch, S.; Starzec, M. Using domestic drain water as affordable and clean energy: Predictive modeling and sensitivity analysis of a horizontal heat exchanger. *Energy* **2025**, *341*, 139326. [CrossRef]
33. Piotrowska, B. Modeling, optimization and sensitivity analysis of integrated photovoltaic, wind turbine, battery storage and drain water heat recovery systems. *Energy* **2026**, *346*, 140191. [CrossRef]
34. Cui, X.; Lee, M.; Koo, C.; Hong, T. Energy consumption prediction and household feature analysis for different residential building types using machine learning and SHAP: Toward energy-efficient buildings. *Energy Build.* **2024**, *309*, 113997. [CrossRef]
35. Ponce-Bobadilla, A.V.; Schmitt, V.; Maier, C.S.; Mensing, S.; Stodtmann, S. Practical guide to SHAP analysis: Explaining supervised machine learning model predictions in drug development. *Clin. Transl. Sci.* **2024**, *17*, e70056. [CrossRef]
36. Yılan, Y.; Beykent, A. Day-Ahead Electricity Price Forecasting Using the XGBoost Algorithm: An Application to the Turkish Electricity Market. *Comput. Mater. Contin.* **2025**, *86*, 1–16. [CrossRef]
37. Natural Gas Price Statistics. Available online: https://ec.europa.eu/eurostat/statistics-explained/index.php?title=Natural_gas_price_statistics (accessed on 30 March 2026).
38. GHG Emission Factors for Local Energy Use. Available online: https://data.jrc.ec.europa.eu/dataset/72fac2b2-aa63-4dc1-ade3-4e56b37e4b7c (accessed on 30 March 2026).
39. Słyś, D.; Stec, A.; Bednarz, K.; Ogarek, P.; Zeleňáková, M. Managing and Optimizing Hybrid Distributed Energy Systems: A Bibliometric Mapping of Current Knowledge and Strategies. *Energies* **2025**, *18*, 2497. [CrossRef]


https://doi.org/10.3390/en19081845

40. Fisher, P.M.J.; Smith, D. The water industry and the decarbonisation of cities: A comprehensive review in the context of Cop26. *PLoS Water* **2022**, *1*, e0000023. [CrossRef]
41. Rożnowski, M.; Żywiec, J.; Szpak, D.; Tchórzewska-Cieślak, B. The Problem of Water Losses in the Visegrad Group (V4) Countries: Challenges and Opportunities. *Water* **2026**, *18*, 640. [CrossRef]
42. Arsenyeva, O.; Biletskyi, I.; Plankovskyy, S.; Kapustenko, P.; Tsegelnyk, Y.; Teliura, N. Heat Recovery Systems From Sewage: The Current State and Development Prospects. In *Sustainable Development and the Evolution of Civil Engineering*, 1st ed.; Neves Lousada, S.A., Ed.; IGI Global Scientific Publishing: Hershey, PA, USA, 2025; pp. 69–106. [CrossRef]
43. Hosseini Dehshiri, S.S.; Hosseini Dehshiri, S.J.; Mostafaeipour, A. Waste heat recovery systems integrated in hotels: An uncertain and reliable planning. *Appl. Therm. Eng.* **2025**, *281*, 128653. [CrossRef]
44. Zanganeh, J.; Seyfaee, A.; Gates, G.; Moghtaderi, B. Thermal Management and Energy Recovery in Commercial Dishwashers: A Theoretical and Experimental Study. *Energies* **2025**, *18*, 2338. [CrossRef]
45. Song, J.; Zhang, Y.; Ma, Z.; Zhang, H. Research progress and application prospect of wastewater heat pump technology in recent decade: A review. *Clean Energy* **2025**, *9*, 123–147. [CrossRef]
46. Ma, C.; Yang, Y.; Jiang, Y.; Liu, K. Control Optimization of Multi-Energy System in Residential Heating and Domestic Hot Water Application. In *Proceedings of the 12th International Conference on Cold Climate HVAC & Energy, CCHVAC 2025; Environmental Science and Engineering*; Ni, L., Wang, P., Li, J., Liu, Y., Eds.; Springer: Singapore, 2025; Volume 1, pp. 351–360. [CrossRef]
47. Amanowicz, Ł.; Turski, M. PCM-Filled Capsules (RT44HC) for Heat Storage—Laboratory Scale Pilot Study. *Energies* **2025**, *18*, 345. [CrossRef]
48. Ogarek, P.; Stec, A.; Słyś, D. Technical and economic analysis of an autonomous hybrid photovoltaic-hydrogen energy system for academic buildings with the focus on energy Independence—A case study for Poland. *Appl. Energy* **2025**, *392*, 125985. [CrossRef]
49. Ahmad, S.; Alsalmah, H.A. Numerical and artificial intelligent analysis of double pipe heat exchanger exposed to solar irradiation for sustainable energy solutions. *Results Eng.* **2025**, *27*, 105994. [CrossRef]
50. Li, W.; Sui, W.; Cheng, L.; Ji, Y.; Guo, Y.; Zhu, J. Quantifying seasonal demand-side flexibility in residential air conditioning under diverse control strategies. *Energy Build.* **2026**, *352*, 116764. [CrossRef]
51. Ciupek, B.; Frąckowiak, A. Review of Thermal Calculation Methods for Boilers—Perspectives on Thermal Optimization for Improving Ecological Parameters. *Energies* **2024**, *17*, 6380. [CrossRef]
52. Serrano-Arévalo, T.I.; Ramírez-Márquez, C.; Ponce-Ortega, J.M. Comprehensive management of energy carriers: A circular economy perspective. *Curr. Opin. Chem. Eng.* **2026**, *51*, 101224. [CrossRef]
53. Borowski, P.F.; Bandala, E.R.; Lubańska, A.; Stein, E. The Effects and Challenges of Climate Change on the Food-Energy-Water Nexus—Poland as a Case Study. In *Optimization in the Agri-Food Supply Chain: Recent Studies*; Koubaa, M., Ammar, M.H., Dhouib, D., Mnejja, S., Eds.; Wiley: Hoboken, NJ, USA, 2024; pp. 17–40. [CrossRef]




https://doi.org/10.3390/en19081845

*energies*

MDPI

*Article*

# Heat Transfer and Thermo-Mechanical Analysis of Plastic-Strain Evolution in Laser-Welded Thin-Walled Laminated Cooling Plates with Non-Uniform Stiffness

**Chengkun Li, Yujia Cai, Han Wang, Zhihang Zhang, Fang Han, Xiaoqing Zhu, Chengcheng Wang and Zhibo Dong ***

National Key Laboratory of Precision Welding & Joining of Materials and Structures, Harbin Institute of Technology, Harbin 150001, China
* Correspondence: dongzhb@hit.edu.cn

**Abstract**

Thin-walled laminated cooling plates integrate internal channels and pin-fin cores, producing reduced and spatially non-uniform stiffness that changes welding restraint and distortion. This study investigates stiffness-controlled plastic-strain evolution in laser butt welding of GH3230 laminated plates, with geometrically identical solid plates as reference. A coupled heat-transfer and thermo-mechanical finite element model was developed in MSC Marc using a composite Gaussian surface–volumetric moving heat source and temperature-dependent properties. The thermal solution was validated against near-weld thermal cycles and fusion geometry; mechanical predictions were evaluated by CMM distortion and residual-stress measurements. Both structures show comparable residual-stress magnitudes and spatial trends, indicating that residual stress is governed mainly by the local weld thermal gradient. In contrast, the laminated plate exhibits larger angular/bending distortion. Simulations show that, although the plastic-strain pattern is similar, the laminated plate develops higher peak plastic strain confined to a narrower band near the weld, with the transverse plastic strain dominating. Plastic strain–temperature paths reveal continued transverse plastic-strain accumulation during cooling with limited recovery, consistent with restraint redistribution induced by stiffness non-uniformity. An equivalent restraint–stiffness spring model explains this "narrower-but-stronger" plastic zone and links stiffness to yielding and residual plastic-strain magnitude, supporting distortion prediction and stiffness-informed control of welded laminated cooling plates.

**Keywords:** heat transfer; thermo-mechanical coupling; laminated cooling plate; laser butt welding; plastic strain evolution

## 1. Introduction

Laminated cooling plates are key thermal-management components in advanced energy systems, where lightweight design must be achieved without compromising heat-spreading and cooling performance. Their internal flow channels and supporting pillars result in significantly reduced and spatially non-uniform stiffness, which can substantially modify the restraint conditions during fabrication. Fusion welding is widely used to manufacture such components because of its high efficiency and material utilization. However, the localized moving heat input generates a strongly non-uniform thermal cycle, together with rapid transient heat conduction through the thin walls and continuous surface heat loss, making residual stresses and welding-induced distortion largely unavoidable. This issue is particularly critical in thin-walled structures for energy equipment, where the low bending stiffness makes them highly susceptible to angular and bending deformation, thereby compromising dimensional accuracy, assembly tolerance, and ultimately service reliability. Consequently, developing predictive and mechanistic approaches for welding distortion in thin-walled laminated cooling plates remains an important topic at the intersection of heat-transfer analysis and thermo-mechanical response. Horajski et al. [1] proposed an integrated structural–process strategy for reducing distortion in thin-walled welded structures and emphasized the continuing practical importance of deformation control in engineering applications. Qiu et al. [2] further showed that welding-induced distortion can amplify stresses and affect fatigue assessment, indicating that distortion is not merely a geometric imperfection, but also a factor that propagates into strength evaluation and safety margins. Experimental and numerical studies [3–5] have also shown that geometric scale, such as plate dimensions, and boundary constraints strongly influence both welding deformation and residual stress. These findings suggest that predictive models should be established on a mechanistic basis rather than relying solely on empirical corrections.

Factors influencing welding deformation mainly include the dimensions of the base metal, plate thickness, clamping conditions, and welding parameters [6]. For properly assembled thin-walled structures, welding deformation is mainly governed by the highly non-uniform temperature field generated during welding and by the intrinsically low stiffness of thin plates, which makes them less resistant to deformation [7]. From a mechanics perspective, welding deformation originates from non-recoverable inelastic strains accumulated under strongly non-uniform thermal cycles. In most fusion-welding processes, plastic strain is the dominant component, which forms the basis of the inherent strain concept [8,9]. Within this framework, the complex thermo-mechanical process is represented by an equivalent inelastic strain field, and the resulting global deformation is obtained by applying this field to the structure [10]. Following this concept, Lee et al. [11] developed a layered shell-element inherent strain method for thin-plate structures, improving computational efficiency for large-scale problems, and further proposed a time-saving workflow for predicting inherent deformation in large welded structures [12], thereby demonstrating its engineering applicability. To improve prediction accuracy, Wang et al. [13] introduced a stepwise loading strategy for inherent strains and validated its effectiveness in stiffened plate structures, showing that the loading scheme can significantly influence the final distortion field. Despite these advances in computational efficiency and predictive capability, a central issue remains insufficiently understood: why plastic strain evolves in a particular manner and how its spatial distribution is governed, especially under conditions of non-uniform stiffness and non-uniform restraint. In practice, the inherent strain field is still often treated as a black-box input, which limits the interpretability of distortion mechanisms and the transferability of distortion-control strategies.

The onset and evolution of plasticity during welding are governed by the combined effects of transient temperature gradients, temperature-dependent material properties, and structural constraints/stiffness [14]. A highly reliable understanding of plastic strain formation therefore typically requires thermo-mechanical coupled analysis [15]. Park et al. [16] numerically investigated the role of transverse restraint in welding distortion and residual stress, demonstrating that boundary conditions can fundamentally alter the final distortion mode and stress level. Chen et al. [17] analyzed residual stress distributions in U-rib stiffened plates, highlighting how stiffeners redistribute local restraint and, consequently, the stress/strain fields. Xu et al. [18] conducted experimental and numerical studies on residual stress and distortion in welded "I"-section curved beams, showing that section non-uniformity and curvature modify the deformation response. Collectively, these studies

https://doi.org/10.3390/en19061536

indicate a strong coupling between structural stiffness/constraint and welding response. However, most of the existing literature still evaluates outcomes mainly through final-state distortion or residual stress, while systematic and transferable explanations are scarce for how stiffness influences the restraint path and thereby governs the temporal evolution and spatial distribution of plastic strain.

This limitation becomes more pronounced for thin-walled laminated structures. Compared with conventional plates, internal channels and supporting pillars introduce pronounced spatially non-uniform stiffness, which may change how thermal strain is constrained and how the plastic zone develops. In contrast, the current research on laminated configurations has largely focused on thermal management and cooling performance [19,20], and an explicit framework is still lacking for interpreting plastic strain formation mechanisms, dominant components, and the conditions under which the spatial distribution transitions in welded joints with non-uniform stiffness.

To address these gaps, this work establishes an experimentally validated thermo-mechanical finite element model that explicitly accounts for the non-uniform stiffness characteristics of GH3230 laminated plates and systematically analyzes the initiation and evolution of plastic strain during the welding thermal cycle. By comparison with a conventional plate joint, the influence of non-uniform stiffness on plastic-strain evolution is identified and summarized. On this basis, an equivalent restraint–stiffness spring model is further proposed to parameterize how stiffness reduction and changes in the restraint path affect plastic-strain distributions, thereby providing actionable structural and process-oriented guidance for distortion control of thin-walled laminated welded components.

## 2. Experiments and Measurements

### 2.1. Welding Experiments

A TRUMPF TruDisk 6001 laser (From TRUMPF Photonics Components GmbH, Ulm, Germany) was used to perform a butt weld on two laminated plates with dimensions of 200 mm × 358 mm. The welding parameters are listed in Table 1. For comparison, a butt-welding experiment on a conventional (solid) plate with the same dimensions was carried out using the identical process parameters.

**Table 1.** The parameters of laser welding process.

| Power (W) | Welding Speed (mm/s) | Heat Input (J/mm) | Spot Diameter (mm) | Shielding Gas |
|---|---|---|---|---|
| 2700 | 30 | 90 | 0.3 | Argon |

Note: Specimens were tack-welded at both ends before welding. The defocus distance was 0 mm, and the focal plane was located on the top surface of the workpiece. The beam was perpendicular to the workpiece surface.

During welding, K-type thermocouples were employed to record thermal cycles at locations 3, 4, and 5 mm away from the weld center, as shown in Figure 1. The measured temperature histories were used to validate the thermo-mechanical coupled finite element model. The validation results are presented in Section 4.1.

https://doi.org/10.3390/en19061536

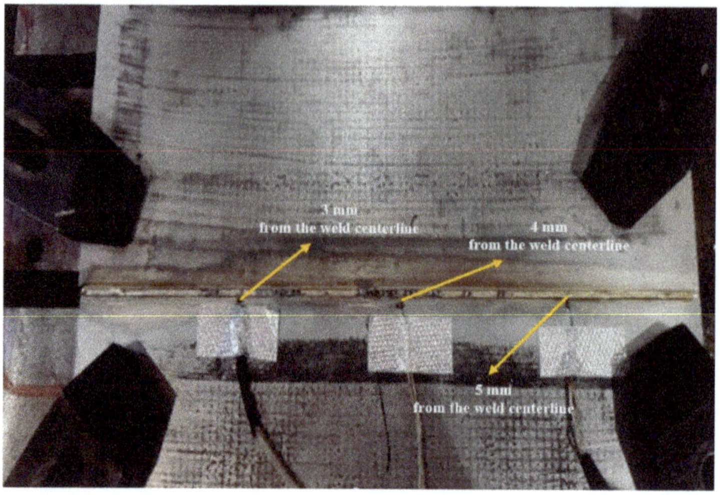


**Figure 1.** Schematic of thermocouple arrangement for thermal cycle measurement during welding.

### 2.2. Measurement of Welding Residual Stress and Distortion

Welding-induced distortion was measured using a coordinate measuring machine (CMM, Hexagon GLOBAL SILVER CLASSIC SR 09.20.08) (From Hexagon Metrology (Qingdao) Co., Ltd., Qingdao, China) with a measurement accuracy of 3 μm, as shown in Figure 2. Welding residual stresses were measured using X-ray diffraction (XRD, $\sin^2\psi$ method) and the hole-drilling method, as shown in Figure 3. Due to the size limitation of the XRD measurement stage, the central region of the welded plate was cut to a size of 90 mm × 90 mm by wire electrical discharge machining prior to XRD measurement. The associated stress relaxation induced by cutting was additionally considered in the finite element simulations. Specifically, the elements in the removed region were progressively deactivated to reproduce the material removal process and the associated stress redistribution. The corresponding simulation results after cutting are presented in Section 4.2, where the residual stress state used for comparison with the XRD measurements is discussed.

**Figure 2.** Measurement of welding-induced distortion using a coordinate measuring machine.

https://doi.org/10.3390/en19061536

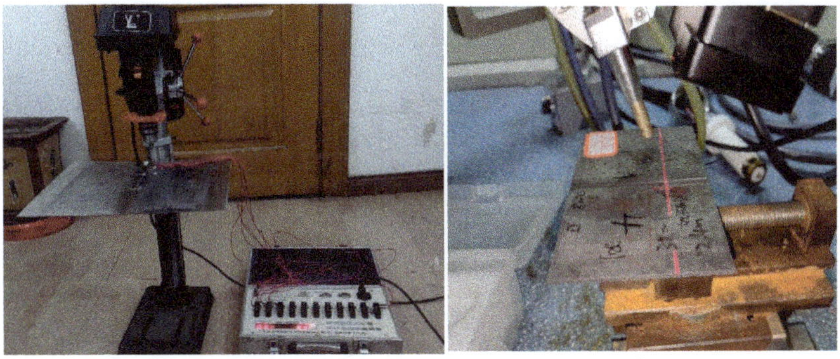

(**a**) (**b**)

**Figure 3.** Experimental setup for welding residual stress measurement: (**a**) hole-drilling method; (**b**) X-ray diffraction (XRD).

Before residual-stress measurements, the specimens were electropolished to remove the surface layer affected by machining and welding spatter. The material removal depth was 50 μm. The measured residual stresses and distortions are reported in Section 4.2, where they are compared with the corresponding finite element results.

## 3. Theory and Modeling

### *3.1. Governing Equations*

Welding is a typical transient thermo-mechanical coupling process involving heat transfer, thermal expansion, elastic–plastic deformation, and constraint-induced stress redistribution. In the present work, the governing equations consist of the transient heat conduction equation and the mechanical equilibrium equations, which are solved in a coupled manner to capture the initiation and evolution of plastic strain during the welding thermal cycle.

The governing equation for transient heat conduction is given in Equation (1):

$$c\rho\frac{\partial T}{\partial t} = \frac{\partial}{\partial x}\left(k_x\frac{\partial T}{\partial x}\right) + \frac{\partial}{\partial y}\left(k_y\frac{\partial T}{\partial y}\right) + \frac{\partial}{\partial z}\left(k_z\frac{\partial T}{\partial z}\right) + Q(x,y,z,t) \tag{1}$$

where $Q$ is the internal heat source generated during welding; $T$ is the temperature; $k_x$, $k_y$, and $k_z$ are the thermal conductivities in three directions (for an isotropic material, $k_x = k_y = k_z = k(T)$); $c(T)$ is the specific heat capacity; $\rho(T)$ is the density; $t$ is time. It should be noted that $c(T)$, $\rho(T)$, and $k(T)$ are temperature-dependent.

In addition to heat conduction inside the welded component, heat dissipation to the surroundings occurs during welding and subsequent cooling. The heat loss consists of convection $q_c$ and radiation $q_r$, which follow Newton's cooling law and the Stefan–Boltzmann law, respectively, as expressed in Equations (2) and (3):

$$q_c = -h_f(T_S - T_0) \tag{2}$$

$$q_r = -\varepsilon\sigma(T_S^4 - T_0^4) \tag{3}$$

where $T_s$ is the surface temperature, $T_0$ is the ambient temperature, $h_f$ is the convection coefficient, whose value depends on the local thermal boundary condition. $E = 0.8$ is the emissivity, and $\sigma = 5.67 \times 10^{-8}$ W/(m$^2$·K$^4$) is the Stefan–Boltzmann constant.

https://doi.org/10.3390/en19061536

Considering that the material remains at elevated temperatures only for a short duration during laser welding, creep effects were neglected [21]. The total strain increment is expressed in Equation (4) [22]

$$\Delta\varepsilon^{total} = \Delta\varepsilon^{e} + \Delta\varepsilon^{p} + \Delta\varepsilon^{th} \tag{4}$$

where $\Delta\varepsilon^{total}$ is the total strain increment, $\Delta\varepsilon^{e}$ is the elastic strain increment, $\Delta\varepsilon^{p}$ is the plastic strain increment, and $\Delta\varepsilon^{th}$ is the thermal strain increment. The thermal strain increment $\Delta\varepsilon^{th}$ is expressed in Equation (5):

$$\Delta\varepsilon^{th} = \alpha(T) \cdot \Delta T \tag{5}$$

where $\alpha(T)$ is the temperature-dependent coefficient of thermal expansion and $\Delta T$ is the temperature change.

To account for possible geometric nonlinearity in the thin-plate structure, the large-strain formulation was adopted in this work [23].

### 3.2. Welding Heat Source

As illustrated in Figure 4, a composite heat source model was adopted in the present laser welding simulation [24]. This model combines a Gaussian surface heat flux with a Gaussian volumetric heat source to describe energy deposition at the surface and within the penetration zone, respectively.

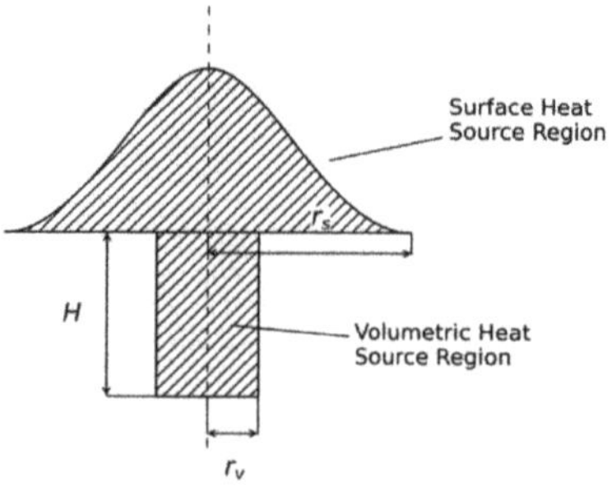


**Figure 4.** Schematic illustration of the composite heat source model for laser welding.

The surface heat flux distribution is expressed as

$$q_s(x,y) = \frac{\alpha Q_s}{\pi r_s^2} \exp\left[-\frac{\alpha(x^2+y^2)}{r_s^2}\right] \tag{6}$$

where $\alpha$ is the heat concentration coefficient, $Q_s$ is the power of the surface heat source, and, $r_s$ is the effective radius of the surface heat source.

The volumetric heat source distribution is expressed in Equation (7):

$$q_v(x,y,z) = \frac{6Q_V(H-\beta z)}{\pi r_v^2 H^2(2-\beta)} \exp\left[\frac{-3(x^2+y^2)}{r_v^2}\right] \tag{7}$$

where $\beta$ is the decay coefficient, $Q_v$ is the power of the volumetric heat source, $r_v$ is the effective radius of the volumetric heat source, and $H$ is the effective penetration depth.

The total welding power satisfies $\eta P = Q_s + Q_v$, where $P$ is the laser power and $\eta$ is the absorption efficiency.

https://doi.org/10.3390/en19061536

### 3.3. Material Properties

The chemical composition of GH3230 is listed in Table 2. Material nonlinearity was considered in the finite element model. The temperature-dependent thermophysical and mechanical properties of GH3230 were adopted from the relevant literature [25], as shown in Figure 5. The elastic behavior of the material was assumed to follow isotropic Hooke's law, while the plastic behavior after yielding was described by the von Mises yield criterion with isotropic strain hardening. The technical requirements for GH3230 comply with GB/T 14992-2025 [26] and this alloy is internationally equivalent to HAYNES® 230 alloy (From Junmu Steel Group, Shanghai, China).

**Table 2.** The chemical composition of GH3230.

| Composition | Ni | Cr | Co | Mo | Al | Ti |
|---|---|---|---|---|---|---|
| Percentage | 48% | 25% | 5% | 2% | 2.5% | 1.5% |

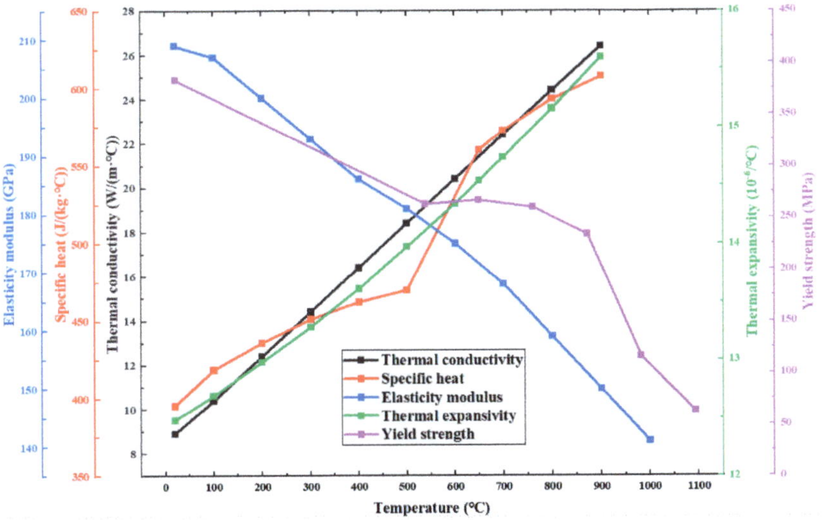


**Figure 5.** Temperature-dependent thermophysical and mechanical properties of GH3230.

### 3.4. Finite Element Mesh and Boundary Conditions

The three-dimensional finite element models of the laminated plate and the conventional solid plate are shown in Figure 6. The laminated plate model explicitly represents the outer skin, inner skin, pin-fin core, and base material, whereas the solid plate model adopts a homogeneous geometry with the same external dimensions. To reduce computational cost, a graded transition meshing scheme was employed. The mesh was refined in the vicinity of the weld seam with an element size of 0.5 mm, while a larger element size was used in regions far from the weld. All simulations were performed in MSC Marc. The models were discretized using hexahedral elements throughout. A coupled thermo-mechanical element type (Element 7) was adopted to enable fully coupled heat transfer and mechanical analysis. The extraction paths used in the subsequent result analysis (Lines 1–3) are also indicated in Figure 6. The total numbers of elements and nodes for each model are labeled in the same figure.

https://doi.org/10.3390/en19061536

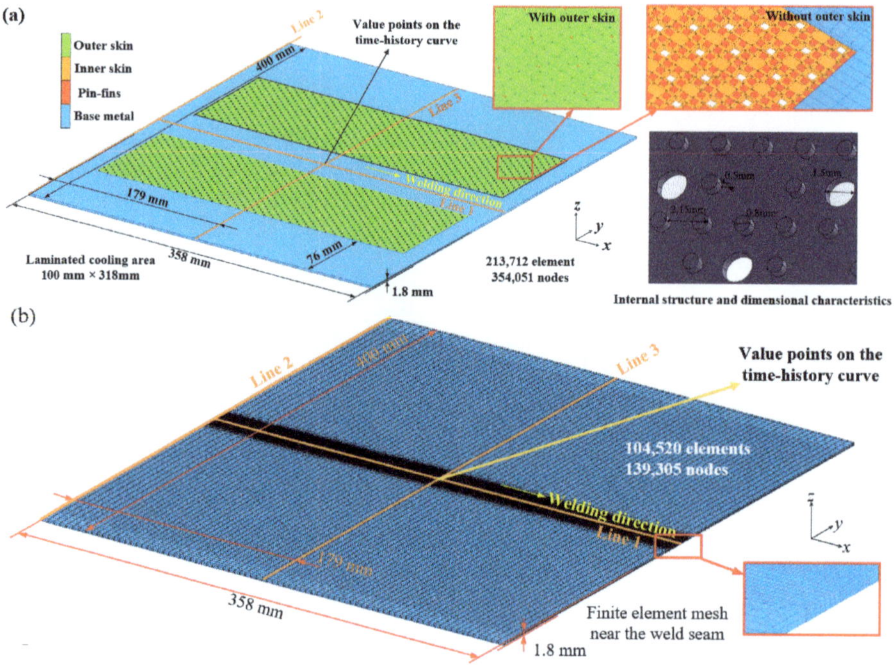


**Figure 6.** Finite element models and mesh configurations of the welded plates. The predefined weld path (Line 1) and reference lines for result extraction (Lines 2 and 3) are also indicated. (**a**) Finite element model of the laminated plate with internal pin-fin structure; (**b**) finite element model of the conventional solid plate.

The clamping conditions in the welding experiments, with the clamp locations shown in Figure 1, were reproduced in the FE model by constraining the corresponding nodes [27]. Since the clamps were positioned far from the weld seam, their influence on heat transfer was neglected. This assumption was later supported by the agreement between the simulated and measured thermal cycles.

To improve the accuracy of the thermal analysis, the additional heat exchange induced by the shielding gas during welding was taken into account. Specifically, an enhanced convection coefficient was applied in the weld-adjacent region exposed to the flowing shielding gas, whereas the regions without forced gas flow were assigned a natural convection coefficient [28]. Thermal radiation was incorporated by means of a temperature-dependent equivalent heat transfer coefficient [29]. The effect of solid-state phase transformation was neglected in the present model [30]. Geometric nonlinearity was included in the finite element analysis to improve computational accuracy [30].

## 4. Results and Discussion

### *4.1. Validation of Thermal Analysis*

Figure 7 compares the simulated transient temperature field with the macrograph of the experimental weld cross-section. The fusion-zone geometry provides an integral validation of the underlying heat-transfer description, because it reflects the balance between the imposed moving heat input and the subsequent heat dissipation/spreading through the plate thickness and along the surface. The results confirm that the adopted heat-source model can reproduce the experimentally observed fusion zone of the laser weld. Full penetration was achieved in the experiment, and the fusion width on the top surface was

https://doi.org/10.3390/en19061536

approximately 2.2 mm; the simulation predicts an essentially identical weld-pool profile under the same heat-source parameters and heat-transfer boundary conditions. Therefore, the calibrated thermal model offers a consistent and reliable basis for subsequent thermo-mechanical analysis, and for isolating the effect of structural stiffness non-uniformity under a fixed thermal (heat-transfer) baseline.

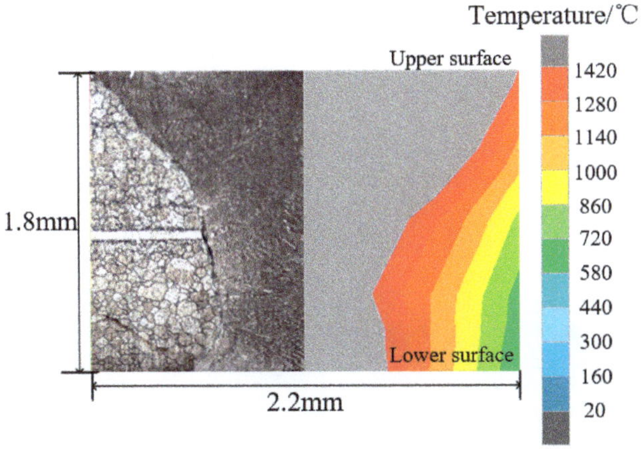


**Figure 7.** Comparison of simulated temperature distribution and weld cross-section.

To ensure reliable thermal predictions, the simulated thermal cycles near the weld were compared with the experimentally measured results. The comparison is shown in Figure 8. Both the experimental measurements and the finite element results were taken at the material surface.

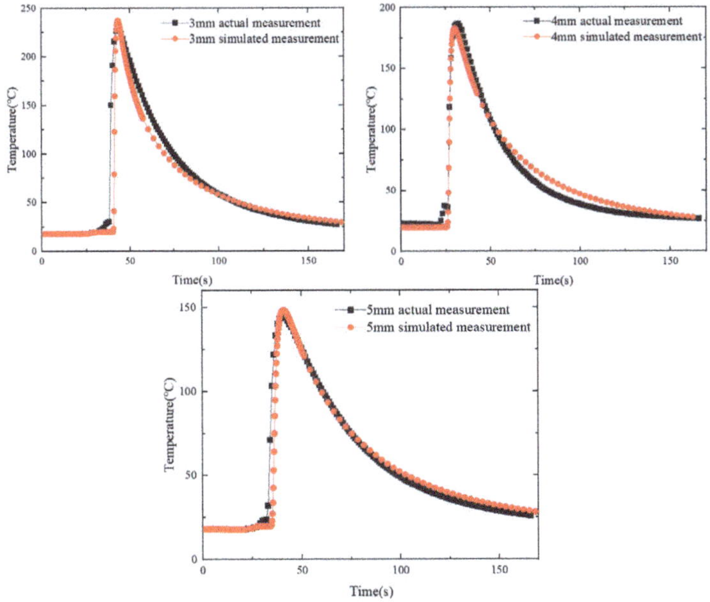


**Figure 8.** Comparison of simulated and measured thermal cycles at different distances from the weld centerline.

https://doi.org/10.3390/en19061536

The thermal histories measured at 3, 4, and 5 mm from the weld centerline exhibit a clear spatial attenuation of the peak temperature with increasing distance, reflecting the combined effects of localized heat input and subsequent heat spreading through conduction and surface heat loss.

Figure 9 further compares the transient temperature fields of the solid and laminated plates during the steady welding stage. In both cases, the isotherms exhibit a characteristic elongated teardrop-shaped pattern, which is associated with the highly concentrated energy input of the laser heat source and its high travel speed. The high-temperature zone is confined to the vicinity of the weld, and the isotherms are stretched along the welding direction. Owing to rapid heat spreading/conduction in thin plates and surface heat losses, the transverse temperature gradient remains very steep. Under identical heat-source parameters and heat-transfer boundary conditions, the overall heat-affected region and the melting-boundary isotherm are comparable for the two configurations, indicating that the subsequent differences in plastic-strain localization and distortion are mainly attributed to stiffness-dependent restraint redistribution.

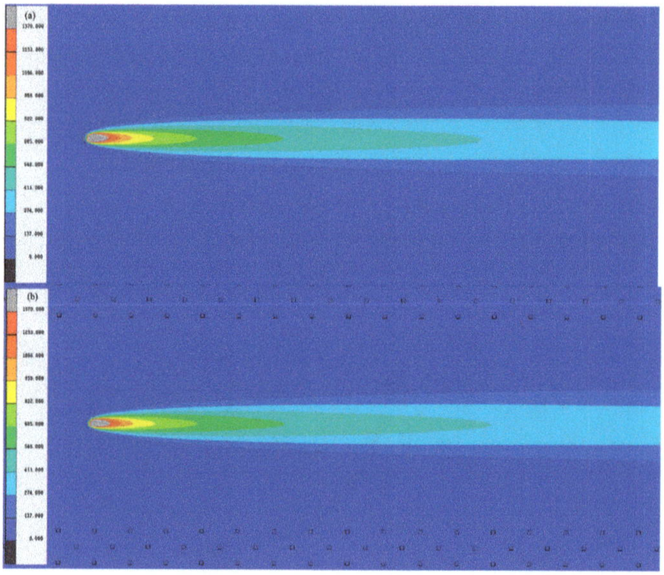

**Figure 9.** Transient temperature contours during the steady welding stage under identical heat input and heat-transfer boundary conditions: (**a**) solid plate; (**b**) laminated plate.

The numerical results reproduce the full heating–peak–cooling sequence with good agreement, capturing both the rapid temperature rise during heat-source passage and the post-heating decay governed by heat dissipation. This agreement indicates that the adopted moving heat-source representation, together with the applied thermal boundary conditions, can reliably describe the transient heat-transfer process in laser welding. The validated thermal model is therefore taken as a consistent heat-transfer baseline for the subsequent thermo-mechanical coupled analysis and welding-distortion prediction.

### *4.2. Welding Residual Stress and Distortion*

The simulated stress-relief effect induced by cutting is shown in Figure 10. Both the longitudinal and transverse residual stresses are released to different extents after cutting, while boundary effects remain evident. As a result, the peak longitudinal residual

https://doi.org/10.3390/en19061536

stress decreases after cutting, whereas the maximum transverse residual stress shows a moderate increase.

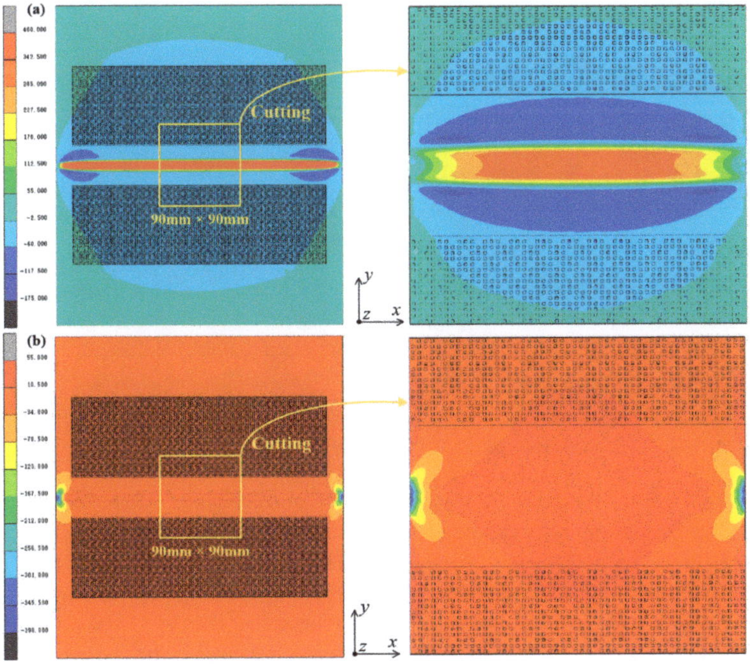


**Figure 10.** Simulated stress redistribution caused by cutting: (**a**) longitudinal residual stress; (**b**) transverse residual stress.

The simulated and measured welding residual stresses of the solid plate and laminated plate are presented in Figures 11 and 12. Owing to the spatial symmetry of the residual stress distribution, only half of the measurement results for each plate are shown.

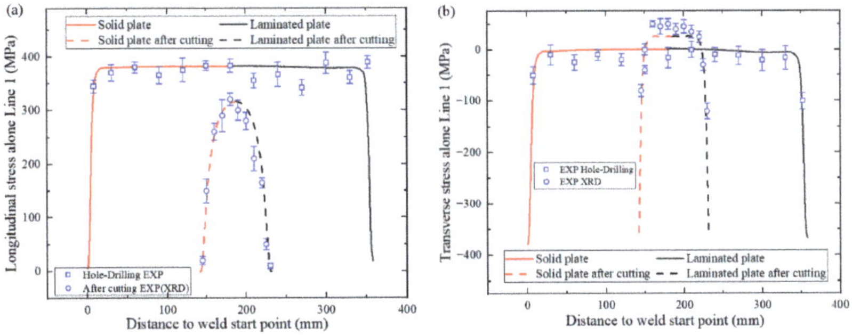


**Figure 11.** Simulated and measured residual stress distributions along Line 1: (**a**) longitudinal residual stress; (**b**) transverse residual stress.

Figure 11a shows the longitudinal residual stress extracted along Line 1. A stress-relaxation zone is observed near both ends of the weld seam, located approximately 12 mm from the plate edges, which is associated with edge effects. The longitudinal stress decreases to nearly zero at the ends, while it remains relatively stable in the central region at about 380 Mpa. Figure 11b presents the transverse residual stress along Line 1. A pronounced

https://doi.org/10.3390/en19061536

compressive region appears within roughly 20 mm from both seam ends; outside this zone, the transverse stress is tensile but remains below 5 Mpa. Overall, the transverse residual-stress field is nearly self-equilibrated, and the integral of transverse stress along the coordinate direction approaches zero.

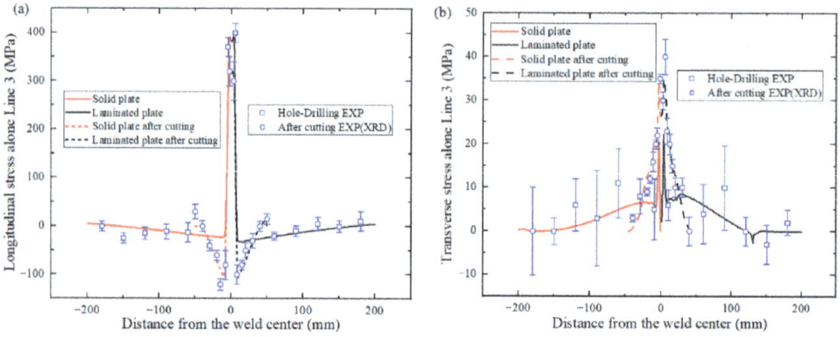


**Figure 12.** Simulated and measured residual stress distributions along Line 3: (**a**) longitudinal residual stress; (**b**) transverse residual stress.

No significant difference is observed between the solid and laminated plates in terms of the residual-stress distributions along Line 1. This indicates that, along the weld direction, residual-stress formation is governed primarily by the localized, highly non-uniform thermal cycle (i.e., the near-field temperature gradients and cooling conditions) imposed by the moving heat source and subsequent heat dissipation, rather than by global stiffness variations introduced by the laminated architecture. In other words, under identical heat-source parameters and heat-transfer boundary conditions, the thermal driving field dominates the residual-stress level along the seam, while stiffness non-uniformity plays a secondary role for this specific stress component and extraction path.

Figure 12a presents the longitudinal residual-stress distribution along Line 3. The tensile longitudinal residual stress is mainly concentrated within the weld seam and within approximately 15 mm on both sides, while compressive stresses develop in the remaining regions to satisfy force equilibrium. The integral of longitudinal residual stress along the transverse direction is approximately zero, indicating a self-equilibrated stress state. Owing to the highly localized laser heat input and the resulting steep thermal gradients, the thermally affected zone is narrow and the residual-stress field is correspondingly concentrated. The maximum longitudinal residual stress reaches approximately 385 Mpa for both the solid and laminated plates, and the overall distribution patterns are similar. Although the laminated plate shows slightly higher compressive levels in certain regions, the difference between the two plates is limited.

Figure 12b shows the transverse residual-stress distribution along Line 3. Both the XRD and hole-drilling measurements exhibit relatively large scatter, mainly because the transverse residual stress is much smaller than the longitudinal component, whereas the experimental uncertainty of both methods in the present study remains on the order of 10 Mpa. In addition, the presence of columnar grains in the weld region makes accurate XRD measurement more difficult. Nevertheless, the overall experimental trends are consistent with the simulation results, and the absolute deviations are generally within 10 Mpa, indicating acceptable agreement between experiment and simulation. The transverse component is much smaller than the longitudinal one, and a noticeable stress drop occurs at the weld center. This behavior can be attributed to the relatively weaker transverse restraint and to partial stress relaxation associated with welding-induced angular/bending

https://doi.org/10.3390/en19061536

distortion, which modifies the constraint state during cooling. The solid and laminated plates exhibit consistent distribution trends, with only minor numerical differences.

Although the solid and laminated plates exhibit highly similar residual-stress distributions, their welding-induced distortions differ markedly, as shown in Figure 13. Figure 13 compares the post-weld out-of-plane (Z-direction) displacement profiles extracted along two representative lines (Line 1 and Line 2). The finite-element predictions agree well with the experimental measurements (EXP), supporting the credibility of the coupled analysis.

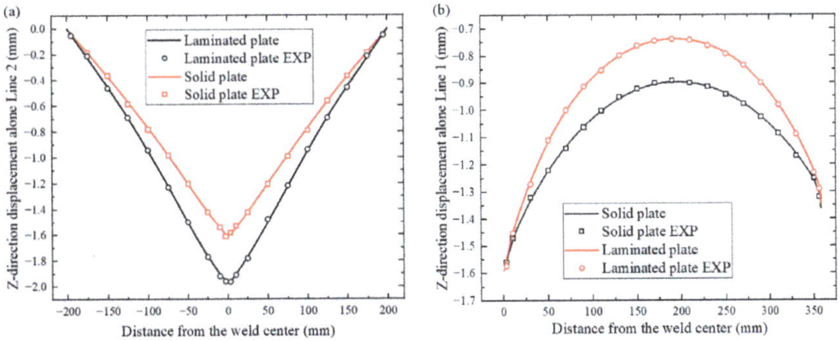


**Figure 13.** Comparison of simulated and measured out-of-plane (Z-direction) displacements: (**a**) displacement distribution along Line 2; (**b**) displacement distribution along Line 1.

Figure 13a shows the Z-direction displacement along Line 2. The deformation profile is approximately symmetric about the weld center and presents a characteristic "V" shape [31]. The maximum displacement occurs near the weld center and decays rapidly with increasing distance, approaching zero in the far field. This deformation mode is typical for thin-plate welding and can be interpreted as the macroscopic manifestation of the localized thermal cycle: steep temperature gradients and heat-flow-driven non-uniform expansion/contraction generate an imbalance between compressive and tensile plastic strains across the weld region, which accumulates as angular distortion during cooling. Under identical heat input and overall dimensions, the laminated plate exhibits significantly larger angular distortion than the solid plate; the maximum displacement at the weld center increases from approximately 1.6 mm to about 2.0 mm.

Figure 13b presents the Z-direction displacement along Line 1, reflecting the overall bending deformation of the plate. Line 1 corresponds to the weld zone, where plastic strain is concentrated. Owing to the difference in weld-pool width between the top and bottom surfaces, the longitudinal contraction during cooling becomes non-uniform through the thickness, leading to the maximum bending deformation in this region. Under the same boundary conditions, the laminated plate shows more pronounced deflection than the solid plate. These results indicate that welding distortion is not determined by residual stress alone. Even when the thermally induced residual-stress patterns are similar, differences in structural stiffness and the associated restraint redistribution during the heating and cooling process can still result in distinct deformation responses. Therefore, the stiffness non-uniformity introduced by the laminated configuration plays a dominant role in the observed distortion difference.

### 4.3. Plastic Strain Distribution

According to inherent strain theory, welding distortion is primarily governed by the spatial distribution of non-recoverable inelastic strain, with plastic strain being the dominant contributor in fusion welding. Because plastic strain is generated under a highly

https://doi.org/10.3390/en19061536

localized thermal cycle, the final deformation pattern can be viewed as an integrated response to the heat-transfer-driven temperature gradients and the associated restraint conditions. Figure 14 presents the post-weld out-of-plane (Z-direction) displacement contours of the laminated and solid plates, with a magnification factor of 20 applied for clarity.

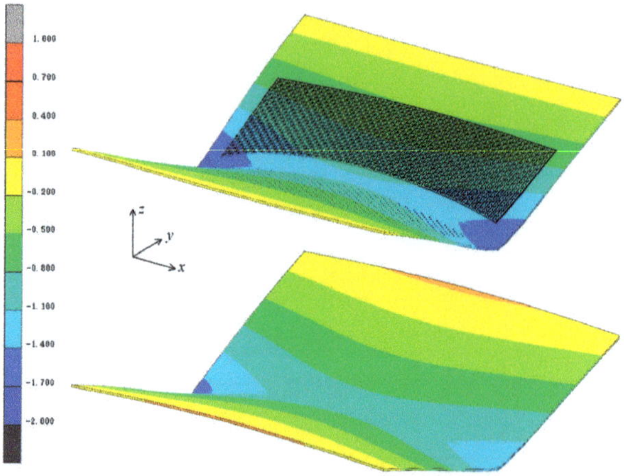


**Figure 14.** Contour plots of welding-induced out-of-plane deformation (20× magnification).

The global deformation modes of the laminated and solid plates are highly similar, and both exhibit the characteristic angular-distortion pattern typical of thin-plate laser welding. The representative displacement profiles extracted along selected lines (Figure 13) further confirm the similarity in deformation shape between the two plates. This consistency suggests that the overall spatial pattern of the plastic-strain field is also broadly comparable, while differences are expected mainly in magnitude and local concentration due to the stiffness non-uniformity introduced by the laminated configuration.

To further clarify the specific differences in plastic strain under identical heat transfer conditions, the distributions of plastic strain were extracted, as shown in Figures 15 and 16.

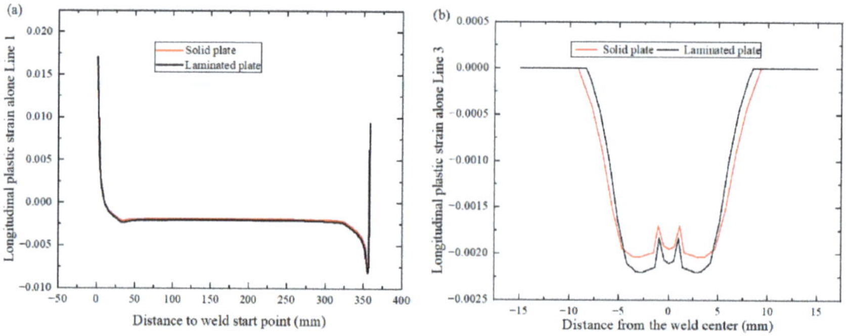


**Figure 15.** Longitudinal plastic strain distributions of the solid and laminated plates: (**a**) along Line 1; (**b**) along Line 3.

Figure 15a presents the post-weld longitudinal plastic strain extracted along Line 1. The laminated and solid plates exhibit similar overall trends. A pronounced positive peak appears near the weld start, followed by a rapid drop into a relatively stable plateau region; another abrupt variation occurs close to the weld termination. These end effects

https://doi.org/10.3390/en19061536

are consistent with the transient thermal boundary at the weld start/stop, where the local thermal cycle (and thus thermal expansion–contraction mismatch) differs from that in the steady-travel segment. Within the plateau region, the laminated plate shows slightly larger absolute longitudinal plastic strain than the solid plate.

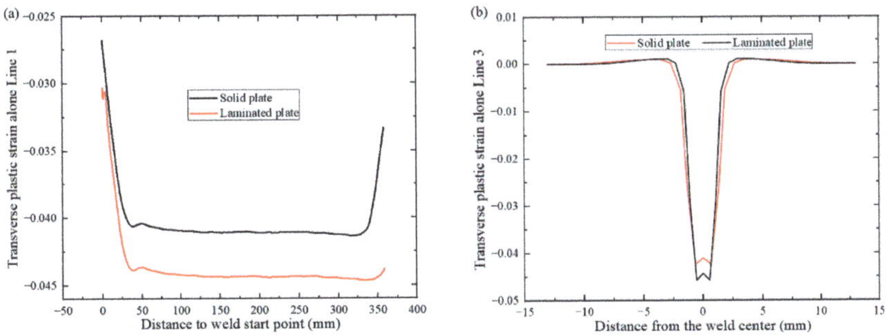


**Figure 16.** Transverse plastic strain distributions of the solid and laminated plates: (**a**) along Line 1; (**b**) along Line 3.

Figure 15b shows the local longitudinal plastic-strain distribution along Line 3 with the weld center as the origin. The plastic strain is strongly localized around the weld center and decays rapidly with distance, approaching zero beyond approximately 10 mm from the centerline. The narrow affected region reflects the highly concentrated laser heat input and the resulting steep temperature gradients. Although the curve shapes for the two plates are similar, differences appear in the depth of the concave region and in the local details near the weld center. The laminated plate exhibits larger absolute plastic-strain values, while the affected zone is relatively narrower, indicating a more localized longitudinal plastic response under the non-uniform stiffness condition.

Figure 16a presents the transverse plastic-strain distribution along Line 1. Both the laminated and solid plates show rapid variations near the weld start and termination, followed by a relatively stable plateau over the long central segment. This start/stop behavior is consistent with the non-steady thermal boundary at the beginning and end of welding, where the transient heat input and heat dissipation conditions differ from those in the steady-travel region. In the plateau region, the laminated plate exhibits larger absolute transverse plastic-strain values than the solid plate.

Figure 16b shows the local transverse plastic-strain distribution along Line 3 with the weld center as the origin. A sharp negative valley appears near the weld center, and the strain rapidly recovers toward zero away from the weld, becoming negligible beyond approximately 12 mm from the centerline. The strong localization reflects the highly concentrated laser heat input and the resulting steep temperature gradients. Although the overall curve shapes of the two plates are similar, differences are evident in the valley depth and local curvature near the weld center. The laminated plate develops a larger peak magnitude of transverse plastic strain, while the affected zone is relatively narrower.

Overall, the transverse plastic strain is significantly larger than the longitudinal component, consistent with the different restraint levels in the transverse and longitudinal directions during welding. By comparing Figures 15 and 16, it is evident that the laminated and solid plates share broadly similar spatial patterns of plastic strain, whereas clear differences exist in magnitude and effective distribution width. The laminated plate generally exhibits larger plastic-strain magnitudes within a narrower affected region. Under an identical thermal loading baseline (same heat-source parameters and heat-transfer boundary conditions), these differences can be attributed to the stiffness-dependent restraint

https://doi.org/10.3390/en19061536

redistribution, which modifies how the heat-transfer-driven thermal strains are constrained and hence alters the evolution and spatial distribution of plastic strain.

### *4.4. Plastic Strain Evolution During Thermo-Mechanical Coupling*

To further clarify the roles of heat transfer and structural stiffness in governing plastic strain evolution, the temperature-dependent evolution paths of plastic strain during a representative thermal cycle were extracted, as shown in Figure 17.

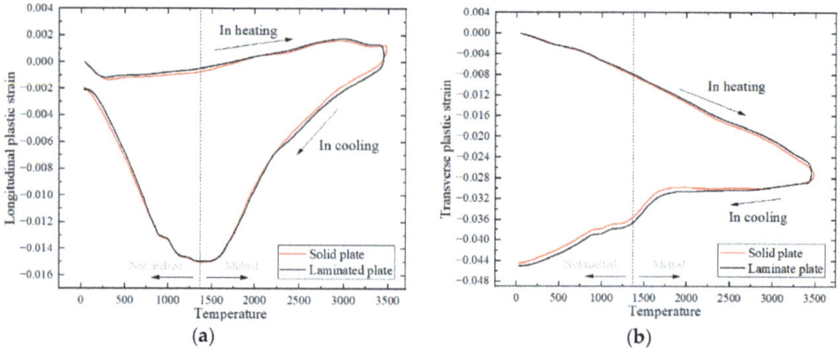


**Figure 17.** Evolution of plastic strain with temperature during a representative thermal cycle: (**a**) longitudinal plastic strain; (**b**) transverse plastic strain.

Figure 17 presents the plastic strain–temperature evolution for the solid and laminated plates during a representative thermal cycle. The arrows indicate the heating and cooling stages. The vertical dashed line denotes the melting boundary temperature (1370 °C), separating the non-melted and melted regimes.

Figure 17a shows the evolution of longitudinal plastic strain. During heating, the longitudinal plastic strain is slightly negative at low temperatures and increases gradually with temperature, becoming positive at elevated temperatures. During cooling, the minimum longitudinal plastic strain appears near the melting boundary temperature; as temperature decreases further, the strain partially recovers and approaches a small negative value at room temperature. The laminated and solid plates exhibit similar trajectories, while the laminated plate maintains slightly higher strain levels throughout the cycle.

Figure 17b illustrates the transverse plastic-strain evolution. During heating, the transverse plastic strain increases monotonically with temperature. During cooling, it continues to increase and reaches its maximum magnitude after complete cooling. The transverse component is markedly larger than the longitudinal one, and no sign reversal or apparent recovery is observed.

These directional differences arise from the distinct constraint paths under a highly localized welding thermal cycle. Along the longitudinal direction, each material point undergoes a sequential heating–cooling history as the moving heat source passes, and the shrinkage of previously heated regions imposes tensile effects on the subsequently heated material, enabling partial redistribution and recovery of longitudinal plastic strain during cooling. In contrast, the transverse direction is predominantly constrained by the adjacent unmelted material on both sides of the weld, and the lateral restraint under steep temperature gradients promotes the accumulation of compressive plastic strain. In addition, welding-induced angular distortion modifies the transverse constraint state during cooling, further intensifying transverse plastic-strain accumulation and leading to larger transverse magnitudes.

https://doi.org/10.3390/en19061536

### 4.5. Influence of Structural Stiffness on the Thermo-Mechanical Process

To provide a more quantitative interpretation of the plastic strain evolution described above, a stiffness–bar–spring model was developed, as illustrated in Figure 18. The model is intended to describe the evolution of longitudinal plastic strain during heating by representing the restraint effect of the surrounding base material with an equivalent spring.

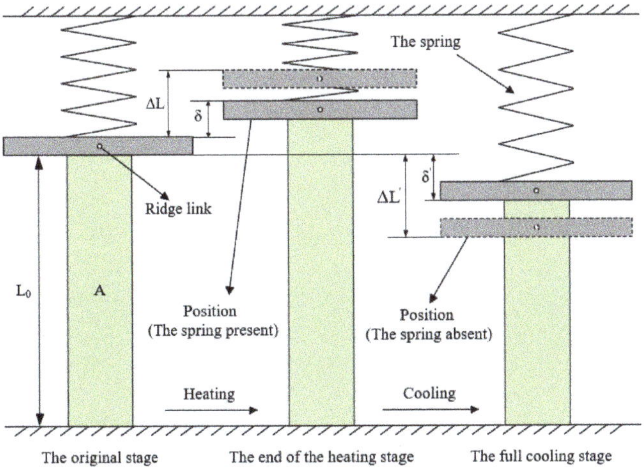


**Figure 18.** Schematic illustration of the proposed stiffness–bar–spring model.

In the proposed stiffness–bar–spring model, the two ends of the bar are assumed to be rigidly connected through an end link, which enforces displacement compatibility. The material is assumed to be temperature-independent for analytical simplicity, and an elastic–perfectly plastic constitutive law is adopted, such that the stress remains constant once the yield strength is reached. Owing to the sufficiently large longitudinal restraint, the bar ends are further assumed to remain horizontal during deformation. The coefficients of thermal expansion, yield strength, and elastic modulus are denoted by $\alpha$, $\sigma_s$, and $E$, respectively. The stiffness of the equivalent spring is denoted by K.

As illustrated in Figure 18, during the heating stage, if no spring restraint is present, the free thermal elongation of the bar is

$$\Delta L = L_0 \alpha \Delta T \quad (8)$$

where $L_0$ is the initial length of the bar, $\alpha$ is the coefficient of thermal expansion, and $\Delta T$ is the temperature increment.

When the spring restraint is introduced, force equilibrium must be satisfied between the bar and the spring. The spring force equals the internal axial force in the bar, leading to:

$$\delta K = \frac{(\Delta L - \delta)EA}{L_0 + \Delta L} \approx \frac{(\Delta L - \delta)EA}{L_0} \quad (9)$$

where $K$ is the stiffness of the equivalent spring, $E$ is the elastic modulus, $A$ is the cross-sectional area of the bar, and $\delta$ is the restrained displacement induced by the spring.

Solving for $\delta$, the restrained displacement is obtained as

$$\delta = \frac{EAL_0 \alpha \Delta T}{KL_0 + EA} \quad (10)$$

https://doi.org/10.3390/en19061536

The compressive strain in the bar under restrained thermal expansion is:

$$\varepsilon = \frac{\Delta L - \delta}{L_0} \tag{11}$$

When the temperature increase is sufficiently large such that yielding occurs in the bar, the compressive strain reaches the yield strain. Therefore,

$$\varepsilon = \frac{\Delta L - \delta}{L_0} = \frac{\alpha \Delta T K L_0}{K L_0 + EA} = \varepsilon_s \tag{12}$$

from which the temperature rise required to initiate yielding is obtained as

$$\Delta T_Y = \frac{\varepsilon_s (K L_0 + EA)}{\alpha K L_0} \tag{13}$$

Equation (13) indicates that as the stiffness $K$ increases, the required temperature increment for yielding decreases. In other words, stronger restraint promotes earlier yielding. This explains why the solid plate, which exhibits higher effective stiffness, develops a wider plastic zone, consistent with the differences observed in Figures 15 and 16.

Compared with the solid plate, the laminated plate contains numerous non-uniformly distributed internal channels, which reduce its overall structural stiffness and consequently lower its resistance to deformation. In the model shown in Figure 18, this stiffness reduction is reflected by a decrease in the equivalent external restraint parameter $K$. During the analysis, both the laminated plate and the solid plate were subjected to the same thermal input and surface heat-flux distribution. According to Equation (13), a smaller $K$ leads to a higher critical temperature for yielding during the thermal expansion stage. In other words, weaker restraint delays yielding and tends to confine plastic deformation to a narrower region, whereas stronger restraint promotes yielding over a wider zone. This explains why the plastic strain in the laminated plate is more concentrated near the weld seam, as shown in Figures 15b and 16a.

After complete cooling, the stress in the weld region is assumed to approach the yield level. From force equilibrium:

$$\delta' K = \varepsilon_Y EA \tag{14}$$

The corresponding elastic recovery of the bar is

$$\varepsilon_Y L_0 \tag{15}$$

The residual compressive plastic strain is therefore expressed as:

$$\varepsilon' = \frac{\varepsilon_Y EA}{K} - \varepsilon_Y L_0 \tag{16}$$

Equation (16) shows that the peak residual compressive plastic strain decreases with increasing stiffness $K$. Since the solid plate has a larger effective stiffness than the laminated plate, its final compressive plastic strain remains lower. This theoretical prediction is consistent with the evolution results shown in Figure 17. Overall, the proposed model successfully captures the relationship between welding-induced plastic strain and external stiffness, and qualitatively clarifies how the external stiffness parameter $K$ influences the magnitude of plastic strain. It therefore provides a useful mechanistic interpretation and reference for understanding the evolution of plastic strain during the thermo-mechanical process.

https://doi.org/10.3390/en19061536

## 5. Conclusions

1. A coupled thermo-mechanical FE model with a composite Gaussian surface–volumetric moving heat source was developed and validated using near-weld thermal cycles and fusion geometry.
2. Under identical laser welding parameters, laminated and solid plates exhibit similar residual-stress distributions, suggesting residual stress is governed primarily by the local weld thermal gradient.
3. In contrast to residual stress, the laminated plate shows larger angular/bending distortion, highlighting the effect of spatially non-uniform stiffness on restraint redistribution and deformation.
4. Plastic-strain distributions are similar in pattern but differ in magnitude and localization: the laminated plate develops higher peak plastic strain within a narrower zone, dominated by transverse plastic strain.
5. An equivalent restraint–stiffness spring model links stiffness to yielding and plastic-zone characteristics, providing a mechanistic basis for distortion prediction and stiffness-informed control of welded laminated cooling plates.

**Author Contributions:** Conceptualization, C.L.; Methodology, Z.D.; Software, Y.C., H.W., Z.Z., F.H., X.Z. and C.W.; Validation, C.L., Y.C., H.W., Z.Z., F.H., X.Z. and C.W.; Data curation, C.L.; Writing—original draft, C.L.; Writing—review & editing, Z.D.; Supervision, Z.D.; Funding acquisition, Z.D. All authors have read and agreed to the published version of the manuscript.

**Funding:** This work is supported by Heilongjiang Chunyan Innovation Team Program (CYCX24008).

**Data Availability Statement:** The original contributions presented in this study are included in the article. Further inquiries can be directed to the corresponding author.

**Conflicts of Interest:** The authors declare that they have no known competing financial interests or personal relationships that could have appeared to influence the work reported in this paper.

## References

1. Horajski, D.; Cieślik, J.; Cieślik, M. Advanced Structural and Technological Method of Reducing Distortion in Thin-Walled Welded Structures. *Materials* **2021**, *14*, 504. [CrossRef] [PubMed]
2. Qiu, Y.; Yan, R.; Wang, N.; Shen, W.; Xu, S.; Li, M.; Qin, K. Stress Amplification Effect and Fatigue Strength Evaluation of Marine Thin Plate Welded Structure Considering Welding Deformation: Theoretical and Experimental Analysis. *Thin-Walled Struct.* **2023**, *188*, 110871. [CrossRef]
3. Ruíz, H.; Sanchez, A. Influence of the Plate Dimensions of a Fillet Weld Joint on Angular Deformation and Residual Stresses. *Pol. Marit. Res.* **2025**, *32*, 124–131. [CrossRef]
4. He, Y.; Yang, Z.; Yang, D.; Duan, R.; Xu, J.; Wang, W.; Shi, M.; Xu, G.; Chen, S. Numerical Simulation of Local Cooling on Residual Stress and Deformation of Welded Joints of Unequal Thickness Plates. *J. Mater. Res. Technol.* **2025**, *36*, 8410–8420. [CrossRef]
5. Liu, Y.; Ma, N.; Lu, F.; Fang, H. Measurement and Analysis of Welding Deformation in Arc Welded Lap Joints of Thin Steel Sheets with Different Material Properties. *J. Manuf. Process.* **2020**, *61*, 507–517. [CrossRef]
6. Mao, Z.; Zhou, C.; Wang, C.; Li, Y.; Huang, Z.; Deng, D. Investigation on Residual Stress and Welding Deformation in a Ultra-High-Strength Quenched Steel Thin-Plate Butt Joint. *China Weld.* **2026**, *35*, 100032. [CrossRef]
7. Wang, J.; Ma, N.; Deng, D. Progress in Welding Distortion Prediction and Control Technology for Advanced Manufacturing. *J. Manuf. Process.* **2025**, *152*, 1012–1036. [CrossRef]
8. Liang, X.; Chen, Y.; Cheng, L.; Yang, Q.; To, A. A Modified Inherent Strain Method for Fast Prediction of Residual Deformation in Additive Manufacturing of Metal Parts. In Proceedings of the 2017 Solid Freeform Fabrication Symposium, Austin, TX, USA, 7–9 August 2017.
9. Chino, T.; Kunugi, A.; Kawashima, T.; Watanabe, G.; Can, C.; Ma, N. Fast Prediction for Resistance Spot Welding Deformation Using Inherent Strain Method and Nugget Model. *Materials* **2021**, *14*, 7180. [CrossRef] [PubMed]

https://doi.org/10.3390/en19061536

10. Ma, N.; Nakacho, K.; Ohta, T.; Ogawa, N.; Maekawa, A.; Huang, H.; Murakawa, H. Inherent Strain Method for Residual Stress Measurement and Welding Distortion Prediction. In *Proceedings of the ASME 2016 35th International Conference on Ocean, Offshore and Arctic Engineering*; American Society of Mechanical Engineers: New York, NY, USA, 2016; Volume 50008, p. V009T13A001. [CrossRef]
11. Lee, C.H.; Perrera, D.; Chung, H. Welding Deformation Analysis by Inherent Strain Method Based on Layered Shell Element. *J. Mar. Sci. Eng.* **2021**, *9*, 632. [CrossRef]
12. Yu, L.; Deng, D.; Xie, L.; Murakawa, H. Prediction of deformation for large welded structure based on inherent strain. In Proceedings of the International Institute of Welding (IIW), Osaka, Japan, 11–16 July 2004.
13. Wang, L.; Miao, Y.; Hu, G.; Zhao, Y.; Deng, Q. Numerical Simulations and Experimental Verification of T-Structure Welding Deformation Using the Step-by-Step Loading Inherent Strain Method. *Metals* **2023**, *13*, 1312. [CrossRef]
14. Muránsky, O.; Hamelin, C.J.; Patel, V.I.; Luzin, V.; Braham, C. The Influence of Constitutive Plasticity Models on Accumulated Plastic Strains in Finite Element Weld Analyses. *Int. J. Solids Struct.* **2015**, *66–67*, 518–530. [CrossRef]
15. Han, Y. Multiphysics Study of Thermal Profiles and Residual Stress in Welding. *Materials* **2024**, *17*, 886. [CrossRef] [PubMed]
16. Park, J.; An, G.; Ma, N.; Kim, S. Effect of Transverse Restraint on Welding Residual Stress in V-Groove Butt Welding. *Metals* **2022**, *12*, 654. [CrossRef]
17. Chen, L.; Wang, T.; Pan, J.; Ma, N.; Wang, R. Welding Residual Stress Distribution of U-Rib Stiffened Plates of Steel Box Girders and Its Influence on Structural Natural Frequencies. *Front. Mater.* **2022**, *9*, 876111. [CrossRef]
18. Xu, F.; Zhao, Y.; Chen, J. Experimental and Numerical Investigation on Residual Stress and Deformation of Welded I-Section Curved Steel Beams. *Thin-Walled Struct.* **2024**, *199*, 111821. [CrossRef]
19. Chen, Z.; Chen, X.; Yang, X.; Yu, B.; Wang, B.; Zhu, J.; Chen, Y.; Cai, W. Numerical Study on Cooling Characteristics of Turbine Blade Based on Laminated Cooling Configuration with Clapboards. *Energy* **2024**, *299*, 131372. [CrossRef]
20. Li, H.; Xie, F.; Zhuang, Y.; Wang, Y.; Yan, Y.; Xu, G.; Cui, J. Numerical Investigation of Cooling Performance of 3D-Printable Cooling Structure for Aero-Engine Combustor. *Appl. Therm. Eng.* **2024**, *238*, 122196. [CrossRef]
21. Liu, C.; Luo, Z.; He, Y.; Huang, S.; Tuo, J.; Ding, H. Numerical simulation of welding residual stresses in 6082-T6 thin aluminum alloy. *Measurement* **2024**, *234*, 114800. [CrossRef]
22. Służalec, A. *Theory of Thermomechanical Processes in Welding*; Springer: Dordrecht, The Netherlands, 2005. [CrossRef]
23. Wu, C.; Kim, J.-W. Numerical prediction of deformation in thin-plate welded joints using equivalent thermal strain method. *Thin-Walled Struct.* **2020**, *157*, 107033. [CrossRef]
24. Wang, S.; Shi, G.; Zhao, L.; Dai, Y.; Hou, T.; He, Y.; Chen, P.; Qin, F. Modelling and Optimization of Continual Laser Joining Processes for Silicon Aluminum Alloy in Microwave Devices. *Crystals* **2023**, *13*, 631. [CrossRef]
25. Li, C.K.; Dong, Z.B.; Wang, H.; Han, F.; Teng, J.F.; Lü, Y.L. Research on Service Life Prediction of Closely Spaced Array Hole Column Laminated Cooling Structure. *Trans. China Weld. Inst.* **2022**, *43*, 101–106. [CrossRef]
26. *GB/T 14992-2025*; Superalloys and Intermetallic Compounds—Designation and Chemical Composition. China Iron & Steel Association: Beijing China, 2025.
27. Alghamdi, A.; Saleh, M. Finite Element Simulation of the Effect of Phase and Thermal Boundary Conditions on Temperature Distribution and Residual Stress. *Crystals* **2022**, *12*, 1422. [CrossRef]
28. Bannour, S.; Abderrazak, K.; Mhiri, H.; Palec, G.L. Effects of Temperature-Dependent Material Properties and Shielding Gas on Molten Pool Formation During Continuous Laser Welding of AZ91 Magnesium Alloy. *Opt. Laser Technol.* **2012**, *44*, 2459–2468. [CrossRef]
29. Chiumenti, M.; Cervera, M.; Salmi, A.; Agelet de Saracibar, C.; Dialami, D.; Matsui, K. Finite Element Modeling of Multi-Pass Welding and Shaped Metal Deposition Processes. *Comput. Methods Appl. Mech. Eng.* **2010**, *199*, 2343–2359. [CrossRef]
30. Liu, Y.; Yu, Y.; Wang, P.; Fang, H.; Ma, N. Analysis and Mitigation of the Bending Deformation in Girth-Welded Slender Pipes with Numerical Modelling and Experimental Measurement. *J. Manuf. Process.* **2022**, *78*, 278–287. [CrossRef]
31. Zubairuddin, M.; Vasudevan, M.; Das, P.K.; Alam, M.M.; Kumar, K.S.; Prabhakar, S. FEM Based Thermal and Mechanical Analysis of Comparative Study of TIG and A-TIG Welding on P91 Steel. *Sci. Rep.* **2025**, *15*, 10271. [CrossRef] [PubMed]




https://doi.org/10.3390/en19061536

*energies*

MDPI

*Article*

# Evaluation of the Intensity of Heat and Mass Transfer Processes in Cavitation Environments

**Anatoliy Pavlenko**

Department of Building Physics and Renewable Energy, Kielce University of Technology, Aleja Tysiąclecia Państwa Polskiego, 7, 25-314 Kielce, Poland; apavlenko@tu.kielce.pl

**Abstract**

This study investigates the impact of cavitation phenomena on heat and mass transfer in working fluids. To quantify the intensity of transport processes within cavitation bubble clusters, a numerical analysis of bubble dynamics was carried out with explicit consideration of fluid compressibility. The results demonstrate that physicochemical transformations induced by cavitation are governed not only by shock waves and pressure pulses generated during bubble collapse, but also by extreme thermal effects arising within collapsing cavitation clouds. Under conditions of maximum bubble compression, the vapor inside the bubbles and the surrounding liquid may undergo a transition to a supercritical state. The developed model elucidates the structure of microflows in the interbubble region and provides a quantitative evaluation of local velocity, pressure, and heat flux fields. The systematic assessment of cavitation-enhanced heat and mass transfer offers valuable insights for the advancement of conventional heat and mass transfer technologies and the design of innovative devices in mechanical and chemical engineering.

**Keywords:** cavitation bubble; impact pressure; bubble dynamics; bubble model; mass transfer; phase transitions

## 1. Introduction

Cavitation phenomena have long attracted considerable scientific interest owing to the unique effects associated with the collapse of bubble clusters, including the generation of extreme temperature fields and intense pressure fluctuations in the surrounding fluid. During bubble oscillations, heat and mass exchange occur across the bubble–liquid interface, while heat transfer in the bulk liquid may be substantially enhanced, for example, through an increase in the critical heat flux and overall heat transfer intensity [1]. A variety of cavitation-based mechanisms for heat transfer enhancement in liquids have been proposed and investigated [2,3].

Despite these advances, the influence of cavitation on heat and mass transfer remains highly complex, involving multiple interacting mechanisms that are difficult to generalize across different flow regimes and applications. Consequently, a detailed analysis of cavitation cluster dynamics and the conditions governing the onset and development of cavitation effects is essential for a fundamental understanding of cavitation-enhanced transport processes.

In the early stages of cavitation research, efforts were focused primarily on its destructive effects. Subsequently, however, researchers began to explore and exploit its unique properties. Hydrodynamic cavitation is associated with the formation and collapse of cavitation bubbles caused by pressure pulsations, leading to the creation of localized regions of extremely high temperature and pressure. These effects can be successfully utilized in various applications, such as water purification, biomass conversion and pretreatment, and substance extraction [4].

Both acoustic and hydrodynamic cavitation enhance anaerobic digestion and fermentation processes, thereby increasing biogas production [5]. In addition to large-scale industrial applications, cavitation is also employed in smaller-scale technical processes, including homogenization, emulsification, and extraction [6,7]. Ultrasonic treatment is widely used in mineral flotation due to its ease of operation, absence of secondary pollutant formation, and high level of safety [8].

The technical applications of cavitation effects are diverse, and hydrodynamic cavitation is considered a promising technology for process intensification owing to its high energy efficiency and ability to induce chemical reactions [9]. Cavitation occurs through several stages; during bubble compression, extremely high temperatures and pressures are generated within the liquid, resulting in localized heating of the surrounding medium.

In oil and gas production technologies, gas hydrates may form under conditions of high pressure and low temperature. If the temperature remains above a critical value due to cavitation effects, gas hydrate formation can be inhibited [10,11]. As cavitation technologies continue to gain widespread application, the development of cavitation generators is also progressing rapidly [12,13]. Therefore, detailed investigation and a deeper understanding of the physical processes of cavitation—particularly those occurring within cavitation clusters—are essential for further advancement in this field of research.

In this work, we consider the patterns of bubble behaviour within a cluster, taking into account the mutual influence of the effects accompanying the cavitation development process. Regardless of whether a dynamic effect at a local point in the fluid is a consequence of the collective influence of a large number of bubbles or is solely caused by the action of a single bubble (for example, the formation of a cumulative microjet or the emission of a spherical shock wave by a collapsing bubble), all bubbles participate directly or indirectly in its creation. The difference between the behaviour of a single bubble within a bubble cloud and the behaviour of an isolated bubble in an infinite volume of liquid is primarily due to the influence of its nearest neighbours on the bubble's evolution. As they develop, these neighbours create their own pressure field in the vicinity of the bubble, which differs from the external pressure currently acting on the system. Therefore, the behaviour of an individual bubble within a cluster under variation of the external pressure is more complex than that represented in the single-bubble dynamics model.

Compared to the large volume of work dedicated to the study of single bubble dynamics, publications related to the study of the collective behaviour of bubbles and the evolution of an individual bubble within a dynamically developing ensemble are less prevalent. These studies can be broadly divided into three main groups.

The first group focuses on phenomena related to the dynamic effects of bubbles, for example, [14,15]. The authors show that hydrostatic pressure is the dominant factor in controlling cavitation activity, but due to the inertia of the cavitation cluster, dynamic effects are difficult to predict for clustered bubbles. A cavitation system possesses a specific bubble density, which can be manipulated by changing the hydrostatic pressure to enhance dynamic effects. This could serve as a promising approach for developing technologies, such as material synthesis or processing [16,17].

The core idea of the second group of studies focuses on the fact that bubbles can accumulate into clusters and thereby intensify the cavitation effect. However, many authors point to the ambiguity of such effects [18,19]. The behaviour of bubbles in clusters differs significantly from the dynamics of single bubbles. Their dynamics are mutually influenced

https://doi.org/10.3390/en19051120

by neighboring bubbles, leading to attraction or repulsion. These effects can be explained by the presence of the secondary Bjerknes force.

Some authors [20,21] indicate that the radial oscillations of a cavitation bubble are suppressed by larger neighbors when they are in close proximity, and the magnitude and direction of the secondary Bjerknes force change, leading to a transition between mutual attraction and repulsion of the bubbles. Authors in [22] assert that the suppression of bubble pulsations increases with increasing bubble density. If the bubble density exceeds a critical value, a repulsive region emerges, preventing their coalescence. For example, at a pressure of 3 MPa and a bubble density up to $1 \times 10^{13}$ m$^{-3}$, distinct zones of attraction appear, and as the hydrostatic pressure increases, the attraction zone gradually expands towards a higher density region. In fact, increasing hydrostatic pressure can increase the stable bubble density of the cluster, which will intensify the cavitation effect. It can be argued that while there are numerous publications devoted to experimental studies of single bubbles [23,24], there are significantly fewer studies on the mutual influence of double bubbles [25,26], and very few experimental reports focus on clusters, which constitute the third approach to research. Furthermore, the question of what causes the dynamic effect—whether it is the cumulative effect of the cluster or the influence of individual bubbles surrounding the cluster—remains unexplained. Modelling of dynamic effects [27–29] is more widely applied. Among the approaches presented in publications for describing cavitation processes, several well-established types of models can be distinguished.

1. Multiscale models [30–33].

This approach can describe a compressible multiscale cavitation region where the dynamics of large vapour cavities and microbubbles are modeled, taking into account the average compressibility. The cavitation cluster is treated as a liquid–vapor mixture continuum in the Eulerian sense. These models are developed by coupling an Eulerian medium and functional dependencies, expressed in terms of velocities and radii, which are tracked in the Lagrangian sense using various approaches. For a stable transition between the Eulerian and Lagrangian frameworks, the transition criterion can be defined based on a parameter that accounts for the non-uniformity of the domain discretization. In hybrid models, the onset of cavitation is first estimated in low-pressure regions, which limits predictions using Eulerian mass transfer models.

2. Models relating fluid characteristics to thermodynamic parameters [34–37].

This method, based on equations of state that link fluid characteristics to thermodynamic parameters, is limited only to very specific problems unrelated to either bubble growth dynamics or the dynamic effects of cavitation. In general, these models represent a simplified representation of real processes and are usually solved for an averaged substance whose properties reflect the weighted average of the volume fraction of the two media. Another approach involves modelling cavitation using real gas polynomial equations of state to accurately predict the properties of water in any state. Modelling complex dynamic effects relies on numerical methods which, although accurate, require significant computational resources and databases, and are not applicable for rapid prediction in practical scenarios.

3. Phase boundary modelling [38,39].

4. Modelling using transport equations [40–43].

Such models accurately predict the dynamics of the bubble cloud in the initial stage by considering bubble interaction. However, in the final stage, the inertia effect, represented as a second-order derivative, plays a more significant role and shows a higher collapse speed.

5. Predicting bubble dynamics using the Rayleigh–Plesset equation [44–46].

The Rayleigh–Plesset equation accurately describes bubble oscillations in incompressible liquids but neglects energy losses due to acoustic frequencies, as it was based on

https://doi.org/10.3390/en19051120

incompressible fluids. Therefore, this equation becomes inapplicable in cases where energy loss cannot be neglected. Nevertheless, this approach to describing cavitation processes is fundamental in the development of cavitation modelling, including the description of bubble cluster dynamics.

Still, the idea of modelling the impulse of a bubble cluster has not received significant attention in the literature, despite the fact that the mathematical interpretation of cavitation phenomena may be highly attractive for a large number of problematic cases. A critical question remains unanswered by existing models: how do neighbouring bubbles affect the dynamic effects within a cluster?

The principles of the theoretical approach to solving this problem are outlined in various works, e.g., in [47–49]. One simplified approach to modelling cavitation clusters involves describing the dynamics of an aggregate of bubbles using the cell model approximation, which allows the evolution of a monodisperse ensemble to be linked to the behaviour of a single bubble. Despite the small number of publications, the available information on the results of experimental and theoretical studies indicates the complexity and extraordinary nature of the processes being studied. These models are based on the fundamental assumption that at all stages of cloud evolution, the bubble sizes are much smaller than the distance between the nearest neighbors.

The results of theoretical calculations of pressure fields and flow velocities, even for two interacting bubbles, show poor agreement with experiments [50,51]. According to researchers, this is explained by the fact that the bubble dynamics equations are derived with the assumption of potential flow in its vicinity, although the real velocity field is neither symmetrical nor potential. Accounting for all the features of bubble cluster dynamics will likely remain an unresolved problem, but the development of simplified methods for assessing the behaviour of a cavitation cloud is a promising task.

## 2. Bubble Dynamics Within a Cluster Using the Cell Model Approximation

Without detailing the complex spatial configurations of velocity and pressure fields within the inter-bubble regions of a dynamically evolving cluster, the general nature of the fluid flow can be conceptualized as follows. As the bubbles expand or contract within an elementary volume of the two-phase system, an unsteady liquid flow is induced in the space between the bubbles. Relative to any given point within this volume, the liquid flow exhibits inherent asymmetry, consequently leading to a non-symmetrical pressure field. This scenario is further complicated by the continuous temporal changes in the distribution of bubble sizes and their mutual distances throughout the volume. The microflows observed in the inter-bubble space are fundamentally vortical, which causes additional localized pulsations in the fluid's velocity and pressure. When the amplitude and frequency of these pulsations are sufficiently large, a specific collective effect emerges, whereby the bubble cloud intensifies hydromechanical and heat and mass transfer processes.

The intricate pressure distribution surrounding each bubble complicates the analytical description of its dynamics. Under these conditions, the fundamental assumption of spherical symmetry for the velocity and pressure fields in the bubble's vicinity becomes inaccurate, necessitating the inclusion of higher-order effects. In contrast to a single, isolated bubble, where developing reliable and consistent models presents no significant challenges, describing the behaviour of an individual bubble within a cloud, and the cluster as a whole, is fraught with difficulties of a fundamental nature.

In the review section of this article, attention is drawn to the dependence of bubble behaviour on their concentration. The use of cell models reflects this dependency by adopting two distinct approaches: one for dilute systems [52,53] and another for concentrated clusters [54,55].

https://doi.org/10.3390/en19051120

These foundational principles of the cell model enable the utilization of established single-bubble dynamics equations as a first approximation for calculating the averaged ensemble parameters. Subsequent calculation steps then involve sequentially incorporating higher-order effects. Crucially, when a bubble is surrounded by other growing bubbles, the pressure fields generated by its nearest neighbours mutually overlap. Consequently, the character of the pressure distribution in the vicinity of each bubble, $p_b = f(r)$, deviates significantly from the distribution described by the widely recognized single-bubble equations [56,57].

$$p_b(\Delta,\tau) = p_f + \left[p_b - p_f + \frac{1}{2}\cdot\rho_b\cdot u_R^2\cdot\left(1-\frac{R^3}{r^3}\right) - \frac{2\sigma}{R} - \frac{4\mu\cdot u_R}{R}\right]\cdot\frac{R}{r} \quad (1)$$

where $\Delta$ is the distance between the bubbles.

The dynamic evolution of each bubble is therefore not governed by the external pressure $p_f$, acting on the cluster as a whole, but rather by an average pressure within the inter-bubble space, denoted as $p_b$, which is characteristically greater than the external pressure $p_b > p_f$. In the approximation utilized by the cell model, the external boundary condition for the pressure of an individual bubble should be applied not at infinity, but at a distance corresponding to the average spacing between adjacent bubbles—specifically, at the radius of the spherical cell within which the bubble is centrally located (Figure 1).

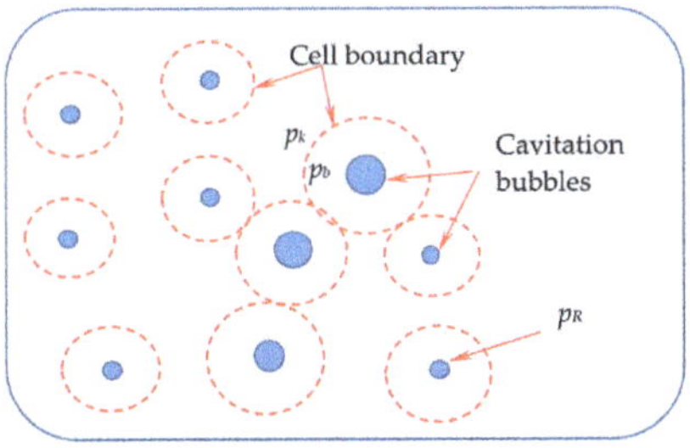


**Figure 1.** Computational scheme.

Analysing the available literature, the fundamental principles of the cell model, as applied to the dynamics of vapour bubbles within a cluster, can be summarized. If the volumetric vapour content (void fraction) of the two-phase bubble medium, $\gamma(\tau)$, and the mean bubble radius, $R(\tau)$, are known, then the volume allocated per bubble is given by $4/3\pi R^3\gamma(\tau)$. This volume can be conceptually treated as an elementary spherical cell containing a single bubble at its center. The outer radius of this cell, $\Psi$, is defined as $\Psi = R/\gamma^{0.3}$, and the thickness of the liquid layer within the cell is $\Psi = R + R(1-\gamma^{0.3})/\gamma^{0.3}$. It is postulated that the liquid motion within the boundaries of the cell, induced by the expansion or compression of the central bubble, is strictly radial. Crucially, there is no mass exchange across the external boundary of the cell with the rest of the cluster. Consequently, the mass of the substance contained within the cell is conserved (remains constant). Within the framework of the cell model, it is further postulated that the following conditions are satisfied at the outer boundary:

$$\left.\frac{dm}{dr}\right|_{r=\psi} = 0; \quad \left.\frac{dT}{dr}\right|_{r=\psi} = 0. \quad (2)$$

This condition facilitates the incorporation of temperature variations within the cell's liquid due to heat and mass transfer with the central bubble. The temperature at the external surface of the cell only begins to change once the thickness of the thermal layer surrounding the bubble, $\theta(\tau)$, exceeds the radius of the cell [49].

https://doi.org/10.3390/en19051120

The behaviour of each bubble is treated as that of an isolated single bubble confined within the volume of a closed cell. As the cell expands or contracts, it undergoes a force interaction with the surrounding portion of the two-phase system, which dictates the pressure exerted on the cell's external surface. This pressure, $p(\Psi) = p_k$, is then employed as the external boundary condition during the derivation of the equation governing the fluid motion within the cell. The resulting pressure distribution in the liquid within the confines of the cell, according to the Lagrange–Cauchy integral [58], is given by the expression:

$$p(\theta) = p_k - \rho_f \cdot \left( \frac{\partial \varphi}{\partial \tau} - \frac{\partial \varphi}{\partial \tau}\bigg|_{r=\psi} + \frac{u^2}{2} - \frac{u^2}{2}\bigg|_{r=\psi} \right). \tag{3}$$

By applying the boundary condition for the pressure at the bubble interface, $p(R) = p_R$, and subsequently defining the parameter values on the right-hand side of Equation (3) using the following formulas:

$$u(r) = \frac{u_R R^2}{r^2}; \quad \varphi(r) = -\frac{u_R R^2}{r}; \quad \frac{\partial \varphi}{\partial \tau} = -\frac{1}{r}\left( \frac{du_R}{d\tau} \cdot R^2 + 2R \cdot u_R \cdot \frac{dR}{d\tau} \right). \tag{4}$$

Consequently, by applying the condition for pressure at $r = R$ (the bubble interface) and at r = Ψ (the cell boundary), we derive the equation for the radial motion of the fluid within the cell:

$$\frac{du_R}{d\tau} = \frac{p_R - p_k - 0.5\rho_b u_R^2 (3 - 4\gamma^{0.3} + \gamma^{1.3})}{\rho_f R (1 - \gamma^{0.3})}. \tag{5}$$

This resulting equation differs significantly from the standard equation of motion developed for a single bubble in an infinite liquid volume [49], due to the inclusion of the cell boundary effects.

$$\frac{du_R}{d\tau} = \frac{p_R - p_f - 1.5\rho_b \cdot u_R^2}{\rho_b \cdot R} \tag{6}$$

This difference stems from two main factors: the presence of correction terms on the right-hand side of Equation (5) that account for the non-negligible volume occupied by the bubbles, and the replacement of the external pressure $p_f$ with the currently unknown pressure at the cell boundary, $p_k$. As the vapor volume fraction $\gamma \to 0$, the cell radius $\Psi$ tends towards infinity $\Psi \to \infty$, and the pressure at the cell's outer boundary $p_k$ converges to the external pressure $p_k \to p_f$. Consequently, Equation (5) simplifies to Equation (6) under dilute conditions. It is worth noting that for the initial growth stages of small bubbles or the final collapse phase, $\gamma$ is small, rendering the correction terms in Equation (5) of secondary importance. However, the cell model approximation becomes inherently flawed for comparatively high values of the two-phase system's vapour content ($\gamma > 0.5$).

The principal distinction between Equations (5) and (6) lies in their boundary conditions: the behaviour of the isolated bubble in Equation (6) is determined by the known externally applied pressure $p_f$. Conversely, the behaviour of a bubble surrounded by others is dependent on the pressure $p_k(\tau)$ at the external cell boundary, a value that is not known a priori and is, in turn, determined by the intensity of expansion or compression across the entire bubble cluster. Finally, the precise magnitude of the pressure in the liquid phase $p_R$, as well as the quantitative influence of capillary and viscous forces on this pressure, remain open questions within this modelling framework.

## 3. Modified Model

We consider an alternative approach to modelling bubble behaviour within a cluster, one that is based on averaging the pressure solely within the liquid's inter-bubble space. We assume that within a liquid volume, initially at temperature $T_{f0}$, there are uniformly

https://doi.org/10.3390/en19051120

distributed vapor bubbles of identical size, with an initial radius $R_0$. These bubbles are considered to be in thermodynamic equilibrium with the liquid. The initial vapour pressure inside each bubble is equal to the saturated vapor pressure at the liquid temperature, $p_{sat}(T_{f0})$. This assumption is used solely as the initial condition for the calculation. It is assumed that the liquid contains only vapor nuclei of uniform size, and their number can be estimated using established methods for various temperatures and pressures. Subsequent calculation steps account for changes in bubble size depending on their thermodynamic state and the influence of neighboring bubbles. The initial pressure in the liquid is uniform throughout the entire volume and is determined by the condition: $p_{f0} = p_{sat}(T_{f0}) - 2\sigma/R$. The number of bubbles per unit mass of the system, $k_b$, is assumed to be known and remains constant throughout the system's subsequent evolution, thereby excluding the effects of bubble coalescence and breakup. At time $\tau = 0$, the external pressure acting on the system, initially $p_f = p_{f0}$, begins to change according to a known law, initiating the growth or compression of all bubbles. For a monodisperse system, the volumetric vapour content (void fraction) is related to the size and concentration of the bubbles and is determined by the expression:

$$\gamma = \frac{4\pi R^3 k_b \rho_b}{3 + 4\pi R^3 k_b \rho_b}. \quad (7)$$

Within the vicinity of each bubble, specifically within the boundaries of the elementary cell of radius $\psi(\tau) = R(\tau)/\gamma^{0.3}$, a spherical pressure distribution, $p_f(r,\tau)$, is established. This pressure field can be determined using Equation (8):

$$\frac{\partial \varphi}{\partial \tau} + \frac{u^2}{2} + \frac{p}{\rho_b} = f(\tau), \quad (8)$$

under the condition $p_f(R,\tau) = p_R(\tau)$.

When determining the function $f(\tau)$ on the right-hand side of Equation (8) using the boundary condition $p(\psi) = p_k$, the pressure distribution obtained takes the form of Equation (3). Conversely, if we utilize the boundary condition $p(R) = p_R$ to determine $f(\tau)$, the pressure distribution inside the cell adopts the following form:

$$p_f(r) = p_R - \rho_f \cdot \left( \frac{\partial \varphi}{\partial \tau} - \frac{\partial \varphi}{\partial \tau}\bigg|_{r=\psi} + \frac{u^2}{2} - \frac{u^2}{2}\bigg|_{r=\psi} \right). \quad (9)$$

By utilizing the relationships defined in Equation (4) for the calculation of the quantities on the right-hand side of Equation (9), we then reduce this equation to the final form (Equation (10)):

$$p_f(r,\tau) = p_R - \rho_b \left( R\frac{du_R}{d\tau} + 2u_R^2 \right) \frac{1 - R^4}{r^4} + 0.5\rho_b u_R^2 \frac{1 - R^4}{r^4}. \quad (10)$$

By setting the radial coordinate equal to the cell radius, $r = \Psi$. Equation (10) can be re-expressed as follows:

$$p_f(r,\tau) = p_R - \rho_b \left( R\frac{du_R}{d\tau} + 2u_R^2 \right) \cdot \left(1 - \gamma^{0.3}\right) + 0.5\rho_b u_R^2 \cdot \left(1 - \gamma^{1.3}\right). \quad (11)$$

Considering the right-hand side of Equation (5), we can derive the equation for calculating the pressure in the vicinity of the bubble:

$$p_b(r,\tau) = p_R - \rho_b \left( \frac{p_R - p_k + 0.5\rho_b u_R^2 (1 - \gamma^{1.3})}{1 - \gamma^{0.3}} \right) \cdot \left(1 - \frac{R}{r}\right) + 0.5\rho_b u_R^2 \left(1 - \frac{R^4}{r^4}\right). \quad (12)$$

https://doi.org/10.3390/en19051120

The expression for the averaged pressure inside the cell is as follows:

$$p_m(\tau) = p_k + \frac{\rho^{0.3}\left[p_R - p_k + 0.5\rho_b u_R^2\left(1-\gamma^1\right)\right]}{1-\gamma^{0.3}} \cdot \left[\frac{1.5\left(1-\gamma^{0.6}\right)}{1-\gamma^{0.6}} - 1\right] - 0.5\rho_b u_R^2 \cdot \left(\frac{3\gamma\cdot\left(1-\gamma^{0.3}\right)}{1-\gamma^{0.3}} - \gamma^{1.3}\right). \quad (13)$$

The following algorithm is employed to determine the pressure at the external cell boundary $p_k$, the pressure at the bubble boundary $p_R$, the mean pressure in the liquid $p_m$, as well as the bubble radius $R$ and the velocity of its surface $u_R$ at each stage of the two-phase system's evolution.

Algorithm for dynamic parameter calculation. At the initial time $\tau = 0$, the external pressure acting on the system instantaneously changes from its initial equilibrium value $p_f = p_{f0}$ to some constant prescribed value $p_f < p_{f0}$. At this precise moment, the pressure within the inter-bubble space, including the external cell boundary, is equal to this external pressure. By substituting this value of the pressure into Equation (5) as the parameter $p_k$, we determine the magnitude of the liquid acceleration at the bubble interface within the cell $du_R/d\tau$. Subsequently, we calculate the new values for the system's governing parameters, including $R$, $u_R$, $\gamma$, and $p_R$. Substituting these newly calculated quantities into the right-hand side of Equation (13) allows us to compute the value of the liquid pressure averaged over the cell volume, $p_m$. The pressure value $p_m$ found in this manner is then utilized in the next calculation step as the cell boundary pressure $p_k$. This value of $p_k$ is then inserted back into Equation (5) to subsequently determine the new values for $du_R/d\tau$, $u_R$ and $R$. Other governing system equations are solved simultaneously with Equation (5) to compute the new values of the remaining parameters. This procedure is repeated iteratively at each time step of the numerical solution for the bubble dynamics equations. If the external pressure $p_f$ acting on the system does not change instantaneously but over a finite time interval $\Delta\tau$ according to some known function, for example:

$$p_f = p_{f0} - \left(p_{f0} - p_b\right) sin(0.5\pi\tau/\Delta\tau) \quad (14)$$

the change in external pressure, $dp_f$, within the time interval $0 < \tau < \Delta\tau$ is determined by the expression:

$$dp_f = -0.5\left(p_{f0} - p_b\right) \cdot cos(0.5\pi\tau/\Delta\tau) \cdot \pi d\tau/\Delta\tau \quad (15)$$

At each calculation step, the pressure $p_f$ is computed according to the specified scheme, and the obtained value is then adjusted by the increment $dp_f$, substituting the time step interval as $d\tau$. After the change in external pressure is completed $\tau > d\tau$, $p_f$ remains constant $p_f = const$ and $dp_f = 0$. This iterative procedure allows for the calculation of the time-dependent pressure at the cell boundary, $p_k$, and the mean pressure within the inter-bubble space of the ensemble, $p_m$, for any variation in the external pressure.

When the vapor content of the bubble system changes, the liquid temperature must vary due to the release or absorption of heat associated with phase transitions occurring at the expanded interfacial surface. The mean temperature of the liquid phase in the bubble ensemble, $T_m$, can be identified with the mean temperature of the liquid within the elementary cell. In the vicinity of each bubble, within the confines of the non-stationary thermal layer, a temperature distribution $T_f(r,\tau)$ exists.

The mean liquid temperature within the cell, $T_m$, at any given time is determined by integrating the functional expression $T_f(r,\tau)$ over the volume of the liquid in the cell and subsequently dividing by that volume. As previously noted, the cell model inherently assumes that each elementary cell is thermally isolated. Therefore, the amount of heat transferred through the bubble surface due to phase transition and heat exchange, $4\pi R^2(JQ + q)$ [16], uniquely determines the change in the heat content of the liquid in the cell and, consequently, its mean temperature. Thus, the energy balance can be formulated as:

$$\frac{d}{d\tau}\left[\frac{4}{3}\pi\rho_b c\left(\psi^3 - R^3\right)\cdot T_m\right] = 4\pi R^2(JQ + q). \tag{16}$$

By substituting the expression for the void fraction $\gamma$ from Equation (7) into Equation (16) and performing the necessary mathematical transformations, we arrive at a simplified equation for determining the mean temperature of the liquid in the inter-bubble space of the ensemble:

$$\frac{dT_m}{d\tau} = \frac{k_b(JQ + q)}{c}. \tag{17}$$

A new initial parameter introduced into this model is the concentration of bubbles per unit mass, $k_b$. The system of equations, transformed in this manner and incorporating the proposed iterative scheme for determining the pressure within the inter-bubble space, can be effectively utilized to predict the dynamic behaviour of a cluster of monodisperse bubbles when subjected to variations in external pressure.

## 4. Liquid Flow in the Inter-Bubble Space of a Cluster

Cell models offer a capability to estimate the average pressure and void fraction within the inter-bubble space of the system at any given moment. This, in principle, allows for the investigation of steady and unsteady two-phase bubble flows in pipes and channels, as well as tracking the evolution of an individual bubble within such a system. However, these models are fundamentally incapable of detailing the hydrodynamic and heat and mass transfer phenomena that occur at the micro-level within the inter-bubble space. The model achieves pressure or void fraction averaging over a volume that significantly exceeds the volume of a single bubble. Consequently, the cluster is effectively treated as a homogeneous medium, with bubbles represented merely as singularities.

To accurately describe the behaviour of bubbles at the micro-level and to model the velocity and pressure fields in the inter-bubble space, several determining factors influencing the hydrodynamic environment in the liquid phase during intense bubble surface motion must be considered.

1. The bubbles within the cluster vary in geometrical size, shape, velocity, and the acceleration of their radial surface motion. This disparity necessitates an individualized description of each bubble's evolution.

2. The force interaction between dynamically developing bubbles results in their chaotic translational movement. This, in turn, causes continuous changes in the velocity and direction of the microflows over time, leading to pulsations of local liquid pressure.

3. The dynamic interaction between neighbouring bubbles causes their continuous deformation. This further induces oscillations on the surface and significantly complicates the hydrodynamic environment in the vicinity of each bubble.

In essence, the physical model that accounts for these complex factors is sufficiently straightforward and apparent. For instance, the non-uniformity of the bubbles within the ensemble can be addressed by specifying their initial size and spatial distribution. In estimating the force interaction between bubbles, the expression for the Bjerknes force, which arises between two oscillating bubbles [59–61], can be used as a first approximation. The nature of the surface deformation is managed by the condition of pressure equality at the interfacial surface, allowing the calculation of the bubble's radius of curvature in any direction [62]. However, the major modeling difficulty lies in the excessive duration of numerical computations, even for a single time step. Consequently, a detailed analysis of the behaviour of even two interacting bubbles proves to be a highly complex and labor-intensive task [63].

To analyze the nature of microflows, pressure fields, and velocity fields in the liquid phase of a dynamically evolving bubble system, we consider a model as a first approximation based on the following key assumptions:

(1) the influence of viscosity is manifested only at the interfacial surface;

(2) bubbles retain a spherical shape during their evolution, this is justifiable for relatively small bubbles, where capillary pressure counteracts the external forces from other bubbles, provided that $\gamma < 0.55$.

(3) the force interaction between neighbouring bubbles, caused by the radial motion of their surfaces, is accounted for, which results in the mutual translational movement of bubbles within the ensemble;

(4) the liquid flow in the inter-bubble space is potential.

We treat the expanding or compressing bubbles as discretely distributed dynamic and thermal sources (or sinks) within the liquid. Based on assumption (4) (potential flow), we utilize the superposition principle in accordance with potential theory: the resultant flow is the sum of flows from each source, and the magnitude of the scalar potential of the velocity field at an arbitrary point in the liquid is the sum of the potentials generated at that point by all sources.

Consider a cluster of vapour bubbles comprising $k_b$ elements. We place the origin of the Cartesian coordinate system at some point in the center of the bubble system. If the center coordinates (x, y, z), radii $R_k$, and surface expansion velocity $u_R$, of each of the $k_b$ bubbles are known at a specific instant in time, the total field potential at an arbitrary point in the liquid with chosen coordinates, according to the superposition rule of potentials, is determined by the formula:

$$\varphi(x,\, y,\, z) = \sum_{i=1}^{\mathrm{k}} -\frac{u_{Ri}R_i^2}{\overline{\omega}_i}, \tag{18}$$

where $\overline{\omega}_i = \sqrt{\Delta x^2 - \Delta y^2 - \Delta z^2}$.

The components of the resultant velocity vector at the point (x, y, z) are determined by the following expressions:

$$u_x = \sum_{i=1}^{k} \frac{u_{Ri}R_i^2\Delta x}{\overline{\omega}_i^3}; \quad u_y = \sum_{i=1}^{k} \frac{u_{Ri}R_i^2\Delta y}{\overline{\omega}_i^3}; \; u_z = \sum_{i=1}^{k} \frac{u_{Ri}R_i^2\Delta z}{\overline{\omega}_i^3}, \tag{19}$$

The spatial direction of the velocity vector is subsequently determined using the standard formulas of analytical geometry, based on the known values of its projections:

$$\frac{\partial\varphi}{\partial\tau} = \sum_{i=1}^{k} -\left(\frac{du_{Ri}}{d\tau}\cdot\frac{R_i^2}{\overline{\omega}_i} + \frac{2R_i u_{Ri}^2}{\overline{\omega}_i}\right), \tag{20}$$

where $du_{Ri}/d\tau$ is the acceleration of the radial surface motion of the individual bubble, as determined by Equation (6).

By substituting the value of the partial derivative of the potential with respect to time, $\partial\varphi/\partial\tau$, from Equation (20) and the velocity vector from Equation (19) into the Lagrange–Cauchy integral (Equation (8)), the magnitude of the pressure at a local point (x, y, z) of the microflow can be determined. The function $f(\tau)$ on the right-hand side of Equation (8) is determined by imposing the condition that at a sufficiently large distance from the bubble cloud, the pressure in the liquid is constant and equal to $p_f$, and both the

https://doi.org/10.3390/en19051120

flow velocity and the potential are zero. Under these far-field conditions, the local pressure within the inter-bubble space inside the cloud is determined by the following equation:

$$p_b(x,y,z) = p_f - \rho_b\left(\frac{\partial\varphi}{\partial\tau} + \frac{\vec{u}^2}{2}\right) \tag{21}$$

by substituting the values of $\vec{u}$ and $\partial\varphi/\partial\tau$ from Equations (19) and (20), respectively, into its right-hand side.

Since the point (x, y, z) is arbitrarily chosen, this scheme allows for the estimation of the velocity and pressure magnitudes at any point within the inter-bubble space. Consequently, the instantaneous distribution of velocities and pressures throughout the two-phase system under consideration can be determined. It is essential to ensure that the point where $\vec{u}$ and $p_b$ are determined does not fall within the region currently occupied by the bubble's gas phase. Determining the pressure field in the liquid between the bubbles makes it possible to average the pressure over the volume of each individual cell (which are not necessarily identical). This averaged pressure is then used in the next time step to solve the single-bubble dynamics equation for each of the $k_b$ bubbles, thereby determining the new values for $R$, $u_R$, $du_R/d\tau$, and the new spatial coordinates x, y, z. Subsequently, the entire aforementioned procedure for calculating the velocity and pressure fields is performed again.

In this model, the correct calculation of the dynamics of each individual bubble within the finite-volume ensemble is of fundamental importance. As previously indicated, Equation (5) must be used as the equation of motion to determine the value of the radial acceleration $du_R/d\tau$ on the right-hand side of Equation (20).

The main question centers on the appropriate values for the parameters $p_k$ and $\gamma$ to be substituted into these equations. During the derivation of Equation (5), it was established that the void fraction $\gamma$ in this equation characterizes the cell radius $\Psi = R/\gamma^{0.3}$ surrounding the specific bubble, while $p_k$ represents the liquid pressure averaged over the volume of that cell. These bubbles located on the periphery of the cloud are in constant contact with an unlimited volume of the liquid phase, where the pressure $p_f$ = const. For these peripheral bubbles, the cell radius $\Psi$ is assumed to be half the distance from their centers to the center of their nearest neighbours, which determines the corresponding value of $\gamma$. The pressure calculated via Equation (13) is substituted as $p_k$ into Equation (5). These peripheral bubbles essentially behave as isolated bubbles in an infinite liquid volume throughout the process. However, the pressure that these bubbles create in the outer layer of the cluster, calculated using Equation (13), is hypothesized to be transmitted as the cell boundary pressure $p_k$ for all bubbles situated in the next adjacent spherical layer—the layer bordering the outer cloud layer. This pressure $p_k$, created by the bubbles in the second layer, is then passed on as the boundary condition for the bubbles in the third spherical layer, and so on. Consequently, the averaged pressure within a spherically shaped bubble cloud will vary from the periphery to the center, and the pressure gradient along the cloud's radius depends on the intensity of expansion or compression of the bubbles within each spherical layer. This macro-level procedure for determining pressures within a finite ensemble allows for the calculation of the current parameters of each bubble as a function of its distance from the cloud center and its initial size. Subsequently, at each computational time step, the known values of the radial acceleration $du_R/d\tau$ and velocity $u_R$ for every bubble enable the estimation of the velocity, pressure magnitudes, and direction of the microflow at any point in the inter-bubble space, utilizing the field Equations (18)–(21). In summary, the presented equations for both unlimited and finite numbers of vapour bubbles, combined with the single-bubble dynamics equations, provide a framework for predicting the behaviour

https://doi.org/10.3390/en19051120

of bubbles within a cluster and the cluster's overall behaviour under varying external pressure, as well as for determining the velocity and pressure fields in the inter-bubble space and their temporal evolution under any operational parameters.

## 5. Discussion

Based on the modified model, an analysis of the behaviour of vapor bubbles in water was performed under different operating modes and various external pressure change conditions. Below, some of the prediction results are presented as examples to illustrate the model's capabilities

### *5.1. Dynamics of Bubble Growth and Collapse in a Cluster*

Within the framework of the modified model, the evolution of a monodisperse vapour bubble cloud was considered for known initial values of the liquid temperature $T_{f0}$ and initial pressure и $p_{f0}$. At the initial instant, all bubbles with radius $R_0$ and concentration $k_b$ are in equilibrium with the liquid. At the initial time τ = 0, the external pressure acting on the system begins to change. In analyzing various experimental methodologies related to the rapid change of pressure in a liquid volume [64–66], we established that, regardless of the method used to change the pressure in the system, the nature of the actual curve $p_f$ = f(τ) is well approximated by a cosinusoidal dependence, taking the form:

$$p_f(\tau) = P + \frac{p_{f0} - P}{2} \cdot \left[1 + cos\left(\frac{\pi\tau}{\Delta\tau}\right)\right] \qquad \text{at } 0 < \tau < \Delta\tau;$$

$$p_f(\tau) = P \qquad \text{at } \tau \geq \Delta\tau, \tag{22}$$

where $P$ represents the constant pressure established in the liquid far from the bubble as a result of the change in external pressure, and $\Delta\tau$ is the duration of this change. After the specified time interval $\Delta\tau$, the external pressure assumes a new constant value $P$.

Figure 2 illustrates how the mean pressure in the liquid within the inter-bubble space of the cluster, $p_m = p_k$, changes following an instantaneous drop in external pressure from an initial equilibrium value of 60 kPa to 15 kPa. The pressure in the liquid rapidly increases, asymptotically approaching the saturated vapour pressure $p_{sat}(T_f)$, which corresponds to approximately 30 kPa under the conditions presented in the figure. The figure also includes a graph showing the pressure pulsations within a cavitation cloud, based on the data presented in the study [67–70].

The presented data suggests that the pressure within the cluster undergoes an averaging process, but the dynamic effects originating from the bubbles at the cloud boundaries can be substantially greater than those observed inside the cluster. These internal effects within the cavitation cloud become more prominent as the bubble concentration increases. Figure 3 illustrates the change in the liquid pressure inside a monodisperse cloud of collapsing bubbles following an instantaneous increase in the external pressure from the equilibrium value of 70 kPa to 150 kPa. Due to the intense initial compression of the bubble ensemble and the associated stretching of the liquid, the pressure in the inter-bubble space initially drops sharply to nearly 70 kPa. However, the rapid compression leads to a significant increase in the vapour pressure inside the bubble, which in turn causes the pressure to rise in the liquid layers adjacent to the surface. This elevated pressure then propagates throughout the entire cell volume. Similarly to the previous case of bubble growth, the pressure inside the cell is controlled by the vapour pressure in the bubble, and this control is more noticeable as the cell size decreases, meaning the higher the bubble concentration. Consequently, as shown in the figure, at high concentrations, the mean pressure in the liquid substantially exceeds the external pressure until the bubbles completely collapse.

https://doi.org/10.3390/en19051120

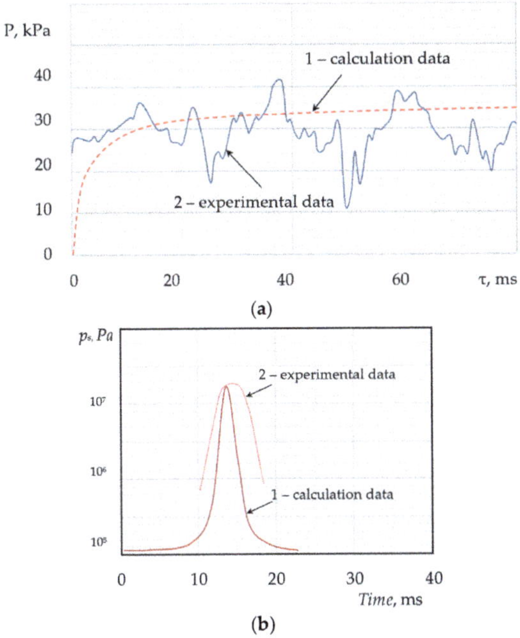


**Figure 2.** Comparison of bubble cluster and single bubble dynamics: (**a**) change in averaged pressure within the cluster during vapor bubble growth in water following a sudden reduction in external pressure. The calculation is based on the modified model using the following conditions: $T_{f0}$ = 380 K; $p_{f0}$ = 60 kPa; $p_f$ = 15 kPa; $R_0$ = 20 μm; $k_b$—$10^4$ kg$^{-1}$; (**b**) change in the pressure of a single bubble during its maximum compression stage in the process of collapse, under the same conditions. Curve 1 represents the model calculation, and curve 2 represents the data reported in: (**a**)—[67]; (**b**)—[68].

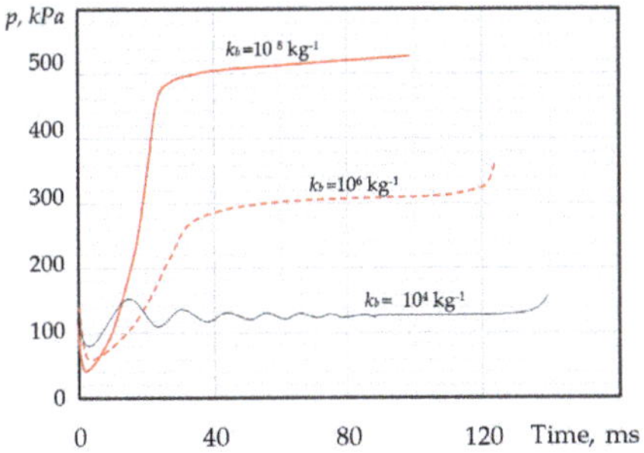


**Figure 3.** Change in averaged pressure $p_m = p_k$ in the liquid phase during the cavitation collapse of vapor bubbles in water following an increase in external pressure. The calculations were performed using the following initial conditions $T_{f0}$ = 380 K; $p_{f0}$ = 70 kPa; $p_f$ = 130 kPa; $R_0$ = 100 μm; and for various values of the bubble concentration $k_b$.

The elevated pressure established within the cell acts to suppress the bubble's expansion following maximum compression and its subsequent oscillation, which would be characteristic of the collapse of a single, isolated bubble. However, at comparatively low concentrations (curve 1 in Figure 3), the behaviour of the bubbles closely approaches that of a single bubble. During the collapse process, the pressure inside the cell exhibits only weak oscillation, tending toward the external pressure value, $p_f$ = 130 kPa. The collapse of bubbles at low concentrations is accompanied by oscillation of their size, as evident from Figure 3 (curve 1), whereas at high concentrations, the bubble radius decreases mono-

https://doi.org/10.3390/en19051120

tonically until complete collapse. Interestingly, despite the qualitative difference in the compression process, the lifetime of the bubbles in the cluster does not depend on the concentration and is practically identical to the duration of the single bubble collapse under the same conditions. Furthermore, for identical values of $k_b$, the initial bubble size $R_0$ has only a minor influence on their behaviour. A study of the growth process at various $R_0$ values, ranging from 5 μm to 100 μm, demonstrated that shortly after the onset of bubble expansion (the time depending on concentration), the magnitude of the current pressure in the inter-bubble space and the rate of vaporization are practically independent of $R_0$.

The magnitude of the pressure drop or the degree of liquid superheating exerts a similarly minor influence on the dynamics of bubble growth for identical values of $T_{f0}$ and $k_b$. For example, when the pressure in a monodisperse ensemble is reduced from an equilibrium value of 60 kPa to 20 kPa, the corresponding time dependencies for the averaged pressure $p_m$ = f(τ) essentially coincide. This congruence also applies to the dependencies of $R(\tau)$, $u_R(\tau)$ and $\gamma(\tau)$. However, during the short initial growth phase, before the quasi-equilibrium pressure is established in the liquid phase, the intensity of bubble expansion is higher when the pressure differential is larger. The duration of the external pressure change, $\Delta\tau$, also has only a weak effect on the characteristics of the dependencies $p_m(\tau)$, $R(\tau)$. In the inertial stage, the rate of vapor phase growth is higher when the rate of pressure reduction $dp_f/d\tau$ is higher. At the same time, intensive bubble growth facilitates a more rapid approach to the quasi-equilibrium value $p_m$. Regardless of the duration of the pressure drop, the same averaged pressure $p_m$ is established inside the cluster after a very short period. This pressure is what governs the subsequent evolution of the bubbles, meaning that after only approximately 80 μs following the onset of the bubble ensemble expansion, the recorded parameter values for $R$, $u_R$ or $\gamma$ no longer allow one to discern the cluster's development history or determine whether the change in external pressure has ceased or is still ongoing.

### *5.2. Velocity and Pressure Fields in the Inter-Bubble Space*

The modified model allows for the individualization of each bubble, taking into account its current size, location within the cluster, kinematic and thermodynamic characteristics, and other properties. Information obtained through the cell model approximation is also incorporated into the description of the bubble cloud behaviour. Figure 4 is presented as an illustration, showing the instantaneous distribution of liquid microflow velocities inside the cluster at a specific moment in time. The microflows, which exhibit a complex nature, are caused by the expansion of the bubbles during the cavitation process. The arrows in the figure indicate the direction of the velocity at local points but not its magnitude. Clusters containing between 10 and 1500 bubbles were considered in the analysis. At any given time, the local values of the velocity vector $\vec{u}$ (magnitude and direction), the velocity field potential $\varphi$, and the pressure magnitude $p$ are determined at various points in the liquid. To visualize the results, the projections of the velocity vector onto a specified plane passing through the volume occupied by the bubble cloud were calculated. It is characteristic that the velocity field exhibits a truly cellular structure. Directly at the surface of the bubbles, the flow is close to radial. However, as the distance from the surface increases, the microflow pattern becomes more complex, and at the boundaries of bubble interaction, the velocity vectors change direction sharply. This figure demonstrates that bubble growth occurs more intensely at the edges of the cloud, which is attributed to the non-uniform distribution of the averaged pressure $p_k$ throughout the cloud volume: in the central part, $p_k >> p_f$, whereas at the periphery, $p_k \rightarrow p_f$.

https://doi.org/10.3390/en19051120

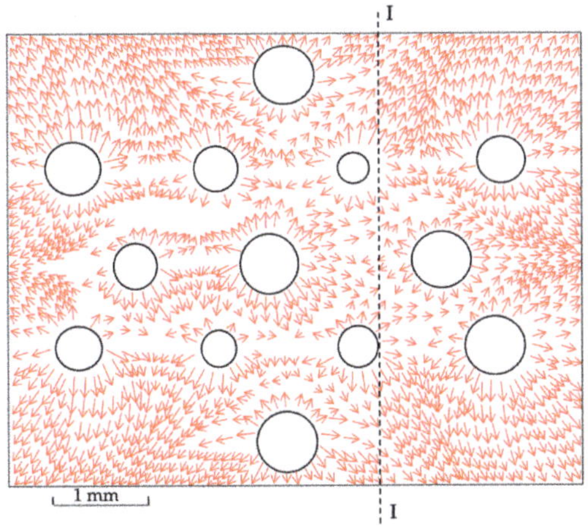


**Figure 4.** Directions of liquid velocity vectors inside a cavitation cloud composed of 30 bubbles growing as a result of a pressure drop in the system. The figure displays the velocity projections onto the XY plane, which intersects the center of the cluster.

Figure 5 illustrates how the magnitudes of the local pressure $p$ and local velocity $\varphi$ change along the direction I-I (as defined in Figure 4). Significant velocity fluctuations are observed within the cluster along any direction, whereas outside the cluster, even near its boundaries, the velocity magnitude changes monotonically, maintaining a practically constant direction. Despite such abrupt velocity changes in the inter-bubble space, the pressure pulsations are, as evident from Figure 5, relatively weak. Nevertheless, the magnitude of the pressure gradient reaches up to 20 kPa/mm at local points, and the velocity gradients exceed 20 m/s per 1 mm. The figure clearly indicates that the pressure within the cloud volume is higher than at the periphery. This increased internal pressure leads to a situation where the bubble size and, consequently, the void fraction are higher in the peripheral zone of the cloud than in its interior.

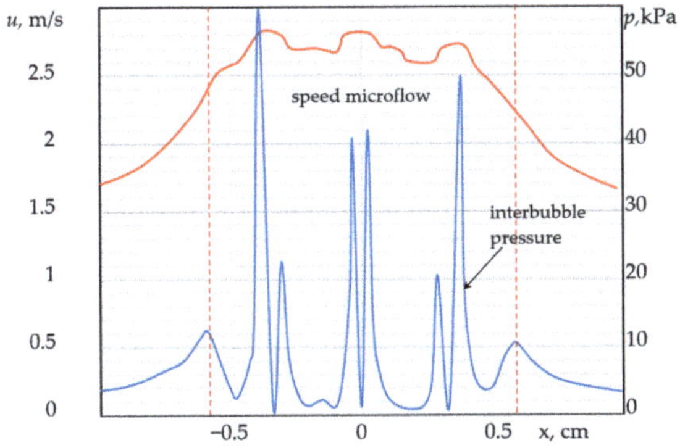


**Figure 5.** Variations in local flow velocity and local pressure magnitude in the liquid within the inter-bubble space along the direction I-I (Figure 4). The calculation was performed using the same conditions specified for Figure 4.

The dashed lines in Figure 5 demarcate the region occupied by the cluster. The data obtained indicates that within the inter-bubble space of the cloud, even during the monotonic expansion of the bubbles, there are sharp jumps in pressures and velocities characteristic of turbulent flow. This nature of flow contributes to the intensification and

https://doi.org/10.3390/en19051120

stimulation of heat and mass transfer and hydromechanical processes within the liquid phase of the bubble system. When cavitation conditions are realized in a liquid, a multitude of vapour-gas bubbles grow and then simultaneously collapse, forming a cavitation cluster. The effectiveness of the cavitation impact is determined by the collective influence of all bubbles within the cluster. Cluster sizes typically range from 0.1 to 5 cm. The concentration of nuclei in the cluster depends on the liquid's temperature, gas content, and the presence of impurities. The superposition of shock waves from each individual bubble forms a single destructive shock wave. However, this conventional concept does not account for the destruction of micro-objects inside the cluster itself. Furthermore, when modelling the behaviour of a cavitation cluster and developing analytical methods for cavitation reactors, temperature is often not even included as one of the model parameters [59,60]. The following section provides an assessment of the main characteristics of heat and mass transfer across the bubble surface.

### 5.3. Heat and Mass Transfer Across the Interfacial Surface During Bubble Collapse

As demonstrated earlier, the dynamic effects of individual bubbles at the cluster boundary during the collapse stage significantly exceed the average effects of the cluster itself. The most intense heat and mass transfer across the interfacial surface, and the most vigorous mass transfer processes (evaporation and condensation), occur during the stage of maximum bubble compression, as illustrated in Figures 6 and 7. Although the amplitude values of the heat flux impulse $JQ + q$ are extremely high during the maximum compression stage, the quantity of thermal energy transferred through the bubble wall during this time is very small. Due to the extremely short duration of this impulse, this part of the energy constitutes only a negligible fraction of the total thermal energy accumulated in the bubble after the transformation of kinetic energy into potential energy.

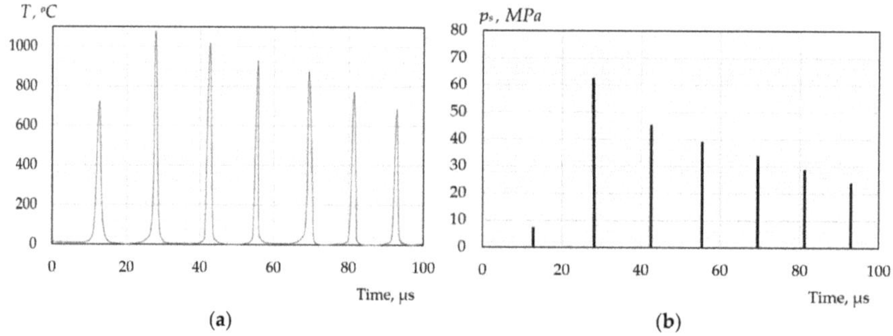


**Figure 6.** Change in vapor temperature (**a**) and vapor pressure (**b**) during the collapse of a vapor bubble in a subcooled liquid. A modified model was used for the calculation under the following conditions: $T_{l0}$ = 20 °C; $T_{s0}$ = 130 °C; $p_{l0}$ = 100 kPa; $R_0$ = 100 μm.

The intensity of mass transfer, $J$, also increases to an extreme degree during the stage of maximum compression, but this fractional change in acquired or lost vapours mass is negligibly small in the overall mass balance. For the remainder of the bubble's development time, the inherently low rates of heat and mass transfer can still exert a relatively large influence on the bubble's dynamics. Figure 8a,b illustrates how the vapours mass in the bubble, $m_s$, and the quantity of heat accumulated in the bubble, $q$, change during the oscillation process. Despite the strong fluctuations in the bubble size, both quantities, $m_s$ and $q$, decrease practically monotonically, with the largest reduction in vapours mass and thermal energy occurring during the first period of oscillations. The absolute quantities of energy and mass transferred across the bubble surface per unit time are not very large. However, considering the minute dimensions of the bubble at the moment of maximum

https://doi.org/10.3390/en19051120

compression, the specific fluxes across a unit surface area are substantial. As shown in Figures 8 and 9, the magnitude of the specific mass flux of vapours condensing into the liquid or evaporating into the bubble is estimated to be on the scale of 100 $kg/m^2s$, and the specific energy flux is on the scale of 100 $MW/m^2$.

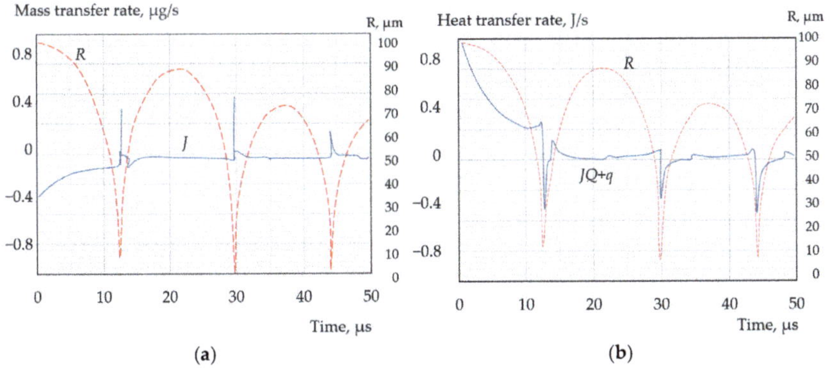


**Figure 7.** Variation in the intensity of heat flux (**a**) and mass flux (**b**) across the surface of a vapor bubble during its collapse in subcooled liquid. The calculation conditions are: $T_{f0}$ = 20 °C; $p_{f0}$ = 100 kPa; $R_0$ = 100 μm).

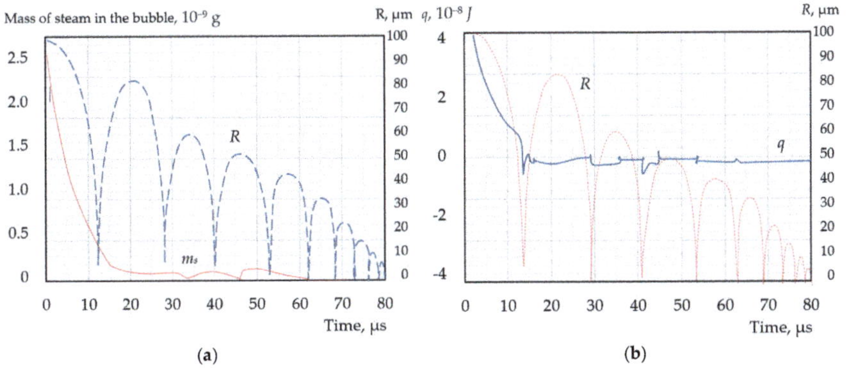


**Figure 8.** Vapor mass (**a**) and quantity of heat (**b**) in a vapor bubble during its collapse in subcooled liquid. The dashed curve indicates the simultaneous change in the bubble radius.

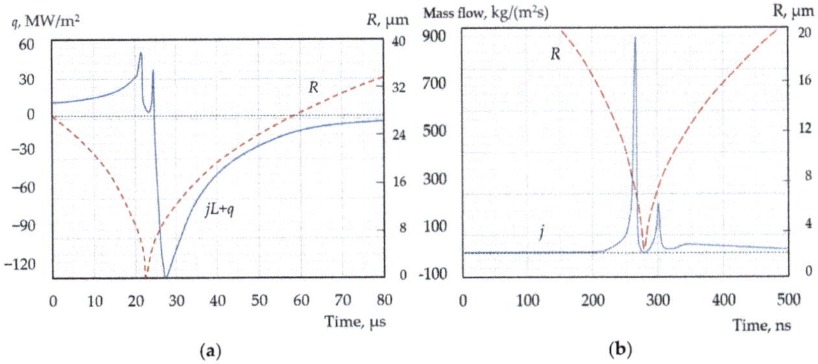


**Figure 9.** Intensity of Heat Flux $JQ + q$ (**a**) and Mass Flux $J$ (**b**) across the interfacial surface during the maximum compression stage of a bubble's collapse in subcooled liquid. The dashed curve indicates the simultaneous change in the bubble radius.

https://doi.org/10.3390/en19051120

## 6. Conclusions

Based on the conducted analysis, a mathematical model for bubble cluster dynamics has been developed, which describes the evolution of a collection of vapour bubbles when thermodynamic equilibrium is disrupted in the vapour-liquid system. Computational experiments were performed to investigate the patterns of bubble evolution resulting from a sharp decrease or increase in external pressure. When the external pressure is released in a vapour-liquid bubble medium, the pressure in the liquid phase rapidly increases, asymptotically approaching the saturated vapour pressure corresponding to the liquid temperature. The rate of approach to saturation pressure is proportional to the bubble concentration. he degree of superheat, the rate of pressure release, and the initial size of the bubble nuclei only influence the initial stage of cavitation development.

The computational experiments provided quantitative information regarding the local values of pressure and velocity at any point in the liquid phase of the developing bubble ensemble. It was established that even during the monotonic expansion of bubbles in the thermal regime, the magnitudes of the velocity and pressure gradients reach extremely high values.

The dynamic action of cavitation is attributed to the following factors:

- during bubble growth and compression, complex non-stationary microflows are formed inside the cluster due to the superposition of radial flows towards each bubble; these flows feature extremely high values of shear rates and shear stresses, the effect of which is present throughout the entire lifetime of the cluster;
- calculations indicate that in the inter-bubble space, shear rates at local points on the boundaries of flow interaction can reach $10^6$ $s^{-1}$, which corresponds to a shear stress of approximately 1 kPa when water is the dispersed phase;
- during collapse, each bubble in the cluster acts as the center of a spherical shock wave with a pressure amplitude that can reach 100 MPa.

The non-stationary fields of velocities and pressures in the liquid inside the cluster are extremely complex. During bubble compression, the volume-averaged pressure inside the cluster rapidly increases and can exceed 10 MPa in the final stage. The simultaneous collapse of bubbles causes the pressure within the cluster volume to drop sharply, and a high-amplitude resultant impulse is radiated into the surrounding medium.

A significant finding is that even during the stage of monotonic cluster expansion, when the bubble growth rate does not exceed 2 m/s, the shear rates in local liquid zones are quite high—on the order of $5 \times 10^4$ $s^{-1}$. Thus, while a high level of thermal impact is only realized during the nanosecond-scale stage of maximum cluster compression, the effective influence of shear stresses is evident throughout the entire existence of the cluster.

The question of the total dynamic effect of the cluster remains a subject for discussion. If one assumes that dynamic effects are determined only by the collapse of the outer peripheral bubbles, theoretical models are likely to predict overestimated dynamic parameter values. This is because real processes exhibit effects of mutual compensation of counter-propagating impulses directed from different bubbles, a finding confirmed by certain experimental studies.

The developed modified mathematical model is designed for the analysis of pressure and temperature fields, as well as microflows, in the inter-bubble space of the ensemble during boiling and cavitation processes.

The model is applicable to both hydrodynamic and acoustic cavitation conditions and can be used for low and high bubble concentrations. In the former case—for example, at a bubble concentration of up to $1 \times 10^{13}$ $m^{-3}$ and a pressure of 3 MPa—the bubbles can be treated as isolated local volumes that do not significantly influence one another. The initial

https://doi.org/10.3390/en19051120

input for the mathematical model specifies the bubble size distribution and concentration, while the calculation algorithm itself remains unchanged.

**Funding:** This research received no external funding.

**Data Availability Statement:** The original contributions presented in this study are included in the article. Further inquiries can be directed to the corresponding author.

**Conflicts of Interest:** The authors declare no conflicts of interest.

## Nomenclature

| | |
|---|---|
| $p_b$ | pressure in the vicinity of the bubble, Pa |
| $R$ | bubble radius, m |
| $\Delta$ | distance between bubbles, m |
| $r$ | radial coordinate, m |
| $p_f$ | external pressure, Pa |
| $\rho_b$ | gas density inside the bubble, kg/m$^3$ |
| $u_R$ | velocity of the bubble boundary motion, m/s |
| $\mu$ | dynamic viscosity, Pa s |
| $\sigma$ | interfacial surface tension, N/m |
| $\gamma(\tau)$ | volumetric vapour content (void fraction), |
| $\Psi$ | cell radius, m |
| $\theta$ | thickness of the thermal layer, m |
| $p_k$ | pressure at the cell boundary, Pa |
| $p_R$ | pressure at the bubble boundary, Pa |
| $T$ | temperature, K |
| $k_b$ | number of bubbles, |
| $p_m$ | average pressure, Pa |
| $Q$ | latent heat of vaporization, J/kg |
| $q$ | specific heat flux, W/m$^2$ |
| $c$ | specific heat capacity, J/(kg K) |
| $\vec{u}$ | modulus of the velocity vector. |

## References


1. Zhang, D.; Guan, J.; He, Z.; Shen, C.; Cao, H. Experimental investigation on heat transfer and flow patterns of pulsating heat pipe assisted by ultrasonic cavitation. *Int. J. Heat Mass Transf.* **2022**, *183*, 122187. [CrossRef]
2. Li, C.; Luo, Z.; Shao, Y.; Qian, Y.; Du, S.; Yang, Q.; Chen, Z.; Chen, H.; Zha, Y.; Fang, X. A review on ultrasound-enhanced heat transfer. *Ultrason. Sonochemistry* **2025**, *121*, 107570. [CrossRef] [PubMed]
3. Alenezi, A.; Ali, H.M.; Phelan, P.E. Experimental investigation of the effect of ultrasonic waves on heat transfer enhancement and pressure drop for propylene glycol/water mixtures in a rectangular minichannel heat sink. *Appl. Therm. Eng.* **2025**, *267*, 125717. [CrossRef]
4. Song, Y.; Guan, S.; Wang, Y.; Zhang, C.; Liu, J.; Zhang, L. The application of hydrodynamic cavitation technology and the synergistic effect of hybrid advanced oxidation processes: A review. *Water Sci. Technol.* **2025**, *92*, 301–325. [CrossRef]
5. Askarniya, Z.; Sun, X.; Wang, Z.; Boczkaj, G. Cavitation-based technologies for pretreatment and processing of food wastes: Major applications and mechanisms—A review. *Chem. Eng. J.* **2023**, *454*, 140388. [CrossRef]
6. Pavlenko, A.M. Change of emulsion structure during heating and boiling. *Int. J. Energy A Clean Environ.* **2019**, *20*, 291–302. [CrossRef]
7. Galloni, M.G.; Fabbrizio, V.; Giannantonio, R.; Falletta, E.; Bianchi, C.L. Applications and applicability of the cavitation technology. *Curr. Opin. Chem. Eng.* **2025**, *48*, 101129. [CrossRef]
8. Zhang, H.; Du, M.; Hu, H.; Zhang, H.; Song, N. A Review of Ultrasonic Treatment in Mineral Flotation: Mechanism and Recent Development. *Molecules* **2024**, *29*, 1984. [CrossRef] [PubMed]
9. Wang, M.; Qiu, J.; Chen, W. Towards the development of cavitation technology for gas hydrate prevention. *R. Soc. Open Sci.* **2021**, *8*, 202054. [CrossRef] [PubMed]


https://doi.org/10.3390/en19051120

10. Huang, H.; Zhang, Y.; Wu, X.; Hui, C.; Tan, Y.; Li, G. Study on the Decomposition of Natural Gas Hydrate Induced by Cavitation Damage. In Proceedings of the 57th U.S. Rock Mechanics/Geomechanics Symposium, Atlanta, GA, USA, 25–28 June 2023. [CrossRef]
11. Pavlenko, A.M. Thermodynamic Features of the Intensive Formation of Hydrocarbon Hydrates. *Energies* **2020**, *13*, 3396. [CrossRef]
12. Li, J.; Li, D.; Li, T. Progress in the Understanding and Modeling of Cavitation and Related Applications. *Fluid Dyn. Mater. Process.* **2025**, *21*, 445–470. [CrossRef]
13. Koshlak, H.; Pavlenko, A. A New Mathematical Model for the Features of Bubble Collapse in Steam Cavitation Processes. *Appl. Sci.* **2025**, *15*, 9948. [CrossRef]
14. Fattahi, K.; Boffito, D.C.; Robert, E. Quantifying the chemical activity of cavitation bubbles in a cluster. *Sci. Rep.* **2024**, *14*, 7978. [CrossRef] [PubMed]
15. Li, P.; Wang, J.; Liao, Z.; Ueda, Y.; Yoshikawa, K.; Zhang, G. Microbubbles for effective cleaning of metal surfaces without chemical agents. *Langmuir* **2022**, *38*, 769–776. [CrossRef]
16. Tao, F.; Zhao, G.-Y.; Chen, W.-Z.; Tao, D. 1/2 order subharmonic waves of two cavitation bubbles. *Ultrason. Sonochemistry* **2024**, *110*, 107022. [CrossRef]
17. Kenzhaliyev, B.; Surkova, T.; Berkinbayeva, A.; Baltabekova, Z.; Smailov, K.; Abikak, Y.; Saulebekkyzy, S.; Tolegenova, N.; Omirbek, T.; Dosymbaeva, Z. Innovative Methods for Intensifying the Processing of Zinc Clinker: Synergy of Microwave Treatment and Ultrasonic Leaching. *Metals* **2025**, *15*, 246. [CrossRef]
18. Shen, X.; Wu, P.; Lin, W. A new model for bubble cluster dynamics in a viscoelastic media. *Ultrason. Sonochemistry* **2024**, *107*, 106890. [CrossRef] [PubMed]
19. Pavlenko, A.; Szwaba, R. Behaviour of vapour bubbles in an acoustic field. *Appl. Therm. Eng.* **2025**, *280*, 128495. [CrossRef]
20. Zou, Q.; Zhong, X.; Zhang, B.; Gao, A.; Wang, X.; Li, Z.; Qin, D. Bubble pulsation characteristics in multi-bubble systems affected by bubble size polydispersity and spatial structure. *Ultrasonics* **2023**, *134*, 107089. [CrossRef] [PubMed]
21. Liu, J.; Wang, X.; Liang, J.; Qiao, Y. Refined secondary Bjerknes force equation for double bubbles with pulsation, translation, and deformation. *Ultrason. Sonochemistry* **2024**, *102*, 106756. [CrossRef]
22. Liu, R.; Hu, J.; Wu, Y.; Chen, S.; Wang, C.; Mo, R.; Guo, J.; Lin, S. Manipulation mechanisms of bubble aggregation and evolution in inertial cavitation fields. *Ultrason. Sonochemistry* **2025**, *119*, 107384. [CrossRef]
23. Wang, X.; Hu, S.; Wang, Y.; Guo, Z.; Zhang, Y. Collapse dynamics of a single cavitation bubble in proximity to a wall with two protrusions. *Phys. Fluids* **2025**, *37*, 063334. [CrossRef]
24. Pavlenko, A.; Koshlak, H. Study of the Dynamics of a Single Bubble. *Energies* **2024**, *17*, 4236. [CrossRef]
25. Zhao, Y.; Jiang, S.; Wang, X.; Guo, W.; Zhao, G.; Zhao, F.; Xu, W. Study on the collapse process of double cavitation bubble based on coarse-grained force field. *J. Mol. Liq.* **2025**, *429*, 127517. [CrossRef]
26. Xu, W.; Zhou, Z.; Zhao, Y.; Zhao, G.; Guo, W.; Zhao, F.; Wang, X. Multifactorial Analysis of the Collapse Process of Double Bubbles Based on the Coarse-Grained Force Field in Free Domain. *Langmuir* **2025**, *41*, 3706–3715. [CrossRef]
27. Folden, T.S.; Aschmonei, F.J. A classification and review of cavitation models with an emphasis on physical aspects of cavitation. *Phys. Fluids* **2023**, *35*, 081301. [CrossRef]
28. Stavropoulos-Vasilakis, E.; Kyriazis, N.; Jadidbonab, H.; Koukouvinis, P.; Gavaises, M. Chapter 1—Review of Numerical Methodologies for Modeling Cavitation. In *Cavitation and Bubble Dynamics*; Koukouvinis, P., Gavaises, M., Cavitation, Dynamics, D., Eds.; Academic Press: Cambridge, MA, USA, 2021; pp. 1–35. [CrossRef]
29. Fuster, D.A. Review of Models for Bubble Clusters in Cavitating Flows. *Flow. Turbul. Combust.* **2019**, *102*, 497–536. [CrossRef]
30. Madabhushi, A.; Mahesh, K. A compressible multi-scale model to simulate cavitating flows. *J. Fluid Mech.* **2023**, *961*, A6. [CrossRef]
31. Ghahramani, E.; Ström, H.; Bensow, R.E. Numerical simulation and analysis of multi-scale cavitating flows. *J. Fluid Mech.* **2021**, *922*, A22. [CrossRef]
32. Li, J.; Carrica, P.M. A population balance cavitation model. *Int. J. Multiph. Flow.* **2021**, *138*, 103617. [CrossRef]
33. Zhang, A.-M.; Li, S.-M.; Cui, P.; Li, S.; Liu, Y.-L. Interactions between a central bubble and a surrounding bubble cluster. *Theor. Appl. Mech. Lett.* **2023**, *13*, 100438. [CrossRef]
34. Hanimann, L.; Mangani, L.; Casartelli, E.; Darwish, M.; Kclaramunt, K. Cavitation modelling using real gas state equation: A conceptual study. In *Proceedings of the 30th IAHR Symposium on Hydraulic Machinery and Systems, Lausanne, Switzerland, 21–26 March 2021*; IOP Conference Series: Earth and Environmental Science; IOP Publishing: Philadelphia, PA, USA, 2021; Volume 774, p. 012080. [CrossRef]
35. Huang, M.; Yao, C.; Wang, P.; Cheng, L.; Ying, W. Physics-informed data-driven cavitation model for a specific Mie–Grüneisen equation of state. *J. Comput. Phys.* **2025**, *524*, 113703. [CrossRef]
36. Hong, F.-W.; Zhai, S.-C.; Zheng, C.-S.; Liu, D.-C. Cavitation model of non-equilibrium between phase transition and hydrodynamics. *J. Hydrodyn.* **2025**, *37*, 489–496. [CrossRef]


https://doi.org/10.3390/en19051120

7. Li, W.; Yu, Z.; Kadam, S. An improved cavitation model with thermodynamic effect and multiple cavitation regimes. *Int. J. Heat Mass Transf.* **2023**, *205*, 123854. [CrossRef]
3. Jin, W. Phase change mechanism of periodic cavitation evolution in turbulent cavitation boundary layer. *Ocean Eng.* **2023**, *289*, 116230. [CrossRef]
9. Wang, Z.; Li, L.; Li, X.; Zhu, Z.; Yang, S.; Yang, G. Investigation on multiscale features of cavitating flow in convergent-divergent test section using Eulerian–Lagrangian method. *Int. J. Mech. Sci.* **2023**, *238*, 107853. [CrossRef]
). Budich, B.; Schmidt, S.J.; Adams, N.A. Numerical simulation and analysis of condensation shocks in cavitating flow. *J. Fluid Mech.* **2018**, *838*, 759–813. [CrossRef]
1. Asnaghi, A.; Svennberg, U.; Bensow, R.E. Numerical and experimental analysis of cavitation inception behaviour for high-skewed low-noise propellers. *Appl. Ocean Res.* **2018**, *79*, 197–214. [CrossRef]
2. Shi, Y.; Luo, K.; Chen, X.; Li, D.; Jia, L. A new cavitation model considering inter-bubble action. *Int. J. Nav. Archit. Ocean Eng.* **2021**, *13*, 566–574. [CrossRef]
3. Wang, K.; Li, Z.; Zhang, R.; Ma, R.; Huang, L.; Wang, Z.; Jiang, X. Computational fluid dynamics-based ship energy-saving technologies: A comprehensive review. *Renew. Sustain. Energy Rev.* **2025**, *207*, 114896. [CrossRef]
4. Pavlenko, A. Energy conversion in heat and mass transfer processes in boiling emulsions. *Therm. Sci. Eng. Prog.* **2020**, *15*, 100439. [CrossRef]
5. Zhang, A.-M.; Li, S.-M.; Cui, P.; Li, S.; Liu, Y.-L. A unified theory for bubble dynamics. *Phys. Fluids* **2023**, *35*, 033323. [CrossRef]
6. Abu-Nab, A.K.; Hakami, A.M.; Abu-Bakr, A.F. Charged Cavitation Multibubbles Dynamics Model: Growth Process. *Mathematics* **2024**, *12*, 569. [CrossRef]
7. Cao, P.; Hao, C.; Li, B.; Jiang, H.; Liu, Y. Effect of ruptured cavitated bubble cluster on the extent of the cell deformation by ultrasound. *Ultrason. Sonochemistry* **2021**, *80*, 105843. [CrossRef]
8. Pavlenko, A. Self-preservation Effect of Gas Hydrates. *Rocznik Ochrona Środowiska* **2021**, *23*, 346–355. [CrossRef]
9. Pavlenko, A. Numerical Modeling of the Behavior of Bubble Clusters in Cavitation Processes. *Energies* **2024**, *17*, 1741. [CrossRef]
0. Zhang, L.; Chen, W.; Shen, Y.; Wu, Y.; Zhao, G. The nonlinear characteristics of the pulsations, translations and the secondary Bjerknes force. *Chaos Solitons Fractals* **2021**, *152*, 111322. [CrossRef]
1. Wu, Y.; Lei, Z.; Liu, R.; Wang, C. The oscillations of non-spherical bubbles in liquid. *Ultrason. Sonochemistry* **2025**, *114*, 107262. [CrossRef]
2. Fuster, D.; Colonius, T. Modelling bubble clusters in compressible liquids. *J. Fluid Mech.* **2011**, *688*, 352–389. [CrossRef]
3. Nguyen, V.T.; Phan, T.H.; Park, S.H.; Trong-Nguyen, D.; Quang-Thai, N.; Warn-Gyu, P. Numerical study of shock waves and supersonic jets triggered by cavitation bubble collapse in different pressurized ambiences. *J. Mech. Sci. Technol.* **2024**, *38*, 4167–4173. [CrossRef]
4. Singhal, A.K.; Athavale, M.M.; Li, H.; Jiang, Y. Mathematical Basis and Validation of the Full Cavitation Model. *ASME J. Fluids Eng.* **2002**, *124*, 617–624. [CrossRef]
5. Lester, T.; Kyungjun, C.; Hyunji, K.; Chongam, K. Generalized physics-based cavitation model encompassing multiple cavitation regimes. *Int. J. Heat Mass Transf.* **2025**, *244*, 126898. [CrossRef]
6. Pavlenko, A.M.; Koshlak, H. Application of Thermal and Cavitation Effects for Heat and Mass Transfer Process Intensification in Multicomponent Liquid Media. *Energies* **2021**, *14*, 7996. [CrossRef]
7. Pavlenko, A. Dispersed phase breakup in boiling of emulsion. *Heat Transf. Res.* **2018**, *49*, 633–641. [CrossRef]
8. Pavlenko, A.M.; Koshlak, H. Intensification of Gas Hydrate Formation Processes by Renewal of Interfacial Area between Phases. *Energies* **2021**, *14*, 5912. [CrossRef]
9. Pavlenko, A. Analysis of Methods for Intensifying Heat and Mass Transfer in Liquid Media. *Energies* **2025**, *18*, 1419. [CrossRef]
0. Zhang, X.; Li, F.; Wang, C.; Mo, R.; Hu, J.; Guo, J.; Lin, S. Effects of translational motion on the Bjerknes forces of bubbles activated by strong acoustic waves. *Ultrasonics* **2022**, *126*, 106809. [CrossRef]
1. Radchenko, M.; Radchenko, A.; Trushliakov, E.; Pavlenko, A.; Radchenko, R. Advanced Method of Variable Refrigerant Flow (VRF) Systems Designing to Forecast On-Site Operation—Part 1: General Approaches and Criteria. *Energies* **2023**, *16*, 1381. [CrossRef]
2. Margulis, I.M.; Polovinkin, V.N.; Yashin, A.I. Modern Approaches to the Description of the Dynamics of Cavitation Bubbles and Cavitation Clouds. *Tech. Phys. Lett.* **2024**, *50*, 276–281. [CrossRef]
3. Radchenko, M.; Radchenko, A.; Trushliakov, E.; Koshlak, H.; Radchenko, R. Advanced Method of Variable Refrigerant Flow (VRF) Systems Designing to Forecast Onsite Operation—Part 2: Phenomenological Simulation to Recoup Refrigeration Energy. *Energies* **2023**, *16*, 1922. [CrossRef]
4. Deng, J.; Huang, T.; Zhu, L.; Pan, L.; Sun, W.; Zhang, L.; Ma, Z. Study on bubble growth interphase heat transfer of rapid decompression. *Prog. Nucl. Energy* **2025**, *180*, 105578. [CrossRef]
5. Jukanti, V.G.; Srinivasan, V. Modeling the chaotic dynamics of bubble growth under hydrodynamic interactions in pool boiling. *Int. J. Heat Mass Transf.* **2024**, *234*, 126114. [CrossRef]

66. Yan, S.; Chen, Y.; Lyu, K.; Qin, H.; Zhang, A.-M.; Li, S. Numerical and experimental benchmark study of nonspherical cavitation bubble dynamics near a rigid wall. *Appl. Ocean Res.* **2025**, *154*, 104423. [CrossRef]
67. Sun, T.; Zhang, X.; Zhang, J.; Wang, C. Experimental Study on the Unsteady Natural Cloud Cavities: Influence of Cavitation Number on Cavity Evolution and Pressure Pulsations. *J. Mar. Sci. Eng.* **2021**, *9*, 487. [CrossRef]
68. Mnich, D.; Reuter, F.; Denner, F.; Ohl, C.-D. Single cavitation bubble dynamics in a stagnation flow. *J. Fluid Mech.* **2024**, *979*, A18. [CrossRef]
69. Kansal, D.; Gururani, P.; Joshi, N.C.; Gaurav, P.; Avnish, C. A Comprehensive Review on Utilization of Cavitation Technology for Industrial Waste Water Treatment: A Step Toward Sustainability. *Water Air Soil Pollut.* **2025**, *236*, 438. [CrossRef]
70. Yogesh, P.; Shirish, H.S.; Perugu, S.; Xun, S.; Sivakumar, M. Hybrid hydrodynamic cavitation (HC) technique for the treatment and disinfection of lake water. *Ultrason. Sonochemistry* **2023**, *97*, 106454. [CrossRef] [PubMed]



https://doi.org/10.3390/en19051120

energies

MDPI

Article

# Numerical Analysis of the Flow Downstream of the Exhaust Nozzle of a Miniature Turbojet Engine During Co-Combustion of Kerosene and Hydrogen

**Łukasz Brodzik [1,*], Bartosz Ciupek [1,*], Grigore Cican [2,3], Łukasz Semkło [1] and Dominik Schroeder [4]**

1 Faculty of Environmental Engineering and Energy, Institute of Thermal Energy, Poznan University of Technology, 5 M. Sklodowska-Curie Square, 60-965 Poznan, Poland; lukasz.semklo@put.poznan.pl
2 Faculty of Aerospace Engineering, National University of Science and Technology Politehnica Bucharest, 1–7 Polizu Street, District 1, 011061 Bucharest, Romania; grigore.cican@upb.ro
3 National Research and Development Institute for Gas Turbines COMOTI, 220D Iuliu Maniu Boulevard, 061126 Bucharest, Romania
4 Faculty of Environmental Engineering and Energy, Poznan University of Technology, 5 M. Sklodowska-Curie Square, 60-965 Poznan, Poland; dominik.schroeder@student.put.poznan.pl
* Correspondence: lukasz.brodzik@put.poznan.pl (Ł.B.); bartosz.ciupek@put.poznan.pl (B.C.)

**Abstract**

Research related to the use of hydrogen is crucial because it can be a significant factor in reducing exhaust emissions into the environment. This work represents the second stage of research related to the analysis of exhaust gases from a miniature GTM 400 MOD turbojet engine. Based on the parameters obtained in the experiment, a numerical study of the flow downstream of the exhaust nozzle was performed in ANSYS software (version R1, 2025). The main goal of the calculations was to obtain the temperature distribution and compare the values at checkpoints where actual measurements were taken. The study was conducted in two stages. In the first stage, the engine operated conventionally, burning kerosene. In the second stage, co-combustion of kerosene and hydrogen occurred in the engine. The numerical analysis enabled the visualization of the flow phenomenon for these stages within the considered rotational speeds. The analysis examined flow in domains where the gap was included and excluded. This gap was created by the different diameters of the container opening and the engine nozzle outlet. The study showed that ignoring the gap allows for temperatures closer to the experimental values. Results deemed satisfactory were less likely to be found in vortex areas, but more likely in areas closer to the container walls and the engine exhaust stream itself. The results of the conducted tests showed that the Root Mean Square Error RMSE was within the range of 9.5–22.8% relative to the average temperature values obtained from experimental measurements. The highest turbulence intensity occurred for hydrogen co-combustion at a higher rotational speed, reaching 284% for Variant 2 and 300% for Variant 1. The largest standard deviation for both fuel types at a distance of 0.083 m from the container wall was 68.5 K for variant 2 and 76 K for variant 1. At a distance of 0.166 m from the container wall, the deviation was 76 K for variant 2 and 83.8 K for variant 1.

**Keywords:** co-combustion; hydrogen; numerical analysis; turbojet engine; CFD; combustion diagnostics; RANS

## 1. Introduction

Engine research is currently one of the most important fields of study in the development of aviation technology. This involves continuous engine improvement, as well as its optimization in terms of fuel composition changes and the possible future use of alternative fuels, such as hydrogen. Such research can result in newly observed phenomena, which may lead to new design solutions. The existence of these phenomena can be proven primarily experimentally, but also using modern mathematical tools that solve the scientific problem through approximation, depending on the model used.

The integration of pressure gain combustion into aeroengine systems has gained increasing attention as a promising pathway to significantly enhance thermal efficiency. In this context, the study by Varatharajulu and Stathopoulos [1] presents a thermodynamic and exergy-based comparison of turbojet and turbofan engines equipped with a Rotating Detonation Combustor (RDC). Although their findings confirm potential performance benefits, the analysis relies on simplified cycle modeling and does not consider the complex flow interactions downstream of the nozzle, leaving unresolved questions regarding exhaust dynamics and structural loading.

In pursuit of alternatives to complex vectoring systems for Short Take-Off and Landing STOL applications, Zhang et al. [2] demonstrated the potential of a bypass dual throat nozzle with a bending transition duct. While their results highlight thrust vectoring feasibility, the absence of structural and unsteady flow assessments limits understanding of how such concepts would affect nozzle outlet turbulence and long-term operability. Similarly, Meng et al. [3] investigated serpentine-configured multi-stream supersonic nozzles, identifying stealth advantages at the expense of thrust. However, the lack of experimental validation or optimization studies leaves a gap in understanding how such flow distortions translate to thermomechanical impacts on nozzle walls.

Ensuring stable combustion under unconventional conditions was addressed by He et al. [4], who examined water injection in Turbine-Based Combined Cycle TBCC engines. Their combined experimental and numerical study shows improved stability, but the research remains restricted to specific regimes and does not explore how altered exhaust properties propagate through downstream nozzle sections. In parallel, Aygun et al. [5] compare metaheuristic optimization strategies for small turbojet cycles, revealing trade-offs in thrust and fuel consumption. However, these optimizations assume idealized operation, neglecting secondary effects such as structural loads or turbulent mixing in exhaust jets.

Efforts to integrate hydrogen into aviation fuels have also exposed critical challenges. Niszczota and Gieras [6] demonstrated that hydrogen inlet injection can reduce $CO_2$ emissions but increases $NO_x$ formation, while long-term nozzle durability remains unaddressed. Ayaz et al. [7] provide comparative insights into kerosene, diesel, and JP-4, identifying environmental trade-offs but without assessing how fuel chemistry alters exhaust flow and structural constraints. At the combustor level, Kang et al. [8] proposed a combined injection scheme that improves detonation efficiency but also increases pulsating exhaust pressures, which could amplify nozzle thermal and mechanical stresses, yet this aspect is not explicitly analyzed.

The off-design behavior of micro gas turbines has been studied by Wu et al. [9], who achieved improved accuracy with corrected maps. While effective for performance prediction, their method does not provide detailed flow-field resolution at the nozzle exit. Zhang et al. [10] further contributed with a multi-modal combustor design, but their system-level simulations lack downstream validation, limiting their applicability for nozzle thermomechanical studies. Xu et al. [11] introduced a rotor-motion-coupled CFD framework, reducing prediction errors under crosswind conditions, but its high computational cost constrains systematic nozzle load assessments under fuel variation.

https://doi.org/10.3390/en19040938

On the broader energy landscape, Boretti [12] emphasizes hydrogen's potential in aviation while acknowledging persistent barriers, particularly in combustion stability and $NO_x$ mitigation. Yet, as highlighted, much of the literature remains conceptual, with limited integration of fuel chemistry effects on exhaust nozzle flow, temperature fields, and material stresses.

Addressing these gaps more directly, Brodzik et al. [13] conducted a detailed thermomechanical analysis of the GTM 400 MOD turbojet nozzle operating on kerosene and kerosene–hydrogen co-combustion, supported by coupled CFD/FEM modeling. Their results confirmed that hydrogen addition did not significantly increase nozzle wall temperature or deformation, with peak temperatures below 1100 K and stresses remaining within the material yield limit. These values are consistent with the present findings for comparable engine speeds, indicating that the temperature rise induced by hydrogen enrichment is moderate and does not exceed thermal safety margins. However, Brodzik et al. [13] focused primarily on structural aspects, without analyzing the spatial evolution of the exhaust flow field beyond the nozzle exit.

Bogoi et al. [14] offered complementary insight into flow and acoustic behavior in micro turbojet exhausts using chevron and ejector nozzles. Their Schlieren imaging revealed strong local gradients and vortex formation near the nozzle exit, corresponding to flow structures similar to those observed in the present numerical results. Nevertheless, their work addressed noise attenuation rather than thermal or pressure field coupling, leaving the thermofluid interaction effects unexplored.

Related experimental investigations by Catana and Badea [15] demonstrated that variations in exhaust nozzle geometry significantly affect the operating line and thrust characteristics of gas turbine engines, indirectly confirming the strong sensitivity of downstream flow structures to geometric modifications. However, their analysis focused on global performance parameters and did not resolve local temperature or turbulence distributions in the exhaust region.

Similarly, Villarreal-Valderrama et al. [16] investigated thrust augmentation through a variable exhaust nozzle using active disturbance rejection control, showing that controlled nozzle geometry variations can strongly influence engine thrust and stability. While their results highlighted the dynamic coupling between nozzle configuration and engine performance, detailed thermofluid field analysis downstream of the nozzle was beyond the scope of their study.

In contrast, the present work explicitly quantifies temperature, pressure, and turbulence fields in a confined exhaust environment, thereby complementing these studies by providing spatially resolved thermofluid insight into the effects of exhaust geometry and operating conditions.

Ciupek et al. [17] previously conducted an experimental investigation of the GTM400 MOD turbojet engine operating under kerosene–hydrogen co-combustion, focusing on the measurement of exhaust gas temperature, emission composition ($CO_2$, CO, $NO_x$), and thrust variation at two rotational speeds (31,630 rpm and 47,110 rpm). Their results demonstrated a temperature increase proportional to the hydrogen share in the fuel mixture, reaching up to 760 K, and confirmed corresponding reductions in $CO_2$ and CO emissions alongside an increase in $NO_x$ due to thermal effects. The present study therefore represents the second stage of this research, extending the earlier experimental findings through numerical modeling of the flow downstream of the exhaust nozzle, where pressure distribution, turbulence intensity, and aerodynamic structures were analyzed for the same engine configurations and operating parameters established in [17].

Overall, compared to these studies, the present numerical analysis advances understanding by correlating temperature and pressure field development behind the nozzle

https://doi.org/10.3390/en19040938

with fuel composition effects, bridging the gap between thermomechanical modeling [13], acoustic investigations [14], and emission-oriented experiments [17].

Taken together, these works underline both the potential and the limitations of current research on alternative fuels and nozzle configurations. Despite recent advances in thermodynamic modeling, cycle optimization, and partial hydrogen integration in aviation propulsion, a significant research gap remains in the field of high-fidelity characterization of flow structures and thermomechanical interactions downstream of miniature turbojet nozzles operating under kerosene–hydrogen co-combustion. Filling this gap is essential for understanding how alternative fuels modify the exhaust-jet behavior of temperature, and turbulence generation factors that directly influence emission control, structural loads, and the durability of next-generation propulsion systems.

A review of the literature indicates that very limited experimental or numerical information exists on the detailed representation of exhaust-jet development behind turbojet nozzles, particularly for engines operating at small scales. Where such studies exist, they typically address conventional kerosene combustion, while comprehensive analyses for kerosene–hydrogen mixtures are absent. This lack of data complicates validation of numerical models and hinders the development of generalized predictive tools for hydrogen-enriched fuels.

In this context, the primary objective of the present study is to clearly characterize the exhaust-flow behavior downstream of a miniature turbojet nozzle operating under kerosene and kerosene–hydrogen co-combustion. The scientific significance of this work lies in its integration of dedicated experimental measurements with validated numerical modeling, enabling a detailed analysis of temperature distribution, total pressure field evolution, and turbulence intensity in regions where no such data have previously been available. By providing this combined framework, the study contributes to establishing a foundation for future development of generalized models describing exhaust-jet structure under alternative-fuel operation.

This work constitutes the second part of a broader research effort devoted to the characterization of the exhaust-gas plume of a miniature turbojet engine and its diagnostic interpretation. While previous studies in the field of turbojet engine combustion diagnostics have primarily focused on radiative heat transfer modeling in combustion environments [18,19] or on advanced non-contact diagnostic techniques applied to engine components and dynamic responses [20]. The present study provides a novel contribution by delivering a quantitative, experimentally validated numerical description of the flow and temperature field downstream of the exhaust nozzle.

The originality of this work lies in bridging the gap between combustion-oriented diagnostics and downstream exhaust analysis by demonstrating that key thermofluid characteristics of the exhaust plume can be accurately reproduced without explicit combustion modeling, using experimentally derived boundary conditions. In this way, the study complements existing combustion diagnostics approaches by extending their applicability to confined exhaust regions, offering new insights that support the validation and refinement of future theoretical and computational models of miniature turbojet engines.

## 2. Materials and Methods

The numerical analysis was performed using ANSYS Fluent software (version R1, 2025). It was intended to illustrate the exhaust flow in a specially designed container attached to a turbojet engine (Figure 1a). The test stand was equipped with connection points for the temperature measurements (1), an exhaust gas outlet from the engine (2), a GTM400 turbojet engine stand (3), the GTM400 turbojet engine (4), a fuel connection (5), a kerosene tank (6), a hydrogen tank (7) and a fuel mixing and dosing system (8). The

https://doi.org/10.3390/en19040938

probing temperatures stand (Figure 1b) was equipped with thermoelectric temperature sensors type K (1–2), measuring the temperature transversely to the direction of exhaust gas flow (3).

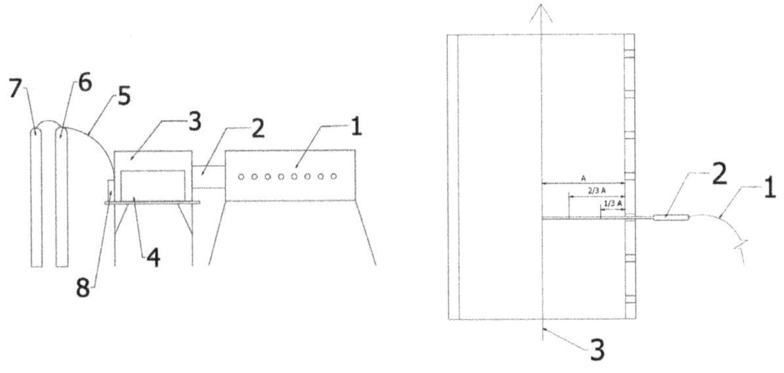


(a) (b)

**Figure 1.** The test stand: (**a**) engine stand with container; (**b**) scheme of probing temperatures in the exhaust gas (A = ±250 mm) [17].

In reality, during the test, the engine was placed in a container whose diameter on the engine side was 0.13 m. In the case of the engine, the nozzle end diameter was 0.12 m. Due to this difference, there was a gap where air could freely flow between the container interior and the test stand environment. Therefore, two calculation variants were adopted in the analysis:

- Variant 1—the existence of a gap related to the difference in the mentioned diameters was taken into account, Figure 2.
- Variant 2—a simplification was applied, consisting of eliminating the gap, then the diameter of the nozzle end cross-section was equal to the diameter of the container.

The container's dimensions for length, width, and height (excluding the stand) were 1 m, 0.496 m, and 0.5 m, respectively. Its outlet was unobstructed. The dimensions of the air-filled space behind the container were 3 m, 1.5 m, and 1.5 m.

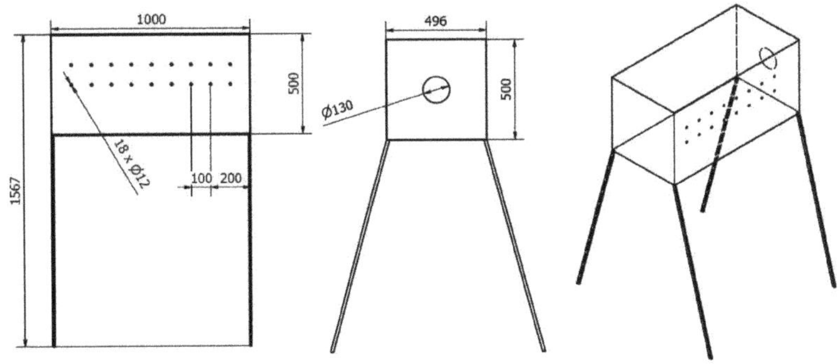


**Figure 2.** Location of checkpoints in the container [17].

## 2.1. GTM400 MOD Turbojet Engine

The calculations concern the GTM400 MOD turbojet engine manufactured by JetPol (Poznań, Poland). This is a miniature single-flow engine, modernized to enable combustion of a kerosene–hydrogen mixture in the combustion chamber. The engine specifications are shown in Table 1. Hydrogen supply was provided by a hydrogen system with hydrogen

https://doi.org/10.3390/en19040938

cylinders manufactured by Linde Gaz Polska (Poznań, Poland). The engine uses Jet A-1 kerosene and 99.955% pure hydrogen. The numerical analysis was based on experimental measurements, in which the flow conditions were determined as follows: For the co-combustion of kerosene and hydrogen, the kerosene flow was reduced to 85% each time. At this time, kerosene mass flow rate was 228 mL/min for the lower speed and 306 mL/min for the higher speed. Simultaneously, the hydrogen flow was started, which was 149 L/min for the lower speed and 393 L/min for the higher speed. Hydrogen was injected into the return duct near the inlet to the vaporizers. The engine control panel was responsible for kerosene control. A 2000 SLPM flow regulator from Alicat Scientific (Tucson, AZ, USA) was installed at the gas system outlet. The pressure regulators were designed to reduce the system pressure to 8 bar. The flow regulator was calibrated for this pressure and maintained the highest accuracy there. For pressure and temperature, the accuracy was 0.5% and 0.75 °C, respectively. The measured flux range for this device is 2000 SLPM (standard liters per minute). In this way, after feeding the kerosene and hydrogen mixture, the same rotational speed was achieved as when burning kerosene alone. More detailed data related to the gas system and the engine are described in the literature [17,21]. Initial conclusions regarding engine operation with kerosene–hydrogen co-combustion were also formulated there.

**Table 1.** The engine specifications.

| Parameter | Value |
|---|---|
| Thrust [N] | 15–300 |
| Rotational speed [rpm] | 30,000–81,000 |
| Maximum air mass flow [kg/s] | 0.59 |
| Maximum fuel consumption [kg/min] | 1.03 |

### 2.2. *Experimental Studies Behind the Nozzle*

The numerical simulation was performed based on experimental data presented in Table 2. The measured volumetric flow rate at the nozzle takes into account both the mass of the supplied air and the mass of the fuel burned in the combustion chamber. Tests were performed for two engine speeds: 31,630 rpm and 47,110 rpm. At the lower speed, the kerosene-fueled engine produced 14.7% less thrust than when fueled with kerosene and hydrogen. At the higher speed, the difference was similar, 8.4% less thrust. During engine operation, temperature was measured behind the nozzle using thermoelectric sensors. To measure the flue gas temperature, 18 K-type (NiCr–Ni) thermocouples were used, with a measurement range of 73 K to 1274 K. Based on the calibration of the entire measurement system, the temperature measurement uncertainty in the range of 373 K to 773 K did not exceed ±1.2 K. Temperature measurements were taken in three zones (distances) from the engine nozzle. In each zone, the measurement was taken along the engine's axis of symmetry and closer to the container housing wall. A more precise arrangement of the measurement points, hereinafter referred to as checkpoints, is presented in Figure 2. More detailed information about the experiment is provided in the literature [17].

To confirm the reliability of temperature measurements, all thermocouples were calibrated before the experimental campaign using a certified dry-block temperature calibrator with an accuracy of ±0.5 K and cross-checked against a platinum resistance reference thermometer. Calibration was carried out at three temperature points: 373 K, 573 K, and 773 K, covering the full range expected during testing. Each measurement series was repeated three times under the same conditions to assess repeatability, with variations between repetitions remaining within ±1.2 K. Calibration and verification were performed both

https://doi.org/10.3390/en19040938

prior to and after the engine tests to ensure sensor stability and to minimize the influence of possible signal drift.

**Table 2.** Experimental data.

| Fuel Type | Kerosene | Kerosene and Hydrogen | Kerosene | Kerosene and Hydrogen |
|---|---|---|---|---|
| Rotation speed [rpm] | 31,630 | 31,630 | 47,110 | 47,110 |
| Mass flow rate [kg/s] | 0.151 | 0.158 | 0.275 | 0.333 |
| Temperature in the outlet cross-section of the engine nozzle [K] | 650 | 750 | 660 | 760 |

*2.3. Input Data for Numerical Analysis*

The engine operated for an extended period at the same speed, fueled with kerosene and hydrogen at constant, set values. Simultaneously, temperature measurements were taken at various checkpoints at the same time. Therefore, the steady flow condition was used for the calculations. Flow modeling requires solving the basic equations of fluid mechanics. The first is the continuity equation:

$$\frac{\partial \rho}{\partial t} + \frac{\partial}{\partial x_i}(\rho \mathrm{v}_i) = 0 \tag{1}$$

where $\rho$ is the fluid density, kg/m$^3$, $t$ is time, s, and $\mathrm{v}_i$ $(i = 1, 2, 3)$ are velocity vectors in the Cartesian coordinate system, m/s.

The second is the momentum equation:

$$\rho\left(\frac{\partial \mathrm{v}_i}{\partial t} + \mathrm{v}_j \frac{\partial}{\partial x_j}\mathrm{v}_i\right) = \frac{\partial}{\partial x_j}\sigma_{ij} + \rho \mathrm{F}_i \tag{2}$$

where $\sigma_{ij}$ $(i = 1, 2, 3)$, $j = (1, 2, 3)$ is the stress tensor $\mathrm{F}_i$ is the external force, N.

The third is the energy equation:

$$\frac{\partial}{\partial t}(\rho c_p T) + \frac{\partial}{\partial x_i}(\rho c_p T \mathrm{v}_i) = \frac{\partial}{\partial x_i}\left(K \frac{\partial T}{\partial x_i}\right) \tag{3}$$

where $c_p$ is the specific heat, J/(kgK), $T$ is the temperature, K, a $K$ s the thermal conductivity coefficient, W/mK.

Additionally, the real combustion process requires solving the diffusion equation that takes into account the combustion reaction. This formula, in simplified form, is as follows:

$$\frac{\partial c}{\partial t} + \frac{\partial c \mathrm{v}_i}{\partial x_i} = \frac{\partial}{\partial x_i}\left(D \frac{\partial c}{\partial x_i}\right) + R \tag{4}$$

where $D$ is the diffusion coefficient, m$^2$/s, $c$ is the concentration of the reacting substance, kg/m$^3$, $R$ is the term describing the combustion reaction, kg/(m$^3$s).

For calculations of real flows, i.e., those that take viscosity into account, RANS (Reynolds Averaged Navier–Stokes) models are used, among others. In such an analysis, the time-averaged Navier–Stokes equations are solved. This simplification shortens the computational time. These models simulate the behavior of disturbed, vortex flow in various ways, often depending on their complexity. The SST k–$\omega$ model was used for flow analysis. Using a blending function, it combines the k–$\omega$ model near the wall and the k-$\varepsilon$ model further from the wall, where free flow occurs. The standard k–$\omega$ model and the transformed k-$\varepsilon$ model are multiplied by the mixing function and added [13,22]. In this case, the turbulent viscosity accounts for the transport of turbulent shear stress. Combining

https://doi.org/10.3390/en19040938

both models reduces sensitivity to turbulence in the free inlet flow and allows for more versatile calculation options [23,24]. Appropriate calculations in both layers (models) are possible due to the presence of a damped term of the cross-diffusion derivative in the $\omega$ equation. The k–$\omega$ Shear Stress Transport (SST) model contains two equations [25,26]:

(a) the specific turbulent kinetic energy for $k$ [m$^2$s$^{-2}$]

$$\frac{\partial}{\partial t}(\rho k) + \frac{\partial}{\partial x_i}(U_i \rho k) = \frac{\partial}{\partial x_j}\left(\mu_k \frac{\partial}{\partial x_j} k\right) + \overline{P}_k - \beta * \rho \omega k \tag{5}$$

(b) the specific turbulent dissipation rate (also specific turbulent frequency) for $\omega$ [s$^{-1}$]

$$\frac{\partial}{\partial t}(\rho \omega) + \frac{\partial}{\partial x_i}(U_i \rho \omega) = \frac{\partial}{\partial x_j}\left(\mu_\omega \frac{\partial}{\partial x_j} \omega\right) +$$

$$+ P_\omega - \beta \rho \omega^2 + 2\rho(1 - F_1)\frac{1}{\omega}\frac{1}{\sigma_{\omega,2}}\frac{\partial}{\partial x_j} k \frac{\partial}{\partial x_j} \omega \tag{6}$$

where the effective viscosity $\mu_\omega$ [kgm$^{-1}$s$^{-1}$]

$$\mu_k = \mu_t \frac{1}{\sigma_k} \tag{7}$$

and the effective viscosity $\mu_\omega$ [kgm$^{-1}$s$^{-1}$]

$$\mu_\omega = \mu_t \frac{1}{\sigma_\omega} \tag{8}$$

$\mu_t$ is modified eddy (or turbulent) viscosity [kgm$^{-1}$s$^{-1}$] and $\sigma_k$ and $\sigma_\omega$ are diffusion constants of the model. $P_\omega$ is the rate of production of $\omega$ [kgm$^{-3}$s$^{-2}$] and is given by the equation

$$P_\omega = \gamma \left[ 2\rho S_{ij} S_{ij} - \frac{2}{3}\rho\omega \left(\frac{\partial}{\partial x_j} U_j\right) \delta_{ij} \right] \tag{9}$$

where $\delta_{ij}$ is the Kronecker delta function and $\gamma$ is a model constant. $S_{ij}$ is the mean rate of deformation component [s$^{-1}$]

$$S_{ij} = \frac{\tau_{ij}}{2\mu_t} + \frac{1}{3}\frac{\rho k \delta_{ij}}{\mu_t} \tag{10}$$

The first blending function $F_1$ is given by the equation

$$F_1 = \tanh\left\{ \left[ min\left( \max\left( \frac{L_t}{y}; \frac{500 v_t}{y^2 \omega} \right); \frac{4\omega\sigma_{\omega,2} k}{X y^2} \right) \right]^4 \right\} \tag{11}$$

$X$ is the cross-diffusion term [kgm$^{-1}$s$^{-2}$]

$$X = \max\left( 2\rho \frac{1}{\omega}\frac{1}{\sigma_{\omega,2}} \frac{\partial}{\partial x_i} k \frac{\partial}{\partial x_i} \omega; 10^{-10} \right) \tag{12}$$

$L_t$ is turbulent length scale [m]

$$L_t = \frac{\sqrt{k}}{0.09\omega} \tag{13}$$

The function $F_1$ is equal to 0 outside the boundary layer and equal to 1 inside the boundary layer. Its values depend, among other things, on $k$ i $\omega$.

https://doi.org/10.3390/en19040938

$v_t$ is the kinematic eddy viscosity $[m^2s^{-1}]$

$$v_t = \frac{\mu_t}{\rho} = \frac{a_1 k}{max(a_1\omega,\ SF_2)} \tag{14}$$

where $a_1$ is constant, $S$ is the strain norm, and $F_2$ is the second mixing function

$$F_2 = \tanh\left\{\left[\max\left(\frac{\sqrt{k}}{0.09\omega y};\ \frac{500v_t}{y^2\omega}\right)\right]^2\right\} \tag{15}$$

In places with large shear gradients, the overestimation of the $v_t$ value limits the expression in the denominator $max(a_1\omega,\ SF_2)$.

$\overline{P}_k$ is the effective rate of production $[kgm^{-1}s^{-3}]$

$$\overline{P}_k = \min(P_k; 10\beta^*\rho k\omega) \tag{16}$$

and prevents the formation of turbulence in stagnation areas, where

$$P_k = 2\mu_t S_{ij} S_{ij} - \frac{2}{3}\rho k\left(\frac{\partial}{\partial x_j}U_i\right)\delta_{ij} \tag{17}$$

According to the above, the specific turbulent kinetic energy $k$ and the specific turbulent dissipation rate $\omega$ are coupled to each other via the modified eddy viscosity $\mu_t$, the production $P_k$ and $P_\omega$, the dissipation and the cross-diffusion term. The k-$\epsilon$ model constants used in the calculations were $C_{2\epsilon} = 1.9$, TKE $Pr = 1$, TDR $Pr = 1.2$, Energy $Pr = 0.85$, Wall $Pr = 0.85$, where $Pr$ is the Prandtl number.

The energy model was used in the calculations, and thermal radiation was included in the flow. The conservation of the radiative energy along a ray path of direction $\hat{s}$ can be written as

$$\hat{s}\nabla I(r,\hat{s}) = -\kappa I(r,\hat{s}) + \kappa_a n_I^2 \frac{\sigma T^4}{\pi} + \frac{\kappa_s}{4\pi}\int_{4\pi} I(r,\hat{s}')P(\hat{s}'\cdot\hat{s})d\omega' \tag{18}$$

where $I(r,\hat{s})$—radiative intensity at a point with coordinates r $(r \in \Omega_2)$, and direction determined by the unit vector $\hat{s}$, $\kappa$—extinction coefficient, $\kappa_a$—coefficient of absorption, $\kappa_s$—coefficient of scattering, $P(\hat{s}'\cdot\hat{s})$—scattering phase function, $n_I$—refractive index, $d\omega'$—solid angle.

The radiation equation was solved by approximation, using discrete ordinates. A pressure-based solver was used, and the air density was given a constant. During the analysis, the domain was described by boundary conditions. At the inlet, the experimentally obtained mass flow and temperature were assumed, which, in simplified terms, was equal to the value measured at the closest distance from the nozzle, within the container. Atmospheric pressure of 101,325 Pa and temperature of 300 K were assumed for the slot inlet. The same values were determined at the container outlet. The container walls were assumed to be made of steel, the density of which was 8030 $kg/m^3$. Air density was 1.225 $kg/m^3$. Other material properties of the container and air are presented in Table 3. The assumed material properties are assumed and adopted as for 316 steel. The adiabatic condition was used for the outer walls of the container. The numerical analysis undertaken did not take into account the fuel combustion process in a turbojet engine. Therefore, the flow stream studied was a mixture of air, kerosene, and hydrogen. Therefore, the paper does not present any information regarding the composition of the exhaust gases. The analysis was therefore a simplification aimed at reproducing the temperature field downstream of the engine nozzle as faithfully as possible.

https://doi.org/10.3390/en19040938

**Table 3.** The container material and air properties.

| Properties of Steel | | | |
|---|---|---|---|
| Temperature [K] | Thermal Conductivity Coefficient [W/(m·K)] | Specific Heat at Constant Pressure [J/(kg·K)] | Thermal Emissivity Coefficient [-] |
| 300 | 16.0 | 477 | 0.20 |
| 400 | 17.0 | 515 | 0.25 |
| 500 | 18.0 | 536 | 0.30 |
| 600 | 18.5 | 557 | 0.35 |
| 700 | 19.0 | 570 | 0.40 |
| 800 | 19.5 | 582 | 0.45 |
| 900 | 20.0 | 597 | 0.50 |
| 1000 | 20.5 | 611 | 0.55 |
| 1100 | 21.0 | 626 | 0.60 |

| Properties of Air | | |
|---|---|---|
| Temperature [K] | Thermal Conductivity Coefficient [W/(m·K)] | Specific Heat at Constant Pressure [J/(kg·K)] |
| 300 | 0.0263 | 1005 |
| 400 | 0.0338 | 1009 |
| 500 | 0.0411 | 1026 |
| 600 | 0.0469 | 1047 |
| 700 | 0.0521 | 1068 |
| 800 | 0.0569 | 1093 |
| 900 | 0.0613 | 1114 |
| 1000 | 0.0654 | 1135 |
| 1100 | 0.0691 | 1156 |

An unstructured mesh filled with polyhedral elements was used in the calculations. The number of mesh elements for variant 1 was 815,117, while for variant 2 it was 804,054. These grids were adopted after the validation stage described in the next section. An example mesh is presented in Figure 3. An example parameter for assessing mesh quality, i.e., orthogonal quality, for variant 1 was 0.693, while for variant 2 it was 0.615. These were the minimum values for both parameters. The average y+ value along the container walls was 4.62. The number of layers installed was 5 and the growth rate was 1.15.

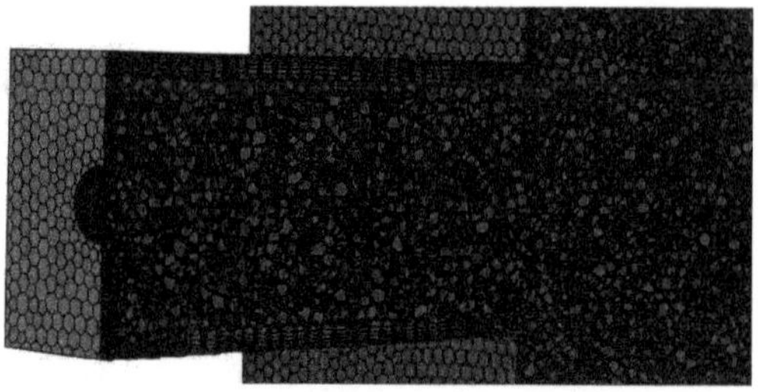

**Figure 3.** The domain Mesh—Variant 4, close-up of the container area.

## 3. Domain Validation

Before conducting the main analysis, a model validation was performed based on five sample meshes. The tests included simulations for a test engine operating at 47,110 rpm with a kerosene–hydrogen co-combustion. The domain was constructed according to the assumptions of Variant 2. In this case, its dimensions were as follows:

(a) D-1: container dimensions, i.e., 1 m, 0.496 m, and 0.500 m, variant 2;

https://doi.org/10.3390/en19040938

(b) D-2: container dimensions and the space behind its outlet: 3 m, 1.5 m, 1.5 m, variant 2;
(c) D-3: container dimensions and the space behind its outlet: 4 m, 1.5 m, 1.5 m, variant 2;
(d) D-4: container dimensions and the space behind its outlet: 6 m, 1.5 m, 1.5 m, basic mesh, variant 2;
(e) D-5: container dimensions and the space behind its outlet: 6 m, 1.5 m, 1.5 m, dense mesh, variant 2;
(f) D-6: container dimensions and the space behind its outlet: 6 m, 1.5 m, 1.5 m, basic mesh, variant 1;
(g) D-7: container dimensions, the space behind its outlet: 6 m, 1.5 m, 1.5 m, and the space around the slot 0.5 m long with external limits 1.5 m wide and 1.5 m high, variant 1.

In each case, the dimensions are given in the following order: length, width, and height. Each subsequent case considered involved a more complex computational domain. For variant 2, the domain dimensions were increasingly increased, and in case D-5, additional mesh refinement was performed. The selected parameters for each mesh are presented in Table 4. The minimum orthogonality was not less than 0.51 and the maximum skewness did not exceed 0.489. These values were considered acceptable taking into account the degree of simplification of the domain and boundary conditions. These are also parameters generally considered acceptable or good. An additional analysis involved comparing two models, namely the Realizable k-ε and the SST k–ω. In this case, a domain with 1,031,894 finite volumes was used. The minimum orthogonality was 0.57 and the minimum skewness was 0. The domain consisted of a portion of the container and the space behind the container, each measuring 3 m, 1.5 m, and 1.5 m. The Realizable k-ε model differs from the k-ε model in two respects. First, it provides an alternative formulation for turbulent viscosity. Second, it has a modified transport equation for epsilon dissipation. For these reasons, this model can better simulate regions of highly sheared flow.

**Table 4.** Selected parameters for mesh evaluation.

| | D-1 | D-2 | D-3 | D-4 | D-5 | D-6 | D-7 |
|---|---|---|---|---|---|---|---|
| Orthogonality (min/mean/max) | 0.573/ 0.958/1 | 0.600/ 0.959/1 | 0.582/ 0.959/1 | 0.615/ 0.960/1 | 0.614/ 0.960/0.9999 | 0.693/ 0.992/1 | 0.510/ 0.960/1 |
| Skewness (min/mean/max) | $1.1 \times 10^{-13}$/ 0.041/0.42 | $1.4 \times 10^{-6}$ 0.040/0.39 | $6.4 \times 10^{-6}$/ 0.041/0.42 | $2.7 \times 10^{-6}$/ 0.039/0.385 | $2.7 \times 10^{-6}$/ 0.039/0.385 | $5 \times 10^{-6}$/ 0.008/0.307 | $4.3 \times 10^{-14}$/ 0.039/0.489 |
| Number of finite mesh volumes | 531,264 | 640,504 | 733,907 | 804,054 | 1,961,564 | 815,117 | 1,098,361 |
| Turbulence intensity | 398 | 274 | 215 | 284 | 284 | 300 | 260 |

The validation of the computational domain was performed based on the determination of the mean absolute error according to the equation

$$MAE = \frac{1}{N}\sum_{i=1}^{N}\left|T_{i'} - \hat{T}_i\right| \quad (19)$$

where $N$ is the number of numbers in the set, $T_{i'}$ is the temperature measured during the experiment at the checkpoint [K], and $\hat{T}_i$ is the temperature calculated during simulation at the checkpoint [K].

The values of mean absolute errors for individual domains were presented in Figure 4.

https://doi.org/10.3390/en19040938

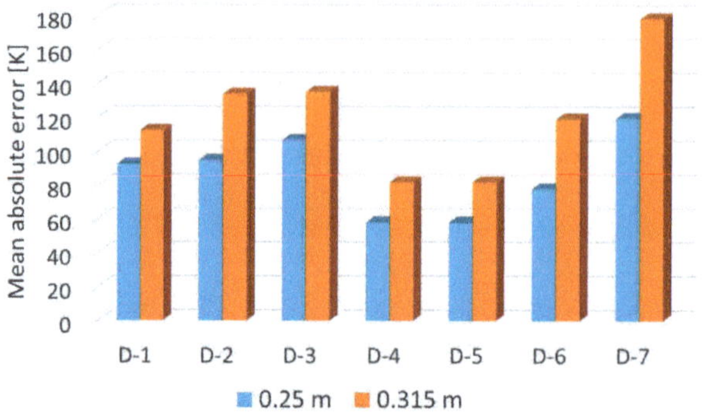


**Figure 4.** The mean absolute errors for domains, for the two considered measurement heights, a rotational speed of 47,110 rpm and kerosene–hydrogen co-combustion.

The largest errors with respect to experimental values occurred for domain D-1, and the smallest for domain D-4. Of these defined cases, only two were selected for further, more detailed, and fundamental research analysis. For variant 2, increasing the domain size led to increasingly smaller mean absolute errors. In variant 1, the presence of additional space outside the gap (D-7) increased the mean absolute error. It is worth noting that in domain D-3 the turbulence intensity had the lowest value, equal to 215. However, for domain D-4 much larger mean absolute errors occurred. For the domain variant including the gap, an inlet-air pressure inlet with ambient pressure (101,325 Pa) was used. Turbulence was left at the default settings: Turbulent Intensity = 5% and Turbulent Viscousity Ratio = 10. Analysis of the simulations taking into account the gap showed that the separated external area around the gap causes air to be directed into the container during engine operation, Figure 5. The presence of a gap without the external area, but using the inlet condition, does not account for this phenomenon. However, vertical flows are observed from the inside of the gap. Flows with different directions are observed in the corner of the container. In the simulation, both in domain 7 and domain 6, there are no back flows near the gap. A different image of the velocity distribution is presented by the simulation for domain 4. The visible flow directions near the gap are oriented towards the main stream and the container outlet. Ultimately, domains D-4 and D-6 were selected for further calculations. For domains 6 and 7, the first case was selected as variant 1 despite obtaining a higher turbulence intensity value. The reason for this choice was the temperature results, which contained significantly larger errors compared to domain 6.

In the comparison of the SST k–ω and Realizable k-ε models, the same regions were analyzed for Variant 2, i.e., control points at distances of 0.25 m and 0.315 m from the container wall. Calculations were also performed for a speed of 47,110 rpm in the kerosene–hydrogen co-combustion mode. Mean absolute errors were calculated for both regions. In the first case, for the Realizable k-ε model, this error was 93.8 K, which was 20.9 K greater than for the SST k–ω model. In the second case, for the Realizable k-ε model, the error was 122.6 K, which was 38.3 K greater than for the SST k–ω model. The reason for this may be that the Realizable k-ε model predicts temperature distributions in near-wall regions worse. The turbulence intensity for the SST k–ω model was equal to 289, while for the Realizable k-ε it was 280. The analysis results showed that the SST k–ω model better predicted the temperature field and was therefore used in the further part of the study.

https://doi.org/10.3390/en19040938

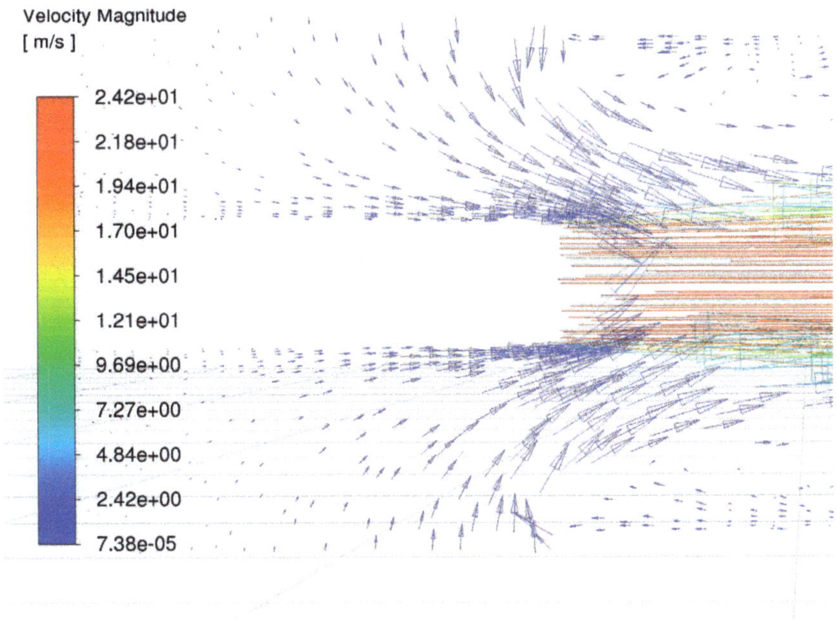


**Figure 5.** Velocity distribution—domain 7 (variant 1).

## 4. Results

The analysis of the results included a comparison of experimental measurements and the results of numerical calculations. Considering the two domain variants under consideration, parameters such as total pressure, temperature, and turbulence intensity were analyzed. In each case, speeds of 31,630 rpm and 47,110 rpm were considered, both in the kerosene combustion mode and in the kerosene and hydrogen co-combustion mode. The presented figures of the distributions of selected parameters inside the container primarily concern variant 2, for which the calculation results differed significantly less from the experimental results. This was described in Section 4.4.

### *4.1. Discussion Related to the Total Pressure Distribution*

Combustion of a kerosene and hydrogen mixture generates a smaller area of total pressure with higher values. However, the maximum value is observed for kerosene–hydrogen co-combustion, at 31,630 rpm, at 101,388 Pa. This is an increase of 6 Pa compared to engine operation, where only kerosene is burned. The results for 47,110 rpm present a similar pattern. In this case, however, the gradient values obtained were significantly larger. For kerosene combustion, the maximum total pressure value is 101,515 Pa, while for kerosene–hydrogen co-combustion it is 101,604 Pa. For this rotational speed, the distribution area of this parameter is more similar, although the pressure gradients are stronger. The changes described above concern the domain for variant 2. The total pressure distribution was presented in Figures 6–9.

The results obtained for variant 1 differ significantly from those for variant 2. Including a gap in the domain resulted in increased pressure gradients, primarily in its immediate vicinity. At a speed of 31,630 rpm for kerosene combustion, the total pressure was 101,396 Pa, while for kerosene–hydrogen co-combustion it was 101,404 Pa. At a speed of 47,110 rpm, the total pressure was 101,560 Pa and 101,681 Pa, respectively. Pressure gradients result directly from the increased dynamic pressure. For variant 1, an increase in this parameter

https://doi.org/10.3390/en19040938

was observed for speeds of 31,630 rpm and 47,110 rpm during kerosene combustion by 25.4% and 13.7%, respectively, compared to variant 2. Similarly, for both rotational speeds during kerosene–hydrogen co-combustion, there was an increase of 24.6% and 13.9%.

**Figure 6.** The total pressure distribution behind the nozzle for a speed of 31,630 rpm during kerosene combustion.

**Figure 7.** The total pressure distribution behind the nozzle for a speed of 31,630 rpm during kerosene and hydrogen combustion.

**Figure 8.** The total pressure distribution behind the nozzle for a speed of 47,110 rpm during kerosene combustion.

https://doi.org/10.3390/en19040938

**Figure 9.** The total pressure distribution behind the nozzle for a speed of 47,110 rpm during kerosene and hydrogen combustion.

### *4.2. Discussion Related to Temperature Distribution*

The shape and pattern of temperature gradient for both analysis variants are similar. The temperature field for variant 2 is presented in Figures 10–13. However, the simulation showed that including the gap in the calculations leads to a significant temperature drop at the checkpoints. The maximum temperature for both variants, however, was the same. For the lower speed and kerosene combustion, it was 650 K, and for kerosene and hydrogen combustion, it was 750 K. For the higher speed and kerosene combustion, the temperature was 660 K, and for kerosene and hydrogen combustion, it was 760 K.

The analysis showed that in both variants, there are checkpoints where the temperature differs only slightly or not at all from the experimentally measured values. There are also points where the temperature gradients were very large. Examining the area halfway up the container at a speed of 31,630 rpm, larger temperature gradients appear much more frequently during kerosene–hydrogen co-combustion for both variants considered. A complete temperature comparison for this case was presented in Table 5.

**Table 5.** The temperature comparison at altitude checkpoints (0.25 m container height) for a rotational speed of 31,630 rpm.

| Distance from Side Wall [m] | Distance from the Nozzle [m] | Temperature Obtained from the Kerosene Experiment [K] | Temperature Obtained from the Simulation for Kerosene [K]—Variant 2 | Temperature Obtained from the Simulation for Kerosene [K]—Variant 1 | Temperature Obtained from the Experiment for Kerosene and Hydrogen [K] | Temperature Obtained from the Simulation for Kerosene and Hydrogen [K]—Variant 2 | Temperature Obtained from the Simulation for Kerosene and Hydrogen [K]—Variant 1 |
|---|---|---|---|---|---|---|---|
| 0 | 0.1 | 387 | 385 | 370 | 461 | 418 | 424 |
| | 0.2 | - | 386 | 375 | 440 | 420 | 428 |
| | 0.3 | 378 | 387 | 371 | 375 | 422 | 429 |
| | 0.4 | 393 | 387 | 374 | 467 | 422 | 426 |
| | 0.5 | - | 417 | 400 | 445 | 453 | 438 |
| | 0.6 | 384 | 422 | 403 | 432 | 461 | 428 |

https://doi.org/10.3390/en19040938

**Table 5.** *Cont.*

| Distance from Side Wall [m] | Distance from the Nozzle [m] | Temperature Obtained from the Kerosene Experiment [K] | Temperature Obtained from the Simulation for Kerosene [K]—Variant 2 | Temperature Obtained from the Simulation for Kerosene [K]—Variant 1 | Temperature Obtained from the Experiment for Kerosene and Hydrogen [K] | Temperature Obtained from the Simulation for Kerosene and Hydrogen [K]—Variant 2 | Temperature Obtained from the Simulation for Kerosene and Hydrogen [K]—Variant 1 |
|---|---|---|---|---|---|---|---|
| 0.083 | 0.1 | 519 | 392 | 363 | 521 | 423 | 398 |
| | 0.2 | - | 392 | 370 | 508 | 422 | 402 |
| | 0.3 | 457 | 397 | 377 | 466 | 430 | 407 |
| | 0.4 | 498 | 412 | 390 | 549 | 447 | 419 |
| | 0.5 | - | 430 | 404 | 529 | 469 | 430 |
| | 0.6 | 423 | 441 | 418 | 474 | 484 | 447 |
| 0.166 | 0.1 | 681 | 517 | 460 | 572 | 582 | 519 |
| | 0.2 | - | 496 | 432 | 569 | 558 | 476 |
| | 0.3 | 681 | 478 | 444 | 564 | 534 | 492 |
| | 0.4 | 540 | 495 | 463 | 548 | 556 | 518 |
| | 0.5 | - | 527 | 504 | 557 | 597 | 516 |
| | 0.6 | 534 | 497 | 496 | 560 | 557 | 569 |
| 0.249 | 0.1 | 636 | 649 | 649 | 747 | 749 | 749 |
| | 0.2 | - | 645 | 646 | 747 | 744 | 745 |
| | 0.3 | 647 | 641 | 632 | 749 | 738 | 728 |
| | 0.4 | 579 | 634 | 616 | 629 | 730 | 709 |
| | 0.5 | - | 625 | 602 | 633 | 718 | 693 |
| | 0.6 | 566 | 618 | 591 | 636 | 710 | 678 |

The results are similar for a speed of 47,110 rpm. For this case, the full temperature summary was provided in Table 6. For both the lower and higher speeds, the solution approximated using variant 2 proved to be closer to the actual measured results.

**Table 6.** The temperature comparison at altitude checkpoints (0.25 m container height) for a rotational speed of 47,110 rpm.

| Distance from Side Wall [m] | Distance from the Nozzle [m] | Temperature Obtained from the Kerosene Experiment [K] | Temperature Obtained from the Simulation for Kerosene [K]—Variant 2 | Temperature Obtained from the Simulation for Kerosene [K]—Variant 1 | Temperature Obtained from the Experiment for Kerosene and Hydrogen [K] | Temperature Obtained from the Simulation for Kerosene and Hydrogen [K]—Variant 2 | Temperature Obtained from the Simulation for Kerosene and Hydrogen [K]—Variant 1 |
|---|---|---|---|---|---|---|---|
| 0 | 0.1 | 442 | 376 | 351 | 484 | 403 | 373 |
| | 0.2 | - | 379 | 361 | 433 | 404 | 394 |
| | 0.3 | 393 | 383 | 365 | 396 | 409 | 391 |
| | 0.4 | 420 | 388 | 362 | 486 | 414 | 392 |
| | 0.5 | - | 418 | 364 | 460 | 450 | 417 |
| | 0.6 | 398 | 421 | 394 | 439 | 454 | 417 |

https://doi.org/10.3390/en19040938

**Table 6.** *Cont.*

| Distance from Side Wall [m] | Distance from the Nozzle [m] | Temperature Obtained from the Kerosene Experiment [K] | Temperature Obtained from the Simulation for Kerosene [K]—Variant 2 | Temperature Obtained from the Simulation for Kerosene [K]—Variant 1 | Temperature Obtained from the Experiment for Kerosene and Hydrogen [K] | Temperature Obtained from the Simulation for Kerosene and Hydrogen [K]—Variant 2 | Temperature Obtained from the Simulation for Kerosene and Hydrogen [K]—Variant 1 |
|---|---|---|---|---|---|---|---|
| 0.083 | 0.1 | 675 | 386 | 351 | 561 | 410 | 372 |
| | 0.2 | - | 389 | 357 | 550 | 413 | 386 |
| | 0.3 | 538 | 400 | 365 | 489 | 427 | 386 |
| | 0.4 | 542 | 418 | 374 | 536 | 449 | 412 |
| | 0.5 | - | 459 | 379 | 529 | 464 | 426 |
| | 0.6 | 520 | 438 | 407 | 489 | 475 | 442 |
| 0.166 | 0.1 | 661 | 517 | 460 | 668 | 578 | 511 |
| | 0.2 | - | 492 | 432 | 616 | 545 | 469 |
| | 0.3 | 692 | 470 | 443 | 608 | 517 | 491 |
| | 0.4 | 551 | 490 | 467 | 640 | 543 | 507 |
| | 0.5 | - | 525 | 461 | 578 | 588 | 511 |
| | 0.6 | 565 | 494 | 507 | 576 | 548 | 561 |
| 0.249 | 0.1 | 665 | 659 | 659 | 761 | 759 | 759 |
| | 0.2 | - | 655 | 656 | 761 | 754 | 755 |
| | 0.3 | 650 | 651 | 642 | 770 | 748 | 738 |
| | 0.4 | 567 | 643 | 626 | 649 | 739 | 717 |
| | 0.5 | - | 635 | 612 | 655 | 729 | 699 |
| | 0.6 | 576 | 628 | 601 | 650 | 720 | 686 |

Comparing the results at the control points in the upper plane (0.315 m), both for lower and higher speeds, they are similar to those at the midpoint of the container. All temperatures were presented in Tables 7 and 8.

**Table 7.** The temperature comparison at altitude checkpoints (0.315 m container height) for a rotational speed of 31,630 rpm.

| Distance from Side Wall [m] | Distance from the Nozzle [m] | Temperature Obtained from the Kerosene Experiment [K] | Temperature Obtained from the Simulation for Kerosene [K]—Variant 2 | Temperature Obtained from the Simulation for Kerosene [K]—Variant 1 | Temperature Obtained from the Experiment for Kerosene and Hydrogen [K] | Temperature Obtained from the Simulation for Kerosene and Hydrogen [K]—Variant 2 | Temperature Obtained from the Simulation for Kerosene and Hydrogen [K]—Variant 1 |
|---|---|---|---|---|---|---|---|
| 0 | 0.1 | 366 | 384 | 376 | 438 | 416 | 419 |
| | 0.2 | - | 393 | 348 | 425 | 428 | 429 |
| | 0.3 | 360 | 396 | 374 | 381 | 432 | 434 |
| | 0.4 | 362 | 401 | 379 | 435 | 435 | 434 |
| | 0.5 | - | 415 | 397 | 422 | 452 | 423 |
| | 0.6 | 357 | 416 | 399 | 426 | 453 | 419 |

https://doi.org/10.3390/en19040938

**Table 7.** *Cont.*

| Distance from Side Wall [m] | Distance from the Nozzle [m] | Temperature Obtained from the Kerosene Experiment [K] | Temperature Obtained from the Simulation for Kerosene [K]—Variant 2 | Temperature Obtained from the Simulation for Kerosene [K]—Variant 1 | Temperature Obtained from the Experiment for Kerosene and Hydrogen [K] | Temperature Obtained from the Simulation for Kerosene and Hydrogen [K]—Variant 2 | Temperature Obtained from the Simulation for Kerosene and Hydrogen [K]—Variant 1 |
|---|---|---|---|---|---|---|---|
| 0.083 | 0.1 | 512 | 395 | 375 | 482 | 424 | 392 |
| | 0.2 | - | 405 | 381 | 473 | 438 | 402 |
| | 0.3 | 419 | 409 | 382 | 441 | 445 | 414 |
| | 0.4 | 456 | 416 | 390 | 479 | 452 | 419 |
| | 0.5 | - | 428 | 405 | 470 | 467 | 430 |
| | 0.6 | 398 | 435 | 410 | 439 | 476 | 433 |
| 0.166 | 0.1 | 584 | 407 | 375 | 560 | 439 | 400 |
| | 0.2 | - | 444 | 387 | 544 | 487 | 413 |
| | 0.3 | 587 | 439 | 425 | 533 | 480 | 458 |
| | 0.4 | 496 | 472 | 429 | 542 | 524 | 469 |
| | 0.5 | - | 466 | 436 | 534 | 516 | 482 |
| | 0.6 | 474 | 476 | 454 | 528 | 529 | 504 |
| 0.249 | 0.1 | 582 | 555 | 516 | 640 | 630 | 559 |
| | 0.2 | - | 527 | 519 | 637 | 594 | 593 |
| | 0.3 | 583 | 576 | 488 | 638 | 657 | 595 |
| | 0.4 | 535 | 524 | 496 | 516 | 590 | 557 |
| | 0.5 | - | 546 | 525 | 507 | 618 | 564 |
| | 0.6 | 536 | 544 | 504 | 505 | 615 | 600 |

The largest differences between the calculated and measured values at the same points occur at a distance of 0.083 m and 0.166 m. This applies to both considered analysis variants.

**Table 8.** The temperature comparison at altitude checkpoints (0.315 m container height) for a rotational speed of 47,110 rpm.

| Distance from Side Wall [m] | Distance from the Nozzle [m] | Temperature Obtained from the Kerosene Experiment [K] | Temperature Obtained from the Simulation for Kerosene [K]—Variant 2 | Temperature Obtained from the Simulation for Kerosene [K]—Variant 1 | Temperature Obtained from the Experiment for Kerosene and Hydrogen [K] | Temperature Obtained from the Simulation for Kerosene and Hydrogen [K]—Variant 2 | Temperature Obtained from the Simulation for Kerosene and Hydrogen [K]—Variant 1 |
|---|---|---|---|---|---|---|---|
| 0 | 0.1 | 409 | 374 | 368 | 469 | 396 | 386 |
| | 0.2 | - | 396 | 371 | 444 | 425 | 406 |
| | 0.3 | 370 | 396 | 369 | 416 | 424 | 401 |
| | 0.4 | 392 | 406 | 364 | 461 | 437 | 399 |
| | 0.5 | - | 414 | 385 | 437 | 445 | 408 |
| | 0.6 | 365 | 414 | 388 | 416 | 445 | 409 |

https://doi.org/10.3390/en19040938

**Table 8.** *Cont.*

| Distance from Side Wall [m] | Distance from the Nozzle [m] | Temperature Obtained from the Kerosene Experiment [K] | Temperature Obtained from the Simulation for Kerosene [K]—Variant 2 | Temperature Obtained from the Simulation for Kerosene [K]—Variant 1 | Temperature Obtained from the Experiment for Kerosene and Hydrogen [K] | Temperature Obtained from the Simulation for Kerosene and Hydrogen [K]—Variant 2 | Temperature Obtained from the Simulation for Kerosene and Hydrogen [K]—Variant 1 |
|---|---|---|---|---|---|---|---|
| 0.083 | 0.1 | 585 | 395 | 365 | 551 | 420 | 388 |
| | 0.2 | - | 403 | 373 | 537 | 431 | 401 |
| | 0.3 | 543 | 409 | 377 | 472 | 438 | 400 |
| | 0.4 | 504 | 416 | 377 | 511 | 447 | 417 |
| | 0.5 | - | 426 | 392 | 504 | 459 | 423 |
| | 0.6 | 467 | 431 | 398 | 463 | 466 | 428 |
| 0.166 | 0.1 | 586 | 402 | 364 | 687 | 430 | 385 |
| | 0.2 | - | 441 | 379 | 608 | 480 | 409 |
| | 0.3 | 596 | 436 | 409 | 597 | 473 | 457 |
| | 0.4 | 537 | 471 | 418 | 623 | 518 | 465 |
| | 0.5 | - | 464 | 428 | 548 | 509 | 473 |
| | 0.6 | 517 | 475 | 449 | 543 | 523 | 497 |
| 0.249 | 0.1 | 583 | 559 | 487 | 682 | 632 | 548 |
| | 0.2 | - | 533 | 518 | 686 | 597 | 582 |
| | 0.3 | 582 | 584 | 527 | 693 | 663 | 584 |
| | 0.4 | 529 | 527 | 495 | 608 | 590 | 539 |
| | 0.5 | - | 554 | 500 | 630 | 624 | 548 |
| | 0.6 | 528 | 551 | 531 | 624 | 620 | 590 |

Measurements were not taken at all analyzed locations during the experiment. Therefore, in such cases, the symbol "-" was used in Tables 5–8.

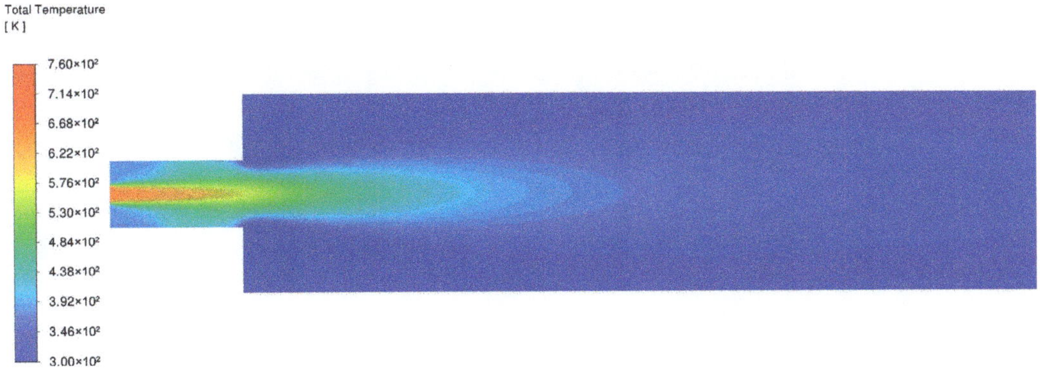


**Figure 10.** The temperature distribution behind the nozzle for a speed of 31,630 rpm during kerosene combustion.

https://doi.org/10.3390/en19040938

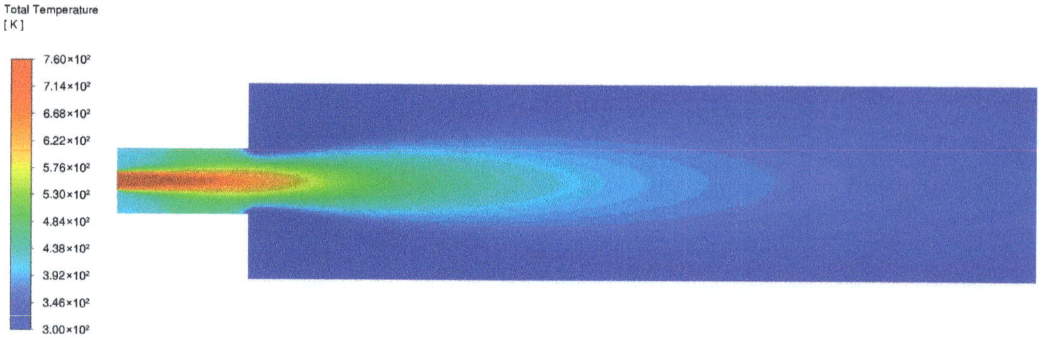


**Figure 11.** The temperature distribution behind the nozzle for a speed of 31,630 rpm during kerosene and hydrogen combustion.

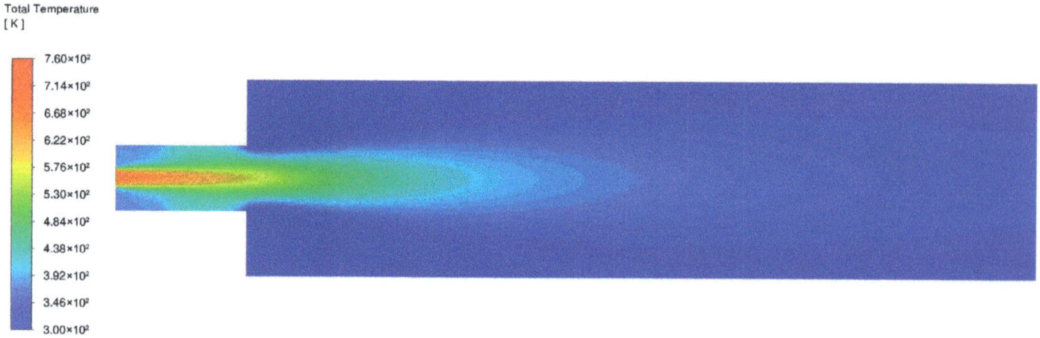


**Figure 12.** The temperature distribution behind the nozzle for a speed of 47,110 rpm during kerosene combustion.

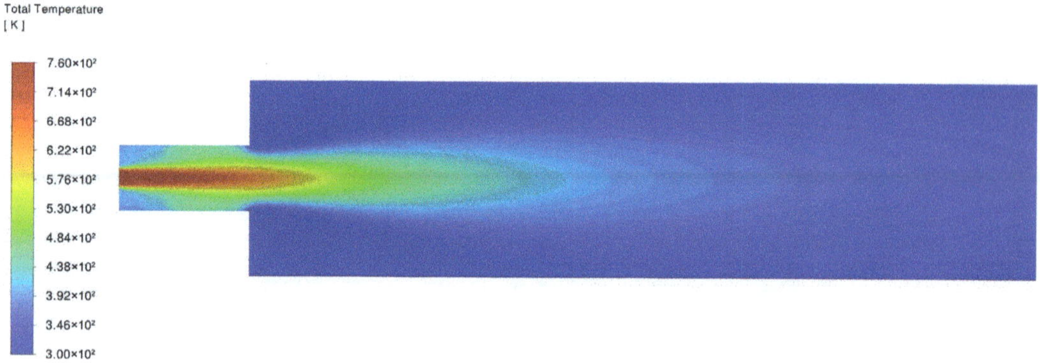


**Figure 13.** The temperature distribution behind the nozzle for a speed of 47,110 rpm during kerosene and hydrogen combustion.

### *4.3. Discussion Related to Turbulence Intensity*

The next parameter examined in the analysis was turbulence intensity. The distribution field of this parameter for both speeds was presented in Figures 14–17. A turbulence intensity below 1% indicates that the flow exhibits very weak characteristics of disturbed flow. The analysis showed that such a flow occurs near the walls, primarily in the front part of the container, i.e., on the engine nozzle side. This parameter increases slightly in the analysis that includes the gap, i.e., for variant 1.

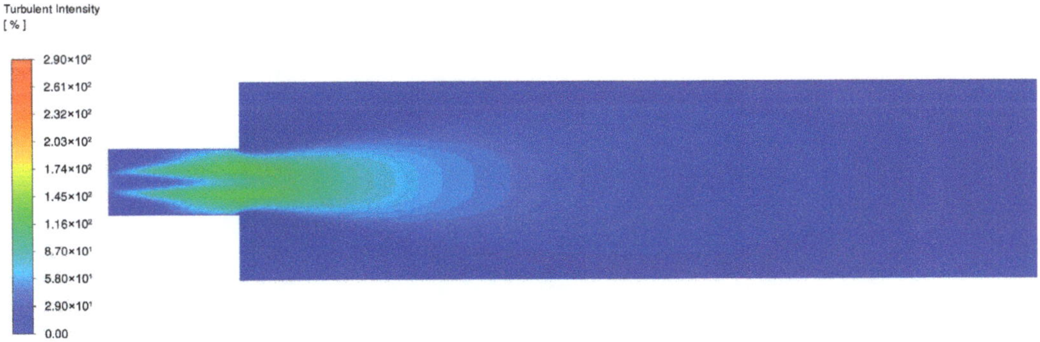


**Figure 14.** The turbulence intensity distribution behind the nozzle for a speed of 31,630 rpm during kerosene combustion.

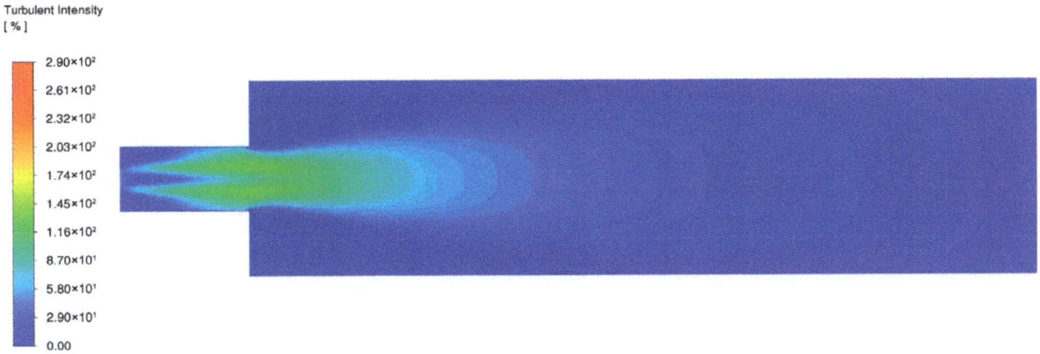


**Figure 15.** The turbulence intensity distribution behind the nozzle for a speed of 31,630 rpm during kerosene and hydrogen combustion.

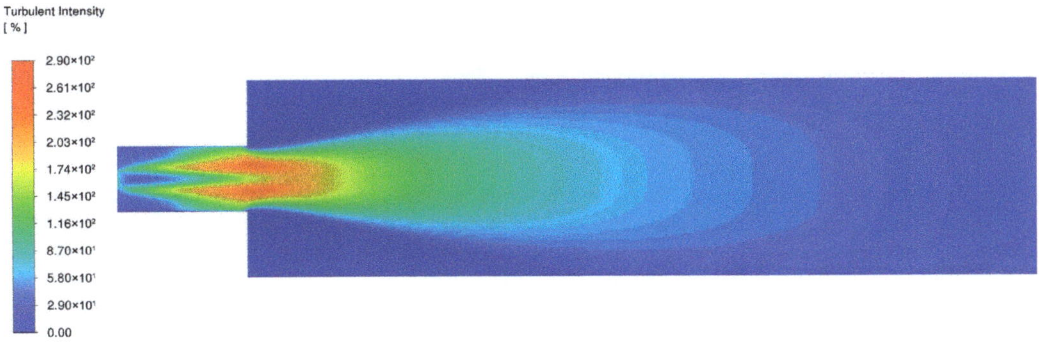


**Figure 16.** The turbulence intensity distribution behind the nozzle for a speed of 47,110 rpm during kerosene combustion.

The maximum turbulence intensity in analysis variant 1 at a speed of 31,630 rpm during kerosene combustion was 136% and for kerosene–hydrogen co-combustion 140%. For higher rotational speeds, this parameter was 244% for kerosene combustion and 300%

https://doi.org/10.3390/en19040938

for kerosene–hydrogen co-combustion. For analysis variant 2 at a lower speed, the turbulence intensity was 130% for kerosene combustion and 136% for kerosene–hydrogen co-combustion. For higher speeds, this parameter was 235% and 284%, respectively. Therefore, the use of the domain in variant 2 resulted in a decrease in this parameter for the lower and higher speeds in the case of using kerosene by 6% and 9%. Similarly, for the kerosene–hydrogen co-combustion, a decrease of 4% and 16% occurred, respectively.

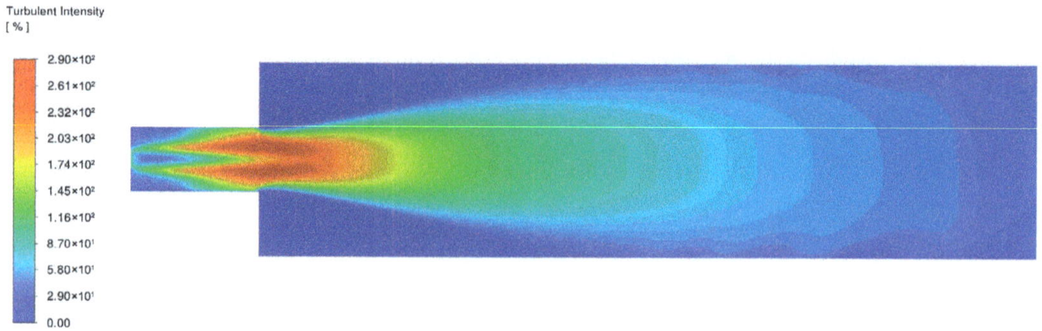


**Figure 17.** The turbulence intensity distribution behind the nozzle for a speed of 47,110 rpm during kerosene and hydrogen combustion.

### *4.4. Discussion Related to the Comparison of Experimental Measurements and Numerical Calculation Results*

The temperature differences obtained at the checkpoints using numerical simulations were analyzed using several parameters. The first of these, also used during domain validation, was the mean absolute error. At a height of 0.25 m for variant 2, for kerosene, in the case of a speed of 31,630 rpm, the mean absolute error of all obtained results was 57.6 K, and for a speed of 47,110 rpm, 100.9 K. For the kerosene–hydrogen co-combustion, for a lower speed this parameter was 40.1 K, and for a higher speed 57.8 K. For variant 1, for kerosene, in the case of a speed of 31,630 rpm, the mean absolute error of all obtained results was 67.1 K, and for a speed of 47,110 rpm, 124.8 K. For the kerosene–hydrogen co-combustion, for a lower speed this parameter was 50.2 K, and for a higher speed 77.8 K.

In the case of kerosene at a measurement height of 0.315 m for variant 2, at a speed of 31,630 rpm, the mean absolute error of all obtained results was 47.5 K, and for a speed of 47,110 rpm – 67.2 K. For the kerosene–hydrogen co-combustion, for a lower speed, this parameter was 38.8 K, and for a higher speed, 58.9 K. For variant 1, for kerosene, in the case of a speed of 31,630 rpm, the mean absolute error of all obtained results was 64.1 K, and for a speed of 47,110 rpm, 91.2 K. For the kerosene–hydrogen co-combustion, for a lower speed, this parameter was 52.3 K, and for a higher speed, 94.5 K. A more accurate distribution of this parameter in the context of the considered distances from the wall is presented in Figures 18 and 19.

The distribution of the mean absolute error confirmed that its highest values occur in two intermediate areas, i.e., around 0.083 m and 0.166 m. For further evaluation of these two areas, the standard deviation was used, according to the following formula

$$\sigma = \sqrt{\frac{1}{N}\sum_{i=1}^{N}(\Delta T_{i'} - \mu)^2} \quad (20)$$

where $N$ is the number of numbers in the set, $\Delta T_{i'}$ is the difference between the measured and calculated temperature for both rotational speeds [K], $\mu$ is the arithmetic mean of all values [K].

https://doi.org/10.3390/en19040938

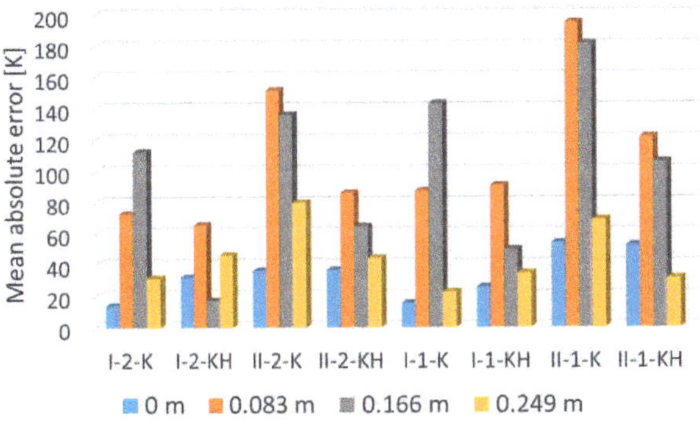


**Figure 18.** The mean absolute error at the checkpoints at a height of 0.25 m: I—the rotational speed of 31,630 rpm, II—the rotational speed of 47,110 rpm, 2—variant 2, 1—variant 1, K—kerosene, KH—the kerosene–hydrogen.

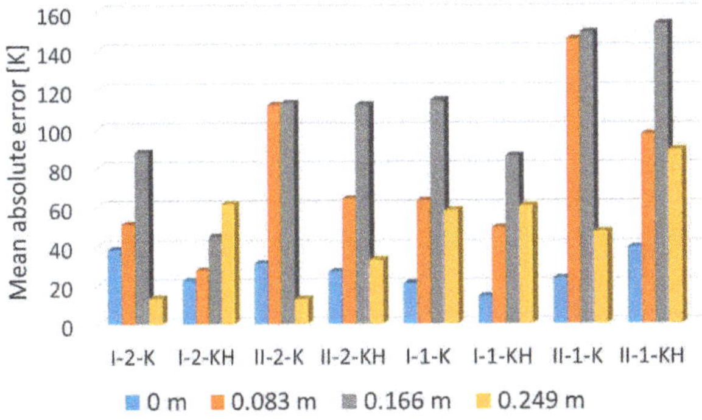


**Figure 19.** The mean absolute error at the checkpoints at a height of 0.315 m: I—the rotational speed of 31,630 rpm, II—the rotational speed of 47,110 rpm, 2—variant 2, 1—variant 1, K—kerosene, KH—the kerosene–hydrogen.

The analysis assumed that the standard deviation, as a certain component of uncertainty, can indirectly indicate the states of uncertainty at individual checkpoints. Therefore, at a height of 0.25 m in variant 2—for kerosene—the area with higher measurement uncertainty was located at a distance of approximately 0.166 m from the tank wall. For the kerosene–hydrogen co-combustion, this area had lower measurement uncertainty. For variant 1, in both cases, the greater measurement uncertainty was around a distance of 0.166 m from the container wall. At a height of 0.315 m, for both variants and both fuel types, the greater uncertainty was always around a distance of 0.166 m from the container wall. The standard deviation values were presented in Table 9. To improve clarity and provide a direct graphical interpretation of the deviations, the standard deviation values listed in Table 9 are additionally presented in Figure 19 in a bar-chart form. The figure enables a direct comparison of uncertainty levels between fuel types, measurement heights, and distances from the container wall. As shown in Figure 20, for both measurement heights (0.25 m and 0.315 m), the standard deviation is consistently higher at a distance of 0.166 m compared to 0.083 m, regardless of the variant considered. This trend confirms that the region located farther from the exhaust axis and closer to the container wall is characterized by stronger temperature fluctuations, which can be attributed to enhanced mixing, recirculation zones, and increased turbulence intensity. Furthermore, the graphical representation highlights that kerosene–hydrogen co-combustion generally exhibits lower

https://doi.org/10.3390/en19040938

standard deviation values at 0.083 m, indicating a more stable temperature field near the exhaust core, while at 0.166 m the differences between fuels become less pronounced, especially at the height of 0.315 m. This observation is consistent with the discussion above and confirms that uncertainty increases with distance from the jet core and with vertical elevation, where flow confinement effects dominate.

**Table 9.** The standard deviation for the cases considered.

| **Measurement Height Level [m]** | **0.25** | | | | **0.315** | | | |
|---|---|---|---|---|---|---|---|---|
| Type of variant | 1 | | 2 | | 1 | | 2 | |
| Fuel type | kerosene | kerosene–hydrogen | kerosene | kerosene–hydrogen | kerosene | kerosene–hydrogen | kerosene | kerosene–hydrogen |
| $\sigma$ [K] (for 0.083 m) | 76.0 | 44.2 | 68.5 | 41.9 | 65.8 | 43.0 | 57.0 | 38.1 |
| $\sigma$ [K] (for 0.166 m) | 83.8 | 49.1 | 76.0 | 49.1 | 69.9 | 75.6 | 69.7 | 69.7 |

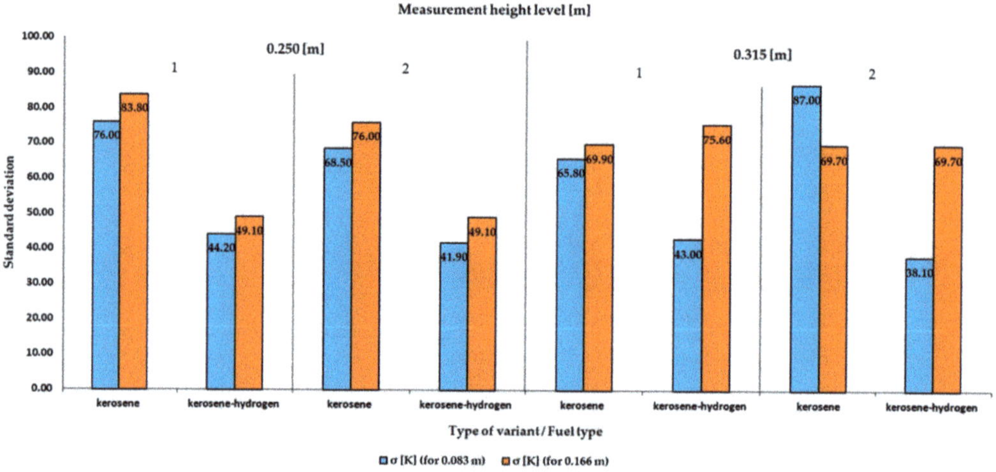


**Figure 20.** Standard deviation of temperature measurements for kerosene and kerosene–hydrogen co-combustion at measurement heights of 0.25 m and 0.315 m, shown for two distances from the container wall (0.083 m and 0.166 m) and for both analyzed variants.

The errors related to the difference between the experimental results and the simulation results for variant 2 were presented in the form of RMSE in Table 10. It was calculated according to the formula

$$RMSE = \sqrt{\frac{1}{N}\sum_{i=1}^{N}\left(T_{i'} - \hat{T}_i\right)^2} \quad (21)$$

and in percentage

$$nRMSE = \frac{RMSE}{\mu} \cdot 100\% \quad (22)$$

The RMSE values obtained in this work (9.5–22.8%) are within the ranges commonly reported for RANS-based temperature predictions in high-gradient jet exhaust environments. For miniature gas turbines, typical discrepancies between numerical and experimental temperature fields range an RMSE of 15% to 30%, as documented in recent small-scale turbine, afterburner and exhaust mixing studies. This confirms that the error levels observed here are consistent with limitations inherent to RANS modeling in strongly sheared, thermally stratified flows.

https://doi.org/10.3390/en19040938

**Table 10.** The RMSE in percent with respect to the mean experimental temperature values.

| Measurement Height Level [m] | 0.25 | | | | 0.315 | | | |
|---|---|---|---|---|---|---|---|---|
| Rotational speed [rpm] | 31,630 | | 47,110 | | 31,630 | | 47,110 | |
| Fuel type | kerosene | kerosene–hydrogen | kerosene | kerosene–hydrogen | kerosene | kerosene–hydrogen | kerosene | kerosene–hydrogen |
| nRMSE [%] | 15.7 | 9.5 | 22.8 | 12.4 | 14.7 | 10.4 | 18.1 | 15.1 |

It should be noted that the very high turbulence intensity values obtained in the analysis (up to 300%) do not indicate physically unrealistic behavior but arise from the definition of turbulence intensity used in ANSYS Fluent. The parameter is computed as the ratio of the root-mean-square fluctuation of the velocity to the local mean velocity. In regions where the mean velocity drops sharply, such as in the shear layers surrounding the exhaust jet or inside recirculation zones, velocity fluctuations remain high while the mean velocity approaches low values. This combination can produce turbulence intensities well above 100%. Similar effects have been reported in confined exhaust jets, under expanded plumes and micro gas turbine mixing layers. The peaks observed in this study therefore correspond to strongly sheared, highly vortical regions at the jet boundary rather than to the core flow.

Figures 21 and 22 present vector distributions of gas velocity in the container for both speed variants and fuel types used. The obtained results reveal differences in the form of local velocity development as well as the formation of vortex regions.

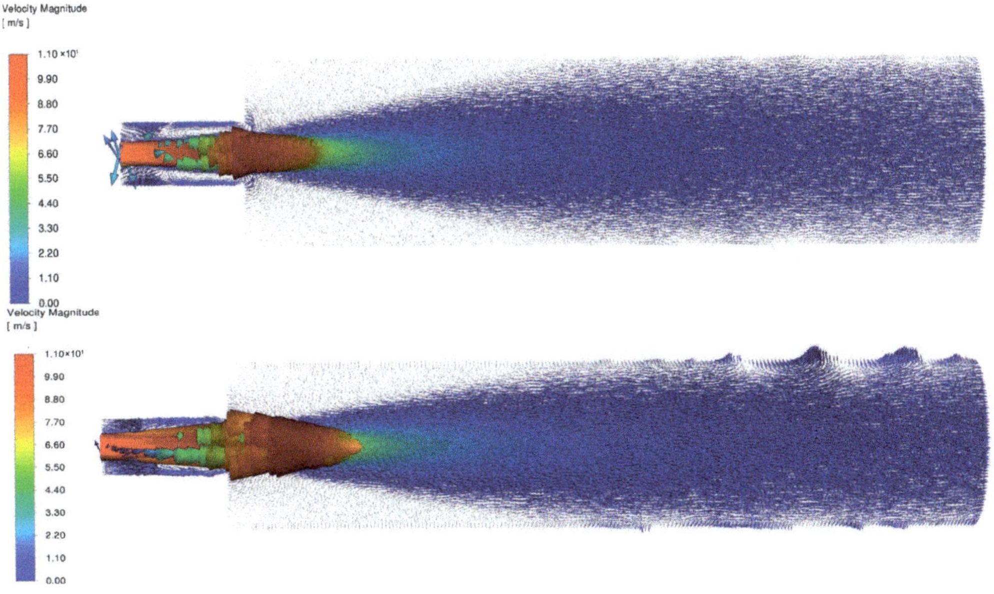


**Figure 21.** The velocity distribution downstream of the nozzle for a speed of 31,630 rpm during kerosene combustion: variant 2 at the (**top**); variant 1 at the (**bottom**).

The same differences are observed for both lower and higher engine speeds. Local gas velocities outside the main exhaust stream are correspondingly higher for a speed of 47,110 rpm. Due to the limited space of the container on the engine side, a more intense gas flow is observed towards the exhaust stream, where the pressure is reduced. In this case, this is the only source of motion, apart from the particle motion in the exhaust cross-section.

https://doi.org/10.3390/en19040938

This leads to an increase in turbulent kinetic energy in the shear layers of the gas flow. In the presence of a gap, this flow is more pronounced. For variant 1, the maximum turbulent kinetic energy was 8.96 $m^2/s^2$, and for variant 2 it was 8.28 $m^2/s^2$.

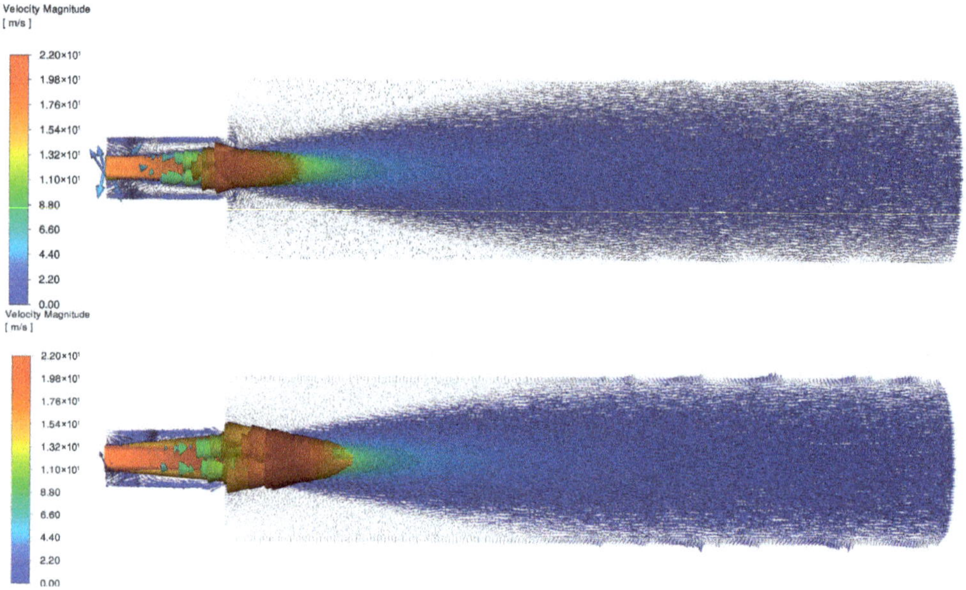


**Figure 22.** The velocity distribution downstream of the nozzle for a speed of 47,110 rpm during kerosene and hydrogen combustion: variant 2 at the (**top**); variant 1 at the (**bottom**).

### 4.5. Model Universality and Comparison with Literature

To assess the universality of the proposed CFD model and confirm its predictive reliability, a comparative analysis was conducted with published numerical and experimental investigations on miniature and small-scale turbojet engines operating under various fuel conditions. The goal was to determine whether the thermal and flow-field distributions obtained in this study are consistent with the general physical trends observed in similar engine configurations.

The obtained RMSE range (9.5–22.8%) aligns with the deviations typically reported for Reynolds-Averaged Navier–Stokes (RANS)-based simulations of small turbojet exhausts. Comparable levels of accuracy were documented by Badami et al. (2014) in their combined experimental–numerical study on small turbojets fueled by Jet-A and biofuel blends, where temperature prediction errors ranged between 10% and 25% depending on operating regime [27].

Similarly, Gao et al. (2014) investigated the flow and liner wall temperature field of a reverse-flow combustor for a miniature turbojet engine, validating CFD results against experimental data with an average temperature deviation of 12–20% [28]. Their findings confirmed that SST k–ω and standard k–ε models provide physically consistent turbulence structures downstream of the combustor, an observation mirrored by the present results.

In turn, Suchocki et al. (2014) analyzed the GTM-140 engine using a coupled non-premixed combustion and k–ε turbulence model, reporting exhaust temperatures between 870 and 950 K, which is consistent with the 890–1000 K outlet temperature range used as boundary conditions in the current simulation [29].

Furthermore, He et al. (2021) optimized a microturbojet combustor numerically and experimentally, demonstrating that turbulence intensity distributions within 200–300%

correspond to flow instabilities localized in the shear layers—closely matching the present observations [30].

Finally, Brodzik et al. (2025) conducted a comprehensive thermomechanical analysis of the same GTM 400 MOD miniature turbojet engine using a coupled CFD–FEM framework, demonstrating that nozzle wall temperatures and stress distributions remained within experimentally verified material limits under both kerosene combustion and kerosene–hydrogen co-combustion conditions [13]. Their results confirmed the predictive robustness of steady-state RANS simulations implemented in ANSYS for evaluating thermal loads and structural response in hydrogen-assisted propulsion systems.

These findings are consistent with broader numerical investigations of alternative fuel combustion in turbojet engines. For instance, Alabas [31] showed that CFD-based modeling of hydrogen- and ammonia-enriched jet fuels yields realistic temperature levels and emission trends when appropriate boundary conditions and turbulence–radiation coupling are applied, further supporting the reliability of RANS-based approaches for hydrogen-containing fuel mixtures.

Moreover, the exhaust temperature ranges and operating conditions adopted in the present study align well with the characteristic parameters reported by Gieras [32] for miniature turbojet engines, who documented typical combustion chamber outlet temperatures and turbine loading levels for small-scale propulsion systems.

Taken together, these studies provide indirect yet consistent validation of the present numerical framework, confirming that the applied modeling assumptions lead to physically realistic predictions of thermal and mechanical behavior in miniature turbojet engines operating with both conventional and hydrogen-enriched fuels.

Taken together, these comparisons indicate that the present numerical model reproduces both the magnitude and spatial character of exhaust temperature fields and turbulence structures reported by other authors (Table 11). Hence, the model can be considered universal within the domain of small-scale turbojet exhaust simulations involving hydrocarbon and hydrogen-based fuels.

**Table 11.** Comparison of study results and literature.

| No. | Source | Engine Type/Fuel | Methods | Key Results | Comparison with Present Study |
|---|---|---|---|---|---|
| 1. | Badami et al., 2014, *Energy Conv. & Management* [27] | 80 N Microturbojet/Jet-A, GTL, JME | Experimental + CFD (k–ε) | CFD–exp. deviation: 10–25%, exhaust T: 700–900 K | Present model within same deviation band; similar T field at nozzle outlet |
| 2. | Gao et al., 2014, *IHTC15 Conf.* [28] | Reverse-flow combustor/ Kerosene | Experimental + CFD (SST k–ω) | T deviation ± 20%, high turbulence near liner wall | Similar turbulence and near-wall gradients validated by SST k–ω model |
| 3. | Suchocki et al., 2014, *Journal of KONES Powertrain and Transport* [29] | GTM-140/ Kerosene | CFD (k–ε) + DPM | Exhaust T: 870–950 K, verified combustor model | Agrees with boundary T (890–1000 K) and pressure gradients in this study |
| 4. | He et al., 2021, Int. J. *Aerospace Eng.* [30] | Microturbojet/ Jet-A | CFD (SST k–ω) + Experiment | Turbulence intensity 200–300% in shear zones, stable combustion | Confirms realistic turbulence peak (284–300%) from current simulations |
| 5. | Brodzik et al., 2025, *Energies* [13] | GTM400 MOD/Kerosene–$H_2$ | CFD + FEM (ANSYS) + Experimental | Max wall T: 1090 K, stress < 261 MPa, ΔT < 12% vs. exp. | Confirms consistent exhaust T range (650–760 K) and mechanical stability under $H_2$ co-combustion |

https://doi.org/10.3390/en19040938

## 5. Conclusions

This study investigated the exhaust flow downstream of a GTM 400 MOD miniature turbojet engine operating under kerosene and kerosene–hydrogen co-combustion conditions. The primary objective was to assess the influence of geometric confinement and fuel composition on the thermal and flow characteristics of the exhaust jet and to verify whether a simplified numerical approach can reliably reproduce experimentally observed temperature fields.

The applied methodology combined experimental temperature measurements with steady-state CFD simulations performed using the SST k–ω turbulence model. Two numerical domain configurations were analyzed: one including the physical gap between the exhaust nozzle and the container (Variant 1) and one neglecting this gap (Variant 2). Model validation was conducted using temperature measurements collected at multiple distances from the exhaust axis and container wall, enabling quantitative assessment through normalized RMSE values.

The results demonstrated that the largest temperature deviations occurred in regions characterized by intensified turbulence and mixing, particularly at intermediate distances from the container wall. The highest turbulence intensities were observed for kerosene–hydrogen co-combustion at higher rotational speeds, reaching approximately 300% in Variant 1 and 284% in Variant 2. The analysis further showed that the simplified configuration without the gap (Variant 2) provided better agreement with experimental data, yielding RMSE values in the range of 9.5–22.8%, which is consistent with accuracy levels reported for RANS-based simulations of confined exhaust flows.

The scientific contribution of this work lies in demonstrating that the temperature field downstream of a miniature turbojet exhaust nozzle can be accurately reproduced without explicitly modeling the combustion process, provided that experimentally derived boundary conditions are applied. Additionally, the study quantifies the effect of geometric confinement on flow and temperature distributions and confirms, through comparison with independent experimental and numerical studies, the robustness and universality of the adopted numerical framework for small-scale turbojet exhaust analysis.

From an application perspective, the developed and validated model can be used for the thermomechanical assessment of miniature turbojet nozzles, optimization of confined exhaust and test chamber geometries, and preliminary evaluation of hydrogen-assisted combustion concepts. The approach is particularly suited for low- and medium-speed operating regimes, where reliable prediction of downstream temperature fields is required without excessive computational cost.

**Author Contributions:** Conceptualization, Ł.B. and B.C.; Methodology, Ł.B., B.C. and G.C.; Software, Ł.B. and B.C.; Validation, Ł.B., B.C., Ł.S., D.S. and G.C.; Formal analysis, Ł.B. and G.C.; Investigation, Ł.B.; Resources, Ł.B.; Data curation, Ł.B.; Writing—original draft preparation, Ł.B., B.C. and Ł.S.; Writing, review and editing, Ł.B., B.C., Ł.S. and G.C.; Visualization, Ł.B. and D.S.; Supervision, Ł.B. and G.C.; Project administration, Ł.B. and B.C.; Funding acquisition, Ł.B. and B.C. All authors have read and agreed to the published version of the manuscript.

**Funding:** This research was funded by the Polish Ministry of Science and Higher Education and the Poznan University of Technology's financial resources for statutory activity. Grant numbers 0712/SBAD/5297 and 0712/SBAD/5298.

**Data Availability Statement:** The original contributions presented in this study are included in the article. Further inquiries can be directed to the corresponding authors.

**Conflicts of Interest:** The authors declare no conflicts of interest.

## Nomenclature

| | |
|---|---|
| $D$ | diffusion coefficient |
| $I$ | radiative intensity |
| $L_t$ | turbulent length scale |
| $F_1$ | first mixing function |
| $F_2$ | second mixing function |
| $\mathrm{F}_i$ | external force |
| $K$ | thermal conductivity coefficient |
| $N$ | number of numbers in the set |
| $P$ | scattering phase function |
| $P_k$ | rate of production of turbulent kinetic energy |
| $\overline{P}_k$ | effective rate of production of turbulent kinetic energy |
| $Pr$ | Prandtl number |
| $P_\omega$ | rate of production of specific turbulent dissipation rate |
| $R$ | term describing the combustion reaction |
| $S_{ij}$ | mean rate of deformation component |
| $T$ | temperature |
| $T_{i'}$ | temperature measured during the experiment at the checkpoint |
| $\hat{T}_i$ | temperature calculated during simulation at the checkpoint |
| $U_i$ | mean flow velocity |
| $X$ | cross-diffusion term |
| $a_1$ | constant |
| $c$ | concentration of the reacting substance |
| $c_p$ | specific heat |
| $k$ | specific turbulent kinetic energy |
| $n_I$ | refractive index |
| $\mathrm{r}$ | coordinate |
| $\hat{\mathrm{s}}$ | unit vector |
| $t$ | time |
| $\mathrm{v}$ | velocity |
| $x$ | direction |
| $y$ | direction |
| CFD | Computational Fluid Dynamics |
| FEM | Finite Element Method |
| MAE | Mean Absolute Errors |
| RDC | Rotating Detonation Combustor |
| RMSE | Root Mean Square Error |
| TBCC | Turbine-Based Combined Cycle |
| TKE | Turbulent Kinetic Energy |
| TDR | Turbulent Diffusivity Ratio |
| $\beta$ | constant |
| $\beta*$ | constant |
| $\gamma$ | constant |
| $\sigma$ | standard deviation |
| $\sigma_{ij}$ | stress tensor |
| $\delta_{ij}$ | Kronecker delta function |
| $\kappa$ | extinction coefficient |
| $\kappa_a$ | coefficient of absorption |
| $\kappa_s$ | coefficient of scattering |
| $\mu$ | arithmetic mean |
| $\mu_k$ | effective viscosity of specific turbulent kinetic energy |
| $\mu_t$ | modified eddy viscosity |
| $\mu_\omega$ | effective viscosity of specific turbulent dissipation rate |

https://doi.org/10.3390/en19040938

| | |
|---|---|
| $\rho$ | density |
| $\sigma_k$ | diffusion constant |
| $\sigma_\omega$ | diffusion constant |
| $d\omega'$ | solid angle |
| $\omega$ | specific turbulent dissipation rate |

## References


1. Varatharajulu Purgunan, G.R.; Stathopoulos, P. Performance and Exergy Analysis of TurboJet and TurboFan Configurations with Rotating Detonation Combustor. *Int. J. Thermofluids* **2024**, *23*, 100739. [CrossRef]
2. Zhang, Y.; Xu, J.; Pan, R.; Li, Y.; Ma, Z.; Huang, S. Numerical Investigation of Short Takeoff and Landing Exhaust System Using Bypass Dual Throat Nozzle. *Aerosp. Sci. Technol.* **2023**, *138*, 108316. [CrossRef]
3. Meng, Y.; Shi, J.; Zhou, L.; Wang, Z. Numerical Investigation on Flow Characteristics of Multi-Stream Supersonic Nozzle with Serpentine Configuration. *Aerosp. Sci. Technol.* **2024**, *144*, 108799. [CrossRef]
4. He, P.; Yue, C.; Fan, Y. The Stable Combustion and Blowout Characteristics Investigation of the Water Injection Pre-Cooling on the Afterburners. *Fuel* **2025**, *397*, 135422. [CrossRef]
5. Aygun, H.; Kirmizi, M.; Kilic, U.; Turan, O. Multi-Objective Optimization of a Small Turbojet Engine Energetic Performance. *Energy* **2023**, *271*, 126983. [CrossRef]
6. Niszczota, P.; Gieras, M. Inlet Injection as an Alternative, Simplified Method of Hydrogen Application in a Miniature Gas Turbine. *Appl. Therm. Eng.* **2024**, *257*, 124175. [CrossRef]
7. Ayaz, S.K.; Caliskan, H.; Altuntas, O. Environmental and Second Law Analysis of a Turbojet Engine Operating with Different Fuels. *Energy* **2023**, *285*, 129402. [CrossRef]
8. Kang, J.; Song, F.; Chen, X.; Wu, Y.; Zhang, D.; Wang, J.; Yang, Z.; Pan, W.; Wang, Z. Performance Enhancement of Rotating Detonation Afterburner through Combined Injection Scheme. *Acta Astronaut.* **2025**, *235*, 639–652. [CrossRef]
9. Wu, X.; Hu, X.; Xiang, X.; Lin, S.; You, J.; Tian, F. An Analysis Approach for Micro Gas Turbine Engine's Performance by Experiment and Numerical Simulation. *Case Stud. Therm. Eng.* **2023**, *49*, 103305. [CrossRef]
10. Zhang, K.; Yu, X.; Zhang, D.; Guo, J.; Qin, F. Evaluation of Multi-Mode Shared Combustor with Different Flame Holder Designs for an Innovative Turbo-Ramjet Combined Cycle Engine. *Energy* **2025**, *323*, 135875. [CrossRef]
11. Xu, Y.; Deng, W.; Liu, X.; Xu, H.; Li, Z. Full Engine 3D Simulation Scheme Considering Rotor Motion for Turbojet Engine under Crosswind Conditions. *Appl. Therm. Eng.* **2025**, *269*, 125951. [CrossRef]
12. Boretti, A. Towards Hydrogen Gas Turbine Engines Aviation: A Review of Production, Infrastructure, Storage, Aircraft Design and Combustion Technologies. *Int. J. Hydrogen Energy* **2024**, *88*, 279–288. [CrossRef]
13. Brodzik, Ł.; Ciupek, B.; Frąckowiak, A.; Schroeder, D. Thermomechanical Analysis of the GTM 400 MOD Turbojet Engine Nozzle During Kerosene and Hydrogen Co-Combustion. *Energies* **2025**, *18*, 4382. [CrossRef]
14. Bogoi, A.; Cican, G.; Gall, M.; Totu, A.; Crunțeanu, D.E.; Levențiu, C. Comparative Study of Noise Control in Micro Turbojet Engines with Chevron and Ejector Nozzles Through Statistical, Acoustic and Imaging Insight. *Appl. Sci.* **2025**, *15*, 394. [CrossRef]
15. Catana, R.M.; Badea, G.P. Experimental Analysis on the Operating Line of Two Gas Turbine Engines by Testing with Different Exhaust Nozzle Geometries. *Energies* **2023**, *16*, 5627. [CrossRef]
16. Villarreal-Valderrama, F.; Zambrano-Robledo, P.; Hernandez-Alcantara, D.; Amezquita-Brooks, L. Turbojet Thrust Augmentation through a Variable Exhaust Nozzle with Active Disturbance Rejection Control. *Aerospace* **2021**, *8*, 293. [CrossRef]
17. Ciupek, B.; Brodzik, Ł.; Frąckowiak, A. Research on Carbon Footprint Reduction During Hydrogen Co-Combustion in a Turbojet Engine. *Energies* **2024**, *17*, 5397. [CrossRef]
18. Mazumder, S.; Roy, S.P. Modeling Thermal Radiation in Combustion Environments: Progress and Challenges. *Energies* **2023**, *16*, 4250. [CrossRef]
19. Beer, J.M. *Methods for Calculating Radiative Heat Transfer from Flames in Combustors and Furnaces, Chapter 2 in Heat Transfer in Flames*; Afgan, N.H., Beer, J.M., Eds.; Scripta Book Company: Washington, WA, USA, 1974.
20. Prokopowicz, W.; Ciupek, B.; Maciąg, A.; Gajewski, T.; Sielicki, P.W. Advanced Spectral Diagnostics of Jet Engine Vibrations Using Non-Contact Laser Vibrometry and Fourier Methods. *Energies* **2025**, *18*, 4837. [CrossRef]
21. Brodzik, Ł. Gas temperature distribution in the combustion chamber of a GTM400 MOD turbojet engine powered by JET A-1 Fuel and Hydrogen. *Energies* **2024**, *17*, 745. [CrossRef]
22. Karim, M.M.; Rahman, M.M.; Alim, M.A. Performance of SST k–ω turbulence model for computation of viscous drag of axisymmetric underwater bodies. *Int. J. Eng.* **2011**, *24*, 2.
23. Liu, J.; Guo, W.; Yin, M.; Xi, W.; Sunden, B. Flow and heat transfer characteristic of regenerative cooling channels using supercritical $CO_2$ with circular tetrahedral lattice structures. *Case Stud. Therm. Eng.* **2025**, *71*, 106204. [CrossRef]


https://doi.org/10.3390/en19040938

24. Liu, J.; Guo, W.; Zhao, G.; Ma, J.; Xi, W.; Sunden, B. Study on Heat Transfer Characteristics of Supercritical CO2/n-decane Composite Double Triangular Regenerative Cooling for Scramjets. *Aerosp. Sci. Technol.* **2026**, *172*, 111725. [CrossRef]
25. Goldberg, U.C.; Batten, P. A wall-distance-free version of the SST turbulence model. *Eng. Appl. Comput. Fluid Mech.* **2015**, *9*, 1. [CrossRef]
26. Rocha, P.C.; Rocha, H.B.; Carneiro, F.M.; da Silva, M.V.; Bueno, A.V. k–ω (shear stress transport) turbulence model calibration: A case study on a small scale horizontal axis wind turbine. *Energy* **2013**, *65*, 412–418. [CrossRef]
27. Badami, M.; Nuccio, P.; Pastrone, D.G.; Signoretto, A. Performance of a Small-Scale Turbojet Engine Fed with Traditional and Alternative Fuels. *Energy Convers. Manag.* **2014**, *82*, 219–228. [CrossRef]
28. Gao, X.; Duan, F.; Lim, S.C.; Yip, M.S. Experimental and Numerical Study of Flow Structure and Liner Wall Temperature in Reverse Flow Combustor. In Proceedings of the 15th International Heat Transfer Conference (IHTC-15), Kyoto, Japan, 10–15 August 2014. [CrossRef]
29. Suchocki, T.; Lampart, P.; Klonowicz, P. Numerical Investigation of a GTM-140 Turbojet Engine. *J. KONES Powertrain Transp.* **2014**, *21*, 331–338. [CrossRef]
30. He, W.; Guo, Z.; Xu, S.; Jin, Y. Numerical and Experimental Investigations on Performance Optimization of a Microturbojet Engine Combustor. *Int. J. Aerosp. Eng.* **2021**, *2021*, 6616537. [CrossRef]
31. Alabas, H.A. CFD study of jet fuel with ammonia—Hydrogen mixtures in a jet engine: Emissions and combustion characteristics analysis. *Int. J. Hydrogen Energy* **2025**, *142*, 140–150. [CrossRef]
32. Gieras, M. *Miniaturowe silniki turboodrzutowe*; Wydawnictwo Politechniki Warszawskiej: Warszawa, Poland, 2016.



https://doi.org/10.3390/en19040938

*energies*

MDPI

*Article*

# Experimental Verification of a Method for Improving the Efficiency of an Evaporative Tower Using IEC

**Bartosz Jagieła [1] and Magdalena Jaremkiewicz [2,*]**

[1] Doctoral School, Cracow University of Technology, 31-155 Cracow, Poland; bartosz.jagiela@doktorant.pk.edu.pl

[2] Faculty of Environmental and Energy Engineering, Cracow University of Technology, 31-155 Cracow, Poland

* Correspondence: magdalena.jaremkiewicz@pk.edu.pl

**Abstract**

This paper analyses the impact of inlet air precooling on the efficiency and electricity consumption of an open-type evaporative cooling tower. An Indirect Evaporative Cooler (IEC) was used to reduce the inlet air temperature, and its influence on system efficiency was experimentally evaluated. Although IEC units and the Maisotsenko cycle are increasingly discussed in the literature, no research to date has considered their effect on evaporative tower efficiency under actual operating conditions. For this purpose, a test stand was constructed comprising an open cooling tower and an IEC unit. The system operated automatically for 2952 h, corresponding to a full cooling season in Poland. Two sets of data collected during cooling tower operation were analysed: without precooling (Stage I) and with precooling using IEC (Stage II). Measurements were recorded every 10 s. Additionally, tests were conducted at elevated thermal loads and peak ambient temperatures. The comparative analysis concluded that air precooling using IEC reduced the cooling tower's electricity consumption by approximately 15% and increased the SCOP of the cooling tower by 30%. This demonstrates the significant potential of the proposed solution.

**Keywords:** evaporative cooling tower; indirect evaporative cooling; inlet air precooling; Maisotsenko cycle; energy efficiency; HVAC system

## 1. Introduction

Meeting electricity demand in the 21st century is becoming a challenge due to climate change and a growing population. Especially since, due to the average ambient temperature, each subsequent year is usually warmer than the previous one. The Heating, Ventilation, and Air Conditioning (HVAC) market is constantly evolving to provide customers with a comfortable environment through cooling, heating, and ventilation. In certain developed nations, including Japan, the USA, and Korea, more than 90% of households have individual air conditioning units. The electricity consumption of these devices accounts for approximately 10% of total global electricity consumption [1]. Consumers frequently overlook the most energy-efficient air conditioning units, instead basing their decisions solely on cost. Investing in more efficient HVAC systems could cut future energy demand by 45% by 2050 [1].

Most modern HVAC systems are Vapour-Compression Refrigeration Systems (VCRSs), in which the refrigerant undergoes a phase change and are characterised by significant electricity consumption. They also harm the environment through the refrigerants they use. Leading HVAC manufacturers are now exploring more efficient, environmentally friendly solutions in response to climate change. Among the known refrigerants, environmentally friendly ones, i.e., those with low ozone depletion potential (ODP) and low global warming potential (GWP), include ammonia, $NH_3$, known as R717 (ODP = 0 and GWP = 0); carbon dioxide, $CO_2$, or R744 (ODP = 1, GWP = 0); and propane, $C_3H_8$, or R290 (ODP = 3, GWP = 0) [2]. All of them represent the future of modern HVAC systems on an industrial scale. Unfortunately, for devices with lower cooling capacities (below 20 kW), the use of these refrigerants is limited due to their flammability, toxicity, and high operating pressure [3–5]. A refrigerant that does not have these disadvantages and is neutral and clean is water, $H_2O$, designated as R718 (ODP < 1, GWP = 0). Water can be used as a primary or secondary refrigerant. Due to its high heat of evaporation (approximately 2200 kJ/kg at 0 °C), it is one of the most effective refrigerants. Water is used as a refrigerant in evaporative devices, such as evaporative air conditioners and cooling towers. Cooling air using evaporative technology enables temperatures below the ambient temperature to be achieved, and since it utilises natural processes, energy consumption can be reduced. The environmental and economic issues discussed above are driving evaporative systems to become increasingly popular.

### *1.1. Evaporative Cooling Systems*

Evaporative systems are categorised into two types: Direct Evaporative Coolers (DECs) and Indirect Evaporative Coolers (IECs). In DEC units, the cooled air comes into direct contact with water in the wet channel. During cooling, the water evaporates directly into the stream of cooled air, lowering its temperature. IEC uses a heat exchanger to cool the air, in which heat is transferred from the cooled air in the dry channel to the air flowing through the wet channel. This prevents moisture from being added to the cooled air. The lowest temperature to which air can be cooled in evaporative devices is the wet-bulb temperature, which depends on the dry-bulb temperature and relative humidity of the air. The cooled air obtained in IEC may have similar parameters to those obtained in VCRS, but unlike them, evaporation systems are innovative and can be considered a cheaper cooling technique. A variation of the IEC system is the Maisotsenko cycle (M-cycle), which allows lower temperatures than the classic ICE. Figure 1 illustrates the heat-exchange process in the IEC and diagrams the thermodynamic processes within the unit.

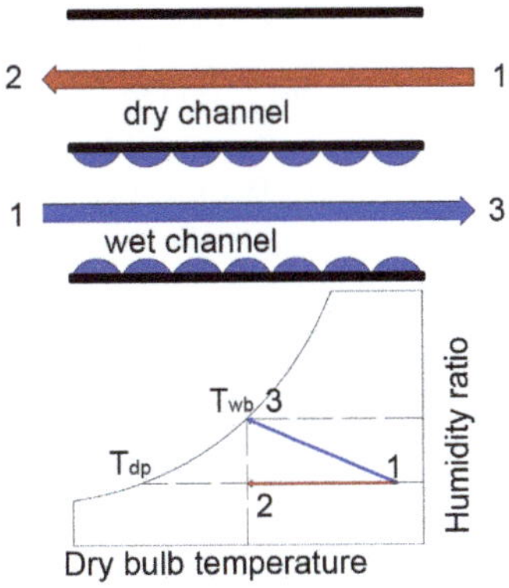


**Figure 1.** Principle of operation of IEC.

The effectiveness of evaporative cooling can be determined based on the wet-bulb efficiency:

$$\varepsilon_{wb} = \frac{T_{in} - T_{out}}{T_{in} - T_{wb}} \quad (1)$$

https://doi.org/10.3390/en19020554

where $T_{in}$ is the inlet temperature, $T_{out}$ is the outlet temperature, and $T_{wb}$ is the wet-bulb temperature.

In the first concept of the M-cycle, the product is the main stream of cooled air, which is obtained through heat exchange with the so-called auxiliary stream of air flowing through the wet channel. The inlet air is divided into two parts, one of which, i.e., the main stream, flows through the dry channel and is used, for example, to cool rooms. The second part, i.e., the auxiliary air stream, is first pre-cooled during heat exchange with the air flowing through the wet channel, and then flows through the wet channel, further lowering its temperature. Cooling systems using the M-cycle employ cross-flow heat and mass exchangers (HMXs). As in IEC systems, the cooled air (product) is not moisturised. Theoretically, the lowest temperature achieved in the M-cycle is the dew point temperature. The efficiency of the M-cycle can be determined based on the dew point efficiency:

$$\varepsilon_{dp} = \frac{T_{in} - T_{out}}{T_{in} - T_{dp}} \quad (2)$$

where $T_{dp}$ is the dew point temperature.

The M-cycle concept is continually being developed by its creator and other researchers and scientists, as evidenced by the increasing number of publications on topics such as improving cycle efficiency, the technologies used, and potential applications.

The M cycle is primarily used in cooling systems, but it can also be applied in water desalination and distillation processes [6,7] as well as in power plants, including NOx reduction [8,9]. The presented applications of the M-cycle have been analysed in terms of optimisation for various locations around the world, characterised by both low and high air humidity, e.g., cities in the Arabian Gulf [10], Riyadh in Saudi Arabia [11], Beijing in China [12], Bushehr in Iran [13], northern Italy [14], and Martos in Spain [15]. The research was conducted both numerically [10–14] and experimentally [14,15].

Research conducted on M cycles includes HMXs. In [16], a mathematical model of heat and mass transfer was developed for two selected HMX designs using a modified epsilon-NTU method. The authors of [17] also developed an HMX model in Python 3.10, based on a spatially discretised differential equation solver and a psychrometric chart. In both cases, the models were validated using experimental data. Research on heat exchangers also concerns technology. For example, in [18], a nanoporous membrane covering the surface of a wet channel was tested to improve heat and mass transfer. The study experimentally confirmed the beneficial effect of the coating on the heat exchanger's efficiency. Another solution studied, as described in [19], involved the use of aluminium fins in the heat exchanger's dry ducts to increase heat transfer efficiency. This increased the cooling capacity by 18%.

Evaporative cooling systems can operate in various configurations, including those with desiccant wheels, ejectors, or as hybrid systems that incorporate freon devices. The operating conditions and limitations of individual solutions are summarised in [20]. A desiccant air conditioning system uses a moisture-absorbing wheel to dehumidify air drawn from the surroundings. Then, in a typical system, the air is pre-cooled in a rotary heat exchanger, then humidified in a spray chamber, and cooled simultaneously. The air returning from the cooled space flows through the subsequent parts of the system in the opposite direction, but is heated before the desiccant wheel, allowing the desiccant wheel to be regenerated. In [21], indirect cooling using the M cycle was proposed. A numerical analysis of this system was performed, demonstrating its higher temperature efficiency. Research on such systems also concerns moisture-absorbing materials used in the desiccant wheel, as in [22]. Hybrid systems combine IEC and VCRS. Research on such a system, where IEC uses an M-cycle, is presented in [23]. The report shows

https://doi.org/10.3390/en19020554

that the tested system consumes 80% less energy than a conventional air conditioning system. In turn, a numerical simulation of two hybrid systems was conducted in [24], where preliminary cooling was achieved using a conventional IEC and an IEC with an M cycle. In this case, it was also demonstrated that systems of this type enable reduced electricity consumption in locations with a humid climate. In [25], tests were carried out on three air conditioning systems: a refrigeration unit, a refrigeration unit coupled with evaporative cooling, and a refrigeration unit coupled with regenerative evaporative cooling. An analysis of annual electricity consumption showed savings of 17% and 22% for the second and third solutions, respectively, compared to the first. In [26], an analysis of the operation of a two-stage evaporative cooling system consisting of IEC followed by DEC in public buildings in Tehran was presented. It was shown that such a system could replace VCRS while simultaneously reducing electricity consumption; however, it would consume 55% more water than a stand-alone DEC.

### *1.2. Cooling Towers*

In cooling towers, waste heat is dissipated into the environment through a stream of coolant (usually water), which undergoes partial evaporation, causing the cooling air to reach a temperature close to that of a wet-bulb thermometer. Cooling towers can be divided into two types: natural draft and forced draft (utilising a fan). Another classification, based on the circulation of the cooled medium, is into open and closed evaporative towers. Cooling towers are commonly used across various industries, including chemical and petrochemical plants, refineries, and both conventional and nuclear power stations. The heat transfer principle in the cooling tower and the thermodynamic process occurring within the unit are illustrated in Figure 2.

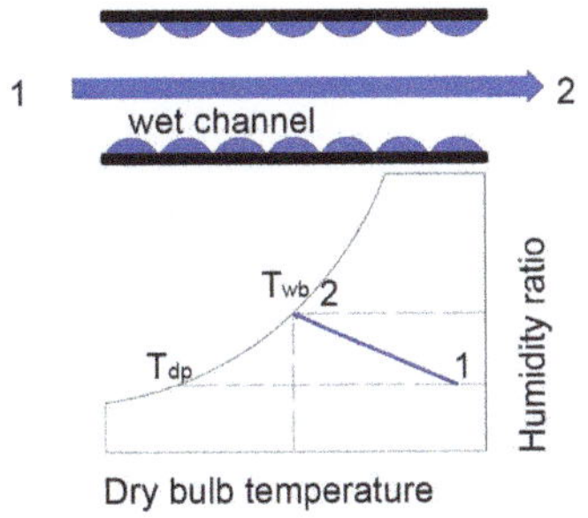


**Figure 2.** Principle of operation of DEC.

Research on cooling towers primarily focuses on increasing wet-bulb efficiency. In [27], the cooling tower capacity was improved by pre-cooling and dehumidifying the inlet air by using a reverse Brayton cycle. The proposed system was thermodynamically modelled and optimised for the operating conditions using a genetic algorithm to obtain the highest power plant output. The M-cycle can be used in cooling towers. A comparison between a traditional cooling tower and an M-cycle cooling tower was conducted in [28]. The comparison was carried out using a one-dimensional heat and mass transfer model based on the NTU method. It was found that using an M-cycle in a cooling tower allows the water temperature to be lowered below the wet-bulb temperature. This result cannot be achieved with a traditional approach. In [29], on the other hand, optimal dry duct lengths were sought for a cooling tower with an M-cycle, where the fill is arranged in parallel counterflow. The calculations were performed using the NTU method and based on experimental data. Determining the optimal duct length increased the efficiency of the wet thermometer.

https://doi.org/10.3390/en19020554

The IEC and the evaporation tower are considered the most efficient (low electricity consumption) and environmentally friendly (no F-gases and low water consumption). The first device cools the air to the dew point, and the second allows subcooling the process water to the wet-bulb temperature. This paper proposes a solution in which the air is pre-cooled in the IEC and then routed to the cooling tower. A key premise of the research was to demonstrate that using IEC to precool the inlet air improves the cooling tower's efficiency. To test the proposed solution, an experimental stand was constructed to compare a cooling system consisting of a cooling tower and a cooling tower with pre-cooling of the air in the IEC. The comparison focused on electricity consumption and the Seasonal Coefficient of Performance (SCOP). Based on the literature review, no studies to date have analysed the efficiency of this solution. Examples of research on cooling towers carried out so far have focused on the use of a reverse Brayton cycle to cool and dehumidify the inlet air and, above all, the use of the M-cycle in them. These studies have mainly based on simulation results, whereas this paper is based on a two-year experimental study. In addition, studies on new cooling tower solutions focused on improving wet-bulb temperature efficiency and not the SCOP, as in this paper. An increase in SCOP means lower electricity consumption in relation to the cooling energy produced by the cooling tower. The use of an IEC in combination with a cooling tower would be an advantageous solution not only because of lower operating costs, but also because of the environmental benefits of lower electricity consumption and the use of water as a refrigerant (elimination of fluorinated gases). Increasing the efficiency of cooling towers by adding an IEC could lead to greater interest in and use of evaporative units for process cooling or room air conditioning in hotels, hospitals, and office buildings. With the proposed solution, the user will achieve reduced energy consumption or more stable cooling system operation despite higher ambient temperatures.

## 2. Description of the Experimental Stand and Research Methodology

The study was carried out in two stages. In Stage I, the system operated in the reference (baseline) configuration, i.e., in which water heated by a heat pump was next cooled in a cooling tower using ambient air intake. In Stage II, the inlet air to the cooling tower was precooled by means of an indirect air-conditioning unit. The experimental site was located in the village of Zwrócona in the Ząbkowice district, situated in the Sudetic Foreland region of Poland. One of the most important factors influencing the selection of an appropriate cooling device is the local climate, as it directly affects the achievable final temperature of the liquid or air after cooling. Cooling equipment is designed to provide maximum cooling capacity under the least favourable ambient conditions. These conditions correspond to the highest dry-bulb and wet-bulb temperatures, both of which depend on air humidity. Using the software from the cooling tower manufacturer MITA Cooling Technologies from Siziano, Italy, based on the Meteonorm climate database [30], a suitable unit can be selected, and its operating costs analysed. This ensures the chosen unit operates optimally and is not oversized.

An analysis of climate data for Ząbkowice Śląskie was conducted to determine the most suitable cooling and heating solutions. The maximum and minimum ambient temperatures $T_{amb}$ are 32.5 °C (occurring on 9 July at 15:00) and −17.4 °C (recorded on 12 January at 23:00), respectively. The average dry-bulb temperature is 8.6 °C. The maximum wet-bulb temperature $T_{wb}$ is 22.7 °C on 21 July at 9:00, corresponding to an ambient temperature of $T_{amb}$ = 26.8 °C and coupled with a relative humidity of $RH$ = 71.3%, while the minimum wet-bulb temperature $T_{wb}$ is −17.48 °C on 12 January at 23:00, corresponding to $T_{amb}$ = −17.4 °C and $RH$ = 99.3%. The average wet-bulb temperature is 6.5 °C. The maximum ambient humidity $RH$ is 100%, occurring for approximately 276 h per year; the minimum $RH$ was recorded on 29 May at 13:00 and amounted to 29.5%, with $T_{amb}$ = 23.3

https://doi.org/10.3390/en19020554

°C and $T_{wb}$ = 13.1 °C, and the average ambient humidity is 79.1%. Sorting the wet-bulb temperatures from highest to lowest, it can be seen that the sixteen consecutive highest wet-bulb temperatures correspond to dry-bulb temperatures below 30 °C. The highest wet-bulb temperature recorded for Ząbkowice Śląskie is 22.7 °C. The MITA cooling tower selection software recommends a design wet-bulb temperature of 21 °C for this location. This means that the cooling tower will not achieve the required capacity at wet-bulb temperatures above this value. In the case of Ząbkowice Śląskie, such wet-bulb temperatures only occur for about 35 h a year.

### 2.1. Stage I

Stage I involved the installation of the cooling device, i.e., an open-circuit cooling tower, and the heating device, represented by an air-source heat pump. The two units were hydraulically separated by a plate heat exchanger. The cooling tower circuit is the secondary circuit, while the heat pump circuit is the primary circuit. The secondary circuit was equipped with a circulation pump and the necessary fittings. The entire system was additionally equipped with an automation unit and measurement sensors. The schematic of the laboratory stand used for the Stage I tests is shown in Figure 3.

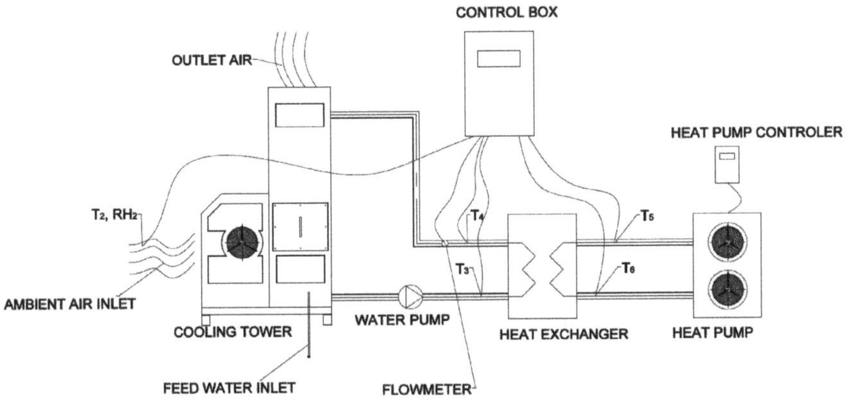


**Figure 3.** Schematic of the Stage I test stand.

The cooling tower was delivered as a complete solution. The delivery scope included the centrifugal fan, fill pack, spray nozzles, and a float valve for replenishing the circuit with treated water. The tower was not equipped with its own control system, which should include a water temperature sensor at the tower outlet, and the fan speed was modulated based on the setpoint. The fan can be modulated within a frequency range of 25 Hz to 60 Hz. The control system consists of a frequency inverter that adjusts the tower fan's rotational speed based on the temperature sensor's setpoint, which measures the temperature $T_3$ of the cooled water (i.e., the temperature of the water entering the heat exchanger). Due to water evaporation and associated losses, makeup water must meet specific quality parameters, and circulating water must be refreshed to ensure its properties, such as pH, conductivity, and hardness, do not exceed permissible limits. Because the water volume in the system is small, the refreshing process is performed manually after preliminary water-quality measurements. These measurements were performed at least once every two weeks and no more frequently than once per week.

The parameter that defines the size of a cooling tower is the hydraulic load, expressed as the ratio of the water flow rate ($m^3/h$) to the fill surface area ($m^2$). The appropriate flow rate and pressure of water in the cooling tower ensure proper distribution of water throughout the filling pack. Correct water dispersion within the filling pack is essential

https://doi.org/10.3390/en19020554

for achieving efficient operation and maintaining the required outlet water temperature. The filling pack is a wet heat exchanger in which heat transfer occurs within counterflow air–water channels. Water is introduced at the top of the exchanger and subsequently flows downward through the channels as droplets under gravity. Ambient air is supplied from the bottom of the exchanger. Evaporation occurs when the water comes into contact with the air stream. The selected cooling tower model is equipped with a single centrifugal fan providing a nominal airflow rate of 2.770 $m^3/h$. The unit also includes a float valve for replenishing evaporated water, a water-distribution system with spray nozzles, and a fill pack. The cooling capacity of the evaporative tower is 17 kW at a temperature difference of $\Delta T = 5$ K. The fan power is 0.4 kW. The air inlet is located at the bottom of the unit, where the fan forces air upwards through the entire height of the tower. The manufacturer, MITA Cooling Technologies, indicates that the total pressure drop across the cooling tower is $270 \pm 20$ Pa at the maximum airflow rate through the device. This value corresponds to the maximum available fan static pressure in the CW3 cooler.

The use of a heat pump was essential because, for the cooling system to operate continuously throughout the study period (from the beginning of May to the end of August), it was necessary to maintain a sufficiently high and constant water temperature at the cooling tower inlet. A monoblock heat pump is an optimal solution, offering high efficiency and a compact design. In the selected air-source heat pump, 1 kilowatt of electrical power can deliver up to 6 kilowatts of thermal output, enabling efficient operation with low electrical energy consumption. The heat pump is equipped with its own control system, which modulates the operation of the refrigeration compressor and the condenser fans. The monoblock air-source heat pump was selected based on the cooling tower's capacity. The unit's circulation pump should ensure a heating-fluid (water) flow rate of approximately 3 $m^3/h$, and the heat pump's heating capacity should be approximately 17 kW at a temperature difference ($\Delta T = T_{in} - T_{out}$) of 5 K. A Kaysun heat pump, model KHP-MO 12 DTN, was selected. The refrigerant used in the unit is R410A, an HFC-based blend consisting of R32 and R125 in equal (50/50) mass fractions. R410A is non-flammable, has an ozone depletion potential (ODP) of 0, and a global warming potential (GWP) of 2088. This refrigerant is considered environmentally unfriendly, and as of 1 January 2027, the sale of monoblock heat pumps operating with R410A will be prohibited within the European Union.

It is recommended to hydraulically separate the open cooling tower circuit from the closed heat-pump circuit. This is necessary because the water in the cooling tower circuit is exposed to contaminants present in the air and in the tower's surroundings. The heat exchanger was selected based on the water temperatures and flow rates on both the secondary and primary sides of the research station. A plate heat exchanger manufactured by ONDA from Mussolente, Italy, model S 12-26, was chosen. The ONDA plate heat exchanger with installed sensors is shown in Figure 4. The primary function of the circulation pump is to maintain a constant flow rate and pressure of the cooled water passing through the cooling tower and the plate heat exchanger. The Yonos MAXO-Z 25/0.5–7 PN 10 pump manufactured by WILO in Warsaw, Poland, was selected for this purpose.

https://doi.org/10.3390/en19020554

**Figure 4.** View of the plate heat exchanger hydraulically separating the heat pump circuit (left) from the cooling tower circuit (right).

At the core of the control system is a frequency inverter, which regulates the speed of the cooling tower fan motor group based on measurements obtained from the cooling tower outlet water temperature sensor $T_3$. The primary function of the system is to maintain the outlet temperature of the circulating water leaving the cooling tower at a predefined setpoint. The setpoint can be adjusted manually using a potentiometer that is synchronised with the PLC controller. Cooling towers are selected based on design conditions, which are always defined for the least favourable operating scenario. The maximum thermal load occurs under the most demanding ambient conditions, when the tower fan operates at its maximum rotational speed (60 Hz). For the remainder of the operating time throughout the year, the fan speed is modulated between 20 and 60 Hz. Measurement data from all sensors are recorded at 10 s intervals. The control and data-logging system is equipped with the following: a MIM 1205HG4C3T0 electromagnetic flow meter manufactured by KOBOLD in Settimo Milanese, Italy; PT100 temperature sensors manufactured by Limatherm Sensor in Limanowa, Poland, with a measurement range of −50 to 200 °C; an integrated TH2E thermo-hygrometer manufactured by Papouch in Prague, Czech Republic, enabling relative humidity measurements from 0% to 100%, temperature measurements from −40 to +123.8 °C, and dew-point calculation; and an ME300 frequency inverter adapted for a 0.37 kW fan, as well as an AS218 PLC equipped with a PID controller that is capable of direct data logging to a USB drive, both manufactured by Delta in Taoyuan City, Taiwan. The proper operation of the heat pump is managed by a wall-mounted controller on which the desired outlet water temperature must be set. The circulation pump of the heating circuit, installed inside the unit's housing, offers three-stage flow regulation. The controller activates the water pump when the measured temperature drops by 2 K relative to the setpoint. It also allows monitoring of the unit's electrical energy consumption, which is essential for comparing energy use between Stage I and Stage II of the experimental study.

On the laboratory bench, measurements are taken of the water temperature at the inlet ($T_4$) and outlet ($T_3$) of the cooling tower, at the inlet ($T_6$) and outlet ($T_5$) of the heat pump, as well as the air temperature ($T_2$) at the inlet to the cooling tower. In addition, the relative humidity ($RH_2$) of the air at the cooling tower inlet and the water flow rate through the tower ($\dot{V}_{cool}$) and heat pump ($\dot{V}_{hp}$) are measured, as well as the frequency and current ($I$) of the fan motor. The resolutions, measurement accuracies, and descriptions of the sensors used are shown in Table 1.

https://doi.org/10.3390/en19020554

**Table 1.** Measuring instruments and their accuracy.

| Measured Parameter | Symbol | Location/Description | Resolution | Accuracy |
|---|---|---|---|---|
| Cooled water temperature | $T_3$, °C | Cooling tower outlet | 0.1 °C | 1.5% |
| Heated water temperature | $T_4$, °C | Cooling tower inlet | 0.1 °C | 1.5% |
| Heated water temperature | $T_5$, °C | Heat pump outlet | 0.1 °C | 1.5% |
| Cooled water temperature | $T_6$, °C | Heat pump inlet | 0.1 °C | 1.5% |
| Ambient air temperature | $T_2$, °C | Cooling tower air inlet | 0.1 °C | 1.5% |
| Relative humidity | RH, % | Cooling tower air inlet | 1 °C | 1.5% |
| Water flow rate | $\dot{V}_{cool}$, m$^3$/h | Cooling tower water loop | 0.1 m$^3$/h | 0.1% |

### 2.2. *Stage II*

The research station for Stage II of the study was expanded to include a precooling system for the cooling tower inlet air. The optimal solution is the indirect evaporative cooling system Coolerado$^{TM}$, manufactured by Seeley International form Lonsdale, Australia, which provides a complete configuration and enables the supply-air temperature to be reduced below the wet-bulb temperature. Since this system is initially designed to cool indoor spaces by delivering conditioned air, it is equipped with fans, a water distribution system, and an integrated control unit. The complete schematic of the Stage II setup is shown in Figure 5.

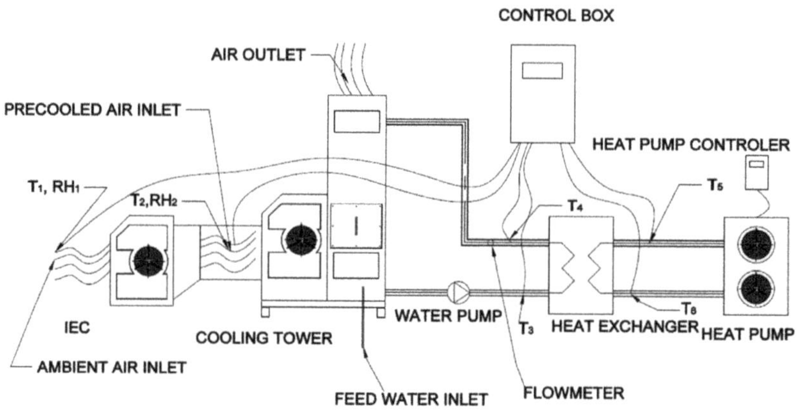


**Figure 5.** Scheme of Stage II of the research station.

An IEC CW3 model by Seeley International, Lonsdale, Australia, was used. This unit is a complete solution with an indirect regenerative heat exchanger. The maximum available fan static pressure is 270 Pa, which should be sufficient to overcome the highest expected pressure drops in the ventilation ducts and within the cooling tower itself. The unit has a cooling capacity of 17 kW and an electrical input of 1.75 kW. The nominal airflow rate of the fan is 5360 m$^3$/h, and the water consumption is 13 L/min. The IEC connection to the cooling tower is shown in Figure 6.

https://doi.org/10.3390/en19020554

**Figure 6.** Connection of evaporative cooler (1) to a cooling tower (2).

The CW3 IEC is equipped with its own controller, connected to the unit via a communication cable. The cooler operates in automatic cooling mode with a minimum allowable supply air temperature set point of +16 °C.

Measurements in Stage II were extended from those in Stage I to include measurement of temperature ($T_1$) and relative humidity ($RH_1$) at the inlet to the evaporative cooler.

### 2.3. Methodology

To demonstrate the impact of precooling the inlet air on energy consumption and the evaporative tower's cooling capacity, experimental tests were conducted. These tests involved comparing the operation of a cooling installation equipped solely with a cooling tower to that of an installation in which the tower was supplemented with the IEC, providing precooling of the air entering the tower. The study was divided into two stages. The measured values and calculation results obtained in Stages I and II were compiled into tables and graphs.

The Stage I measurements were carried out from 1 May 2023 to 31 August 2023. In turn, the Stage II measurements were conducted from 1 May 2024 to 31 August 2024. During Stage I research, the system operated for a total of 2952 h, and during Stage II, it operated for 2951 h, with a constant heat load supplied by the heat pump. Measurements were recorded every 10 s throughout the system's operating period in Stages I and II.

Measurements taken during the operation of the experimental research station in both Stage I and Stage II did not include the data necessary to perform complete analyses. The additional quantities calculated for both research stages were as follows: the cooling capacity of the cooling tower, the thermal capacity of the heat pump, the electrical power of the fan motor in the tower, the ambient air wet-bulb temperature, and the cooling tower's Coefficient of Performance (COP).

In addition to the primary performance evaluation, supplementary analyses were conducted, including an economic assessment of the proposed system. The economic review accounted for the electricity consumption of the IEC auxiliary fans and the cooling tower fan. These analyses confirmed that the financial performance is strongly influenced by the auxiliary fans' configuration and their electricity demand; however, they did not have a significant impact on the thermodynamic performance trends or the main conclusions of

https://doi.org/10.3390/en19020554

this study. Therefore, the results of these analyses are not presented in detail, while their execution is explicitly noted here for methodological completeness and transparency.

To calculate the cooling capacity of the cooling tower $\dot{Q}_{cool}$, an energy balance equation on the cooling water side was used:

$$\dot{Q}_{cool} = \dot{m}_{w,cool} C_{w,cool} (T_4 - T_3), \tag{3}$$

where $\dot{m}_{w,cool}$ is the mass flow rate of cooling water, $C_{w,cool}$ is the specific heat capacity of water, $T_4$ is the inlet water temperature, and $T_3$ is the outlet water temperature.

The mass flow rate of the cooled water in the tower is determined on the basis of the measured volumetric flow rate $\dot{V}_{cool}$:

$$\dot{m}_{w,cool} = \rho_{w,\ cool} \dot{V}_{cool} \tag{4}$$

where $\rho_{w,\ cool}$ is the density of the cooled water in the cooling tower.

The thermal capacity of the heat pump $\dot{Q}_{hp}$ was also calculated on the basis of an energy balance, defined in this case on the heated-water side of the device:

$$\dot{Q}_{hp} = \dot{m}_{w,hp} C_{w,hp} (T_5 - T_6) \tag{5}$$

where $\dot{m}_{w,hp}$ is the mass flow rate of water in the heat pump loop, $C_{w,hp}$ is the specific heat capacity of water, $T_5$ is the inlet water temperature, and $T_6$ is the outlet water temperature.

Similarly to the cooling tower, in the case of the heat pump, the water mass flow rate was determined on the basis of the known volumetric water flow rate $\dot{V}_{hp}$:

$$\dot{m}_{w,hp} = \rho_{w,\ hp} \dot{V}_{hp} \tag{6}$$

where $\rho_{w,\ hp}$ is the density of the water heated in the heat pump.

In the case of the heat pump, the volumetric flow rate of the heated water $\dot{V}_{hp}$ is constant and equal to 2.6 $m^3/h$.

The density in Equations (4) and (6) and the specific heat in Equations (3) and (5) are determined from polynomial functions for the mean water temperature flowing through the cooling tower or heat pump. The water density as a function of temperature is given by the following:

$$\rho = 1003.871389 - 0.49355774\, T + 0.006389001\, T^2 - 8.7267 \times 10^{-5}\, T^3 + 3.4308 \times 10^{-7}\, T^4 - 4.6788 \times 10^{-10}\, T^5 \tag{7}$$

where $T$ is the fluid temperature (in °C). The standard deviation of the water density is $\sigma = \pm 4.0586\ \mathrm{kg/m^3}$, and the coefficient of determination is ($r^2$ = 0.999).

The relationship for the specific heat capacity of water as a function of temperature is given by

$$\begin{aligned} C_w = {} & 4220.853492 + 3.969000719\, T - 0.35459395\, T^2 + 0.009487507\, T^3 - 0.00011820\, T^4 \\ & + 7.87319 \times 10^{-7}\, T^5 - 2.8613 \times 10^{-9}\, T^6 + 5.33092 \times 10^{-12}\, T^7 - 3.9578 \times 10^{-15}\, T^8 \end{aligned} \tag{8}$$

where $T$ is the fluid temperature (in °C). The standard deviation of the specific heat capacity of water is $\sigma = \pm 39.99\ \mathrm{J/(kgK)}$, and the coefficient of determination is $r^2$ = 0.999.

Equations (7) and (8) are valid for the temperature range from 0 to 370 °C.

To determine the active electrical power of the three-phase fan motor of the cooling tower, the following equation was used:

$$P_{fan} = \sqrt{3} U I cos(\varphi) \tag{9}$$

https://doi.org/10.3390/en19020554

where $U$ is the voltage, $I$ is the electric current, and $cos(\varphi)$ is the power factor.

The ambient air wet-bulb temperature was calculated using the formula proposed by Stull [31]:

$$T_{wb} = T_{amb}\, arctan\left(0.151977(RH + 8.313659)^{\frac{1}{2}}\right) + arctan(T_{amb} + RH) - arctan(RH - 1.676331) + 0.00391838\, RH^{\frac{3}{2}}\, arctan(0.023101\, RH) - 4.686035 \quad (10)$$

where $T_{amb}$ is the ambient temperature (in °C), and $RH$ is the relative humidity of the air (in %).

According to [32], Stull's formula provides the highest accuracy for the calculated wet-bulb temperature within the range from −1 °C to 0.65 °C, for ambient temperatures between −20 °C and 50 °C, and relative humidity from 5% to 99%.

The measurements performed also made it possible to determine the theoretical SCOP of the cooling tower, expressed by the following equation:

$$\mathrm{SCOP_{CT}} = \frac{\dot{Q}_{cool}}{P_{fan}} \quad (11)$$

where $\dot{Q}_{cool}$ is the heat flux dissipated by the cooling tower, and $P_{fan}$ is the power consumed by the fan.

Based on the cooling tower capacity $\dot{Q}_{cool}$ calculated for each measuring point (the time step was 10 s), the amounts of heat transferred to the air during each successive 10 s were calculated. Their sum for the entire operating period of the system represents the amount of cooling energy produced by the cooling tower $Q_{cool}$. Similarly, the electricity consumption of the cooling tower fans $W_{fan}$ was determined on the basis of their power $P_{fan}$ and the amount of heat generated by the heat pump $Q_{hp}$ on the basis of its capacity $\dot{Q}_{hp}$. These calculations were performed separately for Stage I and Stage II.

For the performed calculations of indirect measurements, their uncertainties were determined. In accordance with the law of error propagation, the precision limits, assuming a 95% confidence level, are calculated using the following equation [33]:

$$U = \sqrt{P^2 + B^2}, \quad (12)$$

where $P$ is the precision limit of measurement, and $B$ is the bias limit of measurement, which was omitted in these studies.

The method of determining measurement uncertainty is illustrated by the example of a cooling tower's capacity. The formula for the precision limits for cooling tower capacity is as follows:

$$\begin{aligned} P^2_{\dot{Q}_{cool}} &= \left(\frac{\partial \dot{Q}_{cool}}{\partial \dot{V}_{cool}}\right)^2 P^2_{\dot{V}_{cool}} + \left(\frac{\partial \dot{Q}_{cool}}{\partial \rho_{w,cool}}\right)^2 P^2_{\rho_{w,cool}} + \left(\frac{\partial \dot{Q}_{cool}}{\partial C_{w,cool}}\right)^2 P^2_{C_{w,cool}} + \left(\frac{\partial \dot{Q}_{cool}}{\partial T_4}\right)^2 P^2_{T_4} + \left(\frac{\partial \dot{Q}_{cool}}{\partial T_3}\right)^2 P^2_{T_3} \\ &= \left(\rho_{w,cool} \cdot C_{w,cool}(T_4 - T_3)\right)^2 P^2_{\dot{V}_{cool}} + \left(\dot{V}_{cool} \cdot C_{w,cool}(T_4 - T_3)\right)^2 P^2_{\rho_{w,cool}} \\ &+ \left(\dot{V}_{cool} \cdot \rho_{w,cool}(T_4 - T_3)\right)^2 P^2_{C_{w,cool}} + \left(\dot{V}_{w,ch} \cdot \rho_{w,cool} \cdot C_{w,cool}\right)^2 P^2_{T_4} \\ &+ \left(-\dot{V}_{cool} \cdot \rho_{w,cool} \cdot C_{w,ch}\right)^2 P^2_{T_3} \\ &= \left(\frac{\dot{Q}_{cool}}{\dot{V}_{cool}}\right)^2 P^2_{\dot{V}_{cool}} + \left(\frac{\dot{Q}_{cool}}{\rho_{w,cool}}\right)^2 P^2_{\rho_{w,cool}} + \left(\frac{\dot{Q}_{cool}}{C_{w,ch}}\right)^2 P^2_{C_{w,cool}} + \left(\frac{\dot{Q}_{cool}}{T_4 - T_3}\right)^2 P^2_{T_4} + \left(\frac{-\dot{Q}_{cool}}{T_4 - T_3}\right)^2 P^2_{T_3} \end{aligned} \quad (13)$$

where $P_{\dot{V}_{cool}}$, $P_{T_3}$, and $P_4$ are the measurement uncertainties (type B) determined from the class of measurement sensors, while $P_{\rho_{w,cool}}$ and $P_{C_{w,cool}}$ are the standard deviations of the approximated densities and specific heat, respectively (type A).

Hence, the measurement uncertainty of the cooling tower capacity is expressed by the following formula:

$$U_{\dot{Q}_{cool}} = \sqrt{\left(\frac{\dot{Q}_{cool}}{\dot{V}_{cool}}\right)^2 P^2_{\dot{V}_{cool}} + \left(\frac{\dot{Q}_{cool}}{\rho_{w,cool}}\right)^2 P^2_{\rho_{w,cool}} + \left(\frac{\dot{Q}_{cool}}{C_{w,ch}}\right)^2 P^2_{C_{w,cool}} + \left(\frac{\dot{Q}_{cool}}{T_4 - T_3}\right)^2 P^2_{T_4} + \left(\frac{-\dot{Q}_{cool}}{T_4 - T_3}\right)^2 P^2_{T_3}}$$
$$= \sqrt{\dot{Q}^2_{cool}\left[\left(\frac{P_{\dot{V}_{cool}}}{\dot{V}_{cool}}\right)^2 + \left(\frac{P_{\rho_{w,cool}}}{\rho_{w,cool}}\right)^2 + \left(\frac{P_{C_{w,cool}}}{C_{w,cool}}\right)^2 + \left(\frac{P_{T_4}}{T_4 - T_3}\right)^2 + \left(\frac{-P_{T_3}}{T_4 - T_3}\right)^2\right]} \quad (14)$$

In summary, the study used both qualitative and quantitative research methods. An overview of the applied research methods is presented in Table 2.

**Table 2.** Summary of research methods.

| Research Methods | Description |
|---|---|
| Document examination method | A literature review was conducted, and existing research on evaporative cooling devices was analysed. |
| Experimental method | Measurements on the experimental stand were carried out and recorded. An analysis of the measured values was carried out. |
| Mathematical modelling | Based on the measurements, calculations were performed for quantities that could not be directly measured, such as $SCOP_{CT}$. An analysis of the results obtained was carried out. |

## 3. Results

This chapter presents the measurement results for Stages I and II. The dataset was supplemented with calculations of the evaporative tower's cooling capacity $\dot{Q}_{cool}$ (from Equation (3)), the heat pump's heating capacity $\dot{Q}_{hp}$ (from Equation (5)), the ambient air wet-bulb temperature $T_{wb}$ (from Equation (10)), and the SCOP of the cooling tower (from Equation (11)). In this chapter, ambient temperature $T_{amb}$ is equal to temperature $T_2$ in Stage I and temperature $T_1$ in Stage II. The measurement data also include the dew-point temperature, calculated by the controller based on the dry-bulb temperature and relative humidity. Table 3 summarises the average values of selected measured and computed parameters for Stages I and II. They were calculated as arithmetic means over the entire operational period of the system in Stages I and II.

**Table 3.** Average values of the parameters measured and calculated in Stages I and II.

| Parameter | Stage I | Stage II |
|---|---|---|
| $T_3$, °C | 30.1 ± 1.5% | 30.6 ± 1.5% |
| $T_4$, °C | 25.1 ± 1.5% | 24.8 ± 1.5% |
| $T_5$, °C | 36.0 ± 1.5% | 35.3 ± 1.5% |
| $T_6$, °C | 30.0 ± 1.5% | 29.7 ± 1.5% |
| $\dot{Q}_{cool}$, kW | 17.3 ± 1.9% | 19.9 ± 1.9% |
| $\dot{Q}_{hp}$, kW | 18.2 ± 2.0% | 19.3 ± 1.7% |
| $T_{amb}$, °C | 19.0 ± 1.5% | 19.9 ± 1.5% |
| $T_{wb}$, °C | 15.9 ± 1.5% | 17.1 ± 1.6% |
| Frequency, Hz | 31.7 ± 5.0% | 29.0 ± 5.0% |
| $I$, mA | 380.4 ± 5.0% | 327.6 ± 5.0% |
| $\dot{V}_{cool}$, m$^3$/h | 2.9 ± 0.1% | 2.9 ± 0.1% |

To compare all stages, the total amount of cooling energy produced by the cooling tower $Q_{cool}$, the total thermal energy delivered by the heat pump $Q_{hp}$, and the electrical energy consumed by the cooling tower fan motor $W_{fan}$ were calculated for the respective

https://doi.org/10.3390/en19020554

measurement periods. From these results, the SCOP of the cooling tower was determined for each stage. All corresponding data are presented in Table 4.

**Table 4.** A summary of the cooling energy produced $Q_{cool}$ and the electrical energy consumed $W_{fan}$ by the cooling tower, the thermal energy delivered by the heat pump $Q_{hp}$, and the SCOP of the cooling tower.

| | Stage I | Stage II |
|---|---|---|
| Quantity of cooling energy generated by the cooling tower $Q_{cool}$, kWh | 51,050 ± 3.5% | 58,550 ± 3.7% |
| Electricity consumption of the cooling tower $W_{fan}$, kWh | 1115 ± 3.4% | 980 ± 3.1% |
| Quantity of heat energy generated by the heat pump $Q_{hp}$, kWh | 53,460 ± 3.8% | 56,659 ± 3.9% |
| $SCOP_{CT}$, kWh/kWh | 45.8 ± 2.2% | 59.7 ± 2.4% |

### *3.1. Test Under Increased Thermal Load—16 August 2024*

On 16 August 2024, a test of the cooling system was conducted during the period of the highest ambient temperatures. The purpose of this test was to evaluate how maximum precooling of the inlet air, combined with an increased thermal load, would affect the performance of the evaporative tower. During the test period (10:45–14:45), the ambient temperature ranged from 25 °C to 32 °C, and the relative humidity ranged from 45% to 75%.

The test consisted of turning off the automatic operation of both the CW3 cooler and the MCT 25 cooling tower. Both devices were operated at maximum fan speed and with lowered setpoints for the precooled air temperature ($T_2$) and the water temperature ($T_3$). The cooling tower inverter frequency was set to 60 Hz, while the cooler's controller was switched to manual mode with the minimum allowable temperature setpoint of 18 °C and maximum supply air flow. Additionally, the outlet water temperature $T_5$ from the heat pump was increased to 45 °C. Table 5 presents hourly average values of selected measured and calculated parameters obtained during the test. In Table 3, the symbols $T_{amb}$ and $RH_{amb}$ represent the temperature $T_1$ and relative humidity $RH_1$ of the air measured at the inlet to the IEC, while $T_{precool}$ and $RH_{precool}$ represent $T_2$ and $RH_2$ measured at the cooling tower air inlet.

**Table 5.** Average values of the parameters measured and calculated in Stage II.

| Parameter | 1st Hour 10:45/11:45 | 2nd Hour 11:45/12:45 | 3rd Hour 12:45/13:45 | 4th Hour 13:45/14:45 |
|---|---|---|---|---|
| $T_{amb}$, °C | 25.1 ± 1.5% | 31.1 ± 1.5% | 30.9 ± 1.5% | 31.1 ± 1.5% |
| $RH_{amb}$, % | 71.6 ± 5.0% | 52.4 ± 5.0% | 51.1 ± 5.0% | 48.3 ± 5.0% |
| $T_{precool}$, °C | 20.7 ± 1.5% | 21.2 ± 1.5% | 21.3 ± 1.5% | 21.6 ± 1.5% |
| $RH_{precool}$,% | 90.1 ± 5.0% | 90.6 ± 5.0% | 89.7 ± 5.0% | 86.9 ± 5.0% |
| $T_3$, °C | 25.5 ± 1.5% | 25.3 ± 1.5% | 25.3 ± 1.5% | 25.3 ± 1.5% |
| $T_4$, °C | 34.9 ± 1.5% | 35.1 ± 1.5% | 35.1 ± 1.5% | 34.8 ± 1.5% |
| $T_5$, °C | 44.8 ± 1.5% | 45.4 ± 1.5% | 45.1 ± 1.5% | 45.1 ± 1.5% |
| $T_6$, °C | 34.7 ± 1.5% | 35.3 ± 1.5% | 35.1 ± 1.5% | 34.9 ± 1.5% |
| $\dot{Q}_{cool}$, kW | 32.8 ± 3.5% | 34.2 ± 3.5% | 35.1 ± 3.5% | 33.4 ± 3.5% |
| $\dot{Q}_{hp}$, kW | 30.7 ± 3.1% | 33.5 ± 3.1% | 32.3 ± 3.1% | 31.9 ± 3.1% |

### *3.2. Comparison of Measurements for Selected Days of Stage I and Stage II*

From the available measurement data, two representative days were selected: one from Stage I (17 August 2023) and another from Stage II (11 August 2024). The selection criterion was a similar daily profile of ambient dry-bulb temperature and wet-bulb temperature.

https://doi.org/10.3390/en19020554

Table 6 presents the measured and calculated average values, while Table 7 summarises the amounts of cooling energy $Q_{cool}$ produced by the cooling tower, thermal energy $Q_{hp}$ produced by the heat pump, the electrical energy consumed by the fans of the cooling tower $W_{fan}$, and the resulting $SCOP_{CT}$ values. Figure 7 illustrates the daily distribution of the measured dry-bulb temperatures and the calculated wet-bulb temperatures.

**Table 6.** Average values of the parameters measured and calculated in Stages I and II.

| Parameter | Stage I | Stage II |
|---|---|---|
| $T_3$, °C | 25.0 ± 1.5% | 25.0 ± 1.5% |
| $T_4$, °C | 30.2 ± 1.5% | 30.1 ± 1.5% |
| $T_5$, °C | 30.0 ± 1.5% | 29.9 ± 1.5% |
| $T_6$, °C | 35.0 ± 1.5% | 35.1 ± 1.5% |
| $\dot{Q}_{cool}$, kW | 17.4 ± 1.5% | 19.6 ± 1.9% |
| $\dot{Q}_{hp}$, kW | 18.2 ± 1.7% | 19.9 ± 2.1% |
| $T_{amb}$, °C | 20.1 ± 1.5% | 20.5 ± 1.5% |
| $T_{wb}$, °C | 16.6 ± 1.6% | 16.9 ± 1.6% |
| Frequency, Hz | 32.5 ± 5.0% | 29.3 ± 5.0% |
| $I$, mA | 390.7 ± 5.0% | 378.1 ± 5.0% |
| $\dot{V}_{cool}$, $m^3/h$ | 2.9 ± 0.1% | 2.9 ± 0.1% |

**Table 7.** A summary of the cooling energy produced $Q_{cool}$ and the electrical energy consumed $W_{fan}$ by the cooling tower, the thermal energy delivered by the heat pump $Q_{hp}$, and the $SCOP_{CT}$ of the cooling tower.

| | Stage I | Stage II |
|---|---|---|
| Quantity of cooling energy generated by the cooling tower $Q_{cool}$, kWh | 417.6 ± 3.4% | 470.4 ± 3.7% |
| Electricity consumption of the cooling tower $W_{fan}$, kWh | 9.4 ± 3.4% | 9.1 ± 3.1% |
| Quantity of heat energy generated by the heat pump$Q_{hp}$, kWh | 436.8 ± 3.2% | 477.6 ± 3.3% |
| $SCOP_{CT}$, kWh/kWh | 44.4 ± 2.2% | 51.7 ± 2.3% |

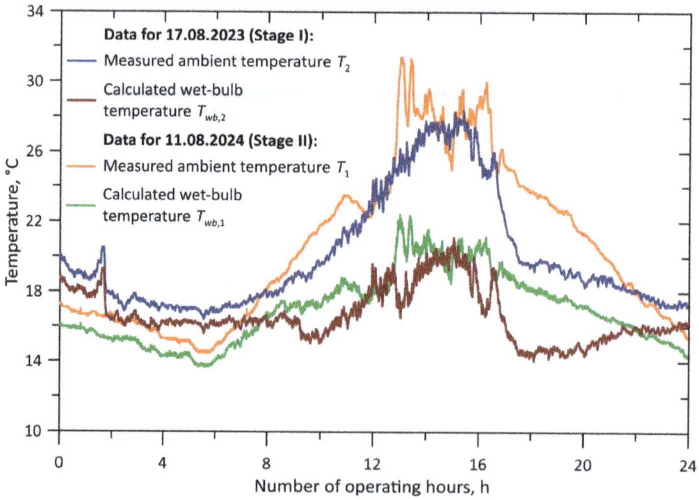


**Figure 7.** Ambient (dry-bulb) and wet-bulb temperatures for 17 August 2023 (Stage I) and for 11 August 2024 (Stage II).

https://doi.org/10.3390/en19020554

## 4. Analysis of Results and Discussion

This paper presents the results of experimental investigations (Stages I and II) conducted over two cooling seasons (from early May to the end of August) in 2023 and 2024. The statistical climate data included in the MITA software assume a wider range of dry-bulb temperatures. The minimum and maximum values are 1.5 °C and 32 °C, respectively. However, measurements from Stages I and II showed that during the analysed period, the ambient temperature did not fall below 9 °C. Figure 8 illustrates the temperature distribution for Stage I, while Figure 9 presents the corresponding distribution for Stage II. The maximum dry-bulb temperature recorded in Stage I was 32.9 °C, and in Stage II it reached 35.6 °C. The average dry-bulb temperature in Stage II was 0.9 °C higher than in Stage I.

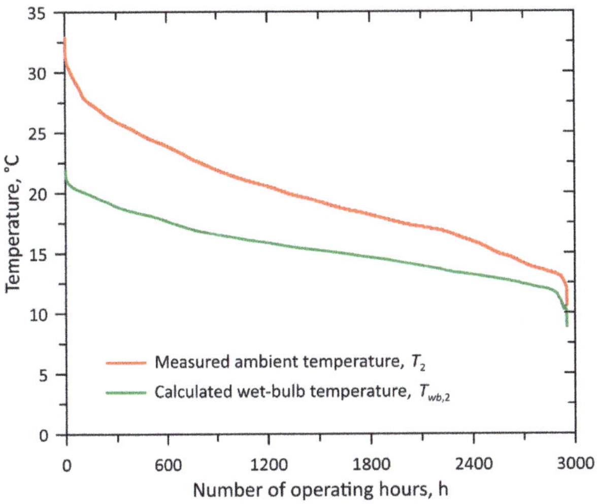


**Figure 8.** Hourly profiles of the ambient temperature ($T_2$) and the wet-bulb temperature ($T_{wb,2}$) at the inlet of the cooling tower for Stage I.

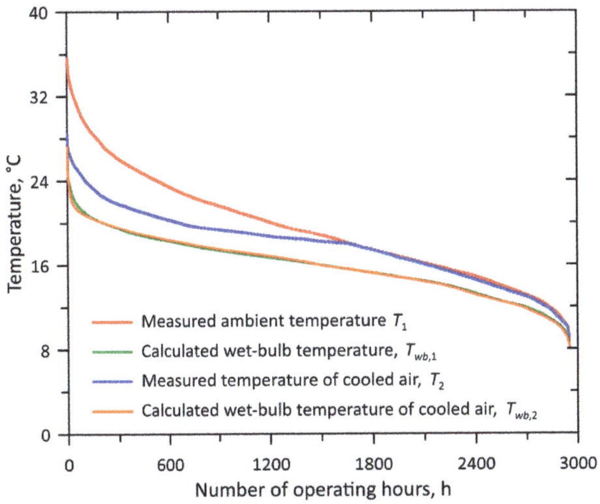


**Figure 9.** Hourly profile of ambient temperature $T_1$ and wet-bulb temperature $T_{wb,1}$ at the inlet to the IEC and air temperature $T_2$ and wet-bulb temperature $T_{wb,2}$ after pre-cooling at the inlet to the cooling tower for Stage II.

https://doi.org/10.3390/en19020554

The wet-bulb temperatures also differed between the two stages. Once again, the statistical data values are lower than the actual measurements. The lowest calculated wet-bulb temperature was 9.1 °C for Stage I and 8.3 °C for Stage II. The average wet-bulb temperature in 2024 was 1.2 °C higher than in 2023.

Higher ambient dry-bulb and wet-bulb temperatures reduce the available driving potential for heat and mass transfer, thereby reducing cooling tower capacity. In Stage II, the higher maximum and average dry-bulb temperatures increased the average wet-bulb temperature, further increasing the outlet water temperature from the cooling water. As a result, operation under such conditions generally leads to reduced cooling capacity, higher outlet temperatures, and increased electricity demand. Consequently, the improved capacity observed in Stage II, despite less favourable ambient conditions, highlights the beneficial impact of the air at the cooling tower inlet on the cooling tower's operation.

In Figure 9, a decreasing difference between the ambient air temperature and the cooled air temperature can be observed. This effect is associated with the control process of the CW3 cooler. The fan speed in the CW3 unit automatically adjusts to the set temperature, which in this case was 18 °C. The cooler's water consumption, which supplies the wet channels, is constant and not subject to regulation. Consequently, the closer the ambient temperature is to the set point, the lower the fan speed.

The calculated $SCOP_{CT}$ for Stage I is consistent with MITA's software's theoretical assumptions. Although the statistical dataset assumes a broader range of minimum and maximum dry-bulb and wet-bulb temperatures, these values are representative of the actual climatic conditions that may occur at the studied location. The cooling tower using precooled air from the CW3 unit achieved a $SCOP_{CT}$ approximately 30% higher than in Stage I. This improvement demonstrates that the proposed solution enables higher cooling capacity while reducing electricity consumption compared with the stand-alone cooling tower. Moreover, the cooling tower operating in combination with the IEC exhibited significantly higher efficiency, despite experiencing higher peak dry-bulb and wet-bulb temperatures than those recorded during the Stage I measurements.

The test conducted on 16 August 2024 demonstrated that the cooling tower with a precooled inlet air temperature could achieve substantially higher cooling capacity. This effect was observed under high ambient temperatures and increased thermal load from the heat pump. Under these conditions, the cooling tower maintained the outlet water temperature ($T_3$) below 25.5 °C, while the maximum difference between the ambient air temperature and the cooled air temperature reached 9.9 K. The cooling capacity of the evaporative tower increased from its nominal value of 17 kW to approximately 35 kW, and the heat pump's thermal output likewise increased to a maximum of about 35 kW. This enhancement was a direct result of the increased water-side temperature difference $\Delta T$.

The comparison confirms that precooling the inlet air to the cooling tower using an IEC reduces electricity consumption and increases the thermal efficiency of the open-circuit cooling tower. The selected day of Stage II (11 August 2024) was warmer than the corresponding day of Stage I (17 August 2023), with an average dry-bulb temperature higher by 0.5 °C and an average wet-bulb temperature higher by 0.3 °C. Despite these less favourable ambient conditions, the average electricity consumption by the tower fan motor in Stage II was approximately 4% lower than in Stage I. The increased water temperature difference observed in the cooling tower of the Stage II system is a direct result of inlet air precooling, which improves heat and mass transfer by reducing the inlet air's effective wet-bulb temperature. Consequently, the higher thermal load on the heat pump is enabled by the improved air-side conditions and is not an independent effect.

https://doi.org/10.3390/en19020554

## 5. Conclusions

This study demonstrated that precooling of the inlet air to the cooling tower using an IEC significantly improves system capacity under both nominal and extreme operating conditions. Compared to the reference system without precooling, the proposed configuration increased cooling capacity by 14.7% and reduced the cooling tower's electricity consumption by 135 kWh, while enabling lower outlet water temperatures.

The main scientific achievement of this work is the experimental confirmation that IEC-based inlet air precooling leads to a substantial improvement in the Seasonal Coefficient of Performance of the cooling tower ($SCOP_{CT}$), which increased by approximately 30% compared to operation without precooling.

The key assumption of the research was that using IEC to precool the inlet air improves the cooling tower's efficiency, and this thesis was proven.

Unlike previous studies, which predominantly focused on improving wet-bulb efficiency, the present work introduces SCOP as a comprehensive, seasonally representative performance indicator for cooling towers. Based on an extensive literature review, no experimental studies were identified that evaluate the impact of IEC-based inlet air precooling on the SCOP of an open-circuit cooling tower, which highlights the novelty of the present research.

An increase in SCOP directly indicates lower electricity consumption relative to the useful cooling energy generated, providing a more application-oriented assessment of cooling tower performance.

Additional tests conducted at high ambient temperatures and elevated thermal loads confirmed that the integrated cooling tower and IEC system ensures reliable operation even under extreme conditions, which is particularly important for industrial facilities lacking backup cooling capacity or undergoing periodic expansions.

Although including auxiliary IEC fans increases total electricity consumption, the improved operational reliability may outweigh the associated increase in operating costs in many practical applications.

The results further indicate that additional energy savings may be achieved through alternative system configurations, particularly those that integrate only an indirect heat exchanger rather than a complete IEC with auxiliary fans.

Possible alternative configurations include the following:

- Standalone HMX or MicroCore (CW3) modules installed at the cooling tower air inlet;
- A single fan driving the ambient air stream, which is then precooled and finally exhausted from the cooling tower, overcoming total pressure losses in the system.

The proposed solution applies not only to open-circuit cooling towers but also to closed-circuit systems, where even greater efficiency improvements and extended dry-operation (free-cooling) periods can be expected. Furthermore, inlet air precooling can compensate for reduced cooling capacity associated with the transition to environmentally friendly, low-GWP refrigerants, enabling system modernisation without performance degradation.

The study confirms that precooling the inlet air using the IEC significantly enhances the capacity of open-circuit evaporative cooling towers, particularly in climates characterised by a large dry-bulb to wet-bulb temperature difference. The experimental results demonstrate that precooling the inlet air reduces cooling tower electricity consumption by approximately 15%, accompanied by a noticeable increase in cooling capacity and a lower outlet water temperature.

The benefits of precooling the inlet air to the cooling tower are most pronounced in hot, dry, or semi-arid climates, where the high evaporative potential enables effective reduction in air temperature before it enters the cooling tower. In temperate climates,

https://doi.org/10.3390/en19020554

significant electrical energy savings can still be achieved during peak summer conditions. In contrast, in hot, humid climates, the effect is more limited and is primarily suitable for precooling air-inlet applications. Overall, the study's results indicate that IEC-based inlet air precooling is an effective strategy for reducing electricity consumption and improving operational reliability of open evaporative cooling towers, particularly under high ambient temperatures.

**Author Contributions:** Conceptualization, B.J.; methodology, B.J.; validation, B.J.; formal analysis, B.J.; investigation, B.J.; resources, B.J. and M.J.; writing—original draft preparation, B.J.; writing—review and editing, M.J.; visualisation, B.J. and M.J.; supervision, M.J. All authors have read and agreed to the published version of the manuscript.

**Funding:** This research was funded by the Implementation Doctorate programme, contract No. DWD/4/25/2020.

**Data Availability Statement:** The data presented in this study are available on request from the corresponding author due to the know-how of producers of cooling devices.

**Conflicts of Interest:** The authors declare no conflicts of interest.

## Abbreviations

The following abbreviations are used in this manuscript:

| | |
|---|---|
| COP | Coefficient of Performance |
| DEC | Direct Evaporative Cooler |
| GWP | Global Warming Potential |
| HMX | Heat and Mass Exchanger |
| HVAC | Heating, Ventilation, and Air Conditioning |
| IEC | Indirect Evaporative Cooler |
| M-cycle | Maisotsenko Cycle |
| ODP | Ozone Depletion Potential |
| RH | Relative Humidity |
| SCOP | Seasonal Coefficient of Performance |
| VCRS | Vapour-Compression Refrigeration System |

## References


1. International Energy Agency. The Future of Cooling. Opportunities for Energy-Efficient Air Conditioning, OECD/IEA, 2018. Available online: https://www.iea.org (accessed on 3 December 2025).
2. BITZER Kuehlmaschinenbau GmbH. Refrigerant Report 20; A-501-20 EN. Available online: https://www.bitzer.de (accessed on 3 December 2025).
3. Yang, Z.; Feng, B.; Ma, H.; Zhang, L.; Duan, C.; Liu, B.; Zhang, Y.; Chen, S.; Yang, Z. Analysis of lower GWP and flammable alternative refrigerants. *Int. J. Refrig.* **2021**, *126*, 12–22. [CrossRef]
4. Artusoa, P.; Marinetti, S.; Minetto, S.; Del Col, D.; Rossetti, A. Modelling the performance of a new cooling unit for refrigerated transport using carbon dioxide as the refrigerant. *Int. J. Refrig.* **2020**, *115*, 158–171. [CrossRef]
5. Longo, G.; Mancin, S.; Righetti, G.; Zilio, C. Hydrocarbon refrigerants HC290 (Propane) and HC1270 (Propylene) low GWP long-term substitutes for HFC404A: A comparative analysis in vaporisation inside a small-diameter horizontal smooth tube. *Appl. Therm. Eng.* **2017**, *124*, 707–715. [CrossRef]
6. Dizaji, H.S.; Mrabet, B.M. Development and experimental sensitivity analysis of a Maisotsenko-cycle water desalination system under diverse climate conditions. *Appl. Therm. Eng.* **2025**, *281*, 128730. [CrossRef]
7. Pandelidis, D.; Cichoń, A.; Pacak, A.; Drąg, P.; Drąg, M.; Worek, W.; Cetin, S. Water desalination through the dewpoint evaporative system. *Energy Convers. Manag.* **2021**, *229*, 113757. [CrossRef]
8. Saghafifar, M.; Gadalla, M. Innovative inlet air cooling technology for gas turbine power plants using integrated solid desiccant and Maisotsenko cooler. *Energy* **2015**, *87*, 663–677. [CrossRef]
9. Saghafifar, M.; Gadalla, M. Analysis of Maisotsenko open gas turbine power cycle with a detailed air saturator model. *Appl. Energy* **2015**, *149*, 338–353. [CrossRef]


https://doi.org/10.3390/en19020554

10. Baakeem, S.; Orfi, J.; Mohamad, A.; Bawazeer, S. The possibility of using a novel dew point air cooling system (M-Cycle) for A/C application in Arab Gulf Countries. *Build. Environ.* **2019**, *148*, 185–197. [CrossRef]
11. Sohani, A.; Sayyaadi, H.; Mohammadhosseini, N. Comparative study of the conventional types of heat and mass exchangers to achieve the best design of dew point evaporative coolers at diverse climatic conditions. *Energy Convers. Manag.* **2018**, *158*, 327–345. [CrossRef]
12. Badiei, A.; Akhlaghi, Y.G.; Zhao, X.; Li, J.; Yi, F.; Wang, Z. Can whole building energy models outperform numerical models, when forecasting performance of indirect evaporative cooling systems? *Energy Convers. Manag.* **2020**, *213*, 112886. [CrossRef]
13. Delfani, S.; Karami, M. Transient simulation of solar desiccant/M-Cycle cooling systems in three different climatic conditions. *J. Build. Eng.* **2020**, *29*, 101–152. [CrossRef]
14. Zanchini, E.; Naldi, C. Energy saving obtainable by applying a commercially available M-cycle evaporative cooling system to the air conditioning of an office building in North Italy. *Energy* **2019**, *179*, 975–988. [CrossRef]
15. Comino, F.; González, J.C.; Navas-Martos, F.J.; De Adana, M.R. Experimental energy performance assessment of a solar desiccant cooling system in Southern Europe climates. *Appl. Therm. Eng.* **2020**, *165*, 114579. [CrossRef]
16. Pandelidis, D.; Anisimov, S.; Worek, W.M. Performance study of the Maisotsenko Cycle heat exchangers in different air-conditioning applications. *Int. J. Heat Mass Transf.* **2015**, *81*, 207–221. [CrossRef]
17. Pokorný, J.; Madejski, P.; Fišer, J. Sustainable air conditioning with a focus on evaporative cooling and the Maisotsenko cycle. *Arch. Thermodyn.* **2025**, *46*, 111–121. [CrossRef]
18. Lv, J.; Zhou, B.; Zhu, M.; Xi, W.; Hu, E. Experimental Study on the Performance of a Dew-Point Evaporative Cooling System with a Nanoporous Membrane. *Energies* **2022**, *15*, 2592. [CrossRef]
19. Ali, M.; Ahmad, W.; Sheikh, N.A.; Ali, H.; Kousar, R.; Rashid, T. Performance enhancement of a cross flow dew point indirect evaporative cooler with circular finned channel geometry. *J. Build. Eng.* **2021**, *35*, 101980. [CrossRef]
20. Cuce, P.M.; Riffat, S. A state of the art review of evaporative cooling systems for building applications. *Renew. Sustain. Energy Rev.* **2016**, *54*, 1240–1249. [CrossRef]
21. Pandelidis, D.; Anisimov, S.; Worek, W.; Drag, P. Numerical analysis of a desiccant system with cross-flow Maisotsenko cycle heat and mass exchanger. *Energy Build.* **2016**, *123*, 136–150. [CrossRef]
22. Chen, C.; Hsu, C.; Chen, C.; Chen, S. Silica gel polymer composite desiccants for air conditioning systems. *Energy Build.* **2015**, *101*, 122–132. [CrossRef]
23. Kozubal, E.; Slayzak, S. *Coolerado 5 Ton RTU Performance: Western Cooling Challenge Results*; NREL: Golden, CO, USA, 2010.
24. Cui, X.; Chua, K.J.; Islam, M.R.; Ng, K.C. Performance evaluation of an indirect pre-cooling evaporative heat exchanger operating in hot and humid climate. *Energy Convers. Manag.* **2015**, *102*, 140–150. [CrossRef]
25. Khandelwal, A.; Talukdar, P.; Jain, S. Energy savings in a building using regenerative evaporative cooling. *Energy Build.* **2011**, *43*, 581–591. [CrossRef]
26. Heidarinejad, G.; Bozorgmehr, M.; Delfani, S.; Esmaeelian, J. Experimental investigation of two-stage indirect/direct evaporative cooling system in various climatic conditions. *Build. Environ.* **2009**, *44*, 2073–2079. [CrossRef]
27. Karmakar, A.; Salomon, B.C.; Acharya, S.; Kozlov, A.; Chudnovsky, Y. Utilizing Reversed Brayton Cycle for Enhanced Cooling Tower Technology. *Heat Transf. Eng.* **2025**, *46*, 928–942. [CrossRef]
28. Pandelidis, D.; Drag, M.; Drag, P.; Worek, W.; Cetin, S. Comparative analysis between traditional and M-Cycle based cooling tower. *Int. J. Heat Mass Transf.* **2020**, *159*, 120–124. [CrossRef]
29. Fan, X.; Lu, X.; Wang, J.; Li, Z.; Wang, Q.; Dong, Z.; Zhang, R. Performance Evaluation of a Maisotsenko Cycle Cooling Tower with Uneven Length of Dry and Wet Channels in Hot and Humid Conditions. *Energies* **2021**, *14*, 8249. [CrossRef]
30. MITA Cooling Technologies S.r.l.,Via del Benessere, 13, Siziano 27010, Pavia, IT. Available online: https://www.mitacoolingtechnologies.com/ (accessed on 3 December 2025).
31. Stull, R. Wet-bulb temperature from relative humidity and air temperature. *J. Appl. Meteorol. Climatol.* **2011**, *50*, 2267–2269. [CrossRef]
32. Lee, C.; Wang, Y. A novel method to derive formulas for computing the wet-bulb temperature from relative humidity and air temperature. *Measurement* **2018**, *128*, 271–275. [CrossRef]
33. Taler, D. *Calculation and Experimental Research on Heat Exchangers*; Wydawnictwo PK: Cracow, Poland, 2016. (In Polish)




https://doi.org/10.3390/en19020554

*energies*

MDPI

*Article*

# Analysis of Heat Transfer Characteristics in a Latent Heat Storage Module Using Circular-Finned Tubes

**Ji-Woon Ko [1], Tae Hwan Song [1], Jong-Hoon Lee [1,2], Jong Hyeon Peck [1] and Seung Jin Oh [1,2,*]**

[1] Celan Energy Transition Group, Korea Institute of Industrial Technology, 102 Jejudaehak-ro, Jeju-si 63242, Jeju Special Self-Governing Province, Republic of Korea; jiwoonko@kitech.re.kr (J.-W.K.); xoghks4551@kitech.re.kr (T.H.S.); rrl@kitech.re.kr (J.-H.L.); pjh6240@kitech.re.kr (J.H.P.)

[2] Department of Convergence Manufacturing System Engineering, University of Science and Technology (UST), 217 Gajeong-ro, Yuseong-gu, Daejeon 34113, Republic of Korea

* Correspondence: ohs8680@kitech.re.kr

**Abstract:** Latent heat thermal energy storage (LHTES) using inorganic salt hydrates is a promising technology for buffering renewable energy fluctuations; however, phase-dependent heat transfer remains insufficiently understood for design optimization. In this study, a shell-and-tube storage module with a circular-finned tube was constructed and filled with 13.17 kg of barium hydroxide octahydrate (BHO). Discharge tests were conducted with heat transfer fluid (HTF) inlet temperatures ranging from 20 °C to 50 °C and flow rates of 10–25 L/min, while charging was performed at 90 °C. The overall heat transfer coefficient ($U_o$) was derived using the logarithmic mean temperature difference method, the inside coefficient ($h_i$) was calculated by the Petukhov correlation, and the outside coefficient ($h_o$) was obtained via thermal-resistance network. Results show that the average discharge energy was approximately 1.027 kWh (except 0.859 kWh at 50 °C inlet), with a mean utilization efficiency of 79.25%. The $U_o$ was consistently highest in the liquid phase, followed by the latent and solid phases, with ranges of 0.257–0.863, 0.025–0.072, and 0.015–0.044 kW/m$^2$·°C, respectively. Sensitivity analysis revealed that the HTF flow rate strongly influenced $U_o$ across all phases, whereas inlet temperature played only a minor role. The outside coefficient $h_o$ was 0.033–0.162 kW/m$^2$·°C in the latent regime and 0.018–0.064 kW/m$^2$·°C in the solid regime, with a notable peak around Reynolds number $1.3 \times 10^4$ in the latent phase. These findings provide detailed phase-resolved $U_o$ and $h_o$ data for inorganic salt hydrate storage and highlight design insights such as the diminishing returns of flow rate increase beyond a threshold, offering valuable guidelines for sizing and operation of LHTES in Power-to-Heat applications.

**Keywords:** latent heat thermal energy storage (LHTES); phase change material (PCM); finned-tube heat exchanger; barium hydroxide octahydrate; overall heat transfer coefficient; Power-to-Heat (P2H)

## 1. Introduction

The global transition toward carbon-neutral energy systems, driven by the urgent need to address climate change and enhance energy security, has led to a rapid expansion of renewable energy sources such as solar and wind power. The International Energy Agency (IEA) projects that by 2030, the world's installed renewable energy capacity will increase by over 90% compared to 2023 levels. In response to this trend, nations are implementing ambitious long-term energy strategies. For instance, South Korea's "11th Basic Plan for Long-term Electricity Supply and Demand (2024–2038)" aims to increase renewable energy capacity from 31.4 GW in 2023 to 125.9 GW by 2038, constituting approximately 47% of the total power generation infrastructure [1].

This fundamental shift in the energy landscape introduces significant technical challenges. The inherent intermittency of renewable energy sources, which fluctuate with weather conditions and time of day, creates a temporal mismatch between energy generation and demand. To ensure grid stability and reliably utilize this variable output, energy storage systems are indispensable [2,3]. Among the various enabling technologies, Power-to-Heat (P2H) has emerged as an effective strategy for sector coupling. P2H converts surplus renewable electricity into thermal energy, which can be stored to alleviate grid congestion and improve the overall efficiency of renewable energy utilization. The viability of P2H is critically dependent on an associated Thermal Energy Storage (TES) technology. TES systems are broadly categorized as sensible, latent, or thermochemical storage. Latent Heat Thermal Energy Storage (LHTES) is particularly attractive due to its high energy storage density and its ability to store and release heat at a nearly constant temperature during the phase transition of a Phase Change Material (PCM).

LHTES systems operate on a simple principle: during the charging process, a high-temperature Heat Transfer Fluid (HTF) melts the solid PCM, storing latent heat [4]. During discharging, the liquid PCM releases this stored heat to a low-temperature HTF, causing it to solidify. However, a major impediment to the widespread application of PCMs is their typically low thermal conductivity, which limits the rates of heat charging and discharging. To overcome this limitation, various thermal enhancement techniques have been investigated, including the dispersion of highly conductive nanoparticles and the use of porous media. Among these, the integration of fins into the heat exchanger is one of the most practical and effective methods due to its simple structure, low cost, high durability, and proven reliability [5,6]. Agyenim et al. demonstrated through extensive experiments that various fin structures can significantly improve heat transfer rates compared to unfinned tubes [7]. Moreover, Esapour et al. numerically confirmed that fin structures accelerate the movement of the phase-change interface and markedly improve the system's response time [8]. In addition to these methods, microencapsulation strategies, the incorporation of nanofillers, and the use of metal foams represent other promising approaches for enhancing the effective thermal conductivity of PCM. Rolka et al. suggested adding 1 wt% graphene nanoparticles to PCM, which increased its thermal conductivity by approximately threefold in the solid stability reduction of about 6–9% [9]. Chen et al. reported that a PCM integrating SA and MF reduced the melting time by 61.42% compared with pure PCM [10]. Anfas Mukram et al. evaluated the performance of microencapsulated PCM embedded in specially fabricated cement bricks. Their results indicated that, relative to conventional bricks, the modified bricks effectively reduced heat ingress and lowered indoor temperatures by 1.2 °C [11].

Despite these advancements, inorganic salt hydrates, which offer attractive properties such as high latent heat capacity, superior thermal conductivity, and lower cost, have faced delayed practical implementation due to inherent material challenges, including supercooling, phase segregation, and corrosiveness [12,13]. To address these limitations, numerous studies have focused on developing effective strategies for mitigating supercooling, stabilizing phase behavior, and preventing corrosion. Zhang et al. found that 4 wt% NaCl reduced the supercooling degree to 1.1 °C, while adding 0.8 wt% nano-Cu particles further lowered it to approximately 0.5 °C [14]. Cui et al. demonstrated that the incorporation of 3 wt% carboxymethyl cellulose (CMC) into nano-CU/SAT composite phase change materials effectively suppressed phase separation and preserved the latent heat of fusion even after 50 repeated heating-cooling cycles [15]. Moreno et al. investigated the corrosion behavior of eleven salt hydrates on four metallic substrates (copper, aluminum, carbon steel, stainless steel) classifying the PCMs according to their applicability in cooling or heating systems [16]. Barium Hydroxide Octahydrate (BHO), the PCM utilized in the present study, has likewise attracted research interest aimed at resolving issues such as supercooling, phase segregation, and corrosion. Nevertheless, investigations that implement BHO in actual thermal storage modules and rigorously evaluate their performance are still limited. Advancing the practical deployment of BHO in thermal energy storage applications therefore requires precise experimental determination of its thermo-physical characteristics, accompanied by structured methodologies for system design and performance optimization.

In particular, most finned-tube LHTES studies to date have examined paraffin-based PCMs combined with straight or longitudinal fin geometries, whereas the present work focused on a single circular-finned tube module filled with the inorganic salt hydrate BHO operating in a high-temperature range relevant to P2H applications. Furthermore, previous studies on BHO and similar salt hydrates have mainly reported material-level thermophysical properties and bulk charging/discharging behavior, rather than detailed module-scale heat transfer characteristics. In contrast, the present study experimentally determines phase-resolved overall and outside heat transfer coefficients ($U_o$ and $h_o$) for the solid, latent, and liquid regimes in a practical finned-tube LHTES module, thereby addressing this specific deficiency in the existing literature.

This study investigated the heat transfer characteristics of an LHTES module specifically designed for an inorganic PCM. A shell-and-tube module featuring circular fins was designed and fabricated to maximize the heat transfer area between the HTF and the PCM. BHO, an inorganic salt hydrate with a phase change temperature suitable for medium-temperature applications, was selected as the storage medium. A dedicated experimental apparatus was constructed to perform a quantitative analysis of the module's charging and discharging characteristics under a range of operating conditions. The primary objectives are to determine the phase-dependent heat transfer performance by calculating the overall and outside heat transfer coefficients and to identify the key operational parameters influencing the module's efficiency, thereby providing essential data for the design and optimization of LHTES systems for P2H and other industrial applications.

## 2. Materials and Methods

### *2.1. LHTES Module and Phase Change Material*

The LHTES module for this study was designed as a shell-and-tube heat exchanger with a target thermal energy storage capacity of approximately 1 kWh, based on established heat transfer correlations and prior research [17]. The module consists of a single heat transfer tube with external circular fins, all enclosed within an outer shell. A detailed schematic and a photograph of the manufactured module are shown in Figure 1, and its specifications are summarized in Table 1.

The central heat transfer tube, made of SUS304 stainless steel, had an outer diameter of 48.6 mm, an inner diameter of 44.0 mm, and a total length of 1450 mm. To enhance heat transfer, 165 circular fins made of SUS316 stainless steel were installed on the tube's outer surface using an interference fit. Each fin had a height of 20.0 mm, and the fins were spaced with a pitch of 8.46 mm. This finned-tube assembly is housed within a shell with an outer diameter of 101.6 mm. The effective internal volume within the shell available for the PCM is 0.0076 $m^3$, which was filled with 13.17 kg of the selected PCM.

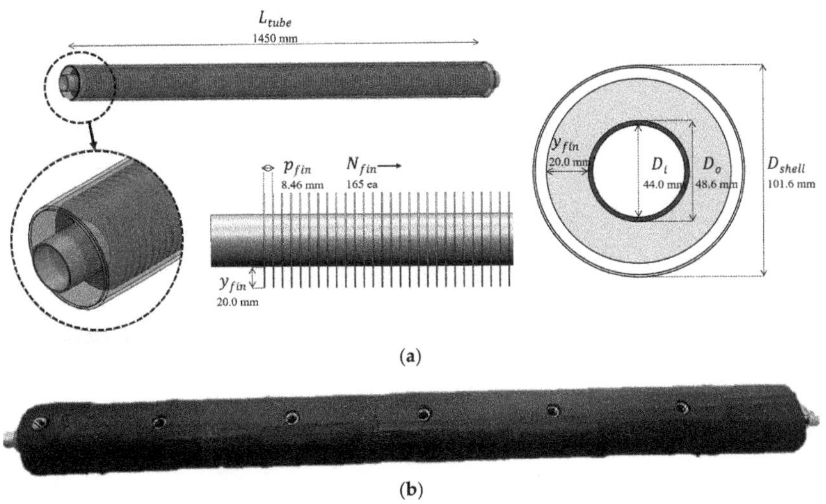


**Figure 1.** Latent heat thermal storage module with PCM mass of 13.17 kg: (**a**) Design drawing including a cross-sectional view and dimensions, and (**b**) photograph of the manufactured module.

**Table 1.** Specifications of latent heat thermal storage module.

| Item | | Specification | |
|---|---|---|---|
| Tube side | Tube material | SUS304 | [-] |
| | Fin material | SUS316 | [-] |
| | Outer diameter, $D_o$ | 48.6 | [mm] |
| | Inner diameter, $D_i$ | 44.0 | [mm] |
| | Tube thickness, $\delta_{tube}$ | 2.3 | [mm] |
| | Tube length, $L_{tube}$ | 1450 | [mm] |
| | Number of tubes, $N_{tube}$ | 1 | [ea] |
| | Fin height, $y_{fin}$ | 20.0 | [mm] |
| | Fin pitch, $p_{fin}$ | 8.46 | [mm] |
| | Number of fins, $N_{fin}$ | 165 | [-] |
| | Heat transfer area, $A_o$ | 0.2214 | [m$^2$] |
| Shell side | Shell diameter, $D_{shell}$ | 101.6 | [mm] |
| | External volume, $V_{ext}$ | 0.0118 | [m$^3$] |
| | Effective internal volume, $V_{int}$ | 0.0076 | [m$^3$] |

The PCM selected in this study was barium hydroxide BHO, an inorganic salt hydrate. The thermophysical properties of BHO and the phase-change regimes illustrated in the above schematic are summarized in Table 2, indicating that BHO is suitable for high-temperature thermal energy storage applications. The temperature–stored-energy (enthalpy) behavior of the PCM during heat charging and discharging can be described as follows: starting from the solid state, sensible heating increases the stored energy approximately in proportion to temperature. When the phase-change temperature is reached, a nearly isothermal latent region forms around 78 °C. In this regime, melting (during charging) or freezing (during discharging) proceeds while a large amount of energy is absorbed or released at almost constant temperature. Once the phase change is completed, the PCM enters the liquid sensible region, where the stored energy again increases (or decreases) almost linearly with temperature. With a phase-change temperature of about 78 °C and a high latent heat of fusion in the range of 233–332 kJ/kg, BHO provides a substantially higher energy storage density than many organic PCMs [18]. BHO was selected for this study due to its high density, making it suitable for high-density thermal storage applications. Furthermore, its specific heat capacities in both solid and liquid states contribute to its overall effective sensible heat storage.

**Table 2.** Thermophysical Properties of Barium Hydroxide Octahydrate (BHO) * [18,19].

| Property | Value | |
|---|---|---|
| Phase change temperature | ≈78 | [°C] |
| Latent heat capacity | 233~332 | [kJ/kg] |
| Solid specific heat capacity | 1.14~2.11 (for 25~78.5 °C) | [kJ/kg·°C] |
| Liquid specific heat capacity | 2.56~2.61 (for 78.5~150 °C) | [kJ/kg·°C] |
| Density | 2.18 | [kg/L] |

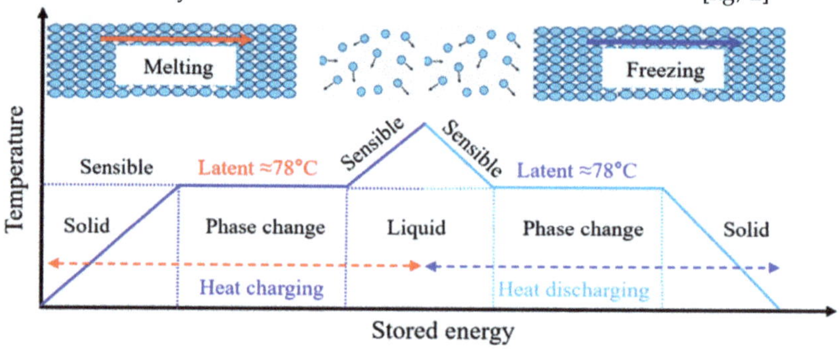


* Note: BHO properties are taken from Refs. [18,19] and were not measured in this study.

## 2.2. Experimental Apparatus and Procedure

To quantitatively evaluate the thermal performance of the LHTES module, a dedicated experimental apparatus was constructed, as shown schematically in Figure 2. The setup was designed to precisely control and monitor the charging and discharging processes. The HTF (water) temperature was regulated using high- and low-temperature constant-temperature baths connected via a plate heat exchanger. An inverter pump and a volumetric flow meter (accuracy: ±0.03%) were used to control and measure the HTF flow rate circulating through the module.

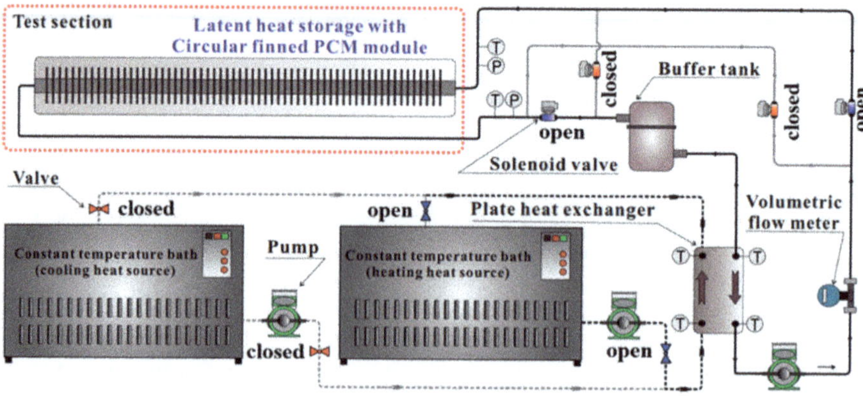


**Figure 2.** Schematic diagram of experimental setup.

To capture the thermal behavior and phase change dynamics of the PCM, T-type thermocouples (±0.1 °C) were strategically placed within the module. The HTF inlet and outlet temperatures were measured to determine the heat transfer rate. Inside the shell, thermocouples were installed at five equidistant points along the module's length. At each point, sensors were positioned at the tip of a fin and in the space between fins to monitor the temperature distribution and track the solid–liquid interface during phase change. All sensor signals were collected and recorded in real-time by a data acquisition system for subsequent analysis.

The experiments were conducted under various operating conditions, as detailed in Table 3, to simulate a range of real-world scenarios. For all tests, the charging process was performed with a fixed HTF inlet temperature of 90 °C and continued until the average PCM temperature reached 85 °C. The discharging process was investigated by varying both the HTF inlet temperature (20, 30, 40, 50 °C) and the HTF flow rate (10, 15, 20, 25 L/min). The discharging tests were terminated when the average PCM temperature dropped to 45 °C, except for the case with a 50 °C HTF inlet, where the termination temperature was set to 55 °C.

**Table 3.** Experimental Conditions for Charging and Discharging Tests.

| HTF Inlet Temperature, $T_{HTF,i}$ | | HTF Flow Rate, $\dot{m}_{HTF}$ |
|---|---|---|
| **Charging** | **Discharging** | |
| **[°C]** | **[°C]** | **[L/min]** |
| 90 | 20<br>30<br>40<br>50 | 10<br>15<br>20<br>25 |

*2.3. Procedure Data Reduction and Analysis*

The experimental data were processed using a set of established thermodynamic and heat transfer equations to quantify the module's performance. The instantaneous heat transfer rates for charging ($\dot{Q}_c$) and discharging ($\dot{Q}_d$) were calculated from the HTF's mass flow rate ($\dot{m}_{HTF}$), specific heat ($c_{p,HTF}$), and the temperature difference between the inlet ($T_{HTF,i}$) and outlet ($T_{HTF,o}$), as shown in Equation (1). To address measurement accuracy and uncertainty, the HTF and PCM temperatures were measured using thermocouples calibrated against a reference sensor prior to the experiments, yielding an accuracy of approximately ±0.1 °C for each sensor. Since the data reduction in this study is based on the HTF-side heat transfer rate, the uncertainty of the HTF temperature difference was evaluated using the concept of uncertainty propagation. For a representative operating condition with an HTF inlet temperature of 20 °C, the resulting relative uncertainty of the HTF-side heat quantity was found to be on the order of 10.52% on average. In the late part of discharging, during the solid phase, the HTF inlet–outlet temperature difference becomes very small, so including this region in the evaluation would further increase the relative uncertainty. However, in the liquid and latent regimes, the relative uncertainty remains on the order of a few percent, which we considered sufficient for discussing the phase-dependent heat transfer characteristics.

$$\dot{Q}_c = \dot{m}_{HTF} c_{p,HTF} (T_{HTF,i} - T_{HTF,o}) \text{ or } \dot{Q}_d = \dot{m}_{HTF} c_{p,HTF} (T_{HTF,i} - T_{HTF,o}) \quad (1)$$

The total accumulated heat stored ($Q_{c,acc}$) or released ($Q_{d,acc}$) over a process was determined by integrating the instantaneous heat transfer rate from the start time ($t_s$) to the final time ($t_f$), as given by Equation (2).

$$Q_{c,acc} = \int_{t_s}^{t_f} \dot{Q}_c dt \text{ or } Q_{d,acc} = \int_{t_s}^{t_f} \dot{Q}_d dt \tag{2}$$

The thermal utilization efficiency ($\varepsilon_{eff}$), representing the ratio of total energy discharged to total energy charged, was calculated using Equation (3).

$$\varepsilon_{eff} = \frac{Q_{d,acc}}{Q_{c,acc}} \times 100(\%) \tag{3}$$

For practical comparison, the total accumulated energy was converted to an hourly energy rate in kilowatt-hours (kWh) using Equation (4).

$$Q_{hr} = \frac{1}{3600} \int_{t_s}^{t_f} \dot{Q} dt \tag{4}$$

The overall heat transfer coefficient based on the outer surface area ($U_o$) was calculated using the Log Mean Temperature Difference (LMTD) method, as shown in Equations (5) and (6), where $A_o$ is the total outer heat transfer area and $T_{PCM}$ is the average PCM temperature.

$$U_o = \frac{\dot{Q}_d}{A_o \Delta T_{lm}} \tag{5}$$

$$\Delta T_{lm} = \frac{(T_{PCM} - T_{HTF,o}) - (T_{PCM} - T_{HTF,i})}{\ln\left(\frac{T_{PCM} - T_{HTF,o}}{T_{PCM} - T_{HTF,i}}\right)} \tag{6}$$

The outside heat transfer coefficient ($h_o$) was determined from the total thermal resistance ($1/U_o$) by subtracting the resistances due to internal convection and tube wall conduction, as expressed in Equation (7), and should therefore be interpreted as an effective, lumped outside coefficient that includes the effects of fin efficiency and fin–tube contact resistance. Here, $A_i$ and $A_o$ (0.2214 $m^2$) are the inner and outer surface areas, $h_i$ is the inside heat transfer coefficient, and $k_{tube}$ is the thermal conductivity of the tube material.

$$h_o = \left[\frac{1}{U_o} - \frac{A_o}{A_i h_i} - \frac{A_o \ln\left(\frac{D_o}{D_i}\right)}{2\pi k_{tube} L}\right]^{-1} \tag{7}$$

The inside heat transfer coefficient ($h_i$) was calculated using the Nusselt number ($Nu_{HTF}$), as shown in Equation (8).

$$h_i = \frac{Nu_{HTF} k_{HTF}}{D_i} \tag{8}$$

For the turbulent flow regime within the tube, the Nusselt number was calculated using the Petukhov correlation (Equation (9)), which is highly regarded for its accuracy.

$$Nu_{HTF} = \frac{(f/8) Re_{HTF} Pr_{HTF}}{1.07 + 12.7 (f/8)^{1/2} \left(Pr_{HTF}^{2/3} - 1\right)} \tag{9}$$

The friction factor (f) required for the Petukhov correlation was calculated using the Gnielinski correlation for fully developed turbulent flow in smooth tubes, given by Equation (10) [20].

$$f = (0.790 \ln Re_{HTF} - 1.64)^{-2} \text{ for } 3000 \le Re_{HTF} \le 5 \times 10^6 \tag{10}$$

# 3. Results

## 3.1. Transient Thermal Dynamics of Charging and Discharging

The dynamic thermal behavior of the LHTES module was characterized through detailed time-series analysis of temperature and energy data. Figure 3 illustrates a representative experimental cycle, specifically for a discharge process with an HTF inlet temperature of 30 °C and a flow rate of 10 L/min, followed by a charging process. The discharge phase lasted 47 min, while the subsequent charging phase required 105 min. This difference in duration reveals an inherent asymmetry in the heat transfer dynamics. Charging is driven by a larger and more stable temperature difference between the 90 °C HTF and the cooler PCM, leading to a relatively consistent heat transfer rate. In contrast, during discharging, the temperature of the PCM continuously decreases, which in turn reduces the LMTD and slows the rate of heat extraction over time.

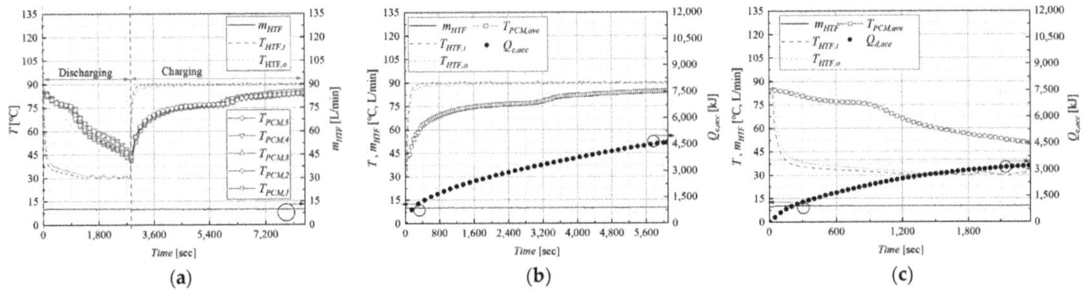


**Figure 3.** Heat transfer characteristics during charging and discharging of the latent heat thermal storage module: (**a**) Temperature and flow rate profiles during heat charging and discharging process (at: $\dot{m}_{HTF} = 10$ L/min, $\dot{T}_{HTF,i} = 90$ °C (Charging), $\dot{T}_{HTF,i} = 30$ °C (Discharging)). (**b**) Heat charging process (at: $\dot{m}_{HTF} = 10$ L/min, $\dot{T}_{HTF,i} = 90$ °C). (**c**) Heat discharging process (at: $\dot{m}_{HTF} = 10$ L/min, $\dot{T}_{HTF,i} = 30$ °C).

During the charging process, the average PCM temperature rises steadily. This rise is characterized by three distinct stages: a rapid increase when the PCM is in its solid state (sensible heating), followed by a much slower temperature rise or plateau near 78 °C as the PCM absorbs latent heat during melting, and finally another period of rapid temperature increase once the PCM is fully in the liquid state. Correspondingly, the accumulated stored energy increases almost linearly with time, indicating a relatively stable heat charging rate.

Conversely, during the discharging process, the average PCM temperature decreases. A similar three-stage pattern is observed: a period of sensible heat release from the liquid PCM, a prolonged period of nearly isothermal heat release at the phase-change temperature (latent heat), and a final stage of sensible heat release from the solid PCM. The temperature difference between the HTF inlet and outlet diminishes as the PCM cools, causing the rate of heat discharge to decrease over time. This results in a non-linear, decelerating curve for the accumulated discharged energy. For this specific case, the total discharged energy was 3330 kJ, compared to a stored energy of 4665 kJ, yielding a thermal utilization efficiency of 71.37%.

## 3.2. Heat Transfer Performance and Efficiency

As shown in Figure 4a, across all experimental conditions, excluding those with a 50 °C HTF inlet temperature, the module demonstrated consistent performance, delivering an average heat discharge of 1.027 kWh. The average thermal utilization efficiency for these conditions was calculated to be 79.25%, and under the same discharging end condition, it reached a maximum of 89.97%. This value represents a practical, system-level efficiency that accounts not only for the thermodynamic processes but also for real-world factors such as ambient heat loss from the module's exterior and piping.

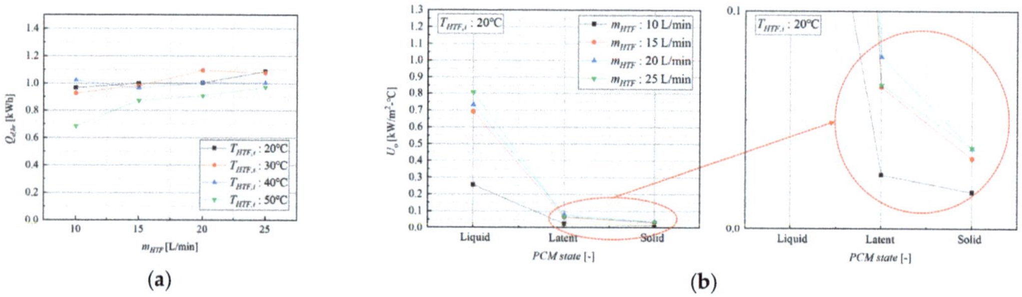

**Figure 4.** Heat transfer rate and overall heat transfer coefficient of a latent heat storage module: (**a**) Heat transfer rate (discharging case), and (**b**) overall heat transfer coefficient (at: $\dot{T}_{HTF,i}$ = 20 °C).

A notable deviation was observed for tests conducted with an HTF inlet temperature of 50 °C, where the average discharged energy was significantly lower at 0.859 kWh. This reduction is directly attributable to the experimental termination condition for this specific case, which was set at a PCM temperature of 55 °C instead of 45 °C. This higher cut-off temperature meant that a substantial portion of the PCM's latent heat, and some of its sensible heat in the solid phase, was not extracted before the test concluded. This finding underscores that the operational cut-off temperature is not merely a procedural detail but a critical control parameter that directly dictates the achievable energy recovery and round-trip efficiency of the system. Setting this temperature too high strands a significant portion of the stored energy within the module, while setting it too low may be impractical if the application requires a certain minimum output temperature.

A central finding of this study is the profound dependence of the overall heat transfer coefficient ($U_o$) on the physical state of the PCM. As shown in Figure 4b and summarized in Table 4, the heat transfer performance varies dramatically across the three phases. The highest $U_o$ values, ranging from 0.257 to 0.863 kW/m$^2$·°C, were observed when the PCM was in the liquid state. In the latent phase, where both solid and liquid coexist, the $U_o$ values were an order of magnitude lower, ranging from 0.025 to 0.072 kW/m$^2$·°C. The lowest performance was recorded in the solid state, with $U_o$ values between 0.015 and 0.044 kW/m$^2$·°C.

This significant variation is governed by the dominant heat transfer mechanisms in each phase. In the liquid phase, natural convection currents develop within the molten PCM, creating fluid motion that significantly enhances heat transfer from the finned surface, resulting in a high $U_o$. During solidification, a layer of solid PCM begins to form on the cooler tube and fin surfaces. Since heat transfer through this solid layer is dominated by pure conduction, which is limited by the PCM's low thermal conductivity, it introduces a substantial thermal resistance. As this solid layer thickens, the overall heat transfer coefficient progressively decreases. Once the PCM is entirely solid, conduction becomes the sole mechanism for heat transfer on the shell side, leading to the lowest observed $U_o$ values. This phase-dependent behavior is a critical characteristic of LHTES systems; it implies that performance is not static but changes dynamically throughout an operational cycle. Consequently, traditional heat exchanger design methods that assume a constant U-value are inadequate for LHTES modules and can lead to significant errors in performance prediction and system sizing. A dynamic or phase-averaged approach is essential for accurate engineering design.

**Table 4.** Heat transfer rate and overall heat transfer coefficient during discharge process.

| $\dot{m}_{HTF}$ | $T_{HTF,i}$ | $Q_{c,hr}$ | $\varepsilon_{ff}$ | $U_{o,liquid}$ Liquid | $U_{o,latent}$ Latent | $U_{o,solid}$ Solid |
|---|---|---|---|---|---|---|
| L/min | °C | kWh | % | kW/m²-°C | kW/m²-°C | kW/m²-°C |
| 10 | 20 | 0.966 | 74.54 | 0.285 | 0.034 | 0.027 |
| 10 | 30 | 0.925 | 71.37 | 0.257 | 0.025 | 0.017 |
| 10 | 40 | 1.023 | 78.90 | 0.320 | 0.027 | 0.015 |
| 10 | 50 | 0.686 | 52.93 | 0.409 | 0.031 | 0.019 |
| 10 | 20 | 0.999 | 77.08 | 0.395 | 0.030 | 0.018 |
| 15 | 30 | 0.983 | 75.85 | 0.693 | 0.065 | 0.032 |
| 15 | 40 | 1.166 | 89.97 | 0.763 | 0.067 | 0.033 |
| 15 | 50 | 0.872 | 67.26 | 0.603 | 0.060 | 0.037 |
| 20 | 20 | 0.992 | 76.52 | 0.863 | 0.063 | 0.038 |
| 20 | 30 | 1.095 | 84.47 | 0.795 | 0.072 | 0.034 |
| 20 | 40 | 1.006 | 77.62 | 0.395 | 0.062 | 0.037 |
| 20 | 50 | 0.909 | 70.12 | 0.675 | 0.070 | 0.040 |
| 25 | 20 | 1.087 | 83.87 | 0.670 | 0.062 | 0.044 |
| 25 | 30 | 1.079 | 83.26 | 0.808 | 0.066 | 0.037 |
| 25 | 40 | 1.006 | 77.59 | 0.489 | 0.057 | 0.044 |
| 25 | 50 | 0.971 | 74.92 | 0.803 | 0.067 | 0.041 |

### *3.3. Analysis of the Outside Heat Transfer Coefficient*

To isolate the performance on the PCM side, the outside heat transfer coefficient ($h_o$) was calculated. An analysis of the heat capacity distribution during the discharge process from 85 °C to 45 °C showed that the latent phase accounted for the majority of the heat transfer (72.94%), followed by the solid phase (22.62%), while the liquid phase contribution was minimal (4.44%). Therefore, the analysis of $h_o$ focused on the dominant latent and solid phases.

As shown in Figure 5, $h_o$ varied with the HTF Reynolds number ($Re_{HTF}$). In the latent phase, ranged from 0.033 to 0.162 kW/m$^2$·°C. For all discharging inlet temperatures considered (20–50 °C), a similar non-monotonic trend was observed: $h_o$ increased with $Re_{HTF}$ in the low-Re region, reached a local maximum in the range $Re_{HTF} \approx 1.2 \times 10^4$–$1.4 \times 10^4$, and then tended to level off or slightly decrease at higher $Re_{HTF}$. This behavior reveals a critical trade-off. Initially, as the HTF flow rate and Reynolds number increase, the internal convective heat transfer ($h_i$) improves, reducing the dominant thermal resistance on the tube side and allowing more heat to be extracted, which increases the calculated $h_o$. However, beyond the optimal point, the higher heat flux accelerates the formation of an insulating layer of solid PCM on the finned surface. The conductive thermal resistance of this growing solid layer eventually becomes the dominant limiting factor, causing the effective outside heat transfer performance to decline. This non-monotonic behavior indicates that simply maximizing the HTF flow rate is not an optimal strategy for enhancing the discharge rate; an optimal flow rate exists that balances internal convection with external solidification dynamics.

In the solid phase, $h_o$ values ranged from 0.018 to 0.064 kW/m$^2$·°C, which is, on average, 56.4% lower than in the latent phase. This confirms that once the PCM is fully solidified, the system's performance is severely limited by the poor thermal conduction through the solid PCM layer.

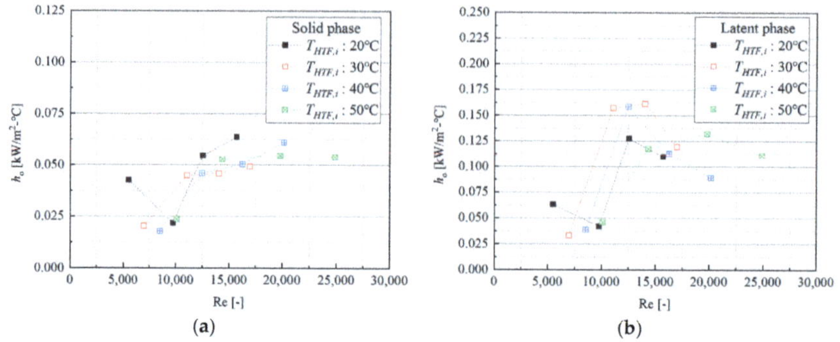


**Figure 5.** Outside heat transfer coefficient of a latent heat storage module: (**a**) latent phase, and (**b**) solid phase.

### *3.4. Statistical Analysis of Influencing Parameters*

In this study, response surface methodology (RSM) was applied using Minitab (Release 14.12.0) to analyze key response variables obtained from the discharging experiments, which characterize the thermal performance of the module, with the HTF flow rate and inlet temperature treated as the main factors. In particular, the overall heat transfer coefficient $U_o$ for each of the three PCM phases (solid, latent, and liquid) was taken as a primary response. Main effects plots and Pareto charts were generated to assess the influence of HTF flow rate and HTF inlet temperature on the overall heat transfer coefficient ($U_o$) of each of the three PCM phases.

As shown in Figure 6, the HTF flow rate had a large, positive, and statistically significant effect on $U_o$ for all phases. This is visually represented by the steep upward slope in the main effects plots and the large bars for flow rate in the Pareto charts, which far exceeded the threshold for statistical significance. This is because increasing the flow rate enhances turbulence inside the tube, reducing the internal convective thermal resistance and thereby improving the overall heat transfer. In contrast, the HTF inlet temperature was found to have a minimal and statistically insignificant effect on $U_o$ across all phases. This is evident from the nearly flat lines for temperature in the main effects plots and the very small bars in the Pareto charts. This finding is further supported by the contour plots in Figure 7, where the contour lines for constant $U_o$ are nearly vertical, indicating strong dependence on the x-axis (flow rate) and weak dependence on the y-axis (temperature). This robust statistical confirmation has a critical implication for system design and control. It is far more effective to modulate the HTF flow rate (i.e., pumping power) to control the charge and discharge rates than it is to alter the operating temperature of the thermal source or sink. This is particularly relevant for applications such as waste heat recovery, where the source temperature may be fixed, as it demonstrates that performance can still be effectively controlled via the HTF flow.

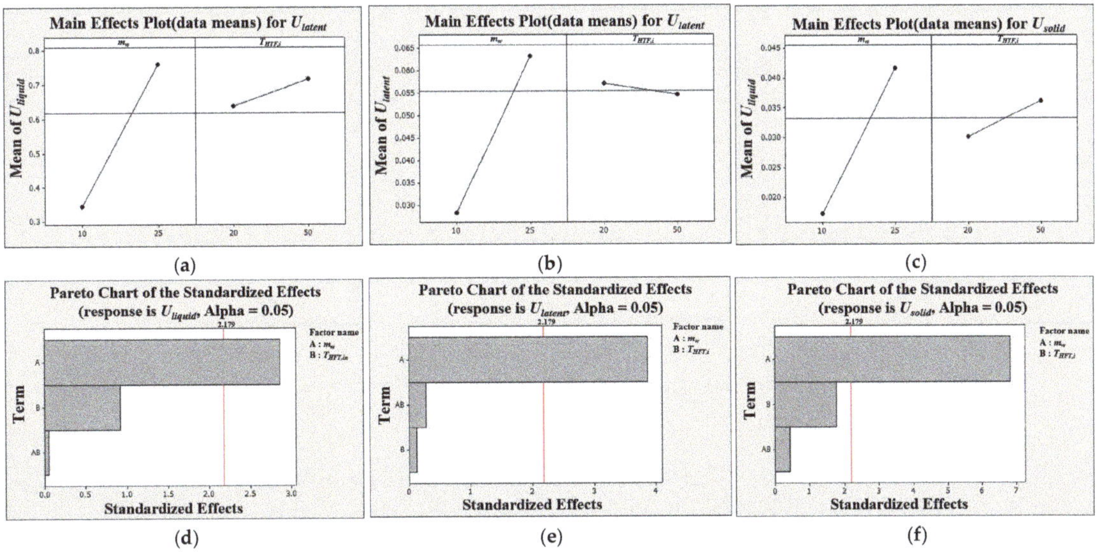


**Figure 6.** Main effects and pareto charts of overall heat transfer coefficient for different PCM states: (**a**) main effects of $U_{liquid}$, (**b**) main effects of $U_{latent}$, (**c**) main effects of $U_{solid}$, (**d**) Pareto chart of $U_{liquid}$, (**e**) Pareto chart of $U_{latent}$, and (**f**) Pareto chart of $U_{soild}$.

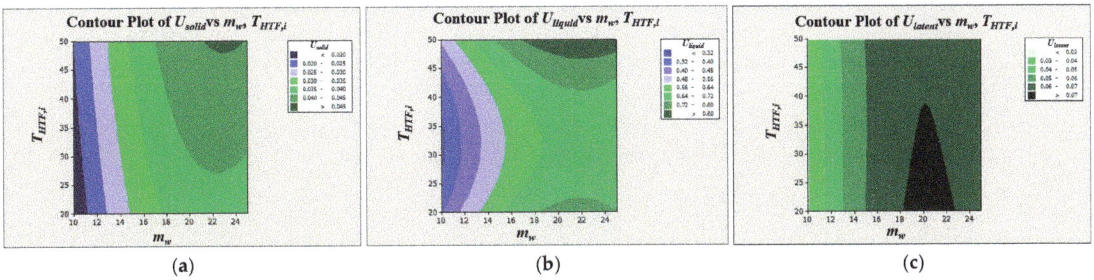


**Figure 7.** Contour plots of overall heat transfer coefficient according to PCM phase state: (**a**) $U_{solid}$, (**b**) $U_{liquid}$, and (**c**) $U_{latent}$.

## 4. Conclusions

In this study, an LHTES module using circular-finned tubes and an inorganic PCM (BHO) was fabricated and experimentally tested to analyze its phase-dependent heat transfer characteristics. Based on the experimental results and statistical analysis, the following conclusions were drawn:

(1) The LHTES module demonstrated a reliable average heat discharge capacity of 1.027 kWh and a thermal utilization efficiency of 79.25% under the tested conditions, confirming the viability of the design.

(2) The discharge termination temperature was identified as a critical operational parameter that directly influences the total amount of energy recovered and the overall system efficiency. Incomplete phase change due to a high cut-off temperature can lead to significant reductions in performance.

(3) Statistical analysis confirmed that the HTF flow rate is the dominant factor influencing the overall heat transfer coefficient across all PCM phases (liquid, latent, and solid). In contrast, the effect of the HTF inlet temperature was found to be statistically insignificant.

(4) The overall heat transfer coefficient is highly dependent on the PCM's phase, following the order: Liquid > Latent > Solid. The coefficient in the liquid state (up to 0.863 kW/m$^2$·°C) can be more than an order of magnitude greater than in the solid state (as low as 0.015 kW/m$^2$·°C), highlighting the necessity of using phase-aware models for system design.

Analysis of the outside heat transfer coefficient revealed that the latent and solid phases dominate the discharge process. The coefficient in the solid phase was, on average, 56.4% lower than in the latent phase, indicating that conduction through the solid PCM is the primary performance bottleneck. These two phases should be the priority for design optimization.

The experimental data and derived heat transfer coefficients obtained in this work provide a crucial foundation for the design, modeling, and operational optimization of LHTES modules. These findings are particularly valuable for integrating such storage systems into P2H applications to support the transition to renewable energy. Future studies will extend this work to investigate the heat transfer characteristics of modules connected in series and parallel, as well as the long-term cycling stability and safety of BHO under practical operating conditions.

**Author Contributions:** Conceptualization, S.J.O. and J.H.P.; methodology, J.-W.K.; software and validation, T.H.S., J.-H.L. and J.H.P.; formal analysis, J.-W.K.; investigation, J.-W.K.; resources, S.J.O.; data curation, T.H.S.; writing—original draft preparation, J.-W.K.; writing—review and editing, S.J.O.; visualization, J.-W.K.; supervision, J.H.P.; project administration, S.J.O.; funding acquisition, S.J.O. All authors have read and agreed to the published version of the manuscript.

**Funding:** This research was supported by the Korean Institute of Energy Technology Evaluation and Planning (KETEP), funded by the Ministry of Trade, Industry, and Energy (MOTIE) (grant No. 20226210100050).

**Data Availability Statement:** The original contributions presented in this study are included in the article. Further inquiries can be directed to the corresponding author.

**Conflicts of Interest:** The authors declare no conflicts of interest.

## References

1. Ministry of Trade, Industry and Energy. *The 11th Basic Plan for Long-Term Electricity Supply and Demand (2024–2038)*; Public Notice No. 2025-169; MOTIE: Sejong, Republic of Korea, 2025; pp. 1–82.
2. Wang, W.; Xie, X. Flexible Heat and Power Load Control of Heating Units Based on Energy Demand-Supply Balance. *SSRN Electron. J.* **2024**, *4940421*, 1–25. [CrossRef]
3. Kauko, H.; Braekken, A.; Askeland, M. Flexibility through Power-to-Heat in Local Integrated Energy Systems with Renewable Electricity Generation and Seasonal Thermal Energy Storage. *Energy* **2024**, *313*, 134017. [CrossRef]
4. Farid, M.M.; Khudhair, A.M.; Razack, S.A.K.; Al-Hallaj, S. A Review on Phase Change Energy Storage: Materials and Applications. *Energy Convers. Manag.* **2004**, *45*, 1597–1615. [CrossRef]
5. Sharma, A.; Tyagi, V.V.; Chen, C.R.; Buddhi, D. Review on Thermal Energy Storage with Phase Change Materials and Applications. *Renew. Sustain. Energy Rev.* **2009**, *13*, 318–345. [CrossRef]
6. Jegadheeswaran, S.; Pohekar, S.D. Performance Enhancement in Latent Heat Thermal Storage System: A Review. *Renew. Sustain. Energy Rev.* **2009**, *13*, 2225–2244. [CrossRef]
7. Agyenim, F.; Hewitt, N.; Eames, P.; Smyth, M. A Review of Materials, Heat Transfer and Phase Change Problem Formulation for Latent Heat Thermal Energy Storage Systems (LHTESS). *Renew. Sustain. Energy Rev.* **2010**, *14*, 615–628. [CrossRef]
8. Chueh, C.-C.; Hung, S.-C. Numerical Investigation of Melting of a Phase-Change Material in H-type Shell Tubes. *Appl. Therm. Eng.* **2024**, *236*, 121470. [CrossRef]
9. Rolka, P.; Przybylinski, T.; Kwidzinski, R.; Lackowski, M. Investigation of low-temperature phase change material (PCM) with nano-additives improving thermal conductivity for better thermal response of thermal energy storage. *Sustain. Energy Technol. Assess.* **2024**, *66*, 103821. [CrossRef]

10. Chen, T.; Liu, S.; Khan, S.-Y.; Shen, Y.; Zhang, S.; Wang, Y.; Kumar, M.; Li, Y.; Li, X. Investigation and optimal design of partially encapsulated metal foam in a latent heat storage unit for buildings. *J. Energy Storage* **2024**, *84*, 110979. [CrossRef]
11. Mukram, T.-A.; Daniel, J. Performance evaluation of a novel cement brick filled with micro-PCM used as a thermal energy storage system in building walls. *J. Energy Storage* **2024**, *77*, 109910. [CrossRef]
12. Purohit, B.K.; Sistla, V.S. Inorganic Salt Hydrate for Thermal Energy Storage Application: A Review. *Energy Storage* **2021**, *3*, e212.
13. Kumar, N.; Ness, R.V.; Chavez, R., Jr.; Banerjee, D.; Muley, A.; Stoia, M. Experimental Analysis of Salt Hydrate Latent Heat Thermal Energy Storage System with Porous Aluminum Fabric and Salt Hydrate as Phase Change Material with Enhanced Stability and Supercooling. *J. Energy Resour. Technol.* **2021**, *143*, 041801.
14. Zhang, Z.; Duan, Z.; Chen, D.; Xie, Y.; Cao, X.; Wang, J. Sodium acetate trihydrate-based composite phase change material with enhanced thermal performance for energy storage. *J. Energy Storage* **2021**, *34*, 102186. [CrossRef]
15. Cui, W.; Yuan, Y.; Sun, L.; Cao, X.; Yang, X. Experimental studies on the supercooling and melting/freezing characteristics of nano-copper/sodium acetate trihydrate composite phase change materials. *Renew. Energy.* **2016**, *99*, 1029–1037. [CrossRef]
16. Moreno, P.; Miro, L.; Sole, A.; Barreneche, C.; Sole, C.; Martorell, I.; Cabeza, L.-F. Corrosion of metal and metal alloy containers in contact with phase change materials (PCM) for potential heating and cooling applications. *Appl. Energy.* **2014**, *125*, 238–245. [CrossRef]
17. Collar-Serradilla, V.; Silva, R.; Satchi, C.-S.; Lopez-Pedrajas, D.; Pato-Doldan, B. Multiphysics modeling and economic design of a high-temperature phase-change-material thermal energy storage, for saturated steam generation industrial applications. *Appl. Therm. Eng.* **2025**, *279*, 127678. [CrossRef]
18. Wang, Q.; Wang, J.; Chen, Y.; Zhao, C.Y. Experimental Investigation of Barium Hydroxide Octahydrate as Latent Heat Storage Materials. *Sol. Energy* **2019**, *177*, 99–107. [CrossRef]
19. Kenisarin, M.; Mahkamov, K. Salt Hydrates as Latent Heat Storage Materials: Thermophysical Properties and Costs. *Sol. Energy Mater. Sol. Cells* **2016**, *145*, 255–286. [CrossRef]
20. Incropera, F.P.; DeWitt, D.P. *Fundamentals of Heat and Mass Transfer*, 5th ed.; John Wiley & Sons: New York, NY, USA, 2003; pp. 550–553.

energies

MDPI

Article

# Conjugate Study on Thermal–Hydraulic Performance of Topology-Optimized Lattice-Filled Cooling Channel for Thermal Management of Solid-Oxide Fuel Cells

**Kirttayoth Yeranee [1], Yuli Cheng [1] and Yu Rao [1,2,*]**

1 Institute of Turbomachinery, School of Mechanical Engineering, Shanghai Jiao Tong University, Shanghai 200240, China; nifzahunt@sjtu.edu.cn (K.Y.); chengyuli_sjtu@sjtu.edu.cn (Y.C.)
2 Shenzhen Research Institute, Shanghai Jiao Tong University, Shenzhen 518000, China
* Correspondence: yurao@sjtu.edu.cn

**Abstract:** Integrated additional cooling channels offer precise thermal management for solid-oxide fuel cells (SOFCs), mitigating temperature gradients. This research studies the thermal–hydraulic performance of cooling channels integrated between SOFC interconnectors, including a Diamond-type triply periodic minimal surface (TPMS), a conventional topology-optimized structure, and a topology-optimized lattice-filled structure. A conjugate heat transfer analysis is employed to investigate the influences of flow rate within the range of Reynolds numbers from 300 to 5000, and the effects of coolant type, including air and liquid metals, as well as the impacts of structural material. The results demonstrate that the topology-optimized lattice-filled structure, generating high turbulence mixing, achieves superior temperature uniformity, especially at high flow rates, despite having higher thermal resistance and pressure loss than the conventional topology-optimized design. The coolant types show the largest influence on thermal–hydraulic performance, and the use of liquid gallium in the conventional optimized design obtains the best temperature uniformity, yielding differences between the maximum and minimum temperatures of less than 5 K. Moreover, the higher-thermal-conductivity material improves temperature uniformity, even at low flow rates. Overall, the optimized-baffle designs in the conventional topology-optimized model, utilizing high-conductivity coolant and structural materials, could be the most suitable for thermal management of the SOFC.

**Keywords:** solid-oxide fuel cell; temperature uniformity; thermal–hydraulic performance; topology optimization; triply periodic minimal surface

## 1. Introduction

Solid oxide fuel cells (SOFCs) are an advanced technology for the future of sustainable energy, capable of directly converting chemical reactions into electricity with high efficiencies exceeding 70% [1]. Their high operating temperatures, ranging from 873 K to 1273 K, enable the favorable reaction kinetics, the use of non-precious metal catalysts, and the recovery of valuable high-grade waste heat [2]. However, the heat generated from exothermic electrochemical reactions, Joule heating, and resistive losses causes substantial temperature within the cell. These non-uniform thermal gradients induce thermomechanical stresses, leading to performance degradation through mechanisms such as electrode sintering and interconnect delamination [1–3]. Aguiar et al. [4] have suggested that a maximum local temperature gradient of 10 K/cm is critical to mitigate such structural damage, emphasizing that precise temperature control is not merely beneficial but essential for the commercial applicability and durability of SOFC systems.

Effective thermal management is therefore crucial for the SOFCs to minimize temperature gradients in the positive electrode–electrolyte–negative electrode (PEN) and interconnector structures. Enhancing thermal conduction in the PEN materials and improving convective heat transfer in gas channels are the most effective strategies to obtain high temperature uniformity. Different thermal management approaches have been investigated, including optimized air/fuel channels to increase surface area and uniform gas distribution [5,6]; flow arrangements, where adopting a co-flow arrangement generally yields a smaller temperature gradient compared to other configurations [7–9]; heat pipes [10,11], which can significantly lower gradients yet face challenges of heavy stainless-steel envelopes and long startup times; and fuel composition control [12], since endothermic reforming creates severe gradients, while adjusted air-to-fuel ratios help mitigate them.

Conventional thermal management also relies heavily on excess air circulation through cathode channels [13–16]. Experimental and numerical studies have confirmed that regulating air flow rate significantly influences thermal gradients. Wang et al. [17] demonstrated that increasing the airflow from 1.5 m/s to 7.5 m/s reduced the peak temperature gradient by around 17%. A similar increase in fuel velocity also enhances convective heat transfer and improves thermal uniformity. While increasing the air flow rate can moderate temperatures, the low volumetric heat capacity of air consumes significant pumping power for compression, accounting for up to 10% of the net output [18]. Furthermore, excessive air cooling can lower the average stack temperature, weakening electrochemical performance and reducing power density. These inherent limitations have encouraged the investigation of advanced cooling strategies [2,3].

Another innovation in thermal management for SOFCs is the use of high-performance coolants in an additional cooling channel integrated within the SOFC stack, as shown in Figure 1b–d. Promsen et al. [16] proposed a saturated water-cooled SOFC stage. They found that the water-based cooling method reduced temperature differences in planar stacks from 185 K to 26 K. This strategy also reduced the parasitic losses from the air blower, increasing the overall system efficiency. Li et al. [14] proposed a supercritical fuel-cooled SOFC scheme. They observed that this design significantly reduced the temperature gradient from 22.7 K/cm in a conventional air-cooled SOFC to 8.35 K/cm, resulting in a substantial reduction in air usage and thereby lowering parasitic power consumption, improving overall efficiency. Fan et al. [15] revealed that an additional liquid tin cooling channel achieved a highly uniform temperature of 15 K. Later, Hasbi et al. [19] demonstrated that liquid gallium flowing through additional cooling slots of SOFC interconnectors significantly improved temperature uniformity over air cooling, though its high flow rates caused a larger pumping power penalty. These methods decouple the cooling function from the oxidant supply, allowing for a drastic reduction in the excess air ratio and its associated parasitic losses.

Despite these promising results, a critical gap remains in the design and optimization of the additional cooling structures for the SOFCs. Advanced architectures, such as triply periodic minimal surface (TPMS) lattices, offer superior thermal performance and structural integrity in various thermal management systems [20]. However, their use in cooling channels often causes a high-pressure loss, which again demands excessive pumping power [21]. To address this trade-off, researchers have turned to optimization techniques [22,23]. One promising method combines density-based topology optimization with lattice infills [24]. This process first computes an optimal channel layout under defined constraints and objectives. The obtained intermediate densities are then filled with lattice structures. This hybrid approach has been shown to significantly reduce pressure loss while maintaining excellent thermal performance, offering a potential pathway to more efficient thermal management of the SOFCs.

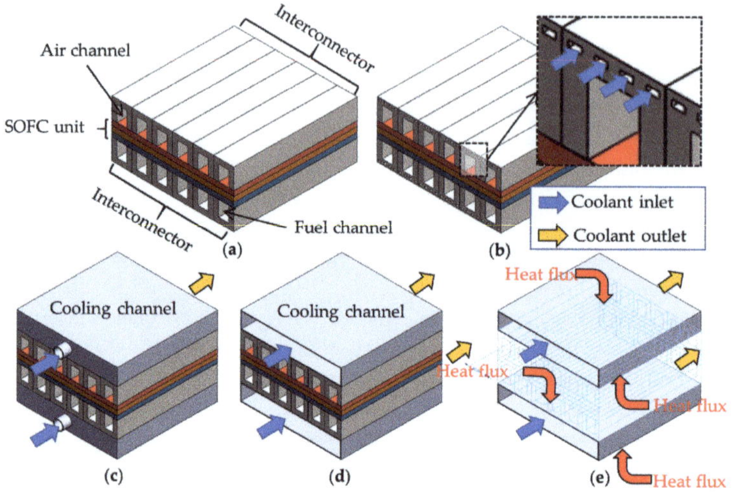


**Figure 1.** Different cooling strategies for SOFC stack: (**a**) excessive air; (**b**) coolant flowing through the interconnectors; (**c**,**d**) coolant passing through an additional cooling channel; and (**e**) concept of simplified boundary conditions in this study.

Hence, this study proposes high-performance cooling channel designs integrated between SOFC interconnectors [14–16], focusing on thermal–hydraulic performance and temperature uniformity of three advanced cooling structures: (1) a Diamond-type TPMS network, known for its high surface-area-to-volume ratio and efficient flow mixing [25]; (2) conventional topology-optimized structures, designed through computational methods to achieve optimal heat transfer performance under given constraints [26,27]; and (3) a hybrid topology-optimized-filled lattice structure, which combines the tailored flow paths of topology optimization with the enhanced heat transfer of an infilled lattice [24].

This investigation employs a simplified but practical approach by modeling cell heat generation as a constant heat flux boundary condition, as demonstrated in Figure 1d, allowing a comparative evaluation of these complex geometries. This methodology, successfully applied in studies on proton-exchange membrane fuel cells (PEMFCs) [28,29], enables the isolation and detailed analysis of the thermal–hydraulic performance of the cooling structures. This study further examines three critical factors that govern thermal management performance in the SOFCs. The analysis evaluates the thermal–hydraulic performance and temperature uniformity of the proposed cooling structures within the Reynolds numbers of 300 to 5000 [19], compares different coolants (air, liquid tin [15], and liquid gallium [19]), and assesses the impacts of commonly used metallic materials [30], including ferritic stainless steel [15], high-thermal conductivity metal [31], and nickel–chromium superalloy [32]. Through this comprehensive analysis, this work aims to serve as a guideline for implementing advanced cooling channels between the SOFC stacks to improve efficiency and durability.

## 2. Methodology

This study employs conjugate heat transfer (CHT) analysis to investigate the thermal–hydraulic performance, evaluating the impact of flow rate, coolant type, and structural material on heat transfer efficacy, pressure drop, and temperature uniformity. A comparative analysis covers three cooling architectures, including the Diamond-TPMS channel, a high-performance cooling structure generated via density-based topology optimization to maximize heat transfer, and the topology-optimized lattice-filled geometry, where the optimized density fields ($\gamma$) are interpreted into a manufacturable circular pin-fin array. This lattice structure is selected for its compatibility with traditional manufacturing and integration within narrow SOFC stack interconnects.

It should be noted that this study employs 2D topology optimization for computational efficiency, justified by the narrow channel gap. While a full 3D optimization generates complex geometries requiring advanced additive manufacturing, the 2D method produces extrudable designs, which are suitable for conventional processes [33–36]. This simplification is also validated by the comparable cooling performance achieved between the 2D-derived and full 3D models [37]. The topology optimization procedure is elucidated in the Appendix A, while subsequent sections detail the numerical models, boundary conditions, mesh sensitivity, and validation.

### 2.1. Numerical Models

The numerical models investigated in this study are presented in Figure 2. Figure 2a shows the overall dimensions of the cooling structure, with a characteristic length ($D$) of 20 mm. The channel has a cross-section of 20 mm $\times$ 20 mm, attached by heated plates with a thickness of 5 mm, consistent with the previous study [16]. The length of the cooling structure ($L$) is 120 mm. Figure 2b shows the Diamond TPMS structure, which is generated from a mathematical equation [21]. This structure has a wall thickness of 1 mm and a unit cell size of 20 mm. The conventional topology-optimized design, shown in Figure 2c, is obtained by extracting the density field ($\gamma$) from the density-based topology optimization with an iso-contour threshold of 0.5. In Figure 2d, the optimized layout is interpreted as an array of circular pin-fins, where the pin diameter ($D_p$) varies consistently with the local density field ($0.1 \leq \gamma \leq 0.9$). The $D_p$ = 1 mm corresponds to $\gamma$ = 0.9, increasing consistently to $D_p$ = 2 mm where $\gamma$ = 0.1. A summary of all design parameters is provided in Table 1.

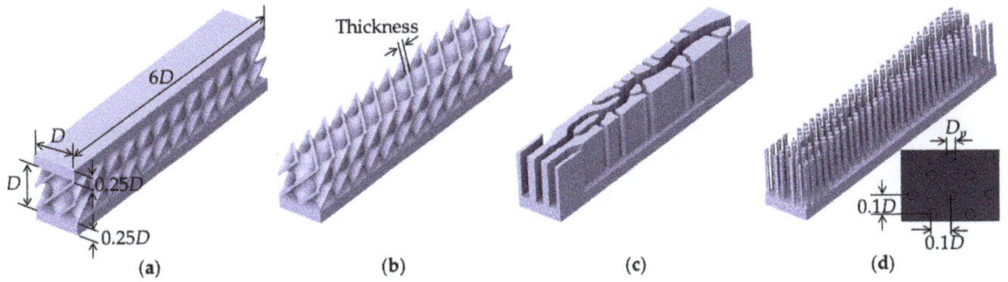


**Figure 2.** Numerical models in this study: (**a**) Dimension of cooling structure; (**b**) Diamond-type TPMS structure; (**c**) Conventional topology-optimized design, and (**d**) Topology-optimized lattice-filled geometry.

**Table 1.** Details of numerical models in this study.

| Models | Details | Surface Area (mm$^2$) | Porosity (%) |
|---|---|---|---|
| Diamond TPMS | The unit cell size of 20 mm is arranged in the channel, and the wall thickness of TPMS is 1 mm | 23,082.2 | 80 |
| Conventional topology-optimized | The final optimized structure is directly extracted from the density field with a threshold value of 0.5 | 17,310.2 | 50 |
| Topology-optimized lattice-filled | Arrays of circular pin-fin structures with a $D_p$ of 2 mm are filled in the density field of 0.1, and the $D_p$ decreases consistently to 0.5 mm in the density field of 0.9. | 28,958.9 | 80 |

## 2.2. Computational Domain and Boundary Conditions

The conjugate heat transfer simulations are conducted using Ansys Fluent v. 2021R1, Ansys Inc., Canonsburg, PA, USA. Figure 3 depicts the computational domain for the cooling channel configuration. Spanwise periodic boundary conditions are applied to the Diamond TPMS model, while the symmetry conditions are used for the topology-optimized and lattice-filled models to minimize computational cost. The bottom and top walls of the solid structure are treated by a uniform heat flux of 3200 $W/m^2$, and this value was obtained based on an operating voltage of 0.75 V by Zhang et al. [38]. The interfaces between the solid and fluid domains are modeled using conjugate boundaries, and the rest surfaces are set as adiabatic.

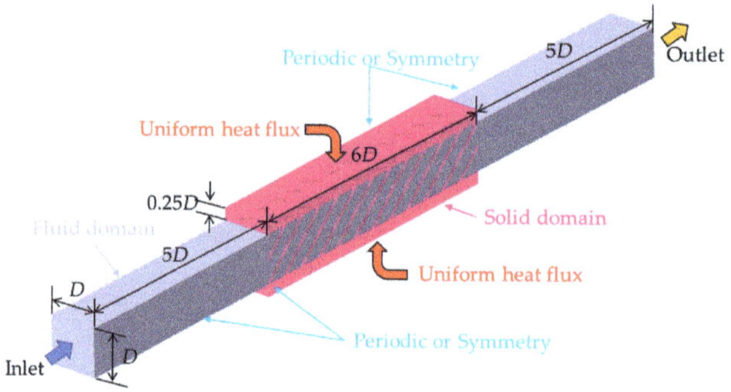


**Figure 3.** Computational domain and boundary conditions.

It is important to note that this study employs a uniform heat flux boundary condition to facilitate a direct cooling performance comparison of each case. In the SOFC, the current density and electrochemical reactions typically result in non-uniform heat generation within the PEN layer. While this simplification does not capture the uneven temperature distributions, it provides a consistent method for evaluating the cooling performance and temperature uniformity of each design. Future work will incorporate coupled electrochemical–thermal models to investigate the interaction between non-uniform stack generation and advanced cooling.

Three coolants, including compressible air, liquid tin, and liquid gallium, are studied. The inlet is defined with a uniform velocity profile and a constant temperature of 1073 K, while the outlet is set to a zero-gauge pressure [15,19]. The influence of the flow rate is investigated for the air-cooled channel at the Reynolds numbers of 300, 1000, and 5000 [19]. For the subsequent studies on coolant type and structural material, a Reynolds number of 300 is maintained. The solid structures are modeled with the properties of ferritic stainless steel, a high-thermal-conductivity metal, and a nickel–chromium superalloy. The corresponding thermophysical properties are detailed in Table 2, and all studied cases are summarized in Table 3.

**Table 2.** Material properties of the coolant and solid structure.

| | Materials | Dnesity ($kg/m^3$) | Specific Heat Capacity (J/kg·K) | Thermal Conductivity (W/m·K) | Viscosity (Pa·s) |
|---|---|---|---|---|---|
| Fluid | Air | Compressible | Linear interpolation | Linear interpolation | Linear interpolation |
| | Liquid tin [15] | 6330 | 240 | 33.8 | 0.00101 |
| | Liquid gallium [19] | 5600 | 374 | 61.67 | 0.00175 |
| Solid | Ferritic stainless steel [15] | 8900 | 444 | 30 | - |
| | High-thermal-conductivity metal [19,31] | 8900 | 450 | 72 | - |
| | Nickel–chromium superalloy [32] | 8330 | 640 | 22.8 | - |

**Table 3.** Studied cases in this research.

| Study Case | Reynolds Number | Coolant | Structure |
|---|---|---|---|
| Coolant flow rate | 300, 1000, and 5000 | Air | Ferritic stainless steel |
| Coolant type | 300 | Air, liquid tin, and liquid gallium | Ferritic stainless steel |
| Structural material | 300 | Air | Ferritic stainless steel, high-thermal-conductivity metal, and nickel–chromium superalloy |

### 2.3. Parameter Definition

The Reynolds number ($Re$) is calculated based on the velocity ($V_{in}$) at the inlet channel, as follows:

$$Re = \frac{\rho V_{in} D_h}{\mu} \tag{1}$$

where $\rho$ and $\mu$ are the density and dynamic viscosity of the coolant, defined at the inlet.

Generally, the hydraulic diameter ($D_h$), applied to highly complex structures, such as porous media or two-fluid heat exchangers, is calculated as $4\psi/S_v$, where $\psi$ is the porosity and $S_v$ is the specific surface area [21]. However, for this study, which focuses on channel-based heat sinks, the $D_h$ is defined using the conventional channel-flow definition [19,25]:

$$D_h = \frac{4A_{in}}{L_p} \tag{2}$$

where $A_{in}$ indicates the cross-sectional area at the inlet of the cooling channel, and $L_p$ is the wetted perimeter. Using this definition, all designs are created with a consistent $D_h$ = 20 mm. Meanwhile, the $D_h$ values calculated using the former porous media definition are provided in the Appendix A.

The Nusselt number ($Nu$), which is generally used to characterize the convective heat transfer in the channel, is calculated as follows:

$$Nu = \frac{hD_h}{k} \tag{3}$$

where $k$ is the thermal conductivity of the cooling material. $h$ is the local heat transfer coefficient, which is defined by

$$h = \frac{q}{(T_w - T_b)} \tag{4}$$

where $q$ is the wall heat flux at the observed area. $T_w$ and $T_b$ are the endwall and bulk temperatures. $T_b$ is evaluated by linearly interpolating between inlet ($T_{in}$) and outlet ($T_{out}$) temperatures.

The heat exchange capability for all cases is evaluated based on the total thermal resistance ($R_{tot}$), defined as follows [33]:

$$R_{tot} = \frac{T_{heat} - T_{in}}{Q} \tag{5}$$

where $T_{heat}$ is the area-averaged temperature on the heated wall. $Q$ is the total heat input.

For the hydraulic performance, the friction factor ($f$) and pumping power ($P_e$) of the flow along the channel are expressed as follows:

$$f = \frac{2\Delta p D_h}{\rho V L} \tag{6}$$

$$P_e = \Delta p V A_{in} \tag{7}$$

where $\Delta p$ is the average pressure difference between the inlet and outlet of the cooling structure.

The performance evaluation criterion (PEC) is calculated to quantify the trade-off between heat transfer enhancement and the associated pressure loss:

$$\text{PEC} = \frac{Nu_{avg}/Nu_0}{(f/f_0)^{1/3}} \tag{8}$$

where $Nu_{avg}$ indicates the area-averaged Nusselt number. The baseline Nusselt number ($Nu_0$) and friction factor ($f_0$) are obtained from the Dittus-Boelter equation and the Blasius correlation, respectively, as follows:

$$Nu_0 = 0.023Re^{0.8}Pr^{0.4} \tag{9}$$

$$f_0 = 0.316Re^{-0.25} \tag{10}$$

where the Prandtl number ($Pr$) for air is assigned a value of 0.7. It should be noted that Equations (9) and (10) are valid for the turbulent regimes ($Re > 4000$).

The temperature uniformity index (TUI) is calculated to quantify the temperature uniformity on the heated surface. The TUI is defined by the following equation:

$$\text{TUI} = 1 - \frac{\sum_{i=1}^{n} |T_i - T_{heat}| A_i}{2T_{heat}A} \tag{11}$$

where $T_i$ is the local temperature, $A_i$ is the local area, and $T_{heat}$ is the area-averaged temperature of the heated wall.

### 2.4. Mesh Independence Check and Numerical Scheme

The computational domain is discretized using poly-hexcore and prism meshes generated in Fluent Meshing. Figure 4 shows the mesh system for the conventional topology-optimized design. The "Body of Influence" is activated over the cooling structure to control high-quality surface mesh. Moreover, to capture complex structures, especially the Diamond TPMS and lattice-filled models, the value of the curvature normal angle and cell per gap is set at 9 degrees and 1. The boundary layers are generated using the smooth-transition offset method with the transition ratio of 0.272. Overall, the solid and fluid domains primarily consist of hexagonal elements, while prism layers are densely generated near all no-slip walls to ensure the $y^+$ value below 1.0.

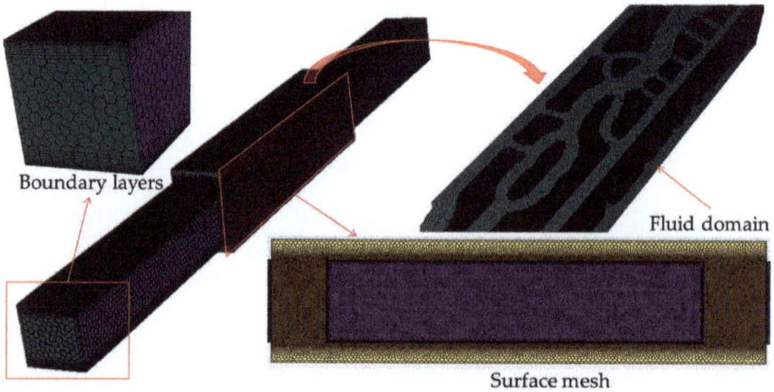


**Figure 4.** Overview of the mesh system in the conventional topology-optimized model.

The realizable $k$-$\varepsilon$ turbulence model is selected for this study based on its demonstrated accuracy in predicting the thermal–hydraulic performance of complex TPMS channel geometries in previous work [25], where it showed excellent agreement with experimental transient liquid crystal thermography (TLC) data at higher Reynolds numbers ($Re$ = 10,000–30,000). While the Reynolds number range in the present study is ranging from $Re$ = 300 to 5000, encompassing both laminar and transitional regimes, the use of turbulence models throughout the entire range is a common approach for comparative analysis of complex geometries, as seen in various air- and liquid-cooled heat sink studies [34–37]. Since this study also focuses on a relative comparison of cooling structures under a consistent numerical setting, the use of the realizable $k$-$\varepsilon$ turbulence model is valid due to its dampening function, which provides a reasonable approximation in transitional flows, enabling an efficient evaluation within the studied Reynolds numbers.

The mesh independence analysis is conducted by comparing the temperature distribution within three mesh systems of $4.21 \times 10^6$ (Low), $4.92 \times 10^6$ (Medium), and $7.21 \times 10^6$ (High) elements. The results in Figure 5 show that all plots are almost identical within the range of 0 mm < $X$ < 100 mm, while small deviations can be observed near the outlet ($X \geq 100$ mm). The area-averaged Nusselt number at the endwall in the low and medium meshes is 3.3% and 0.3%, respectively, deviating from the highest mesh resolution. Therefore, the mesh with $4.92 \times 10^6$ elements is selected for the conventional topology-optimized model, due to a balance between computational cost and accuracy. Similar mesh settings are applied to all other models.

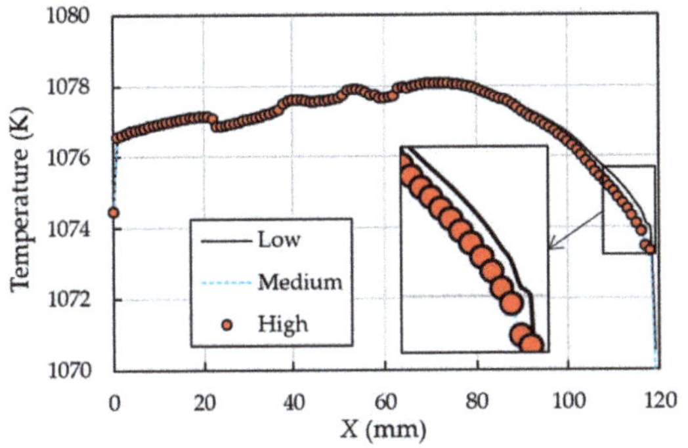


**Figure 5.** Plots of temperature distributions on the endwall with different meshes at $Re$ = 5000.

## 2.5. Validation

To validate the present numerical models, the average Nusselt number correlation established by Xu et al. [25] from their experimental study of a Diamond TPMS structure is employed. Since their experimental data are acquired via the TLC under adiabatic conditions, a corresponding numerical model is constructed without a solid domain, as shown in Figure 6. In this model, a uniform heat flux is applied directly to the bottom endwall to emulate the TLC measurement technique. The model is validated across a range of $Re$ = 10,000, 15,000, and 20,000, with all other boundary conditions maintained as specified in Figure 3.

Figure 7 shows the agreement between the present numerical results and the experimental correlation ($Nu_{avg}/Nu_0 = 135.1Re^{-0.34}t^{0.16}$) established by Xu et al. [25]. The data points from this numerical study closely follow the trend of the empirical correlation within $Re$ = 10,000–20,000. The average deviation in all validated $Re$ is 12.5%, which can be explained by the implemented periodic boundary conditions. By design, this setup excludes the side impingement flow, as found in the experiments, which is a key mechanism for high heat transfer, leading to the observed discrepancy. Therefore, this validation ensures the ability to simulate more complex conjugate heat transfer scenarios accurately.

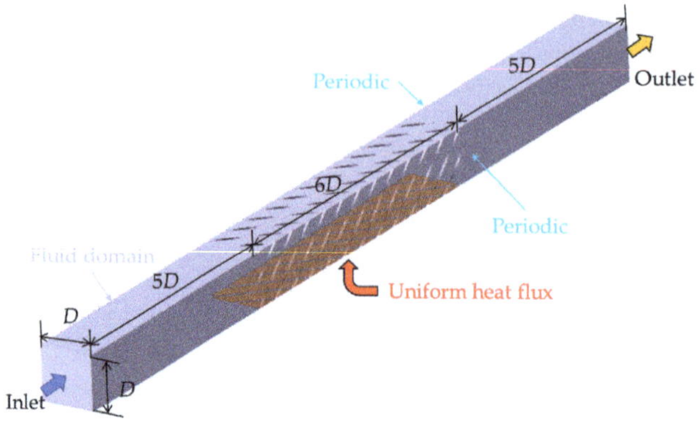


**Figure 6.** Numerical model of the Diamond-type TPMS structure for the validation.

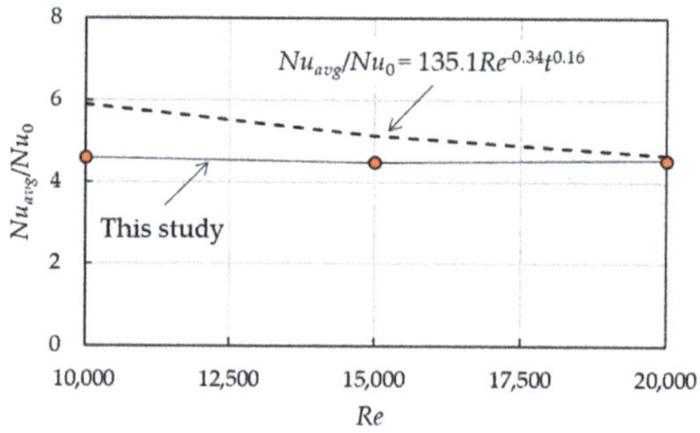


**Figure 7.** Plots of the average Nusselt number ratio between the present study and the correlations from experimental investigation within the Reynolds number of 10,000 to 20,000.

## 3. Results and Discussion

The thermal–hydraulic performance is characterized by plotting heat transfer and pressure loss to illustrate the effects of coolant flow rate, coolant type, and structural material. The flow and temperature distributions are used to explain the variations in each geometry.

### 3.1. Influences of Coolant Flow Rate

In this section, to effectively evaluate the influence of coolant flow rate within the range of $Re$ = 300 to 5000, air and ferritic stainless steel are selected as the working fluid and structural material, respectively. Their thermal properties serve to clearly observe the thermal–hydraulic results.

#### 3.1.1. Thermal–Hydraulic Performance

Figure 8 provides the thermo-hydraulic performance within the studied $Re$ for the Diamond-type TPMS structure (DiamondTPMS), conventional topology-optimized structure (ConTopOpt), and topology-optimized lattice-filled structure (TopOptLattice). In Figure 8a, the area-averaged Nusselt number ($Nu_{avg}$) monotonically increases as $Re$ increases due to higher flow momentum and mixing, resulting in a thinner thermal boundary layer. All models show a competitive and steadily increasing trend at $Re$ = 300 and 1000.

The Diamond TPMS geometry, due to its intense flow accumulation [25], achieves the highest $Nu_{avg}$ at higher $Re$, followed by the topology-optimized lattice-filled model. At $Re$ = 5000, the $Nu_{avg}$ value in the Diamond is 66.1, which is higher than in the conventional topology-optimized and lattice-filled models by 58.6% and 22.4%, respectively.

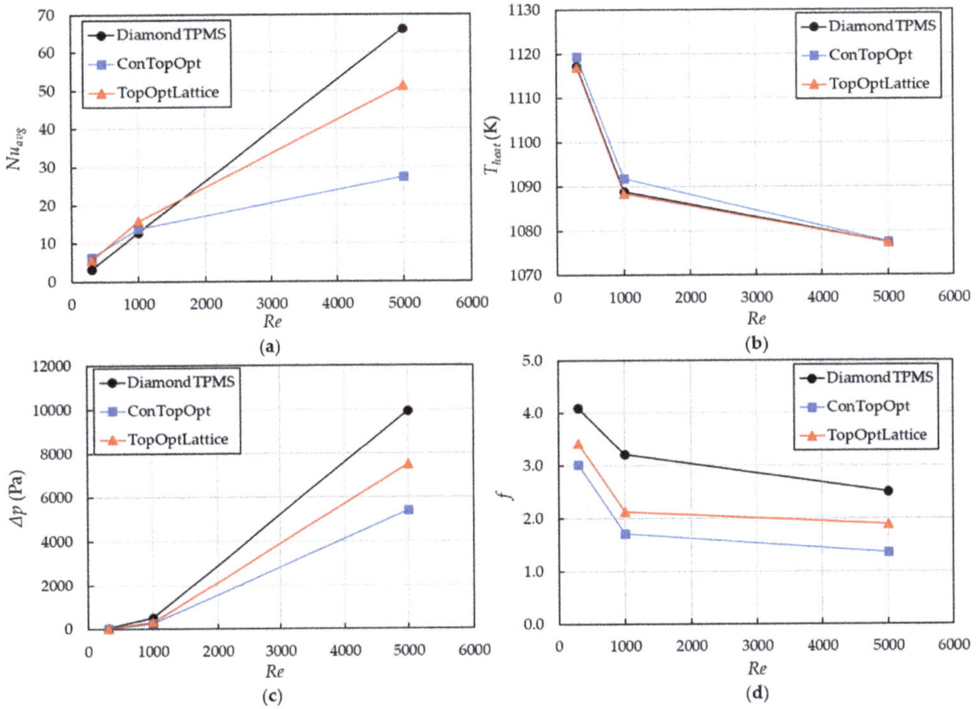


**Figure 8.** Thermal–hydraulic performance of all models within the range of the Reynolds numbers of 300 to 5000: (**a**) area-averaged Nusselt number on the endwall inside the cooling channel; (**b**) area-averaged temperature on the heated wall; (**c**) pressure loss; and (**d**) friction factor.

This enhanced heat removal inside the channel directly reflects the temperature removal from the heated wall in Figure 8b. All models sharply decrease wall temperatures when $Re$ increases from 300 to 1000. At the highest coolant flow rate, the Diamond TPMS model shows the best cooling structure, followed by the topology-optimized lattice-filled model. However, the cooling performance in the Diamond TPMS structure causes the highest pressure loss and friction factor, as shown in Figure 8c,d. A direct consequence of the high pressure is that its topology blocks the flow passing through the cooling channel.

Meanwhile, the conventional topology-optimized and lattice-filled models demonstrate a significant advantage, as they exhibit a substantially lower pressure penalty. Compared to the Diamond TPMS model that blocks the through-flow path, the pressure loss in the conventional topology-optimized and lattice-filled models is 43.2 Pa to 5417 Pa and 48.9 Pa to 7500 Pa, which decreases by 26.2% to 45.6% and 16.4% to 24.6%, respectively, when the $Re$ increases from 300 to 5000. Moreover, the friction factor, particularly in the conventional topology-optimized model, which allows for a smoother flow and reduces strong vortices, exhibits the lowest values within the studied flow rates and decreases as $Re$ increases, indicating a significantly more hydraulically efficient design.

Figure 9 plots the $Nu/Nu_0$, $f/f_0$, and PEC of all geometries at $Re$ = 5000 to evaluate the trade-off between heat transfer enhancement and flow resistance. The Diamond TPMS geometry yields the highest $Nu/Nu_0$ but also incurs the largest $f/f_0$. For all geometries, the PEC values exceed 1.0, confirming a net performance gain. Among them, the conventional topology-optimized model achieves the highest PEC of 5.48, outperforming the Diamond TPMS and lattice-filled models by 22.5% and 11.5%, respectively. This demonstrates that while the Diamond TPMS is most effective at augmenting heat transfer, the conventional topology-optimized design offers a superior overall balance by mitigating the friction penalty.

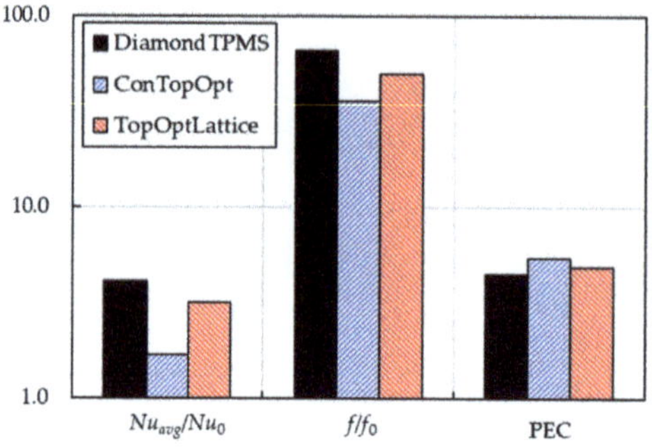


**Figure 9.** Plots of the heat transfer enhancement, friction factor ratio, and the performance evaluation criterion of all models at $Re$ = 5000.

Figure 10 presents the trade-off between thermal performance and energy use by plotting thermal resistance ($R_{tot}$) against pumping power ($P_e$) for all models. The trends are similar to those in Figure 9, where the Diamond TPMS structure consistently requires the highest $P_e$ to achieve a low $R_{tot}$, indicating that this cellular-lattice model is less effective at removing heat for a given pumping input. The conventional topology-optimized and lattice-filled models obtain a lower $R_{tot}$ at a given $P_e$. In this comparison, the topology-optimized lattice-filled model shows the best performance at low $P_e$, outperforming the Diamond TPMS and conventional topology-optimized geometries. Meanwhile, the conventional topology-optimized geometry obtains a similar $R_{tot} \approx 0.55$ K/W but requires a lower $P_e$ compared to the other models. Beyond overall heat transfer efficiency, the temperature uniformity is a critical requirement for SOFC system integrity. The following section evaluates the flow pattern, heat transfer distribution, and temperature uniformity for all designs.

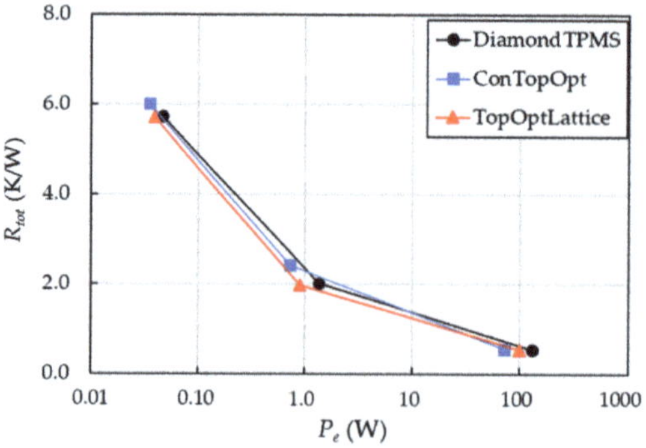


**Figure 10.** Plots of the thermal resistance on the heated walls against pumping power in all models.

### 3.1.2. Flow Characteristics

The results in the flow and heat transfer distributions at *Re* = 300 are excluded because the thermal–hydraulic performance of all geometries is insignificant, making it difficult to distinguish the flow effect. Therefore, the internal flow characteristics of all studied models are compared at *Re* = 1000 and 5000, as shown in Figure 11. The velocity contours and flow trajectories provide detailed insights into the internal flow characteristics that directly affect the cooling efficiency and temperature uniformity of the SOFC stack. As seen in Figure 11a, the Diamond TPMS lattice exhibits exceptional flow uniformity at both *Re* values [25]. Its diagonal topology provides consistent flow with minimal recirculation and promotes uniform heat transfer. These flow characteristics imply low thermal gradients across the SOFC components.

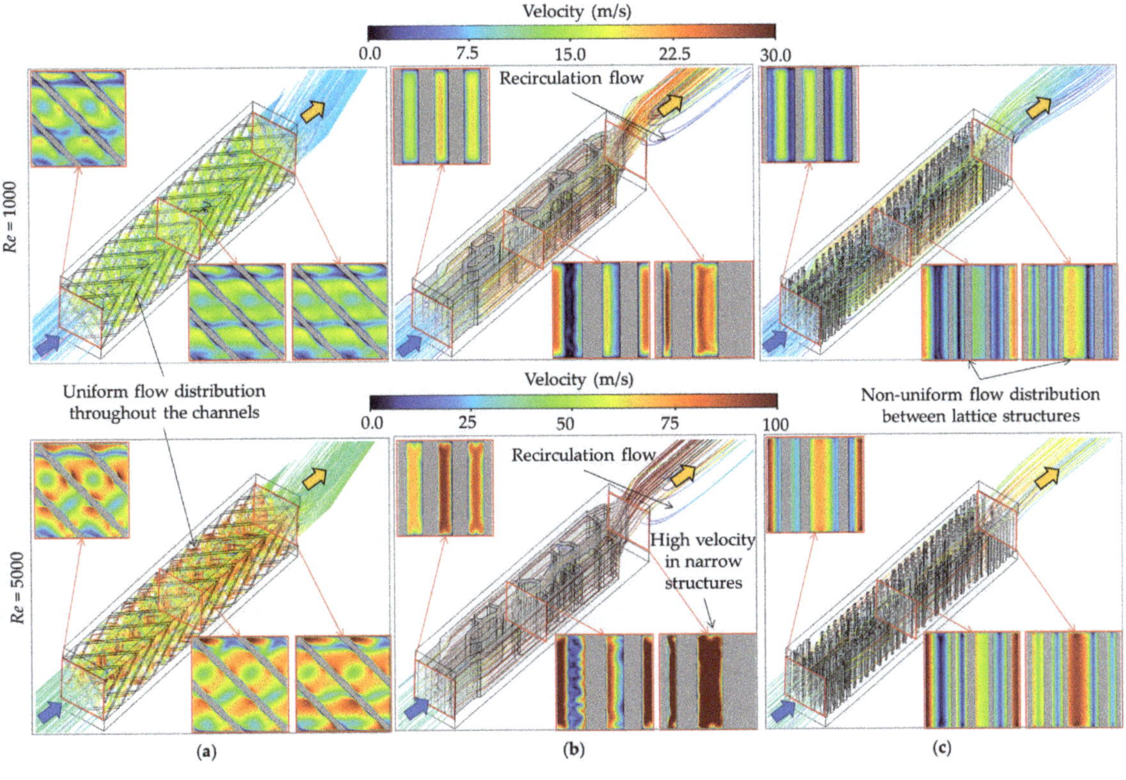


**Figure 11.** Flow characteristics of all models at the Reynolds numbers of 1000 and 5000: (**a**) Diamond TPMS model. (**b**) Conventional topology-optimized model. (**c**) Topology-optimized lattice-filled model.

In contrast, the conventional topology-optimized design in Figure 11b generates high-velocity jets between the baffles, leading to intense local heat transfer enhancement. However, this model generates strong recirculation zones at the outlet and a highly non-uniform flow distribution within the channel, particularly at *Re* = 5000. On the other hand, the topology-optimized lattice-filled structure in Figure 11c emerges as a hybrid solution. Although its flow distribution is not perfectly uniform as in the Diamond TPMS model, the integrated lattice structures with large surface areas induce local turbulence and mixing, enhancing overall cooling uniformity more effectively than the conventional topology-optimized model.

### 3.1.3. Heat Transfer Distributions

Figure 12 demonstrates the Nusselt number distributions at *Re* = 1000 and 5000 for all models. The contours indicate local heat transfer intensity, demonstrating the effectiveness and uniformity of convective heat transfer within each geometry. Based on the flow analysis, the Diamond TPMS structure demonstrates a correspondingly uniform and stable Nusselt number distribution at high *Re*, as seen in Figure 12a. The high flow uniformity translates into consistent heat transfer, confirming its capability for balanced thermal management of the SOFC system.

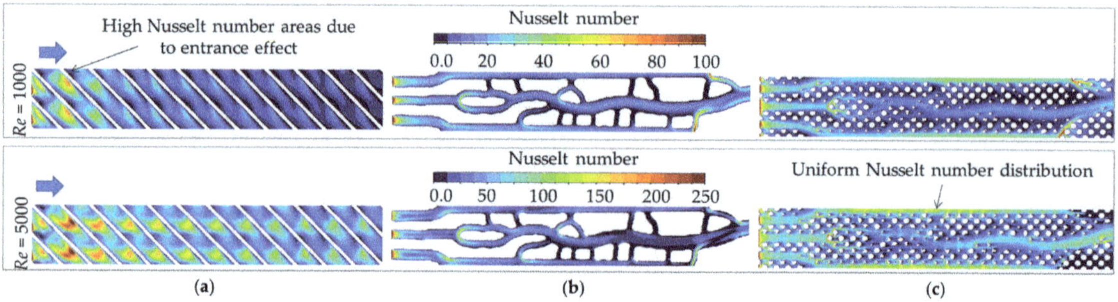


**Figure 12.** Nusselt number distributions on the endwall of all models at the Reynolds numbers of 1000 and 5000: (**a**) Diamond TPMS model. (**b**) Conventional topology-optimized model. (**c**) Topology-optimized lattice-filled model.

The jet impingement on the baffles and intense flow recirculation at the outlet in the conventional topology-optimized design cause non-uniform Nusselt number distribution, as observed in Figure 12b, particularly at high coolant flow rates, which can lead to high thermal stresses. Positively, the topology-optimized lattice-filled structure shows a more significant improvement with increasing flow rate. Although the non-uniform flow distribution at *Re* = 1000 results in a lower and uneven Nusselt number, the complex lattice interacts to create a uniform and enhanced heat transfer field at *Re* = 5000. This demonstrates that the lattice-filled design utilizes higher flow rates to improve heat transfer within its large surface area, thereby overcoming initial limitations.

Figure 13 shows the temperature distributions in the solid domain for all models at *Re* = 1000 and 5000. The temperature contours demonstrate the coolant flow behavior that influences heat removal and temperature uniformity in the cooling channel of the SOFC system. As shown in Figure 13a, the Diamond TPMS structure, consistent with its uniform flow, provides excellent temperature uniformity. However, at *Re* = 5000, a significant thermal gradient develops between the cooled structure and the heated walls, indicating a high thermal stress when integrating into the SOFC system. The conventional topology-optimized model in Figure 13b, which aligns with its non-uniform flow and Nusselt number distributions, exhibits uneven temperature contours and sharp gradients at both *Re* values, demonstrating a clear risk for thermal fatigue.

While the non-uniform flow characteristics in the topology-optimized lattice-filled structure at *Re* = 1000 lead to significant temperature variations, the high flow rate enhances mixing within the lattice, generating a uniform and lower overall temperature field. This structure also reduces the temperature imbalance found in the Diamond model and the high temperature in the conventional topology-optimized design. By combining a topologically guided flow path with a high-surface-area lattice, this complex structure shows a superior temperature uniformity at high coolant flow rates, ensuring effective heat dissipation and minimizing thermal variation, thereby enhancing SOFC reliability.

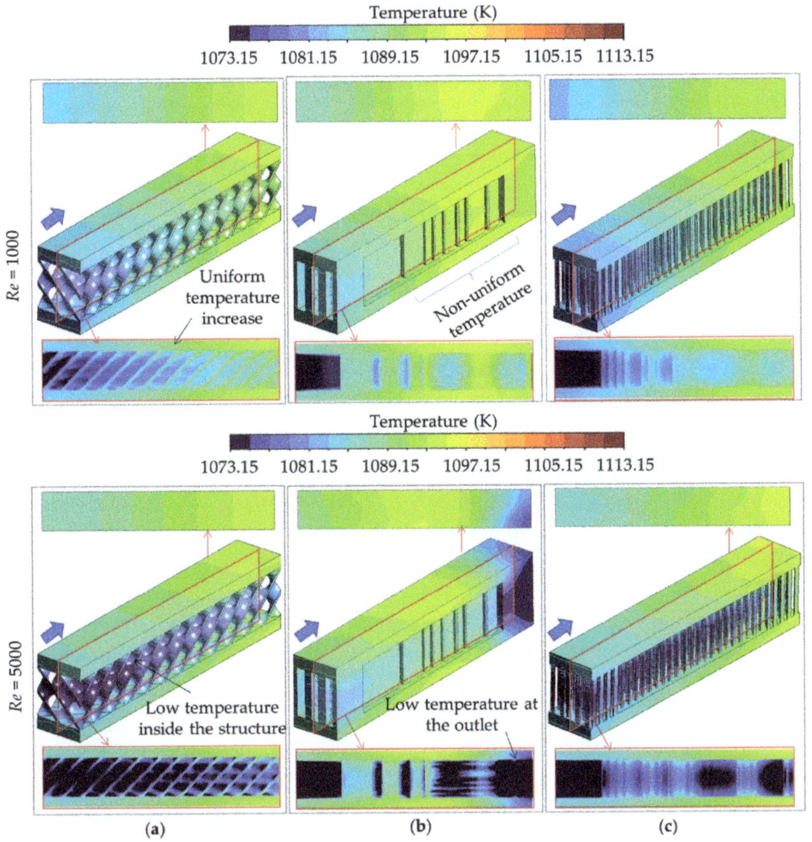


**Figure 13.** Temperature distributions on the solid domain of all models at the Reynolds numbers of 1000 and 5000: (**a**) Diamond TPMS model. (**b**) Conventional topology-optimized model. (**c**) Topology-optimized lattice-filled model.

3.1.4. Temperature Uniformity

Figure 14 shows the temperature uniformity ($T$-$T_{avg}$) on the heated wall for all models within the range of $Re$ = 300–5000. The color scale, ranging from −18 K to +12 K, indicates that values closer to zero signify superior temperature uniformity. Among them, the Diamond TPMS structure in Figure 14a exhibits minimal temperature deviation from $Re$ = 300 to $Re$ = 5000, confirming its inherent design provides highly uniform temperature on the heated wall. Meanwhile, the conventional topology-optimized structure in Figure 14b causes a lower temperature uniformity, particularly at $Re$ = 5000. Large and contiguous zones of positive and negative deviations are evident, revealing inherent non-uniform flow distribution, which is due to recirculation and stagnant zones. This temperature pattern fails to achieve the reliability for thermal management of the SOFC system at high coolant flow rates.

The topology-optimized lattice-filled structure in Figure 14c improves the temperature uniformity with increasing flow rate. At $Re$ = 5000, this model achieves temperature uniformity similar to that of the Diamond TPMS model, demonstrating its advanced design effectively leverages higher flow inertia to activate uniform cooling across a complex structure. The topology-optimized lattice-fill structure could be an excellent candidate in the cooling channel between the SOFC stack, especially for high cooling air supplies.

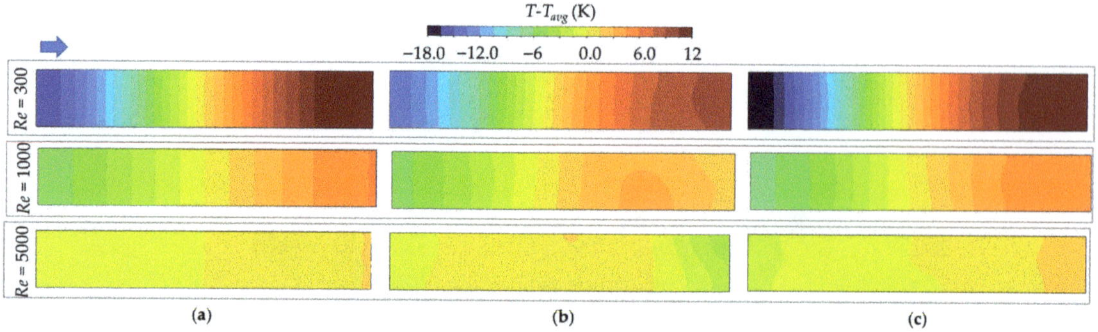


**Figure 14.** Contours of the temperature variation on the heated wall of all models at the Reynolds numbers of 300, 1000, and 5000: (**a**) Diamond TPMS model. (**b**) Conventional topology-optimized model. (**c**) Topology-optimized lattice-filled model.

Figure 15 provides quantitative values of the temperature range ($T_{max}$–$T_{min}$) and a temperature uniformity index (TUI) against *Re*. In Figure 15a, the conventional topology-optimized model exhibits the lowest temperature range ($T_{max}$–$T_{min}$) value of 23 K and 8.6 K when *Re* = 300 and 1000, while the Diamond TPMS and topology-optimized lattice-filled structures show a smaller temperature range of around 2 K at *Re* = 5000. Here, the smaller ($T_{max}$–$T_{min}$) value indicates a more uniform temperature field by minimizing the difference between the hottest and coldest spots, which is critical for reducing thermal stresses.

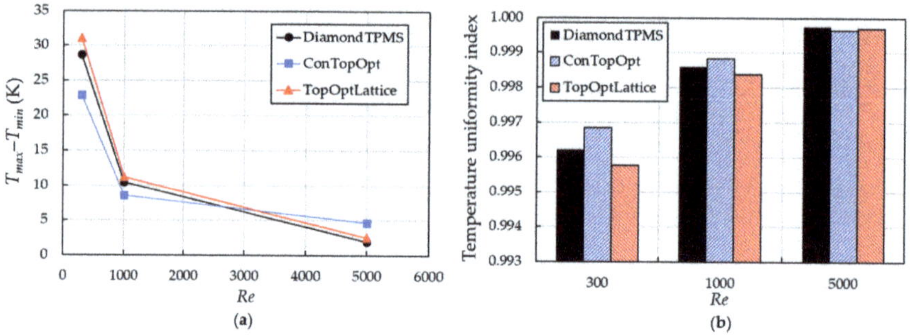


**Figure 15.** Evaluation of temperature uniformity on the bottom heated wall of all models at the Reynolds numbers of 300, 1000, and 5000: (**a**) $T_{max}$–$T_{min}$ and (**b**) temperature uniformity index.

This superior performance is further validated in Figure 15b by the TUI, where a value closer to 1.0 represents perfect uniformity. The Diamond TPMS structure achieves the highest index values, confirming its excellent thermal distribution at *Re* = 5000. While the conventional topology-optimized models perform well at low *Re* values, the topology-optimized lattice-filled structure slightly exceeds its uniformity at higher flow rates. Overall, the analyses in this section demonstrate that the topology-optimized lattice-filled structure offers a highly uniform thermal environment on the heated wall of the air-cooled SOFC system, indicating that it is the optimal design for mitigating thermal stress and ensuring reliable operation at high cooling air flow rates in this study.

### 3.2. Effect of Coolant Type

To isolate the influence of coolant flow rate on thermal performance, the thermal–hydraulic performance, heat transfer distribution, and temperature uniformity are analyzed at $Re = 300$. The following analysis utilizes a ferritic stainless steel structure to compare three coolants: air, liquid tin, and liquid gallium.

3.2.1. Thermal–Hydraulic Performance

Figure 16 compares the thermal–hydraulic performance of all models under three coolant types at $Re = 300$. The conventional topology-optimized and lattice-filled structures achieve $Nu_{avg}$ approximately twice that of the Diamond TPMS with air cooling, as shown in Figure 16a, due to their enhanced flow redistribution and mixing. While liquid metals yield lower Nusselt number values of $Nu_{avg} < 4$ [19], the topology-optimized lattice-filled model slightly outperforms others, as its complex lattice promotes mixing even in low-flow-rate liquid metals. This translates directly to the temperature on the heated wall in Figure 16b, where liquid metals reduce wall temperatures by up to 43 K compared to air, with the gallium-cooled channel obtaining the lowest values.

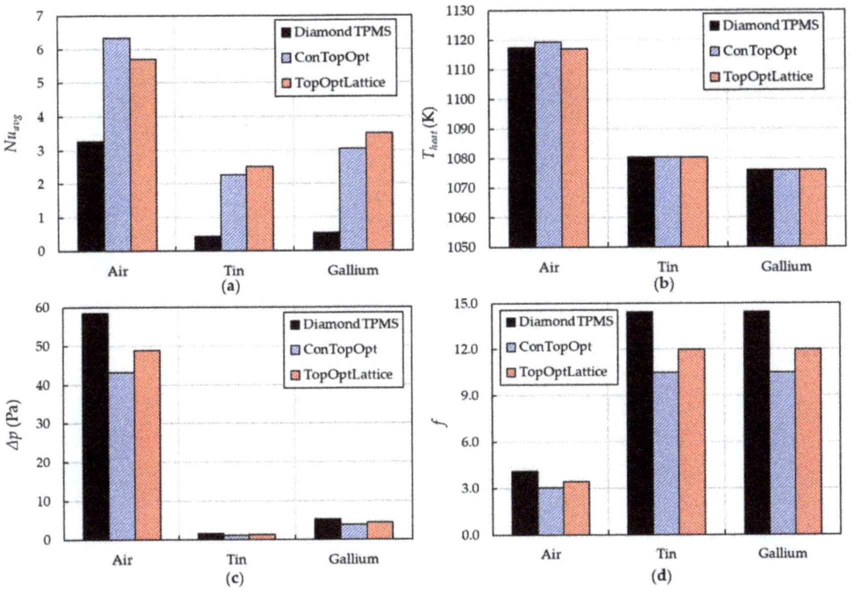


**Figure 16.** Thermal–hydraulic performance of all models with different coolants at Reynolds numbers of 300: (**a**) area-averaged Nusselt number on the endwall inside the cooling channel; (**b**) area-averaged temperature on the heated wall; (**c**) pressure loss; and (**d**) friction factor.

Figure 16c reveals that pressure loss is governed by both fluid properties and channel topology. For the air-cooled channel, the Diamond TPMS structure exhibits the highest pressure loss of 58.5 Pa, while the conventional topology-optimized model achieves the lowest. For liquid tin and gallium, the pressure losses are significantly lower than those of the air-cooled channel due to their higher density and lower dynamic viscosity, which drastically reduces fluid flow resistance at an equivalent $Re$. The $f$ in Figure 16d further elucidates this hydraulic behavior, where the Diamond TPMS model exhibits the highest $f$ value, ranging from 4.1 to 14.2, which is 26.2% to 27.2% higher than the conventional topology-optimized model, and from 16.4% to 16.9% higher than the lattice-filled design for all coolant types. The topology-optimized lattice-filled model exhibits an obviously higher $f$ value than the conventional topology-optimized model, by 13.2% to 14.2%, because of its finer internal structures, which strongly interact with the high density of the liquid metals.

Figure 17 compares the $R_{tot}$ and $P_e$ for all designs with different coolants at $Re = 300$. As shown in Figure 17a, the $R_{tot}$ decreases significantly when the coolant is changed from air to liquid metals. The air-cooled channel exhibits the highest $R_{tot}$ in all geometries due to its inherently low thermal conductivity and heat capacity. Meanwhile, the liquid tin and gallium achieve higher performance, with gallium offering the lowest $R_{tot} \approx 0.25$ K/W, a reduction of 93%. A noticeable difference of approximately 60% between the liquid metals is due to variations in coolant conductivity. Notably, the geometric differences cause only minor changes in the $R_{tot}$ with liquid metals, confirming that the coolant properties dominate heat transfer under these flow conditions. Overall, for the air-cooled channel, the topology-optimized lattice-filled structure exhibits the best performance. Meanwhile, for liquid metals, the conventional topology-optimized geometry is the most effective, indicating that its baffle structures are more efficient in removing heat at low coolant flow rates.

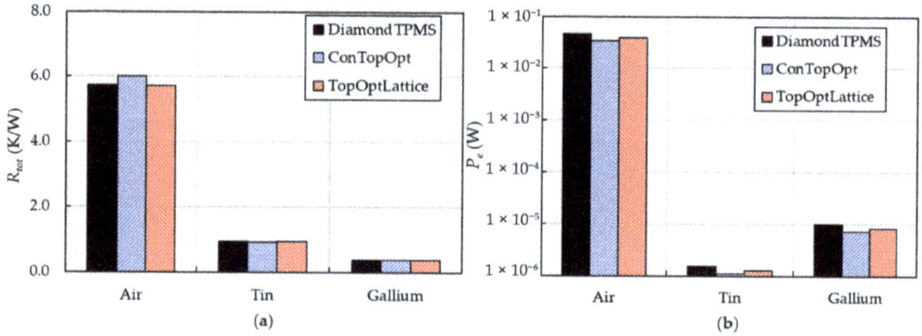


**Figure 17.** Influence of coolant type on: (**a**) thermal resistance and (**b**) pumping power for all models at Reynolds numbers of 300.

Figure 17b reveals that air cooling requires the highest $P_e$ of 0.047 W, despite its low heat transfer performance, which is attributed to its low density requiring high volumetric flow rates [18]. The liquid metals significantly reduce pumping requirements, with liquid tin exhibiting the lowest $P_e$ of up to 100% compared to the air-cooled channel, due to its viscosity–density balance. The higher viscosity and density of the gallium-cooled channel result in a 5.6% increase in its $P_e$ compared to the tin-cooled channel. Overall, the geometric design influences on the $P_e$ are minor for liquid metals but more pronounced for air, where the Diamond TPMS model shows higher losses.

### 3.2.2. Heat Transfer Distributions

Figure 18 shows the effect of coolant types on the Nusselt number distributions for all models at $Re = 300$. The liquid metal-cooled channels exhibit lower Nusselt numbers than the air-cooled channel despite their superior performance. It is essential to clarify that the Nusselt number is a measure of the convective relative to the conductive heat transfer. The relatively high thermal conductivity of liquid metals is the primary mechanism for their effectiveness, enabling rapid heat transfer through conduction rather than convection. This results in a lower temperature difference between the wall and the bulk fluid, which leads to a lower Nusselt number. While the air-cooled channel exhibits high Nusselt numbers, the tin- and gallium-cooled channels maintain a more uniform Nusselt number along the flow direction, even at this low flow rate. Furthermore, the Diamond TPMS structure creates a high heat transfer only at the channel entrance, whereas the conventional topology-optimized and lattice-filled models maintain a more uniform distribution.

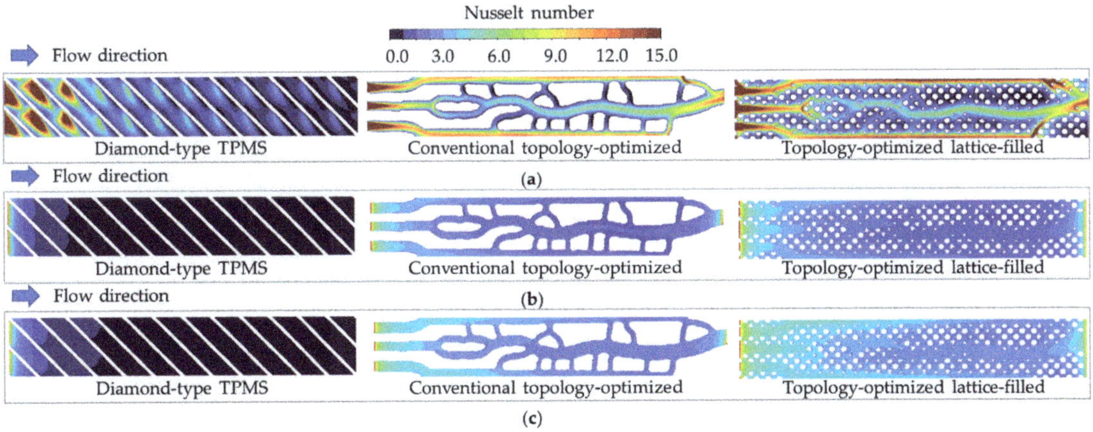


**Figure 18.** Nusselt number distributions on the endwall of all models at the Reynolds numbers of 300: (**a**) Air-cooled channel. (**b**) Liquid tin-cooled channel. (**c**) Liquid gallium-cooled channel.

A comparison of the liquid-metal coolants reveals that the gallium-cooled channel in Figure 18c generates a higher magnitude and a more uniform Nusselt number distribution than the tin in Figure 18b. It is noticeable that the conventional topology-optimized structures consistently demonstrate the most uniform distributions when using liquid gallium. This optimal performance arises from the combination of high thermal conductivity in the liquid gallium and the baffle topology, which provides a simple and large surface area for heat exchange while maintaining smooth flow pathways that enable efficient heat removal.

The temperature fields on the solid domain for the three coolant types at $Re = 300$ are compared in Figure 19. The cross-sectional views show that although the constant flow rate offers a similar temperature trend in all geometries, the liquid tin and gallium generate substantially lower and more uniform temperature distributions than the air-cooled channel. Specifically, the gallium-cooled channel in Figure 19c exhibits the most uniform temperature and the lowest peak temperature, closely followed by the tin-cooled channel in Figure 19b. Meanwhile, the air cooling in Figure 19a results in non-uniform and high peak temperatures, indicating that it is unsuitable for the cooling channel of the SOFC without substantial geometric or operational adjustments.

### 3.2.3. Temperature Uniformity

Figure 20 shows the impact of coolant type on the temperature uniformity for all designs at $Re = 300$. A dominant trend of all structures is the profound improvement in temperature uniformity when using liquid tin and gallium coolants compared to air. The air-cooled channels in Figure 20a exhibit the largest temperature change, indicating poor heat spreading and a high risk of thermal stress. On the other hand, both liquid metal coolants provide more uniform temperature fields, with significantly smaller deviations concentrated near zero. Among the liquid metals, the gallium-cooled channel in Figure 20c yields the most uniform temperature distributions, slightly outperforming the liquid tin-cooled channel in Figure 20b. This is also consistent with the higher thermal conductivity of the liquid gallium [19]. Regarding the geometries, the conventional topology-optimized structure consistently demonstrates the best uniformity with the liquid-metal coolants. Its optimized-baffle design integrates well with high-conductivity liquid metals to create a uniform temperature contour.

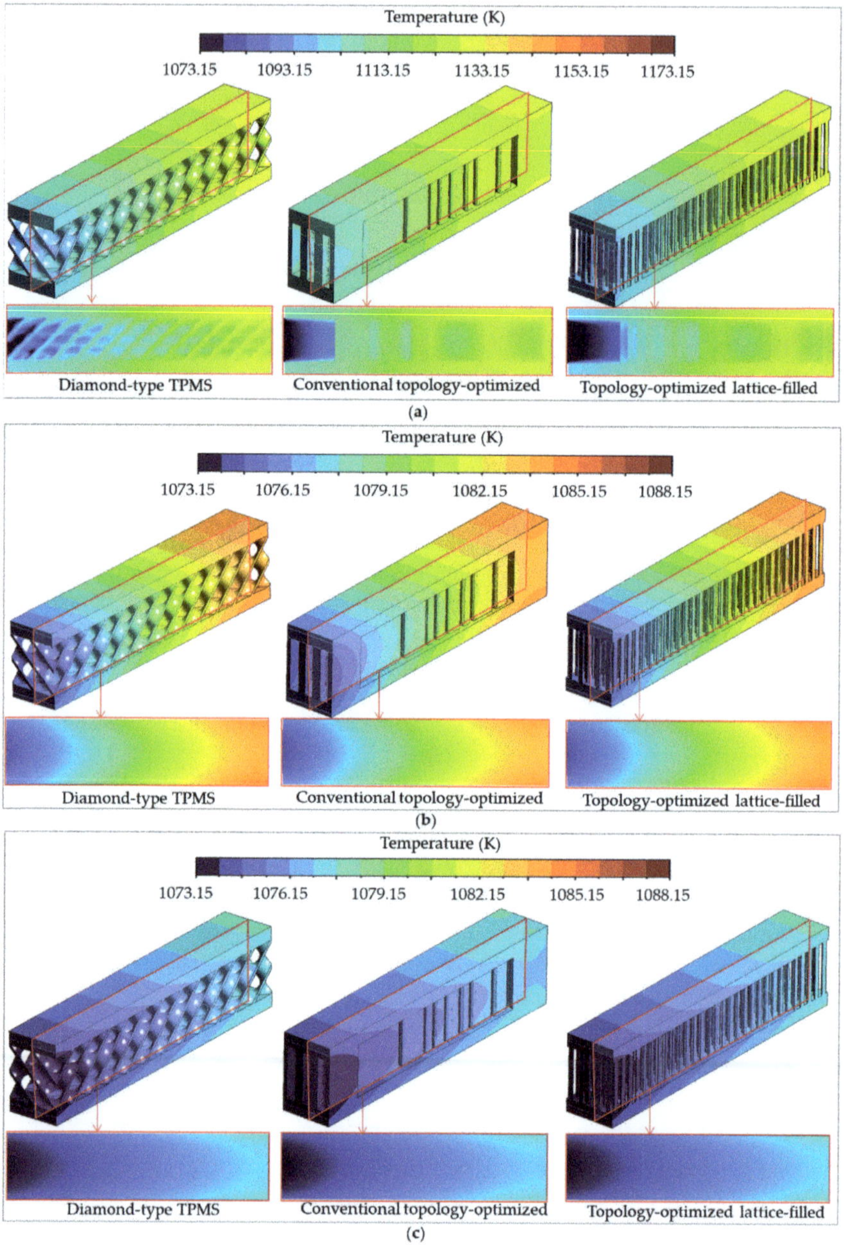


**Figure 19.** Temperature distributions on the solid domain of all models at the Reynolds numbers of 300: (**a**) Air-cooled channel. (**b**) Liquid tin-cooled channel. (**c**) Liquid gallium-cooled channel.

Figure 21 quantitatively evaluates the temperature uniformity on the heated wall for all models using air, liquid tin, and liquid gallium as coolants at $Re$ = 300. With the air-cooled channel, all geometries exhibit significant temperature differences ($T_{max}$–$T_{min}$) up to 30 K and lower temperature uniformity index values, indicating limited heat spreading capability. Among the designs, the conventional topology-optimized model achieves the best air-cooled performance due to its simple baffle cooling structures, which enable smooth flow at low *Re*. In contrast, the liquid metals drastically improve temperature uniformity, reducing the temperature difference value to below 10 K and achieving temperature uniformity index values close to 1.0. The gallium-cooled channel performs slightly better than the tin, yielding differences below 5 K due to its superior thermal conductivity. These results confirm that while the geometry influences convective heat transfer, the coolant properties dominate temperature uniformity. Consequently, these findings affirm the promise of liquid-metal coolants for thermal management in high-power-density SOFCs, though practical integration issues remain to be improved.

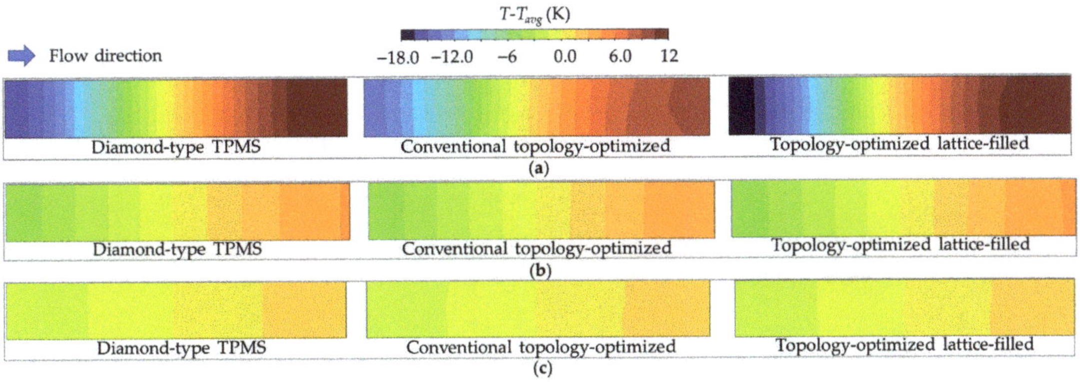


**Figure 20.** Contours of the temperature variation on the heated wall of all models at the Reynolds numbers of 300: (**a**) Air-cooled channel. (**b**) Liquid tin-cooled channel. (**c**) Liquid gallium-cooled channel.

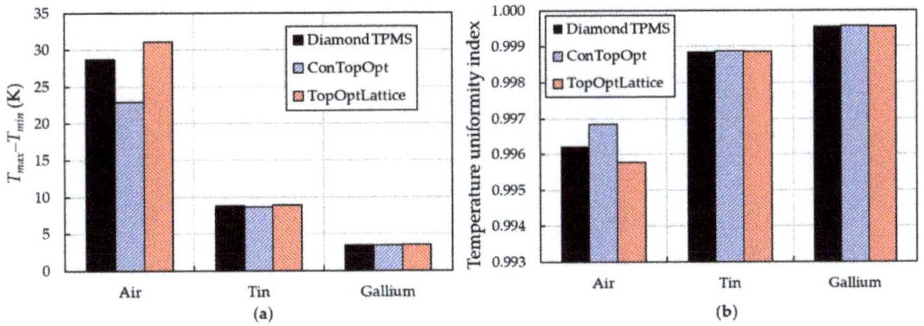


**Figure 21.** Evaluation of temperature uniformity on the bottom heated wall of all models with different coolants at the Reynolds numbers of 300: (**a**) $T_{max}$–$T_{min}$ and (**b**) Temperature uniformity index.

Although these findings affirm the excellence of liquid-metal coolants for thermal management in high-power-density SOFCs, their practical implementation in the SOFCs requires careful consideration of material compatibility, corrosion, and cost. The liquid tin is highly corrosive to ferritic stainless steel under high operating temperatures of the SOFCs, necessitating the use of specialized nickel-based superalloys or protective ceramic coatings [15]. The liquid gallium could cause embrittlement in many structural metals and is significantly more expensive than liquid tin. Moreover, these liquid metals face wetting issues in highly complex channels. Thus, their thermal benefits must be balanced against higher costs, material problems, and potential long-term degradation.

### 3.3. *Effect of Structural Material*

This section evaluates the impact of various structural materials on the thermal performance of the air-cooled channel. A comparative analysis of pressure loss shows insignificant variation between different structural materials at $Re = 300$; consequently, the discussion is confined to thermal characteristics.

#### 3.3.1. Heat Transfer Performance

Figure 22 illustrates the comparison of thermal performance parameters for all air-cooled geometries, fabricated with three different solid materials at $Re = 300$. The $Nu_{avg}$ is substantially higher in the conventional topology-optimized and lattice-filled configurations than in the Diamond TPMS structure across all structural materials, as seen in Figure 22a. The improvement is significant, with the $Nu_{avg}$ increases of 56–89% over the Diamond TPMS model. The stainless steel and nickel–chromium structures provide similar value, which is slightly higher than that of the high-thermal-conductivity metal.

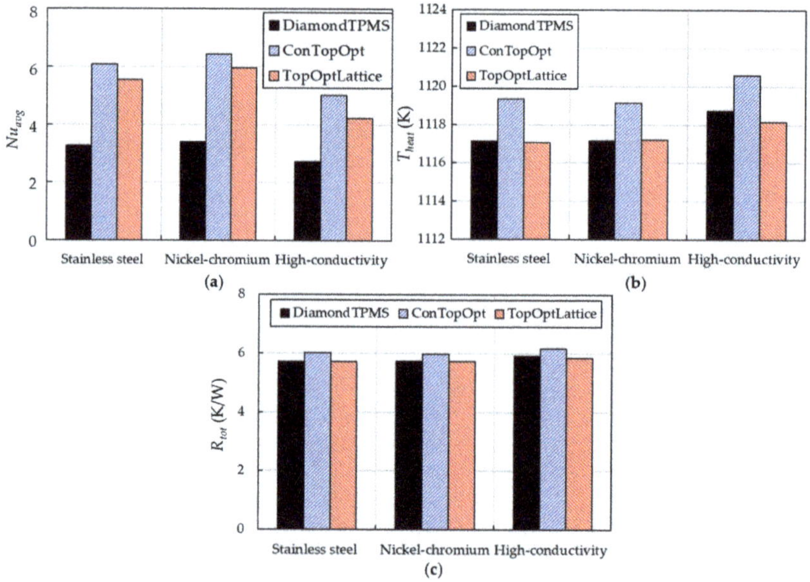


**Figure 22.** Thermal performance of all air-cooled geometries at the Reynolds numbers of 300: (**a**) area-averaged Nusselt number on the endwall inside the cooling channel; (**b**) area-averaged temperature on the heated wall, and (**c**) Thermal resistance.

Despite these convective heat transfer increases, the conventional topology-optimized model exhibits higher $T_{heat}$ than the Diamond TPMS structure, as found in Figure 22b. This indicates that convective enhancement alone does not dictate the final wall temperature. The high-thermal-conduction metal causes a slightly higher $T_{heat}$ value than the others. Lastly, the $R_{tot}$ shows a similar value in all designs in Figure 22c, indicating that the high convective heat transfer does not translate proportionally into lower thermal resistance with different structural materials. It could also be deduced that the performance of the flow and heat transfer conditions in this study is governed by a complex balance between convection and conduction.

3.3.2. Heat Transfer Distributions

Figure 23 illustrates the Nusselt number distributions on the endwall of all designs for different structural materials at *Re* = 300. In the ferritic stainless-steel and nickel–chromium superalloy cases, as shown in Figure 23a,b, the conventional topology-optimized geometry generates high Nusselt number values along the flow direction, indicating improved convective mixing at low coolant flow rate. In contrast, the Diamond TPMS model exhibits alternating high- and low-Nusselt number values, resulting from periodic flow recirculation, which leads to more pronounced local hot spots [25]. The topology-optimized lattice-filled structure exhibits intermediate behavior, maintaining relatively continuous high-Nusselt number regions near flow passage intersections.

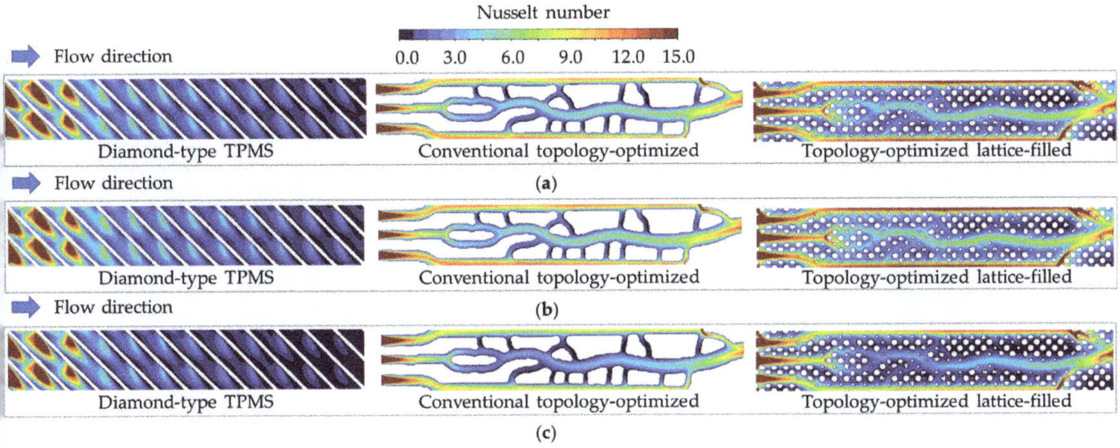


**Figure 23.** Nusselt number distributions on the endwall of all air-cooled geometries at the Reynolds numbers of 300: (**a**) Ferritic stainless steel. (**b**) Nickel–chromium superalloy. (**c**) High-thermal-conductivity metal.

When replacing the structural material with a high-thermal-conductivity metal in Figure 23c, the Nusselt number gradients become clearly non-uniform. Although the increased thermal conductivity promotes lateral heat spreading within the solid, a non-uniform endwall temperature distribution can be observed. Consequently, the distinction between high- and low-Nusselt number zones becomes more prominent. Among all designs, the topology-optimized lattice-filled maintains the high uniform Nusselt number regions, confirming that the topology optimization effectively tailors the internal flow paths to maximize thermal utilization of the coolant under air-cooling conditions.

Figure 24 presents the temperature distributions on the solid domain of all air-cooled geometries at *Re* = 300 for different structural materials. In the ferritic stainless-steel and nickel–chromium superalloy cases shown in Figure 24a,b, all designs exhibit noticeable temperature gradients along the flow direction, with higher temperatures near the outlet region, particularly in the topology-optimized lattice-filled structure. The conventional topology-optimized structure shows the most uniform temperature distribution, although it demonstrates slightly lower temperatures, followed by the Diamond TPMS model.

As the material changes to the high-thermal-conductivity metal in Figure 24c, the overall temperature levels increase, but the temperature variations become smaller, which can be attributed to lateral heat spreading within the solid wall, significantly reducing the thermal nonuniformity between the inlet and outlet. In this material condition, the temperature contours in all geometries become more uniform, indicating that material thermal conductivity has a significant influence on the wall temperature distribution.

Nevertheless, the conventional topology-optimized configuration consistently provides the lowest temperature variation among all designs.

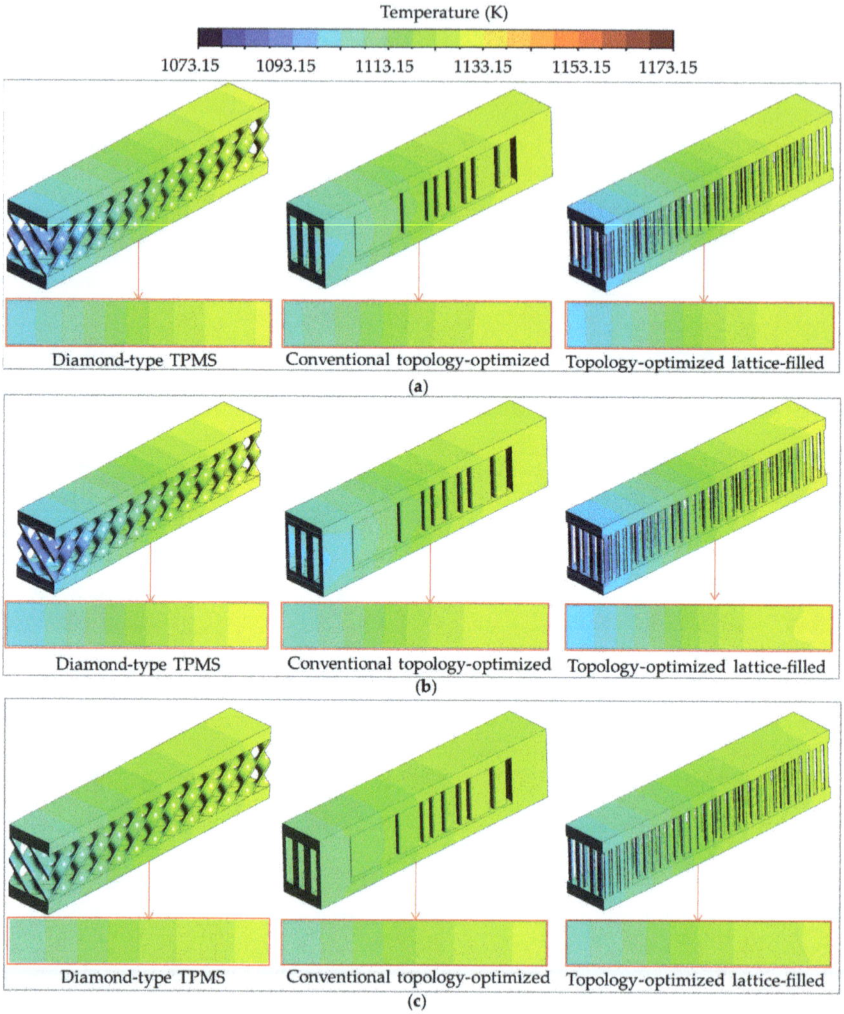


**Figure 24.** Temperature distributions on the solid domain of all air-cooled geometries at the Reynolds numbers of 300: (**a**) Ferritic stainless steel. (**b**) Nickel–chromium superalloy. (**c**) High-thermal-conductivity metal.

### 3.3.3. Temperature Uniformity

Figure 25 illustrates the contours of temperature variation ($T$-$T_{avg}$) on the heated walls of all air-cooled geometries at $Re$ = 300 for different materials. Similar to the previous discussion, the ferritic stainless steel in Figure 25a and the nickel–chromium superalloy in Figure 25b cause significant temperature gradients. The conventional topology-optimized model exhibits the most non-uniform temperature distribution, a consequence of its less complex structure, which allows smooth flow between the baffles at low $Re$. The transition to the high-thermal-conductivity metal in Figure 25c shows a more uniform temperature contour. This improvement signifies that heat is being effectively conducted from hot spots, creating a more isothermal heated surface.

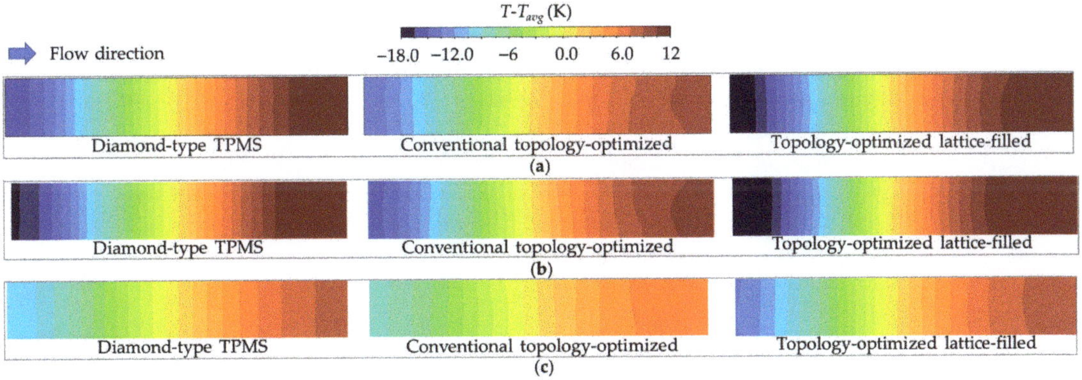


**Figure 25.** Contours of the temperature variation on the heated wall of all air-cooled geometries at the Reynolds numbers of 300: (**a**) Ferritic stainless steel. (**b**) Nickel–chromium superalloy. (**c**) High-thermal-conductivity metal.

Figure 26 evaluates the temperature uniformity on the heated wall of all air-cooled designs, fabricated from different materials, at $Re = 300$. For all materials, the conventional topology-optimized configuration exhibits the smallest temperature range ($T_{max}$–$T_{min}$) of 16 K to 30 K, as shown in Figure 26a, indicating superior temperature uniformity compared to the other geometries. Meanwhile, the topology-optimized lattice-filled structure causes the largest temperature range values, spanning 40 K to 51 K, due to complex flow recirculation behind each lattice, particularly at low *Re*. As the thermal conductivity of the wall material increases, from ferritic stainless steel to high-thermal-conductivity metals, the overall temperature range decreases across all designs, and the conventional topology-optimized model achieves the lowest temperature range of 16 K.

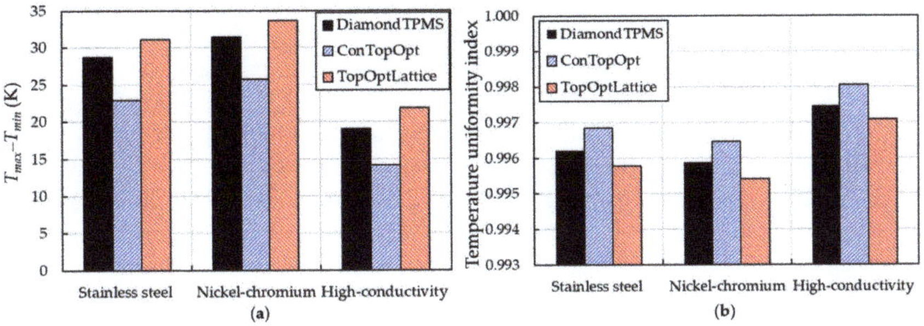


**Figure 26.** Evaluation of temperature uniformity on the bottom heated wall of all air-cooled geometries with different structural materials at the Reynolds numbers of 300: (**a**) $T_{max}$–$T_{min}$ and (**b**) temperature uniformity index.

A similar trend is observed in the temperature uniformity index in Figure 26b, where the conventional topology-optimized model achieves the highest index for all materials. The topology-optimized lattice-filled model slightly reduces uniformity despite its enhanced surface area, implying that excessive internal complexity may hinder uniform cooling, particularly at low coolant flow rates. The Diamond TPMS model maintains moderate uniformity across studied materials, benefiting from its periodic structure but limited by reduced internal mixing.

The superior temperature uniformity observed with high-conductivity materials can also be quantitatively explained by the Biot number. For all geometries at $Re = 300$, the Biot number is significantly less than 0.1. This occurs because the intricate structures have a very small characteristic length, defined by the ratio of solid volume to wetted surface area. Among the materials, the high-thermal-conductivity metal yields the lowest Biot number, confirming its minimal internal conductive resistance relative to the external convective resistance and thus the most uniform temperature field.

Overall, these findings indicate that for the air-cooled system utilizing high-conductivity materials and low coolant flow rates, the baffle-shaped designs (the conventional topology-optimized model) that enable low pressure loss and a smooth flow field are the most suitable for thermal management of the SOFC stack.

## 4. Conclusions

This study presents a conjugate analysis of cooling channel designs, including the Diamond TPMS, conventional topology-optimized, and topology-optimized lattice-filled structures, which integrate between the SOFCs to improve thermal management. The investigation assesses the impact of coolant flow rate, coolant type, and structural material on thermal–hydraulic performance and temperature uniformity. The conclusions can be drawn as follows:

(1) The Diamond TPMS structure provides excellent flow and heat transfer distribution, but causes a significant pressure penalty of 45.6% and 24.6% at *Re* = 5000, compared to the conventional topology-optimized and lattice-filled designs. The conventional optimized model minimizes pressure loss but can lead to non-uniform flow and temperature distributions at high flow rates. The topology-optimized lattice-filled structure emerges as an excellent compromise, especially at high *Re*. By combining the tailored flow paths from topology optimization with the high surface area of a lattice-filled array, this design in the air-cooled system achieves temperature uniformity comparable to that of the Diamond TPMS at *Re* = 5000, while incurring a significant reduction in pressure loss penalty.

(2) While both topology-optimized designs can improve convective heat transfer, the choice of coolant is found to be the highest dominant factor, governing the thermal–hydraulic performance. The transition coolants from air to liquid metals resulted in a significant reduction in thermal resistance, up to 93%, and a relatively high temperature uniformity index at *Re* = 300. The liquid gallium, with its superior thermal conductivity, outperformed liquid tin, achieving the most uniform temperature distributions on the heated wall, where the differences between the maximum and minimum temperatures ($T_{max}$–$T_{min}$) are less than 5 K in both topology-optimized designs. However, the excellent cooling performance of liquid metals might come with a higher cost and greater engineering challenges.

(3) A higher thermal conductivity material improves heat spreading within the solid structure, creating a more uniform temperature in all geometries. Under low coolant flow rates and air-cooled conditions, the baffle-like structures in the conventional topology-optimized model exhibit preferable thermal–hydraulic performance, facilitating a smooth flow field, and achieve the lowest temperature range ($T_{max}$–$T_{min}$) of 16 K at *Re* = 300.

(4) For maximum performance, employing a liquid metal coolant within the optimized-baffle design channel is highly recommended for the precise thermal management of the SOFC. This configuration leverages their balanced thermal–hydraulic performance and the high thermal conductivity of the liquid metals. Additionally, fabricating the channel with a high thermal conductivity material for a conventional topology-optimized design further achieves better temperature uniformity while minimizing parasitic power.

The design strategies outlined in this work offer an alternative approach to enhancing the thermal management of SOFC systems, providing significantly improved temperature uniformity. To transition these concepts from simplified simulations to real-world applications, future investigations could establish full-system computational modeling and experimental validation to quantify the realistic performance improvements.

**Author Contributions:** Conceptualization, K.Y. and Y.R.; methodology, K.Y. and Y.C.; software, K.Y.; validation, K.Y. and Y.C.; formal analysis, K.Y. and Y.C.; investigation, K.Y. and Y.C.; resources, K.Y. and Y.R.; data curation, K.Y. and Y.C.; writing—original draft preparation, K.Y.; writing—review and editing, K.Y., Y.R. and Y.C.; visualization, K.Y. and Y.C.; supervision, Y.R.; project administration, Y.R.; funding acquisition, K.Y. and Y.R. All authors have read and agreed to the published version of the manuscript.

**Funding:** This work was supported by grants from the National Natural Science Foundation of China (Nos. 11972230, 24Z033103800), the National Science and Technology Major Project of China (No. 2017-III-0009-0035), and the Science and Technology Commission of Shanghai Municipality (No. 24110712700).

**Data Availability Statement:** The original contributions presented in this study are included in the article. Further inquiries can be directed to the corresponding author.

**Acknowledgments:** Yu Rao gratefully acknowledges support from the Shandong Provincial Industrial Leading Talents Project.

**Conflicts of Interest:** The authors declare no conflicts of interest.

## Nomenclature

| | |
|---|---|
| $A_i$ | Local area ($m^2$) |
| $A_{in}$ | Inlet area of the cooling channel (m) |
| $D$ | Characteristic length (m) |
| $D_h$ | Hydraulic diameter of the cooling channel (m), defined in Equation (2) |
| $D_p$ | Diameter of the circular pin-fin structure (m) |
| $f$ | Friction factor, defined in Equation (6) |
| $f_0$ | Blasius correlations, defined in Equation (10) |
| $h$ | Heat transfer coefficient ($W/(m^2 \cdot K)$), defined in Equation (4) |
| $T_b$ | Bulk flow temperature (K) |
| $T_{heat}$ | Temperature of the heated wall (K) |
| $T_{in}$, $T_{out}$ | Inlet and outlet coolant temperatures (K) |
| $T_i$ | Local temperature (K) |
| $T_{max}$, $T_{min}$ | Maximum and minimum temperature on the heated wall (K) |
| $T_w$ | Endwall temperature(K) |
| $L$ | Length of the cooling structure (m) |
| $L_p$ | Wetted perimeter (m) |
| $Nu$ | Nusselt number, defined in Equation (3) |
| $Nu_{avg}$ | Area-averaged Nusselt number |
| $Nu_0$ | Dittus-Boelter equation, defined in Equation (9) |
| $q$ | Wall heat flux ($W/m^2$) |
| $k$ | Thermal conductivity of the fluid ($W/m \cdot K$) |
| $P_e$ | Pumping power (W), defined in Equation (7) |
| $p_{in}$ | Coolant inlet pressure (Pa) |
| $Re$ | Reynolds number, defined in Equation (1) |
| $R_{tot}$ | Total thermal resistance (K/W), defined in Equation (5) |
| $S_v$ | Specific area surface ($m^2/m^3$) |
| $V_{in}$ | Velocity at the inlet (m/s) |
| $X, Y, Z$ | Cartesian coordinate |

| | |
|---|---|
| Greek letter | |
| $\gamma$ | Density field in the topology optimization |
| $\mu$ | Dynamic viscosity of the coolant (Pa·s) |
| $\rho$ | Density of the coolant ($kg/m^3$) |
| $\Delta p$ | Different pressure between the inlet and outlet boundaries (Pa) |
| $\Gamma_{in}$ | Inlet boundary |
| $\psi$ | Porosity (%) |
| Abbreviations | |
| CHT | Conjugate heat transfer |
| PEC | Performance evaluation criterion, defined in Equation (8) |
| PEMFCs | Proton-exchange membrane fuel cells |
| PEN | Positive electrode-electrolyte-negative electrode |
| TLC | Transient liquid crystal thermography |
| TPMS | Triply periodic minimal surface |
| TUI | Temperature uniformity index, defined in Equation (11) |
| SOFC | Solid-oxide fuel cell |

## Appendix A

The density-based topology optimization is used to maximize heat transfer within the internal cooling channel. This method solves a material distribution problem, defining an optimal layout of fluid and solid phases within the design domain ($\Omega$). A continuous design variable ($\gamma$) represents the material distribution, ranging from 0 (solid) to 1 (fluid) at any cell in the design domain. A review of this approach for cooling channels is available in Refs [26,27].

To reduce the high computational cost required for convergence, a simplified 2D model is employed, as shown in Figure A1. The symmetry boundary conditions are applied to the side walls to further accelerate the optimization process. The inlet velocity ($V_{in}$) is typically used to determine the Reynolds number ($Re$). However, during topology optimization, the channel geometry changes, which alters the inlet velocity even when the input power is held constant. To maintain a consistent basis for comparison at fixed input power, the following integral equation for inlet pressure is derived [27], as follows:

$$\int_{in} p_{in} u d\Gamma_{in} = \frac{\mu Re^3}{\rho^3 D_h} \tag{A1}$$

where $p_{in}$ is the coolant inlet pressure. $u$ is the velocity component in the $X$-axis.

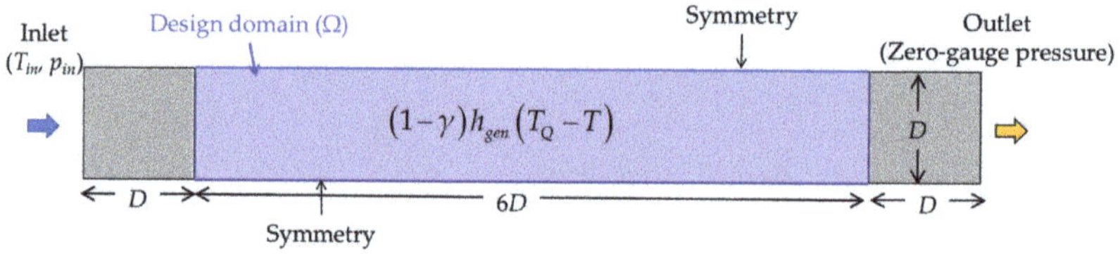


**Figure A1.** Design domain and boundary conditions of the optimization model.

In this topology optimization, the Reynolds number, based on the coolant hydraulic diameter, is maintained at 10,000 to facilitate the simulation. The heat generation coefficient ($h_{gen}$), which governs the rate of heat generation, is set to $1.3 \times 10^6$ W/($m^3$K) to promote the formation of a complex structure. $T_Q$ is a constant temperature applied at the heat source. A volume constraint with a volume fraction of 0.5 is also imposed. The governing equations, material interpolations, problem formulation, filtering, projection, and optimization procedures follow established methods from previous studies.

The intermediate densities ($0.1 \leq \gamma \leq 0.9$) from the topology optimization are interpreted as an array of circular pin-fins, as shown in Figure A2. This optimized 2D layout is extruded to a 20 mm height to form a 3D lattice structure. The pin diameter ($D_p$) varies consistently with the local density field, ranging from $D_p$ = 1 mm for $\gamma$ = 0.9 to $D_p$ = 2 mm for $\gamma$ = 0.1. The pin-fin configuration is selected for its compatibility with traditional manufacturing and its suitability for integration within the narrow interconnects of an SOFC stack.

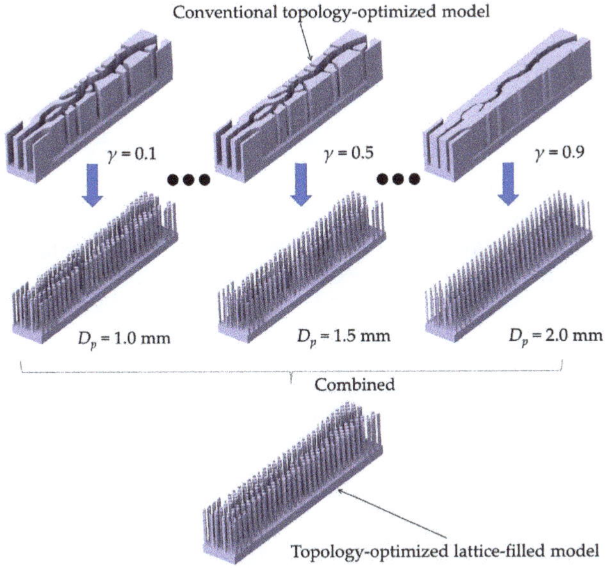


**Figure A2.** Procedure to obtain the cooling channel with a topology-optimized lattice-filled structure.

The hydraulic diameter of all geometries, calculated by $D_h = 4\psi/S_v$, is shown in Table A1.

**Table A1.** The hydraulic diameter calculated based on the porous media definition.

| Geometry | Diamond TPMS Model | Conventional Topology-Optimized Model | Topology-Optimized Lattice-Filled Model |
|---|---|---|---|
| Hydraulic diameter (mm) | 6.65 | 5.55 | 5.3 |

## References

1. Lin, C.; Kerscher, F.; Spliethoff, H. Thermal Gradient Management in Solid Oxide Fuel Cells: Mechanisms, Strategies, and Future Directions. *J. Power Sources* **2025**, *656*, 238017. [CrossRef]
2. Veza, I. Fuel-Cell Thermal Management Strategies for Enhanced Performance: Review of Fuel-Cell Thermal Management in Proton-Exchange Membrane Fuel Cells (PEMFCs) and Solid-Oxide Fuel Cells (SOFCs). *Hydrogen* **2025**, *6*, 65. [CrossRef]
3. Zeng, Z.; Qian, Y.; Zhang, Y.; Hao, C.; Dan, D.; Zhuge, W. A Review of Heat Transfer and Thermal Management Methods for Temperature Gradient Reduction in Solid Oxide Fuel Cell (SOFC) Stacks. *Appl. Energy* **2020**, *280*, 115899. [CrossRef]
4. Aguiar, P.; Adjiman, C.S.; Brandon, N.P. Anode-Supported Intermediate-Temperature Direct Internal Reforming Solid Oxide Fuel Cell: II. Model-Based Dynamic Performance and Control. *J. Power Sources* **2005**, *147*, 136–147. [CrossRef]
5. Manglik, R.M.; Magar, Y.N. Heat and Mass Transfer in Planar Anode-Supported Solid Oxide Fuel Cells: Effects of Interconnect Fuel/Oxidant Channel Flow Cross Section. *J. Therm. Sci. Eng. Appl.* **2015**, *7*, 041003. [CrossRef]
6. Gonçalves, J.F.; Sá, L.F.N.; Lopes, T.; Meneghini, J.R.; Silva, E.C.N. Design of Solid Oxide Fuel Cell (SOFC) Channel Layout Using the Topology Optimization Method with a Design Variable Propagation Approach. *Struct. Multidiscip. Optim.* **2024**, *67*, 216. [CrossRef]

7. Ren, K.; Su, Y.; Zhong, Z.; Jiao, Z. Microstructure-Insight Topology Optimization for Efficient Interconnect Flow Channels in Solid Oxide Fuel Cells. *Int. J. Heat Mass Transf.* **2025**, *242*, 126823. [CrossRef]
8. Lee, W.; Lang, M.; Costa, R.; Lee, I.S.; Lee, Y.S.; Hong, J. Enhancing Uniformity and Performance in Solid Oxide Fuel Cells with Double Symmetry Interconnect Design. *Appl. Energy* **2025**, *381*, 125178. [CrossRef]
9. Gong, C.; Luo, X.; Tu, Z.; Chan, S.H. A Novel Flow Channel Design to Achieve High Temperature Homogenization in Solid Oxide Fuel Cell. *Int. J. Hydrogen Energy* **2024**, *52*, 442–453. [CrossRef]
10. Dillig, M.; Plankenbühler, T.; Karl, J. Thermal Effects of Planar High Temperature Heat Pipes in Solid Oxide Cell Stacks Operated with Internal Methane Reforming. *J. Power Sources* **2018**, *373*, 139–149. [CrossRef]
11. Dillig, M.; Meyer, T.; Karl, J. Integration of Planar Heat Pipes to Solid Oxide Cell Short Stacks. *Fuel Cells* **2015**, *15*, 742–748. [CrossRef]
12. Dokmaingam, P.; Irvine, J.T.S.; Assabumrungrat, S.; Charojrochkul, S.; Laosiripojana, N. Modeling of IT-SOFC with Indirect Internal Reforming Operation Fueled by Methane: Effect of Oxygen Adding as Autothermal Reforming. *Int. J. Hydrogen Energy* **2010**, *35*, 13271–13279. [CrossRef]
13. Dillig, M.; Leimert, J.; Karl, J. Planar High Temperature Heat Pipes for SOFC/SOEC Stack Applications. *Fuel Cells* **2014**, *14*, 479–488. [CrossRef]
14. Li, C.; Cheng, K.; Li, B.; Liu, H.; Qin, J.; Wei, L. Feasibility Study on Supercritical Fuel Cooled Solid Oxide Fuel Cell Stack with Internal Reforming. *Int. J. Hydrogen Energy* **2024**, *50*, 312–322. [CrossRef]
15. Fan, J.; Wang, Y.; Tian, R.; Shi, J.; Shi, Y.; Cao, H.; Cai, N. Thermal Management of Solid Oxide Fuel Cells with Liquid Metal. *Int. J. Hydrogen Energy* **2023**, *48*, 10659–10670. [CrossRef]
16. Promsen, M.; Komatsu, Y.; Sciazko, A.; Kaneko, S.; Shikazono, N. Feasibility Study on Saturated Water Cooled Solid Oxide Fuel Cell Stack. *Appl. Energy* **2020**, *279*, 115803. [CrossRef]
17. Wang, M.; Wang, K.; Wang, Y.; Chen, J.; An, B.; Tu, S. Effect of Operation Parameters on the Thermal Characteristics in a Planar Solid Oxide Fuel Cell. *J. Therm. Sci.* **2024**, *33*, 1974–1989. [CrossRef]
18. Peters, R.; Blum, L.; Deja, R.; Hoven, I.; Tiedemann, W.; Küpper, S.; Stolten, D. Operation Experience with a 20 KW SOFC System. *Fuel Cells* **2014**, *14*, 489–499. [CrossRef]
19. Hasbi, S.; Amber, I.; Hossain, M.; Leong, K.Y.; Saharudin, M.S. Feasibility Study of Liquid Metal-Based Thermal Management for Solid Oxide Electrolyzer Cell (SOEC). *Int. J. Green Energy* **2025**, 1–22. [CrossRef]
20. Wang, H.; Zhao, C.; Liu, W.; Liu, Z.; Bian, H.; Zhang, K. Advances in Triply Periodic Minimal Surface Structures for Thermal Management Systems: A Comprehensive Review. *Appl. Therm. Eng.* **2025**, *279*, 127481. [CrossRef]
21. Yeranee, K.; Rao, Y. A Review of Recent Investigations on Flow and Heat Transfer Enhancement in Cooling Channels Embedded with Triply Periodic Minimal Surfaces (TPMS). *Energies* **2022**, *15*, 8994. [CrossRef]
22. Cheng, L.; Liu, J.; Liang, X.; To, A.C. Coupling Lattice Structure Topology Optimization with Design-Dependent Feature Evolution for Additive Manufactured Heat Conduction Design. *Comput. Methods Appl. Mech. Eng.* **2018**, *332*, 408–439. [CrossRef]
23. Jiang, Y.; Hu, J.; Wang, S.; Lei, N.; Luo, Z.; Liu, L. Meshless Optimization of Triply Periodic Minimal Surface Based Two-Fluid Heat Exchanger. *CAD Comput. Aided Des.* **2023**, *162*, 103554. [CrossRef]
24. Yeranee, K.; Xu, C.; Rao, Y.; Zhang, Y. Experimental and Numerical Study of Improving Flow and Heat Transfer in a Serpentine Cooling Channel with Topology-Optimized TPMS Porous Structures. *Int. J. Heat Mass Transf.* **2024**, *231*, 125873. [CrossRef]
25. Xu, C.; Rao, Y.; Yeranee, K. Experiments and Numerical Analysis of Turbulent Flow Heat Transfer and Pressure Loss of a Channel with Triply Periodic Minimal Surface Structures. *Int. J. Therm. Sci.* **2026**, *220*, 110334. [CrossRef]
26. Li, H.; Ding, X.; Jing, D.; Xiong, M.; Meng, F. Experimental and Numerical Investigation of Liquid-Cooled Heat Sinks Designed by Topology Optimization. *Int. J. Therm. Sci.* **2019**, *146*, 106065. [CrossRef]
27. Matsumori, T.; Kondoh, T.; Kawamoto, A.; Nomura, T. Topology Optimization for Fluid-Thermal Interaction Problems under Constant Input Power. *Struct. Multidiscip. Optim.* **2013**, *47*, 571–581. [CrossRef]
28. Baek, S.M.; Yu, S.H.; Nam, J.H.; Kim, C.J. A Numerical Study on Uniform Cooling of Large-Scale PEMFCs with Different Coolant Flow Field Designs. *Appl. Therm. Eng.* **2011**, *31*, 1427–1434. [CrossRef]
29. Shahsavari, S.; Desouza, A.; Bahrami, M.; Kjeang, E. Thermal Analysis of Air-Cooled PEM Fuel Cells. *Int. J. Hydrogen Energy* **2012**, *37*, 18261–18271. [CrossRef]
30. Fergus, J.W. Metallic Interconnects for Solid Oxide Fuel Cells. *Mater. Sci. Eng. A* **2005**, *397*, 271–283. [CrossRef]
31. Ghorbani, B.; Vijayaraghavan, K. 3D and Simplified Pseudo-2D Modeling of Single Cell of a High Temperature Solid Oxide Fuel Cell to Be Used for Online Control Strategies. *Int. J. Hydrogen Energy* **2018**, *43*, 9733–9748. [CrossRef]
32. Liu, Y.; Tan, J.; Zhang, D.; Kuai, Z. Modeling and Performance Analysis of Solid Oxide Fuel Cell Power Generation System for Hypersonic Vehicles. *Aerospace* **2024**, *11*, 846. [CrossRef]
33. Haertel, J.; Engelbrecht, K.; Lazarov, B.; Sigmund, O. Topology Optimization of a Pseudo 3D Thermofluid Heat Sink Model. *Int. J. Heat Mass Transf.* **2018**, *121*, 1073–1088. [CrossRef]

4. Zeng, S.; Lee, P. Topology Optimization of Liquid-Cooled Microchannel Heat Sinks: An Experimental and Numerical Study. *Int. J. Heat Mass Transf.* **2019**, *142*, 118401. [CrossRef]
5. Zeng, S.; Kanargi, B.; Lee, P. Experimental and Numerical Investigation of a Mini Channel Forced Air Heat Sink Designed by Topology Optimization. *Int. J. Heat Mass Transf.* **2018**, *121*, 663–679. [CrossRef]
6. Zeng, S.; Sun, Q.; Lee, P.S. Thermohydraulic Analysis of a New Fin Pattern Derived from Topology Optimized Heat Sink Structures. *Int. J. Heat Mass Transf.* **2020**, *147*, 118909. [CrossRef]
7. Marshall, S.D.; Lee, P.S. 3D Topology Optimisation of Liquid-Cooled Microchannel Heat Sinks. *Therm. Sci. Eng. Prog.* **2022**, *33*, 101377. [CrossRef]
8. Zhang, M.; Wang, E.; Ni, M.; Zheng, K.; Ouyang, M.; Hu, H.; Wang, H.; Lu, L.; Ren, D.; Chen, Y. A Numerical Analysis of Metal-Supported Solid Oxide Fuel Cell with a Focus on Temperature Field. *Heliyon* **2024**, *10*, e37271. [CrossRef] [PubMed]

*energies*

MDPI

*Review*

# Pathways for Greenhouse Thermal Management's Contribution to Net-Zero Food Production

**Samson Sogbaike, Celestina Ezenwajiaku, Amir Badiee, Chris Bingham and Aliyu M. Aliyu ***

School of Engineering and Physical Sciences, University of Lincoln, Bayford Pool, Lincoln LN6 7TS, UK; abadiee@lincoln.ac.uk (A.B.)
* Correspondence: aaliyu@lincoln.ac.uk

**Abstract**

Decarbonising greenhouse food production requires improvements in thermal management, energy efficiency, and system integration. Greenhouse energy demand is shaped by coupled heat and mass transfer processes, particularly envelope performance, ventilation, and latent heat associated with humidity control. This article synthesises recent advances in greenhouse microclimate control with emphasis on heat transfer, low-carbon heating and cooling, thermal storage, renewable and waste heat integration, and advanced modelling and control approaches. The review shows that humidity control and latent load management are primary drivers of winter energy use, as moisture removal through ventilation and dehumidification directly increases the sensible heating required to maintain indoor temperature setpoints. When assessed using realistic psychrometric relationships, ventilation and dehumidification can dominate peak heating demand and seasonal consumption. The performance of heat pumps, storage systems, semi-closed greenhouse concepts, and renewable heat pathways depends on how thermal loads are defined, how system boundaries are set, and how technologies are integrated in operation. Digital twins, predictive control, and hybrid physics-data models are increasingly used to manage variability in weather, energy prices, and infrastructure constraints. Greenhouse decarbonisation cannot be treated as a simple substitution of energy sources. System performance depends on coordinated design and operation, including heat recovery, moisture removal, and integration of supply technologies. Semi-closed and heat recovery-based configurations can reduce the ventilation–heating penalty and lower primary energy demand compared with vent-to-dry approaches. Long-term market projections suggest that the commercial greenhouse sector could expand substantially by 2050 under plausible growth scenarios, reflecting increased capital investment rather than a proportional rise in global food output. Net-zero greenhouse production is achievable through combined improvements in thermal management, electrification, and renewable energy integration. However, large-scale deployment depends on consistent modelling assumptions, credible economic assessment, and alignment with heat and $CO_2$ supply infrastructure. The transition is therefore shaped by system integration and planning as much as by individual technologies.

**Keywords:** controlled-environment agriculture; greenhouse thermal management; net-zero greenhouse systems; low-carbon heating technologies; greenhouse energy systems

## 1. Introduction

Global food production is subject to growing pressure as a result of population growth, urbanisation, dietary change, climate change, and increasingly long and exposed supply

 https://doi.org/10.3390/en19081975

chains. A meta-analysis of global projections suggests a 35–56% increase in food demand between 2010 and 2050, depending on socioeconomic pathways and assumptions [1]. Climate-related extremes (e.g., heat, droughts and floods) have already affected agricultural productivity and are expected to intensify risks to food availability, prices, and livelihoods across regions (The IPCC AR6 Working Group II [2]).

Protected cultivation, including the use of greenhouses and the wider controlled-environment agriculture (CEA), has expanded as a strategy to stabilise food production, improve yield predictability, and extend growing seasons. By protecting crops from outdoor variability, greenhouse systems enable targeted management of microclimate variables such as temperature, humidity, radiation and $CO_2$, which improves food productivity and quality consistency. Recent reviews of greenhouse automation and CEA emphasise the roles of sensing, actuation, and integrated environmental control in this transition [3]. Benke and Tomkins [4] likewise discuss how vertical farming and other controlled-environment models are positioned as emerging food production pathways responding to land constraints, urban demand, and environmental pressures.

The benefit of protected cultivation comes with a clear trade-off, particularly the higher energy requirements for thermal regulation. As food production systems transition from open-field agriculture and conventional greenhouses towards semi-closed or fully closed structures and plant factories, the energy required for heating, cooling, dehumidification and sometimes artificial lighting will increase sharply. Mills's [5] study contrasts low-tech open greenhouses (with relatively minor energy uses) with closed/fully controlled facilities that require substantial energy for climate control and lighting, which leads to higher capital and operating costs than open-field cultivation. In temperate and colder regions, heating remains the dominant energy usage for commercial greenhouses. Vadiee and Martin [6] review energy management strategies for commercial greenhouses, with emphasis on Nordic climates, and show that fossil fuel-based heating still plays a major role, making combined conservation (energy savings) and supply strategies central to decarbonisation. In high-latitude greenhouses, energy requirements can increase further because year-round cultivation often combines heating with supplemental lighting [7].

This high level of energy use places greenhouse horticulture at the centre of the food–energy climate nexus. According to the Food and Agriculture Organization of the United Nations (FAO), agrifood systems emitted approximately 16.2 Gt $CO_2$-eq in 2022, essentially unchanged from 2021, corresponding to 29.7% of total anthropogenic greenhouse gas emissions. They also note that energy use along supply chains is an important and growing contributor [8]. The Sixth Assessment Report (AR6) of the Intergovernmental Panel on Climate Change (IPCC) note that climate risks to food systems grow as temperatures rise, and that mitigation and adaptation must progress together to avoid increasing food insecurity and wider socio-economic vulnerability [9,10].

From an engineering perspective, greenhouse microclimates arise from coupled heat and mass transfer processes within a semi-transparent structure that contains a crop canopy. Incoming short-wave solar radiation usually supplies most of the energy, while losses occur through long-wave radiative exchange with the sky, conduction through the cover material, and convection driven by infiltration and ventilation [11,12]. Crop evapotranspiration adds substantial latent loads, coupling temperature regulation to humidity control [13]. This coupled behaviour is well established in greenhouse climate modelling. Katzin et al. [14] developed a process-based greenhouse model and identified latent heat as a core component of the greenhouse energy balance. Other modelling studies use similar energy and mass balance formulations to capture these interactions and to evaluate energy demand and climate control performance under real operating conditions [15,16].

https://doi.org/10.3390/en19081975

Recent work on greenhouse decarbonisation focuses on cutting thermal demand, switching to low-carbon sources of heat and cooling, and improving operation through modelling and control. On the demand side, measures such as improved envelopes, thermal screens and operational strategies aim to reduce heat losses and stabilise conditions with lower energy input [6,17]. On the supply side, electrification and the integration of renewable sources are becoming more attractive as electricity systems decarbonise. The International Energy Agency (IEA) identifies heat pumps as an important option for reducing heating-related emissions, with their advantages growing as grid carbon intensity decreases [18]. Thermal energy storage, ranging from short-term buffering to seasonal storage, can capture surplus heat and improve both efficiency and flexibility in greenhouse systems [19].

Modelling, simulation, and optimisation in greenhouse research has shifted from design evaluation toward operational decision support. Alongside process-based models and computational fluid dynamics (CFD) tools, reduced-order models are now used more frequently because they allow faster optimisation and make it easier to evaluate how design decisions, ventilation strategies, and climate control actions influence energy demand [14]. At the operational level, climate control is moving beyond rule-based or simple feedback systems toward more advanced methods such as model predictive control (MPC) and AI-driven approaches [20,21]. Comparative studies that evaluate MPC against reinforcement learning (RL) methods reflect the growing interest in applying AI-based control to improve energy efficiency and overall greenhouse operation [22].

Despite these advances, research in this area is still dispersed across different climates, greenhouse designs, and technology scales. Many studies report results using metrics that are hard to compare (for example, energy savings without carbon intensity, or thermal improvements without considering cost or scalability assessment), limiting comparability and slowing translation to deployable solutions. Studies also often evaluate individual devices or strategies on their own, even though a change in one part of the system (such as lighting) can shift heat demand, electricity import or export, and $CO_2$ supply requirements at the system level [7]. These gaps point to the need for an integrated assessment that links heat transfer processes, technology performance, techno-economic factors, infrastructure constraints, and food security implications within a single framework.

Therefore, this review brings together recent developments in greenhouse thermal management and examines practical pathways toward net-zero food production. Rather than treating technologies, control strategies, or energy-saving measures in isolation, it considers heat transfer processes, load structure, techno-economic constraints, infrastructure requirements, and food security implications within a combined framework. It shows that greenhouse decarbonisation depends not only on the technologies adopted, but also on how thermal demand is characterised, how system boundaries are defined, and how interventions are implemented over time. Based on this, the review outlines a staged pathway linking demand reduction, electrification, renewable and waste heat integration, thermal storage, and advanced control under realistic operating conditions. It also covers heat transfer mechanisms, heating and cooling options, storage strategies, energy sources, modelling approaches, and deployment constraints (including cost, infrastructure readiness, and scalability). The aim is to clarify the state of the art, identify the main technical and deployment barriers and outline research directions most likely to accelerate low-carbon and resilient greenhouse food production.

The terms net-zero greenhouse (or net-zero energy greenhouse), low-carbon and decarbonisation are used with defined boundaries. Net-zero greenhouse may refer to balancing annual energy demand with on-site or purchased renewable energy, or to reducing operational emissions from energy use to net zero over a defined period. A broader

https://doi.org/10.3390/en19081975

interpretation includes upstream emissions from energy supply, infrastructure, inputs, and $CO_2$ provision. These definitions are not equivalent and can lead to different assessments of system performance. In this review, net-zero is interpreted mainly in operational terms, focusing on emissions from greenhouse energy use and related services, including heating, electricity consumption, and $CO_2$ supply, unless stated otherwise. Low-carbon describes approaches that reduce emissions compared with conventional fossil-based systems without eliminating them entirely. Decarbonisation refers to the reduction in emissions through demand reduction, electrification, integration of renewable or recovered heat, storage, and improved control. Where relevant, a distinction is made between operational emissions and broader life cycle impacts, as outcomes depend on factors such as grid carbon intensity, auxiliary electricity use, infrastructure configuration, and $CO_2$ sourcing.

### *1.1. Greenhouse Energy Use, Decarbonisation, and Food Security*

Greenhouse horticulture is increasingly positioned to reduce food-system vulnerability under climate change, offering the ability to stabilise production, extend growing seasons, and reduce exposure to extreme weather. However, the same mechanisms that provide this stability (such as active control of temperature, humidity, radiation, and $CO_2$ concentration) make production highly dependent on energy availability, energy prices, and the carbon intensity of supply. As a result, greenhouse food production sits at the intersection of food security, energy security, and climate mitigation.

#### 1.1.1. Scale and Structure of Greenhouse Energy Use

Energy use in greenhouses varies widely depending on climate, crop, structural design, and the level of environmental control. In cold and temperate regions, space heating often dominates annual demand, and in warm and humid climates, cooling and dehumidification can become the main loads. In high-latitude year-round systems, supplemental lighting adds a major electrical load and interacts with heating requirements because lighting technologies differ in their waste heat output [7].

Paris et al. [23] distinguish low-energy greenhouse systems (often relying on passive solar gains and basic ventilation) from high-energy systems where tight climate control requires substantial external energy input. Heating and cooling account for the majority of operational energy consumption, and fossil fuels remain a common heat source in many regions [23].

The Dutch greenhouse sector is a frequently cited example of intensive but highly integrated greenhouse production. Sector reports indicate that total energy consumption in Dutch greenhouse horticulture was about 106 PJ in 2019, dominated by natural gas, with extensive use of combined heat and power (CHP) systems to supply both heat and electricity [7,24]. A focused transition study shows the rapid diffusion of CHP in Dutch greenhouse horticulture, growing from about 1000 $MW_{el}$ in 2003 to over 3000 $MW_{el}$ in 2009, reshaping the sector's energy and $CO_2$ profile [25].

These arrangements also demonstrate why decarbonisation needs to be assessed at the system level. Katzin et al. [7] report that moving from high-pressure sodium (HPS) to LED lighting can lower overall energy use by about 10–25%, but often increases heating demand because LEDs release less heat. The net benefit depends on how heat is supplied (fossil or low carbon) and the carbon intensity and price of electricity [7].

#### 1.1.2. The Food Security Lens: Why Energy and Emissions Matter

Food security is commonly described through availability, access, utilisation, and stability, as shown in Figure 1 [26]. Greenhouse production can enhance availability and stability by buffering climate variability, but dependence on external energy can introduce

https://doi.org/10.3390/en19081975

vulnerabilities that affect access and long-term stability, especially in cold-climate regions where heating can be a large share of operating cost.

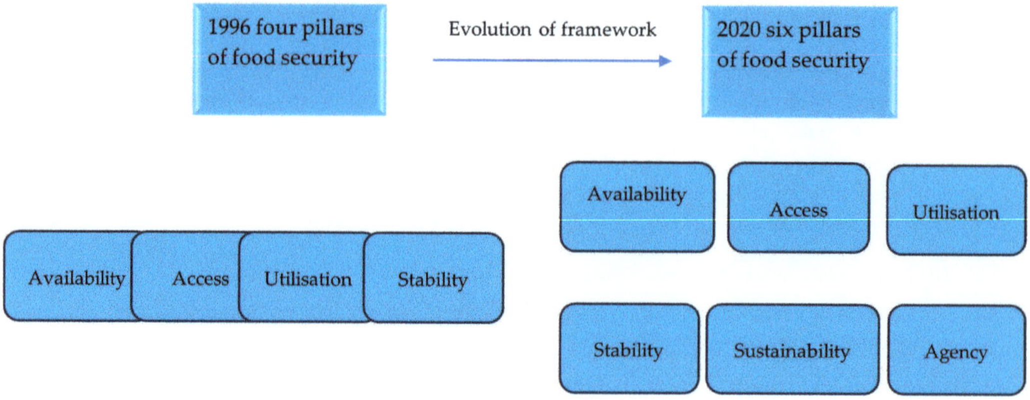


**Figure 1.** Evolution of food security frameworks from the four pillars (availability, access, utilisation, and stability) introduced in 1996 to the six dimensions adopted in 2020, which adds sustainability and agency [27].

Market and policy reports on Dutch horticulture during the energy crisis show how rising energy costs shaped production decisions, including scaling back winter output and reassessing crop choices [24]. From a food security perspective, such responses can reduce off-season supply and increase price volatility, particularly in regions that rely on greenhouse production during winter months. Decarbonisation options that reduce exposure to fossil fuel price swings (such as geothermal heat, waste heat, or electrification with stable tariffs or thermal storage) can therefore support both emissions reduction and greater resilience in the food system.

### *1.2. Summary and Contribution of Reviewed Literature*

The reviewed studies show rapid progress in greenhouse heat transfer analysis, thermal technologies, energy storage, renewable and waste heat integration, and advanced control. However, the synthesis remains uneven: the results are difficult to compare across climates and greenhouse types, and practical deployment factors (such as cost, infrastructure limits, and $CO_2$ supply) are often treated as secondary issues. This review brings these elements together through a heat transfer-centred framework that connects technology performance and operational control with net-zero goals and practical deployment realities. This approach allows system-level trade-offs to be identified more clearly, particularly those related to humidity control, latent load management, auxiliary electricity use, supply temperature, and shared infrastructure.

A key contribution is the development of a staged decarbonisation pathway that reflects both engineering requirements and implementation constraints. The analysis shows that a credible net-zero greenhouse operation depends not only on low-carbon heat supply, but also on the sequence in which demand reduction, electrification, storage, renewable integration, and network coupling are introduced under clearly defined system boundaries. In doing so, it consolidates evidence from multiple disciplines, clarifies interactions between system components, and identifies research directions most likely to support low-carbon, resilient, and scalable greenhouse food production. Figure 2 summarises this framework, linking heat transfer processes with technologies, storage options, heat supply pathways, and the constraints that shape deployment. It also shows

https://doi.org/10.3390/en19081975

how economic factors, infrastructure availability, energy price volatility, and policy support influence decarbonisation pathways.

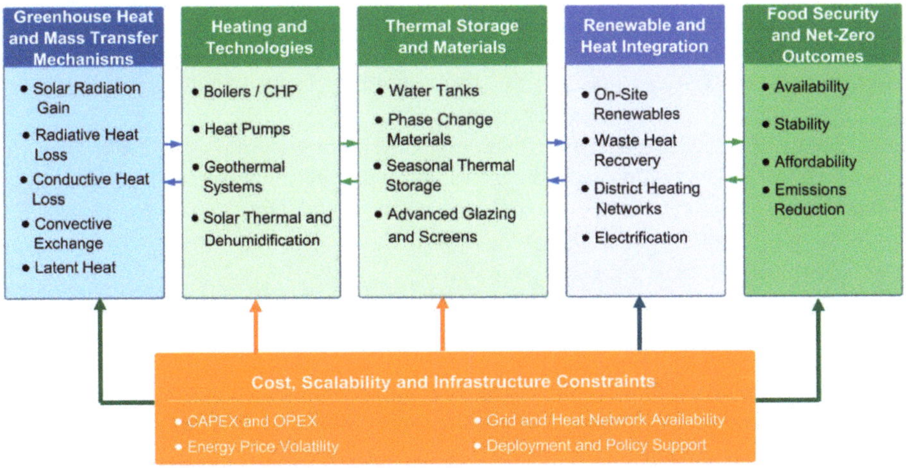


**Figure 2.** Greenhouse energy management framework linking heat transfer processes, heating and cooling technologies, thermal storage options, and renewable integration with system performance constrained by economic, infrastructure, and policy factors.

## 2. Methodology

This study was carried out as a systematic literature review using a framework-guided thematic synthesis. The evidence identification, screening, eligibility and reporting processes followed the PRISMA 2020 [28]. The review process was structured using the SALSA approach (Search, Appraisal, Synthesis, and Analysis) as described by Grant and Booth [29]. Given heterogeneity in study designs, boundary conditions and reported metrics, synthesis followed the SWiM guideline [30].

The review focused on development in greenhouse thermal management and on the feasibility of pathways toward net-zero greenhouse food production. The assessment considered operational energy-related emissions associated with heating, electricity use, and $CO_2$ supply, while broader life cycle impacts were included where relevant and where studies reported clearly defined boundaries. The scope was limited to protected cultivation systems in which thermal loads, heating and cooling, storage and envelope/cover measures, energy supply decarbonisation strategies, modelling and control approaches, and tech-economic constraints can be assessed using extractable quantitative or validated modelling evidence. Seven research questions guided the analysis:

- Covering dominant heat and mass transfer mechanisms (dominate greenhouse thermal loads across climates and typologies);
- Performance of thermal technologies;
- Contributions of storage and material innovations;
- Options for integrating renewable and waste heat;
- Impacts of modelling and control approaches;
- Techno-economic performance and scalability;
- Implications of net-zero feasibility and food-system resilience.

Searches were conducted in Scopus, Web of Science, and ScienceDirect, with Google Scholar used for supplementary retrieval. The primary time window was 2008–2025, with particular attention to post-2015 work reflecting rapid growth in electrified heating, thermal storage integration, and optimisation-based control. Institutional publications such

https://doi.org/10.3390/en19081975

as the Food and Agriculture Organization, Intergovernmental Panel on Climate Change, and International Energy Agency were used for system boundaries (such as emissions accounting and decarbonisation context) but not used as primary sources for device-level performance. Sector guidance and market reports were used only to provide context on commercial practice, policy conditions, and market scale. They were not treated as primary evidence for technical conclusions and were considered alongside peer-reviewed studies, taking account of their scope and limitations.

Studies were considered eligible if they examined greenhouse or protected cultivation in which thermal management (heating, cooling, dehumidification, humidity control or energy modelling) formed part of the design or operation, and if they provided extractable evidence for at least one of the following:

- Quantitative thermal or energy performance;
- Validated modelling (energy balance, CFD, and reduced-order or hybrid/ML surrogate models);
- Quantified outcomes of control or optimisation strategies;
- Explicit emissions or life cycle assessment (LCA) with boundary definition;
- Techno-economic metrics (such as CAPEX, OPEX, payback, NPV, and LCOH).

Pure agronomic studies without thermal or energy relevance, papers lacking methodological detail, and duplicated publications were excluded.

The extracted data fields included system boundaries, climate context, greenhouse type, thermal load representation, technology configuration, control strategy, validation method, and reported energy, carbon and economic metrics. Studies were grouped by system layer: thermal load characterisation, heating and cooling technologies, storage and envelope measures, renewable and waste heat integration architectures, modelling and control approaches, and techno-economic and scalability constraints. Because modelling assumptions and reporting standards vary widely, cross-study averages were not used when system boundaries were not directly comparable. Instead, findings are interpreted within their climatic and operational contexts, with attention on interactions relevant to net-zero pathways, such as how envelope performance affects heat pump sizing, how LED lighting shifts the heat–electricity balance, how storage influences peak demand, and how control strategies shape ventilation-driven latent loads.

Key performance indicators such as U-value, coefficient of performance (COP), energy saving rate, and ventilation rate are interpreted in the context in which they are reported, rather than being standardised across studies. These metrics are presented using different units, boundary conditions, and evaluation methods, depending on climate, greenhouse configuration, control strategy, and modelling approach. Comparisons are therefore limited to studies with comparable assumptions. Where this is not possible, the results are discussed with reference to the conditions under which they were obtained. Table 1 summarises the main indicators, their units, and the factors that influence their interpretation.

**Table 1.** Key performance indicators, reported units, and factors influencing comparability across greenhouse studies.

| Indicator | Typical Units | Definition/Basis | Factors Affecting Comparability | Sections |
|---|---|---|---|---|
| U-value (overall heat transfer coefficient) | $W\ m^{-2}\ K^{-1}$; in some studies, reported as $W\ m^{-2}\ °C^{-1}$ | Heat transfer through the greenhouse envelope per unit area and temperature difference | Wind speed, cover condition, radiative exchange (including sky temperature), structural elements, measurement method, thermal screens | Sections 3.2.1 and 3.5.1 |

https://doi.org/10.3390/en19081975

**Table 1.** *Cont.*

| Indicator | Typical Units | Definition/Basis | Factors Affecting Comparability | Sections |
|---|---|---|---|---|
| Coefficient of performance (COP) | (dimensionless); sometimes reported as instantaneous COP or seasonal COP/COPh | Ratio of useful heating or cooling output to electrical input | Source and sink temperatures, operating mode, part-load behaviour, inclusion of auxiliary electricity, instantaneous vs. seasonal definition, system boundary (electrical vs. primary energy) | Sections 3.4.2, 3.6.1 and 3.8 |
| Energy-saving rate | %; in some sections discussed alongside energy savings in kWh $m^{-2}$, and heat-cost metrics such as $/GJ or $/MWh_{th}$ | Reduction in energy use relative to a reference system | Baseline definition (critical), climate, control strategy, system boundary, inclusion of auxiliary loads and $CO_2$ supply | Sections 3.5, 3.8 and 3.9 |
| Ventilation rate/air exchange | $h^{-1}$ or $s^{-1}$; sometimes expressed as mass flow rate (kg $s^{-1}$) | Rate of air exchange between greenhouse and outdoor air | Ventilation type, wind conditions, vent geometry, discharge coefficients, infiltration, control objective (temperature or humidity driven) | Sections 3.2.3 and 3.3 |
| Latent heat/moisture removal rate | W; kg $s^{-1}$; L $h^{-1}$; sometimes discussed with energy input such as W $m^{-2}$ | Moisture-related heat transfer from evapotranspiration or condensation | Crop transpiration, humidity definition (RH, VPD, or dew point), condensation modelling, dehumidification method | Sections 3.2.4 and 3.3.4 |
| Heating demand/load | W, kW, W $m^{-2}$, kWh·$m^{-2}$ $year^{-1}$, and in some studies MWh·$year^{-1}$ | Heat required to maintain setpoint temperature | Climate, envelope performance, ventilation, latent loads, internal gains, temporal basis (peak vs. annual) | Sections 3.2 and 3.8 |

## 3. Results and Discussion

The search process outlined in Section 2 returned 1427 records in total (Scopus: 1334; Web of Science: 93). After deduplication based on DOI and title matching, 1220 unique entries remained. Screening of title and abstract against the eligibility criteria removed 103 records, mainly papers addressing greenhouse gas/climate-only studies or non-agricultural greenhouse effect. This resulted in 1117 records for full-text review.

### 3.1. Overview of Included Studies

All the included studies ($n$ = 1117) were coded using the system-layer and evaluation-lens framework described in Section 2. Many publications span several analytical layers, for example, combining thermal modelling with techno-economic assessment or evaluating heat pump deployment alongside control optimisation, so theme classifications are non-mutually exclusive. Within the screened dataset ($n$ = 1220), the most commonly represented themes were:

- Heating, cooling, and dehumidification technologies (906 records);
- Renewable and waste heat integration, including geothermal, district heat, and solar thermal systems (733 records);
- Techno-economic and life cycle/emissions assessments (466 records);

- Storage and envelope/material innovations, including phase change materials (PCM), thermal energy storage (TES), glazing systems, and thermal screens (404 records);
- Modelling, optimisation, and advanced control approaches (MPC, reinforcement learning, digital twins, surrogate modelling) (347 records);
- Heat transfer and modelling studies (181 records).

These counts reflect thematic tagging within the screened dataset and therefore exceed the total number of included publications due to multi-category classification. Publication output shows a clear shift in research focus. Early work (late 2000s to early 2010s) focused on thermal characterisation and energy balance modelling. From around 2015 onward, there has been a notable increase in research on electrification, renewable integration, storage coupling, and intelligent control. As shown in Figure 3, annual publication output increases steadily across the review period and accelerates in the most recent years, aligning with wider interest in greenhouse decarbonisation.

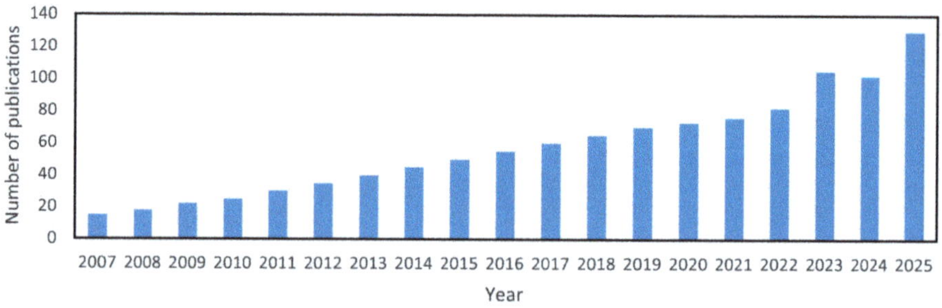


**Figure 3.** Annual number of publications in greenhouse agriculture from 2007 to 2025, showing a steady increase in research activity and accelerated growth in recent years.

Geographically, most publications come from countries with established greenhouse sectors or strong energy-system research infrastructures. As shown in Figure 4, China accounts for the largest number of publications, followed by the United States and several European and Asian countries. This distribution reflects differences in production scale and research investment rather than climatic similarity, so the prevalence of themes does not necessarily align with their climatic relevance, and it follows directly from the material captured in this review.

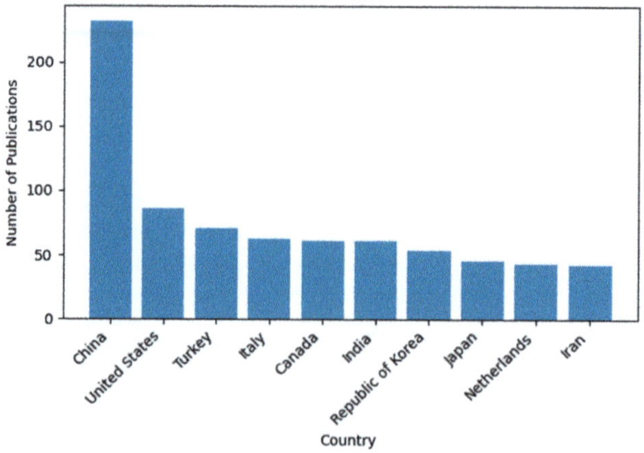


**Figure 4.** Distribution of publications by country within the screened dataset, with China contributing the largest share, followed by the United States, and countries across Europe and Asia.

https://doi.org/10.3390/en19081975

The screened literature shows substantial coverage of technology-specific solutions and growing attention to renewable integration and economic performance. However, thematic dominance does not equate to methodological consistency/comparability. Reporting practices vary widely across studies, particularly with respect to boundary definitions, emission factors, auxiliary energy use, and economic assumptions. The following subsections, therefore, synthesise evidence within clearly defined system layers, emphasising transparency in underlying parameters and comparability across studies, consistent with the review methodology.

### 3.2. Heat Transfer Mechanism and Thermal Load Characterisation

A credible assessment of greenhouse decarbonisation pathways starts with a clear description/account of the thermal loads that low-carbon systems must meet/serve. The performance of heat pumps, geothermal sources, solar thermal collectors, and waste heat integration depends not only on annual energy demand, but also on how the demand is distributed over time (peaks vs. base load) and by type (sensible vs. latent). These demands arise from coupled heat and mass transfer processes across the enclosure and within the crop–air system. Process-based greenhouse climate models, ranging from compact energy balance formulations to fully dynamic simulations, share the same conservation principles and primarily differ in parameter choices, subsystem coupling, and spatial resolution [14].

A common starting point is the energy balance of the greenhouse air volume, which can be expressed as

$$C_{air}\frac{dT_{air}}{dt} = Q_{solar} + Q_{heat} - Q_{trans} - Q_{vent} - Q_{rad} - Q_{lat} \tag{1}$$

where $C_{air}$ is the effective thermal capacitance of greenhouse air (J $K^{-1}$), $T_{air}$ is the indoor air temperature (K), $Q_{solar}$ is the absorbed short-wave radiation (W), $Q_{heat}$ is the heating input (W), $Q_{trans}$ is transmission losses through the envelope (W), $Q_{vent}$ is the sensible exchange due to ventilation and infiltration (W), $Q_{rad}$ is the net long-wave radiative exchange with the sky and surroundings (W), and $Q_{lat}$ is the latent heat associated with evapotranspiration and condensation (W). Variants of this balance (with additional states for cover, soil, and crop) underpin most process-based greenhouse models used for design and control analysis [14]. The formulation describes the main heat transfer pathways and their interactions, rather than serving as a detailed predictive model. In practice, greenhouse thermal behaviour is analysed using coupled energy mass balance models, CFD, or hybrid approaches that resolve spatial variation, transient behaviour, and crop–climate coupling. These are discussed in Section 3.2.5. Figure 5 shows the main heat and mass transfer processes within a greenhouse, corresponding to the terms in the air energy balance.

This decomposition matters for decarbonisation because heating demand is not a fixed input; it is the residual of several interacting fluxes. Changes in cover properties, air exchange, radiative conditions, or moisture dynamics can shift peak power and seasonal energy use. These same shifts can alter the required supply temperature, which directly affects heat pump COP and the feasibility of integrating low-grade heat sources.

#### 3.2.1. Envelope/Cover Transmission and U-Values

In cold (heating-dominated) climates, conductive heat loss through the cover and structural elements can account for a large share of seasonal heating demand. Under quasi-steady conditions, transmission loss is commonly written as

$$Q_{trans} = UA(T_{in} - T_{out}) \tag{2}$$

https://doi.org/10.3390/en19081975

where $U$ is the overall heat transfer coefficient (W m$^{-2}$ K$^{-1}$), $A$ is the envelope/cover area (m$^2$), and $T_{in} - T_{out}$ is the indoor and outdoor temperature difference (K). This formulation remains the basis of most greenhouse heating load calculations [31].

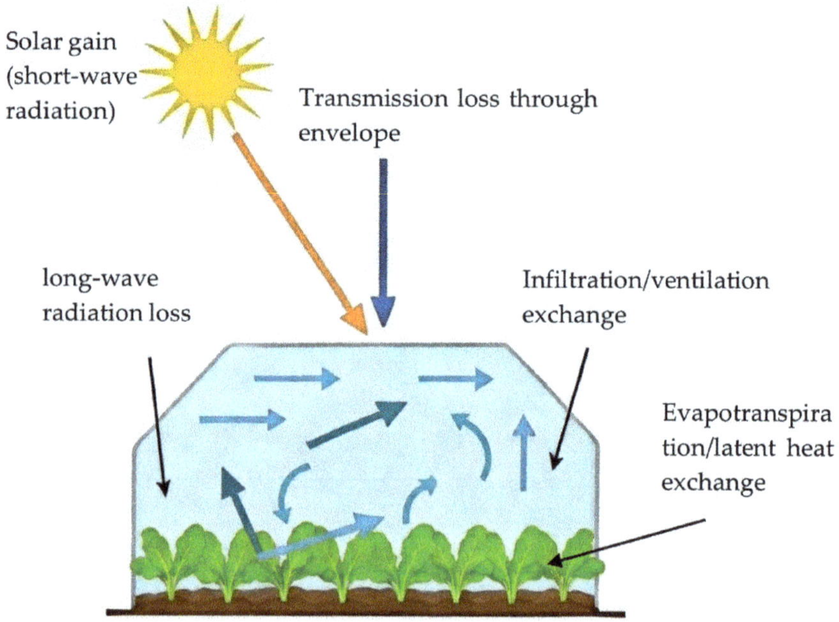


**Figure 5.** Heat and mass transfer processes in a greenhouse, including solar gains, transmission losses, ventilation exchange, radiative losses, and latent heat exchange associated with evapotranspiration.

The reported U-values vary widely with cladding material and external conditions (especially wind speed and whether the cover surface is dry or wet). Stanghellini et al. [32] report indicative maximum U-values at a reference wind speed of 4 m s$^{-1}$ of approximately 8.0–8.8 W m$^{-2}$ K$^{-1}$ for single-glass and single-polyethylene (PE) films, around 5.2–6.0 W m$^{-2}$ K$^{-1}$ for double-glass or double-polyethylene film, and about 4.8 W m$^{-2}$ K$^{-1}$ for twin-wall polycarbonate (PC). These values show why cladding selection has a direct influence on peak heating demand. For the same indoor setpoints, a single-layer film structure loses far more heat than a multi-layer system.

The measured U-values can be lower than these maximum indicative values when specific constructions or test conditions are used. For instance, Zhang [33] reports U-values of around 2.9–3.4 W m$^{-2}$ K$^{-1}$ for double-polyethylene configurations. Hot-box experiments also show that thermal screens can reduce effective U-values by 50–60%, depending on screen type, sealing quality and test/boundary conditions [34]. Cross-study comparison remains difficult because U-values are often reported without consistent reporting/specification of wind conditions, radiative assumptions, wet-cover state, or which envelope components are included. These uncertainties matter because envelope assumptions feed directly into decarbonisation assessments. Lower U-values reduce peak heat demand and enable a lower supply-temperature heating system, improving heat pump performance and reducing the size requirements for storage and distribution networks. However, U-value assumptions can understate peak loads and overstate the feasibility assessment of low-grade heat integration/options [6,35,36].

https://doi.org/10.3390/en19081975

#### 3.2.2. Longwave Radiative Exchange

During night-time and under clear-sky conditions, long-wave radiative exchange between the greenhouse cover and the sky becomes a major source of heat loss. Radiative heat exchange can be represented in simplified form as

$$Q_{rad} = \varepsilon \sigma A \left( T_{cover}^4 - T_{sky}^4 \right) \tag{3}$$

where $\varepsilon$ is surface emissivity (-), $\sigma$ is the Stefan–Boltzmann constant (W $m^{-2}$ $K^{-4}$), and $T_{cover}$ and $T_{sky}$ are the absolute cover and effective sky temperatures (K). Although many greenhouse models linearise this term for computational efficiency, the fourth-power temperature dependence explains why radiative losses intensify during cold, clear nights when the effective sky temperature can fall below ambient air temperature.

Cover conditions further influence radiative behaviour. Condensation on the cover alters surface temperature and can increase the combined convective and radiative losses. Stanghellini et al. [32] report higher total heat loss under wet-cover conditions than under dry surfaces, illustrating the coupling between condensation processes and thermal performance. This interaction is relevant for electrified heating systems, as radiative peaks often coincide with periods of highest heating demand. These peaks shape required electrical capacity, storage sizing/requirement, and heat pump operating conditions.

#### 3.2.3. Air Exchange (Infiltration and Ventilation)

Air exchange contributes to both sensible and latent heat transfer. Sensible ventilation and infiltration losses are commonly expressed as

$$Q_{vent} = \dot{m} c_p (T_{in} - T_{out}) \tag{4}$$

or, expressed in terms of air change rate,

$$Q_{vent} = \rho c_p V N (T_{in} - T_{out}) \tag{5}$$

where $\dot{m}$ is the mass flow rate (kg $s^{-1}$), $c_p$ is the specific heat capacity of air (J $kg^{-1}$ $K^{-1}$), $\rho$ is the air density (kg $m^{-3}$), $V$ is the greenhouse volume ($m^3$), and $N$ is the air change rate ($h^{-1}$ or $s^{-1}$) [37].

Ventilation is important for maintaining both temperature and humidity, but it also becomes a major driver of heating demand during cool periods because warm indoor air is replaced by colder outdoor air. Natural ventilation studies by Boulard and Baille [38] and Papadakis et al. [39] show that wind-driven exchange often dominates buoyancy under moderate wind speeds, and that ventilation rate formulations must account for vent geometry and discharge coefficients. Fatnassi et al. [40] further demonstrate that insect screens and vent design can reduce airflow and change greenhouse microclimate, affecting both energy use and crop conditions.

For decarbonisation analysis, the key point is the link between ventilation and humidity control. During heating seasons, ventilation-driven moisture removal introduces a sensible heating load. As a result, humidity management cannot be independent of heating load estimation. Comparison of technologies that assume a fixed ventilation rate or omit humidity-driven exchange will misrepresent both the magnitude and timing of heating demand.

https://doi.org/10.3390/en19081975

### 3.2.4. Latent Heat, Evapotranspiration, and Condensation

The greenhouse climate cannot be described using sensible heat transfer alone. Crop evapotranspiration requires latent heat, influences humidity and leaf temperature, and directly shapes the need for ventilation. The latent heat flux is typically expressed as

$$Q_{lat} = \lambda \dot{m}_v \quad (6)$$

where $\lambda$ is the latent heat of vaporisation (J $kg^{-1}$) and $\dot{m}_v$ is the vapour mass flux (kg $s^{-1}$). This formulation underlines latent heat representation in greenhouse climate models, whether evapotranspiration is estimated through Penman–Monteith relations or semi-empirical crop functions [14,15].

Condensation on the cover releases latent heat locally, alters cover temperature and changes the balance of convective and radiative losses. Coupled CFD–energy balance studies show that omitting humidity and latent processes can lead to understated heating requirements and misrepresent spatial climate behaviour, particularly in unheated or weakly heated structures where condensation is common [41,42]. This coupling also explains interest in semi-closed greenhouse designs and heat recovery/dehumidification systems, which aim to reduce the energy penalty associated with humidity control through ventilation.

### 3.2.5. Modelling Approaches and Implications for Load Estimation

Three modelling approaches appear frequently in the reviewed literature. Energy balance models (EBMs) remain the standard for seasonal energy analysis and for developing control strategies because they are computationally efficient. Multi-node dynamic formulations typically resolve temperatures for the air, cover, soil, and crop, and are often calibrated against experimental measurements [15]. CFD is used when spatial variability is of primary interest (such as airflow patterns/distributions, thermal stratification, and condensation risk), but its high computational cost generally limits simulations to short time windows. To extend its use, some studies combine CFD with lower-order models. For example, Piscia et al. [42] coupled CFD with an energy balance model to examine humidity control in unheated greenhouses, while Gu and Goto [43] used CFD to investigate airflow within and around plant canopies.

Within CFD studies, turbulence modelling influences predicted airflow and heat transfer. Most applications rely on Reynolds-averaged Navier–Stokes formulations, which provide a balance between accuracy and computational cost. The k–ε model, including standard and renormalisation group (RNG) variants, is widely applied to buoyancy-driven indoor flows and greenhouse ventilation [44,45]. More advanced models, such as Reynolds Stress Models (RSMs), improve the representation of turbulence anisotropy but at a higher computational cost, which limits routine use [46]. Higher fidelity methods, including LES and scale-adaptive simulations, can resolve transient eddies and detailed flow structures, but their computational requirements remain too expensive for full-scale greenhouse studies [47]. In practice, model selection reflects a trade-off between physical detail and computational feasibility.

Reduced-order and surrogate models, including RC networks and hybrid physics-data models, are increasingly applied in optimisation and predictive control studies, reflecting the need for fast, reliable models that can support MPC and digital twin applications, as noted by Katzin et al. [14]. These differences in modelling approaches introduce a comparability issue across the literature. The reported energy savings are highly sensitive to the load-estimation method, and the results from steady-state models, fully dynamic humidity-coupled formulations, and CFD-based analyses are not directly comparable unless boundary conditions and assumptions are clearly defined. This matters for decarbonisation assessments; an intervention that appears favourable under steady-state heating

https://doi.org/10.3390/en19081975

loads may perform very differently once latent-driven ventilation, auxiliary electricity use, or radiative peaks are represented realistically.

#### 3.2.6. Climate-Dependent Load Structure

The contribution of greenhouse heat transfer mechanisms varies with climate and operating strategy. In cold and temperate regions, winter heating demands are shaped by envelope transmission losses and long-wave radiative exchange, with ventilation for humidity control often adding a substantial sensible load. In milder climates, ventilation and moisture regulation can represent a larger share of total energy use even when temperature differences are moderate [15]. In hot climates, cooling demand is driven mainly by solar gains and latent heat processes; evaporative cooling can be effective, but its performance declines as ambient humidity increases, linking cooling performance directly to moisture control [48].

These climate-dependent load structures help explain why technology performance varies across regions. Electrified heating systems are sensitive to peak load magnitude and required supply temperatures, both of which depend on envelope performance, radiative conditions, and ventilation/humidity control strategy [49]. However, cooling and dehumidification options depend on humidity and whether latent energy can be recovered.

Across the reviewed literature, there is broad agreement that greenhouse heating and cooling demands emerge from coupled transmission, radiative, ventilation-driven and latent fluxes, and that humidity control is a primary driver of winter energy use rather than a secondary constraint. At the same time, the reported load magnitudes differ widely across climates, greenhouse typologies, and operating strategies. Direct comparison is limited by inconsistent boundary condition reporting (wind speed, sky temperature handling, infiltration characterisation), varying treatment of auxiliary electricity use, and different representations of evapotranspiration and condensation. For this reason, a single set of typical greenhouse heat-load values by climate zone cannot be defined in a meaningful way without consistent physical and operational boundaries. Heat demand depends not only on external climate, but also on greenhouse geometry, crop type, temperature and humidity setpoints, lighting, and air-exchange strategy. The reported values should therefore be interpreted within clearly stated system assumptions rather than treated as general climate-based benchmarks.

These factors show that decarbonisation evidence must be interpreted within clearly stated modelling assumptions and that technologies cannot be compared without accounting for climate-specific load structures. In heating-dominated climates, where transmission and long-wave losses define the baseline demand, effective approaches focus on envelope improvement, thermal screening, and low-temperature heat supply supported by storage. In warm and humid climates, where solar gains and latent loads dominate, system performance depends more on mitigating overheating, removing moisture, and the interaction between cooling efficiency and water use. These differences show that greenhouse energy systems cannot be ranked using a single technology hierarchy. The same measure may perform well or poorly depending on whether sensible or latent loads dominate, and on how ventilation and humidity are controlled. As a result, technology choice should therefore follow climate-specific load profiles rather than being treated as an isolated optimisation problem. The following sections build on this. Section 3.3 covers cooling and dehumidification in response to thermal loads. Section 3.4 examines heating and low-carbon heat supply options. Section 3.5 considers envelope improvements and thermal storage in shaping demand and peak loads. Section 3.6 focuses on system-level integration, including heat recovery, renewable coupling, and semi-closed greenhouse configurations.

https://doi.org/10.3390/en19081975

### 3.3. Cooling, Dehumidification, and Overheating Mitigation

This section examines cooling and dehumidification as responses to greenhouse thermal loads, focusing on how sensible and latent loads are managed within the indoor climate. While winter heating often receives most attention, in many regions the main challenge is summer overheating and humidity control. This is true in warm, humid, arid, and high-radiation climates, where maintaining temperature and vapour pressure deficit (VPD) within crop-safe limits becomes the factor that determines both yield and plant health. In these environments, cooling and dehumidification are not secondary services; they are central to maintaining a stable microclimate.

Cooling interventions act through three coupled pathways: they remove sensible heat (reducing dry-bulb temperature), redistribute latent fluxes (through evaporation, condensation, or moisture extraction), and change the ventilation rate required for humidity control and disease prevention. As a result, most studies treat greenhouse cooling as a coupled thermodynamic problem rather than a simple temperature control task. Most strategies combine (i) reducing solar gains and long-wave heat accumulation, (ii) enhancing convective exchange when outdoor conditions are favourable, and (iii) regulating water vapour through controlled ventilation or dedicated dehumidification systems [50,51]. In semi-closed and hybrid greenhouse configurations, available cooling capacity affects how the indoor environment can be decoupled from outdoor conditions. This shapes the viability of operating at low ventilation rates, with implications for $CO_2$ enrichment, water recovery potential, and overall climate stability [52].

#### 3.3.1. Overheating as a Load Problem: Metrics Used in the Literature

Overheating is characterised solely by indoor air temperature. Most studies evaluate overheating using temperature exceedance relative to crop setpoints ($\Delta T$), VPD, and cooling degree-hours or sensible/latent loads partitioning. This reflects the practical reality that plant stress and disease risk are governed by the combined thermal-moisture environment in the crop canopy, not dry-bulb temperature alone.

Experimental evidence from natural ventilation augmented cooling (NVAC) demonstrates this point. McCartney et al. [53] report that an NVAC configuration can produce temperature reductions of 1.9–12.6 °C under the tested conditions, while relative humidity increased by 1.4–31.2% and VPD decreased by 0.3–4.9 kPa depending on inlet conditions and misting configuration. These results show that cooling effectiveness should be evaluated as a joint temperature-humidity outcome. In several cases, VPD improved even as absolute humidity increased, indicating that plant-relevant stress can move in a favourable direction despite higher RH. This is consistent with the evaporative cooling literature by Sethi and Sharma [50], which reports 4–12 °C temperature drops under favourable psychrometric conditions.

Modelling studies reach similar conclusions. Yang et al. [54] use simplified heat-balance calculations to evaluate when natural ventilation, fan-assisted ventilation, and evaporative cooling can maintain target conditions across contrasting climates. Therefore, overheating should be framed as a load problem determined by solar gains, the outdoor air state, internally generated moisture, and achievable air exchange, rather than as a generic high temperature condition.

These metrics also have system-level implications that directly affect decarbonisation pathways. In semi-closed operation, maintaining acceptable VPD at low ventilation rates enables $CO_2$ retention for enrichment, supports condensate recovery, and reduces sensitivity to outdoor disturbances. Katsoulas et al. [52] show that increasing cooling capacity lowers ventilation demand and changes the coupling between indoor and outdoor conditions, with benefits for water use efficiency. This introduces a structural trade-off

https://doi.org/10.3390/en19081975

that recurs in decarbonisation studies: high-ventilation regimes dilute $CO_2$ and increase energy and water penalties, while low-ventilation regimes enabled by mechanical cooling and/or dehumidification improve $CO_2$ and water management but shift the burden toward electricity consumption. In effect, cooling demand is a system integration variable that shapes ventilation policy, $CO_2$ strategy, and resource efficiency rather than as a local comfort metric.

#### 3.3.2. Overheating Mitigation

Ventilation remains the default response to overheating because it is mechanically simple and, in many regions, the lowest-cost option to deploy at scale. Its limitation is thermodynamic rather than operational: ventilation can provide cooling only when the incoming outdoor air has lower enthalpy than the indoor air it replaces. Under hot and humid conditions, increased air exchange may remove sensible heat slowly (or not at all) while introducing moisture that raises disease risk and pushes vapour pressure deficit outside crop-safe bounds. The literature, therefore, treats ventilation as a baseline strategy whose effectiveness depends on climate and greenhouse design [50,55].

Because ventilation performance is constrained by outdoor conditions, some studies focus on reducing the cooling load at the source, particularly by limiting solar gains and long-wave heat accumulation. Cover and shading strategies are often treated as first-line measures: they reduce absorbed short-wave radiation, lower air and surface temperatures, and decrease the frequency and intensity of active cooling operations. In arid and high-radiation climates, radiative and shading materials are reviewed as explicit responses to overheating, with a central design tension clearly recognised that any reduction in solar heat input must be balanced against crop light requirements [50].

#### 3.3.3. Evaporative Cooling: Pad-Fan, Fogging/Misting, and NVAC-Type Concepts

Evaporative cooling remains a core strategy in hot and dry climates because it removes sensible heat by converting it to latent heat, often achieving substantial temperature reductions when the outdoor wet-bulb decline is high. Sethi et al. [50] group greenhouse cooling technologies into ventilation (natural and forced), shading/reflective measures, and evaporative systems (pad–fan, fogging/misting, and roof-based cooling), noting that performance depends on climate suitability.

Recent work on NVAC-type configurations demonstrates a design that combines thermal buoyancy with a targeted misting system to enhance evaporative cooling while improving airflow distribution. The naturally ventilated NVAC greenhouse delivered cooling of 1.3–3.6 °C relative to outside conditions while increasing relative humidity by 5.7–17.7% [53,56]. These paired outcomes matter for net-zero assessment because they show that modest temperature reduction can still be agronomically significant if it reduces VPD, and cooling performance cannot be assessed without its corresponding humidity trajectory.

Beyond conventional evaporative approaches, recent studies treat cooling as an integrated energy–water design problem. In subtropical scenarios, Abedrabboh et al. [57] examine renewable-driven hybrid-cooled greenhouse integrated systems and quantify the value of combining passive measures (e.g., shading and geothermal cooling concepts in their configurations) with active cooling cycles. Their results show 34–58% reductions in annual cooling load, 17–34% reductions in evapotranspiration on peak load days, and reduced life cycle cost by 5.6–28.6% (depending on configuration), alongside lower solar-panel area requirements. While specific system layouts may differ from commercial practice, the transferable insight is methodological: once cooling is treated as a coupled load (thermal + moisture + water supply), the best solution becomes context-dependent and must be reported with a clear system boundary definition (cooling target, water source, and electricity supply).

Across the reviewed literature, there is strong agreement that evaporative cooling performance is controlled primarily by wet-bulb depression and by how uniformly cooled air reaches the crop zone. What differs among climates is whether evaporative systems improve plant comfort without pushing RH or VPD into disease-prone ranges. The NVAC studies are informative because they report temperature, RH, and VPD responses under different inlet conditions, metrics that many cooling studies still omit.

Comparability remains limited when water consumption, control regimes, or humidity responses are not reported, or when performance is presented solely as temperature reduction. For evaporative cooling to be meaningfully evaluated across contexts, studies must document both the thermal and moisture outcomes, along with the associated water and energy requirements.

#### 3.3.4. Dehumidification as an Energy Decision: Ventilation–Heating vs. Moisture Extraction

Humidity control is a major driver of winter energy demand because the common vent to dry, then heat to restore temperature strategy exports sensible heat. Reviews note that ventilation-based dehumidification is simple to implement but can impose large heating penalties in cold climates and performs poorly when outside air is humid [51]. Internal moisture extraction systems avoid this penalty. Heat pump-based dehumidifiers cool greenhouse air below its dew point, condense water on the evaporator surface, and then reheat the air, retaining much of the enthalpy within the enclosure [51]. Experimental studies by Migeon [58] report that such systems can achieve equivalent dehumidification with lower energy input than ventilation–heating under comparable conditions, while also reducing condensation risk on crop surfaces. In their study, the dehumidifying heat pump operated at an energy consumption of about 4 W $m^{-2}$ and, depending on outdoor conditions, required about three to four times less energy than the ventilation–heating method to achieve comparable dehumidification. The system also condensed approximately 30 L $h^{-1}$ of moisture, anchoring performance in a measurable moisture balance rather than qualitative claims [58].

However, the reported energy savings depend on the control strategy, the specified moisture target (RH, VPD, or dew point), and the evaluation must include auxiliary electricity use and any reheating requirements [51]. From a net-zero perspective, dehumidification is therefore not simply a humidity control choice but a thermodynamic decision between exporting enthalpy through ventilation or retaining and partially recovering it through internal condensation. Beyond its energy role, internal moisture extraction produces a measurable condensate stream. This water can be recovered and reused, linking humidity control with water management. In low-ventilation systems, particularly semi-closed configurations, recovered volumes can be significant relative to irrigation demand, depending on crop transpiration and operating conditions. This reduces heat loss and improves water use efficiency. Condensate recovery is therefore not only a result of humidity control but also part of system integration, especially in water-constrained settings or intensive greenhouse production.

### *3.4. Heating Technologies and Low-Carbon Heat Supply*

Heating remains the dominant energy demand in commercial greenhouses across temperate and cold regions, especially where year-round production requires tight control of temperature and humidity. The heating requirement is not simply sensible heat provision; it is shaped by the load structure (envelope transmission, infiltration/ventilation, long-wave exchange, and latent-heat coupling). These factors determine peak capacity, feasible

https://doi.org/10.3390/en19081975

supply temperatures, and the scope for heat recovery or storage. They also influence which low-carbon heating pathways are technically viable and economically resilient.

This section reviews major heating supply routes used in greenhouse systems. Conventional boilers and CHP systems serve as the baseline, followed by electrified heat pumps, geothermal and earth-coupled systems, solar thermal and hybrid renewable technologies, and the use of waste heat or district heating networks.

#### 3.4.1. Baseline Systems: Boilers and CHP (And Why They Persist)

Commercial heated greenhouses have traditionally relied on natural gas boilers, with CHP systems widely adopted in some regions as an integrated source of heat and electricity. Beyond their controllability and established infrastructure, CHP units offer horticulture-specific advantages: they supply on-site electricity and produce a flue-gas $CO_2$ stream that (once cleaned and conditioned) can be used for crop enrichment. As low-carbon heating replaces combustion-based CHP, $CO_2$ enrichment shifts from a by-product of heat generation to a separate supply and control requirement. This change affects system design, operation, and cost. The implications for supply routes, storage, safety, and integration with greenhouse operation are discussed in Section 3.6.5. Andrews and Pearce [59] note that biomass combustion is generally unsuitable for direct $CO_2$ enrichment at greenhouse scale because of pollutant risks, whereas CHP exhaust (from natural gas or biogas) can be treated to meet horticultural requirements. This capability has been a key factor in the use of CHP relative to other combustion-based alternatives.

The Netherlands remains the most prominent example of large-scale CHP deployment in greenhouse horticulture. Andrews and Pearce [59] report literature indicating approximately 2500 MW of CHP capacity associated with Dutch greenhouses, describing the sector as an established user of CHP-based heat supply. While CHP can reduce primary fuel consumption relative to separate heat and power generation, its climate benefit depends on system boundary assumptions (particularly the carbon intensity of displaced grid electricity and how exported power is accounted for).

Another reason boilers and CHP persist is that $CO_2$ enrichment practice has been structurally coupled to on-site fossil combustion in many greenhouse systems. Vermeulen [60] reports that $CO_2$ is commonly sourced from combustion gases of natural gas boilers or CHP in Dutch greenhouse horticulture. As efficiency measures expand and fossil-free heat sources (e.g., geothermal) are adopted, natural gas use declines, and with it the supply of on-site $CO_2$. This creates a $CO_2$ shortfall that must be covered through alternative sources. van Tuyll et al. [61] make the same linkage explicit in the context of electrification: historically, most supplemental $CO_2$ has come from natural gas combustion, and as the sector decarbonises, it must increasingly rely on non-combustion sources and/or external supply chains (pipelines or delivered food-grade $CO_2$).

Therefore, the implication is that greenhouse heating decarbonisation is not simply a matter of replacing the heat source. It requires a coordinated redesign of $CO_2$ supply strategies to avoid productivity losses and ensure safe, reliable $CO_2$ delivery under low-carbon energy configurations.

#### 3.4.2. Electrification: Air-Source, Ground-Source, and Dual-Source Heat Pumps

Heat pumps are central to greenhouse heating decarbonisation because they convert electrical input into useful heat with a COP greater than unity. However, in greenhouse applications, performance depends on operating conditions, especially the temperature of the heat source and the supply temperature required for distribution.

Zhou et al. [62] evaluated a dual-source heat pump system designed to transfer surplus thermal energy from Chinese solar greenhouses (CSGs) to a multi-span greenhouse and

https://doi.org/10.3390/en19081975

reported results from a winter field test. Using CSG air as a heat source enhanced heat pump performance relative to outdoor air sourcing, increasing heating capacity by 27% and COP by 23%. Operating in dual-source mode produced a further 10.8% increase in collected heat and a 7.9% improvement in COP relative to the single air-source case. The reported COP values reached 2.8 during heat collection and 2.5 during greenhouse heating, with CSG air-sourced heat contributing between 2.1% and 28.2% of total heat over the test period. These findings are valuable because they reflect performance under realistic control and weather variability rather than idealised COP assumptions.

Ground-coupled and hybrid ground-source concepts are also being explored for greenhouse use, where the seasonal stability of the ground loop can support higher performance. Abir Ahsan et al. [63] assessed a hybrid ground-source heat pump (HGSHP) system integrated with a cooling tower for hot-humid greenhouse operation using a dynamic TRNSYS model. Their simulations show that the HGSHP configuration reduced cooling energy consumption by up to 10.8% compared with a GSHP-only system, while also improving long-term ground thermal stability by rejecting part of the cooling load through the tower. The study further includes a payback period sensitivity analysis, illustrating how boundary-condition assumptions and cost factors can influence the economic performance of electrified systems.

#### 3.4.3. Geothermal and Earth-Coupled Heating: Deep Geothermal, Shallow Systems, and EAHE

Geothermal options for greenhouse heating range from deep geothermal wells to shallow ground-coupled systems such as GSHPs and earth-to-air heat exchangers (EAHEs) [64]. Deep geothermal resources can provide stable, high-temperature heat suitable for direct hydronic supply where geological conditions allow, substantially reducing dependence on fossil boilers. However, deployment is constrained by drilling depth, reservoir characteristics, and high upfront investment.

For shallow systems, the reported case studies show that GSHPs benefit from relatively stable ground temperatures and can reduce conventional energy consumption compared with air-source systems. Their performance depends strongly on ground-loop sizing and achieving seasonal thermal balance (continuous winter extraction without adequate summer recharge can lower ground temperature and reduce efficiency over time) [64]. EAHEs systems are primarily used as air pre-conditioning technologies that moderate inlet temperature and help reduce heating or cooling loads rather than serving as standalone heat sources.

Therefore, geothermal feasibility is shown to depend on required temperature grade, long-term ground energy balance, and site-specific infrastructure constraints [64]. These factors determine whether geothermal systems can deliver stable, low-carbon heat at scales compatible with commercial greenhouse operation.

#### 3.4.4. Solar Thermal and Solar-Integrated Heating (Often with Storage)

Solar energy integration in greenhouse systems has been explored through solar thermal collectors, solar-air heating configurations, and photovoltaic–thermal (PVT) systems [65]. A recurring design constraint in the literature is the balance between energy capture and light transmission. Because crop growth depends on photosynthetically active radiation (PAR), any collector or shading element must be assessed against crop-specific light requirements, as excessive shading can negatively affect plant growth and yield [65].

In heating-dominated climates, the contribution of solar thermal systems is closely linked to the temporal mismatch between solar availability and heating demand. Vadiee and Martin [19] show that solar-based greenhouse concepts often rely on storing surplus summer heat for use during winter, using both short-term and seasonal thermal storage. This introduces additional operational requirements, including the need to manage excess

https://doi.org/10.3390/en19081975

heat and, in some cases, upgrade stored low-temperature heat to reach temperatures suitable for distribution [19,66].

As a result, the performance of solar-integrated heating depends not only on collector efficiency but also on how the available heat is used within the greenhouse, with further discussion provided in Section 3.6.

#### 3.4.5. Waste Heat Recovery and District Heating: Low-Grade Heat as a Decarbonisation Lever

Waste heat and district heating provide low-carbon heat for greenhouse operation where suitable sources are available. Their use depends on the availability, temperature grade, and continuity of the external heat stream, as well as the proximity of the greenhouse to the source. Asgari et al. [67] note that colocation is a fundamental limitation for waste heat deployment in greenhouses. Even when the thermal resource is substantial (particularly in cold seasons), adoption is only possible if the greenhouse is close enough, and if the infrastructure allows heat to be transferred at an acceptable cost.

Case studies illustrate the potential scale of these sources. For a glass-furnace flue-gas system, Andrews & Pearce [59] estimate around 5.1 MW recoverable heat after imposing a non-condensing outlet temperature constraint. Under their assumptions, this output could supply approximately 3.9 acres of greenhouse area, with associated carbon offset calculations depending on how $CO_2$ enrichment is handled. In a low-temperature industrial discharge application, Fguiri et al. [68] report winter-period waste heat of about 145,800 MWh and estimate that it could meet the heating needs of about 45 ha of greenhouses (in southern Tunisia). They also project annual emissions reductions of roughly 746 $tCO_2$-eq and strong financial performance (NPV and IRR) under the assumed conditions.

These studies show that low-grade external heat can replace a substantial share of fossil-based heating where suitable sources exist. However, performance depends on source temperature, continuity, distance to the greenhouse, and the assumptions used in emissions and economic metrics.

#### 3.4.6. Biomass and Renewable Fuel: Decarbonisation Potential with Operational Constraints

Biomass boilers and biogas-based systems are often discussed as renewable substitutes for fossil boilers in greenhouse heating, particularly in regions where electrification is limited or where agricultural residues and livestock waste are locally available. Esen and Yuksel [69] demonstrated this feasibility in a winter greenhouse, maintaining indoor temperatures near 23 °C using a biogas-integrated system that produced 2231.83 L of biogas under mesophilic digestion. Their results show that small-scale greenhouse heating can be maintained under practical operating conditions.

However, renewable combustion introduces operational requirements that extend beyond heat delivery. Fuel logistics, storage volume, moisture variability, ash handling, and emissions control (especially particulate matter and $NO_x$) affect both operating costs and regulatory compliance.

Life cycle analyses show that the environmental performance of pellet-based greenhouse heating depends heavily on upstream system boundaries (including feedstock sourcing and processing intensity, and supply-chain (transport)). Wu et al. [70] report cradle-to-grave emissions of 288 kg $CO_2$eq per tonne of Quebec wood pellets (14.89 kg $CO_2$eq/GJ), with notable contributions from harvesting, densification energy, and transportation. Their comparison across regions also shows considerable variability driven by drying energy sources, transport logistics, and electricity mix characteristics. These show that greenhouse gas reductions relative to natural gas depend on how the biomass supply chain is configured rather than on the combustion process itself. As a result, life cycle performance depends on clearly defined system boundaries and upstream conditions.

https://doi.org/10.3390/en19081975

### 3.5. Storage and Envelope/Material Innovations

Decarbonising greenhouse heating becomes considerably easier when overall thermal demand is reduced and peak loads are smoothed. The literature points consistently to two main points. The first is the use of envelope and material measures that lower transmission and long-wave radiative losses without imposing unacceptable penalties on crop light availability. The second is the use of thermal energy storage (TES) to shift heat supply in time, reduce peak heating capacity requirements, and improve compatibility with low-carbon heat sources.

#### 3.5.1. Envelope and Material Innovations: Reducing UA and Radiative Loss Without Over-Shading the Crop

Thermal screens are among the most established envelope measures for reducing winter energy use because they act on two major night-time loss pathways (overall transmission and long-wave radiative exchange). Using hot-box measurements, Geoola et al. [34] report that deploying a thermal screen to greenhouse films can lower the effective overall heat transfer coefficient by about 50–60%, with reduction depending on thermal screen type.

However, screen performance cannot be captured by a single U-value. It depends on the radiative properties and air exchange through the screen layer. Andersson [71] shows that aluminised energy screens reduce radiative losses and that a two-layer screen can further reduce radiative exchange compared with a single layer, depending on the material combination.

Rabiu et al. [72] combined hot-box experiments with TRNSYS modelling to quantify overall heat transfer coefficients for energy-saving screens and to evaluate the effect of screen porosity and permeability. Their results show that air leakage through the screen layer has an impact on measured U-values and must be accounted for in both experimental and simulation-based evaluation [72]. Rafiq et al. [73] developed a method to determine total hemispherical long-wave radiative properties of greenhouse energy screens and reported challenges in estimating reflectance. They emphasise that incomplete radiative characterisation limits cross-study comparability. These studies show that screen performance depends on radiative properties, leakage behaviour, and test boundary conditions rather than on nominal material types alone.

Beyond internal screens, attention has moved to advanced covering materials because lower U-values make electrified heating more practical. A recent technical review by Zhang et al. [74] finds that low-emissivity coated covers can reduce total energy consumption, but realised performance is sensitive to operational factors such as condensation and climate conditions. Song et al. [75] analyse low-emissivity envelope configurations and show that larger reductions in thermal loss can coincide with lower visible light transmission. Their optimisation results underline the need to balance insulation gains with crop light requirements rather than assuming a negligible optical penalty.

In warm climates, envelope choices are used to shape cooling demand through spectral selectivity, not only through ventilation or evaporative cooling. Lin et al. [76] evaluate a low-emissivity light-blocking film (LBF) that transmits approximately 85% of photosynthetically active radiation (PAR) by roughly 27% relative to control glazing and produced cultivar and season-dependent yield responses, including significant reductions during the summer experiment. Cooling energy use increased by about 10–12% under the LBF despite its lower nominal solar heat gain coefficient. The authors attribute this to convective re-emission and airflow dynamics within the greenhouse, indicating that thermal outcomes cannot be inferred from optical specifications alone [76].

Therefore, selective spectral filtering and advanced covers must be evaluated as part of a coupled system. Material properties, climate control strategy, airflow configuration,

https://doi.org/10.3390/en19081975

and crop physiology together determine energy use and productivity. As a result, claims of energy savings or yield benefits are transferable only when optical and thermal parameters, boundary conditions, and control settings are reported with sufficient detail.

Table 2 summarises studies that report greenhouse cover materials and thermal screen or curtain configurations used to reduce night-time heat loss. The reported performance depends not only on the nominal material but also on the screen-envelope configuration, including layer arrangement, emissivity, permeability, and humidity interactions. The literature, therefore, indicates that screens and covers should be evaluated as integrated thermophysical systems, accounting for U-value reduction, radiative behaviour, air leakage, and resulting microclimate effects.

**Table 2.** Studies reporting greenhouse cover materials and thermal screen/curtain configurations used to reduce night-time heat loss, overall heat transfer coefficient (U-value), or seasonal heating demand. The table shows that performance depends on both material properties and operational context, including humidity management, leakage, and radiative boundary conditions.

| Study | Greenhouse Cover/Cladding | Screen Type | Method | Reported Thermal or Energy Effect | Main Implication |
|---|---|---|---|---|---|
| Geoola et al. [34] | Greenhouse plastic films | Thermal screens (hotbox tested; screen types varied) | Laboratory hot-box measurements | Adding a thermal screen reduced the effective U-value by about 50–60%, with corresponding energy-saving potential of about 50–60% under the test conditions. | Thermal screens can strongly reduce conductive/convective heat loss. |
| Kim et al. [77] | Double-layer polyethylene greenhouse (0.1 mm PE inner and outer layers) | Internal thermal curtain | Field study in winter, Republic of Korea | U-value decreased from 3.85 to 2.76 W $m^{-2}$ $°C^{-1}$ with the thermal curtain, and night-time use produced heating energy savings of about 28.7%. | Thermal curtains provide measurable field-scale savings, but the realised reduction is lower than the idealised hot-box results. |
| Akpenpuun et al. [17] | Single-layer greenhouse vs. double-layer greenhouse with polyolefin, thermal screen, and inner polyethylene layer | Thermal screen in both systems; additional inner PE in the double-layer system | Winter strawberry greenhouse comparison | The double-layer configuration improved night-time insulation and reduced operating cost relative to the single-layer system but also increased RH and reduced VPD; the single-layer system produced 14% higher yield in this case. | Better insulation can reduce heat demand, but envelope/screen combinations may create humidity and crop-performance trade-offs if ventilation and moisture control are not adjusted. |
| Rasheed et al. [78] | Greenhouse with alternative screen materials assessed in simulation | Polyester, Luxous, Tempa, and multi-layer night screens | Model-based optimisation under Korean winter conditions | Multi-layer night screens reduced heating requirement by 20%, 5.4%, and 13.5% relative to polyester, Luxous, and Tempa, respectively. Lower emissivity and greater thickness contributed to lower heating demand. | Screen performance depends not only on thickness but also on emissivity; multi-layer arrangements can outperform single commercial screen types. |
| Rafiq et al. [73] | Four representative materials: polyethylene film, Luxous-1347, white polyester, and Tempa 8672 energy-saving screen | | Modified radiation-balance method using wide-band radiometers under natural conditions; values checked against spectrophotometer | The study develops and validates a method for estimating long-wave radiative properties of greenhouse screens and films, showing that radiative behaviour must be measured explicitly because simplified methods can overestimate transmission or give unstable results for some material types. | Radiative-loss control at night depends on measured emittance, reflectance, and transmittance. |

https://doi.org/10.3390/en19081975

**Table 2.** *Cont.*

| Study | Greenhouse Cover/Cladding | Screen Type | Method | Reported Thermal or Energy Effect | Main Implication |
|---|---|---|---|---|---|
| Rabiu et al. [72] | Ten commercial greenhouse thermal screens, including white polyester, Luxous, PH-55, PH-super, PH-66, Clima45, New-Lux, and multilayer M1–M3 screens | | TRNSYS plus experimental hotbox, combining thermophysical, radiometric, and aerodynamic characterisation | Screen thermal performance depended strongly on the combined effect of material thickness, radiative properties, and permeability. Low-porosity multilayer screens (M1 and M3) showed the lowest simulated permeable U-values (4.4 W $m^{-2}$ $K^{-1}$), while more open screens showed much poorer retention. | Screen comparison should include air leakage/permeability alongside radiative and conductive properties; otherwise, U-value and energy-retention performance can be misjudged. |

#### 3.5.2. Sensible Thermal Energy Storage: Water Tanks and Rock-Bed Storage

Sensible thermal energy storage (TES) is widely used in greenhouse heating because it is technically simple, scalable, and compatible with low-temperature heat sources. In cold climates, water-based heat buffers are common. They store surplus daytime heat, often produced when gas firing is used for $CO_2$ enrichment, and release it at night. This separates the timing of heat generation from heat delivery and supports more stable operation [79]. Buffering storage also improves system stability by reducing part-load operation and limiting frequent start-stop cycling. Evidence from heat pump systems shows that even modest storage volumes can reduce cycling losses. Lyu et al. [80] report losses of about 12.5% without a buffer tank, falling to below 1% when a tank is added, with substantial benefits achieved at small storage volumes.

Rock-bed thermal energy storage represents an air-based alternative in which greenhouse air acts as the heat transfer medium. In the foil-tunnel application evaluated by Malinowski et al. [81] warm air accumulating near the roof is extracted by fans and forced through a packed stone bed via perforated ducts to charge the store. During discharge, airflow is routed through the bed to recover stored heat and return it to the greenhouse. Amiri et al. [82] identify pressure drop across the bed and limited heat transfer between the air and storage medium as key constraints in rock thermal storage systems. Reduced particle size improves heat transfer rates but increases flow resistance and fan power demand, while larger particles ease airflow but limit charging and discharging effectiveness. These coupled trade-offs directly influence system efficiency and must be considered explicitly in the design of packed bed systems.

#### 3.5.3. Latent Thermal Storage Using PCMs: Promising, but Performance Hinges on Heat Transfer Design

Latent thermal energy storage (LTES) using PCMs is attractive for greenhouses because it can store and release large amounts of heat within a narrow temperature range. When the phase-transition temperature is chosen to match crop-relevant air or root-zone setpoints, PCMs can support passive night-time heating and help smooth daytime peaks. Nishad and Krupa [83] report that PCM-integrated greenhouse systems have achieved diurnal temperature reductions of up to 7 °C and nocturnal temperature increases of up to 9 °C relative to ambient conditions, contributing as much as 30% of heating demand in some configurations. They also note that performance depends strongly on PCM type, quantity, encapsulation, and how the material is incorporated into the greenhouse.

Across controlled-environment agriculture studies, PCM performance is shaped less by the base material than by heat transfer design and encapsulation strategy. Low thermal conductivity, leakage during melting, and cycling stability are recurrent challenges. Zhu et al. [84] show that organic PCMs usually require conductivity enhancement (such as porous matrices or graphite additives), along with encapsulation to prevent leakage and improve heat transfer. Cui et al. [85] identify macro-encapsulated modules, shape-stabilised composites, and capsule-based systems as the main implementation pathways in greenhouse applications due to leakage and corrosion constraints.

Experimental optimisation studies further show the importance of formulation and integration. Liu et al. [86] demonstrate that additive selection and mixing ratios influence supercooling, phase-change temperature, latent heat, and discharge characteristics in a solar greenhouse PCM system. These results show that PCM performance depends on a combination of thermal properties, encapsulation quality, and system-level design rather than on the choice of PCM alone.

#### 3.5.4. Seasonal Thermal Energy Storage (STES): Enabling High Renewable Fractions, but Highly Site-Dependent

Seasonal thermal energy storage addresses the mismatch between summer heat and winter heating demand by storing excess heat for later use. A comprehensive techno-economic review by Yang et al. [87] classifies STES technologies into sensible systems (tank, pit, borehole, aquifer), latent systems, and thermochemical systems, and compares them using metrics such as storage efficiency, energy density, and levelised cost of heat. Their analysis shows that economic viability depends heavily on configuration and site conditions, with pit storage and low-temperature aquifer systems approaching competitiveness under favourable conditions.

Borehole thermal energy storage (BTES) is widely studied at the district scale. Gao et al. [88] describe BTES systems as arrays of boreholes, typically 30 to 100 m deep, that store heat in the surrounding ground via conduction, often integrated with solar collectors and heat pumps. Performance depends on multi-year thermal stabilisation and local geological conditions, including groundwater flow and subsurface conductivity. Sadeghi et al. [89] emphasise that BTES integration depends on how well the system aligns with the heating network it serves, the operating temperature range, and subsurface conditions such as hydraulic conductivity, groundwater flow, and borehole interference. These factors make STES inherently site-dependent, with geology, land availability, drilling cost, and regulatory context determining both technical outcomes and economics.

From a net-zero planning perspective, Baeuerle et al. [90] view seasonal storage as an important infrastructure for achieving high shares of renewable heat in industrial and district-scale systems, particularly where thermal networks are constrained or where direct electrification would create a large winter peak. The literature, therefore, presents STES not as a universal solution but as a context-specific option that becomes effective when geological conditions, infrastructure needs, and scale align.

### 3.6. *Renewable, Waste Heat, and Integration System Configurations*

Decarbonising greenhouse heating is primarily a system-integration challenge rather than a matter of substituting one device for another. Closed and semi-closed greenhouse studies show that reducing ventilation losses and improving the thermal envelope performance are often prerequisites for effective heat recovery and renewable-heat integration. Most recovered heat is low temperature, which means it must be stored and, when necessary, lifted to a usable supply temperature for reuse (in the greenhouse) [66].

Comparative studies depend strongly on the modelling assumptions used. In particular, the results are sensitive to how the cover heat transfer coefficient is represented, how ventilation and humidity control are modelled, and whether radiative and latent fluxes are included explicitly in the energy balance. Clear reporting of these boundary condi-

https://doi.org/10.3390/en19081975

tions is therefore essential if the results are to be interpreted consistently and compared across studies.

#### 3.6.1. PV-Heat Pump Coupling and Electrified Near-Zero Concepts

PV-powered heat pumps are increasingly studied as a pathway for electrifying greenhouse heating, with recent work moving beyond idealised COP assumptions. Instead, heat pump behaviour is embedded within greenhouse load models that represent air exchange, humidity control, and nonideal operating conditions. Asgari et al. [91] develop an open-source quasi-steady-state model that couples a greenhouse energy balance formulation with a non-ideal air-source heat pump (ASHP)/GSHP cycle (including vertical and horizontal ground heat exchangers) and links these to a SAM-Python PV model. A dedicated ventilation sub-model is included to represent cooling-load management.

In their Canadian case study (London, Ontario) [91], technology choice had a direct effect on PV requirements. The air-source heat pump delivered the lowest heating efficiency (COPh = 2.52 on the coldest day), while GSHP configurations performed better (vertical COPh = 3.68; horizontal GSHP COPh = 3.79). These higher efficiencies reduced the PV capacity required across greenhouse sizes. The study also identified a persistent winter mismatch between PV output and heating demand. Even with net metering-based sizing, annual self-consumption and self-sufficiency remained modest (23.5–25.1% and 23.7–26.0%, respectively).

Longer-horizon electrified concepts, therefore, rely on storage and seasonal balancing to stabilise source temperatures and protect heat pump performance. In the broader borehole thermal energy storage literature, Sadeghi et al. [89] report that coupling GSHPs with BTES/seasonal storage can raise entering water temperatures and improve system performance, while also allowing design relaxations such as reduced borehole length in the reported simulations.

#### 3.6.2. Solar Thermal + TES and Seasonal Storage Architecture

Solar thermal integration in greenhouse systems is usually implemented with thermal energy storage to manage the mismatch between solar availability and heating demand. Two configurations dominate the literature: short-term (diurnal) storage for day-night shifting and seasonal thermal energy storage for balancing summer surplus with winter needs. STES is proposed to reduce dependence on fossil backup during winter, although its techno-economic feasibility remains strongly site dependent [87].

Greenhouse system studies increasingly evaluate combined architectures that include solar collectors, seasonal storage, and heat pumps. Kim et al. [92] examine such a configuration and report solar contribution, system COP, operating cost reductions, and emissions impacts relative to a boiler-based reference case, with the results showing strong sensitivity to investment costs and policy incentives.

Low-cost soil-based seasonal storage has also been explored. Zhang et al. [93] propose a seasonal solar soil heat storage to avoid the need for a heat pump. Under controlled indoor temperature operation, their model indicates annual savings in auxiliary electricity use, demonstrating the potential of ground-based STES where installation constraints allow. Comparative simulation studies further show that seasonal solar thermal storage can displace a measurable share of greenhouse heating demand relative to conventional HVAC systems, depending on storage scale and collector sizing [94].

Short-term storage improves nighttime heat coverage, while seasonal storage enables a much larger proportion of renewable heat to be used throughout the year. The reported performance and economic outcomes vary widely because they depend on the modelling

https://doi.org/10.3390/en19081975

assumptions and boundary conditions applied. Clear reporting of these assumptions is essential when comparing different storage configurations.

### 3.6.3. Waste Heat and District Heat Coupling

Industrial waste heat is widely recognised as a valuable resource for district heating networks in locations where spatial and network conditions allow integration. Many industrial processes release substantial quantities of low- to mid-temperature heat, and pinch-based screening methods have been developed to identify streams that match district heating supply temperatures [95]. However, technical compatibility does not ensure deployment. Feasibility depends on source temperature, distance to the network, operational continuity, and the complexity of integrating the heat stream [95].

Analysis of real industrial waste heat projects also shows that non-technical factors can be decisive. Project initiation, contractual clarity, risk-sharing arrangements, and business-model choices strongly influence outcomes (organisational and incentive barriers often prevent utilisation even when the technical potential is clear) [96].

For greenhouse applications, the feasibility of linking to district heating or industrial waste heat sources is shaped mainly by required supply temperature and network configuration. Effective coupling depends on temperature matching: lower greenhouse supply temperatures increase the number of industrial heat streams that can be used directly and reduce the need for temperature boosting. This reflects the same temperature-matching principles formalised in waste heat screening methods for district heating integration [95].

### 3.6.4. Semi-Closed/Closed Greenhouse Configurations

Semi-closed and closed greenhouse systems combine heating, cooling, dehumidification, and ventilation control within a single framework. Their performance depends not only on the heat source, but on how ventilation and humidity are managed through mechanical processes. Closed greenhouses minimise air exchange and rely on mechanical systems to regulate temperature and humidity. Semi-closed systems retain vents but limit their use by using heat exchangers for cooling and moisture control, resorting to ventilation only when mechanical capacity is insufficient [97].

van Beveren et al. [97] describe semi-closed greenhouse operation as a means of reducing energy use by minimising ventilation losses and storing surplus heat, thereby lowering natural gas demand. The system they evaluated integrates short-term thermal buffers, seasonal aquifer storage, and a heat pump that upgrades low-temperature heat for distribution within the greenhouse. This links heat supply, storage, and climate control within a single operating approach.

Work on dehumidification further illustrates this integration. In heated tomato production, Gilli et al. [98] found that heat pump-based dehumidification reduced energy use by 15% and 25% in two consecutive seasons compared with conventional ventilation and heating control, without affecting yield or fruit quality. These results show that reducing reliance on ventilation for moisture removal can lower heating demand and improve compatibility with low-temperature, electrified heat supply, as temperature and humidity are regulated through coordinated mechanical processes.

### 3.6.5. $CO_2$ Supply and Enrichment in Low-Carbon Greenhouse Systems

$CO_2$ enrichment in commercial greenhouses has traditionally relied on combustion-based heating. In boiler-based systems, flue gas can be treated and used for crop dosing, linking heat generation with $CO_2$ supply. As heating shifts toward electrification and other low-carbon sources, this link is reduced, and $CO_2$ provision must be addressed as a separate system requirement. The literature identifies combustion-derived $CO_2$, compressed or liquefied $CO_2$, and industrial or network-based sources as the main supply

https://doi.org/10.3390/en19081975

routes. Maintaining target concentrations remains difficult, as $CO_2$ behaviour is closely linked to ventilation, temperature, humidity, and crop uptake [99–101].

The choice of $CO_2$ source shapes greenhouse design. Delivered or externally supplied $CO_2$ requires infrastructure for storage, pressure regulation, distribution, and controlled injection. Symbiotic or external supply depends on the availability and compatibility of nearby $CO_2$ sources. Studies on enrichment control highlight the need for accurate monitoring and regulation, given the interaction between $CO_2$ and other microclimate variables [99,102]. Work on integrated systems further shows that the feasibility of external $CO_2$ supply depends on how well source characteristics match greenhouse demand.

$CO_2$ enrichment depends on greenhouse operating mode. In ventilation-driven systems, injected $CO_2$ is diluted and lost through air exchange. Semi-closed systems reduce ventilation and rely on mechanical air treatment, allowing temperature, humidity, and $CO_2$ concentration to be regulated more independently. This improves $CO_2$ retention but increases dependence on air treatment, recirculation, and dehumidification systems [52,103].

External $CO_2$ supply introduces constraints related to gas quality and system suitability. Studies of $CO_2$-rich industrial streams show that impurity profiles vary with source and purification method, and that additional treatment may be required before use. In greenhouse applications, $CO_2$ supply cannot be assessed by quantity alone. Source quality, upgrading requirements, storage, transport, and dosing control influence system design, cost, and feasibility. Techno-economic comparisons are therefore less reliable when $CO_2$ provision is excluded from the system boundary, particularly in low-carbon systems that no longer obtain $CO_2$ as a by-product of combustion-based heating [99,101,104]. External $CO_2$ supply also introduces safety and handling requirements, including controlled storage, leak management, and safe dosing within greenhouse environments.

### 3.7. *Modelling, Optimisation and Intelligent Control*

As greenhouse energy systems shift from stand-alone boilers toward integrated, multi-vector architectures, operational strategy becomes as important as the technologies in use. Greenhouse climate evolves through coupled energy and mass balances that change with weather, crop stage, and choices around ventilation or dehumidification. At the same time, renewable electricity and waste heat sources introduce variability on the supply side. For these reasons, greenhouse climate management is increasingly formulated as an optimisation problem under uncertainty, typically implemented through predictive control and hybrid physics-data modelling frameworks.

Recent studies show that controller performance depends on how the underlying model is structured and on the boundary conditions applied. Assumptions about envelope heat loss, ventilation and infiltration rates, radiative boundary conditions, and latent heat coupling directly shape estimated heating and cooling loads and therefore influence the performance of different control strategies.

#### 3.7.1. Modelling Hierarchy: From Energy Balance Models to Hybrid and Multi-Fidelity Stacks

Greenhouse climate control has traditionally been based on lumped energy and mass balance models that treat indoor air as spatially uniform and governed by conservation equations. Katzin et al. [14] review the development of these process-based models and identify recurring structural elements, including explicit long-wave radiative exchange, transpiration-driven latent heat fluxes, simplified canopy energy balances, and varying formulations for infiltration and condensation. They also show that model complexity is linked to the intended application: simplified formulations are often used for control and optimisation, while more detailed formulations are applied in design evaluation and process analysis.

https://doi.org/10.3390/en19081975

Computational fluid dynamics has been combined with energy balance modelling to capture airflow and humidity patterns that cannot be represented in spatially averaged formulations. Piscia et al. [42] coupled CFD-derived ventilation rates and convective coefficients with an energy balance model for night-time humidity analysis, showing that detailed flow simulations can inform reduced-order models while preserving the computational efficiency needed for parametric studies.

Taken together, the recent advances support a layered modelling approach in which reduced order models handle real-time optimisation and constraints, higher fidelity simulations provide structural validation and sensitivity testing, and data-driven elements deliver fast prediction and uncertainty handling.

#### 3.7.2. PID and Rule-Based Control: Still Dominant in Practice, but Limited by Coupling and Disturbances

Proportional-Integral-Derivative (PID) and rule-based controllers remain common in commercial greenhouse climate control because they are straightforward to implement, easy to tune, and transparent in operation [105]. However, greenhouse environments are nonlinear and strongly coupled. Temperature, humidity, and $CO_2$ concentration interact through ventilation, heating and crop transpiration, creating behaviour that is difficult to manage with simple control loops. In these multivariable environments, PID controllers are usually applied as single-loop or in loosely coordinated arrangements, which limit their effectiveness when strong coupling or external disturbances are present [105].

Comparative evaluations illustrate these structural constraints. Mengesha et al. [106] assessed a practical PID arrangement against a feedback linearisation-LQR controller on a nonlinear multi-variable greenhouse model. The feedback linearisation-LQR design managed disturbances more effectively, with lower perturbation peaks and faster recovery, while the PID controller showed stronger setpoint tracking with minimal overshoot.

As greenhouse energy systems evolve toward hybrid configurations that include heat pumps, storage, semi-closed dehumidification, and variable energy supply, it becomes increasingly important to account for multivariable interactions, external disturbances, and operational constraints [105]. This shift has driven growing interest in predictive and optimisation-based supervisory control strategies that operate above the level of conventional feedback loops.

#### 3.7.3. Model Predictive Control: Deterministic, Stochastic, and Economic Approaches

Model predictive control is used increasingly in greenhouse climate management because it handles multivariable interactions, actuator constraints, and forecast disturbances within a single optimisation framework. Bwambale et al. [107] identify greenhouse climate control as a suitable application for MPC because of the strong influence of weather variability and the interactions between temperature, humidity, and $CO_2$ [107].

Recent MPC studies fall into three broad categories. First, robust and data-driven MPC formulations address modelling uncertainty and disturbance variability. Mahmood et al. [21] propose a data-driven robust MPC framework for greenhouse temperature control and evaluate its performance under uncertain dynamics. They report more stable regulation and a clearer assessment of energy use than nominal formulations, reflecting a shift from traditional tracking MPC toward designs that explicitly account for uncertainty.

The second direction involves stochastic MPC, which represents uncertainty probabilistically. Kim and You [108] develop a stochastic MPC scheme based on Gaussian process models that combine probabilistic disturbance modelling with online learning. Their optimisation framework seeks to minimise expected operating costs while maintaining climate constraints, and their results show improved regulation and lower resource use under variable weather.

https://doi.org/10.3390/en19081975

The third direction integrates explicit electricity and $CO_2$ cost terms into the objective function alongside climate tracking penalties. In the formulation by Kim and You [108], the stage cost includes price-weighted energy and $CO_2$ consumption together with quadratic penalties for deviation from climate setpoints. Tracking remains part of the objective, but the explicit monetisation of resource use shifts the optimisation toward economically efficient operation.

Across MPC studies, the reported energy savings depend strongly on the underlying climate model assumptions, particularly ventilation-dehumidification control, latent heat coupling, and envelope heat transfer representation.

#### 3.7.4. AI and Surrogate Models: Forecasting, Fast Optimisation, and Hybrid Physics-Data Control

In greenhouse control, AI is mostly applied in three ways. Short-horizon forecasting models are used to predict near-term disturbances and state trajectories (such as outdoor weather, solar irradiance, and indoor climate dynamics) to improve predictive control performance. Data-driven surrogate models are developed to approximate computationally intensive simulators, including CFD and detailed crop–climate models, thereby enabling rapid optimisation and real-time decision support. In addition, policy-learning approaches, notably reinforcement learning and imitation learning, are employed to derive control policies from interaction data or historical operations.

A recurring challenge is robustness outside the conditions represented in the training data. Work on machine learning for physical systems shows that out-of-distribution inputs can produce unreliable extrapolations even when in-sample fit is excellent, which remains a central barrier to reliable deployment across changing regimes [109].

In greenhouse operation, regime shifts are routine across seasons, weather patterns, and management policies. Liu et al. [110] frame this as a limitation of fixed-parameter mechanistic models under cross-seasonal nonlinear dynamics and propose a hybrid method in which deep neural networks predict time-varying calibration parameters to maintain performance across seasons.

For these reasons, hybrid strategies that keep a mechanistic backbone and use data-driven components for calibration or residual correction are increasingly reported. Zhou et al. [111] present a knowledge-based workflow in which a process-based greenhouse tomato model is first improved using particle-filter parameter estimation and then refined with a deep neural network, achieving substantial error reductions while preserving the structure of the underlying process model.

Recent work in building thermal systems shows that AI is also being used for calibration and data generation. Sun et al. [112] applied virtual samples with an autoencoder to correct sensor errors, achieving over 98.9% calibration accuracy and keeping online errors below 3%. This demonstrates that reliable measurements can be maintained without reference sensors or detailed physical models. A related approach uses knowledge-guided virtual samples to overcome limited data. Sun et al. [113] combined LLM-derived knowledge with diffusion and Monte Carlo sampling to generate synthetic fault data, reaching a diagnostic accuracy of 72.80% with improved generalisation compared to conventional small-sample training. Although developed for building systems, these methods are relevant to greenhouse operation, where sensor drift, sparse fault data, and seasonal variability remain persistent challenges.

From a net-zero perspective, surrogate models are valuable because they make large scenario ensembles computationally feasible. This supports the evaluation of alternative climate years, control strategies, tariff structures, and renewable supply profiles within optimisation and planning studies.

https://doi.org/10.3390/en19081975

### 3.7.5. Digital Twin: From Monitoring to Closed-Loop Optimisation

Digital twin approaches extend greenhouse modelling beyond offline simulation by linking a virtual representation to live sensor data for monitoring, state awareness, diagnosis, and decision support. The longer-term aim is to support supervisory optimisation and automated operation. A recent review of greenhouse horticulture notes that explicit digital twin implementations remain limited, although many IoT-enabled monitoring platforms and model-based decision support tools already exhibit twin-like architecture without using the term [114].

Recent application studies show how digital twin concepts are being deployed in practice. Arshad et al. [115] describe a Simulink-based digital twin integrated with greenhouse sensing and analytics, designed for remote monitoring and operational decision support, with a clear pathway toward automated environmental adjustment. Akintan et al. [116] develop a Unity-based digital twin for microclimate monitoring and adaptive sensor selection, using Thompson-sampling to identify season-specific subsets from a 56-sensor network while preserving representative temperature–humidity trends. Slimani et al. [117] present a real-time management workflow that combines IoT data acquisition, a digital twin layer, and augmented-reality visualisation to support operator-facing interpretation and decision making in an integrated cyber-physical loop.

Taken together, these studies show that the most substantive digital twin contributions are those that integrate the model, data pipeline, and decision layer into a functioning operational system, rather than presenting a static 3D visualisation or a standalone simulation with no feedback to real conditions [114].

### 3.7.6. Reinforcement Learning and Hybrid RL-MPC

Reinforcement learning is being explored for greenhouse climate control because it can learn nonlinear control schemes directly from measured data, which is attractive when disturbance-driven behaviour is difficult to represent with a physics-based model. Recent greenhouse studies identify two main barriers to deployment: reduced reliability when a learned policy is applied outside the operating regime present in the training data, and the difficulty of enforcing hard actuator and climate constraints during learning. Ajagekar et al. [118] show that offline value-function learning can produce out-of-distribution errors when the training data captures only a subset of operating conditions. They address constraint enforcement and disturbance uncertainty by embedding the learned value function within a robust MILP layer.

These issues have encouraged hybrid and hierarchical designs that use optimisation to structure decisions while reserving learning for adaptation and tracking. Mansour et al. [119] implement a two-level scheme in which an economic optimiser generates reference trajectories using weather and energy price forecasts, and a lower-level deep reinforcement learning controller tracks these trajectories in real time. They test the system under environmental disturbances and actuator-related uncertainties, including degraded heating capacity, and report shorter training requirements through staged training and transfer.

Comparative studies further clarify how RL and MPC differ in practice. Morcego et al. [22] contrast the two approaches in terms of prediction, objective formulation, and constraint handling. MPC embeds constraints directly within the optimisation problem, while RL typically requires additional mechanisms to maintain feasibility and operational safety. Building on this idea, Mallick et al. [120] introduce an RL-MPC design in which learning updates the parametrisation of an MPC controller online, adjusting model terms, cost weights, and constraint margins while the MPC structure ensures robust constraint management under prediction uncertainty.

https://doi.org/10.3390/en19081975

For net-zero greenhouse operation, these hybrid RL–MPC schemes are particularly relevant. Integrated systems that combine electrified heating, thermal storage, semi-closed dehumidification, and variable energy prices operate under uncertainty and multiple objectives. In these settings, fixed controller tuning is often insufficient. A design that combines optimisation-based constraint handling with adaptive learning offers a more robust pathway to maintain performance without compromising safety or energy efficiency.

### *3.8. Techno-Economic Performance and Scalability*

#### 3.8.1. Economic Framing of Decarbonisation Pathways

Greenhouse decarbonisation differs from building retrofits because the objective is to maintain a controlled crop microclimate, not simply to heat space. Energy demand depends on envelope performance, ventilation and humidity management, and climate control set-points. Capital and operating costs are therefore shaped by how thermal loads are defined and controlled. This section draws on reported case studies, experimental results, and simulation-based analyses to examine how these factors influence cost and performance.

In heating-dominant climates, peak winter load sets installed capacity and drives capital cost. Envelope losses and air exchange rates affect both annual energy consumption and crop requirements [6,31]. In warmer regions, the economic focus tends to shift toward cooling, ventilation and moisture control, which changes the balance between fuel and electricity use. Humidity control further shapes operating cost. Ventilation-based dehumidification during cool periods expels conditioned air and increases heating demand, while internal moisture extraction retains heat within the system [51,121]. Studies of heat pump-based dehumidification report lower primary energy use than ventilation–heating under comparable moisture targets, provided auxiliary electricity and control boundaries are included [58,121].

Feasibility also depends on financial assumptions. Payback, NPV and levelised cost metrics vary with tariffs, discount rate, project lifetime and incentives. Sector guidance reflects this by prioritising demand reduction before supply substitution, since reduced and stabilised loads improve the economic case for electrification and renewable heat integration [122,123]. For interpretation, techno-economic results are grouped by system boundaries. The first is the greenhouse-only boundary, where costs relate solely to internal heating and climate control. The second is the integrated on-site boundary, which includes components such as heat pumps, storage, solar collectors, dehumidification units, and control systems. The third is the external or networked boundary, covering district heating, waste heat, minewater, and other off-site sources, where infrastructure such as pipework, substations, and network layout forms part of the cost. CAPEX, OPEX, payback period, NPV, and LCOH are, therefore, interpreted within these boundary definitions rather than compared directly across studies.

These relationships are evident in the reported case outcomes. Thermal screens are associated with heating energy savings of around 12–35% and payback periods often below three years in commercial applications. Solar thermal systems with seasonal storage have been reported with levelised heat costs of approximately 30–36 $/GJ under favourable conditions. District and waste heat systems depend strongly on network configuration and connection density, with capital investment and pipeline routing often determining overall feasibility. As discussed in the following subsections, these findings show that techno-economic performance is shaped not only by the technology itself, but also by load characteristics, system design, and local economic conditions.

Carbon-intensity comparisons follow the same boundary constraints. The reported emissions depend on electricity emission factors, auxiliary energy use, $CO_2$ supply, and the infrastructure included within the system boundary. Electrified systems may reduce on-site emissions, but can shift impacts upstream, depending on grid carbon intensity and operating conditions. Auxiliary electricity for dehumidification, pumping, and control, along with $CO_2$ provision in non-combustion systems, can affect total emissions. Life cycle and carbon-intensity results are only comparable when system boundaries and energy inputs are defined consistently. For these reasons, payback figures are not transferable unless assumptions are explicit. Without clear reporting of thermal boundary (envelope, ventilation, humidity strategy), economic parameters (tariffs, discount rate, lifetime), and integration choices, reported CAPEX, OPEX and LCO values cannot be compared in a meaningful way. The techno-economic synthesis in Section 3.8, therefore, evaluates pathways within clearly stated physical and financial boundaries rather than treating individual technologies as universally cost-effective or unviable.

#### 3.8.2. CAPEX and OPEX Structure by Pathway

The pathway-specific studies reviewed below do not share a common costing basis. Some report greenhouse-only retrofit costs, while others include storage, control upgrades, or auxiliary electricity. Networked studies extend the boundary further to external infrastructure and heat supply arrangements. The discussion, therefore, interprets the results within their stated boundary conditions and identifies the assumptions that most strongly affect CAPEX, OPEX, payback, NPV, or LCOH.

##### Envelope/Cover and Thermal Screens Upgrades/Retrofits

Envelope improvements and thermal screens are demand-side measures whose economic value lies in reducing annual heat use and, in many cases, reducing peak heating capacity. Because night-time transmission and radiative losses dominate heating demand in many temperate and cold-climate greenhouses (Section 3.2), reducing effective heat loss at night directly lowers fuel consumption and can also reduce peak heating requirement and associated infrastructure.

Sector guidance for covered cropping notes that thermal screens commonly deliver reductions in annual heat demand under commercial practice, although realised savings depend on screen specification and how screens are controlled alongside humidity management. In the EECA case study [123], reported energy savings range from about 12.5% to 35%, with reductions of roughly 25–35% achievable when screens are properly selected and controlled.

A commercial example from the UK HDC report (Pratt, HDC project PC 227) [124] shows this in practice. In a sweet pepper greenhouse, a moveable thermal screen delivered an additional 52 $kWh \cdot m^{-2}$ gas saving compared with a temporary screen. The same study shows that adjusting screen control setpoints increased savings from 29 to 52 $kWh \cdot m^{-2}$, indicating that operational strategy can be as important as the hardware itself. Installed costs were reported at £4–£5 $m^{-2}$, and when energy savings and avoided temporary screen costs are included, simple payback can be below three years under the stated commercial assumptions [124].

From a cost-structure perspective, CAPEX is associated mainly with screen materials, installation, and integration with the control system. OPEX savings come primarily from reduced fuel use, with minimal or no added recurring energy demand unless additional sensing or control upgrades are introduced. Where such upgrades are required to achieve the reported savings, part of the investment should be attributed to improve control capability rather than to the screen hardware alone [124].

##### Heat Pumps and Electrification (ASHP/WSHP/GSHP)

Electrifying greenhouse heating with heat pumps changes the structure of heat supply costs. Expenditure shifts from fuel to equipment, balance-of-plant, and system integra-

https://doi.org/10.3390/en19081975

tion, while operating costs depend mainly on electricity tariffs and achieved seasonal performance. In practice, heat pump CAPEX extends beyond the unit itself and often includes compatible low-temperature distribution (e.g., larger emitters or piping), buffer storage to limit cycling, electrical upgrades, and integration with existing climate control systems such as screens, ventilation, and dehumidification. The reported capital costs, therefore, vary unless the system boundary is clearly defined. Some studies consider only the heat pump, while others include distribution upgrades, storage, electrical upgrades, and dehumidification-related equipment within the same investment boundary.

Operating cost is driven by electricity price and the seasonal coefficient of performance, which depends on source temperature and required heating water temperature. In a Polish greenhouse study, Nemś et al. [125] reported a maximum greenhouse heat demand of 17.36 kW. A 21.5 kW ground-source heat pump operated with a COP of 5.53–5.83, while a 23.00 kW air-source heat pump operated with a COP of 2.59–4.00. Annual operating costs were €385.60 and €678.64, respectively. However, the ground-source system required a significantly higher initial investment (€14,334 compared to €4952 for the air-source unit), resulting in longer payback periods (18 years vs. 5.5 years). This comparison shows the trade-off between efficiency and capital cost, and the importance of matching system capacity to actual demand.

Differences across studies also arise because electrification is often implemented as part of integrated retrofit packages rather than as a standalone intervention. Li et al. [126] evaluate a cold-region greenhouse that combines ground-source heat pumps, solar collectors, insulation upgrades, and intelligent ventilation and lighting controls. They reported an overall energy saving of 17.8%, with 17.6% attributed to heating. The ground-source heat pump system had an installation cost of 80,000 Yuan, annual savings of 16,000 Yuan, a payback period of 5 years, and a 15-year NPV of 98,450 Yuan. These results show that performance improves when heating is considered within a broader system rather than as a standalone measure.

Humidity management further affects both load and cost. Electrified heating cannot be treated solely as a sensible heat problem because the dehumidification strategy drives ventilation and electricity consumption. Sapounas et al. [103] show that semi-closed greenhouse configurations reduce reliance on window ventilation by using mechanical air handling, allowing more independent control of temperature, humidity, and $CO_2$. This reduces heat losses associated with ventilation but introduces additional electrical loads from fans and air-treatment equipment. The reported case evidence shows this trade-off. In a French study, gas consumption in a semi-closed greenhouse was about 10% lower than in a conventional greenhouse, while electricity use increased substantially, reaching about 510% higher during autumn under non-optimal fan operation. Other reported studies indicate energy savings of around 20% in Belgium and up to 43% improvement in energy-use efficiency in Germany when combined with thermal screens and storage. These outcomes show that air distribution design, control strategy, and humidity management directly influence operating cost.

As a result, the performance of heat pump systems in greenhouses depends on the interaction between thermal load, source conditions, control strategy, and the wider system configuration. Comparisons of CAPEX, OPEX, and payback are only meaningful when these factors are clearly defined.

Solar Thermal and Thermal Energy Storage

For solar-assisted greenhouse heating, capital cost is spread across collector area (€/$m^2$), seasonal storage infrastructure (borehole, tank, or pit), civil works and land preparation, pumps and heat exchangers, and integration with the existing heat distribution

https://doi.org/10.3390/en19081975

system and controls. Operating costs are usually lower and are driven mainly by pumping electricity and maintenance. Economic performance is sensitive to system sizing and operating temperature, which influences both collector efficiency and storage losses [127]. The reported LCOH values are therefore not directly comparable across systems with different configurations, particularly where seasonal storage, temperature upgrading, or large collector fields are included, as these elements significantly affect both capital cost and operating requirements.

Semple et al. [127] present a large-scale tech-economic analysis for a 0.4 ha greenhouse in Southwestern Ontario, comparing low-temperature and high-temperature borehole seasonal thermal storage. Both systems are sized to supply roughly 64% of annual sensible heating demand, using 861 $m^2$ and 2009 $m^2$ of collector area, respectively. The reported levelised heat costs range from about 32.9 to 36.2 \$/GJ (115–130 \$/$MWh_{th}$). The high-temperature option requires greater upfront investment but reduces operating electricity demand. In their analysis, achieving a simple payback below seven years requires strong policy support, including capital subsidies and carbon pricing of around 200 \$/tonne $CO_2e$ under the stated gas price cases [127].

Storage coupling also matters in dehumidification and heat recovery configurations. de Zwart [121] measures about 24% energy savings for an internal moisture removal system and notes that adding seasonal storage can increase savings to around 30% by extending heat pump operation and improving the use of recovered heat over the season.

Across these studies, the techno-economic results remain sensitive to modelling assumptions that affect operating temperature and seasonal availability. Semple et al. [127] show that variations in borehole spacing, soil properties, and insulation thickness alter annual heat delivery by less than about 2%. Although the thermal impact is small, such differences can still affect economic outcomes in marginal cases. More broadly, the results are more credible when load models include diurnal dynamics and humidity-driven ventilation losses, since these factors determine peak conditions and the temperature levels at which systems must operate [121,127].

Waste Heat and District Heat

Waste heat recovery and district heating options reshape the techno-economics of greenhouse heating because viability depends less on on-site conversion efficiency and more on network design, demand density, heat-pricing and contractual structure. Capital cost in these systems is driven by distribution infrastructure and connection assets, while operating costs tend to play a smaller role relative to amortisation and revenue assumptions. These systems are best interpreted within an external infrastructure boundary, where network length, substations, trenching, and long-term financing form part of the cost basis.

A district heating feasibility study for greenhouse crops in Almería illustrates the dominance of network-related costs. The optimised configuration supplies 12 greenhouses via a 1.3464 km network delivering 8621.4 MWh·$year^{-1}$, with a total initial investment of 1.7233 M€ and annual operating costs of 0.0158 M€·$year^{-1}$. Substation hardware accounts for 0.3569 M€, confirming that connection equipment forms a substantial part of upfront expenditure alongside the distribution network [128]. The study also shows that economic outcomes are highly sensitive to routing and spatial clustering. Within the optimised population, profitability is closely linked to network length, with a practical threshold in the range of 1.3–1.5 km beyond which additional distance erodes NPV under the assumed supply and demand conditions [128]. This shows that even small routing extensions can dominate project economics where greenhouse loads are dispersed.

Where waste heat comes from another primary process, the cost logic shifts from combustion efficiency to load matching and electricity pricing. Asgari et al. [67] report

https://doi.org/10.3390/en19081975

that an Antminer unit dissipates roughly 3.24 kW as waste heat, consistent with near-complete conversion of electrical input to low-grade heat. This supports the point that, for co-produced heat, feasibility depends on the temporal fit between available waste heat and greenhouse heating demand, and on the tariff basis of the underlying electrical load.

UK-focused work on minewater district heating adopts a long-horizon discounted cashflow approach rather than simple payback. The Horden study models NPV, IRR, and discounted payback over 40 years and aligns with HNDU and CP1 guidance, showing that network business cases are evaluated as infrastructure investments whose viability hinges on pipework requirements and customer clustering [129].

Geothermal district heating concepts highlight the same cost structure. The Penobsquis feasibility report itemises buried HDPE pipework and trenching costs per metre, and includes well maintenance, heat exchanger servicing, and pump replacement cycles as part of ongoing O&M. In the most favourable open-loop scenario, the report estimates total capital investment of about $11.3 million and a discounted payback of approximately seven years under the stated assumptions [130].

Across the literature, waste heat and district heating shift greenhouse heating from a technology substitution problem to an infrastructure planning problem. Economic feasibility is driven primarily by network length, connection density, routing, tariff structure, and financing assumptions, rather than marginal gains in thermal efficiency. Spatial modelling and long-term financial analysis are therefore essential when evaluating district-based heating options for greenhouse systems.

Semi-Closed/Closed Greenhouse Concepts

Semi-closed configurations add capital cost through air-handling units, ducting, sensors, and advanced control systems. Operating cost is shaped by fan power and the electricity used for active dehumidification and heat recovery. Unlike ventilation-based humidity control, which expels conditioned air, semi-closed systems aim to decouple temperature and moisture regulation via mechanical air treatment. Sapounas et al. [103] show that this architectural shift alters both the greenhouse energy balance and system complexity. de Zwart [121] further demonstrates that internal moisture withdrawal with heat recovery can reduce primary energy use when realistic heat pump performance and storage are included. Therefore, these studies indicate that economic outcomes depend on how latent loads, auxiliary electricity, and operational constraints are represented in the thermal model. For this reason, the techno-economic results for semi-closed systems are only comparable where the system boundary includes fan power, dehumidification loads, auxiliary electricity, and the effect of reduced ventilation on heating demand.

### 3.8.3. Sensitivity Drivers: What Most Strongly Shifts Payback and NPV

Across the reviewed techno-economic studies, only a few parameters consistently drive payback and NPV outcomes. Energy prices and tariff structures are often the dominant drivers because they directly determine operating expenditure and can change the relative competitiveness of electrified heat. Tomson et al. [129] showed that moving from BEIS grid-price assumptions to lower-cost electricity supply changed viability outcomes. Under the gas boiler baseline, the 40-year NPV improved from −£5.25 million to −£1.67 million under a power purchase agreement (PPA), and to between −£0.55 million and £0.08 million when priced at onshore wind LCOE. When the baseline was changed to decentralised ASHPs, the same system became economically viable, with 40-year NPV increasing to £1.86–£7.19 million depending on the pricing scenario and configuration. These results show that tariff assumptions and baseline selection alone can move a project from unviable to viable, even when the physical system remains unchanged.

https://doi.org/10.3390/en19081975

Greenhouse thermal boundary conditions exert a similar influence. Envelope heat transfer characteristics and air-exchange rates determine both annual heating demand and peak capacity requirements, shaping capital sizing and operating cost. Demand-reduction measures, therefore, shift project economics before any change in heat supply. Sector guidance for covered cropping reports that well-specified and -operated thermal screens typically reduce annual heat use by roughly 25–35% [123]. Modelling studies likewise show that modest changes in transparent-envelope properties translate into measurable differences in heating and cooling loads [75].

Latent-load representation is another key source of variation in techno-economic results. When humidity control is simplified as ventilation, the associated heat-loss penalty can be underestimated or omitted, leading to lower predicted operating costs and overstating the benefit of ventilation-based strategies. EECA [123] reports a Southern Belle case where dehumidification reduced thermal fuel use by 25%, increased production by 15%, and achieved a payback of 3–4 years in a 3000 $m^2$ twin-skin greenhouse with a capital cost of about $120,000. These results show that payback depends not only on the dehumidification system itself, but also on how humidity control, fan energy, and $CO_2$ retention are represented in the analysis. For networked heat systems, spatial layout and connection density often dominate the economic outcome. In Tomson et al. [129], the southern network required about 800.5 m of trench for the main DN80 pipeline, with pipe costs exceeding £300,000, while the northern extension involved approximately 5977 m of trench with an estimated cost of about £7.26 million and heat losses of around 75 kW. These figures show that network routing and connection density can outweigh the performance of the heat source in determining project viability, which is why feasibility assessments at the district scale focus on network design, cost structure, and heat pricing.

Financial assumptions and reporting boundaries also shape NPV. Discount rate, project lifetime, and whether crop-revenue or quality effects are included can shift economic outcomes even when underlying energy balances are similar, complicating comparison between simple payback metrics and long-horizon discounted cashflow assessments [129]. A cross-cutting issue in gas-to-electric transitions is $CO_2$ enrichment logistics. Dutch sector analyses indicate that a large share of greenhouse $CO_2$ supply has historically been sourced from CHP exhaust. Reduced CHP operation therefore increases reliance on external $CO_2$ supply chains, which introduces additional infrastructure and procurement costs [131]. In a decarbonised setting, $CO_2$ provision becomes a parallel infrastructure requirement rather than a minor operational detail [61].

In greenhouse systems without combustion-based heat sources, $CO_2$ provision becomes a separate cost component. Costs vary with the supply route, including delivered $CO_2$, industrial recovery, or network-based sources, as well as storage, transport, distribution, and injection requirements. Gas purity, compression, and dosing requirements further influence both capital and operating costs. Techno-economic comparisons are therefore incomplete if $CO_2$ provision is excluded from the system boundary, particularly where enrichment interacts with ventilation and humidity control. As a result, outcomes depend not only on heat and power assumptions, but also on how $CO_2$ supply and distribution are defined. Differences across studies reflect system boundaries, operating assumptions, and financial inputs as much as the underlying technology.

### *3.9. Net-Zero Pathways and Food Security Implications (System-Level Synthesis and Roadmap)*

Greenhouse decarbonisation is not a simple technology substitution but a coordinated pathway that links heat and mass transfer, energy infrastructure, storage, operational control, and capital recovery under uncertainty. This system view is essential because agrifood production is a major source of global emissions. FAO estimates places agrifood

https://doi.org/10.3390/en19081975

emissions at 16.2 Gt $CO_2$-eq in 2022, about 29.7% of total anthropogenic emissions, with supply-chain energy use forming a substantial and growing share [8].

Within protected horticulture, life cycle assessments show that energy use is a leading contributor to environmental impact, and that outcomes depend heavily on how heat and electricity are supplied (e.g., conventional heat vs. cogeneration or geothermal sources) [132]. Resource-efficiency studies further show that the balance between heating and electricity demand varies with production typology and climate. This reinforces that decarbonisation outcomes are shaped as much by demand structure and operating conditions as by the choice of technology [133].

### 3.9.1. Net-Zero Greenhouse Archetypes: Four Repeatable System Architectures

Feasibility studies and greenhouse decarbonisation projects reveal a set of recurring system architectures. These archetypes are best defined not by technologies but by how they reshape thermal demand, required temperature lift, and operational flexibility. In practice, net-zero pathways emerge from different combinations of load reduction, electrification, renewable-heat integration, and infrastructure coupling rather than from a single technology acting alone.

#### Load-First Retrofit Pathway

A recurring net-zero pathway prioritises reducing demand and peak load before changing the heat source. Typical measures include thermal screens and night insulation, targeted envelope upgrades and sealing, infiltration reduction, and control adjustments that avoid unnecessary ventilation during heating periods. Sector guidance for covered cropping identifies this sequence as a practical starting point because it delivers immediate reductions in heat use and improves the feasibility and sizing of low-temperature supply options such as heat pumps or low-grade waste heat [123].

Net-zero implication: Reducing uncontrolled heat losses lowers direct emissions and reduces the size and cost of low-carbon heat systems. Sector guidance reports heat energy savings in the range of 12.5% to 35%, with well-selected and well-operated thermal screens typically achieving around 25% to 35% of annual heat input reduction [123].

#### Electrification Pathway: Heat Pumps + Low-Temperature Distribution + Buffering

Heat pump electrification works best when greenhouse heating can operate at low supply temperatures, since this limits the required temperature lift and improves seasonal performance. Outcomes depend on system boundaries, including auxiliary electricity, distribution efficiency, and the interaction between humidity control and heating demand. de Zwart [121] shows that using ventilation-based moisture removal in cool conditions increases heating requirements, whereas internal dehumidification with heat-pump-assisted condensation can reduce primary energy use by approximately 24%, with potential improvement to around 30% when seasonal storage is included.

Net-zero implication: Electrification tends to perform best when combined with reduced envelope losses, appropriate buffering, and an electricity supply with declining carbon intensity. The IEA reports that heat pumps already reduce emissions relative to fossil boilers under current power mixes and that their advantage increases as the electricity supply decarbonises [134,135].

#### Renewable Thermal Pathway

Solar thermal systems coupled with seasonal thermal storage can supply a significant share of greenhouse heating, but the required collector area and storage capacities are capital-intensive and sensitive to design choices such as operating temperature and field sizing. In a 0.4 ha cold-climate case study, Semple et al. [127] show that both the levelised cost of delivered heat and the payback period shift with system configuration and policy conditions. Their results indicate that simple payback is strongly influenced by fuel prices, carbon pricing, and the level of capital support (subsidy).

Net-zero implication: This pathway is most attractive where land is available, the heating season is long, and financing or policy frameworks can support long-lived infrastructure with multi-year payback horizons.

Networked Heat Pathway: Waste Heat/District Heat/Minewater with Greenhouse as Anchor Demand

Networked heat operates as infrastructure rather than a simple technology substitution. Feasibility depends on spatial layout (including pipe routing, connection density, interface costs, and tariff structure), rather than on the intrinsic efficiency of the heat source. In the Almería greenhouse case study, a graph-theory optimisation was used to determine the network layout and demand-point selection. The results showed that investment cost and NPV were strongly shaped by routing choices and the degree of load clustering [128].

Minewater studies reach similar conclusions within discounted cashflow analyses. The Durham work evaluates NPV, IRR, and discounted payback over multi-decade horizons and finds that capital allocation to distribution infrastructure and electricity price assumptions largely determine viability [129].

Net zero implication: Greenhouses can strengthen low-carbon heat networks where demand is sufficiently concentrated to justify distribution infrastructure. Where loads are spread out, the economics are limited by network length and connection costs rather than by the thermodynamic performance of the heat source.

#### 3.9.2. A Practical Roadmap: Staged Decarbonisation That Matches Both Physics and Finance

Across the reviewed studies, net-zero concepts underperform when the engineering sequence clashes with capital recovery logic. A common failure mode is committing early to high-capacity supply infrastructure before stabilising demand. This raises installed capacity and fixed costs and exposes modelling errors in ventilation, humidity control, and tariff exposure.

A staged roadmap that aligns heat transfer drivers with investment risk is more bankable. The early focus is on demand and controllability (thermal screens, leakage reduction, and sensor and setpoint tuning) to cut peak load and lower the required temperature lift. The next step is to electrify the base thermal load with heat pumps under explicit operating boundaries, including humidity control choices that can shift heating demand in semi-closed operation. de Zwart's [121] shows why ventilation-based moisture removal and internal condensation strategies cannot be treated as an afterthought when estimating energy use and sizing.

With a stabilised and electrified base load, renewable integration and storage can be evaluated against realistic demand profiles. Semple et al. [127] show that solar thermal with seasonal storage is technically viable but economically sensitive to configuration and policy assumptions, which argues for optimising cost-effective useful heat rather than maximum solar fraction. At longer horizons, cluster-scale integrations (district heat, minewater, or waste heat networks) become viable where spatial density supports infrastructure recovery. Feasibility studies in Almería and Durham show that routing, demand aggregation, tariff design, and discounted cashflow structure are decisive economic variables [128,129].

Advanced control and digital modelling support every stage. Predictive control, dynamic load estimation, and tariff-aware scheduling reduce uncertainty around ventilation losses, humidity management, and electricity exposure, improving sizing accuracy and the credibility of financial projections. Without realistic operational modelling, electrification and storage strategies risk misestimating both capacity and payback. The pathway follows

https://doi.org/10.3390/en19081975

thermodynamic sequencing (as shown in Figure 6), but its practical value depends on the scale of energy, emissions, cost, and load effects at each stage. Table 3 provides indicative performance ranges from the reviewed literature to support this analysis.

**Table 3.** Indicative performance ranges associated with staged greenhouse decarbonisation pathways.

| Pathway Stage | Typical Energy Effect | Emissions Effect | Peak-Load Effect | Cost Implication | Applicability/ Scale | Literature Basis |
| --- | --- | --- | --- | --- | --- | --- |
| Demand reduction and controllability | Thermal screens: about 12.5–35% heat reduction; reported payback below 3 years in some commercial cases | Reduced fuel use | Lower peak heating demand | Low to moderate CAPEX | Widely applicable, especially in heating-dominant climates | [123,124] |
| Electrification and humidity-aware operation | Internal moisture removal/heat pump-based dehumidification: about 24–30% energy reduction in reported cases; performance depends on tariff and COP | Depends on grid carbon intensity and auxiliary electricity use | Reduce load where ventilation losses are controlled | Moderate to high CAPEX | Greenhouse-only or integrated on-site systems | [103,121,125] |
| Solar thermal with storage | Around 64% annual sensible heating contribution in reported systems; LCOH about 32.9–36.2 $/GJ under stated assumptions | Depends on system configuration and policy context | Reduces seasonal fossil demand; limited by storage design | High CAPEX, long payback without support | Site-specific; more suitable for large-scale or long-term projects | [127] |
| Networked heat (district and waste heat) | High fossil fuel displacement where source and routing are suitable; performance sensitive to network length and demand density | Depends on source characteristics and accounting method | Supports base-load decarbonisation more than peak-load response | Infrastructure-dominated CAPEX | Cluster-scale, site-specific | [128–130] |
| Semi-closed/ integrated control | Energy use depends on dehumidification strategy and auxiliary electricity; some cases report lower gas use with higher electricity demand | Depends on treatment of electricity and $CO_2$ supply | Reduces ventilation-related heat loss | Moderate to high CAPEX | Advanced commercial systems | [103,121] |

Table 3 provides indicative performance ranges for the stages shown in Figure 6. The values are drawn from the reviewed studies and represent typical orders of magnitude rather than directly comparable benchmarks, as outcomes depend on climate, system configuration, control strategy, and system boundary assumptions.

https://doi.org/10.3390/en19081975

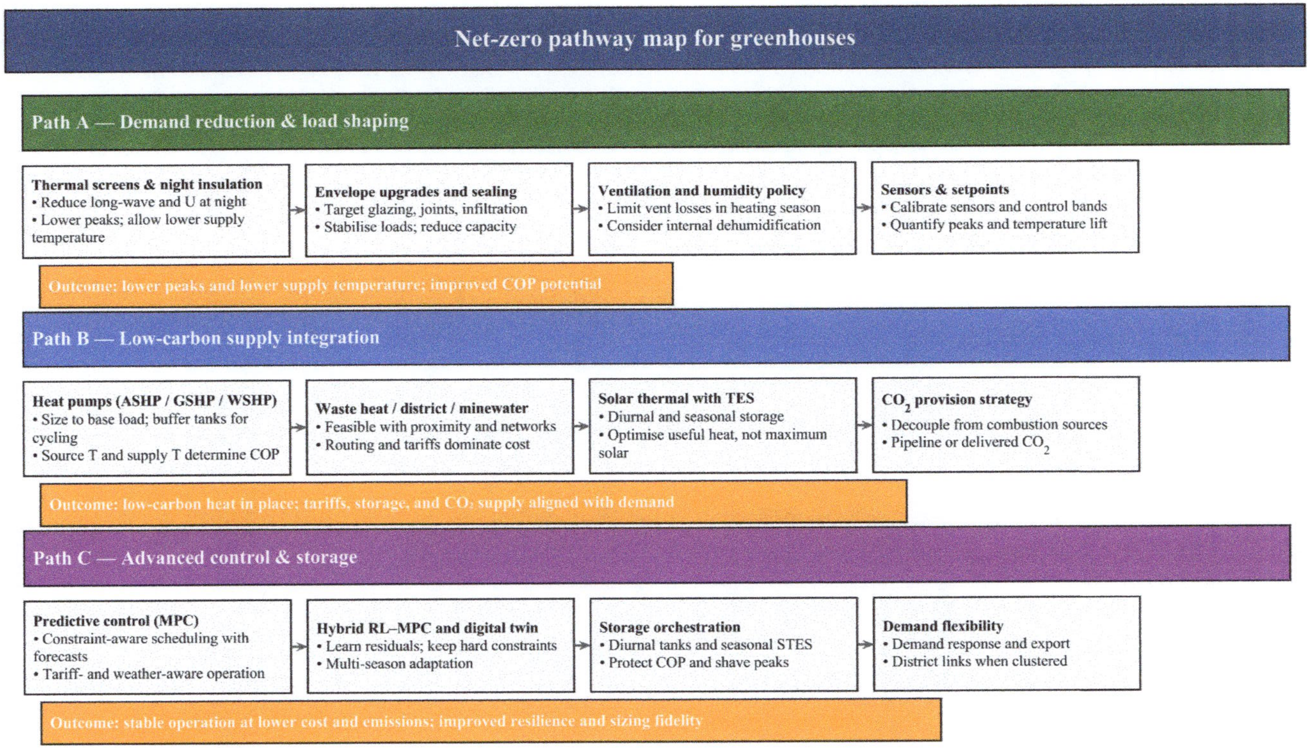


**Figure 6.** Net-zero pathway for greenhouse decarbonisation, showing three staged transitions: (A) reducing demand and improving controllability, (B) integrating low-carbon heat supply, and (C) improving system performance through storage and advanced control.



https://doi.org/10.3390/en19081975

### 3.9.3. Food Security Implications: What "Net-Zero Greenhouse" Changes and What It Risks

Net-zero transitions in protected horticulture matter for food security because they reshape both the reliability and the cost base of year-round supply. At the macro level, the case for decarbonising greenhouse energy is reinforced by the scale of agrifood emissions: FAO estimates 16.2 Gt $CO_2$-eq in 2022, with supply-chain energy a substantial and growing component. At the production-system level, life cycle assessments and resource-efficiency studies consistently show that energy demand is a major driver of environmental impact and that outcomes are sensitive to how heat and power are supplied, as well as to climate and production typology.

Decarbonisation can strengthen food resilience when it lowers exposure to fuel-price volatility and supply disruption. In practice, thermal reliability underpins production continuity: when heating is unaffordable or insecure, growers respond by lowering setpoints, shortening seasons, or temporarily suspending production. The staged pathway developed in this review makes the logic explicit: demand reduction and tighter controllability reduce peak vulnerability, while electrification with buffering and storage helps maintain operation through cold spells and price shocks.

The main equity risk is distributional. Capital-intensive options tend to concentrate in large, well-financed operations, leaving smaller or marginal growers exposed to volatile fossil costs or high electricity tariffs. This is not only a technology problem but also one of infrastructure and finance. Where district or minewater heat is feasible, evidence shows that spatial configuration and financing horizons dominate bankability, which implies that equitable access depends on coordinated planning, transparent connection rules, and tariff design rather than equipment performance alone.

Net-zero claims are credible only when energy sources and boundaries are stated clearly, consistent with the definitions adopted in this review. Energy consumption is among the largest environmental burdens in protected crops, and impacts can fall with a lower-carbon heat and power supply, but the results vary sharply with system definition and energy sourcing. This matters for food security narratives: high-cost green investments that do not reduce life cycle intensity (for example, electrification on a carbon-intensive grid) may raise production costs without delivering a credible climate benefit. $CO_2$ enrichment introduces a parallel constraint. As CHP use declines, supply increasingly depends on external sources, with added infrastructure, quality assurance, and logistics requirements. This affects reliability, as interruptions in supply can limit consistent dosing, and affordability, as $CO_2$ introduces an additional operating cost rather than a by-product of heat generation. It also affects production, as variations in $CO_2$ supply or dosing can affect yield and crop quality, particularly in systems with high ventilation rates where retention is low. $CO_2$ provision also depends on storage, transport, and distribution systems, linking greenhouse operation to wider supply networks. These requirements connect $CO_2$ enrichment to system design and planning, as detailed in Section 3.6.5.

The food security value of net-zero greenhouse pathways is highest when they deliver both operational reliability and credible life cycle emissions reduction, without shifting costs onto growers through poorly structured tariffs, fragmented infrastructure planning, or incomplete accounting of auxiliary electricity and $CO_2$ supply. Figure 7 shows how the sequence of net-zero measures links directly to food security outcomes and the conditions required for credible and equitable deployment.

### *3.10. Long-Term Projections of Commercial Greenhouse Markets and Implications for CEA in Global Food Systems*

Market analyses consistently indicate rapid expansion of commercial greenhouse infrastructure, reflecting growing investment in controlled-environment agriculture. Ac-

https://doi.org/10.3390/en19081975

cording to MarketsandMarkets [136], the commercial greenhouse market is projected to grow from USD 39.6 billion in 2023 to USD 68.7 billion by 2028, equivalent to a compound annual growth rate (CAGR) of 11.6%. Comparable projections are reported by other market analyses. ResearchAndMarkets reports an increase from USD 32.0 billion in 2022 to USD 56.4 billion by 2028 (9.91% CAGR), while Grand View Research projects USD 32.84 billion in 2025, rising to USD 74.10 billion by 2033 (10.9% CAGR). Fortune Business Insights forecasts growth from USD 37.78 billion in 2025 to USD 96.13 billion by 2034 (10.95% CAGR). Despite differences in methodology and market assumptions, these studies converge on a clear finding: commercial greenhouse and associates CEA infrastructure markets are growing at roughly 10–12% per year through the 2020s and into the early 2030s [136–139]. These estimates are used as indicative commercial baselines to inform the scenario analysis, reflecting current market scale and growth trends, and do not represent independently validated forecasts of technology adoption.

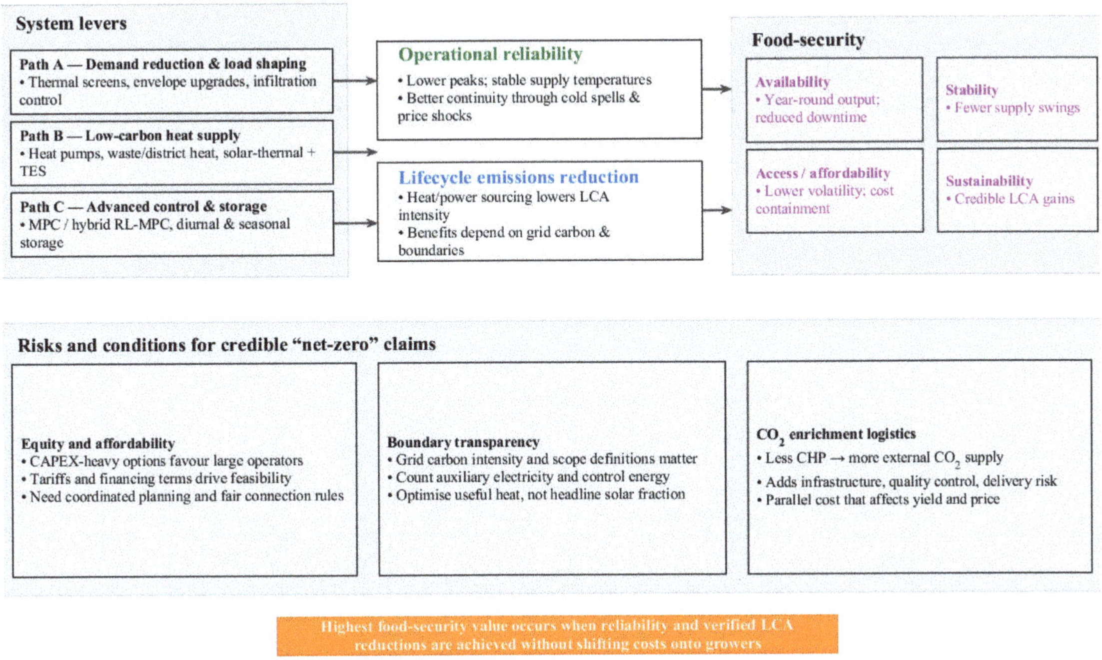


**Figure 7.** Net-zero greenhouse pathway showing system levers, food security benefits, and the conditions required for credible and equitable deployment.

To examine the longer-term implications of this growth for global food systems, the near-term estimates were extended to 2050 using a scenario-based compound-growth model. The analysis uses the USD 68.7 billion market value projected for 2028 as the reference point, with three annual growth rates of 6%, 8%, and 10% (representing conservative, intermediate, and upper-bound pathways within the range suggested by current market reports). These rates also recognise that the present double-digit growth is unlikely to persist indefinitely as the sector matures.

Projections (future market values) were calculated using the standard compound-growth expressions:

$$V_t = V_0(1+g)^n \tag{7}$$

https://doi.org/10.3390/en19081975

where $V_t$ is the market value in year $t$, $V_0$ is the 2028 anchor value, $g$ is the assumed annual growth rate, and $n$ is the number of years after the anchor year. This formulation is widely used in economic and market trend assessment and provides a transparent means of extending short- term forecasts into longer- range- scenario pathways while keeping the underlying assumptions explicit.

A limitation of this approach should be recognised. The projections are based on constant annual growth rates over the entire horizon, whereas the adoption of technology-intensive systems rarely follows such patterns. In practice, adoption develops unevenly, with slower early growth constrained by capital cost and infrastructure, followed by a period of faster expansion and eventual slowdown as markets mature. While this behaviour is often described using logistic diffusion models, the lack of long-term data for the greenhouse sector makes it difficult to define these dynamics in a consistent way. The projections should therefore not be interpreted as a model of market diffusion. Instead, they extend current growth trends under fixed-rate assumptions to illustrate possible long-term trajectories. Early-stage growth may be overstated, while later expansion may be understated. The results are best understood as scenario bounds rather than forecasts of future market behaviour. The use of multiple growth rates partly addresses this limitation by defining a range of plausible outcomes rather than a single estimate.

Applying the three growth scenarios to the 2028 anchor value of USD 68.7 billion yields a projected 2050 market size of approximately USD 247.6 billion at 6%, USD 373.5 billion at 8% and USD 559.2 billion at 10%, as shown in Figure 8. These values remain consistent with the scale and growth implied by current market assessments, while also reflecting the uncertainty associated with extending short-term trends over several decades. The trajectories in Figure 8 are based on constant growth rates and should not be interpreted as realistic diffusion pathways. Over long horizons, adoption is likely to depart from these trends as infrastructure constraints and market maturity begin to limit further expansion.

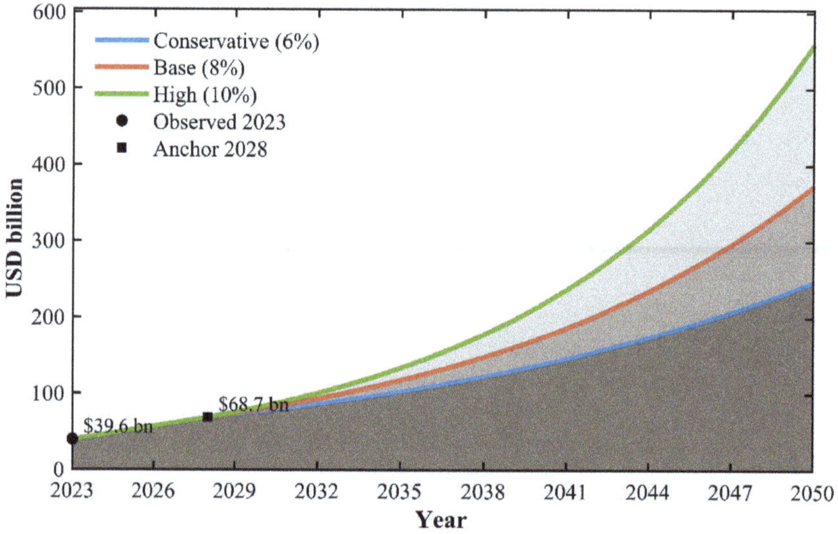


**Figure 8.** Commercial greenhouse market projections (2028–2050) under 6%, 8%, and 10% annual growth scenarios, anchored to the 2028 estimate of USD 68.7 billion.

To contextualise the projected growth of the commercial greenhouse sector within the wider food system, the scenario trajectories were compared against long-run projections of global agricultural production value. FAO statistics indicate that global agriculture value

https://doi.org/10.3390/en19081975

added reached USD 3.8 trillion in 2022, rising to about USD 4.0 trillion in 2023 [140,141]. Extending this series to 2050 using two growth assumptions: 2.6%, reflecting the most recent year-on-year increase, and 2.9%, corresponding to the average growth observed over the past decade. Under these assumptions, global agricultural value increases to approximately USD 8.0–8.7 trillion by 2050, as shown in Figure 9.

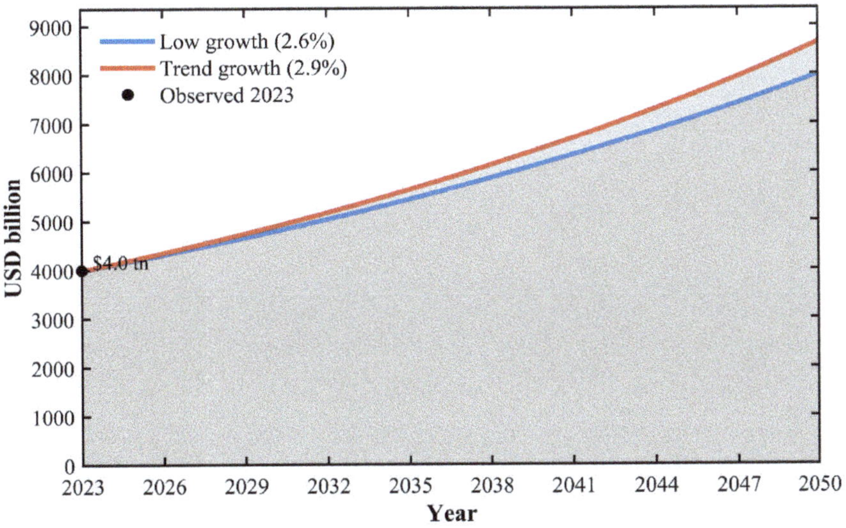


**Figure 9.** Global agricultural value projections for 2023–2050 under 2.6% and 2.9% annual growth scenarios.

These values are broadly aligned with the longer-term trends described in the OECD–FAO Agricultural Outlook, which anticipates a 14% increase in global agricultural production by 2034 when expressed in constant 2014–2016 prices [142]. Moreover, wider food-systems analyses suggest that global food demand will increase by roughly 35–56% between 2010 and 2050, driven by population growth, urbanisation, and dietary change [1]. These benchmarks provide a plausible reference frame for evaluating the potential scale of CEA relative to the broader agricultural economy over the coming decades.

Comparing the two projection sets provides an indication of how the scale of CEA investment may evolve relative to the wider agricultural economy. In 2023, the commercial greenhouse market accounted for approximately 1% of global agricultural value. Under the growth trajectories outlined in Figures 8 and 9, this share rises to around 3.1–6.5% by 2050 (as shown in Figure 10 and Table 4). This should not be interpreted as the proportion of global food output produced in controlled environments. The comparison places a capital-intensive infrastructure market against annual farmgate production value, and the two measure different aspects of the food system.

For this reason, the ratio is best viewed as an investment intensity indicator, signalling the relative weight of CEA infrastructure within global agricultural supply chains rather than its contribution to total food production. The upward trend nevertheless highlights the growing strategic role of controlled-environment systems, particularly in high-value horticulture and climate-sensitive regions where reliability, year-round production, and resource efficiency are increasingly important.

The analysis indicates that CEA technologies are likely to play a far more prominent role in food system resilience by mid-century, while remaining a complement rather than an alternative to open-field agriculture. Their strongest contributions are expected in

https://doi.org/10.3390/en19081975

high-value horticulture, nutrient-dense perishable crops, and regional supply systems that require greater climate tolerance, rather than in the large-scale production of staple grains. Controlled-environment systems can also enhance production stability under climate variability, water scarcity, and energy-price volatility, thereby improving the resilience of food supply chains.

**Table 4.** Summary of data sources, published market estimates, and long-term scenario extensions for the commercial greenhouse sector and global agricultural production value. Values for 2050 represent scenario-based extrapolations from published baselines.

| Item | Source Base | Published Horizon | 2050 Extension |
|---|---|---|---|
| Commercial greenhouse market [136] | USD 39.6 bn (2023) | USD 68.7 bn by 2028, 11.6% CAGR | USD 247.6 bn (6%), USD 373.5 bn (8%), USD 559.2 bn (10%) |
| Global agriculture value [140,141] | USD 4.0 tn (2023) | OECD-FAO: USD 3.96 tn by 2034 constant USD | USD 8.0 tn (2.6%), USD 8.66 tn (2.9%) |
| Greenhouse market/agriculture value | ~1.0% (2023) | | ~3.1% to 6.5% (2050 scenarios) |

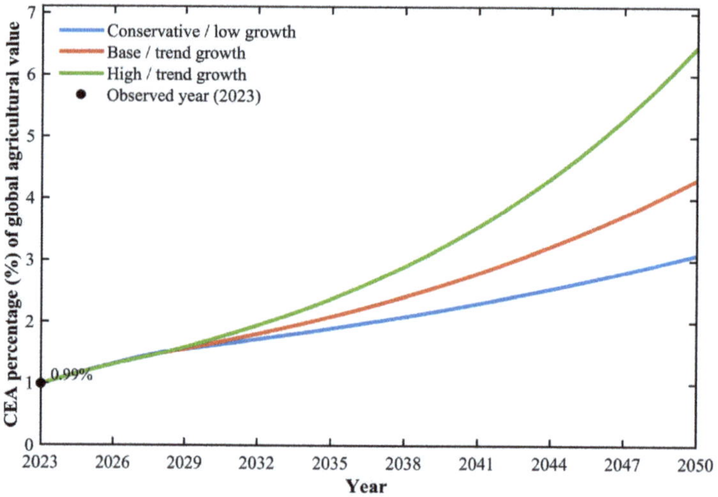


**Figure 10.** Commercial greenhouse industry as a share of global agricultural value, 2023–2050, based on the growth scenarios in Figures 8 and 9. Ratios reflect investment scale, not production share.

These trends have direct implications for Sustainable Development Goal 2 (Zero Hunger), which calls for secure access to safe, nutritious, and sufficient food while promoting sustainable agricultural production systems [143,144]. Greenhouse agriculture can contribute to this objective by supporting year-round production, improving yield reliability, and reducing vulnerability to climate-related disruptions, particularly in regions with extreme weather conditions or high dependence on imported produce. However, the extent to which CEA can influence food system outcomes will depend not only on market growth but also on the policy and institutional arrangements that shape technology uptake.

Several enabling conditions are therefore critical. These include concessional finance for capital-intensive infrastructure; targeted incentives/support for energy-efficient technologies such as thermal screens and heat pumps; investment in thermal storage and low-carbon heat networks; and electricity tariff structures that do not penalise electrified heating during winter peak periods. As reliance on combined heat and power (CHP) declines, $CO_2$ supply logistics will also require deliberate planning. Alongside this, sus-

https://doi.org/10.3390/en19081975

tained public investment in research, knowledge transfer programmes, and workforce development will also be essential for large-scale deployment. Without such measures, controlled-environment agriculture is likely to remain concentrated among larger firms or high-value niche markets; with it, the sector contribution to food system resilience and progress towards SDG 2 could be considerably greater [143–145].

## 4. Future Directions

Advancing greenhouse decarbonisation will require improvements in how systems are analysed, specified, and deployed, rather than the addition of yet more isolated technologies. A persistent limitation across the literature is the lack of comparable system boundaries. Studies differ in how they treat latent loads and humidity control, whether auxiliary electricity is included, the choice of grid emission factors, and the financial assumptions applied (discount rate, project lifetime, and CAPEX scope). To enable comparison, future work should report: envelope performance and infiltration assumptions; the latent load modelling method; seasonal COP including auxiliaries; the scope of capital costing; lifetime and discount rate; and the emissions boundary definition used for greenhouse gas accounting.

A major priority for future work is improving the realism of performance and cost modelling. Many techno-economic assessments still rely on steady state formulations that cannot represent evapotranspiration, condensation heat recovery, or the effects of control strategy on heating and dehumidification demand. A more robust evaluation will require dynamic energy-mass balance models that include these interactions and incorporate weather and tariff variability. Digital twin approaches linking physical simulation with discounted cashflow metrics offer a credible route for exploring scenario uncertainty and testing net-zero pathways under operational constraints.

Electrification pathways also demand closer alignment with power system conditions. The feasibility of heat pump-based systems depends on tariff structure, demand charges, marginal grid carbon intensity, and interactions with on-site photovoltaics and storage. For district heat, minewater, and industrial waste heat options, research priorities lie in spatial and financial optimisation rather than component efficiency: cluster design, network cost normalisation, and shared infrastructure financing determine whether these pathways can be deployed at scale.

Finally, thermal management research must connect more directly to food system outcomes. Emissions intensity per unit of product, resilience to energy price shocks, and the relationship between capital intensity and food affordability remain insufficiently resolved, particularly in emerging-economy contexts. Without such linkage, decarbonisation pathways cannot be assessed adequately in terms of scalability, cost implications, or their contribution to wider food system resilience.

## 5. Conclusions

This review examined greenhouse thermal management in the context of net-zero food production, focusing on the interaction between heat transfer processes, system boundaries, operating strategy, and infrastructure constraints. The findings show that decarbonisation is not determined by individual technologies alone, but by how thermal demand, humidity control, and energy supply are represented and integrated. Envelope losses, ventilation, and latent heat exchange set the baseline demand, while tariffs, capital cost, and infrastructure access determine whether low-carbon solutions are viable in practice.

Humidity control plays a central role in greenhouse energy use. Although temperature remains the primary control variable, ventilation and latent loads often drive both seasonal energy consumption and peak heating demand, particularly in winter. This affects the performance of electrification, dehumidification, and heat recovery systems,

https://doi.org/10.3390/en19081975

whose effectiveness depends on how sensible and latent processes are managed within the control strategy. The reported results are not comparable without a clear definition of system boundaries, auxiliary electricity, and control assumptions. Greenhouse decarbonisation follows a staged pathway rather than a single-technology transition. Demand reduction through envelope improvement and thermal screening is a prerequisite for viable electrification. Electrification systems perform best when supply temperatures, humidity management, and tariff exposure are addressed together. Further reductions depend on the integration of renewable and recovered heat, storage, and, where available, networked heat systems. The sequence of implementation and system integration is therefore as important as the technologies themselves.

Long-term projections indicate substantial growth in the greenhouse sector, with market value potentially increasing from USD 68.7 billion in 2028 to USD 248–559 billion by 2050. Relative to projected global agricultural value, this reflects rising capital intensity rather than a proportional increase in food output. The review has several limitations. The reported results vary widely due to differences in boundary definitions, climate conditions, and modelling approaches, which limit direct comparison across studies. In addition, there is limited empirical evidence on integrated systems that combine heating, dehumidification, $CO_2$ supply, and control, and long-term operational performance is not consistently reported.

From a food security perspective, controlled-environment agriculture is likely to play a growing role in climate-resilient production, but market growth does not directly translate to increased food output. The contribution of low-carbon greenhouse systems depends on affordability, access to infrastructure, and reliable heat and $CO_2$ supply. While greenhouses can support stable year-round production under climate stress, wider adoption depends on coordinated energy systems, financing, and policy support. Progress will require clearer system boundaries, improved treatment of latent loads and auxiliary electricity, and closer integration of modelling, techno-economic analysis, and operational control. Under these conditions, greenhouse systems can contribute meaningfully to both decarbonisation and resilient food production.

**Author Contributions:** Conceptualization, S.S. and A.M.A.; methodology, S.S.; software, S.S.; formal analysis, S.S.; data curation, S.S.; writing—original draft preparation, S.S. and C.E.; writing—review and editing, S.S., C.E., A.B., C.B., and A.M.A.; supervision, A.B., C.B., and A.M.A.; project administration, S.S. and A.M.A.; funding acquisition, A.M.A. All authors have read and agreed to the published version of the manuscript.

**Funding:** This research was funded by the University of Lincoln's Graduate Teaching Fellow Scheme (Ref No: 1AB-2AL-539354).

**Data Availability Statement:** The original contributions presented in this study are included in the article. Further inquiries can be directed to the corresponding author.

**Acknowledgments:** The authors acknowledge the University of Lincoln for supporting Samson Sogbaike through a fully funded PhD scholarship under the Graduate Teaching Fellow Scheme (Ref No: 1AB-2AL-539354).

**Conflicts of Interest:** The authors declare no conflicts of interest.

## Abbreviations

The following abbreviations are used in this manuscript:

| | |
|---|---|
| AI | Artificial Intelligence |
| ASHP | Air-Source Heat Pump |
| BTES | Borehole Thermal Energy Storage |

https://doi.org/10.3390/en19081975

| | |
|---|---|
| CAGR | Compound Annual Growth Rate |
| CAPEX | Capital Expenditure |
| CEA | Controlled-Environment Agriculture |
| CFD | Computational Fluid Dynamics |
| CHP | Combined Heat and Power |
| COP | Coefficient of Performance |
| CSG | Chinese Solar Greenhouse |
| DC | Discount Rate |
| DRL | Deep Reinforcement Learning |
| EAHE | Earth-to-Air Heat Exchanger |
| EECA | Energy Efficiency and Conservation Authority |
| FAO | Food and Agriculture Organization of the United Nations |
| GHG | Greenhouse Gas |
| GSHP | Ground-Source Heat Pump |
| HPS | Heat Pump System |
| HVAC | Heating, Ventilation, and Air Conditioning |
| IEA | International Energy Agency |
| IPCC | Intergovernmental Panel on Climate Change |
| IRR | Internal Rate of Return |
| LCA | Life Cycle Assessment |
| LCO | Levelised Cost of Operation |
| LCOH | Levelised Cost of Heat |
| LED | Light Emitting Diode |
| LTES | Latent Thermal Energy Storage |
| MILP | Mixed-Integer Linear Programming |
| ML | Machine Learning |
| MPC | Model Predictive Control |
| NPV | Net Present Value |
| NVAC | Natural Ventilation and Air Circulation |
| OECD–FAO | Organisation for Economic Co-operation and Development and Food and Agriculture Organization |
| OPEX | Operational Expenditure |
| PAR | Photosynthetically Active Radiation |
| PC | Polycarbonate |
| PCM | Phase Change Material |
| PE | Polyethylene |
| PID | Proportional Integral Derivative (controller) |
| PRISMA | Preferred Reporting Items for Systematic Reviews and Meta-Analyses |
| PV | Photovoltaics |
| PVT | Photovoltaic Thermal |
| RH | Relative Humidity |
| RL | Reinforcement Learning |
| SALSA | Search, Appraisal, Synthesis and Analysis |
| SDG | Sustainable Development Goal |
| STE | Solar Thermal Energy |
| STES | Seasonal Thermal Energy Storage |
| STH | Solar Thermal Heating/Solar Thermal Heat |
| TES | Thermal Energy Storage |
| TRNSYS | Transient System Simulation |
| U-value | Overall heat transfer coefficient ($W \cdot m^{-2} \cdot K^{-1}$) |
| UK | United Kingdom |
| USD | United States Dollar |
| VPD | Vapour Pressure Deficit |
| WSHP | Water-Source Heat Pump |

https://doi.org/10.3390/en19081975

## References


1. van Dijk, M.; Morley, T.; Rau, M.L.; Saghai, Y. A meta-analysis of projected global food demand and population at risk of hunger for the period 2010–2050. *Nat. Food* **2021**, *2*, 494–501. [CrossRef] [PubMed]
2. Intergovernmental Panel on Climate Change (IPCC). *Climate Change 2022—Impacts, Adaptation and Vulnerability*; Cambridge University Press: Singapore, 2023. [CrossRef]
3. Shamshiri, R.R.; Kalantari, F.; Ting, K.C.; Thorp, K.R.; Hameed, I.A.; Weltzien, C.; Ahmad, D.; Shad, Z.M. Advances in greenhouse automation and controlled environment agriculture: A transition to plant factories and urban agriculture. *Int. J. Agric. Biol. Eng.* **2018**, *11*, 1–22. [CrossRef]
4. Benke, K.; Tomkins, B. Future food-production systems: Vertical farming and controlled-environment agriculture. *Sustain. Sci. Pract. Policy* **2017**, *13*, 13–26. [CrossRef]
5. Mills, E. The emergence of indoor agriculture as a driver of global energy demand. *Npj Sustain. Agric.* **2025**, *3*, 52. [CrossRef]
6. Vadiee, A.; Martin, V. Energy management strategies for commercial greenhouses. *Appl. Energy* **2014**, *114*, 880–888. [CrossRef]
7. Katzin, D.; Marcelis, L.F.M.; van Mourik, S. Energy savings in greenhouses by transition from high-pressure sodium to LED lighting. *Appl. Energy* **2021**, *281*, 116019. [CrossRef]
8. Food and Agriculture Organization of the United Nations (FAO). Greenhouse Gas Emissions from Agrifood Systems: Global, Regional and Country Trends, 2000–2022. 2024. Available online: https://openknowledge.fao.org/server/api/core/bitstreams/111b7ee8-282b-42ff-ad95-cccecd90f8ea/content (accessed on 17 February 2026).
9. IPCC Sixth Assessment Report. Agriculture, Forestry and Other Land Uses (AFOLU). In *Climate Change 2022—Mitigation of Climate Change*; Cambridge University Press: Singapore, 2023; pp. 747–860. [CrossRef]
10. IPCC. Food, Fibre and Other Ecosystem Products. In *Climate Change 2022—Impacts, Adaptation and Vulnerability*; Cambridge University Press: Singapore, 2023; pp. 713–906. [CrossRef]
11. Boyacı, S.; Kocięcka, J.; Jagosz, B.; Atılgan, A. Energy Efficiency in Greenhouses and Comparison of Energy Sources Used for Heating. *Energies* **2025**, *18*, 724. [CrossRef]
12. Singh, M.C.; Singh, J.P.; Singh, K.G. Development of a microclimate model for prediction of temperatures inside a naturally ventilated greenhouse under cucumber crop in soilless media. *Comput. Electron. Agric.* **2018**, *154*, 227–238. [CrossRef]
13. Choab, N.; Allouhi, A.; El Maakoul, A.; Kousksou, T.; Saadeddine, S.; Jamil, A. Review on greenhouse microclimate and application: Design parameters, thermal modeling and simulation, climate controlling technologies. *Sol. Energy* **2019**, *191*, 109–137. [CrossRef]
14. Katzin, D.; van Henten, E.J.; van Mourik, S. Process-based greenhouse climate models: Genealogy, current status, and future directions. *Agric. Syst.* **2022**, *198*, 103388. [CrossRef]
15. Reyes-Rosas, A.; Molina-Aiz, F.D.; Valera, D.L.; López, A.; Khamkure, S. Development of a single energy balance model for prediction of temperatures inside a naturally ventilated greenhouse with polypropylene soil mulch. *Comput. Electron. Agric.* **2017**, *142*, 9–28. [CrossRef]
16. Golzar, F.; Heeren, N.; Hellweg, S.; Roshandel, R. A novel integrated framework to evaluate greenhouse energy demand and crop yield production. *Renew. Sustain. Energy Rev.* **2018**, *96*, 487–501. [CrossRef]
17. Akpenpuun, T.D.; Na, W.-H.; Ogunlowo, Q.O.; Rabiu, A.; Adesanya, M.A.; Addae, K.S.; Kim, H.-T.; Lee, H.-W. Effect of Greenhouse Cladding Materials and Thermal Screen Configuration on Heating Energy and Strawberry (*Fragaria ananassa* var. "Seolhyang") Yield in Winter. *Agronomy* **2021**, *11*, 2498. [CrossRef]
18. IEA. The Future of Heat Pumps, Paris. 2022. Available online: https://www.iea.org/reports/the-future-of-heat-pumps (accessed on 8 January 2026).
19. Vadiee, A.; Martin, V. Energy management in horticultural applications through the closed greenhouse concept, state of the art. *Renew. Sustain. Energy Rev.* **2012**, *16*, 5087–5100. [CrossRef]
20. Chen, W.-H.; Mattson, N.S.; You, F. Intelligent control and energy optimization in controlled environment agriculture via nonlinear model predictive control of semi-closed greenhouse. *Appl. Energy* **2022**, *320*, 119334. [CrossRef]
21. Mahmood, F.; Govindan, R.; Bermak, A.; Yang, D.; Al-Ansari, T. Data-driven robust model predictive control for greenhouse temperature control and energy utilisation assessment. *Appl. Energy* **2023**, *343*, 121190. [CrossRef]
22. Morcego, B.; Yin, W.; Boersma, S.; van Henten, E.; Puig, V.; Sun, C. Reinforcement Learning versus Model Predictive Control on greenhouse climate control. *Comput. Electron. Agric.* **2023**, *215*, 108372. [CrossRef]
23. Paris, B.; Vandorou, F.; Balafoutis, A.T.; Vaiopoulos, K.; Kyriakarakos, G.; Manolakos, D.; Papadakis, G. Energy Use in Greenhouses in the EU: A Review Recommending Energy Efficiency Measures and Renewable Energy Sources Adoption. *Appl. Sci.* **2022**, *12*, 5150. [CrossRef]
24. New Zealand Ministry of Foreign Affairs and Trade (MFAT). The Dutch Horticulture sector: An Innovative Sector Facing an Energy Crisis and Rising Costs. 2022. Available online: https://www.mfat.govt.nz/en/trade/mfat-market-reports/the-dutch-horticulture-sector-an-innovative-sector-facing-an-energy-crisis-and-rising-costs-september-2022 (accessed on 8 January 2026).


https://doi.org/10.3390/en19081975

25. van der Veen, R.A.C.; Kasmire, J. Combined heat and power in Dutch greenhouses: A case study of technology diffusion. *Energy Policy* **2015**, *87*, 8–16. [CrossRef]
26. FAO; IFAD; UNICEF; WFP; WHO. *The State of Food Security and Nutrition in the World 2023*; FAO: Rome, Italy, 2023. [CrossRef]
27. Termine, P. Ensuring Food Security: Why Agency and Sustainability Matter, FAO (2024). Available online: https://www.fao.org/cfs/cfs-hlpe/insights/news-insights/news-detail/ensuring-food-security--why-agency-and-sustainability-matter/en (accessed on 17 February 2026).
28. Page, M.J.; McKenzie, J.E.; Bossuyt, P.M.; Boutron, I.; Hoffmann, T.C.; Mulrow, C.D.; Shamseer, L.; Tetzlaff, J.M.; Akl, E.A.; Brennan, S.E.; et al. The PRISMA 2020 statement: An updated guideline for reporting systematic reviews. *BMJ* **2021**, *372*, n71. [CrossRef]
29. Grant, M.J.; Booth, A. A typology of reviews: An analysis of 14 review types and associated methodologies. *Health Inf. Libr. J.* **2009**, *26*, 91–108. [CrossRef]
30. Campbell, M.; McKenzie, J.E.; Sowden, A.; Katikireddi, S.V.; Brennan, S.E.; Ellis, S.; Hartmann-Boyce, J.; Ryan, R.; Shepperd, S.; Thomas, J.; et al. Synthesis without meta-analysis (SWiM) in systematic reviews: Reporting guideline. *BMJ* **2020**, *368*, l6890. [CrossRef]
31. D'Arpa, S.; Colangelo, G.; Starace, G.; Petrosillo, I.; Bruno, D.E.; Uricchio, V.; Zurlini, G. Heating requirements in greenhouse farming in southern Italy: Evaluation of ground-source heat pump utilization compared to traditional heating systems. *Energy Effic.* **2016**, *9*, 1065–1085. [CrossRef]
32. Stanghellini, C.; Ooster, B.V.; Heuvelink, E. Heating in climate-controlled greenhouses. In *Greenhouse Horticulture*; Brill | Wageningen Academic: Gelderland, The Netherlands, 2024; pp. 164–192. [CrossRef]
33. Zhang, Y. Effect of covering materials on energy consumption and greenhouse microclimate. *Agric. For. Meteorol.* **1996**, *82*, 227–244. [CrossRef]
34. Geoola, F.; Kashti, Y.; Teitel, M.; Levi, A.; Brickman, R.; Esquira, I. A study of u value of greenhouse films with thermal screens using the hot box method. *Acta Hortic.* **2011**, *893*, 367–372. [CrossRef]
35. Rasheed, A.; Lee, J.W.; Lee, H.W. A Review of Greenhouse Energy Management by Using Building Energy Simulation. *Prot. Hortic. Plant Fact.* **2015**, *24*, 317–325. [CrossRef]
36. Song, A.; Kim, Y.; Hwang, S.; Shin, M.; Lee, S. A Comprehensive Review of Thermal Transmittance Assessments of Building Envelopes. *Buildings* **2024**, *14*, 3304. [CrossRef]
37. Ahamed, M.S.; Guo, H.; Tanino, K. A quasi-steady state model for predicting the heating requirements of conventional greenhouses in cold regions. *Inf. Process. Agric.* **2018**, *5*, 33–46. [CrossRef]
38. Boulard, T.; Baille, A. Modelling of Air Exchange Rate in a Greenhouse Equipped with Continuous Roof Vents. *J. Agric. Eng. Res.* **1995**, *61*, 37–47. [CrossRef]
39. Papadakis, G.; Mermier, M.; Meneses, J.F.; Boulard, T. Measurement and Analysis of Air Exchange Rates in a Greenhouse with Continuous Roof and Side Openings. *J. Agric. Eng. Res.* **1996**, *63*, 219–227. [CrossRef]
40. Fatnassi, H.; Boulard, T.; Demrati, H.; Bouirden, L.; Sappe, G. SE—Structures and Environment. *Biosyst. Eng.* **2002**, *82*, 97–105. [CrossRef]
41. Piscia, D. Analysis of Night-Time Climate in Plastic-Covered Greenhouses. Ph.D. Thesis, Universitat Politècnica de Catalunya, Barcelona, Spain, 2012.
42. Piscia, D.; Muñoz, P.; Panadès, C.; Montero, J.I. A method of coupling CFD and energy balance simulations to study humidity control in unheated greenhouses. *Comput. Electron. Agric.* **2015**, *115*, 129–141. [CrossRef]
43. Gu, X.; Goto, E. Computational Fluid Dynamics Model with Realistic Plant Structures to Study Airflow in and around a Plant Canopy on a Cultivation Shelf in a Plant Factory with Artificial Light. *Agriculture* **2024**, *14*, 1199. [CrossRef]
44. Bazgaou, A.; Fatnassi, H.; Bouharroud, R.; Tiskatine, R.; Wifaya, A.; Demrati, H.; Bammou, L.; Aharoune, A.; Bouirden, L. CFD Modeling of the Microclimate in a Greenhouse Using a Rock Bed Thermal Storage Heating System. *Horticulturae* **2023**, *9*, 183. [CrossRef]
45. Kibwika, A.K.; Seo, H.-J.; Seo, I.-H. CFD Model Verification and Aerodynamic Analysis in Large-Scaled Venlo Greenhouse for Tomato Cultivation. *Agriengineering* **2023**, *5*, 1395–1414. [CrossRef]
46. Norton, T.; Sun, D.W.; Grant, J.; Fallon, R.; Dodd, V. Applications of computational fluid dynamics (CFD) in the modelling and design of ventilation systems in the agricultural industry: A review. *Bioresour. Technol.* **2007**, *98*, 2386–2414. [CrossRef]
47. Sudirman, M.; van Hooff, T.; Gillmeier, S.; Blocken, B. Validation study of cross-ventilation in a realistic building geometry: RANS, SAS and LES. *Build. Environ.* **2024**, *269*, 112354. [CrossRef]
48. Allali, F.E.; Fatnassi, H.; Demrati, H.; Errais, R.; Wifaya, A.; Aharoune, A. Greenhouse cooling systems: A systematic review of research trends, challenges, and recommendations for improving sustainability. *Clean. Eng. Technol.* **2025**, *26*, 100973. [CrossRef]
49. Tantau, H.-J.; Akyazi, G. The low energy greenhouse—Heat transfer by long wave radiation. *Acta Hortic.* **2017**, *1170*, 847–854. [CrossRef]


https://doi.org/10.3390/en19081975

50. Sethi, V.P.; Sharma, S.K. Survey of cooling technologies for worldwide agricultural greenhouse applications. *Sol. Energy* **2007**, *81*, 1447–1459. [CrossRef]
51. Amani, M.; Foroushani, S.; Sultan, M.; Bahrami, M. Comprehensive review on dehumidification strategies for agricultural greenhouse applications. *Appl. Therm. Eng.* **2020**, *181*, 115979. [CrossRef]
52. Katsoulas, N.; Sapounas, A.; De Zwart, F.; Dieleman, J.A.; Stanghellini, C. Reducing ventilation requirements in semi-closed greenhouses increases water use efficiency. *Agric. Water Manag.* **2015**, *156*, 90–99. [CrossRef]
53. McCartney, L.; Orsat, V.; Lefsrud, M.G. An experimental study of the cooling performance and airflow patterns in a model Natural Ventilation Augmented Cooling (NVAC) greenhouse. *Biosyst. Eng.* **2018**, *174*, 173–189. [CrossRef]
54. Yang, Y.; Guo, S.; Chen, Y. Estimated viability of greenhouse cooling technologies for growing a tomato crop under various climates. *Int. J. Agric. Biol. Eng.* **2021**, *14*, 90–95. [CrossRef]
55. Abdel-Ghany, A.M.; Al-Helal, I.M.; Alzahrani, S.M.; Alsadon, A.A.; Ali, I.M.; Elleithy, R.M. Covering Materials Incorporating Radiation-Preventing Techniques to Meet Greenhouse Cooling Challenges in Arid Regions: A Review. *Sci. World J.* **2012**, *2012*, 906360. [CrossRef]
56. McCartney, L.; Lefsrud, M.G. Field trials of the Natural Ventilation Augmented Cooling (NVAC) greenhouse. *Biosyst. Eng.* **2018**, *174*, 159–172. [CrossRef]
57. Abedrabboh, O.; Koç, M.; Biçer, Y. Comparative thermoeconomic assessment of renewable-driven hybrid-cooled sustainable greenhouses for subtropical regions. *Energy Convers. Manag.* **2024**, *300*, 117990. [CrossRef]
58. Migeon, C.; Pierart, A.; Lemesle, D.; Travers, A.; Chassériaux, G. A dehumidifying heat pump for greenhouses. *Acta Hortic.* **2012**, *952*, 485–492. [CrossRef]
59. Andrews, R.; Pearce, J.M. Environmental and economic assessment of a greenhouse waste heat exchange. *J. Clean. Prod.* **2011**, *19*, 1446–1454. [CrossRef]
60. Vermeulen, P. Alternative Sources of $CO_2$ for the Greenhouse Horticulture. 2014. Available online: https://www.researchgate.net/publication/282006970 (accessed on 17 February 2026).
61. van Tuyll, A.; Graamans, L.; Boedijn, A. *Carbon Dioxide Enrichment in a Decarbonised Future*; Wageningen University & Research: Wageningen, The Netherlands, 2022. [CrossRef]
62. Zhou, B.; Sun, W.; Guo, W.; Zheng, W.; Qu, M. Performance of a greenhouse heating system utilizing energy transfer between greenhouses based on the dual source heat pump. *Appl. Therm. Eng.* **2025**, *261*, 125088. [CrossRef]
63. Ahsan, T.M.A.; Ahamed, M.S.; Hassan, A.A. Hybrid ground source heat pump for sustainable greenhouse climate control in hot and humid region. *Sustain. Energy Technol. Assess.* **2025**, *83*, 104615. [CrossRef]
64. Ahsan, T.M.A.; Rahman, M.S.; Ahamed, M.S. Geothermal energy application for greenhouse microclimate management: A review. *Geothermics* **2025**, *127*, 103209. [CrossRef]
65. Gorjian, S.; Calise, F.; Kant, K.; Ahamed, M.S.; Copertaro, B.; Najafi, G.; Zhang, X.; Aghaei, M.; Shamshiri, R.R. A review on opportunities for implementation of solar energy technologies in agricultural greenhouses. *J. Clean. Prod.* **2021**, *285*, 124807. [CrossRef]
66. Vadiee, A.; Martin, V. Thermal energy storage strategies for effective closed greenhouse design. *Appl. Energy* **2013**, *109*, 337–343. [CrossRef]
67. Asgari, N.; McDonald, M.T.; Pearce, J.M. Energy Modeling and Techno-Economic Feasibility Analysis of Greenhouses for Tomato Cultivation Utilizing the Waste Heat of Cryptocurrency Miners. *Energies* **2023**, *16*, 1331. [CrossRef]
68. Fguiri, A.; Fatnassi, H.; Jeday, M.R.; Marvillet, C. Study of the energetic and the economic feasibility of a heating agricultural greenhouse using industrial waste-heat at low temperatures. *Int. J. Energy Environ. Econ.* **2017**, *25*, 269–282.
69. Esen, M.; Yuksel, T. Experimental evaluation of using various renewable energy sources for heating a greenhouse. *Energy Build.* **2013**, *65*, 340–351. [CrossRef]
70. Wu, T.; Mukhopadhyay, K.; Thomassin, P.J. A life cycle inventory analysis of wood pellets for greenhouse heating: A case study at Macdonald campus of McGill University. *AIMS Energy* **2016**, *4*, 697–722. [CrossRef]
71. Andersson, N.-E. *Properties of Thermal Screens Used for Energy Saving in Greenhouses*; Aarhus University: Aarhus, Denmark, 2010. Available online: https://pure.au.dk/ws/files/4247455/ref403.pdf (accessed on 18 February 2026).
72. Rabiu, A.; Na, W.-H.; Akpenpuun, T.D.; Rasheed, A.; Adesanya, M.A.; Ogunlowo, Q.O.; Kim, H.T.; Lee, H.-W. Determination of overall heat transfer coefficient for greenhouse energy-saving screen using Trnsys and hotbox. *Biosyst. Eng.* **2022**, *217*, 83–101. [CrossRef]
73. Rafiq, A.; Na, W.H.; Rasheed, A.; Lee, J.W.; Kim, H.T.; Lee, H.W. Measurement of longwave radiative properties of energy-saving greenhouse screens. *J. Agric. Eng.* **2021**, *52*, 1209. [CrossRef]
74. Zhang, E.; Elmi, M.; Jahid, A.; Duan, Q.; Wang, J. Assessing greenhouse covering systems for energy efficiency and solar irradiance modulation: A technical review. *Green Technol. Sustain.* **2026**, *4*, 100281. [CrossRef]
75. Song, S.; Jeon, J.; Yoon, S. Optimizing Energy Efficiency and Light Transmission in Greenhouses Using Rotating Low-Emissivity-Coated Envelopes. *Energies* **2025**, *18*, 1613. [CrossRef]


https://doi.org/10.3390/en19081975

76. Lin, T.; Maier, C.R.; Liang, W.; Klause, N.; He, J.; Tissue, D.T.; Lan, Y.-C.; Sethuvenkatraman, S.; Goldsworthy, M.; Chen, Z.-H. A light-blocking greenhouse film differentially impacts climate control energy use and capsicum production. *Front. Energy Res.* **2024**, *12*, 1360536. [CrossRef]
77. Kim, H.-K.; Kang, G.-C.; Moon, J.-P.; Lee, T.-S.; Oh, S.-S. Estimation of Thermal Performance and Heat Loss in Plastic Greenhouses with and without Thermal Curtains. *Energies* **2018**, *11*, 578. [CrossRef]
78. Rasheed, A.; Na, W.H.; Lee, J.W.; Kim, H.T.; Lee, H.W. Optimization of Greenhouse Thermal Screens for Maximized Energy Conservation. *Energies* **2019**, *12*, 3592. [CrossRef]
79. Seginer, I.; van Straten, G.; van Beveren, P.J.M. Day-to-night heat storage in greenhouses: 1 Optimisation for periodic weather. *Biosyst. Eng.* **2017**, *161*, 174–187. [CrossRef]
80. Lyu, W.; Wang, Z.; Li, X.; Deng, G.; Xu, Z.; Li, H.; Yang, Y.; Zhan, B.; Zhao, M. Influence of the water tank size and air source heat pump size on the energy saving potential of the energy storage heating system. *J. Energy Storage* **2022**, *55*, 105542. [CrossRef]
81. Malinowski, M.; Bodziacki, S.; Famielec, S.; Huptyś, D.; Kurpaska, S.; Latała, H.; Basak, Z. Environmental Impact Assessment of Heat Storage System in Rock-Bed Accumulator. *Energies* **2025**, *18*, 3360. [CrossRef]
82. Amiri, L.; Ermagan, H.; Kurnia, J.C.; Hassani, F.; Sasmito, A.P. Progress on rock thermal energy storage (RTES): A state of the art review. *Energy Sci. Eng.* **2024**, *12*, 410–437. [CrossRef]
83. Nishad, S.; Krupa, I. Phase change materials for thermal energy storage applications in greenhouses: A review. *Sustain. Energy Technol. Assess.* **2022**, *52*, 102241. [CrossRef]
84. Zhu, J.; Zhang, X.; Hua, W.; Ji, J.; Lv, X. Current status and development of research on phase change materials in agricultural greenhouses: A review. *J. Energy Storage* **2023**, *66*, 107104. [CrossRef]
85. Cui, Y.; Gulfam, R.; Ishrat, Y.; Iqbal, S.; Yao, F. Recent Progress of Phase Change Materials and Their Applications in Facility Agriculture and Related-Buildings—A Review. *Buildings* **2024**, *14*, 2999. [CrossRef]
86. Liu, W.; Liu, G.; Shi, W.; Tang, X.; Wu, X.; Wu, J.; Xu, Z.; Zhang, F.; Yang, M. Experimental Investigation on Thermal Performance Optimization of Na2HPO4·12H2O-Based Gel Phase Change Materials for Solar Greenhouse. *Gels* **2025**, *11*, 434. [CrossRef] [PubMed]
87. Yang, T.; Liu, W.; Kramer, G.J.; Sun, Q. Seasonal thermal energy storage: A techno-economic literature review. *Renew. Sustain. Energy Rev.* **2021**, *139*, 110732. [CrossRef]
88. Gao, W.; Qadrdan, M.; Chaudry, M.; Wu, J.; Cardiff University. *Seasonal Storage of Heat in Boreholes*; UK Energy Research Centre: London, UK, 2023.
89. Sadeghi, H.; Jalali, R.; Singh, R.M. A review of borehole thermal energy storage and its integration into district heating systems. *Renew. Sustain. Energy Rev.* **2024**, *192*, 114236. [CrossRef]
90. Baeuerle, Y.I.; Arpagaus, C.; Haller, M.Y. A Review of Seasonal Energy Storage for Net-Zero Industrial Heat: Thermal and Power-to-X Storage Including the Novel Concept of Renewable Metal Energy Carriers. *Energies* **2025**, *18*, 2204. [CrossRef]
91. Asgari, N.; Hayibo, K.S.; Groza, J.; Rana, S.; Pearce, J.M. Greenhouse applications of solar photovoltaic driven heat pumps in northern environments. *Renew. Sustain. Energy Rev.* **2025**, *207*, 114920. [CrossRef]
92. Kim, D.-W.; Kim, M.-H.; Lee, D.-W. Economic and Environmental Analysis of Solar Thermal and Seasonal Thermal Energy Storage Based on a Renewable Energy Conversion System for Greenhouses. *Energies* **2022**, *15*, 6592. [CrossRef]
93. Zhang, L.; Xu, P.; Mao, J.; Tang, X.; Li, Z.; Shi, J. A low cost seasonal solar soil heat storage system for greenhouse heating: Design and pilot study. *Appl. Energy* **2015**, *156*, 213–222. [CrossRef]
94. Byon, Y.-S.; Jeong, J.-W. Energy Efficiency of Seasonal Solar Thermal Energy Storage System for Greenhouse Heating. In *5th International High Performance Buildings Conference*; Purdue University: West Lafayette, Indiana, 2018. Available online: https://docs.lib.purdue.edu/ihpbc/299 (accessed on 5 February 2026).
95. Boldyryev, S.; Krajačić, G.; Garafulić, E. Identification of industrial waste heat potential for district heating systems by pinch-based systematic graphical approach. *Energy Convers. Manag.* **2025**, *333*, 119804. [CrossRef]
96. Moser, S.; Jauschnik, G. Using Industrial Waste Heat in District Heating: Insights on Effective Project Initiation and Business Models. *Sustainability* **2023**, *15*, 10559. [CrossRef]
97. van Beveren, P.J.M.; Bontsema, J.; van 't Ooster, A.; van Straten, G.; van Henten, E.J. Optimal utilization of energy equipment in a semi-closed greenhouse. *Comput. Electron. Agric.* **2020**, *179*, 105800. [CrossRef]
98. Gilli, C.; Kempkes, F.; Munoz, P.; Montero, J.I.; Giuffrida, F.; Baptista, F.J.; Stepowska, A.; Stanghellini, C. Potential of different energy saving strategies in heated greenhouse. *Acta Hortic.* **2017**, *1164*, 467–474. [CrossRef]
99. Li, Y.; Ding, Y.; Li, D.; Miao, Z. Automatic carbon dioxide enrichment strategies in the greenhouse: A review. *Biosyst. Eng.* **2018**, *171*, 101–119. [CrossRef]
100. Marchi, B.; Zanoni, S.; Pasetti, M. Industrial Symbiosis for Greener Horticulture Practices: The $CO_2$ Enrichment from Energy Intensive Industrial Processes. *Procedia CIRP* **2018**, *69*, 562–567. [CrossRef]
101. Wang, A.; Lv, J.; Wang, J.; Shi, K. $CO_2$ enrichment in greenhouse production: Towards a sustainable approach. *Front. Plant Sci.* **2022**, *13*, 1029901. [CrossRef] [PubMed]


https://doi.org/10.3390/en19081975

102. Ghiat, I.; Mahmood, F.; Govindan, R.; Al-Ansari, T. $CO_2$ utilisation in agricultural greenhouses: A novel 'plant to plant' approach driven by bioenergy with carbon capture systems within the energy, water and food Nexus. *Energy Convers. Manag.* **2021**, *228*, 113668. [CrossRef]
103. Sapounas, A.; Katsoulas, N.; Slager, B.; Bezemer, R.; Lelieveld, C. Design, Control, and Performance Aspects of Semi-Closed Greenhouses. *Agronomy* **2020**, *10*, 1739. [CrossRef]
104. Kajolinna, T.; Balshakou, S.; Mustonen, J. Assessment of carbon dioxide quality from industrial point sources for carbon utilization. *Int. J. Greenh. Gas Control.* **2025**, *148*, 104527. [CrossRef]
105. Chen, S.; Liu, A.; Tang, F.; Hou, P.; Lu, Y.; Yuan, P. A Review of Environmental Control Strategies and Models for Modern Agricultural Greenhouses. *Sensors* **2025**, *25*, 1388. [CrossRef]
106. Mengesha, K.A.; Jin, G.-G.; Son, Y.-D. Comparison of the feedback linearization plus LQR controller and the PID controller for greenhouse indoor climate. *Int. J. Agric. Biol. Eng.* **2025**, *18*, 74–82. [CrossRef]
107. Bwambale, E.; Wanyama, J.; Adongo, T.A.; Umukiza, E.; Ntole, R.; Chikavumbwa, S.R.; Sibale, D.; Jeremaih, Z. A review of model predictive control in precision agriculture. *Smart Agric. Technol.* **2025**, *10*, 100716. [CrossRef]
108. Kim, J.; You, F. Energy-efficient greenhouse climate control using Gaussian process-based stochastic model predictive control. *Appl. Energy* **2025**, *391*, 125841. [CrossRef]
109. Beucler, T.; Gentine, P.; Yuval, J.; Gupta, A.; Peng, L.; Lin, J.; Yu, S.; Rasp, S.; Ahmed, F.; O'Gorman, P.A.; et al. Climate-invariant machine learning. *Sci. Adv.* **2024**, *10*, eadj7250. [CrossRef]
110. Liu, K.; Ji, T.; Li, M.; Yang, X.; Sun, J.; Liu, H.; Liu, R. Prediction of greenhouse temperature and humidity across growing seasons: Hybridization of process-based model and deep neural networks. *Inf. Process. Agric.* 2025, *in press*. [CrossRef]
111. Zhou, X.; Liu, Q.; Katzin, D.; Qian, T.; Heuvelink, E.; Marcelis, L.F.M. Boosting the prediction accuracy of a process-based greenhouse climate-tomato production model by particle filtering and deep learning. *Comput. Electron. Agric.* **2023**, *211*, 107980. [CrossRef]
112. Sun, Z.; Yao, Q.; Jin, H.; Xu, Y.; Hang, W.; Chen, H.; Li, K.; Shi, L.; Gu, J.; Zhang, Q.; et al. A novel in-situ sensor calibration method for building thermal systems based on virtual samples and autoencoder. *Energy* **2024**, *297*, 131314. [CrossRef]
113. Sun, Z.; Yao, Q.; Shi, L.; Jin, H.; Xu, Y.; Yang, P.; Xiao, H.; Chen, D.; Zhao, P.; Shen, X. Virtual sample diffusion generation method guided by large language model-generated knowledge for enhancing information completeness and zero-shot fault diagnosis in building thermal systems. *J. Zhejiang Univ. A* **2025**, *26*, 895–916. [CrossRef]
114. Ariesen-Verschuur, N.; Verdouw, C.; Tekinerdogan, B. Digital Twins in greenhouse horticulture: A review. *Comput. Electron. Agric.* **2022**, *199*, 107183. [CrossRef]
115. Arshad, J.; Sheheryar, C.A.A.; Rahmani, M.K.I.; Qayyum, A.; Nasir, R.; Chauhdary, S.T.; Almalki, K.J. Simulink-Driven Digital Twin Implementation for Smart Greenhouse Environmental Control. *Egypt. Inform. J.* **2025**, *30*, 100679. [CrossRef]
116. Akintan, O.; Babawale, S.; Olomo, A.; Adeyemo, R.; Opadotun, O.; Ajayi, J.T.; Mba, P.C.; Njoku, J.N.; Chesang, A.; Zahid, A.; et al. A Digital Twin Framework for Sensor Selection and Microclimate Monitoring in Greenhouses. *AgriEngineering* **2025**, *7*, 315. [CrossRef]
117. Slimani, H.; El Mhamdi, J.; Jilbab, A. Real-time greenhouse management using IoT, digital twin, and augmented reality for optimal control and decision-making. *Acta IMEKO* **2025**, *14*, 1–15. [CrossRef]
118. Ajagekar, A.; Mattson, N.S.; You, F. Energy-efficient AI-based Control of Semi-closed Greenhouses Leveraging Robust Optimization in Deep Reinforcement Learning. *Adv. Appl. Energy* **2023**, *9*, 100119. [CrossRef]
119. Mansour, M.; Sathyanarayanan, K.K.; Sauerteig, P.; Streif, S. Adaptive robust greenhouse climate control: Combining deep reinforcement learning and economic optimization. *Smart Agric. Technol.* **2025**, *12*, 101327. [CrossRef]
120. Mallick, S.; Airaldi, F.; Dabiri, A.; Sun, C.; De Schutter, B. Reinforcement learning-based model predictive control for greenhouse climate control. *Smart Agric. Technol.* **2025**, *10*, 100751. [CrossRef]
121. de Zwart, H.F. Energy conserving dehumidification of greenhouses. *Acta Hortic.* **2014**, *1037*, 203–210. [CrossRef]
122. Energy Efficiency and Conservation Authority (EECA). Covered Cropping Pathway. Available online: https://www.eeca.govt.nz/co-funding-and-support/products/covered-cropping-decarbonisation-pathway/ (accessed on 14 February 2026).
123. EECA. What's Right for Your Greenhouse? 2024. Available online: https://www.eeca.govt.nz/assets/EECA-Resources/Co-funding/Sector-Decarb-Files/Reducing-Energy-Emissions-Decision-Trees-for-Covered-Cropping.pdf (accessed on 14 February 2026).
124. Pratt, C.T. Optimising Greenhouse Environment and Energy Inputs for Sweet Pepper Production in the UK—A Commercial Demonstration of the Use of Thermal Screens and Advanced Climate Control. 2006. Available online: https://projectbluearchive.blob.core.windows.net/media/Default/Research%20Papers/Horticulture/PC%20227%20Final%20report%202007.pdf (accessed on 14 February 2026).
125. Nemś, A.; Nemś, M.; Świder, K. Analysis of the Possibilities of Using a Heat Pump for Greenhouse Heating in Polish Climatic Conditions—A Case Study. *Sustainability* **2018**, *10*, 3483. [CrossRef]

https://doi.org/10.3390/en19081975

26. Li, N.; Xu, W.; Zhang, Y.; Ma, W.; Ren, Y. Energy Saving Technologies and Practices in Facility Agriculture in Cold Regions. *Agronomy* **2025**, *15*, 204. [CrossRef]
27. Semple, L.; Carriveau, R.; Ting, D.S.-K. A techno-economic analysis of seasonal thermal energy storage for greenhouse applications. *Energy Build.* **2017**, *154*, 175–187. [CrossRef]
28. Ramos-Teodoro, J.; Álvarez, J.D.; Torres, J.L. A methodology for feasibility analyses of district heating networks: A case study applied to greenhouse crops. *Energy* **2024**, *301*, 131695. [CrossRef]
29. Tomson, F.; Lovegrove, T.; Allen, O.; Honid, D. Feasibility Study Minewater District Heat Network Horden, Durham County Council. 2023. Available online: https://communityenergypathways.org.uk/wp-content/uploads/2023/11/Durham-Feasibility-Study-FINAL.pdf (accessed on 14 February 2026).
30. Banks, V.; Krebs, M. Final Report: Technical Feasibility Study of the Geothermal Capability of the Penobsquis Mine Site, Penobsquis, NB, Fredericton, New Brunswick, Canada. 2018. Available online: https://sussex.ca/media/Phase-1-Final-Report-Geothermal-Feasibility-Study-1.pdf (accessed on 14 February 2026).
31. Mikunda, T.; Neele, F.; Wilschut, F.; Hanegraaf, M. *A Secure and Affordable $CO_2$ Supply for the Dutch Greenhouse Sector*; TNO: The Hague, The Netherlands, 2015.
32. Torrellas, M.; Antón, A.; Ruijs, M.; Victoria, N.G.; Stanghellini, C.; Montero, J.I. Environmental and economic assessment of protected crops in four European scenarios. *J. Clean. Prod.* **2012**, *28*, 45–55. [CrossRef]
33. Graamans, L.; Baeza, E.; van den Dobbelsteen, A.; Tsafaras, I.; Stanghellini, C. Plant factories versus greenhouses: Comparison of resource use efficiency. *Agric. Syst.* **2018**, *160*, 31–43. [CrossRef]
34. IEA. Energy Efficiency 2025, Paris. 2025. Available online: https://www.iea.org/reports/energy-efficiency-2025 (accessed on 15 February 2026).
35. IEA. Heat Pumps. 2024. Available online: https://www.iea.org/energy-system/buildings/heat-pumps (accessed on 15 February 2026).
36. MarketsandMarkets. Commercial Greenhouse Market Trends, Growth, and Forecast. 2023. Available online: https://www.marketsandmarkets.com/Market-Reports/commercial-greenhouse-market-221045451.html (accessed on 17 February 2026).
37. ResearchAndMarkets. Global Commercial Greenhouse Market to 2028: Ability to Provide High Yield Production Drives Growth. 2023. Available online: https://www.businesswire.com/news/home/20230425005608/en/Global-Commercial-Greenhouse-Market-to-2028-Ability-to-Provide-High-Yield-Production-Drives-Growth---ResearchAndMarkets.com (accessed on 17 February 2026).
38. Grand View Research. Greenhouse Market (2026–2033). 2024. Available online: https://www.grandviewresearch.com/industry-analysis/greenhouse-market-report (accessed on 17 February 2026).
39. Insights, F.B. Commercial Greenhouse Market Size. 2026. Available online: https://www.fortunebusinessinsights.com/commercial-greenhouse-market-114996 (accessed on 17 February 2026).
40. FAO. *World Food and Agriculture—Statistical Yearbook 2024*; FAO: Rome, Italy, 2024. [CrossRef]
41. FAO. *Gross Domestic Product and Agriculture Value Added 2014–2023*; FAO: Rome, Italy, 2025. [CrossRef]
42. OECD-FAO. *OECD-FAO Agricultural Outlook 2025–2034*; OECD Publishing: Paris, France, 2025. [CrossRef]
43. United Nations. Goal 2: Zero Hunger, United Nations. Available online: https://www.un.org/sustainabledevelopment/hunger/ (accessed on 17 February 2026).
44. Nations, U. Food Security and Nutrition and Sustainable Agriculture, United Nations. Available online: https://sdgs.un.org/topics/food-security-and-nutrition-and-sustainable-agriculture (accessed on 17 February 2026).
45. FAO. Achieving SDG 2 Without Breaching the 1.5 °C Threshold: A Global Roadmap Part 1. Available online: https://www.fao.org/interactive/sdg2-roadmap/en/ (accessed on 17 February 2026).




https://doi.org/10.3390/en19081975

energies

MDPI

*Review*

# Applications of Methods of Solving Inverse Heat Conduction Problems for Energy-Intensive Industrial Processes and Energy Conversion—Current State of the Art and Recent Challenges

**Magda Joachimiak * and Damian Joachimiak**

Faculty of Environmental Engineering and Energy, Institute of Thermal Energy, Poznan University of Technology, 60-965 Poznań, Poland; damian.joachimiak@put.poznan.pl
* Correspondence: magda.joachimiak@put.poznan.pl

**Abstract**

This paper presents methods and applications of inverse heat conduction problems (IHCPs) that are ill-posed in the Hadamard sense. The IHCP solution allows for the determination of boundary conditions in the form of heat flux or temperature in places where measurement is impossible or difficult to perform. The applications of IHCP solutions to energy-intensive industrial processes, such as heat treatment and thermochemical treatment, are described. Examples are given of determining boundary conditions on the inner surface of the wall of a power boiler and piston machine, as well as on the surface of a gas turbine blade. It is noted that the application of IHCP solutions to the above-mentioned issues often requires simplification of the computational model, in particular, the method of stabilising the inverse problem (IP). For this purpose, quasi-regularisation of IP and machine learning are currently used. Methods with stabilising properties and neural networks were identified as a challenging and interesting direction for the development of IHCP solutions.

**Keywords:** heat transfer; inverse heat conduction problem (IHCP); quasi-regularisation; regularisation; neural networks; boilers; gas turbines; heat treatment; thermochemical treatment

## 1. Introduction

The concept of a well-posed problem was introduced by Jacques Hadamard [1]. According to him, a problem is well-posed when its solution exists, is unique and stable. In such a case, we say that a so-called direct problem (DP) is being solved. This type of analysis involves determining the differential or integral equations that describe the problem, the shape, type, and size of the area, the initial and boundary conditions, the properties of the substances contained in the area, as well as the internal energy sources and external forces acting on the system. If all information regarding the mathematical model is known, then the problem is well-posed [2]. The state of knowledge necessary to pose and solve a direct problem is often impossible to obtain in real-world engineering problems. If any of the above information from the mathematical model is unknown, for example, about the boundary conditions or the shape of the area, then the inverse problem is considered. Inverse problems are ill-posed in the Hadamard sense [1]. The solution to the inverse heat conduction problem (IHCP) involves determining the initial conditions, the boundary conditions, or the properties of the material, such as the thermal conductivity coefficient or the specific heat, based on the observed temperature values within the area [3,4]. The most common inverse problem during the exploitation of industrial machinery and equipment is to determine the boundary condition in the form of temperature or heat flux for a part of the area based on the temperature measurement inside it. This problem is termed the boundary inverse problem [3–6]. The second type of inverse problem used in solving engineering problems is the Cauchy problem. In this case, the boundary condition for part of the area is unknown, while for another part of the area, two boundary conditions are known, most often temperature and heat flux [3–6].

The solution of the IHCP comes down to solving a differential equation of heat conduction of the form [7]

$$\frac{\partial}{\partial x}\left(\lambda\frac{\partial T}{\partial x}\right)+\frac{\partial}{\partial y}\left(\lambda\frac{\partial T}{\partial y}\right)+\frac{\partial}{\partial z}\left(\lambda\frac{\partial T}{\partial z}\right)+q_V=c\rho\frac{\partial T}{\partial t} \tag{1}$$

where $x$, $y$, $z$—coordinates in the Cartesian system [m], $T$—temperature [K], $t$—time [s], $\lambda$—thermal conductivity [W/mK], $c$—specific heat [J/kgK], ρ—density [kg/m$^3$], and $q_V$—power of internal heat sources related to the unit volume [W/m$^3$]. The differential Equation (1) is solved with the known boundary condition on the part of the boundary $\Gamma_1$ and the sought boundary condition on the part of the boundary $\Gamma_2$, taking into account the measurement of the temperature on the curve $\Gamma^*$, as shown in Figure 1. Depending on the problem under consideration, Equation (1) can be simplified in practice. The absence of heat sources will mean $q_V = 0$. When solving IHCPs, it is important to determine the direction of heat transfer in the case under consideration. Knowledge of physical phenomena is the basis for determining whether calculations can be performed for a one-dimensional $\left(\frac{\partial T}{\partial x} \neq 0, \frac{\partial T}{\partial y} = 0, \frac{\partial T}{\partial z} = 0\right)$ or two-dimensional $\left(\frac{\partial T}{\partial x} \neq 0, \frac{\partial T}{\partial y} \neq 0, \frac{\partial T}{\partial z} = 0\right)$ problem. For a boiler wall or gas turbine blade, the problem often comes down to a two-dimensional equation. The question must be asked whether a non-stationary $\left(\frac{\partial T}{\partial t} \neq 0\right)$ or stationary $\left(\frac{\partial T}{\partial t} = 0\right)$ problem should be considered. When temperature changes over time can be omitted, the steady-state problem is analysed. Many studies consider the optimal cooling of gas turbine blades as a steady-state problem. On the other hand, heat treatment and thermochemical treatment problems analysed using the IHCP solution are non-stationary problems. Significant changes in the thermal conductivity coefficient $\lambda$, specific heat $c$, and density ρ also affect the differential equation considered when solving IHCPs. In heat treatment (e.g., hardening) and thermochemical (e.g., nitriding) processes, taking into account the temperature ranges of these processes and the properties of the steels being treated, the dependence of the thermal conductivity coefficient $\lambda$ and specific heat $c$ on temperature must be taken into account when solving IHCPs.

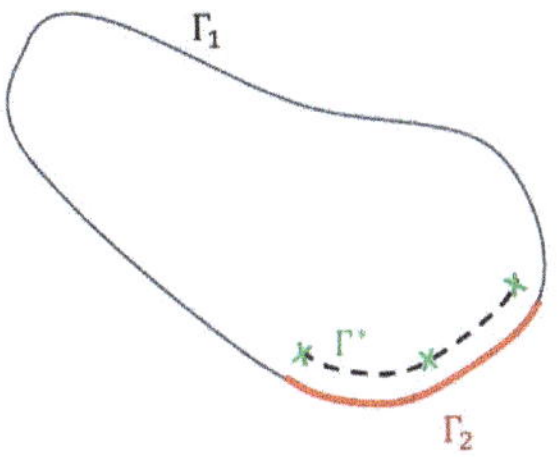


**Figure 1.** Boundaries $\Gamma_1$, $\Gamma_2$ and measurement points on the curve $\Gamma^*$.

To calculate the temperature or heat flux at the boundary $\Gamma_2$, the minimum of the functional of the form should be determined.

$$I_1=\sum_{i=1}^{M}\left(T_i^c-T_i^*\right)^2 \tag{2}$$

https://doi.org/10.3390/en19051291

Functional (2) describes the sum of squares of distances between the calculated temperature $T_i^c$ and the measured temperature $T_i^*$ at the $i$-th measurement point for $i = 1, 2, \ldots,$ M. Using various numerical techniques, the differential Equation (1) is algebraised. Taking into account the information contained in the boundary condition at the boundary $\Gamma_1$ and the temperature measured at the measurement points on the curve $\Gamma^*$, together with the functional (2), the IHCP can be reduced to solving a system of equations

$$A\{x\} = \{b\} \quad (3)$$

where the elements of the vector $\{x\}$ are the requested boundary condition or parameters that allow the aforementioned boundary condition to be determined at the boundary $\Gamma_2$. The elements of matrix $A$ and vector $\{b\}$ usually result from the form of the direct problem.

During the operation of machines and equipment, it is sometimes impossible to measure the temperature of walls with high thermal loads, or the measurement is subject to a very large error [7,8]. Examples of surfaces where measurement is difficult or impossible include: the inner wall of a power boiler, the surface of a gas turbine blade, the body of a gas or steam turbine, the surface of a charge being processed in a furnace for thermochemical treatment, and many others. Determining the temperature or heat flux on such surfaces is based on solving the inverse heat conduction problem. Inverse problems have very widespread technical applications in the analysis of heat flow in machines and devices [6,9].

Inverse problems are often numerically unstable. This is because inverse problems are numerically ill-posed in the Hadamard sense. This means that a small disturbance in the input data can cause a very large disturbance in the result. In such cases, the IHCP solution may be ineffective. In industrial applications, the input data are measurement data. The measured variables are temperature and heat flux. Measurement errors result from the type and accuracy of measuring instruments such as thermocouples and heat flux sensors. When measuring temperature and heat flux, contact resistance occurs at the point of contact between the measured surface and the measuring sensor. This is particularly important for highly dynamic heat transfer processes. In addition, there are thermocouple positioning errors of $\pm 0.2$ mm, which also affect the accuracy of the result. In most cases, the result is a boundary condition in the form of temperature or heat flux on a surface where measurement is impossible or very difficult to perform. In addition, the possibility of placing a limited number of thermocouples and the inability to place thermocouples at points that are advantageous for calculations must be taken into account.

Currently, multiple numerical studies are being conducted on computational methods that allow stable solutions to be obtained for IHCPs [3–6,10]. Computational techniques that stabilise the solution to an inverse problem (IP) are referred to as regularisation.

One of the IHCP regularisation techniques is Tikhonov regularisation [10], which boils down to correcting the functional (2)

$$J = I_1 + \gamma I_2 \quad (4)$$

where the functional $I_1$ is described in Formula (2), $\gamma$ is the regularisation parameter, and $I_2$ is the regularisation functional. The application of regularisation also modifies matrix A in matrix Equation (3). The difficulty lies in selecting the regularisation parameter $\gamma$ in such a way that it stabilises the IP. The Morozov criterion [11–19], the L-curve method [14,20–22], and the energy integral [14] are used for this purpose. In [23], a method for automatic selection of the regularisation parameter is presented. Modifications of the classical Tikhonov regularisation method [24,25] and the iterative Tikhonov regularisation method [26] were investigated. Regularisation of the IHCP was also performed using the

https://doi.org/10.3390/en19051291

SVD algorithm [27,28], discrete Fourier transform [29], and optimal selection of the integral parameter [30].

This paper presents numerous applications of IHCP solutions for energy-intensive industrial processes and energy conversion. It indicates areas of possible application of quasi-regularisation and machine learning for solving IHCPs in terms of the aforementioned industrial applications.

## 2. Methods of Solving IHCPs and Their Quasi-Regularisation for Industrial Applications

To date, many methods have been developed to solve IHCPs. These methods vary greatly depending on the geometry of the area and whether a steady-state or unsteady problem is being considered. For simple geometries such as a rectangle, circle, or ring, numerical studies were conducted to develop analytical–numerical solutions [3–6,31–38]. The solution to the unsteady inverse problem for a circle and a ring is presented in [31,34], respectively. An analytical–numerical calculation method for the stationary inverse problem of heat conduction using Chebyshev polynomials for rectangular geometry was also analysed [36]. In turn, in [37,38], the method of fundamental solutions was also applied to a rectangle. The heat conduction equation was also solved using the Treffetz function [32,33,39]. Numerous research works on methods for solving inverse problems for simple geometries and analysing their stability formed the basis for the development of numerical methods for complex geometries and industrial applications.

In order to be used in industry, the IHCP solution must be numerically stable. The solution is stabilised by IHCP regularisation. Currently, a quasi-regularisation approach is also used to solve IHCPs. In this case, the Tikhonov regularisation with the selection of the regularisation parameter according to Formula (4) is not used. The solution is stabilised by selecting, for example, the time step or the integration method in such a way as to average the input data or values in the calculation process. A properly selected quasi-regularisation method allows for the stabilisation of the IHCP solution while preserving information about the heat transfer process, enabling its analysis. The quasi-regularisation method should be selected individually for each calculation model and industrial application. Quasi-regularisation approaches are often characterised by a simpler mathematical model than models using classical regularisation, which creates great potential for the application of these solutions in industrial applications. Despite its many advantages, quasi-regularisation is a compromise between accuracy and stability of the solution.

### *2.1. Boilers and Heating Devices*

The studies in [40,41] present a solution to the geometric IP in determining the thickness of the stone in a steam boiler pipe. Inverse methods were also used to determine the thermal boundary conditions of the boiler membrane water-wall [42]. Simulations of changing thermal loads on the wall of a heating device using Chebyshev polynomials to solve IHCPs are presented in [43]. The inverse heat transfer method was used to identify the sediment profile in a heat exchanger pipe with two-phase and turbulent flow [44]. Sediments that exist in industrial heat exchangers reduce the thermal efficiency of the exchanger and consequently cause energy losses. To solve the problem, the conjugate gradient method and the volume of fluid method in Open FOAM were used. The inverse problem solution method used is stable and inexpensive, making it suitable for real-world industrial problems. In turn, the estimation of the heat flux and convective heat transfer coefficient in the water membrane of a steam boiler by solving the IHCP using an iterative algorithm is described in [45].

https://doi.org/10.3390/en19051291

It is possible to determine thermal stresses in the boiler pressure components if the fluid temperature and heat transfer coefficient on the inner surface are known. Determining these values under industrial conditions is difficult. However, it is possible to calculate the boundary conditions by solving the IHCP, as described in [46]. The inverse space marching method (ISMM) is presented to determine the heat transfer coefficient on the surface of a thick-walled plate that is sprayed with water (Figure 2a). The experimental stand consists of a steel plate (Figure 2a-1) heated by electric heaters (Figure 2a-2) and insulated on the heater side (Figure 2a-3). The thermocouples are placed on the plate on the side of the heaters (Figure 2a-4). The other surface of the plate is sprayed with water (Figure 2a-5). The plate was divided into control volumes (Figure 2b). Each of the control volume layers is marked with a different colour in Figure 2b. The temperature values measured in the blue layer are used to determine the temperatures in the control volumes of the green layer. In the next step, the temperatures in the yellow layer are calculated, and finally, at the edge of the plate in the red layer.

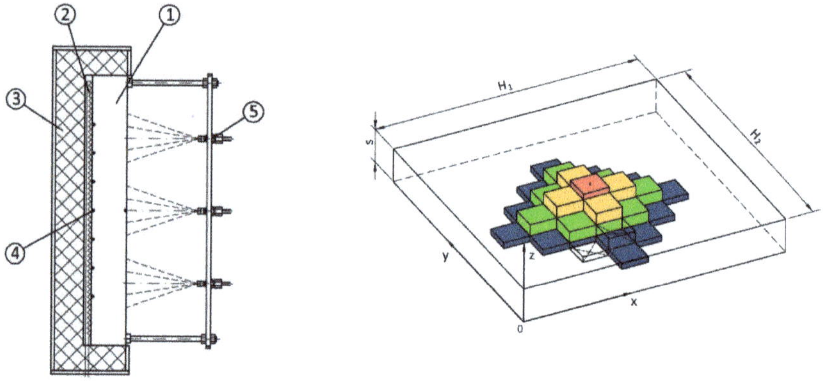


(a) (b)

**Figure 2.** (**a**) Cross-section of the spray cooling stand for thick-walled plate; (**b**) dividing the plate into control volumes [46].

For each control volume, the finite volume method was used to write the energy balance equation. Through step-by-step calculations in the control volumes, the boundary condition is determined. The ISMM method is easy to use and demonstrates numerical stability. The short calculation time for this computational method allows it to be used for online monitoring of thermal stresses in power plant components. The appropriate selection of the number of control volumes allows a stable solution to be obtained without a significant loss of information on the temperature distribution in the wall [46], which gives the regularisation properties of the method. The ISMM method includes a quasi-regularisation approach and has been applied to analyse heat transfer in power boilers.

In the next step, ISMM was used to develop a new technique to measure the temperature of superheated steam in conventional power plants [47] and to calculate the transient thermal stresses on the internal surface of a thick-walled pipe weakened by a hole [48]. The subject of controlling the temperature and thermal stress in the components of the power boiler during start-up, shutdown, and load change to reduce the unit's connection time to the power grid using ISMM, which is a method that uses quasi-regularisation, is presented in [49].

https://doi.org/10.3390/en19051291

## 2.2. Heat Treatment and Thermochemical Treatment

Another area of research worth mentioning where IHCP solutions are applied is heat treatment [50–58] and thermochemical treatment [24,59–62]. In [51], a solution to the one-dimensional IHCP was proposed for heat-treated surfaces. Analyses were performed to determine the temperature of the laser-heated surface and the spray-cooled surface. A mathematical model of the solution was developed using the shifting function method, the least-squares method, and measured temperatures. Quenching of moving metal plates using water jets was analysed using a two-dimensional inverse heat conduction solution [52]. Analyses of the quenching process using IHCPs are also described in [53–56,58].

Machine components such as crankshafts and gears are gas-nitrided. Controlling the surface temperature of thermochemically treated components is difficult to achieve in a closed furnace, where the gas surrounding the load is in constant motion, forced by a fan, and the gas temperature rises to 550 °C. Heat transfer in the furnace occurs through forced convection and, to a large extent, through radiation from the furnace and gas [59,62]. Knowledge of the edge temperature of treated components allows the process to be carried out in such a way that surfaces with the required properties are produced [63]. Controlling this temperature also allows the process time to be reduced without exceeding the thermal stresses [61]. The boundary conditions on elements subjected to thermochemical treatment can be determined by solving the IHCP. This issue has been considered in [24,59–62,64]. The study in [62] presents an analytical and numerical model to solve the IHCP for a cylinder. In industrial applications, it is very important that the method used to solve the IP is stable and does not require regularisation, or that the regularisation method is reliable and automatic. The study presented in [62] describes a quasi-regularisation method consisting of selecting the time step $\Delta t$ in such a way as to eliminate oscillations in the solution. The time step $\Delta t$ was introduced by writing the partial derivative on the right-hand side of the heat conduction differential Equation (1).

$$\frac{\partial T}{\partial t} = \frac{T(x, y, t_k) - T(x, y, t_{k-1})}{\Delta t} \tag{5}$$

The stabilisation method obtained by selecting $\Delta t$ was used to determine the boundary conditions ($T$—temperature, $q$—heat flux, and $h$—heat transfer coefficient) for a cylinder subjected to thermochemical treatment at different heating rates and fan speeds [62]. In numerical tests, the relative error of the boundary condition $\delta BC$ was calculated according to the formula

$$\delta BC = \frac{|BC_{as} - BC_{ip}|}{BC_{as}} \tag{6}$$

where $BC$ is the boundary condition ($T$, $q$, or $h$), $as$ is the analytical solution, and $ip$ is the inverse problem. The ratio of the average relative error for the time step $\Delta t = i$ to the average relative error for the time step $\Delta t = 1$ is described by the formula

$$f(\Delta t) = \frac{av[\delta BC(\Delta t = i)]}{av[\delta BC(\Delta t = 1)]} \tag{7}$$

The ratio of the maximum relative error for the time step $\Delta t = i$ to the maximum relative error for the time step $\Delta t = 1$ is described by the formula

$$g(\Delta t) = \frac{max[\delta BC(\Delta t = i)]}{max[\delta BC(\Delta t = 1)]} \tag{8}$$

Based on the research presented in [62], it can be concluded that changing the time step from $\Delta t = 1$ to $\Delta t = 10$ resulted in a decrease in the ratio of the average relative errors $f$ (Formula (7)) from 1 to 0.5145 for temperature, 0.1441 for heat flux, and 0.1503

https://doi.org/10.3390/en19051291

for the heat transfer coefficient (Table 1). In turn, for the ratio of maximum relative errors $g$ (Formula (8)), the change in the time step resulted in a decrease in the value of the $g$ function to 0.3486 for temperature, 0.0947 for heat flux, and 0.0945 for heat transfer coefficient (Table 1). The data in Table 1 clearly indicate the significant impact of the time step $\Delta t$ on the stability of the solution to the IHCP.

**Table 1.** Values of functions $f(\Delta t)$ and $g(\Delta t)$ for different types of boundary conditions $BC$ (temperature $T$, heat flux $q$, and heat transfer coefficient $h$) for time steps $\Delta t$ = 1, $\Delta t$ = 2, $\Delta t$ = 5, $\Delta t$ = 10 s.

| | Type of $BC$ | $i$ = 1 [s] | $i$ = 2 [s] | $i$ = 5 [s] | $i$ = 10 [s] |
|---|---|---|---|---|---|
| $f(\Delta t)$ | $T$ | 1 | 0.7526 | 0.5921 | 0.5145 |
| | $q$ | 1 | 0.5508 | 0.2298 | 0.1441 |
| | $h$ | 1 | 0.5532 | 0.2362 | 0.1503 |
| $g(\Delta t)$ | $T$ | 1 | 0.6732 | 0.4431 | 0.3486 |
| | $q$ | 1 | 0.6603 | 0.2256 | 0.0947 |
| | $h$ | 1 | 0.6300 | 0.2209 | 0.0945 |

The time step was also a method of stabilising the IHCP solution, where the obtained temperature distribution was used to determine thermal stresses in cylindrical pressure components [65].

In [24], the analysis of heat transfer during the gas nitriding process was extended to include a detailed analysis of the cooling stage. The computational model was supplemented with a modified Tikhonov regularisation method. In turn, in [59], an analysis of experimental studies obtained under industrial conditions was presented and a method using the Broyden–Fletcher–Goldfarb–Shanno algorithm (BFGS) was applied to solve the IHCP. The studies presented in [60,62] were used to analyse heat transfer in an aluminium extrusion die subjected to gas nitriding [64].

A new method for solving the inverse heat conduction problem for the Helmholtz equation using the Trefftz function is presented in [61]. The heat flux sought at the edge $\Gamma_2$ of the thermochemically treated roller is approximated using a step function (Figure 3) and is written using a linear combination

$$q(x) = \sum_{i=0}^{N} q_i(x) = \sum_{i=0}^{N} \varphi_i(x) c_i \tag{9}$$

of the sought coefficients $c_i$ and the functions $\varphi_i$ for $i = 0, 1, 2, \ldots, N$. The function $\varphi_i$ was defined by the formula

$$\varphi_i(x) = \begin{cases} 1 \text{ for x} \in \Gamma_{2,i} \\ 0 \text{ for } \notin \Gamma_{2,i} \end{cases} \tag{10}$$

and shown in Figure 4. For the method under consideration, the functions $\varphi_i(x)$ can be interpreted as an impulse with a value of 1 on the $i$-th fragment of the boundary $\Gamma_2$, which is marked $\Gamma_{2,i}$ (Figure 4). The heat flux $q(x)$ on the boundary $\Gamma_2$ is sought, which is a composite of heat fluxes with a constant value $q_i(x)$ for successive fragments of the boundary. The coefficients $c_i$ are obtained from the IHCP solution. As a result, the flux $q(x)$ is a step function (Figure 3) that is the average of the heat flux on the boundary. The temperature in the analysed area can be written as a combination of the Trefftz functions $\psi_i(x, y)$ added to the value of the function $T_1(x, y)$ obtained from the solution of the direct problem of the form.

$$T(x, y) = T_1(x, y) + \sum_{i=0}^{N} \psi_i(x, y) c_i \tag{11}$$

https://doi.org/10.3390/en19051291

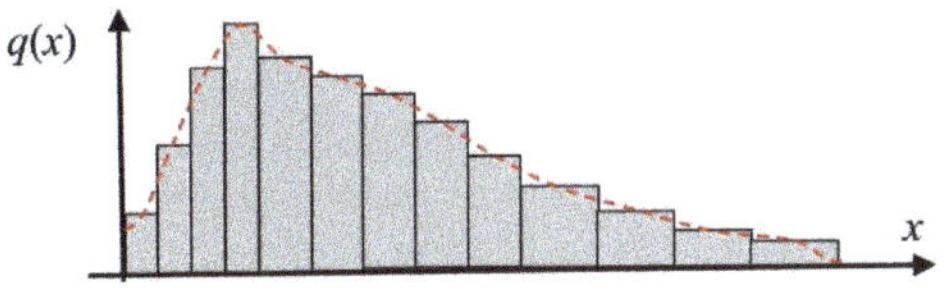


**Figure 3.** Approximation of the heat flux using a step function.

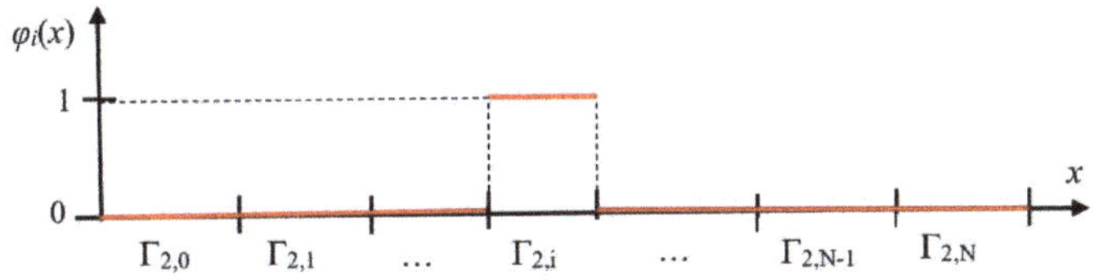


**Figure 4.** Distribution of the step function $\varphi_i(x)$ (red line) at the boundary $\Gamma_2$.

Examples of distributions of the functions $\psi_i(x,y)$ for $i = 0,1,2,\ldots,5$ are shown in Figure 5.

**Figure 5.** Trefftz functions $\psi_0$, $\psi_1$, $\psi_2$, $\psi_3$, $\psi_4$ and $\psi_5$.

This method is described in more detail for steady-state IHCPs in [66,67], and for the unsteady-state problem in [61].

It has been shown that the appropriate selection of the number of functions $\varphi_i$ and the division of the boundary $\Gamma_2$ into segments $\Gamma_{2,0}, \Gamma_{2,1}, \ldots, \Gamma_{2,N-1}, \Gamma_{2,N}$ has a significant impact on the stability of the IP solution. The appropriate selection of step functions allowed for the stabilisation of the inverse problem solution without the need to apply classical Tikhonov regularisation. The appropriate selection of the number of segments and their length on the boundary $\Gamma_2$ constitutes a quasi-regularisation of the IHCP solution method described in [61,66,67]. A dense division of the boundary $\Gamma_2$ allows for a more accurate approximation of the boundary condition. On the other hand, a smaller number of segments on the boundary $\Gamma_2$ results in greater averaging properties and thus stabilises the method. Increasing the density of the division for a fragment of the boundary with greater temperature and heat flux variability allows us to approach a division that is a compromise between the accuracy and stability of the IHCP solution through quasi-regularisation. The method can be successfully applied to the analysis of heat transfer in elements with complex geometry. The calculations are based on variational calculus and are performed in the freeFEM solver [68]. In the initial part of the algorithm, a complex geometry is assumed, which is described parametrically. The known boundary conditions, the division of the boundary where the boundary condition is sought, the coordinates of the temperature measurement points, and the measured values should also be adjusted to the problem under consideration. For geometries such as a gas turbine blade or a cross-section of a real heating device wall, the calculation scheme is analogous. In this case, the functions $\varphi_i(x)$ should be replaced with functions dependent on, for example, two variables $\varphi_i(x,y)$ (Formula (10)).

The presented method was also used to determine the heat flux on the surface of the T-shaped stator cavities of steam turbines [66] and the cross-section of a piston machine body with a cooling channel [69]. The calculation time indicates the possibility of using this method for online temperature control during thermochemical treatment processes. For gas nitriding, the calculation time was approximately 2 s for the assumed time step $\Delta t = 30$ s.

## 2.3. Gas Turbines

A very important challenge in the energy sector is to increase the efficiency of gas turbines. Efficiency increases with the temperature of the gas flowing around the blades. The gas temperature is limited by the properties of the material from which the blades are made. Too high a gas temperature can cause the blades to lose their strength properties and, as a result, damage the turbine. In order for the turbine to operate at a higher gas temperature without risking damage to the blades, cooling systems and protective coatings are used. Optimisation of the gas turbine blade cooling process based on IHCP solutions is discussed in [70–74]. The research area where IHCP solutions were applied to the analysis of thermal loads on gas turbine blades with a protective layer is described in [75–78]. The study in [72] presents an iterative algorithm that allows solving the IHCP for Laplace's equation in a multiply-connected domain. A Cauchy-type problem was solved, where two boundary conditions were known on the outer boundary, while the sought condition was on the inner boundary of an elliptical ring with a shifted inner boundary. The IP solution is based on the DP, where the heat flux on the inner boundary of the domain was assumed and iteratively changed in such a way as to minimise the functional. The aforementioned functional determines the distance between the calculated and specified temperature at the outer boundary and includes a regularisation term. This method, which also uses variational calculus, was developed by the authors and applied to the analysis of the cooling of gas turbine blades [70]. In [70], two iterative algorithms were compared to determine the temperature at the inner limit of the blade. The first one is based on variational calculus, while the second one boils down to solving the least squares approximation problem. The temperature on the outer surface of the blade was assumed to be $T_0 = 600$ K. The IP solution was stabilised using a discrete Fourier transform. The calculation results showed that there was a local loss of stability of the IHCP solution near the leading and trailing boundaries. In these areas, the temperature exceeded the assumed value of $T_0$. This means that the cooling channel is not able to absorb the amount of heat that is transferred from the gas surrounding the blade. In the next step of the investigation, the authors developed an iterative algorithm that allowed the analysis of the cooling of the gas turbine blades using the IHCP solution, where porous material was present in the cooling channels [73]. The temperature of the blade boundary and the porosity distributions in the cooling channel were determined.

https://doi.org/10.3390/en19051291

Another way to increase the temperature of the working medium in a gas turbine is to use ceramic coatings on the surface of the blades. This creates thermal resistance on the surface of the blade. The studies presented in [75–78] describe calculations for an area of metal covered with a layer of ceramic with low thermal conductivity. The issue under consideration is a Cauchy problem. In [77], the influence of thermal conductivity and the thickness of the ceramic layer on the temperature at the metal–ceramic interface was investigated, as exceeding this temperature may cause the blade to lose its mechanical properties. This issue was continued in [75,76], where the stability of the IHCP solution was investigated. This IP was stabilised by means of an energy balance equation for the ceramic layer, which constitutes its quasi-regularisation.

## 3. Applications of Neural Networks for Solving IHCPs

In recent years, a new approach to solving IHCPs has also been proposed. Genetic algorithms [79–81] and machine learning [82–92] are used. Research on the application of machine learning to solve IHCPs is carried out using Artificial Neural Networks (ANNs) [82–92]. ANNs are algorithms whose structure resembles that of the brain [93]. They consist of an input layer $N_{\mathrm{inp}}$, hidden layers $N_{\mathrm{h}}$, and an output layer $N_{\mathrm{out}}$ (Figure 6). The transition between neurons causes the output value of the previous layer to be multiplied by the weight and the assigned bias to be added, and the activation function [82]. The optimisation problem consists of determining the appropriate weights at the network learning stage so that in the next step, the network can be used for data where the expected result is unknown. A special example of a neural network (NN) used to solve IHCPs is Physics-Informed Neural Networks (PINNs) [82,86,87,89,91,92,94–100]. A PINN is a NN that uses fundamental laws of physics, making it a good tool to solve heat transfer problems.

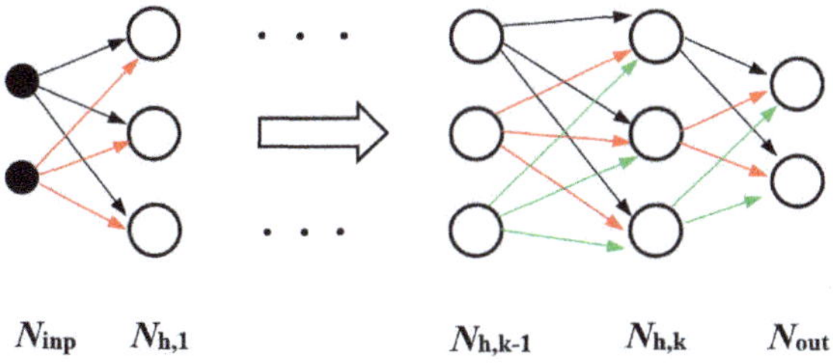


**Figure 6.** Neural network.

In the analysis of boiler and other heating device operation based on the measurement of the temperature inside the wall, the boundary condition in the form of temperature or heat flux on an unavailable surface is calculated. In [91], a PINN model was proposed, which simultaneously identifies the boundary heat flux and thermal conductivity based on temperature values. The mathematical description of the method was prepared for a non-stationary, one-dimensional IHCP. However, it allows the analysis to be extended to two- and three-dimensional problems. The one-dimensional case concerns heat conduction through a flat wall with a thickness of 20 mm, in which the measurement points are located 10 mm and 20 mm from the wall surface where the heat flux is sought. The NN used to predict the solution consists of three blocks: a fully connected NN, a PINN, and a feedback mechanism. The loss function is the sum of functions representing penalty terms for the residuals of the governing equations, boundary conditions, and initial conditions. In the one-dimensional model, transient heat flux $q_{\mathrm{in}}(t)$ was assumed at the left end, while the right end exchanges heat through natural convection. Radiation was taken into account at both ends. The normalised mean squared error (NMSE) was used to evaluate the accuracy

https://doi.org/10.3390/en19051291

of the heat flux sought. In the hidden layers, the activation function was the hyperbolic tangent. The output layer uses a linear activation function. The thermal conductivity of the material was assumed to be k = 5 W/mK. Numerical tests were performed for five different NN architectures (Net-i for i = 1, 2, . . ., 5). The analysed networks have 2, 3, 3, 3, and 4 hidden layers, respectively. The Net-1 has 80 and 60 neurons in its hidden layers, respectively. Accordingly, the Net-2 consists of 80, 60, and 40 neurons; the Net-3 consists of 130, 100, and 70 neurons; the Net-4 consists of 200, 150, and 100 neurons; and the Net-5 consists of 200, 150, 100, and 50 neurons.

For the Net-3 neural network, the NMSE value of heat flux was 0.028, the relative thermal conductivity error was 0.13%, and the training time was approximately 840 s. The authors considered the Net-3 NN to be the best compromise between accuracy of results and training time. The values for the other networks were referenced to Net-3 and are shown in Table 2. During training, the loss function reaches a value below $10^{-3}$ after 20,000 epochs. Training was conducted for up to 50,000 epochs. Numerical tests for Net-3 and Net-4 were also performed for temperature-dependent thermal conductivity described by a linear function using two coefficients. The determination of the heat flux reaches a relatively stable and accurate level after approximately 25,000 epochs (loss function for Net-3 of the order of $10^{-4}$, and for Net-4 of the order of $10^{-5}$). Meanwhile, the identification of two thermal conductivity coefficients converges much more slowly and requires a significantly higher number of iterations.

**Table 2.** NMSE values of heat flux ($q_{NMSE}$), relative error of thermal conductivity (δk), and training time (t) for Net-i referred to Net-3.

| i | $q_{NMSE}$(Net-i)/$q_{NMSE}$(Net-3) | δk(Net-i)/δk(Net-3) | t(Net-i)/t(Net-3) |
|---|---|---|---|
| 1 | 3.1429 | 25.0769 | 0.6212 |
| 2 | 1.1786 | 2.5385 | 0.7657 |
| 3 | 1 | 1 | 1 |
| 4 | 0.9286 | 0.6923 | 1.3457 |
| 5 | 0.6429 | 0.6154 | 1.5295 |

Numerical tests in [91] were performed for measurement data disturbances of 1%, 2%, 4%, and 8%. For the Net-3 test, Table 3 shows the NMSE values for heat flux and the relative error for thermal conductivity, taking into account disturbances in the input data for calculations in relation to values without disturbances. The method described in [91] allows for satisfactory accuracy in heat flux reconstruction despite noise, which is even greater than that encountered in industry. Slightly worse results were obtained for the thermal conductivity coefficient k. However, in many applications, the values of k are known. A two-dimensional heat conduction model was also considered for a rectangular area with dimensions L = 0.05 m and W = 0.02 m at a constant thermal conductivity k = 3 W/mK. The aim of the analysis was to investigate the effect of sensor placement on the result. The number of neurons in the hidden layers was 180, 100, and 150. Simply increasing the number of temperature sensors or improving the sampling resolution in the time–space domain did not lead to a significant improvement in identification results. In the case of the two-dimensional test, taking into account a specific parametric form of heat flow significantly improved accuracy. The PINN method achieved an NMSE value of 0.003 for heat flux and a relative error of 0.3% in identifying thermal conductivity. For this two-dimensional problem, it was found that using four or more sensors yielded satisfactory identification results for both parameters. It is worth noting that the numerical tests were performed for low thermal conductivity values and small areas with simple geometry. In such cases, the wall experiences significantly higher temperature gradients than for thermal conductivity values of 30–50 W/mK, which are typical for steel. In such cases, obtaining an IHCP solution may prove more difficult.

**Table 3.** The $q_{NMSE}$ and δk values with noise level (NL) of input temperature relative to corresponding values without disturbance.

| NL [%] | $q_{NMSE}(NL)/q_{NMSE}(0)$ | δk(NL)/δk(0) |
|---|---|---|
| 0 | 1 | 1 |
| 1 | 1.1429 | 10.46 |
| 2 | 1.3929 | 26.62 |
| 4 | 2.1071 | 55.86 |
| 8 | 3.5714 | 110.46 |

The boundary condition in the form of heat flux for a one-dimensional problem was also determined using a PINN [89]. The NN was trained taking into account physical constraints such as the equation, boundary conditions, initial conditions, and temperature values. The results indicate that the PINN has satisfactory effectiveness, is efficient, and represents a promising direction for the development of methods to solve IP. The studies in [89,91] present the application of PINNs to model the problem of heat transfer through a flat wall. This issue can be developed for future application to the analysis of heat transfer in the walls of thermal devices. However, the cases analysed are theoretical examples at this stage, far from industrial applications.

A comparison of two machine learning-based approaches to solving a nonlinear IHCP in a one-dimensional domain with constant and moving boundary conditions (Case 3 with constant material properties and Case 4 with temperature-dependent material properties) is described in [82]. The issue under consideration corresponds to the analysis of heat transfer through spacecraft panels without and with consideration of the thermal expansion of the plate material. The surface heat flux was determined using ANNs and temperature measurements inside the area. The length of the area was L = 2 cm, and the temperature measurement was taken on one of the surfaces (x = 0) and in the middle of the area, 1 cm from the edge, where the heat flux was determined (x = 1 cm). An Autoregressive-Exogenous Recurrent Neural Network (NARX) and a Feed-Forward Neural Network (FFNN) were used to analyse the four calculation examples. The network consisted of four hidden layers composed of 3, 5, 11, and 21 neurons for the FFNN and one hidden layer composed of 10 neurons for the NARX. Numerical tests were performed for three training cases and three validation cases (parabolic, triangular, and step functions). The training examples were simpler, while the validation examples were more complex.

The training algorithm used was the Levenberg–Marquardt algorithm. Temperature-dependent material (carbon phenolic) properties were taken into account. Thermal conductivity ranged from 0.62 W/mK to 2.77 W/mK. Specific heat ranged from 795.5 J/kgK to 2386.48 J/kgK. In order to compare the FFNN and NARX networks, $r^2$ values (coefficient of determination) and training time were used. The $r^2$ values for Case 3 are higher for the FFNN (from 0.9279 to 0.9891), while for Case 4, they are higher for the NARX (from 0.9458 to 0.9939). For cases with a moving boundary, the FFNN is trained faster. Training times for the FFNN range from 1.8 to 6.75 s, while for the NARX they range from 0.66 to 2.18 s. The analysed networks were shown to show promising properties in solving IHCPs.

Despite favourable $r^2$ values and short learning times, the graphs show significant discrepancies in the heat flow for the stepwise validation function. The paper uses Tikhonov regularisation in the digital filter method. The authors of [82] did not specify the value of the regularisation parameter or the method of its selection.

https://doi.org/10.3390/en19051291

Research on IHCPs through multilayer materials with an insulating layer using neural networks is described in [86,101]. A PINN was used to analyse the heat flow in a multilayer material used as thermal protection for reusable launch vehicles [101]. An improved PINN framework was developed to solve transient nonlinear and inhomogeneous heat conduction problems [86]. These networks were used to optimise the design of thermal protection structures for hypersonic vehicles. Existing numerical methods required the use of a large number of meshes. The use of a NN is a non-mesh method, which has a positive impact on the efficiency of calculations.

In [97], a nonlinear IHCP solution using PINNs was presented to determine the temperature fields, heat flux, and effective thermal conductivity in porous material. The average relative error was 4.5%, 14.4%, and 2.73%, respectively. Compared to numerical calculations using the finite volume method (FVM), a five-order-of-magnitude acceleration of calculations was achieved. The study presented in [102] analysed heat conduction processes in porous materials using machine learning and Fourier transforms. The results presented in [97,102] indicate that PINNs are promising tools for researching heat conduction issues in porous materials.

In [101], the heating of a sample made of carbon fibre–epoxy composite with thermal conductivity k = 0.47 W/mK, density $\rho$ = 1573 kg/m$^3$, and specific heat $c$ = 967 J/kg K was analysed in an oven under convective conditions. The calculations were performed for heat transfer coefficients ranging from 10 W/m$^2$K to 100 W/m$^2$K. The calculations were performed using finite element methods (FEMs) and NNs, including a PINN. Numerical tests for NNs were performed for various activation functions that are exponential functions (ELU, adaptive-ELU) and for the hyperbolic tangent (adaptive-tanh). The activation function for the PINN was selected based on governing heat transfer equations. The network was trained during the first 15 min of heating, which lasted 45 min. The temperature consistency in the central part of the sample was taken into account in the network training process. In the second part of the heating process, the temperature curve was predicted using a NN, and the results were compared with the FEM solution. The curves for ELU, adaptive-ELU, and adaptive-tanh deviated significantly from the assumed function. For adaptive-ELU and adaptive-tanh, the temperature in subsequent heating time units decreased, even reaching zero for adaptive-ELU, which is contrary to the physical phenomena occurring in the oven. The temperature curve obtained for the PINN was almost identical to the curve obtained from the FEM and the assumed function. Calculations using the PINN network were faster than the FEM. The example presented in [101] clearly indicates the usefulness of using PINNs for the analysis of heat transfer processes.

## 4. Conclusions

The difficulties with industrial applications of solutions for IHCPs stem from the fact that these problems are ill-posed in the Hadamard sense. Solving them requires extensive mathematical knowledge, in particular numerical methods, knowledge of computational tools, measurement techniques, and an understanding of the physical phenomena occurring during the operation of machines and equipment. Practical knowledge of the realities of operating these machines and devices in industrial conditions is also important. Therefore, the application of IHCPs in industry is an interdisciplinary issue that is very difficult to implement. There are many scientific papers presenting computational models for stationary and non-stationary one-dimensional simple geometries. They often describe complex IHCP regularisation methods, the application of which in industry could be impractical. Many works also present IHCP solutions for stationary two-dimensional problems for simple and complex geometries, such as gas turbine blades. The difficulty remains in solving unsteady, two-dimensional, and three-dimensional IHCPs for complex

https://doi.org/10.3390/en19051291

geometries. New computational tools such as the freeFEM solver, Python, and COMSOL are creating ever greater opportunities in this field. When controlling the operation of machines and devices using IHCP solutions, we want the calculations to be performed correctly and quickly, and the solution to be numerically stable so that we can obtain results without the need for intervention by a person highly specialised in solving IP. Currently, efforts are being made to automate and simplify methods for solving IHCPs that may have industrial applications. In connection with the above, methods are being sought that allow for online control during the heating or cooling process of walls that are inaccessible for measurement and are most heavily loaded thermally. In order for online control to be possible without human intervention in industrial conditions, the IHCP solution must first undergo a series of numerical tests. These tests cover the impact of temperature and heat flux measurement errors, location, and measurement sensor installation errors on the result, taking into account the heat flow intensity in a given device. It is also very important that the calculation time allows information to be obtained on, for example, the surface temperature, where measurement cannot be performed in real time. Elements with significant thermal loads include the inner wall of a boiler, the casing of a steam and gas turbine, the blades of the gas turbine, and the surfaces of elements subjected to heat or thermochemical treatment. Numerous studies were conducted on IHCP solutions dedicated to these issues and their regularisation for the above-mentioned components, machines, and devices. According to the authors, automation of the control of thermally stressed surfaces may occur when methods for solving inverse heat conduction problems that do not require classical Tikhonov regularisation with the selection of a regularisation parameter are developed and tested. In this case, the IHCP solution can be stabilised using a quasi-regularisation approach, e.g., by selecting the time step, control volume size, or additional energy balance equations. When selecting a large time step, some data may be lost due to high averaging, e.g., of temperature curves, but on the other hand, this is very beneficial in terms of stability. On the other hand, a small time step may mean that the solution obtained will oscillate significantly. The possibility of large oscillations mainly concerns the heat flux and heat transfer coefficient. The quasi-regularisation approach involves finding a compromise between the stability and accuracy of the IHCP solution. Quasi-regularisation methods should be selected individually depending on the technical problem, heat transfer dynamics, and IHCP solution method. A well-chosen quasi-regularisation method does not require human intervention in the IHCP solution during the industrial process. Quasi-regularisation has been applied to control and analyse heat transfer for boilers, gas turbines, and thermochemical treatment. It is also worth mentioning that online control of heat transfer processes using IHCPs and quasi-regularisation was applied to power boiler components and components subjected to gas nitriding. Recently, many papers have been published that describe new approaches to solving IHCPs using neural networks. The use of machine learning makes it possible to simplify the mathematical model and omit the regularisation of the IHCP. Often, the model is trained on the basis of the solution of the DP, which is numerically stable. The trained model then solves the IP, which is no longer an ill-posed problem. Despite many potential benefits, based on the literature presented, there are doubts regarding the effectiveness and applicability of NNs. Calculations using NNs were most often performed for one-dimensional cases. The two-dimensional examples given in the literature concern simple geometries with small dimensions. Numerical tests for heat flow using NNs were performed for materials with a relatively low thermal conductivity coefficient. The maximum value in the analysed papers was 4.4 W/mK. A low thermal conductivity coefficient causes large temperature gradients and facilitates the analysis of the issue. Materials commonly used in industry (e.g., steels) generally have thermal conductivity and specific heat that vary significantly within their

https://doi.org/10.3390/en19051291

applicable temperature range. The analysed studies noted significant difficulties in training networks due to the variability in k with temperature. This is not a major problem for solving IHCPs using classical numerical methods.

In heat treatment and thermochemical treatment, to be able to use neural networks, it would be necessary to perform a large number of experiments to determine the boundary conditions and combine the results with furnace operating parameters such as gas temperature, load weight, load surface properties, etc. Then, surface control could be performed online during the process, without having to solve the ill-posed problem. The possibility of online control springs from shorter computation times for NNs than for classical numerical methods. Therefore, it is very important to train models effectively, acquire training data, and calculate learning time and costs. However, the use of neural networks at their current stage of development for the analysis of very expensive equipment, such as processing furnaces, gas turbines, or power boilers, in industry raises doubts. Currently, there are no methods for certifying these machine learning devices that guarantee their effectiveness. However, quasi-regularisation methods can be successfully applied to them. The application of neural networks to solve IHCPs makes it possible to simplify computational models used to control energy-intensive industrial processes and energy conversion. In numerical tests, the best results were obtained for the PINN. Obtaining a stable solution to the IP allows its industrial application. Nevertheless, in the authors' opinion, at the current stage of research into IHCP solutions using NNs, their effective application to real industrial problems seems unlikely. Currently, the use of neural networks to solve IHCPs is an interesting research direction. Research on solving IHCPs using NNs would need to include larger elements with more complex geometries made from materials commonly used in manufacturing processes.

**Author Contributions:** Conceptualisation, M.J.; methodology, M.J. and D.J.; software, M.J. and D.J.; investigation, M.J. and D.J.; writing—original draft preparation, M.J.; writing—review and editing, M.J. and D.J.; visualisation, M.J. and D.J. All authors have read and agreed to the published version of the manuscript.

**Funding:** This research received no external funding.

**Data Availability Statement:** No new data were created.

**Conflicts of Interest:** The authors declare no conflicts of interest.

## Abbreviations

The following abbreviations are used in this manuscript:

| | |
|---|---|
| ANNs | Artificial neural networks |
| BFGS | Broyden–Fletcher–Goldfarb–Shanno algorithm |
| DP | Direct problem |
| IHCP | Inverse heat conduction problem |
| IP | Inverse problem |
| ISMM | Inverse space marching method |
| NNs | Neural networks |
| PINNs | Physics-informed neural networks |

## References


1. Hadamard, J. Sur les problèmes aux dérivées partielles et leur signification physique. *Princet. Univ. Bull.* **1902**, *13*, 49–52.
2. Grysa, K. *Trefftz Functions and Their Applications in Solving Inverse Problems*; Publishing House of the Kielce University of Technology: Kielce, Polish, 2010; ISSN 1897-2691. (In Polish)
3. Ciałkowski, M. *Selected Methods and Algorithms for Solving Inverse Problems of Heat Conduction Equation*; Poznań University of Technology Publishing House: Poznań, Polish, 1996; ISBN 8371431619. (In Polish)


https://doi.org/10.3390/en19051291

4. Taler, J.; Duda, P. *Solving Direct and Inverse Heat Conduction Problems*, 1st ed.; Springer: Berlin/Heidelberg, Germany, 2006; ISBN 978-3-540-33470-X.
5. Alifanov, O.M. *Inverse Heat Transfer Problems*; Springer: Berlin/Heidelberg, Germany, 1994; ISBN 13:978-3-642-76438-7.
6. Woodbury, K.A.; Najafi, H.; de Monte, F.; Beck, J.V. *Inverse Heat Conduction: Ill-Posed Problems*; John Wiley & Sons: Hoboken, NJ, USA, 2023; ISBN 9781119840190.
7. Taler, J. *Theory and Practice of Identifying Heat Transfer Processes*; Zakład Narodowy imienia Ossolińskich: Wrocław Warszawa Kraków, Polish, 1995; ISBN 83-04-04276-2. (In Polish)
8. Incropera, F.P.; de Witt, D.P. *Fundamentals of Heat and Mass Transfer*; Wiley: New York, NY, USA, 1996.
9. Orlande, H.R.B. *Inverse Heat Transfer: Fundamentals and Applications*, 2nd ed.; CRC Press: Boca Raton, FL, USA, 2021. [CrossRef]
10. Tikhonov, A.N.; Arsenin, V.Y. *Solution of Ill-Posed Problems*; Wiley: New York, NY, USA, 1977.
11. Bedin, L.; Bazán, F.S.V.; Quiroz, J.R. Method for recovering boundary data in a two-dimensional Poisson equation on annular domain. *J. Comput. Appl. Math.* **2018**, *342*, 83–95. [CrossRef]
12. Chen, B.; Sun, Y.; Zhuang, Z. Method of fundamental solutions for a Cauchy problem of the Laplace equation in a half-plane. *Bound. Value Probl.* **2019**, *2019*, 34. [CrossRef]
13. Han, H.; Ling, L.; Takeuchi, T. An energy regularization for Cauchy problems of Laplace equation in annulus domain. *Commun. Comput. Phys.* **2011**, *9*, 878–896. [CrossRef]
14. Joachimiak, M. Choice of the regularization parameter for the Cauchy problem for the Laplace equation. *Int. J. Numer. Methods Heat Fluid Flow* **2020**, *30*, 4475–4492. [CrossRef]
15. Marin, L. An invariant method of fundamental solutions for two-dimensional steady-state anisotropic heat conduction problems. *Int. J. Heat Mass Transf.* **2016**, *94*, 449–464. [CrossRef]
16. Morozov, V.A. *Methods for Solving Incorrectly Posed Problems*; Springer: New York, NY, USA, 1984.
17. Sun, Y. A meshless method based on the method of fundamental solution for solving the steady-state heat conduction problems. *Int. J. Heat Mass Transf.* **2016**, *97*, 891–907. [CrossRef]
18. Sun, Y.; He, S. A meshless method based on the method of fundamental solution for three-dimensional inverse heat conduction problems. *Int. J. Heat Mass Transf.* **2017**, *108*, 945–960. [CrossRef]
19. Wei, T.; Wang, J. A modified quasi-boundary value method for an inverse source problem of the time-fractional diffusion equation. *Appl. Numer. Math.* **2014**, *78*, 95–111. [CrossRef]
20. Jin, B.; Zheng, Y. A meshless method for some inverse problems associated with the Helmholtz equation. *Comput. Methods Appl. Mech. Eng.* **2006**, *195*, 2270–2288. [CrossRef]
21. Marin, L. Numerical solution of the Cauchy problem for steady-state heat transfer in two-dimensional functionally graded materials. *Int. J. Solids Struct.* **2005**, *42*, 4338–4351. [CrossRef]
22. Marin, L.; Munteanu, L. Boundary reconstruction in two-dimensional steady state anisotropic heat conduction using a regularized meshless method. *Int. J. Heat Mass Transf.* **2010**, *53*, 5815–5826. [CrossRef]
23. Pacheco, C.C.; Lacerda, C.R.; Colaço, M.J. Automatic selection of regularization parameter in inverse heat conduction problems. *Int. Commun. Heat Mass Transf.* **2022**, *139*, 106403. [CrossRef]
24. Joachimiak, M.; Joachimiak, D. Stabilization of boundary conditions obtained from the solution of the inverse problem during the cooling process in a furnace for thermochemical treatment. *Int. J. Heat Mass Transf.* **2024**, *224*, 125274. [CrossRef]
25. Yang, F.; Fu, C.; Li, X. A modified Tikhonov regularization method for the Cauchy problem of Laplace equation. *Acta Math. Sci.* **2015**, *35*, 1339–1348. [CrossRef]
26. Ngendahayo, J.P.; Minani, F.; Thim, J.; Berntsson, F. Convergence Analysis of Iterative Tikhonov Regularization Applied to an Inverse Heat Conduction Problem. *Int. J. Appl. Comput. Math.* **2025**, *11*, 149. [CrossRef]
27. Hasanov, A.; Mukanova, B. Relationship between representation formulas for unique regularized solutions of inverse source problems with final overdetermination and singular value decomposition of input-output operators. *J. Appl. Math.* **2015**, *80*, 676–696. [CrossRef]
28. Marin, L.; Karageorghis, A.; Lesnic, D. Regularized MFS solution of inverse boundary value problems in three-dimensional steady-state linear thermoelasticity. *Int. J. Solids Struct.* **2016**, *91*, 127–142. [CrossRef]
29. Frąckowiak, A.; Ciałkowski, M. Application of discrete Fourier transform to inverse heat conduction problem regularization. *Int. J. Numer. Methods Heat Fluid Flow* **2018**, *28*, 239–253. [CrossRef]
30. Joachimiak, M.; Ciałkowski, M. Optimal choice of integral parameter in a process of solving the inverse problem for heat equation. *Arch. Thermodyn.* **2014**, *35*, 265–280.
31. Ciałkowski, M.; Grysa, K. A sequential and global method of solving an inverse problem of heat conduction equation. *J. Theor. Appl. Mech.* **2010**, *48*, 111–134.
32. Grysa, K.; Maciąg, A. Trefftz Method in Solving Inverse Heat Conduction Problems. In *Encyclopedia of Thermal Stresses*; Hetnarski, R.B., Ed.; Springer: Dordrecht, The Netherlands, 2014. [CrossRef]
33. Grysa, K.; Maciag, A.; Pawinska, A. Solving nonlinear direct and inverse problems of stationary heat transfer by using Trefftz functions. *Int. J. Heat Mass Transf.* **2012**, *55*, 7336–7340. [CrossRef]


https://doi.org/10.3390/en19051291

34. Joachimiak, M.; Ciałkowski, M. Non-linear unsteady inverse boundary problem for heat conduction equation. *Arch. Thermodyn.* **2017**, *38*, 81–100. [CrossRef]
35. Joachimiak, M.; Ciałkowski, M. Stable solution to non-stationary inverse heat conduction equation. *Arch. Thermodyn.* **2018**, *39*, 25–37.
36. Joachimiak, M.; Ciałkowski, M.; Frąckowiak, A. Stable method for solving the Cauchy problem with the use of Chebyshev polynomials. *Int. J. Numer. Methods Heat Fluid Flow* **2020**, *30*, 1441–1456. [CrossRef]
37. Mierzwiczak, M.; Kołodziej, J.A. The determination of heat sources in two dimensional inverse steady heat problems by means of the method of fundamental solutions. *Inverse Probl. Sci. Eng.* **2011**, *19*, 777–792. [CrossRef]
38. Mierzwiczak, M.; Kołodziej, J.A. The determination temperature-dependent thermal conductivity as inverse steady heat conduction problem. *Int. J. Heat Mass Transf.* **2011**, *54*, 790–796. [CrossRef]
39. Hożejowska, S.; Piasecka, M. Numerical Solution of Axisymmetric Inverse Heat Conduction Problem by the Trefftz Method. *Energies* **2020**, *13*, 705. [CrossRef]
40. Joachimiak, M.; Ciałkowski, M.; Bartoszewicz, J. Analysis of temperature distribution in a pipe with inner mineral deposit. *Arch. Thermodyn.* **2014**, *35*, 37–49. [CrossRef]
41. Soloviev, A.N.; Shevchenko, M.A.; Germanchuk, M.S. The Inverse Geometric Problem of Thermal Conductivity for Determining the Thickness of Scale in Steam Boiler Pipes. *J. Math. Sci.* **2025**, *287*, 664–672. [CrossRef]
42. Lv, C.; Wang, G.; Chen, H. Estimation of time-dependent thermal boundary conditions and online reconstruction of transient temperature field for boiler membrane water wall. *Int. J. Heat Mass Transf.* **2020**, *147*, 118955. [CrossRef]
43. Joachimiak, M.; Joachimiak, D.; Ciałkowski, M. Investigation on thermal loads in steady-state conditions with the use of the solution to the inverse problem. *Heat Transf. Eng.* **2023**, *44*, 963–969. [CrossRef]
44. Jamail Arand, H.; Sefid, M.; Goudarzi, K. Identification of Fouling Profiles in a Heat Exchanger Pipe with Presence of Two-Phase Flow Using Inverse Heat Transfer Method, a Numerical Study. *Int. J. Eng. Trans. B Appl.* **2026**, *39*, 694–704. [CrossRef]
45. Duda, P.; Konieczny, M. An Iterative Algorithm for the Estimation of Thermal Boundary Conditions Varying in Both Time and Space. *Energies* **2022**, *15*, 2686. [CrossRef]
46. Jaremkiewicz, M.; Taler, J. Online Determining Heat Transfer Coefficient for Monitoring Transient Thermal Stresses. *Energies* **2020**, *13*, 704. [CrossRef]
47. Jaremkiewicz, M.; Taler, D.; Dzierwa, P.; Taler, J. Determination of Transient Fluid Temperature and Thermal Stresses in Pressure Thick-Walled Elements Using a New Design Thermometer. *Energies* **2019**, *12*, 222. [CrossRef]
48. Jaremkiewicz, M. Analysis of the Accuracy of the Inverse Marching Method Used to Determine Thermal Stresses in Cylindrical Pressure Components with Holes. *Energies* **2025**, *18*, 4546. [CrossRef]
49. Taler, D.; Taler, J.; Kaczmarski, K.; Jaremkiewicz, M.; Dzierwa, P.; Trojan, M. Monitoring of thick-walled pressure elements to determine transient temperature and stress distributions using the measured fluid's pressure and wall's temperature. *Energy* **2024**, *309*, 133121. [CrossRef]
50. Duarte Benther, J.; Chalhub, D.J.N.M.; Franklin, E.; Lappas, P.; Rosengarten, G. Semi-analytical solution of transient heat conduction for spray cooling. *Int. J. Therm. Sci.* **2025**, *218*, 110148. [CrossRef]
51. Lee, S.-Y.; Yan, Q.-Z. Inverse analysis of heat conduction problems with relatively long heat treatment. *Int. J. Heat Mass Transf.* **2017**, *105*, 401–410. [CrossRef]
52. Gopalkrishna, S.B.; Kulkarni, G.A.; Ryll, S.; Specht, E. Heat transfer analysis during quenching of moving metal plates using water jets from a mold. *Therm. Sci. Eng. Prog.* **2025**, *64*, 103792. [CrossRef]
53. Barrena-Rodríguez, M.d.J.; Acosta-González, F.A.; Téllez-Rosas, M.M. A Review of the Boiling Curve with Reference to Steel Quenching. *Metals* **2021**, *11*, 974. [CrossRef]
54. Gomez, C.F.; Nieuwenhuizen, R.; van der Geld, C.W.M.; Kuerten, H.G.M.; Bsibsi, M.; van Esch, B.P.M. Inaccuracies in the inverse heat conduction problem solution and their effect on the estimation of heat fluxes during quenching. *Int. J. Heat Mass Transf.* **2022**, *194*, 122953. [CrossRef]
55. Jahedi, M.; Berntsson, F.; Wren, J.; Moshfegh, B. Transient inverse heat conduction problem of quenching a hollow cylinder by one row of water jets. *Int. J. Heat Mass Transf.* **2018**, *117*, 748–756. [CrossRef]
56. Nieuwenhuizen, R. The Inverse Heat Conduction Problem in Quenching of a Hot Steel Plate. MSc. Thesis, Eindhoven University of Technology, Eindhoven, The Netherlands, 2019.
57. Sun, Y.; Jian, H.; Xiong, P.; Zhou, L. An Experimental Study of Boiling Heat Transfer and Quench Front Propagation Velocity During Quenching of a Cylinder Rod in Subcooled Water. *Energies* **2024**, *17*, 5236. [CrossRef]
58. Szénási, S.; Felde, I. Database for Research Projects to Solve the Inverse Heat Conduction Problem. *Data* **2019**, *4*, 90. [CrossRef]
59. Joachimiak, D.; Joachimiak, M.; Frąckowiak, A. Determination of boundary conditions from the solution of the inverse heat conduction problem in the gas nitriding process. *Energy* **2024**, *300*, 131497. [CrossRef]
60. Joachimiak, M. Analysis of Thermodynamic Parameter Variability in a Chamber of a Furnace for Thermo-Chemical Treatment. *Energies* **2021**, *14*, 2903. [CrossRef]

https://doi.org/10.3390/en19051291

1. Joachimiak, M.; Frąckowiak, A. Analysis of heat transfer in a cylinder being treated in a gas nitriding furnace using a new inverse problem solving method. *Energy* **2025**, *333*, 137318. [CrossRef]
2. Joachimiak, M.; Joachimiak, D.; Ciałkowski, M.; Małdziński, L.; Okoniewicz, P.; Ostrowska, K. Analysis of the heat transfer for processes of the cylinder heating in the heat-treating furnace on the basis of solving the inverse problem. *Int. J. Therm. Sci.* **2019**, *145*, 105985. [CrossRef]
3. Małdziński, L. *Thermodynamic, Kinetic and Technological Aspects of Nitrided Layer Production on Iron and Steel in Gas Nitriding Processes*; Poznań University of Technology Publishing House: Poznań, Polish, 2002. (In Polish)
4. Joachimiak, D.; Judt, W.; Joachimiak, M. Numerical analysis of the heating of a die for the extrusion of aluminium alloy profiles in terms of thermochemical treatment. *Arch. Thermodyn.* **2023**, *44*, 159–175. [CrossRef]
5. Jaremkiewicz, M. Analysis of the accuracy of the inverse marching method used to determine thermal stresses in cylindrical pressure components. *Arch. Thermodyn.* **2024**, *45*, 95–105. [CrossRef]
6. Frąckowiak, A.; Spura, D.; Gampe, U.; Ciałkowski, M. Determination of heat transfer coefficient in a T-shaped cavity by means of solving the inverse heat conduction problem. *Int. J. Numer. Methods Heat Fluid Flow* **2020**, *30*, 1725–1742. [CrossRef]
7. Frąckowiak, A.; Wróblewska, A.; Ciałkowski, M. Trefftz numerical functions for solving inverse heat conduction problems. *Int. J. Therm. Sci.* **2022**, *177*, 107566. [CrossRef]
8. Hecht, F. FreeFEM Documentation Release 4.13, 2026. Available online: https://doc.freefem.org/introduction/index.html (accessed on 24 February 2026).
9. Joachimiak, M.; Frąckowiak, A.; Joachimiak, D. Determination of the Temperature on the Inner Wall of the Piston Machine Body by Solving the Inverse Problem. *Heat Transf. Eng.* **2025**, *46*, 1982–1991. [CrossRef]
0. Frąckowiak, A.; Ciałkowski, M.; Wróblewska, A. Application of iterative algorithms for gas-turbine blades cooling optimization. *Int. J. Therm. Sci.* **2017**, *118*, 198–206. [CrossRef]
1. Frąckowiak, A.; Ciałkowski, M.; Wróblewska, A. Iterative algorithms for solving inverse problems of heat conduction in multiply connected domains. *Int. J. Heat Mass Transf.* **2012**, *55*, 744–751. [CrossRef]
2. Frąckowiak, A.; von Wolfersdorf, J.; Ciałkowski, M. An iterative algorithm for the stable solution of inverse heat conduction problems in multiply-connected domains. *Int. J. Therm. Sci.* **2015**, *96*, 268–276. [CrossRef]
3. Frąckowiak, A.; von Wolfersdorf, J.; Ciałkowski, M. Optimization of cooling of gas turbine blades with channels filled with porous material. *Int. J. Therm. Sci.* **2019**, *136*, 370–378. [CrossRef]
4. Frąckowiak, A.; von Wolfersdorf, J.; Ciałkowski, M. Solution of the inverse heat conduction problem described by the Poisson equation for a cooled gas turbine blade. *Int. J. Heat Mass Transf.* **2011**, *54*, 1236–1243. [CrossRef]
5. Ciałkowski, M.; Joachimiak, M.; Mierzwiczak, M.; Frąckowiak, A.; Olejnik, A.; Kozakiewicz, A. Cauchy Type Nonlinear Inverse Problem in a Two-Layer Cylindrical Area. *Heat Transf. Eng.* **2025**, *46*, 2005–2017. [CrossRef]
6. Ciałkowski, M.; Joachimiak, M.; Mierzwiczak, M.; Frąckowiak, A.; Olejnik, A.; Kozakiewicz, A. The analysis of the stability of the Cauchy problem in the cylindrical double-layer area. *Arch. Thermodyn.* **2023**, *44*, 563–579. [CrossRef]
7. Ciałkowski, M.; Olejnik, A.; Frąckowiak, A.; Lewandowska, N.; Mosiężny, J. Cauchy type inverse problem in a two-layer area in the blades of gas turbine. *Energy* **2020**, *212*, 118751. [CrossRef]
8. Ciałkowski, M.; Olejnik, A.; Joachimiak, M.; Grysa, K.; Frąckowiak, A. Cauchy type nonlinear inverse problem in a two-layer area. *Int. J. Numer. Methods Heat Fluid Flow* **2022**, *32*, 313–331. [CrossRef]
9. Allard, D.; Najafi, H. Genetic Algorithm as the Solution of Non-Linear Inverse Heat Conduction Problems: A Novel Sequential Approach. *J. Heat Mass Transf.* **2024**, *146*, 091404. [CrossRef]
0. Liu, W.; Guo, S.; Zhang, S. Temperature field modeling of the plate during hot rolling based on inverse heat conduction problem. *J. Phys. Conf. Ser.* **2019**, *1300*, 1. [CrossRef]
1. Xie, J.; Zhang, H.; Zhu, Z. Analysis of temperature field for rotary friction welding based on genetic-conjugate gradient hybrid inverse algorithm. *Appl. Therm. Eng.* **2025**, *279*, 127508. [CrossRef]
2. Allard, D.; Najafi, H. A systematic study of two machine learning-based approaches for solving non-linear inverse heat conduction problems in one-dimensional domains. *Int. Commun. Heat Mass Transf.* **2024**, *155*, 107494. [CrossRef]
3. Deng, S.; Hwang, Y. Applying neural networks to the solution of forward and inverse heat conduction problems. *Int. J. Heat Mass Transf.* **2006**, *49*, 4732–4750. [CrossRef]
4. Kowsary, F.; Mohammadzaheri, M.; Irano, S. Training based, moving digital filter method for real time heat flux function estimation. *Int. Commun. Heat Mass Transf.* **2006**, *33*, 1291–1298. [CrossRef]
5. Krejsa, J.; Woodbury, K.A.; Ratliff, J.D.; Raudensky, M. Assessment of strategies and potential for neural networks in the inverse heat conduction problem. *Inverse Probl. Eng.* **1999**, *7*, 197–213. [CrossRef]
6. Li, Y.; Liu, W.; Ren, J.; Cui, M.; Zhang, Z.; Nikolaitchik, M.A. An improved physics-informed neural network framework for solving transient nonlinear and inhomogeneous heat conduction problems. *Int. J. Heat Mass Transf.* **2025**, *252*, 127515. [CrossRef]


https://doi.org/10.3390/en19051291

87. Ma, X.; Qiu, L.; Zhang, B.; Wu, G.; Wang, F. Adaptive fractional physics-informed neural networks for solving forward and inverse problems of anomalous heat conduction in functionally graded materials. *Int. J. Heat Mass Transf.* **2025**, *236*, 126393. [CrossRef]
88. Madir, B.-E.; Luddens, F.; Lothodé, C.; Danaila, I. Physics Informed Neural Networks for heat conduction with phase change. *Int. J. Heat Mass Transf.* **2025**, *252*, 127430. [CrossRef]
89. Qian, W.; Hui, X.; Wang, B.; Zhang, Z.; Lin, Y.; Yang, S. Physics-informed neural network for inverse heat conduction problem. *Heat Transf. Res.* **2023**, *54*, 65–76. [CrossRef]
90. Sablani, S.S.; Kacimov, A.; Perret, J.; Mujumdar, A.S.; Campo, A. Non-iterative estimation of heat transfer coefficients using artificial neural network models. *Int. J. Heat Mass Transf.* **2005**, *48*, 665–679. [CrossRef]
91. Shang, Y.; Ban, H.; Liu, D. Simultaneous identification of boundary heat flux and thermal conductivity in inverse heat conduction problems using physics-informed neural networks. *Therm. Sci. Eng. Prog.* **2025**, *65*, 103905. [CrossRef]
92. Sivalingam, S.M.; Govindaraj, V.; Dubey, S. An improved physics informed neural network with theory of functional connections for fractional differential equations. *Eng. Anal. Bound. Elem.* **2025**, *178*, 106281. [CrossRef]
93. Aggarwal, C.C. *Neural Networks and Deep Learning*; Springer Nature Switzerland: Cham, Switzerland, 2023.
94. Ding, S.; Tian, Y.; Qin, L.; Ma, H.; Yang, R. Physics-informed hierarchical neural operator for solving inverse problem of unsteady heat conduction. *Int. J. Heat Mass Transf.* **2026**, *258*, 128335. [CrossRef]
95. Islam, M.S.; Dutta, P. Machine learning assisted inverse heat transfer problem to find heat flux in ablative materials. *Mater. Today Commun.* **2025**, *45*, 112337. [CrossRef]
96. Molchanov, A.M.; Yanyshev, D.S.; Ezhov, A.D.; Bykov, L.V. Solution of the inverse problem of heat conduction using neural networks. *Heat Transf. Res.* **2025**, *56*, 1–12. [CrossRef]
97. Puttock-Brown, M.R.; Mahesh Bindhu, G.K.; Ashby, C.E. A Physics Informed Neural Network for Solving the Inverse Heat Transfer Problem in Gas Turbine Rotating Cavities. *J. Turbomach.* **2025**, *147*, 071010. [CrossRef]
98. Shang, Y.; Tan, C.; Tan, C.; Yu, X.; Hu, X.; Jiang, H.; Ma, W.; Liu, D. Using neural networks: A guidance with application in inverse heat conduction problem. *Eur. J. Phys.* **2025**, *46*, 2. [CrossRef]
99. Xu, J.; Wei, H.; Bao, H. Physics-informed neural networks for studying heat transfer in porous media. *Int. J. Heat Mass Transf.* **2023**, *217*, 124671. [CrossRef]
100. Zobeiry, N.; Humfeld, K.D. A physics-informed machine learning approach for solving heat transfer equation in advanced manufacturing and engineering applications. *Eng. Appl. Artif. Intell.* **2021**, *101*, 104232. [CrossRef]
101. Chen, Y.; Chen, Q.; Ma, H.; Chen, S.; Fei, Q. Transfer machine learning framework for efficient full-field temperature response reconstruction of thermal protection structures with limited measurement data. *Int. J. Heat Mass Transf.* **2025**, *242*, 126785. [CrossRef]
102. Yahya, M.; Saghir, M.Z. Machine Learning and Physics-Informed Neural Networks for Thermal Behavior Prediction in Porous TPMS Metals. *Fluids* **2026**, *11*, 29. [CrossRef]




https://doi.org/10.3390/en19051291

energies

MDPI

*Perspective*

# The Port-Hamiltonian Formulation of Thermodynamics—A New Perspective

Janusz Badur * and Piotr Józef Ziółkowski

Energy Conversion Department, Institute of Fluid Flow Machinery, Polish Academy of Sciences, 80-231 Gdańsk, Poland; pjziolkowski@imp.gda.pl
* Correspondence: jb@imp.gda.pl

**Abstract**

This paper proposes a change in the traditional epistemological paradigm and a look at classical thermodynamics from the point of view of control theory, with the aim of discovering energy state variables. The paper proposes a transition from "causality" to "purposefulness" in nature, which is called port-Hamiltonian thermodynamics. It is unclear whether classical thermodynamics can be incorporated into the formalism of port-Hamiltonian field theory, and it is likely that thermodynamics will need to be expanded or even completely reformulated. The main goal is to satisfy the First and Second Laws of Thermodynamics a priori.

**Keywords:** continuum thermodynamics; Hamiltonian field theory; port-Hamiltonian thermodynamics; theory of control; Poisson brackets; energy dissipation; Rayleighian

## 1. Port-Hamiltonian Formalism and Its Significance in Engineering

Based on Hellenistic science, Western civilization has achieved its goals and exhausted its potential for development. The tasks of humanity related to building, designing, and simply imitating natural phenomena have already been completed. We are now entering a period called the "civilization of sustainable development", so the new tasks are optimization, maintenance, control, and artificial intelligence. The exhaustion of Hellenistic civilization was accompanied by the exhaustion of its main driving force—what we call "Aristotle's thermodynamics". Thermodynamics, as a thoroughly practical science, covered all levels: from quantum thermodynamics, field theory thermodynamics, statistical thermodynamics, phenomenological non-equilibrium thermodynamics, through classical thermodynamics of technical devices, to the thermodynamics of the atmosphere, the sun, the solar system, and the cosmos.

Due to its fundamental motivations, thermodynamics has thus far been a technical art of describing phenomena in which, as Sadi Carnot put it, we are able to obtain mechanical power from a thermal source of power. Thermodynamics thus had anthropological goals, namely, to replace expensive workers and soldiers with cheap mechanical power generated from thermal energy sources. Today, human labor has been completely eliminated, and thermodynamics has new goals, such as maintenance, optimization, control, and AI thermodynamics. These new goals have been taken up by a science called "port-Hamiltonian thermodynamics" [1–3]. Its pioneers are B. Maschke and A.J. van der Schaft [4,5], and the following disciples from their school: R. Ortega, G. Escobar [6], K. Krhac [7], and D. Jeltsema [8]. Among the veterans are also Duindam, Macchelli, Stramigioli nad Bruyncks [9], R. Rashiad [10,11], F. Califano [12], Brugnoli [13], Jäschke [14], Cardoso-Ribeiro [15], R.

Altmann and P. Schulze [16], H. Ramirez and Y. Le Gorrec [17,18], L. Mora [19], and others [1,17] (see reviews).

The importance of the port-Hamiltonian formalism (pH) and its powerful advantages over the Lagrange formalism are evidenced by the extraordinary number of works that have appeared over the last three years (see: van der Schaft [2]). We will mention them in the text, keeping in mind the historical outline of the lecture.

Jean-Francois Pommaret, an eminent theorist and specialist in control theory, elevates the role of control theory above the contemporary foundations of gauge theory and gravitational theory, arguing that when the formalism (pH) is applied to the description of nature, "nothing remains of the mathematical foundations of both general relativity and gauge theory" [20]. He is right in this, because Hamilton's formalism is, by its very nature, a deeper and more comprehensive way of studying the laws of conservation and the equations of evolution of natural phenomena. Pioneers point to the need to rebuild the foundations of the theory of electromagnetism and gravity (Pommaret [21]), rebuild quantum physics (Alicki and Koslof [22]), and rebuild thermodynamics (Beris and Edwards [23], Grmela [3,24]; Sípka et al. [25]; M. Sýkora et al. [26]; Grmela and Pavelka [3], Pavelka, Klika and Grmela [27]). Therefore, if pH thermodynamics is to be based on solid foundations, a reconstruction and modernization of the foundations of both general relativity and gauge theory is needed (Pommaret [21,28], Badur and Dudda [29], Yahalom [30,31]).

Why are both practitioners and theorists involved in the control of natural and technical systems so eager to base their reasoning, tools, and methods of implementation on port-Hamiltonian formalism? This is because this formalism uses the concept of "energy" as a *lingua franca*. This allows two subsystems defined in different physical domains to be coupled by a connection that conserves energy flows. In this sense, within pH, two types of energy-preserving connections are commonly used: a port connection preserving stored energy flows through the Hamiltonian and an energy connection preserving dissipated energy flows through the Rayleighian [23].

The reconstruction of thermodynamics in the new formalism does not have to follow the guidelines set out by the practical principles of thermodynamics. This review investigates the manifestations of thermodynamic laws in the context of thermodynamic pH.

- The Zeroth Law of Thermodynamics establishes empirical methods for determining state parameters and their identities.
- The First Law of Thermodynamics prohibits the creation of energy from nothing (ex nihilo) and mandates the conversion of field energy in the presence of matter.
- The Second Law is a requirement for the perpetual conversion of energy and a prohibition on the reversibility of dynamics: the breaking of time reversal symmetry.
- The Third Law is a requirement for the existence of absolute values of extensive state parameters and a prohibition on the existence of vacuum energy (matter without fields).

The aim of this publication is to show that conventional thermodynamics cannot be directly applied to pH thermodynamics, but must undergo expansion, modernization, and paradigm shifts. It must be based on new experiments and laboratory research.

## 2. Building a Set of Evolutionary Equations Using the Poisson Bracket Apparatus

The motivation for research based on the Poisson brackets came not only from Hamiltonian mechanics, but more generally, from the needs of engineering. Western civilization is based on the undisputed role of mathematical models in recognizing and harnessing natural phenomena. Only harnessed phenomena we can use in technology. This means that only those devices and human products that correctly utilize natural phenomena, which

https://doi.org/10.3390/en19020324

we can control and make the correct predictions. Therefore, most researchers agree that the objects of our civilization possess "harnessed" natural phenomena, described by a set of equations called "governing equations", which are controlled by appropriate boundary and initial conditions. Now is the time for engineers to begin to control all of this.

Researchers who study the construction of the Poisson brackets, understood as a general research tool, are approaching the aforementioned describing equations as a "poor junkyard", viewing the system of equations describing them as a jumble of equations of motion (evolution), conservation laws, and laws of thermodynamics. In other words, the governing equations are not valid for the control equations. Sometimes, it is obvious that the governing equations do not satisfy the conservation laws or one of the laws of thermodynamics. We then say that the system of conservation equations violates the laws, and the simulation of reality performed with this model deviates from the experimental data. These equations cannot be used to achieve control or achieve true goals [32].

Given this state of affairs, researchers in engineering and applied physics as well as chemistry have developed a need to build a research tool that provides more efficient and comprehensive procedures than those derived from the traditional Lagrangean formalism. Reasoning based on the classical Lagrangean apparatus leads to poorer understanding than the more extensive procedures such as GRIMEC, Poisson brackets, or M-Matrix (Grmela and Öttinger [33], Zafferi et al. [34], Grmela [35–37]).

The aim of the Hamiltonian approach, based on the Poisson brackets, is to seek the complete physics of natural phenomena—which practically means that the number of primary variables is many times greater, perhaps even complete, compared to the Lagrangean approach. Furthermore, the structure of the model resulting from the Hamiltonian approach is more flexible, more physical, and more effective in simulations. One of the goals of the Hamiltonian approach is to reinterpret known and worn-out field-theoretic models, for example, extending the fluid model previously known as the "Navier–Stokes–Fourier equation". It is therefore important to understand that the Hamiltonian approach is not simply an approach that re-writes the Navier–Stokes–Fourier equations, but rather an approach that constructs a new physical description for this type of fluid, with new constants, new measurement methods, and new numerical methods. Referring to the well-known and understood name CFD (Computational Fluid Dynamics), which is the research paradigm of the 21st century, it must be said that we are looking for a new science that can be called Hamiltonian-CDF.

The main task of pH thermodynamics can be summarized as follows. The basic structure of equations describing the real world in the Hamiltonian approach using Poisson brackets has the general form, proposed by Grmela [35,36], Morrison [38]:

$$\dot{x} = \{x, H\} + [x, H]. \tag{1}$$

The most important object here is the object named by the letter $x$. This is a state variable that is the fundamental unknown of the control system. It consists of $n$ components with different physical dimensions: $x = (x_1, x_2, x_3, \ldots, x_n)$. Finding and defining these variables for a given physical phenomenon, such as the flow of a Navier–Stokes fluid or a Cosserat fluid, is the most difficult part of control theory—we cannot rely on the simplified knowledge provided by a Lagrangean-type approach. The notation $x$ is not accidental and refers to control theory. Equation (1) contains two brackets, $\{,\}$ and $[,]$, and a Hamiltonian $H$ defining the stored energy of the system.

The equations Equation (1) are evolution equations and fortunately, they can be used in this form appropriate to the control systems. Control practice and theory recognize various variables, but energy variables have been denoted by the letter x for years. Control

https://doi.org/10.3390/en19020324

is most often used in the linear version as (Pommaret [28], Gawroński [32], Ogata [39], Mieloszyk [40]):

$$\dot{x} = Ax + Bu,\ y = Cx + Du, \tag{2}$$

or in the case of field variables

$$\dot{x} = A_{zz}\frac{\partial^2 x}{\partial^2 z} + A_z\frac{\partial x}{\partial z} + Ax + Bu;\ y = Cx + Du, \tag{3}$$

where $A,\ A_z,\ A_{zz}$ are state matrices, $z$ is a spatial variable, $B$ is a control matrix, $u$ and $y$ are the control and the effect—a control output, $C,\ D$ are the output and transmissibility matrices, respectively. Above we have an example of control with a proportional term $y(t) = Cx(t)$ where the output signal is defined in proportion to the state signal. More complex controls are based on a derivative term: $y(t) = k\dot{x}(t)$ or on an integral term: $y(t) = k\int_0^t x(\tau)d\tau$ [40].

Fundamental to the theory of energy control is the system graph (Ogata [39]). Learning how to construct a system graph should become a fundamental element of pH thermodynamics. For example, for the system of Equation (2) consisting of state evolution equations and output equations, the graph is shown in Figure 1, this method of presentation is known from classical monographs [32,39,40].

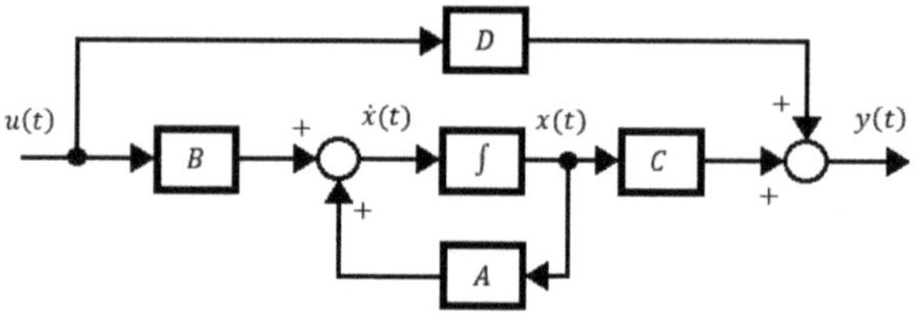


**Figure 1.** Graphical representation of the control system given by Equation (2).

Unfortunately, in control theory, the state variable $x$ is arbitrary defined—frequently uncorrelated with the energy state of the system. However, in the special case where we want to correlate the evolution of the system $\dot{x} = \frac{\partial}{\partial t}x$ with the energy carried by the Hamiltonian, we will call such variables energy variables. This allows us to derive part of the control Equation (2) from the evolution equations placed on Poisson brackets (Equation (1)).

However, let us emphasize that searching for the Hamiltonian is usually a hopeless task and even for a linear system $\dot{x} = Ax + Bu$ is difficult to find. However, if we introduce a function $w$ that is a co-state, then $H(w, x, u) = w^T(Ax + Bu)$ can be a Hamiltonian if we add some adjoint equations to the system of equations [4]. For a system of linear control equations, the most important is the transfer function, which requires the resolvent, i.e., the inverse matrix: $R = (sI - A)^{-1}$. For a system not exceeding $3 \times 3$, it can be found analytically. Modern control theory has also adopted a number of artificial intelligence methods (neural networks, fuzzy logic, genetic algorithms, expert systems). Numerical methods, computational environments such as Matlab (including its Simulink toolkit) or Mathcad, and programming environments such as LabVIEW are also used.

## 3. The Idea of the Hamilton Formalism

In 1834, William Rowan Hamilton showed that Newton's equations: $m_i\dot{q}_i = f_i$ for a set of particles in a conservative force field $f_i = \text{grad } V_i$ with "potential energy"

https://doi.org/10.3390/en19020324

$V_i$ $(i = 1, 2, 3, \ldots, n)$ can be derived from a single function, which he called the "characteristic function",

$$H(p,q,t) = \sum_i \left[ \frac{p_i^2}{2m_i} + V_i(q_1, q_2, \ldots, q_i, \ldots q_n) \right]. \tag{4}$$

Here $q_i$ is the position of the i-th particle, whose mass is $m_i$ and $p_i$ is its canonical momentum $p_i = m_i q_i$. The positions and momenta form the state vector of unknowns $x = (q, p)$. The equations of motion (evolution) are obtained from the formula: $\dot{x} = \mathcal{J}\delta_x H = \mathcal{J}\frac{\delta H}{\delta x}$, which in turn can be transformed to the Newtonian form. The solutions $\dot{x} = \mathcal{J}\delta_x H$ with respect to time form trajectories [28].

At first glance, Hamilton's formula seems to provide only a convenient formulation of Newton's system—a convenience perhaps most obvious, since the scalar function $H(p,q,t)$ encodes all information not in n second-order equations but in 2n first-order equations. Only after the work of Grmela [35], Grmela and Öttinger [33], and Mielke [41] did it become clear that the Hamiltonian formalism offers much more than just this simplification of calculations—it is, in fact, a proper tool for the theory of evolution, control, and optimization. Nature is not inherently optimal and constantly wanders; only humans can teach it how to achieve its goals. The challenges of pH thermodynamics are countless. Never before in the history of the world has man faced such a difficult problem.

Indeed, if we allow for more general functions $H(p,q,t)$ and a more general relation between canonical momenta and position velocities, then practically all previously known models of classical physics fall under the Hamiltonian formalism. Moreover, quantum field theory is formally obtained from Hamiltonian fields by replacing the canonical momentum in the Hamiltonian with a differential operator (Pommaret [42]).

The Hamiltonian formalism, however, imposes strong constraints on the physics of the modeled phenomena. Simply put, when $H(p,q,t)$ is independent of time (it is autonomous), its value is constant along the trajectories: the energy $E = H(q,p)$ is constant. This also implies the fulfillment of the First Law of Thermodynamics, which states that the energy of a system cannot arise from nothing (*ex nihilo*). Similarly, if the Hamiltonian is independent of one of the position variables, e.g., $q_a$ (this variable is negligible), this means that the corresponding canonical momentum $p_a$ is constant—it is an invariant of the evolutionary process. This provides a simple explanation of the relationship between symmetries (e.g., rotational symmetry) and invariants (e.g., angular momentum) and leads to Noether's theorems [43].

One of the stronger constraints imposed by the Hamiltonian structure concerns stability: any trajectory cannot be asymptotically stable in the Hamiltonian system. There is also an even deeper feature: for each equilibrium eigenvalue, there is a corresponding opposite (negative) eigenvalue [28,40].

Another important implication for physics is that knowledge of the invariants of motion (evolution $\dot{x} = \mathcal{J}\delta_x H$) is sufficient to fully characterize the solution of $2n$ equations for a system with $n$ degrees of freedom when the Hamiltonian is integrable. This follows from Liouville's integrability theorem [28]. This strong structural stability of Hamiltonian dynamics was unexpected even in the mid-20th century, when physicists began the first computer simulations of control systems (see: [20,28,31]).

## 4. The Essence of the Poisson Brackets

The Poisson's brackets apparatus gained importance and became "fashionable" only after Dirac used it in quantum mechanics in 1930. Previously, it was used in Hamiltonian mechanics [26]. The starting point is the mechanics of a swarm of points carrying mass $m_i$, $i = 1, 2, 3, \ldots, n$ with position $q_i = q_i(t)$, in which spatial relations play no role and only the changes in positions over time are important: $\partial_t q_i \equiv \dot{q}_i$.

https://doi.org/10.3390/en19020324

It involves reformulating (but with equivalent physics) the Lagrangean approach by introducing new $n$ variables $p_i = \frac{\partial L}{\partial \dot{q}_i}$ called generalized momenta, and transforming the Lagrangean of the action into the Hamiltonian: $H(q, p, t) = \sum p_i \dot{q}_i - L(q, \dot{q}, t)$—this is called the Legendre transformation and allows for the elimination of time derivatives of the fundamental variables and their replacement with generalized momenta. By this operation, we reduce the order of differentiation at the expense of increasing the number of unknowns. In this case, it is a doubling. Instead of the Lagrange-Euler equations of motion, we now have Hamiltonian equations of motion (evolution):

$$\dot{p}_i = -\frac{\partial H}{\partial q_i} + Q \text{ and } \dot{q}_i = \frac{\partial H}{\partial p_i}. \tag{5}$$

The dynamics equations become the evolution equations of a system based on energy variables. They have an interesting structure, called the Dirac structure, and can be expressed by formal operations contained within the Poisson bracket. For example, if we have two functions $f(p, q)$ and $g(p, q)$ in the state space, the Poisson bracket is defined as a commutator [27,35]:

$$\{f, g\} = \frac{\partial f}{\partial q_i}\frac{\partial g}{\partial p_i} - \frac{\partial f}{\partial p_i}\frac{\partial g}{\partial q_i}, \tag{6}$$

in which the indices $i = 1, 2, 3, \ldots, n$ should be traversed. This pattern was used by Dirac to build a quantum commutator on matrices $A, B$ as $[A, B] = AB - BA$. Now, taking as one of the functions the state vector: $x = (q, p)$ and with the other the Hamiltonian $H$, we have the fundamental component of the evolution equation:

$$\dot{x} = \frac{\partial}{\partial t}\begin{pmatrix} q \\ p \end{pmatrix} = \{x, H\} = \frac{\partial x}{\partial q}\frac{\partial H}{\partial p} - \frac{\partial x}{\partial p}\frac{\partial H}{\partial q} = \begin{pmatrix} \frac{\partial H}{\partial p} \\ -\frac{\partial H}{\partial q} \end{pmatrix} = \begin{pmatrix} 0 & 1 \\ -1 & 0 \end{pmatrix}\begin{pmatrix} \partial_q H \\ \partial_p H \end{pmatrix} \equiv \mathcal{J}\partial_x H. \tag{7}$$

We introduced above a dimensionless, skew-symmetric Dirac structure matrix $\mathcal{J} = -\mathcal{J}^T$ and a shortened notation for the generalized derivative of the Hamiltonian $\delta_x H \equiv \partial_x H - \mathrm{div}(\partial_{\nabla x} H) + \cdots$. It is also called "variational derivative" or "functional derivative" in the literature. Due to the physical (thermodynamic) interpretation in the literature over the last 20 years (see review [2,17]), the state variables $x$ are also called energy variables. Now, if the Hamiltonian has a special quadratic form $H = \frac{1}{2}x^T K x$ (for instance, in a case of linear spring, $K = K^T$), then the above evolution system takes the form:

$$\frac{\partial x}{\partial t} = \{x, H\} = \mathcal{J}\frac{\delta H}{\delta x} = \mathcal{J}\delta_x\left(\frac{1}{2}x^T K x\right) = \mathcal{J}Kx \equiv Ax. \tag{8}$$

This control matrix $A = \mathcal{J}K$, known from linear control theory, appears as an object in the fundamental control equation (Equation (2)). This means that only a portion of the control theory tasks—which, as we know, applies to all natural phenomena and phenomena used in technology—is subject to a Hamiltonian approach. We will call such phenomena "Hamiltonian systems". This approach to natural variability has much in common with the Aristotelian thermodynamics, based on the Hamiltonian understanding of the concept of "energy". Therefore, when constructing thermodynamics "a la Hamilton", we call the set of variables $x$ energy variables, which have nothing in common with the state variables defined by the traditional thermodynamics of Clausius, Thomson, and Boltzmann.

Let us recall that when constructing Poisson brackets for field phenomena—that is, when constructing the still very poorly understood Hamiltonian field theory—we must remember that the tool provided by the Poisson bracket is a very blunt one, which often serves to limit rather than expand our knowledge of reality (Sieniutycz and Berris [44]). Therefore, when taking Poisson brackets as a tool for understanding the world, we

https://doi.org/10.3390/en19020324

must remember that they satisfy, for example, the following constraints: asymmetry: $\{f,g\} = -\{g,f\}$; the linearity necessary for the proper evaluation of the phenomenon: $\{\alpha f + \beta g, h\} = \alpha\{f,h\} + \beta\{g,h\}$ for arbitrary real numbers $\alpha, \beta$; and the key identity imposed on nature by Jacobi: $\{f,\{g,h\}\} + \{g,\{h,f\}\} + \{h,\{f,g\}\} = 0$. In practice, to prove that our devised Poisson bracket satisfies this theorem, we need a large piece of paper and a "hot cup of coffee". Then we expand out all 24 terms and watch them cancel one by one [31].

## 5. Origin of the Control Port

This means that the Poisson bracket can be used to write the control system, as follows:

$$\dot{x} = \mathcal{J}\delta_x H(x) + G(x)u(t), \tag{9}$$

when, physically speaking, it represents a skew-symmetric matrix structure allowing energy flows between different regions of the system. If $\mathcal{J}$ is constant, usually $\mathcal{J}(x)$ is a differential operator consisting of skew-symmetric matrices, as follows: $\mathcal{J}(x) = J_0 + a_1 J_1 \partial/\partial\overleftrightarrow{x} + a_2 J_2 \partial/\partial\vec{x}(\partial/\partial\vec{x}) + a_3 J_3 \partial^3_{\vec{x}\vec{x}\vec{x}} + \cdots$; then, it satisfies the Jacobi identity, because for two functions, $f, g$, it is expressed in the form $\{f,g\} = \frac{\partial f}{\partial x}^T \mathcal{J}(x) \frac{\partial g}{\partial x}$, which leads to the following control equations: $\dot{x} = \{x, H\} + G(x)u(t)$.

A further interpretation of the Hamiltonian formalism in the language of control theory is obtained when we exclude from the total degrees of freedom some of the already known degrees $q_a$ with a given momentum $Q$, and we give them a different meaning. When we ask what the change in energy system is, we obtain the following, using Equation (7):

$$\dot{H} = \frac{\partial H}{\partial q}^T \cdot \dot{q} + \frac{\partial H}{\partial p}^T \cdot \dot{p} = -(\dot{p} - Q)^T \cdot \dot{q} + (\dot{q})^T \cdot \dot{p} = Q^T \cdot \dot{q}. \tag{10}$$

Now, by cutting out the degrees of freedom pertaining to the given momentum $Q$ and introducing the notation $u \equiv Q$ and $y = \dot{q}$ for these special degrees of freedom, we have a rearranged system of equations:

$$\dot{q} = \frac{\partial H}{\partial p}, \; \dot{p} = -\frac{\partial H}{\partial q} + Gu, \quad y = G^T \frac{\partial H}{\partial p}, \tag{11}$$

where the output equation appears, which is part of the initial equation $\dot{q}_a = \frac{\partial H}{\partial p_a}$, and the matrix $G$ is generated by the "potential" part of the Hamiltonian. Now calling the "extended" degrees of freedom in the language of control theory, respectively, the control $u$ and the output $y$, we can write the change in the Hamiltonian (Equation (10)) as $\dot{H} = Q^T \cdot \dot{q} = u^T \cdot y$ and express it as follows: the change in the system's energy occurs as a result of the supplied (from the outside) energy $u^T \cdot y$. This reasoning, transforming the control system (Equation (2)) into a problem of the evolution of Hamiltonian field theory, assumes that if the dimensions of the control and output are the same, then the matrix $C = B^T$ in Equation (2) also holds.

## 6. Transition to the Continuum

The above brackets are constructed for "swarm material points" and therefore do not have a field-like character dependent on spatial variables. However, Poisson brackets for the theory of elasticity, Navier–Stokes fluids, and other field models have a differential structure and behave analogously to the "Lie derivative" first used in differential geometry by Wł. Ślebodziński [43]. It is obvious that for any field, relations of the type $\{q_i, q_i\} = 0$; $\{p_i, p_j\} = 0$; $\{q_i, p_j\} = \delta_{ij}$ will hold. For a free function—this is most often an invariant of Hamiltonian motion—$f(q,p,t)$, the complete time change, denoted by the letter $dt$, is equal to [27]:

$$\frac{df}{dt} = \frac{\partial f}{\partial p}\dot{p} + \frac{\partial f}{\partial q}\dot{q} + \frac{\partial f}{\partial t} = -\frac{\partial f}{\partial p}\frac{\partial H}{\partial q} + \frac{\partial f}{\partial q}\frac{\partial H}{\partial p} + \frac{\partial f}{\partial t} = \{f,H\} + \frac{\partial f}{\partial t}. \quad (12)$$

When the function $f$ does not depend directly on time $t$, we have $\frac{df}{dt} = \{f,H\}$. We emphasize this notation of the derivative here, as numerous other notations appear in the literature, leading to unnecessary doubts. Let us emphasize the convention of the "energy control theory" notation, in which $\dot{x} = \frac{\partial}{\partial t}x$. In Hamiltonian thermodynamics, therefore, there are numerous energy invariants of motion. If $\dot{I} = 0$ (invariant), then the condition is that $\{I,H\} = 0$. Conversely, some quantity I is called an energy invariant of motion when the condition $\{I,H\} = 0$ is satisfied [27].

The invariant of motion does not have to be a scalar; it can be any quantity used in Hamiltonian thermodynamics—e.g., if several positions $q_j$ do not appear in the Hamiltonian, then the corresponding object called linear momentum $p_j$ is an invariant of motion $\{p_j,H\} = 0$. This is the case of cyclic variables introduced into Helmholtzian thermodynamics [43]. The most famous invariant in classical Hamiltonian thermodynamics is the "adiabatic invariant": $I = \frac{1}{2\pi}\oint pdq$ and the transition from classical to quantum Poisson brackets $\{,\}_{classical} \rightarrow -\frac{i}{\hbar}[,]_{quantum}$ allows us to find all fundamental laws of Hamiltonian quantum thermodynamics "for free" [20]:

$$\dot{f} = \{f,H\} \qquad \rightarrow \qquad i\hbar\dot{f} = [f,H]. \quad (13)$$

We mentioned above the "problems" associated with the enormous number of transformations when transforming classical control theory into energy-based Hamiltonian control theory, because in practice, the transition from the classical linear control theory formulae $\dot{x} = Ax$ to the energy (thermodynamic) Hamiltonian approach $\frac{\partial x}{\partial t} = \mathcal{J}\frac{\delta H}{\delta x}$ is extremely difficult and usually requires various omissions and "cuts" in the control system. This is because not every control system has a Dirac structure. An even greater limitation of the structure is the Stokes–Dirac structure, which allows the introduction of special ports into the system. Then, the control matrix B in Equation (2) takes on a special (energy) form.

## 7. The Nambu Brackets

Another tool for studying the evolution of natural and technical phenomena is the Nambu bracket [45,46], introduced in 1973. Knowing that natural and technical control systems, $\dot{x} = Ax + Bu$, $y = Cx + Du$, contain larger structures than the Dirac and Stokes–Dirac structures, Nambu proposes formulating the laws of evolution of field and matter physics in the form of brackets based not on one but on two Hamiltonians. Instead of the canonical pair of variables representing positions and momenta $(q_i, p_i)$, Nambu proposes triplets $x = (q_i, p_i, r_i)$ where $i = 1,2,3\ldots,n$. The system has dimension $3n$. It is impossible to determine what these additional state variables are before determining the state of the control system. The Nambu bracket takes the form of a commutator consisting of six terms [47]:

$$\{f,g,h\} = \frac{\partial f}{\partial q}\frac{\partial g}{\partial p}\frac{\partial h}{\partial r} - \frac{\partial f}{\partial q}\frac{\partial g}{\partial r}\frac{\partial h}{\partial p} + \frac{\partial f}{\partial r}\frac{\partial g}{\partial q}\frac{\partial h}{\partial p} - \frac{\partial f}{\partial r}\frac{\partial g}{\partial p}\frac{\partial h}{\partial q} + \frac{\partial f}{\partial p}\frac{\partial g}{\partial r}\frac{\partial h}{\partial q} - \frac{\partial f}{\partial p}\frac{\partial g}{\partial q}\frac{\partial h}{\partial r}. \quad (14)$$

https://doi.org/10.3390/en19020324

This satisfies the asymmetry condition $\{f,g,h\} = -\{g,f,h\} = \{g,h,f\}$, and the Leibniz–Jacobi condition. If we define two energy functional $H$, $G$ (e.g., energy and entropy), then any state function evolves according to the following law:

$$\dot{x} = \{x, H, G\}. \tag{15}$$

Without a doubt, a broader class of control systems can be described using Nambu brackets, and new quantum mechanics can be easily derived. Hence, the Nambu brackets formalism with two Hamiltonians is not a historical curiosity, as evidenced by the works of Nevir and Blender [45], Blender and Badin [46], and Sato [47].

## 8. The Rayleighian Formalism

If the Hamiltonian includes the energetic effects of the dynamics of inertial (completely reversible) phenomena, then the energetic effects of the dynamics of non-inertial (coercive, dissipation-dominated, and completely irreversible) phenomena are included in the Rayleighian, denoted by the letter $\mathcal{R}$. Typically, in natural phenomena, coercion is mixed with inertia, and reversibility with irreversibility; hence, the two functional $\mathcal{H}$ and $\mathcal{R}$ appear together in a single balanced formalism. The undoubted advantage of this formalism is the development of more physically grounded, structure-preserving numerical methods that are less sensitive to energy dissipation phenomena [48]. A possible connection of the Rayleigh functional to pH thermodynamics is presented in Section 15.

## 9. The GENERIC Formalism

Other Hamiltonian formulations of the thermodynamics of irreversible processes require both the conservation of energy and the irreversible entropy creation. Grmela and Öttinger [33] proposed reformulating a conventional thermodynamic system along the lines of a mechanical system in Hamiltonian representation. In this line, the "General Equation for the Non-Equilibrium Reversible Irreversible Coupling", or simply GENERIC, proposes to "complete" the Hamiltonian (reversible) part of a system with a purely dissipative (irreversible) part.

The GENERIC system is expressed with respect to two generating functions, total energy and total entropy, and with respect to two geometric structures, a Poisson bracket and a pseudo-Riemannian metric. This representation is capable to express simultaneously the conservation of energy and the irreversible creation of entropy as a geometric property of the system [34,49].

The reversible part of the evolution is described in terms of the Poisson bracket and the Hamiltonian, which is a function of the total energy of the system. The dissipative part is defined as pseudo-energy and generated by the total entropy as a potential function. The GENERIC formalism proposes the time evolution of the state variables $x$ in the following form [49]:

$$\frac{\partial}{\partial t}x = J(x)\frac{\partial H(x)}{\partial x} + M(x)\frac{\partial S(x)}{\partial x}, \tag{16}$$

where $H, S$ are the total energies and entropy of the control system and the matrices $J, M$ are defined by the Poisson bracket and the "pseudo-Riemannian metric". The matrix $J(x)$ is skew-symmetric and the matrix $M$ is symmetric [27]. They satisfy the following conditions of invariability:

$$J(x)\frac{\partial S(x)}{\partial x} = 0 \text{ and } M(x)\frac{\partial H}{\partial x} = 0. \tag{17}$$

https://doi.org/10.3390/en19020324

If, in analogy to the Poisson bracket, we define the Grmela bracket as $[f,g]_M = \frac{\partial f(x)}{\partial x}^T M(x) \frac{\partial g(x)}{\partial x}$, then using two generators, $H, S$, we can represent the GENERIC formalism using two brackets [27,34,37,41,49]:

$$\frac{\partial}{\partial t} x = \{x, H\}_J + [x, S]_M. \tag{18}$$

This means that the two rules of evolution of functionals are also satisfied:

$$\dot{H} = \{H, H\}_J + [H, S]_M = 0; \ \dot{S} = \{S, H\}_J + [S, S]_M = \sigma_{int} \geq 0, \tag{19}$$

where $\sigma_{int}$ is the irreversible entropy production [35]). This means that the GENERIC formalism satisfies the first and second laws of thermodynamics with an excess.

**Remark 1.** *It must be admitted that the form (Equation (18)) is very attractive for control theory in the Hamilton energy approach. It should be used, but only as a model. Therefore, for practical reasons, it is completely impossible to use the dissipative bracket $[x, S]_M$ based on the entropy functional, because the Port-Hamiltonian approach is based on energy and its modes of dissipation and destruction. Moreover, the introduction of the bracket $[x, S]_M$ limits the class of real systems to a narrow group of problems. Here, one must acknowledge the validity of the control theory approach, which views the problem of energy dissipation and the description of irreversible phenomena in a more general way, devoid of unnecessary assumptions of non-equilibrium thermodynamics acting as a priori imposed simplifications of phenomena. Control theory, arising from the paradigm of the nature purposefulness, has an epistemological advantage over non-equilibrium phenomenological thermodynamics arising from the paradigm of the nature causality.*

## 10. Port-Hamiltonian–Energy Variables

In this article, after introducing port-Hamiltonian systems, the focus is on "connection control" to stabilize the set point of nonlinear physical systems. Most of this theory is well established, see, e.g., [1,7], but new solutions using "energy ports" instead of "power ports" are also considered.

Port-Hamiltonian systems theory emphasizes open systems and their control using additional feedback loops and connections to control systems. This leads to modeling a complex system as an energy-efficient combination of elements responsible for energy storage, dissipation, and "control". Using energy and power as a lingua franca between different physical domains (gravity, electrodynamics, optics, mechanics, electricity, chemistry, and thermodynamics), port-Hamiltonian theory provides a systematic framework for modeling multiphysical systems; see, e.g., [8,9,17]. Moreover, it is not limited to systems with concentrated parameters, but also extends to systems with distributed parameters (with boundary control); see, e.g., [4,10]. In other words, control theory based on port-Hamiltonian thermodynamics allows the achievement of set goals in accordance with the laws of thermodynamics describing "evolutionism" rather than "creationism".

## 11. The Concept of Energy Ports

The essence of port-based modeling and port-Hamiltonian (pH) systems theory is presented in Figure 2. The fundamental element of Hamilton's thermodynamics is that the way the control system is reorganized allows for the modeling of both reversible (energy-storing) and irreversible (energy-dissipating) phenomena. Now let us look at the control system in a new energy-based graphic version. These phenomena can be imagined using ideal energy storage elements $\mathcal{S}$ (capacitors, coils, masses, springs, etc.) and energy-dissipating control elements $\mathcal{R}$ (resistors, dampers, friction elements, etc.),

https://doi.org/10.3390/en19020324

connected to each other by a central energy-converting structure denoted by the letters $\mathcal{S}$-$\mathcal{D}$ (Stokes–Dirac). In some cases, the $\mathcal{S}$-$\mathcal{D}$ structure has physical representations (Krhac, Maschke, and Schaft [7]).

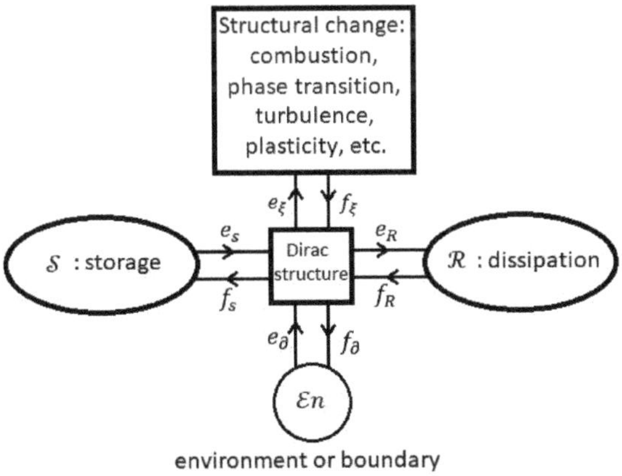


**Figure 2.** Van der Schaft's connection graph for an energy port system containing storage, dissipation, structural change, and environment.

Since control systems are often open systems that interact with their environment, there is an exchange port with the environment $\mathcal{E}n$ from which the control $u$ and the output vector $y$ often originate. The latter, in order to satisfy the conditions of the pH formalism, must be dimensionally correlated and form the so-called supply rate function. Such a function is possible with the interpretation presented in (Equation (9)). It has the form $s(u,y) = u^T \cdot y$ and denotes the power supply to the control system. If we treat the Hamiltonian $H(x)$ as an energy storage function, the ideal situation would be one in which the rate of energy storage would be equal to the rate of its supply through the exchange port: $\dot{H}(x) = s(u,y)$. Practical experience shows that this condition is impossible to achieve. Therefore, if we have a control system measured at two moments $t_0, t_1$ where $t_1 > t_0$, the Hamilton function (stored energy) should satisfy the dissipation inequality [1,2]:

$$H(x(t_1)) \leq H(x(t_0)) + \int_{t_0}^{t_1} s(u(\tau), y(\tau))d\tau. \tag{20}$$

This equation does not say where the dissipated part of the energy that has not been stored "flows away", because in technology it is usually not recovered or controlled. However, adding irreversible ports allows us to find the physical cause of the shortcomings of energy storage and the amount of dissipated energy that is uselessly released into the environment.

## 12. Fluxes and Efforts—The Dual Pairs

The formalism of pH thermodynamics radically changes the foundations of control theory, as it seeks energy pairs in all phenomena. Hence, the connection in the energy control graph is made through a pair of objects with identical dimensions. Let us write them generally as abbreviations of the words *flux* and *effort*—this is the pair $(f,e)$, the first of which is called flux and the second effort. Since these are energy state variables, their product, properly defined, is expressed in units of J/s, i.e., it is energy power. Any pair of vectors $(f,e)$ defines an energy port, and the entire set of variables $f, e$ is also called the set of energy variables of this port [7,9,12].

https://doi.org/10.3390/en19020324

A control system containing three such ports: port $(f_s, e_s)$, which connects to an ideal energy storage device $\mathcal{S}$, port $(f_R, e_R)$, corresponding to the dissipation of available energy $\mathcal{R}$, and external port $(f_\partial, e_\partial)$, through which the system interacts with the environment $\mathcal{E}n$ (including the operation called "control"). The scalar quantities of pH theory form the products $f_s^T \cdot e_s$, $f_R^T \cdot e_R$, $f_\partial^T \cdot e_\partial$, which are instantaneous energy powers transferred by links or "bonds", using the terminology of bond graphs (Ortega et al. [6]).

In particular, the boundary (external) port $(f_\partial, e_\partial)$, $(f_\xi, e_\xi)$ appears in two variants: once as a controllable port (controllable external energy sources) and once as an uncontrollable port (uncontrollable external energy sources). Then the variables are decomposed: $(f_\partial, e_\partial) \to (f_C, e_C) \bigoplus (f_I, e_I)$.

**Remark 2.** *This division of boundary ports results from practice and is best explained using a specific, well-known example of forced vibrations of a solid body $\Omega$ in the shape of a slender rod, one end of which $\partial\Omega_a$ is fixed and at the other end $\partial\Omega_b$. There is an exciting force $\vec{\sigma}_{ext} = \vec{F}_{ext}(t)/A$, which we can consider as control $\vec{u}$ and the corresponding velocity of the rod end as output $\vec{y}$. The remaining sides of the rod $\partial\Omega_c$ are located in atmospheric air, which we can consider to be a constant, zero-force, uncontrollable forcing. The fixed end belongs to the uncontrollable boundary variables in which the value of the state variable (zero velocity and momentum) is set, while the fixing force, called the "reaction", is unknown. No energy flows into the system through this port.*

Considering that practical division of the boundary into three parts $\partial\Omega = \partial\Omega_a \cup \partial\Omega_b \cup \partial\Omega_c$, the Hamiltonian takes the form:

$$\mathcal{H}(x) = \iiint H \, d\Omega + \iint_{\partial\Omega} H_{ext} \, dA, \tag{21}$$

where $H = \frac{1}{2}\rho \dot{\vec{u}}^{\,2} + \rho u(\overset{\leftrightarrow}{\varepsilon})$ is a function of kinetic and internal energy $u(\overset{\leftrightarrow}{\varepsilon})$ dependent on deformation $\overset{\leftrightarrow}{\varepsilon}$ caused by displacement $\vec{u}$. On the other hand, $H_{ext} = \vec{u} \cdot \vec{\sigma}_{ext}(t)$ acts only at the boundary $\partial\Omega_b$. In Lagrange mechanics, vibrations are described by a single variable, which is the displacement $\vec{u}$:

$$\rho \partial_{tt} \vec{u} = \operatorname{div} \overset{\leftrightarrow}{\Sigma} + \rho \vec{b}, \tag{22}$$

where the stress tensor $\overset{\leftrightarrow}{\Sigma} = \frac{\partial u}{\partial \overset{\leftrightarrow}{E}} = \mathcal{L} \overset{\leftrightarrow}{E}$ and the strain tensor is a symmetric object, $2\overset{\leftrightarrow}{E} = \operatorname{grad} \vec{u} + \operatorname{grad}^T \vec{u}$, while $\mathcal{L}$ is a matrix of elastic constants. If we choose the energy state vector as

$$x = \begin{pmatrix} \vec{\varkappa}_u = \rho \dot{\vec{u}} \\ \overset{\leftrightarrow}{\varkappa}_E = \overset{\leftrightarrow}{E} \end{pmatrix} \text{ and } H(x) = \frac{1}{2}\rho^{-1} (\vec{\varkappa}_u)^2 + \frac{1}{2} (\mathcal{L} \overset{\leftrightarrow}{\varkappa}_E) \cdot \overset{\leftrightarrow}{\varkappa}_E, \tag{23}$$

it will obtain the following storage ports:

$$f_s = -\frac{\partial}{\partial t} x = \begin{pmatrix} -\partial_t \vec{\varkappa}_u \\ -\partial_t \overset{\leftrightarrow}{\varkappa}_E \end{pmatrix}; \; e_s = \delta_x H = \begin{pmatrix} \vec{e}_u \\ \overset{\leftrightarrow}{e}_E \end{pmatrix} = \begin{pmatrix} \delta_{\vec{\varkappa}_u} H \\ \delta_{\overset{\leftrightarrow}{e}_E} H \end{pmatrix} = \begin{pmatrix} \dot{\vec{u}} \\ \overset{\leftrightarrow}{\Sigma} = \mathcal{L} \overset{\leftrightarrow}{E} \end{pmatrix}, \tag{24}$$

allowing the momentum evolution to be written as $\dot{x} = \mathcal{J} \delta_x H + Gu$:

$$\frac{\partial}{\partial t} \begin{pmatrix} \vec{\varkappa}_u \\ \overset{\leftrightarrow}{\varkappa}_E \end{pmatrix} = \begin{pmatrix} 0 & \operatorname{div} \\ \operatorname{grad}_s & 0 \end{pmatrix} \begin{pmatrix} \vec{e}_u \\ \overset{\leftrightarrow}{e}_E \end{pmatrix} + \begin{pmatrix} 1 \\ 0 \end{pmatrix} \rho \vec{b}, \tag{25}$$

https://doi.org/10.3390/en19020324

where the control matrix $G$ and mass force control appear: $u_{\vec{b}} = \rho \vec{b}$. The control Equation system (25) consists of 3 + 6 = 9 evolution equations and contains only the first derivatives with respect to time and space. Such low derivatives allow for a completely different discretization than the discretization of the Lagrange approach system (Equation (22)). However, it is important to remember not to spoil the Stokes–Dirac structure with unfortunate discretization; hence, the term "structure preserving discretization" has become common in the literature [5,17,18].

The variation in the Hamiltonian in the region is equal to

$$\dot{\mathcal{H}} = \iiint (\vec{e}_u \cdot \partial_t \vec{x}_u + \overleftrightarrow{e}_E \cdot \partial_t \overleftrightarrow{x}_E) d\Omega = \iiint (\vec{e}_u \cdot \mathrm{div}\, \overleftrightarrow{e}_E + \overleftrightarrow{e}_E \cdot \mathrm{grad}_s \vec{e}_u) d\Omega = \iiint \mathrm{div}(\overleftrightarrow{e}_E \vec{e}_u) d\Omega = \iint_{\partial\Omega} \vec{e}_u \cdot (\overleftrightarrow{e}_E \vec{n})\, dA. \tag{26}$$

By remembering that the boundary $\partial\Omega = \partial\Omega_a \cup \partial\Omega_b \cup \partial\Omega_c$, the following is obtained:

$$\dot{\mathcal{H}} = \vec{e}_u (\partial\Omega_a) \cdot (\overleftrightarrow{e}_E \vec{n}) + \vec{e}_u (\partial\Omega_b) \cdot (\overleftrightarrow{e}_E \vec{n}) + \vec{u} (\partial\Omega_b) \cdot \vec{\sigma}_{ext} (t).$$

By defining boundary ports $f_\partial = u_\partial = \begin{pmatrix} \vec{e}_u(\partial\Omega_a) \\ \overleftrightarrow{e}_E(\partial\Omega_b)\vec{n}_b \end{pmatrix}$ and $e_\partial = y_\partial = \begin{pmatrix} \overleftrightarrow{e}_E(\partial\Omega_a)\vec{n}_a \\ \vec{e}_u(\partial\Omega_b) \end{pmatrix}$, the following can be written:

$$\begin{pmatrix} f_\partial \\ e_\partial \end{pmatrix} = \begin{bmatrix} \mathcal{B}_u\left(\vec{n}\right) \\ \mathcal{B}_y\left(\vec{n}\right) \end{bmatrix} (\delta_x H), \tag{27}$$

The boundary ports use storage-type effort: $e_s$, and $\dot{\mathcal{H}} = (u_\partial)^T y_\partial$.

## 13. Energy-Storing Elements

In the connection graph (Figure 2), the Stokes–Dirac connection structure plays a central role. It is a structure that conserves Hamiltonian energy, i.e., $\dot{H} = \{H, H\} = 0$. This means that the energy flux (power) passing through all ports is conserved. It is said that the Stokes–Dirac structure has the property of power conservation, or the conservation of power built up by flows and forces. It combines three flows $f = (f_s, f_R, f_\partial)$ with efforts $e = (e_s, e_R, e_\partial)$ in such a way that the total power connecting the ports is equal to zero: $f^T \cdot e = 0$. From a mathematical point of view, there is a flow space—let us say $\mathcal{F}$—and an effort space that is energetically dual to it $\mathcal{E} = \mathcal{F}^*$. Holding on to mathematical precision, which, incidentally, is often unnecessary, it is said that the energy port variables $f, e$ form a dual product (duality product) $< f \dagger e >$. It is said that the structure $\mathcal{S}$-$\mathcal{D} \subset$ F×E has a Stokes–Dirac structure if $< f \dagger e > = 0$. The simplest example of a structure $\mathcal{S}$-$\mathcal{D}$ is a skew-symmetric connection matrix $\mathcal{J} = -\mathcal{J}^T$ (Equation (7)).

Now, let us note that the port $(f_s, e_s)$ connecting to the reversible energy storage $\mathcal{S}$ can be treated as "energy-storing elements". It turns out that the flow variables $f_s$ satisfy the condition of state variable evolution, $\dot{x} = -f_s$, and energy can be stored in the Hamiltonian $H(x)$, defining the dual variable $e_s = \delta_x H$. Hence, the evolution equation (Equation (15)) can be written algebraically as follows:

$$\frac{\partial x}{\partial t} = \{x, H\} = \mathcal{J}\frac{\delta H}{\delta x} \quad \rightarrow -f_s = \mathcal{J} e_s, \tag{28}$$

as the relationship between the flux and the effort of the storage port. The change in Hamiltonian (power) is then equal to the following:

$$\frac{\partial}{\partial t} H(x) = \left(\frac{\partial H}{\partial x}\right)^T \cdot \dot{x} = -e_s \cdot f_s \;= e_s^T \mathcal{J} e_s = 0. \tag{29}$$

https://doi.org/10.3390/en19020324

However, $\dot{H} = 0$ only when the power of the other ports is equal to zero.

## 14. Damping in Control Theory

It must be admitted that in control theory, the issue of separating reversible and irreversible phenomena can be resolved by supplementing the initial linear control system (Equation (2)) with additional dissipation matrices (Ogata [39], Gawroński [32]):

$$\dot{x} = (A - R)x + (B + P)u, \tag{30}$$

$$y = (C + P)x + (D + M)u. \tag{31}$$

The condition of the dissipativity of the control system requires that not only the resistance matrix be positive definite, but that the entire "resistance" system is

$$\begin{pmatrix} R & P \\ P & M \end{pmatrix} \geq 0. \tag{32}$$

## 15. The Control System Based on the Irreversible Port-Hamilton System

An open question is the port $(f_R, e_R)$ connecting the $\mathcal{S}$-$\mathcal{D}$ system with the lost energy reservoir $\mathcal{R}$. There is no known way to determine dissipative energy variables, but a one-way flow of dissipated power can be assumed:

$$f_R^T \cdot e_R \leq 0, \qquad \text{for all} \qquad (f_R, e_R) \in \mathcal{R}. \tag{33}$$

From the above arrangement of irreversible phenomena, it follows that Hamilton's thermodynamics elevates energy balances in the form of their power and explicitly defines the power dissipated by the port $(f_R, e_R)$. This distinguishes it from Lagrange's formalism, which adopts the First Law of Thermodynamics as an energy balance that is conserved and in which there are no energy sources (internal or boundary).

A specific class of port-Hamiltonian systems, often used for control, is obtained in such a way as to preserve the skew-symmetric graph called the Stokes–Dirac structure. Only part of this definition is based on Poisson brackets—the rest requires one to impose a model structure in the form of a skew-symmetric matrix (Equation (8)):

$$\begin{pmatrix} f_s \\ f_R \\ f_\partial \end{pmatrix} = \begin{bmatrix} \mathcal{J} & G_R & -G_\partial \\ -G_R^T & 0 & 0 \\ -G_\partial^T & 0 & 0 \end{bmatrix} \begin{pmatrix} e_s \\ e_R \\ e_\partial \end{pmatrix}. \tag{34}$$

The relationship for the dissipated power port can be analogous to Onsager's constitutive equations known in linear irreversible thermodynamics, $e_R = Sf_R$, where similar to the fourth law of thermodynamics, it is assumed that the matrix of coefficients between the effort and the flux is symmetrical and positive definite: $S = S^T$. Furthermore, in Equation (34), by substituting $G_R e_R = -G_R S f_R = -G_R S G^T e_s$ and $e_s = \delta_x H$, it can be reduced (Equations (30) and (31)) to the desired port Hamiltonian of a control system with a specific control and output port:

$$\dot{x} = [\mathcal{J} - R]\delta_x H(x) + Gu; \quad y = G^T \delta_x H(x), \tag{35}$$

where the resistance matrix $R = G_R S G_R^T$ is substituted for the control $u \equiv e_\partial$ and the output $y \equiv f_\partial$. This means that there is a formal possibility of maintaining the Stokes–Dirac structure for a system with dissipation and boundary ports.

https://doi.org/10.3390/en19020324

In general, the definition of $e_R = Sf_R$ irreversible fluxes and efforts can be replaced by Rayleigh's dissipation functional $\mathcal{R}(f_R)$ [9,10,23], which allows for nonlinear relations:

$$e_R = -\frac{\partial^2 \mathcal{R}(f_R)}{\partial f_R (\partial f_R)^T} f_R \cong S f_R. \tag{36}$$

This allows us to write down the evolution of the state (Equation (35)) in a more developed form:

$$\dot{x} = [\mathcal{J} - G_R \frac{\partial^2 \mathcal{R}(f_R)}{\partial f_R (\partial f_R)^T} (G_R)^T] \delta_x H(x) + Gu; \quad y = G^T \delta_x H(x). \tag{37}$$

Note that the introduction of Rayleigh's dissipation functional does not necessarily mean that a model is consistent with any of the extreme variational principles. It should be remembered that most thermodynamicists avoid variational principles and mathematical models based on them [44].

Maintaining the Stokes–Dirac structure guaranteed by Equation (34) is not obvious and should be verified in each case. For solids, this structure is ensured by Betti's theorem [50], while for fluids, this structure is guaranteed only for the Navier–Stokes fluids [8,18]. In a more general case, such as that described in [51], the maintained SD structures of boundary ports must be proven for specific geometries and a specific control.

## 16. Energy-Dissipating Boundary Ports

Another open issue concerns the external (boundary port variables) ports $(f_\partial, e_\partial)$ connecting the system with the environment $\mathcal{E}n$. It is unclear how to approach them. However, it will be desirable to limit their description to algebraic equations that are analogous to $y = Cx + Du$ causal control rather than, for example, differential control: $y = C\dot{x}$. In general, as was pointed out in Equation (34), boundary port variables are not only functions of storage efforts $e_s$ but also of dissipation efforts $e_R$. If the boundary is oriented by the normal vector $\vec{n}$, then according to Cauchy's lemma, boundary variables are defined as follows:

$$\begin{pmatrix} f_\partial \\ e_\partial \end{pmatrix} = \mathcal{B}_{\vec{n}} \begin{pmatrix} e_{s||\partial\Omega} \\ e_{R||\partial\Omega} \end{pmatrix}, \tag{38}$$

where the boundary operator $\mathcal{B}_{\vec{n}}$ has a Stokes–Dirac structure and is skew-symmetric. It provides normal and tangential projections onto the boundary surface $\partial\Omega$. Now, taking into account dissipative and boundary ports, the Hamiltonian increment is expressed as follows:

$$\dot{\mathcal{H}} = \iint_{\partial\Omega} f_\partial \cdot e_\partial \, dA + \iiint_\Omega f_R \cdot e_R d\Omega \leq \iint_{\partial\Omega} f_\partial \cdot e_\partial \, dA. \tag{39}$$

where $\iint_{\partial\Omega} f_\partial \cdot e_\partial \, dA$ denotes the power supplied by the boundary ports and $\iiint_\Omega f_R \cdot e_R d\Omega$ denotes the power dissipated in the form of an internal source. In words, the power flow law can be expressed as follows: the change in energy storage is the result of the power supplied by the boundary ports minus the internal energy dissipation through the dissipative ports. This is, admittedly, a groundbreaking and creative interpretation of the First Law of Thermodynamics. Colloquially speaking, it turns the First Law of Thermodynamics "upside down".

**Remark 3.** *In this paper, there is no focus on known methods of controlling pH systems; however, from a thermodynamic point of view, there is potential interest in this group of methods, which is suitable for controlling Hamiltonian thermodynamics, i.e., power-producing systems. The best-known energy control method is the IDA-PBC method developed by Ortega et al. [6]. A discussion*

https://doi.org/10.3390/en19020324

*of pH system control methods can be found in the mathematical works of Pommaret [42], Mehrmann and Unger [52], and Gawroński [32].*

## 17. Maschke–Kirchhoff Formalism

Maschke and Kirchhoff [50] consider supplementing the equations of evolution with irreversible effects of a mechanical, electrical, or chemical nature. Entropy does not appear in the state vector $x$ they have proposed; hence, this mathematical model has limited value. They do not refer to Rayleigh's dissipative potential, but to an enigmatic function $\mathcal{S}$ with unknown entropy dimensions, almost impossible to define, which they call the "entropy function" satisfying the Poisson bracket conditions, $\{\mathcal{S}, H\} = \delta_x \mathcal{S} J(x)\, \delta_x H$, calculated for the skew-symmetric matrix $J$ and calculable $\delta_x S$. The equations of evolution of the dissipative Dirac system now have the form proposed by Maschke and Kirchhoff:

$$\partial_t x = \gamma(x, \delta_x H)\{\mathcal{S}, H\}\; J\, \delta_x H + \gamma_{port}(x, \delta_x H, u)\{\mathcal{S}, H\}_{port} G u, \tag{40}$$

$$y = \gamma_{port}(x, \delta_x H, u)\{\mathcal{S}, H\}_{port} G^T \delta_x H, \tag{41}$$

where an empirical function $\gamma(x, \delta_x H)$ appears, the content of which is unknown, and its reflecting boundary function $\gamma_{port}(x, \delta_x H, u)$. The proposal by Maschke and Kirchhoff [50] strictly adheres to the research paradigms of control theory, but treats the thermodynamics of phenomena somewhat "neglectfully". The authors' thesis that the product of the bracket $\{\mathcal{S}, H\}$ and the skew-symmetric Dirac structure matrix $J$ "ensures that the second law of thermodynamics is satisfied" is also unconvincing. This may be true, but only in a case simplified to mechanical friction. The formalism presented in Equations (40) and (41) is developed by Ramirez et al. [17,18].

**Remark 4.** *An example of the separation of dissipation port can be found in the Hamiltonian formulation of Fourier heat transport in solids described by the equation (Brugnoli et al. [13], Jäschke et al. [14]):*

$$c_p\, \partial_t T(\vec{x}, t) = \operatorname{div}(\overleftrightarrow{\lambda}\, \operatorname{grad} T) + u_T, \tag{42}$$

*where $c_p$ is heat capacity, $T_0$ is reference temperature, $\overleftrightarrow{\lambda}$ is thermal conductivity, and $u_T$ is internal radiation source. This is represented by the storage variable $x_T = c_p T$, and the following Hamiltonian:*

$$\mathcal{H}(x_T) = \int_\Omega H\, d\Omega = \int_\Omega c_p T_0 \left(\frac{T}{T_0}\right)^2 d\Omega = \int_\Omega x_T \frac{T}{T_0} d\Omega. \tag{43}$$

This allows us to define thermal effort as $e_T = \partial_{x_T} H = T/T_0$ and the flux as $f_T = \dot{x}_T$. In addition to the port pair $(f_T, e_T)$ of the storage type, there is also a dissipation pair $(f_R,\, e_R) \equiv (f_{\vec{q}}, e_{\vec{q}})$ defined as $e_{\vec{q}} = \vec{q}$, $f_{\vec{q}} = -\vec{q}$. Now, both the Fourier conductivity and the constitutive equation of the type $e_R = S f_R = (T_0\, \overleftrightarrow{\lambda})^{-1} f_R$ can be jointly represented as follows:

$$\begin{pmatrix} 1 & 0 \\ 0 & 0 \end{pmatrix} \begin{pmatrix} f_T \\ f_{\vec{q}} \end{pmatrix} = \begin{pmatrix} 0 & \operatorname{div}(\cdot) \\ -\operatorname{grad}(\cdot) & (T_0\, \overleftrightarrow{\lambda})^{-1} \end{pmatrix} \begin{pmatrix} e_T \\ e_{\vec{q}} \end{pmatrix} + \begin{pmatrix} 1 \\ 0 \end{pmatrix} u_T. \tag{44}$$

By taking the fluxes and efforts together, $f = (f_T, f_{\vec{q}})^T$ and $e = (e_T, e_{\vec{q}})^T$, and defining the matrices as follows:

$$P = \begin{pmatrix} 1 & 0 \\ 0 & 0 \end{pmatrix}; \quad J = \begin{pmatrix} 0 & \operatorname{div}(\cdot) \\ -\operatorname{grad}(\cdot) & 0 \end{pmatrix}; \; R = \begin{pmatrix} 0 & 0 \\ 0 & (T_0\, \overleftrightarrow{\lambda})^{-1} \end{pmatrix}; \; G = \begin{pmatrix} 1 \\ 0 \end{pmatrix}, \tag{45}$$



https://doi.org/10.3390/en19020324

the following control system consistent with Equation (30) is obtained:

$$P f = (J + R)e + Gu_T. \tag{46}$$

## 18. Hamilton's Formalism in Physics

Hamiltonian formalism stems from the analytical mechanics of discrete systems and is not suitable for use in classical field theory covering such continua as gravitational fields, electromagnetic fields, and continuous matter fields. There are few attempts in the literature to apply Hamilton's approach to the theory of gravity [28–30], and there are also few Hamiltonian representations of Maxwell's electrodynamics, Infeld–Born, Kauffman, etc. (see: Pavelka [53]; Pavelka et al. [27]). Unsuccessful attempts to find a Hamiltonian may stem from a misunderstanding of Hamilton's concept, from "groping" with blunt mathematical tools such as "contact transformation" or co-tangential bundles. Meanwhile, without the Hamiltonian, there is no correct quantum physics, and the one that exists limits natural phenomena to a reduced minimum.

A completely chaotic situation occurs in the field of fluid or solid material continuum physics. Here, there is no single Hamiltonian approach, and each case has its own mathematical model that is incompatible with others in terms of Hamiltonian and Rayleighian structure. This conceptual chaos stems from the fact that Hamiltonian field theory has virtually no physical foundations and has too many mathematical assumptions and limitations. That is why there are many attempts in the literature to address this topic, which may, after years of struggle, lead to realistic foundations [24,27,54].

The objective epistemological cause of chaos is a certain objective property of continuum physics that discrete systems do not possess. The point is that continuum theories are formulated in different descriptions, i.e., settings of the observer and his measuring instruments. Four descriptions are known: material (Lagrangean), spatial (Eulerian), referential (ALE), and waveform [55]. There are no known general methods in the literature for transitioning between these four descriptions, which is why individual models, such as the model of fluid flow with cavitation, are somewhat trapped within a single description and difficult to transform into another. Meanwhile, as the examples show, a Lagrangean field theory model originally presented in a material description often has its representation as a Hamiltonian field theory in a spatial description. The works [54,56] illustrate this unsatisfactory state of affairs well.

**Remark 5.** *We propose here that a significant reason for the stagnation of pH formalism is the narrow, impoverished mathematical formalism developed in the works of van der Schaft, Maschke, Gilbert, Vanneste, Rashad, Califano, Stramigioli, etc. It is based on differential forms that prefer a notation which, although mathematically correct, deviates in its physical content from the existing classical field theory (Truesdell and Toupin, [55]). For example, bearing in mind the practical computational aspect, we see that the physical content of the three equations of momentum conservation for a viscoelastic fluid (expressed without indices) has its tensor representation and, in differential forms, respectively (Equation (47)-Reynolds (1886) [55], Equation (48)-Califano et al. [20]).*

$$\partial_t(\rho\vec{v}) + \mathrm{div}\,(\rho\vec{v} \bigotimes \vec{v}) = \mathrm{div}[-p\overleftrightarrow{I} + \overleftrightarrow{\tau}] + \rho\vec{b}. \tag{47}$$

$$\dot{v} + \mathrm{d}\left(\frac{1}{2} * (v \wedge *v)\right) + \iota_v \mathrm{d}v = -\frac{1}{*\mu}\,\mathrm{d}p + \frac{1}{*\mu}(*_2\mathrm{d}_\nabla \mathcal{T}) + *b \tag{48}$$

This means that instead of the classic operators div, grad, $\bigotimes$ and tensors $\overleftrightarrow{I} = \delta_{ij}\vec{e}_i \bigotimes \vec{e}_j$, $\overleftrightarrow{\tau} = \tau_{ij}\vec{e}_i \bigotimes \vec{e}_j$ are replaced by the following: the Hodge star-dual $*_2$,

https://doi.org/10.3390/en19020324

two forms of covector-valued $\mathcal{T}$, the $\wedge$ operator, and the exterior derivative operator d, where the fluid velocity vector $\vec{v}$ is uniquely identified with a vector valued zero-form $v$. It is not just about the notation (old wine in a new bottle), but about something more; this apparatus of thinking requires the creation of a new continuum of thermomechanics from scratch.

The geometric formalism of pH thermodynamics forces us to abandon classical notation and classical definitions. But, not only that, it also forces us to conduct new experimental research and broaden our perception of the fundamental concepts of continuum thermodynamics. For example, Cauchy's lemma, well-known in thermodynamics, which defines the way in which the tensor field "comes out" at the edge of the area, $\overleftrightarrow{t}\,\vec{n} \equiv \vec{f}_{ext}$, is challenged by authors such as Califano et al. [20].

## 19. Hamiltonian Formulation of Fluid Mechanics

In this paper, we limit ourselves to the thermodynamics of viscous and conducting gases, which are the mainstay of classical thermodynamics and fluid mechanics. If we ask about a Hamiltonian formulation of classical thermodynamics, focused on the study of changes in a stationary gas that conducts neither heat nor work, the literature is so limited (see Grmela et al. [24,36]) that it can be concluded that Hamiltonian classical thermodynamics does not exist. We will return to the urgent need for Hamiltonian thermodynamics in the conclusion (Section 23)—however, we must begin with a revision of the foundations of quantum thermodynamics [22].

One of the first works on the Hamiltonian formalism of ideal fluid mechanics with energy conversion is the work of Morrison [38]. Here, the problem is the proper formulation of the so-called family of "generalized momenta" and "generalized positions" from which we build the state vector $x = (q, p)$. Unfortunately, the aim of Morrison's work was not to supplement Euler's ideal fluid mechanics with new Hamiltonian variables, but rather to reformulate the classical system of ideal fluid equations consisting of five equations (mass balance, momentum balance, and internal energy) into a Hamiltonian system consisting of five equations (mass balance, momentum balance, and specific entropy), having the state vector $x = (\rho, \vec{v}, s)$, respectively:

$$\begin{aligned} \partial_t \rho &= -\mathrm{div}\,(\rho \vec{v}\,) \\ \partial_t\,(\rho \vec{v}\,) &= -\,\mathrm{div}\,(\rho \vec{v} \otimes \vec{v}\,) + \mathrm{grad}\; p + \rho \vec{b}\cdot \\ \partial_t\,(\rho s) &= -\,\mathrm{grad}\,(\rho s)\,\vec{v} \end{aligned} \tag{49}$$

Following the pioneers, Morrison [38] uses a different representation of the d'Alembert–Euler acceleration, $\vec{a} = \partial_t \vec{v} + (\mathrm{grad}\ \vec{v}\,)\vec{v} \equiv \partial_t \vec{v} + \mathrm{grad}\ (\tfrac{1}{2}\vec{v}^{\,2}) + \vec{v} \times (\mathrm{rot}\ \vec{v})$, sometimes neglects the mass force $\vec{b} = 0$, and transforms the conservative system of equations (Equation (49)) into a system of the (non)conservative form:

$$\begin{aligned} \partial_t \rho &= -\mathrm{div}\,(\rho \vec{v}\,) \\ \partial_t \vec{v} &= -\mathrm{grad}\ (\tfrac{1}{2}\vec{v}^{\,2}) - \vec{v} \times (\mathrm{rot}\ \vec{v}\,) - \tfrac{1}{\rho}\mathrm{grad} p + \vec{b}\cdot \\ \partial_t s &= -\,\mathrm{grad}\,(s)\,\vec{v} \end{aligned} \tag{50}$$

where $\rho$ is the density of the gas, $p$ is the pressure, $\vec{v}$ is its velocity, and $s$ is its specific entropy. Morrison, not knowing the solution to the general problem, puts forward a truncated thesis that Hamiltonian Eulerian fluid mechanics has a mathematical model for which a degenerate Poisson bracket exists. Knowing that this is an invalid bracket, he calls

https://doi.org/10.3390/en19020324

it a "non-canonical Poisson bracket" and no longer attempts to find a canonical (correct) Poisson brace.

It defines an object which, *ex catedra*, is called the Hamiltonian:

$$\mathcal{H}(x) = \iiint_{\Omega} H\, d\Omega = \iiint_{\Omega} \left(\frac{1}{2}\rho \vec{v}^{\,2} + \rho u(\rho,s)\right) d\Omega. \tag{51}$$

This definition has guided blind research for many years. It contains an unusual object: the mass density of internal energy $u = u(\rho,s)$ from which we obtain the definitions of temperature and pressure in classical thermodynamics: $T = \partial_s u$; $p = \rho^2 \partial_\rho u$. This means that Hamiltonian thermodynamics agrees with this unjustified definition.

Morrison [38] proposes the following three-term non-canonical Poisson bracket:

$$\{f,g\} = \iiint_{\Omega} \left\{ \begin{array}{c} \left[\frac{\delta f}{\delta \rho} \operatorname{div}\left(\frac{\delta g}{\delta \vec{v}}\right) + \operatorname{div}\left(\frac{\delta f}{\delta \vec{v}}\right) \frac{\delta g}{\delta \rho}\right] \\ +\left[\rho^{-1}(\operatorname{rot}\vec{v})\cdot\left(\frac{\delta g}{\delta \vec{v}} \times \frac{\delta g}{\delta \vec{v}}\right)\right] + \left[+\rho^{-1}\operatorname{grad}s\cdot\left(\frac{\delta f}{\delta s}\frac{\delta g}{\delta \vec{v}} - \frac{\delta g}{\delta s}\frac{\delta f}{\delta \vec{v}}\right)\right] \end{array} \right\} d\Omega. \tag{52}$$

Let us now show that by taking the state vector $x$ as the function $f$ and the Hamiltonian $H$ as the function $g$ and performing the derivative of the Hamiltonian, we can reduce the system of equations (Equation (50)) to the form $\dot{x} = \{x, H\}$, with the following specific form:

$$\frac{\partial}{\partial t}\begin{pmatrix} \rho \\ \vec{v} \\ s \end{pmatrix} = \begin{pmatrix} \{\rho, H\} \\ \{\vec{v}, H\} \\ \{s, H\} \end{pmatrix} = \begin{pmatrix} -\operatorname{div} H_{\vec{v}} \\ -\operatorname{grad} H_\rho + \rho^{-1}(\operatorname{rot}\vec{v} \times \overleftrightarrow{I})\, H_{\vec{v}} + \rho^{-1}\operatorname{grad}s H_s + 2\operatorname{grad}(\mathrm{g}z) \\ \rho^{-1}\operatorname{grad}s\ H_{\vec{v}} \end{pmatrix}. \tag{53}$$

where definitions and notations consistent with control theory were introduced as follows:

$$\begin{array}{c} \frac{\partial}{\partial \rho} H = \partial_\rho H \equiv H_\rho = \left(\frac{1}{2}\vec{v}^{\,2} + \mathrm{g}z + u(\rho,s)\right) + \rho^{-1}p \\ \frac{\partial}{\partial \vec{v}} H = \partial_{\vec{v}} H \equiv H_{\vec{v}} = \rho\vec{v} \\ \frac{\partial}{\partial s} H = \partial_s H \equiv H_s = \rho \partial_s u = \rho T \end{array}. \tag{54}$$

The extensive parameters of the state $p, T$ are the thermodynamic pressure and temperature. In the momentum evolution Equation $(53)_2$, a mass force $\vec{b} = \operatorname{grad}(\mathrm{g}z) = \mathrm{g}\vec{e}_z$ appears, acting vertically (perpendicular to the Earth's surface), where g is the Galilean gravitational constant. In the momentum equation, a non-canonical term also appears: grads, which comes at the cost of introducing $H_\rho$ into the description and the need to calculate the gradient $\operatorname{grad} H_\rho$.

$$\operatorname{grad} H_\rho = \operatorname{grad}\left(\left(\frac{1}{2}\vec{v}^{\,2} + \mathrm{g}z + u(\rho,s)\right) + \rho^{-1}p\right), \tag{55}$$

where the internal energy gradient is expressed by gradients $\rho, s$, respectively.

$$\operatorname{grad} u(\rho,s) = \partial_\rho u \operatorname{grad}\rho + \partial_s u \operatorname{grad} s. \tag{56}$$

By separating the mass force—treated as an external control in any direction, not only in the z direction—the system of evolution equations (Equation (53)) can be written as a control system:

$$\frac{\partial}{\partial t}\begin{pmatrix} \rho \\ \vec{v} \\ s \end{pmatrix} = \begin{pmatrix} 0 & -\operatorname{div}(\cdot) & 0 \\ -\operatorname{grad}(\cdot) & \overleftrightarrow{m} & \rho^{-1}\operatorname{grad}s \\ 0 & \rho^{-1}\operatorname{grad}s & 0 \end{pmatrix} \begin{pmatrix} H_\rho \\ H_{\vec{v}} \\ H_s \end{pmatrix} + \begin{pmatrix} 0 \\ 2 \\ 0 \end{pmatrix} u_{\vec{b}} = \mathcal{J}_{5\times5}\begin{pmatrix} H_\rho \\ H_{\vec{v}} \\ H_s \end{pmatrix} + G_{1\times3} u_{\vec{b}}. \tag{57}$$

https://doi.org/10.3390/en19020324

where $\overleftrightarrow{m} = \rho^{-1}(\text{rot}\,\vec{v} \times \overleftrightarrow{I})$. Taking full components of the structural matrix $\mathcal{J}$ we are going into the $\mathcal{J}_{5\times5}$ matrix. This matrix is skew-symmetric, so it is consistent with the pH formulation of control theory: $\dot{x} = [\mathcal{J} - R]\delta_x H(x) + Gu., R \equiv 0$.

Going on to the energetic formulation of pH we have the definition of fluxes and efforts $(f_s, e_s)$, given as follows:

$$f_s = \dot{x} = \frac{\partial}{\partial t}\begin{pmatrix}\rho \\ \vec{v} \\ s\end{pmatrix}; \quad e_s = \partial_x H = \begin{pmatrix}H_\rho \\ H_{\vec{v}} \\ H_s\end{pmatrix}, \tag{58}$$

with the help of which the correctness of the energy flux conservation formulation can be demonstrated: $\dot{H} = \{H, H\}_J = 0$.

**Remark 6.** *Let us also note that in the continuum approach, spatial derivative operators appear:* $\text{grad}(\cdot)$, $\text{div}(\cdot)$ *or laplacian:* $\text{lap}(\cdot) = \text{grad}(\text{div}(\cdot))$. *In general, the matrix* $\mathcal{J}(x)$ *becomes an operator with respect to spatial variables, traditionally denoted in fluid mechanics by* $\vec{x}$. *Then, the matrix* $\mathcal{J}(x)$ *can be represented as a composition of dimensionless skew-symmetric matrices* $J_m$, $m = 0, 1, 2, 3\ldots$ *where the first matrix* $J_0$, *traditionally, is denoted without an index. In the case of a compressible Eulerian fluid, we have only the first derivatives* $\mathcal{J}(x) = J_0 + a_1 J_1 \partial_{\vec{x}}$, *where the coefficient* $a_1$ *equalizes the dimensions. Here,* $(\text{div}(\cdot))^T = -\text{grad}(\cdot)$ *also holds, which preserves skew symmetry.*

**Remark 7.** *The convective derivative* $\frac{d}{dt}(\cdot)$, *appearing in the definition of the d'Alembert–Euler acceleration* $\vec{a} = \frac{d}{dt}\vec{v} = \partial_t \vec{v} + (\text{grad}\,\vec{v})\vec{v}$, *appears in all other balances, not only in the momentum balance; hence, the main problem of Hamiltonian fluid mechanics is to find the "fluid displacement field" in the form of scalar, vector, or even tensor gauge potentials [43]. Morrison [38] reviewed and classified these solutions concerning different representations of the velocity field* $\vec{v}$ *in Hamiltonian fluid mechanics. He synthesized the scalar potentials of Euler [57], Helmholtz [58], Clebsch [59], Bateman [60], and Eckart [61] into one general representation:*

$$\vec{v} = A\,\text{grad}\,a + B\,\text{grad}\,b + C\,\text{grad}\,c + \cdots = A_m \text{grad}\,a_m. \tag{59}$$

As an example we have $\vec{v} = \xi_v\,\text{grad}\,v + \xi_s\,\text{grad}\,s + \text{grad}\,\phi$, which is an example of defining the velocity field using three potentials $\xi_v, \xi_s, \phi$ and gradients of specific volume $\text{grad}\,v$ and specific entropy $\text{grad}\,s$. We must agree here that inventing potentials opens the way to gauge field theory (for instance, Helmholtz: $\vec{v} = \text{grad}\,\phi + \vec{A}$ leads toward U(1)-Maxwell continuum). Sieniutycz and Berry [44] believe that these are ad hoc solutions that do not develop the foundations of Hamiltonian field theory. Pommaret even requires the search for new potentials that open up unknown internal symmetries [42].

## 20. Hamiltonian Formulation of Boundary Conditions

Another difficulty is deriving the fluxes $f_s$ and efforts $e_s$ at the boundary of the region $\partial\Omega$ occupied by the fluid, as they do not coincide with the physical conditions for fluids at the boundary known from the literature. The boundary also contains irreversible ports $(f_R, e_R)$ and boundary ports $(f_\partial, e_\partial)$.

Recall that in fluid thermomechanics, there are two types of boundary conditions. They are related to boundary types. The first are conceptual cross-sections through channels (computational regions, most often inlets and outlets), where we impose conditions on the

https://doi.org/10.3390/en19020324

state parameters $x_{inlet}$ and $x_{outlet}$. The second type of conditions are conditions imposed on the constant boundary on the flows:

$$\mathcal{F}\vec{n} = f^{data}, \tag{60}$$

where $\mathcal{F}$ are, respectively, the fluxes of mass, momentum, angular momentum, entropy, volume, current, etc. On this boundary, control can occur by a known quantity $f^{data}$.

When setting boundary conditions for an Eulerian fluid, we must remember that they must have a general structure that easily overlaps with the boundary conditions of a Navier–Stokes fluid and other fluids, such as two-phase fluids and fluids undergoing phase changes, combustion, and turbulence. Therefore, we are dealing with a skew-symmetric Dirac structure that maintains the boundary conditions proposed by du Buite, Navier, Duhem, Stokes, Poisson, Reynolds, Maxwell, Helmholtz and Piotrowski, von Rybczyński, von Smoluchowski, and others (Badur et al. [51]). Also, approaching the interface with the fluid from the solid side creates a complicated physical situation, the existence of which is unknown. This skew-symmetric structure of boundary conditions is called the Stokes–Dirac structure in the literature (Maschke and van der Schaft [4]).

## 21. Energy Picture

In 1866, J.W. Gibbs proposed two equivalent thermodynamic pictures, which, in the case of an Eulerian fluid, lead to two equivalent definitions of energy state variables (Sieniutycz and Berry [44]):

$$\text{Energy picture } x = (\rho, \vec{p}, e), \tag{61}$$

$$\text{Entropy picture } x = (\rho, \vec{p}, s) \tag{62}$$

It should be emphasized that the Eulerian energy picture of fluid mechanics is the oldest, most reliable, and undoubtedly the most important from the point of view of everyday computational practice. It also served as a starting point for other models, for example, solid mechanics and complex fluid mechanics (viscosity and rheology).

This picture is usually used for research on connections with mechanics, kinetic theory, and equilibrium thermodynamics. Almost all branches of classical field theory [55] arose from these studies. Moreover, this picture is considered to have an epistemological connection with Newtonian mechanics, and the ideal Eulerian fluid is treated as a continuation of Newtonian point particle mechanics. During the evolution of $\dot{x} = \{x, H\}$, the constant quantities are mass [kg], momentum [Ns], and energy [J]:

$$M = \int_\Omega \rho d\Omega,\ \vec{P} = \int_\Omega (\rho\vec{v})d\Omega = \int_\Omega \vec{p}\, d\Omega;\ E = \int_\Omega (\rho\epsilon)d\Omega = \int_\Omega e\, d\Omega. \tag{63}$$

And entropy can change [J/K]:

$$S = \int_\Omega (\rho\eta)d\Omega = \int_\Omega s\, d\Omega. \tag{64}$$

Here we use the volumetric densities of mass, momentum, energy and entropy—$\rho, \vec{p}, e, s$ referred to the unit volume, while $\nu$, $\vec{v}$, $\epsilon$, $\eta$ are the proper densities which referred to the unit mass. The system of evolution equations in the energy picture has the implicit (incomplete) form (Grmela [36,37]):

$$\frac{\partial}{\partial t}\rho = \{\rho, H\} = \operatorname{div}(\rho H_{\vec{p}}) = -\operatorname{div}(\vec{J}^{\rho}) + s_\rho, \tag{65}$$

https://doi.org/10.3390/en19020324

$$\frac{\partial}{\partial t}\vec{p} = \{p, H\} = \operatorname{div}(\vec{p} \bigotimes H_{\vec{p}} + ???) = -\operatorname{div}(\overset{\leftrightarrow v}{J}) + \vec{s}_p, \tag{66}$$

$$\frac{\partial}{\partial t}e = \{e, H\} = \operatorname{div}(eH_{\vec{p}}) = -\operatorname{div}(\overset{\to e}{J}) + s_e. \tag{67}$$

where $\overset{\to \rho}{J}$, $\overset{\leftrightarrow v}{J}$, $\overset{\to e}{J}$ are called mass, momentum, and energy fluxes, respectively. Specification of the fluid type requires constitutive equations for the fluxes. The internal sources of mass and energy are equal to zero according to the First Law of Thermodynamics $s_\rho = s_e = 0$, while the internal source of momentum $\vec{s}_p = \rho\vec{b}$ comes from the gravitational field permeating matter. The momentum evolution equation is traditionally treated as a continuum version of Newton's second law ($m_i\dot{\vec{v}}_i = \vec{f}_i + \sum_k \overset{\leftrightarrow}{f}_{ik}$), for the case when there are no dissipative collisions in the swarm of particles $i = 1, 2, 3\ldots, n$). Unfortunately, the Poisson bracket expressing the kinematics of the evolution of $x = (\rho, \vec{p}, e)$ is unknown in the literature, even in the pioneering works of Miroslav Grmela. This creates an obstacle to a formal reconstruction of non-equilibrium thermodynamics that favors and exalts the energy balance; this situation is completely inconsistent with the energetic picture of the Port-Hamiltonian formalism.

The Port-Hamiltonian formalism favors the entropy picture. It tacitly assumes that the two energy and entropy pictures are equivalent. However, such equivalence does not occur in practice.

Thus, by necessity rather than choice, we are left with an entropy picture defining the energy state variables as $x = (\rho, \vec{v}, \eta)$ or, more correctly, as $x = (\rho, \vec{p}, s)$. The Poisson bracket expressing the convective part of the evolution $x = (\rho, \vec{p}, s)$ is as follows (Grmela [24]):

$$\begin{aligned}\{f, g\}_C = &\iiint \rho\,(\operatorname{grad}(f_\rho)\cdot g_{\vec{p}} - \operatorname{grad}(g_\rho)\cdot f_{\vec{p}})\,d\Omega \\ &+ \iiint \vec{p}\cdot(\operatorname{grad} f_{\vec{p}} \otimes g_{\vec{p}} - \operatorname{grad} g_{\vec{p}} \otimes f_{\vec{p}})\,d\Omega \\ &+ \iiint s\,(\operatorname{grad}(f_s)\cdot g_{\vec{p}} - \operatorname{grad}(g_s)\cdot f_{\vec{p}})\,d\Omega.\end{aligned} \tag{68}$$

This bracket allows the formal derivation of the evolution equations without requiring an explicit definition of the Hamiltonian. The following shorthand notations were introduced above:

$$H_\rho = \delta_\rho H = \frac{\partial H}{\partial \rho},\ H_{\vec{p}} = \delta_{\vec{p}} H = \frac{\partial H}{\partial \vec{p}};\ H_s = \delta_s H = \frac{\partial H}{\partial s}. \tag{69}$$

This allows both sides of the equation $\partial_t f = \{f, \mathcal{H}\}$ to be written as follows:

$$\begin{aligned}&\iiint_\Omega (f_\rho \partial_t \rho + f_{\vec{p}} \cdot \partial_t \vec{p} + f_s \partial_t s)\,d\Omega \\ &= \iiint \{f_\rho\,[-\operatorname{div}(\rho H_{\vec{p}})] + f_{\vec{p}}\cdot[-\operatorname{div}(\vec{p} \otimes H_{\vec{p}}) - \vec{p}\operatorname{div}(H_{\vec{p}}) - \rho\operatorname{grad}(H_\rho) - s\operatorname{grad}(H_s)] + \\ &\quad f_s\,[-\operatorname{div}(sH_{\vec{p}})]\}\,d\Omega.\end{aligned} \tag{70}$$

In Equation (70), we used integration by parts. Since the evolution equations hold for every functional, with $\partial_t f = \{f, \mathcal{H}\}$ we have a family of five evolution equations:

$$\partial_t \rho = -\operatorname{div}(\rho H_{\vec{p}}) = -\operatorname{div}(\vec{J}_\rho). \tag{71}$$

$$\partial_t \vec{p} = -\operatorname{div}(\vec{p} \bigotimes H_{\vec{p}}) - \vec{p}\operatorname{div}(H_{\vec{p}}) - \rho\operatorname{grad}(H_\rho) - s\operatorname{grad}(H_s)\,] = -\operatorname{div}(\overset{\leftrightarrow p}{J}). \tag{72}$$

$$\partial_t s = -\operatorname{div}(sH_{\vec{p}}) = -\operatorname{div}(\vec{J}_s). \tag{73}$$



https://doi.org/10.3390/en19020324

The terms in the momentum evolution equation can be written as the divergence of a scalar $p$ called the Hamiltonian pressure:

$$p = -H + \rho H_\rho + sH_s + \vec{p}\cdot H_{\vec{p}}. \tag{74}$$

Since the identity holds, grad $H = H_\rho \mathrm{grad}\rho + H_{\vec{p}}\mathrm{grad}\,\vec{p} + H_s \mathrm{grad}s$, Equation (36) can be replaced by the following:

$$-\vec{p}\,\mathrm{div}\,(H_{\vec{p}}) - \rho\mathrm{grad}(H_\rho) - s\mathrm{grad}(H_s)] \equiv \mathrm{div}\,[-p\overset{\leftrightarrow}{I}\,] = -\mathrm{grad}p. \tag{75}$$

The uniform derivation of the evolution equations based on Poisson brackets (Equation (68)) has the advantage of opening a potential path to generalizations of continuum models. If evolution (motion) is a group of Lie-type continuum transformations, then there is always a Poisson bracket that is canonically associated with the Lie group. This opens the possibility of describing the continuum within the framework of a gauge field theory based on symmetries and their Lie algebras (Pommaret [20], Grmela [24]).

The advantage of this approach is the strict definition of the mass, momentum, and entropy fluxes ($\vec{J}_\rho$, $\overset{\leftrightarrow}{J}^{p}$, $\vec{J}_s$) in terms of the pH effort, $e_s = \delta_x H$, which allows us to write the system of Equations (71)–(73) in terms of the storage ports of the pH system:

$$\partial_t \begin{pmatrix} \rho \\ \vec{p} \\ s \end{pmatrix} = \begin{pmatrix} 0 & -\mathrm{div}(\rho\ldots) & 0 \\ -\rho\mathrm{grad}\ldots & -\mathrm{div}\,(\vec{p}\otimes\ldots) - \vec{p}\mathrm{grad}(\ldots) & \mathrm{sgrad}\ldots \\ 0 & -\mathrm{div}(s\ldots) & 0 \end{pmatrix} \begin{pmatrix} H_\rho \\ H_{\vec{p}} \\ H_s \end{pmatrix} = \mathcal{J}(x) \begin{pmatrix} H_\rho \\ H_{\vec{p}} \\ H_s \end{pmatrix}. \tag{76}$$

which maintains the Dirac structure, $f_s = \mathcal{J}e_s$. Comparing the previous result (57) with (76), we note that using the specific entropy $\eta$ (per kilogram) and the volume entropy density $s$ leads to slightly different formulations of pH.

**Remark 8.** *Now calculating the change in the Hamiltonian* $\dot{\mathcal{H}} = \{\mathcal{H},\mathcal{H}\}_C = 0$, *we can determine the form of the boundary ports* ($f_\partial$, $e_\partial$). *Remembering the definitions,* $e_s = \delta_x H = \begin{pmatrix} H_\rho & H_{\vec{p}} & H_s \end{pmatrix}^T$ *and* $f_s = -\partial_t x = \begin{pmatrix} -\partial_t\rho & -\partial_t\vec{p} & -\partial_t s \end{pmatrix}$, *we calculate the change in the Hamiltonian using (Equations (71)–(73)) as follows:*

$$\begin{gathered} \dot{\mathcal{H}} = \iiint_\Omega e_s^T f_s\, d\Omega = -\iiint_\Omega (H_\rho\dot{\rho} + H_{\vec{p}}\dot{\vec{p}} + H_s\dot{s})\, d\Omega = \iiint_\Omega (H_\rho \mathrm{div}\,(\rho H_{\vec{p}}) + H_{\vec{p}}\cdot \\ [-\mathrm{div}\,(\vec{p}\otimes H_{\vec{p}}) - \vec{p}\mathrm{div}\,(H_{\vec{p}}) - \rho\mathrm{grad}(H_\rho) - \mathrm{sgrad}(H_s)\,] + H_s\mathrm{div}\,(sH_{\vec{p}}))\, d\Omega \\ \dot{\mathcal{H}} = \iiint_\Omega[-\mathrm{div}\,(\rho H_\rho H_{\vec{p}}) + \mathrm{grad}(H_\rho)\cdot(\rho H_{\vec{p}}) - \mathrm{div}\,((\vec{p}\cdot H_{\vec{p}})H_{\vec{p}}) + \mathrm{grad}\,(H_{\vec{p}})\cdot \\ (\vec{p}\otimes H_{\vec{p}}) - H_{\vec{p}}\cdot\mathrm{grad}p - \mathrm{div}\,(sH_sH_{\vec{p}}) + \mathrm{grad}\,(H_s)\cdot(sH_{\vec{p}})]d\Omega = \iiint_\Omega\{-\mathrm{div}[(\rho H_\rho + \\ \vec{p}\cdot H_{\vec{p}} + sH_s)H_{\vec{p}}]\,\}d\Omega = -\iiint_\Omega\{-\mathrm{div}\,[(p+H)H_{\vec{p}}]\,\}d\Omega. \end{gathered} \tag{77}$$

According to Stokes' theorem on the conversion of a volume integral into a surface integral, we obtain the following:

$$\dot{\mathcal{H}} = \iint_{\partial\Omega} \{\,(\rho H_\rho + \vec{p}\cdot H_{\vec{p}} + sH_s)H_{\vec{p}}\cdot\vec{n}_{\partial\Omega}\,\}dA. \tag{78}$$

This formula means that the boundary ports ($f_\partial$, $e_\partial$) can be determined by the efforts $e_s = \delta_x H$ going to the boundary $\partial\Omega$. Unfortunately, if we do not allow any simplifications, the boundary ports should also be a function of the energy state variables $x = (\rho, \vec{p}, s)$. This important fact is not reflected in the literature; see [1,2,62].

**Remark 9.** *Including a state variable such as entropy is not an easy matter and has its unsuccessful attempts in the literature [2,8,10,16,18,62].*

**Remark 10.** *The control system of an ideal Eulerian fluid maintains its Dirac structure if we introduce internal friction of the Navier–Stokes type. To obtain this, we must follow the procedure shown in Section 13 above, where we first define the energetic variables* $(f_R, e_R)$*, which will be "added" to the port structure of the Eulerian fluid. In this case, there are no problems with the identification of the variables—they are the viscous fluid stresses and the deformation velocity,* $(f_R, e_R) = (\overleftrightarrow{\tau}, \overleftrightarrow{d})$*, which are related to each other by the dissipative constitutive equation,* $\overleftrightarrow{\tau} = 2\mu \overleftrightarrow{d} + \lambda(\mathrm{tr}\,\overleftrightarrow{d})\overleftrightarrow{I}$*, where* $\mu, \lambda$ *are the Navier–Lamé viscosity coefficients. This equation can be written in matrix form as follows:*

$$e_{R(6\times1)} = S^{NS}_{(6\times6)} f_{R(6\times1)}. \tag{79}$$

We supplement the evolution equations with fluid viscosity effects as follows:

$$\partial_t x = Euler + \begin{bmatrix} 0 \\ \mathrm{div} \\ 0 \end{bmatrix}_{(5\times6)} e_{R(6\times1)} = \{x, H\}_C + Ge_R, \tag{80}$$

where (Equation (35)) is used and the operator introducing the structure is the matrix $G = G_{5\times6}$. As the structure (Equation (34)) shows, dissipative port variables can be kept in play, but then the Dirac structure must be extended:

$$\begin{pmatrix} \partial_t x \\ f_R \end{pmatrix} = \begin{pmatrix} \mathcal{J} & G \\ G^* & 0 \end{pmatrix} \begin{pmatrix} \delta_x H \\ e_R \end{pmatrix} \quad \text{with} \quad e_R = S^{NS} f_R. \tag{81}$$

Then, however, the dissipative ports extend to the boundary of the region and make the definition of the boundary variables $(f_\partial, e_\partial)$ difficult. The geometric relations $f_{r(6\times1)}$ with the effort $e_s = \partial_x H_{(5\times1)}$ is as follows:

$$f_{R(6\times1)} = \begin{bmatrix} 0 & \mathrm{grad}_s & 0 \end{bmatrix}_{6\times5} \begin{pmatrix} \delta_\rho H \\ \delta_{\vec{p}} H \\ \delta_s H \end{pmatrix} = G^* \delta_x H. \tag{82}$$

This allows the evolution of the Navier–Stokes fluid (Equation (80)) to be written as follows:

$$\partial_t x = \{x, H\}_C + Ge_R = [\mathcal{J} - G(S^{NS})G^*]\delta_x H. \tag{83}$$

Let us emphasize that the constitutive relation (Equation (81)) can be replaced by a more general, nonlinear formula using the Rayleigh dissipative functional. In the case of the Navier–Stokes fluid, remembering that $f_R \equiv \overleftrightarrow{d} = \frac{1}{2}\left(\mathrm{grad}\,\vec{v} + \mathrm{grad}^T \vec{v}\right) \equiv \mathrm{grad}_s \vec{v}$, we obtain the following (Badur et al. [63]):

$$\mathcal{R}(f_R) = \iiint_\Omega \left( \frac{1}{2}\mu \overleftrightarrow{d} \cdot \overleftrightarrow{d} + \frac{1}{2}\lambda (\mathrm{tr}\,\overleftrightarrow{d})^2 \right) d\Omega. \tag{84}$$

**Remark 11.** *From the above, it follows that the use of the Rayleigh functional is possible directly if the flux variables* $f_R$ *are easily expressible in terms of the storage efforts* $e_s = \delta_x H$ *(see Equation (82)). When this is not possible (it is not possible to write the Rayleighian), we must*

https://doi.org/10.3390/en19020324

*extend the control system to the form given by (Equation (40)). The question is, how will the second law of thermodynamics be satisfied in this case? Note that the dissipative part is expressed here entirely in terms of* $\delta_{\vec{p}}H$, *which does not mean that extra definitions of the dissipative ports* $f_R$ *are necessary. However, Cardoso-Ribeiro et al. [15] decompose the definition (Equation (82)) into a divergent part* div $H_{\vec{p}}$ *and a rotational part* rot $H_{\vec{p}}$ *to explicitly maintain the Dirac structure.*

**Remark 12.** *In 1830, Fourier presented a system of simultaneous flow equations in a viscous and conducting fluid, consisting of a mass balance, a momentum balance, and a temperature field evolution. This system of equations is today the basis of the computer science known as Computational Fluid Dynamics (CFD). In the literature, it is also referred to as the Navier–Stokes–Fourier fluid. It imposes a choice of state energy variables as* $x = (\rho, \vec{p}, T)$, *or, worse, as* $x = (\rho, \vec{v}, T)$. *From the point of view of the Hamiltonian approach, the Fourier temperature field evolution equation is the specific entropy evolution equation, while from the point of view of the Lagrangean approach, in contrast, it is a total energy balance in which the internal energy contained in the Hamiltonian has the simplified form:* $u = u(\rho, s) = c_v T$.

$$\partial_t(\rho c_v T) + \mathrm{div}\,(\rho c_v T\vec{v}\,) = \mathrm{div}(\,\kappa\,\mathrm{grad}\,T) + \mathrm{div}\,(-p\vec{v} + \overleftrightarrow{\tau}\,\vec{v}\,) + \rho s_e, \tag{85}$$

where the contribution to the energy flux comes from both the temperature field and the mechanical field. This setting of energy variables was described in Remark 5 as an example for dissipative conduction in a solid. It means that it is possible to find the dissipative port variables corresponding to the heat conduction as $(f_R, e_R) = (\,\vec{g}, \vec{q}\,)$ where $\vec{q}$ is the Carnot heat flux and $\vec{g} = \mathrm{grad}T$ the temperature gradient. The variables $x = (\rho, \vec{v}, T)$ are the basis of the following works: [12,13,18,56].

**Remark 13.** *In thermodynamics, the thermal energy flux* $\vec{q}$ *is related to the entropy flux by the relation:* $\vec{q} = T\vec{h}_s$ *and the mechanical energy flux to the momentum flux. What is the meaning of the mechanical entropy flux* $\vec{h}_m = T^{-1}\vec{q}_m$? *In the work of Mora et al. [19], a vector of state variables was proposed:* $x = (v, \vec{v}, s)$ *consisting of specific volume, velocity, and specific entropy. This state vector is defined under the assumption that the fluid density is a passive quantity and equal to* $\rho = 1$. *The evolution on* $\partial_t s$ *proposed by these authors does not have a reversible term (recoverable)* $\partial_t s = \{s, H\} \equiv 0$ *and has only a dissipative term in the form of the following equation:*

$$\partial_t(\rho c_v T) + \mathrm{div}\,(\rho c_v T\vec{v}\,) = \mathrm{div}(\,\kappa\,\mathrm{grad}\,T) + \mathrm{div}\,(-p\vec{v} + \overleftrightarrow{\tau}\,\vec{v}\,) + \rho s_e, \tag{86}$$

*where the dissipation matrix* $R = (\partial_s H)^{-1} S^{NS} G^T \delta_{\vec{v}} H$ *has a mechanical origin. This is represented by the equation* $\partial_t s = -div\,(\vec{h}_m)$, *where the mechanical term of the entropy flux is defined by* $\vec{h}_m = T^{-1}\vec{q}_m = T^{-1}(\overleftrightarrow{\tau}\,\vec{v}\,)$. *In the pH notation of the control theory,* $\partial_{\vec{v}} H = \vec{v}$, $\overleftrightarrow{\tau} = e_R$ *and* $\partial_s H = T$. *The thermodynamics proposed here cannot be accepted because the potential* $H$ *is not used in the evolution equation for* $\partial_t s$. *Nevertheless, there is a proposal here for a thermodynamic temperature rather than an empirical one. There is also a mechanical entropy flux* $\vec{h}_m$, *which has no deeper justification and is formally incorrect. There is also no reference in (Equation (86)) to the Fourier coefficient of diffusive heat transport.*

**Remark 14.** *Recall that it is more reasonable to choose the variables* $x = (\rho, \vec{p}, s)$. *This is performed by several authors, including Grmela, Klika, and Pavelka [28,46,64].*

Seeking the correct form of the evolution equation for entropy s, Grmela, Klika, and Pavelka [44] propose a "reduction path" from the fine scale, which makes some physical sense by referring to the fundamental phenomena of coercive motion. In ideal gases, these phenomena were studied by Boltzmann, Grad, Chapman, Enskog, Fokker, and Planck. By separating the phenomena into those with different time scales, one can finally obtain the level equations of gas dynamics of turbulent flows, in which the evolution of entropy takes the following form (Grmela, Klika and Pavelka [64]):

$$\partial_t s = -\mathrm{div}\,(sH_{\vec{p}}) - \mathrm{div}\left(\int \eta \frac{\partial H_f}{\partial H_{\vec{p}}}\, dv\right) + \frac{1}{H_s}\int f \frac{\partial H_f}{\partial H_{\vec{p}}} \frac{\partial H_f}{\partial H_{\vec{p}}}\, dv. \quad (87)$$

In this equation, it is easy to isolate the convective term, the diffusion flux of entropy, and its production. The proposal by Grmela, Klika, and Pavelka is extremely attractive—it opens the door to Hamiltonian thermodynamics. Unfortunately, in nature, there are no easy discoveries. Here, although we achieve an extremely suggestive and physically justified entropy evolution equation, we must close the four-state parameter system $x = (\rho, \vec{p}, s, f)$ with an additional evolution equation for the scattering function $f$, e.g., in the form of the Fokker–Planck equations [28].

The great advantage of these evolutionary equations $\partial_t x$, proposed by the aforementioned authors, is the first thermodynamically consistent (coherent) system of turbulent fluid evolution in the literature. By "coherent", we mean consistent with a uniform time scale in which, after exceeding a critical turbulent energy threshold, turbulent fluxes of specific volume, specific momentum, and specific entropy naturally emerge. The most valuable term is the entropy production term, about which thermodynamics is usually silent or told fairy tales. This term, recalled by Clausius in 1854 as "uncompensated heat transformation" (unkompensierte Verwandlung), is the most rotten root in the foundations of thermodynamics.

## 22. Water Flow Undergoing a "Flashing" Phase Transition

Flashing is the rapid evaporation of hot water ($t \geq 100$ °C) and, like cold water cavitation, is induced by a local decrease in pressure below the saturation pressure. Under equilibrium conditions, knowing the pressure and temperature, one can determine the equilibrium degree of filling $\xi_{eq}$ from the internal energy, while the non-equilibrium of the phenomenon and its depth of entry into a metastable state is determined by the non-equilibrium degree of filling with steam $\xi \in (0, 1)$. It is a function of spatial variables (position in the channel) and time. The moment of the beginning of the evolution $\partial_t \xi \rightarrow 0$ is determined by a special closure resulting from the critical overheating of water, determined empirically (Bilicki and Kestin [65]). The Hamiltonian of the system has the following form:

$$\begin{gathered}\mathcal{H} = \iiint_\Omega H(\rho, \xi_{eq}, s, \vec{p}, \xi, \nabla\xi)\, d\Omega = \\ \iiint_\Omega \left(H(\rho, s, \vec{p}) + \alpha(\xi - \xi_{eq})^2 + \beta(\xi - \xi_{eq})^4 + \kappa(\nabla\xi)^2\right)\, d\Omega,\end{gathered} \quad (88)$$

where, alongside the classical Hamiltonian, there is a contribution to the stored energy, which is controlled by three constants: $\alpha, \beta,$ and $\kappa$. This means that the energy parameter of the state is the vector $x = (\rho, \vec{p}, s, \xi)$. The evolution of the parameter $\xi$ (from water to steam) is described by three elements: convection, diffusion, and relaxation to equilibrium, as follows:

$$\partial_t \xi = -\mathrm{div}\,(\xi H_{\vec{p}}) - \mathrm{div}\,(\vec{e}_\xi) - \Gamma H_\xi, \quad (89)$$

which should be corresponding to the composition of three Poisson brackets:

$$\partial_t \xi = \{\xi, H\}_C + \{\xi, H\}_D + \{\xi, H\}_{Relax}. \quad (90)$$

https://doi.org/10.3390/en19020324

The following definitions occur:

$$H_\xi = \delta_\xi H = \frac{\delta H}{\delta \xi} = \frac{\partial H}{\partial \xi} - \text{div}\left(\frac{\partial H}{\partial(\nabla\xi)}\right) \tag{91}$$

where, due to the presence of the gradient $\xi$, defined as grad$\xi \equiv \nabla\xi$ in the Hamiltonian, the derivative with respect to $\xi$ takes the form of a Volterra variational derivative. In the literature, the quantity $H_\xi$ is called affinity $H_\xi \equiv \mathcal{A}$. The quantity $\vec{e}_\xi$ in Equation (89) is called the diffusion effort of the pair $(e_R, f_R)$ (see Equation (79)) and corresponds to its diffusion flux $\vec{f}_\xi = \text{grad}(\mathcal{A}) = \text{grad}H_\xi$:

$$\vec{e}_\xi = \frac{\partial \mathcal{R}(\vec{f}_\xi)}{\partial \vec{f}_\xi} \equiv \frac{\partial^2 \mathcal{R}(\vec{f}_\xi)}{\partial \vec{f}_\xi \otimes \partial \vec{f}_\xi} \vec{f}_\xi = \overset{\leftrightarrow\xi}{S} \vec{f}_\xi. \tag{92}$$

It is based on diffusive Rayleighian $\mathcal{R}(\vec{f}_\xi) = \frac{1}{2}\overset{\leftrightarrow\xi}{S} \cdot (\vec{f}_\xi \otimes \vec{f}_\xi)$. The corresponding dissipative bracket (Equation (89)) has the following form:

$$\{f, g\}_D = \iiint_\Omega \left( -\overset{\leftrightarrow\xi}{S} \cdot \text{grad}(f_\xi) \otimes \text{grad}(g_\xi) \right) d\Omega, \tag{93}$$

which leads to dissipative evolution of the following type:

$$\partial_t \xi = \{\xi, H\}_D = -\text{div}(\vec{e}_\xi) = -\text{div}\left( \overset{\leftrightarrow\xi}{S} \text{ grad}(H_\xi) \right). \tag{94}$$

This structure is known in the literature as the Cahn–Hilliard evolution. The phenological constants $\overset{\leftrightarrow\xi}{S}$ are analogous to the diffusion constants of the Onsager diffusion model, which means a symmetric and positive definite matrix.

As for the relaxation bracket, it has a simple structure, as follows:

$$\{f, g\}_{Relax} = \iiint_\Omega (-\Gamma f_\xi g_\xi) d\Omega. \tag{95}$$

which allows us to calculate the relaxation evolution as follows:

$$\partial_t \xi = \{\xi, H\}_{Relax} = -\Gamma H_\xi = -\Gamma\left( \frac{\partial H}{\partial \xi} - \text{div}\left( \frac{\partial H}{\partial(\nabla\xi)} \right) \right). \tag{96}$$

The parameter $\Gamma > 0$ is the inverse of the relaxation time of the phenomenon (Bilicki and Kestin [65]) and is determined based on the relaxation Rayleighian, $\mathcal{R}(H_\xi) = \frac{1}{2}\Gamma(H_\xi)^2$, through derivatives $\Gamma = \frac{\partial^2 \mathcal{R}}{\partial H_\xi \partial H_\xi}$.

The convective bracket, in the part corresponding to the parameter $\xi$, has the following form:

$$\{f, g\}_C = \iiint_\Omega (\xi \text{grad}(f_\xi) \cdot (g_{\vec{p}})) d\Omega. \tag{97}$$

which allows to calculate convective evolution as follows:

$$\partial_t \xi = \{\xi, H\}_C = -\text{div}(\xi H_{\vec{p}}). \tag{98}$$

The evolution of the entire state vector $x = (\rho, \vec{p}, s, \xi)$ is described by a family of variational derivatives $e_s = (H_\rho \quad H_{\vec{p}} \quad H_s \quad H_\xi)$. The contribution to the structure matrix

https://doi.org/10.3390/en19020324

and resistance matrix derived from the evolutionary Equations (94), (96) and (98) and from the appearance of the variational force $H_\xi$ (Equation (91)) has the following form:

$$\partial_t \begin{pmatrix} \rho \\ \vec{p} \\ s \\ \xi \end{pmatrix} = \begin{bmatrix} 0 & 0 & 0 & 0 \\ 0 & 0 & 0 & J^*_{\xi\vec{p}} \\ 0 & 0 & 0 & J^*_{\xi s} \\ 0 & J_{\xi\vec{p}} & J_{\xi s} & 0 \end{bmatrix}_{6\times 6} \begin{pmatrix} H_\rho \\ H_{\vec{p}} \\ H_s \\ H_\xi \end{pmatrix} - \begin{bmatrix} 0 & 0 & 0 & 0 \\ 0 & 0 & 0 & R_{\xi\vec{p}} \\ 0 & 0 & 0 & R_{\xi s} \\ 0 & R_{\xi\vec{p}} & R_{\xi s} & R_{\xi\xi} \end{bmatrix}_{6\times 6} \begin{pmatrix} H_\rho \\ H_{\vec{p}} \\ H_s \\ H_\xi \end{pmatrix}. \tag{99}$$

This corresponds to the structure (Equation (35)) $\dot{x} = [\mathcal{J} - R]\delta_x H(x)$ in which the structure matrix $\mathcal{J}$ and the resistance matrix $R$ have the dimensions $6 \times 6$. Additionally, as follows from Equations (93) and (95), the resistance matrix has the traditional form $R = G_R S G_R^T$. Obviously, the complete set of evolutionary equations, taking into account convection, resistance, and relaxation, is a full combination of Equations (76), (83), (87) and (99). It is important that the introduction of the variational effort $H_\xi$, despite spatial derivatives reaching the fourth order, allows for discretization that will not destroy the structure of the problem and will preserve the identity of the matrices $\mathcal{J}$ and $R$.

**Remark 15.** *A few remarks are necessary here. The structure of Equation (99) shows the power of pH thermodynamics formalism. In order to preserve the essence of pH thermodynamics, i.e., the skew symmetry of the Stokes–Dirac matrix* $\mathcal{J}$ *and the symmetry of the resistance matrix* $R$*, it was necessary to introduce couplings* $J^*_{\xi\vec{p}}$ *in the momentum evolution equation. Classical non-equilibrium thermodynamics knows nothing about this fact. It can be explained as follows: affinity* $H_\xi$*, determined by the variational derivative (91), contributes to the energy storage* $H$ *(Equation (88)) through the constant* $\kappa$ *interpreted as the surface tension of a vapor bubble. The continuous contribution of millions of bubbles per unit volume can be expressed as elastic shear stress in a fluid medium:*

$$\overset{\leftrightarrow}{\pi} = (\text{grad}\ \xi) \bigotimes \frac{\partial H}{\partial(\nabla\xi)}, \tag{100}$$

*which leads, by supplementing the thermodynamic pressure in Equation (72), to the divergence of pressure being the divergence of the pressure dyad, as follows:* $div(-p\overset{\leftrightarrow}{I}) \rightarrow div(-p\overset{\leftrightarrow}{I} + \overset{\leftrightarrow}{\pi})$*. The induced stress* $\overset{\leftrightarrow}{\pi}$ *has a certain analogy here with the capillary stress introduced by Korteweg in 1901, but on completely different assumptions. In the present work, we rely on gradients* $\nabla\xi$*, while Korteweg relied on* $\nabla\rho$*.*

Similarly, a dissipative term $R_{\xi\vec{p}}$, about which little is known in non-equilibrium thermodynamics, can be introduced into the system. It can be interpreted as turbulent momentum transport induced by phase transition—that is, it would be an additional Reynolds turbulent stress tensor.

However, not much is known about the contribution of convective heat transport $J_{\xi s}$ induced by phase transition $H_\xi$ and turbulent heat transport $R_{\xi s}$ induced by the transition. Both of these quantities have physical meaning, although they have not yet been discovered by phenomenological nonequilibrium thermodynamics. The pH thermodynamics formalism immediately addresses this issue, demanding new experimental research.

The contribution $R_{\xi\xi}$, as stated in Equations (94) and (96) consists of dissipative elements whose magnitude is determined by the matrix $\overset{\leftrightarrow\xi}{S}$ and the relaxation coefficient $\Gamma$.

To summarize point 22, the formalism of pH thermodynamics, applied to a specific example of a well-known phase transition flow, opens up a "thermodynamic treasure trove" of new issues to be solved. First, we need to completely rebuild the entropy evolution equation, and the Grmela Equation (87) provides a certain clue.

https://doi.org/10.3390/en19020324

## 23. Applications of pH Thermodynamics Formalism in Power Systems

The question remains whether pH control is used in engineering and technology. It is well-known that technical thermodynamics is the basis for the design, monitoring, and regulation of various types of power plants (Szewalski [66]). However, a new challenge for control and regulation theory has recently emerged in connection with the dominance of renewable and decentralized energy sources on the market. New necessities have also emerged, such as electric vehicles and AI electronics. In formulating the principles of control theory, a new situation is arising in which, due to the use of energy storage in renewable power plants, it is necessary to rebuild the foundations of control theory and control methods. Port-Hamiltonian systems are most widely used in power plants, where they are the best tool for optimizing operations and maximizing financial profits.

An overview of the applications of pH systems for systems including wind turbines, fuel cells, photovoltaic modules, as well as electrical, thermal, chemical, and other energy storage systems is presented in the work of Tönso, Kapain and Belikov [67], which demonstrates the practical potential of the pH formalism and its advantages in terms of approaches to control theory. This is not surprising, as the pH energy formalism is dedicated to this part of the economy.

However, studies of these individual applications lead to an unpleasant conclusion. The state of development of pH models in specific applications raises concerns and fears, as each of these models has an apparatus created for its own limited needs. This means that the physical foundations of pH formalism are insufficient, inconsistent, limited, opaque, non-universal, non-interdisciplinary, and perhaps also non-thermodynamic. However, works dedicated to specific situations (for instance, [48,68–72]) allow for the development of practical tools and contribute to the development of theoretical foundations.

## 24. Prospects and Open Issues

This paper is summarized in Table 1 and Figure 2. Table 1 lists the efforts, fluxes, energy variables, and control equations for the three fluid models discussed in the paper: Euler fluid, Navier fluid, and flashing flow.

**Table 1.** List of the efforts, fluxes, energy variables, and control equations for the three fluid models discussed in the paper.

| | Effort | Flux | Energetic Variable | Control Equation |
|---|---|---|---|---|
| Euler fluid | $e_s = \begin{pmatrix} H_\rho \\ H_{\vec{v}} \\ H_s \end{pmatrix}$ | $f_s = \partial_t \begin{pmatrix} \rho \\ \vec{v} \\ s \end{pmatrix}$ | $x = \left(\rho,\ \vec{p}\right)$ | Equation (57) |
| Navier fluid | $e_s = \begin{pmatrix} H_\rho \\ H_{\vec{v}} \\ H_s \end{pmatrix}$ | $f_s = \partial_t \begin{pmatrix} \rho \\ \vec{v} \\ s \end{pmatrix}$ | $x = \left(\rho,\ \vec{p}, s\right)$ | Equation (80) |
| Flashing flow | $e_s = \begin{pmatrix} H_\rho \\ H_{\vec{v}} \\ H_s \\ H_\xi \end{pmatrix}$ | $f_s = \partial_t \begin{pmatrix} \rho \\ \vec{v} \\ s \\ \xi \end{pmatrix}$ | $x = \left(\rho,\ \vec{p}, s, \xi\right)$ | Equation (99) |

Figure 2 shows control system containing four ports: port $(f_s, e_s)$, which connects to an ideal energy storage device $\mathcal{S}$, port $(f_R,\ e_R)$, corresponding to the dissipation of available energy $\mathcal{R}$, port $(f_\xi, e_\xi)$, corresponding structural changes such as combustion, phase transition, turbulence, plasticity, etc., and external port $(f_\partial, e_\partial)$, through which the system interacts with the environment $\mathcal{E}n$ (including the operation called "control").

https://doi.org/10.3390/en19020324

This paper provides an answer to the following question: is it possible to formulate thermodynamics in a way that follows from Hamilton and Rayleigh's formalism based on two new concepts: "available energy storage" and "energy lost" during flow between energy ports. From technical practice and the observation of nature, it is known that ideal energy stores are not possible, nor are systems that only destroy available energy. Technical reality is an unrecognized composition of reversible and irreversible phenomena governed jointly by inertia and coercion. The thermodynamics of port-Hamiltonian formalism attempts to combine these two phenomena, but gives priority to reversible phenomena in mathematical formalism. The most popular model is the additive combination of phenomena through the difference between the $\mathcal{J} - R$ matrices. This model is accepted by control theory, but is not necessarily correct from the point of view of pH thermodynamics.

In other words, the additive combination of "reversibility" and "irreversibility" is a temporary solution that completely misses the needs of pH thermodynamics. It is important to remember that we live in an age dominated by additively manufactured materials, which are purpose-built materials. Their design is controlled to meet the new goals facing technology. When designing these materials, there is a constant need to refine models, including thermodynamic ones. This opens up opportunities for the pH thermodynamics approach, which shows promise in this field. We need to look for new, general Poisson brackets that will not only repeat rusty patterns, but will also reach for qualitatively new elements (e.g., see Edwards and Beris [73]). The latest literature review contained in Cardoso-Ribeiro et al. [74] confirms the dynamic development towards new applications and completely new "structure-preserving" numerical methods. However, this review is not conducted from the point of view of thermodynamics, which is why our article is a modest attempt to supplement this state of affairs and point out existing gaps and inconsistencies.

The recent revision of thermodynamics by Arjan van der Schaft [2] exposes the fundamental concepts of classical thermodynamics (technical and non-equilibrium), thereby enabling a new framework for integrating thermodynamics with system modeling and control. This thermodynamics integrated with control theory is called pH thermodynamics. Van der Schaft also shows that systems and control theory is a natural requirement for a radically new formulation and understanding of classical thermodynamics. In particular, pH thermodynamics places the theory of dissipativity, as it were, at the level of the First and Second Laws of Thermodynamics from the outset. It is pH thermodynamics that allows, for the first time in the literature, a field-theoretical approach to the properties of the state and dynamics of control systems.

On the other hand, thanks to pH thermodynamics, it turns out that not everything in the current control theory is consistent. It turns out that pH thermodynamics provides premises and arguments for radical paradigm shifts in systems and control; in particular, for the use of non-minimal state space formulations and a geometric view of them. Moreover, while systems and control theory was based primarily on linear systems with quadratic cost criteria, in accordance with basic models of systems in electrical and mechanical engineering (RLC circuits, mass-spring-damper systems, etc.) [28,32,39,40], pH thermodynamics requires research tools that go beyond this linear–quadratic paradigm.

**Author Contributions:** Conceptualization, J.B.; methodology, J.B.; formal analysis, J.B.; investigation, J.B. and P.J.Z.; writing—original draft preparation, J.B.; writing—review and editing, J.B. and P.J.Z.; supervision, J.B.; project administration, J.B.; funding acquisition, J.B. All authors have read and agreed to the published version of the manuscript.

**Funding:** This research received no external funding.

https://doi.org/10.3390/en19020324

**Data Availability Statement:** The original contributions presented in the study are included in the article, further inquiries can be directed to the corresponding author.

**Conflicts of Interest:** The authors declare no conflicts of interest.

## References


1. Rashad, R.; Califano, F.; van der Schaft, A.J.; Stramigioli, S. Twenty years of distributed port-Hamiltonian systems: A literature review. *IMA J. Math. Control Inform.* **2020**, *37*, 1400–1422.
2. van der Schaft, A.J. Classical thermodynamics revisited: A systems and control perspective. In *IEEE Control Systems Magazine*; IEEE: New York, NY, USA, 2021; Volume 41, pp. 32–60.
3. Grmela, M.; Pavelka, M. Classical and Multiscale Non-equilibrium Thermodynamics. *arXiv* **2025**, arXiv:2510.00610v1. [CrossRef]
4. Maschke, B.; van der Schaft, A.J. Port-controlled Hamiltonian systems: Modelling origins and system theoretic properties. In *Nonlinear Control Systems (NOLCOS 2004), Proceedings of the 2nd IFAC Symposium, Bordeaux, France, 4–26 June 1992*; Fliess, M., Ed.; Elsevier: Amsterdam, The Netherlands, 1993; pp. 359–365.
5. van der Schaft, A.J.; Maschke, B.M. Hamiltonian formulation of distributed-parameter systems with boundary energy flow. *J. Geom. Phys.* **2002**, *42*, 166–194. [CrossRef]
6. Ortega, R.; van der Schaft, A.J.; Maschke, B.; Escobar, G. Interconnection and damping assignment passivity-based control of port-controlled Hamiltonian systems. *Automatica* **2002**, *38*, 585–596. [CrossRef]
7. Krhac, K.; Maschke, B.; van der Schaft, A.J. Port-Hamiltonian systems with energy and power ports. In Proceedings of the 8th IFAC Symposium on Lagrangian and Hamiltonian Methods in Nonlinear Control (LHMNC 2024), Besancon, France, 10–12 June 2024.
8. van der Schaft, A.J.; Jeltsema, D. On energy conversion in port-Hamiltonian systems. In Proceedings of the 60th IEEE Conference on Decision and Control (CDC), Austin, TX, USA, 13–17 December 2021; pp. 2421–2427.
9. Duindam, V.; Macchelli, A.; Stramigioli, S.; Bruyninckx, H. *Modeling and Control of Complex Physical Systems: The Port-Hamiltonian Approach*; Springer: Berlin/Heidelberg, Germany, 2009; pp. 1–423.
10. Rashad, R.; Califano, F.; Schuller, F.P.; Stramigioli, S. Port-hamiltonian modeling of ideal fluid flow: Part I. foundations and kinetic energy. *J. Geom. Phys.* **2021**, *164*, 104201. [CrossRef]
11. Rashad, R.; Califano, F.; Schuller, F.P.; Stramigioli, S. Port-hamiltonian modeling of ideal fluid flow: Part II. compressible and incompressible flow. *J. Geom. Phys.* **2021**, *164*, 104199. [CrossRef]
12. Califano, F.; Rashad, R.; Schuller, F.P.; Stramigioli, S. Geometric and energy-aware decomposition of the Navier–Stokes equations: A port-Hamiltonian approach. *Phys. Fluids* **2021**, *33*, 047114.
13. Brugnoli, A.; Alazard, D.; Pommier-Budinger, V.; Matignon, D. A Port-Hamiltonian formulation of linear thermoelasticity and its mixed finite element discretization. *J. Therm. Stress.* **2021**, *44*, 643–661. [CrossRef]
14. Jäschke, J.; Ehrhardt, M.; Günther, M.; Jacob, B. A Two-Dimensional port-Hamiltonian Model for Coupled Heat Transfer. *Mathematics* **2022**, *10*, 4635.
15. Cardoso-Ribeiro, F.L.; Matignon, D.; Pommier-Budinger, V. Port-Hamiltonian model of two-dimensional shallow water equations in moving containers. *IMA J. Math. Control Inform.* **2020**, *37*, 1348–1366.
16. Altmann, R.; Schulze, P. A port-Hamiltonian formulation of the Navier–Stokes equations for reactive flows. *Syst. Control Lett.* **2017**, *100*, 51–55.
17. Ramirez, H.; Le Gorrec, Y. An overview on irreversible port-Hamiltonian systems. *Entropy* **2022**, *24*, 1478. [CrossRef] [PubMed]
18. Ramirez, H.; Le Gorrec, Y.; Maschke, B. Boundary controlled irreversible port-Hamiltonian systems. *Chem. Eng. Sci.* **2022**, *248*, 117107. [CrossRef]
19. Mora, L.A.; Le Gorrec, Y.; Matignon, D.; Ramirez, H. Irreversible port-Hamiltonian modelling of 3D compressible fluids. *IFAC Papers OnLine* **2023**, *56*, 6394–6399. [CrossRef]
20. Pommaret, J.-F. From Kalman to Einstein and Maxwell: The Structural Controllability Revisited. *Adv. Pure Math.* **2025**, *15*, 570–628. [CrossRef]
21. Pommaret, J.-F. Nonlinear Conformal Gravitation. *J. Mod. Phys.* **2023**, *14*, 1464–1496. [CrossRef]
22. Alicki, R.; Koslof, R. Introduction to Quantum Thermodynamics: History and Prospects. *arXiv* **2019**, arXiv:1801.08314.
23. Beris, A.N.; Edwards, B.J. Dissipation in nonequilibrium thermodynamics and its connection to the Rayleighian functional. *Phys. Fluids* **2024**, *36*, 013102. [CrossRef]
24. Grmela, M. Roles of energy and entropy in multiscale dynamics and thermodynamics. *J. Phys. Commun.* **2024**, *8*, 072001. [CrossRef]
25. Sípka, M.; Pavelka, M.; Esen, O.; Grmela, M. Direct Poisson neural networks: Learning nonsymplectic mechanical systems. *J. Phys. A Math. Theor.* **2023**, *56*, 495201.


https://doi.org/10.3390/en19020324

26. Sýkora, M.; Pavelka, M.; Peshkov, I.; Minakowski, P.; Klika, V.; Romenski, E. Comparison of the symmetric hyperbolic thermodynamically compatible framework with Hamiltonian mechanics of binary mixtures. *Contin. Mech. Thermodyn.* **2024**, *36*, 539–559. [CrossRef]
27. Pavelka, M.; Klika, V.; Grmela, M. *Multiscale Thermo-Dynamics*; de Gruyter: Berlin, Germany, 2018.
28. Pommaret, J.-F. *Partial Differential Control Theory*; Kluwer: Dordrecht, The Netherlands, 2001; Volume VIII, p. 957.
29. Badur, J.; Dudda, W. From the Cosserat continuum backgrounds to modern field theory. *Tech. Sci.* **2024**, *27*, 211–264.
30. Yahalom, A. Gravitational Hamiltonian Systems and the Retarded Gravity Inequality. *Entropy* **2024**, *26*, 986. [CrossRef]
31. Tong, D. Classical Dynamics, University of Cmbridge Part II Mathematical Tripos. 2004. Available online: http://www.damtp.cam.ac.uk/user/tong/dynamics.html (accessed on 28 October 2025).
32. Gawroński, W. *Advances Structural Dynamics and Active Control of Structures*; Springer: Berlin/Heidelberg, Germany, 2004; p. 396.
33. Grmela, M.; Öttinger, H.C. Dynamics and thermodynamics of complex fluids. I. Development of a general formalism. *Phys. Rev. E* **1997**, *56*, 6620–6632. [CrossRef]
34. Zafferi, A.; Peschka, D.; Thomas, M. GENERIC framework for reactive fluid flows. *Z. Angew. Math. Mech.* **2023**, *103*, e202100254.
35. Grmela, M. Bracket formulation of diffusion-convection equations. *Phys. D* **1986**, *21*, 179–212.
36. Grmela, M. Thermodynamics and Rate Thermodynamics. *J. Stat. Phys.* **2024**, *191*, 75. [CrossRef]
37. Grmela, M. Rheological modeling with GENERIC and with the Onsager principle. *arXiv* **2025**, arXiv:2507.22144. [CrossRef]
38. Morrison, P.J. Hamiltonian Description of the Ideal Fluid. *Rev. Mod. Phys.* **1998**, *70*, 467–521. [CrossRef]
39. Ogata, K. *Modern Control Engineering*; Prentice Hall: Hoboken, NJ, USA, 1997.
40. Mieloszyk, E. *Rachunek Operatorów w Zagadnieniach Uogólnionych Układów Dynamicznych*; Wyd. Politechniki Gdańskiej: Gdańsk, Poland, 1999.
41. Mielke, A. Formulation of thermoelastic dissipative material behavior using GENERIC. *Contin. Mech. Thermodyn.* **2011**, *23*, 233–256. [CrossRef]
42. Pommaret, J.-F. Minimum Parametrization of the Cauchy Stress Operator. *arXiv* **2023**, arXiv:2101.03959. [CrossRef]
43. Badur, J. *Eternal Noether's Symmetries*; IMP Press: Gdańsk, Poland, 2021; pp. 1–514.
44. Sieniutycz, S.; Berry, R.S. Conservation laws from Hamilton's principle for nonlocal thermodynamic equilibrium fluids with heat flow. *Phys. Rev. A* **1989**, *40*, 348–361. [CrossRef] [PubMed]
45. Nevir, P.; Blender, R. A Nambu representation of incompressible hydrodynamics using helicity and enstrophy. *J. Phys. A Math. Gen.* **1993**, *26*, 1189–1193. [CrossRef]
46. Blender, R.; Badin, G. Hydrodynamic Nambu Brackets derived by Geometric Constraints. *J. Phys. A* **2015**, *48*, 105501.
47. Sato, N. Generalization of Hamiltonian mechanics to a three-dimensional phase space. *Prog. Theor. Exp. Phys.* **2021**, *2021*, 063A01. [CrossRef]
48. Brugnoli, A. A Port-Hamiltonian Formulation of Flexible Structures. Ph.D. Thesis, Universite de Toulouse, Toulouse, France, 2020.
49. Mielke, A.; Peletier, M.A.; Zimmer, J. Deriving a GENERIC system from a Hamiltonian system. *arXiv* **2024**, arXiv:2404.09284. [CrossRef]
50. Maschke, B.M.; Kirchhhoff, J. Port maps of irreversible port-Hamiltonian systems. *IFAC-Papers OnLine* **2023**, *56*, 6796–6800.
51. Badur, J.; Ziółkowski, P.; Ziółkowski, P.J. On angular velocity slip in nanoflows. *Microfluid. Nanofluid.* **2015**, *19*, 191–199.
52. Mehrmann, V.; Unger, B. Control of port-Hamiltonian differential-algebraic systems and applications. *Acta Numer.* **2023**, *32*, 395–515.
53. Pavelka, M. Multiscale Geometric Approach to Continuum Mechanics and Thermodynamics. Habilitation Thesis, Faculty of Mathematics and Physics, Charles University, Prage, Czech Republic, November 2021.
54. Szücs, M.; Pavelka, M.; Kovács, R.; Fülöp, T.; Ván, P.; Grmela, M. A case study of non-Fourier heat conduction using internal variables and GENERIC. *J. Non-Equilib. Thermodyn.* **2021**, *47*, 31–60. [CrossRef]
55. Truesdell, C.; Toupin, R. The classical field theories. In *Handbuch der Physik*; Flügge, S., Ed.; Springer: Berlin/Heidelberg, Germany, 1960; Volume III/1, pp. 226–793.
56. Bendimerad-Hohl, A.; Haine, G.; Matignon, D.; Maschke, B. Structure-preserving discretization of a coupled Allen-Cahn and heat equation system. *IFAC Papers OnLine* **2022**, *55*, 99–104.
57. Euler, L. Principia motus fluidorum, Opera Omnia Series 2. *Novi Comment. Acad. Sci. Petropolitanae* **1752**, *12*, 133–168.
58. von Helmholtz, H. Over integrals of the hydrodynamic equations, which correspond to eddy movements. *J. Reine Angew. Math.* **1858**, *55*, 22–55.
59. Clebsch, A. Ueber die Integration der hydrodynamischen Gleichunge. *J. Reine Angew. Math.* **1859**, *56*, 1–10.
60. Bateman, H. Notes on a Differential Equation Which Occurs in the Two-Dimensional Motion of a Compressible Fluid and the Associated Variational Problems. *Proc. R. Soc. London* **1929**, *125*, 598–618.
61. Eckart, C.H. The electrodynamics of material media. *Phys. Rev.* **1938**, *54*, 920–923.


https://doi.org/10.3390/en19020324

62. Sanders, J.W.; DeVoria, A.C.; Washuta, N.J.; Elamin, G.A.; Skenes, K.L.; Berlinghieri, J.C. A canonical Hamiltonian formulation of the Navier–Stokes problem. *J. Fluid Mech.* **2024**, *984*, A27. [CrossRef]
63. Badur, J.; Ochrymiuk, T.; Kowalczyk, T.; Dudda, W.; Ziółkowski, P. From fluid mechanics backgrounds to modern field theory. *Acta Mech.* **2022**, *223*, 3453–3465. [CrossRef]
64. Grmela, M.; Klika, V.; Pavelka, M. Gradient and GENERIC time evolution towards reduced dynamics. *Phil. Trans. R. Soc. A* **2020**, *378*, 20190472.
65. Bilicki, Z.; Kestin, J. Physical aspects of the relaxation model in two-phase flow. *Proc. R. Soc. London Ser. A.* **1990**, *428*, 379–397.
66. Szewalski, R. Zjawisko Drobnych Drgań Regulatora i Jego Znaczenie dla Dynamiki Sprawności Regulacji. Ph.D. Thesis, Wydział Mechaniczny, Politechnika Lwowska, Lwów, Ukraine, 1935.
67. Tõnso, M.; Kapain, V.; Belikov, J. Port-Hamiltonian framework in power systems domain: A survey. *Energy Rep.* **2023**, *10*, 2918–2930. [CrossRef]
68. Wang, Y.; Sun, Z.; Li, P.; Ai, J.; Wu, C.; Shen, Z.; Deng, F. A Port-Hamiltonian Perspective on Dual Active Bridge Converters: Modeling, Analysis, and Experimental Validation. *Energies* **2025**, *18*, 5197. [CrossRef]
69. Rettberg, J.; Wittwar, D.; Buchfink, P.; Brauchler, A.; Ziegler, P.; Fehr, J. Port-Hamiltonian fluid–structure interaction modelling and structure-preserving model order reduction of a classical guitar. *Math. Comput. Model. Dyn. Syst.* **2023**, *29*, 116–148. [CrossRef]
70. Lasarzik, R. Analysis of a thermodynamically consistent Navier-Stokes-Cahn-Hilliard model. *Nonl. Anal.* **2021**, *213*, 112526. [CrossRef]
71. Pavelka, M.; Peshkov, I.; Klika, V. On Hamiltonian continuum mechanics. *J. Phys. D Nonlinear Phenom.* **2020**, *408*, 132510.
72. Kekana, M.; Badur, J. An analysis of a piezo-elastic restrictors using a capacitive shunt. *TASK Quaterly* **2004**, *8*, 377–384.
73. Edwards, J.; Beris, A.N. Unified view of transport phenomena based on the generalized bracket formulation. *Ind. Eng. Chem. Res.* **1991**, *30*, 873–881. [CrossRef]
74. Cardoso-Ribeiro, F.L.; Haine, G.; Le Gorrec, Y.; Matignon, D.; Ramirez, H. Port-Hamiltonian formulations for the modelling, simulations and control of fluids. *Comput. Fluids* **2024**, *283*, 106407. [CrossRef]




https://doi.org/10.3390/en19020324

MDPI AG
Grosspeteranlage 5
4052 Basel
Switzerland
Tel.: +41 61 683 77 34

*Energies* Editorial Office
E-mail: energies@mdpi.com
www.mdpi.com/journal/energies

www.ingramcontent.com/pod-product-compliance
Lightning Source LLC
LaVergne TN
LVHW070459061126
851858LV00011B/267
* 9 7 8 3 7 2 5 8 8 4 0 3 2 *

water

MDPI

*Editorial*

# Groundwater Depletion: Current Trends and Future Challenges to Mitigate the Phenomenon

**Nerantzis Kazakis [1,*], Micol Mastrocicco [2] and Konstantinos Chalikakis [3]**

1 Laboratory of Hydrogeology, Department of Geology, Faculty of Natural Sciences, University of Patras, 26504 Rion, Greece
2 Department of Environmental, Biological and Pharmaceutical Sciences and Technologies, University of Campania "Luigi Vanvitelli", Via Vivaldi 43, 81100 Caserta, Italy; micol.mastrocicco@unicampania.it
3 UMR 1114 EMMAH (AU-INRAE), Avignon Université, 84029 Avignon, France; konstantinos.chalikakis@univ-avignon.fr
* Correspondence: nkazakis@upatras.gr

Groundwater constitutes the primary source of drinking water for more than two billion people across the globe. Additionally, groundwater contributes to food and energy security, as well as human health and ecosystem preservation. However, groundwater depletion is a common phenomenon worldwide. This phenomenon appears when extraction surpasses the recharge of the aquifer, and it is triggered by the socioeconomic framework and controlled by the aquifer regime. Groundwater depletion can lead to seawater intrusion in coastal aquifers, a reduction in river base flow, upconing of deep geothermal water, and increased pumping costs. Groundwater depletion has been reported on both a global and regional scale. This Special Issue, entitled "Groundwater Depletion: Current Trends and Future Challenges to Mitigate the Phenomenon", attempts to highlight the efficiency and adequacy of global and regional strategies to mitigate the phenomenon and to promote the transition from regional to aquifer scale. Additionally, it is critically important to appraise the role of climate variability, management practices, and the monitoring of groundwater reserves. The Special Issue also provides the state-of-the-art of tools and methods to map, model, and forecast the phenomenon.

Particularly, this Special Issue includes 11 articles, providing all aspects of the groundwater depletion phenomenon in several different countries incorporating aquifers with different hydrogeological conditions. In an article by Papadopoulos et al. (Contribution 1), the fuzzy logic approach was used to develop a multi-criteria model for the selection of groundwater recharge sites. The model was developed for the aquifer system in the agricultural plain of the Prefecture of Xanthi in northeast Greece. In the unconfined aquifers of the Campania region in southern Italy, a simulation model was obtained in order to determine the dynamic and the impact of boundary conditions on groundwater budget. The non-uniform extinction depth is highlighted as the most influential in aquifer dynamics, while both sea level and rivers scarcely affect groundwater budgets (Contribution 2). Contribution 3 estimated the groundwater storage changes/anomalies (GWCs) in the Indus River Basin, using data from GRACE and the WaterGAP Hydrological Model. Groundwater storage is mainly triggered in the study area by climatic variations and increased pumping for irrigation.

Contribution 4 suggested the use of a closed-surface coal mine as lake to store water and use it for energy production, irrigation, and as a recreation park. Tzampoglou and Loupasakis (Contribution 5) used satellite and in situ data to provide evidence of the overexploitation of the aquifer in the Amyntaio basin, both by dewatering wells of an open-pit coal mine and irrigation wells. Contribution 6 introduced a comprehensive methodology for water resource management, including "framing", "mapping", "involvement", "co-production", and "sharing" phases. Contribution 7 suggested a nature-based solution for the restoration of groundwater level decline in the Gansu Province of China. Contribution 8 provided the results of the APLIS model for the estimation of groundwater recharge in Karst aquifers in southern Greece. The occurrence of microplastics was identified in the groundwater of SE Attica, Greece (Contribution 9). Contribution 10 highlighted the complementarity of long-term piezometric data combined with hydrogeophysics and fracture analysis to understand complex aquifer structure and groundwater dynamics. Groundwater flow dynamics were estimated using a numerical model (FEFLOW) in the Moghra desert of Egypt (Contribution 11).

We wish to extend our gratitude to the authors who contributed to this Special Issue, to the reviewers for their invaluable support, and to the organizers and staff of MDPI for their dedication in bringing this issue to completion and publication.

We believe that this work will stimulate researchers to quantify groundwater depletion at several sites worldwide and suggest strategies to mitigate the phenomenon at the aquifer scale.

**Conflicts of Interest:** The authors declare no conflict of interest.

**List of Contributions:**


1. Papadopoulos, C.; Spiliotis, M.; Pliakas, F.; Gkiougkis, I.; Kazakis, N.; Papadopoulos, B. Hybrid Fuzzy Multi-Criteria Analysis for Selecting Discrete Preferable Groundwater Recharge Sites. *Water* **2022**, *14*, 107. https://doi.org/10.3390/w14010107.
2. Gaiolini, M.; Colombani, N.; Busico, G.; Rama, F.; Mastrocicco, M. Impact of Boundary Conditions Dynamics on Groundwater Budget in the Campania Region (Italy). *Water* **2022**, *14*, 2462. https://doi.org/10.3390/w14162462.
3. Mehmood, K.; Tischbein, B.; Flörke, M.; Usman, M. Spatiotemporal Analysis of Groundwater Storage Changes, Controlling Factors, and Management Options over the Transboundary Indus Basin. *Water* **2022**, *14*, 3254. https://doi.org/10.3390/w14203254.
4. Louloudis, G.; Roumpos, C.; Louloudis, E.; Mertiri, E.; Kasfikis, G. Repurposing of a Closed Surface Coal Mine with Respect to Pit Lake Development. *Water* **2022**, *14*, 3558. https://doi.org/10.3390/w14213558.
5. Tzampoglou, P.; Loupasakis, C. Hydrogeological Hazards in Open Pit Coal Mines–Investigating Triggering Mechanisms by Validating the European Ground Motion Service Product with Ground Truth Data. *Water* **2023**, *15*, 1474. https://doi.org/10.3390/w15081474.
6. Malamataris, D.; Chatzi, A.; Babakos, K.; Pisinaras, V.; Hatzigiannakis, E.; Willaarts, B.A.; Bea, M.; Pagano, A.; Panagopoulos, A. A Participatory Approach to Exploring Nexus Challenges: A Case Study on the Pinios River Basin, Greece. *Water* **2023**, *15*, 3949. https://doi.org/10.3390/w15223949.
7. Liu, M.; Nie, Z.; Cao, L.; Wang, L.; Lu, H. Nature-Based Solutions for the Restoration of Groundwater Level and Groundwater-Dependent Ecosystems in a Typical Inland Region in China. *Water* **2024**, *16*, 33. https://doi.org/10.3390/w16010033.
8. Nanou, E.-A.; Perdikaris, K.; Tserolas, P.; Zagana, E. Recharge Assessment in Greek Karst Systems: Methodological Considerations and Implications. *Water* **2024**, *16*, 568. https://doi.org/10.3390/w16040568.
9. Perraki, M.; Skliros, V.; Mecaj, P.; Vasileiou, E.; Salmas, C.; Papanikolaou, I.; Stamatis, G. Identification of Microplastics Using μ-Raman Spectroscopy in Surface and Groundwater Bodies of SE Attica, Greece. *Water* **2024**, *16*, 843. https://doi.org/10.3390/w16060843.
10. Bertrand, M.; Bertrand, C.; Mazzilli, N.; Gigleux, S.; Denimal, S.; Valois, R.; Girod, L.-M.; Cinkus, G.; Busquet, V.; Chalikakis, K. Advancing the Understanding of Complex Piezometric Information: A Methodological Approach Integrating Long-Term Piezometry, Surface Nuclear Magnetic Resonance, and Fracture Analysis Using Insights from the "Calcaires du Barrois" Series, France. *Water* **2024**, *16*, 1700. https://doi.org/10.3390/w16121700.
11. Shalby, A.; Zeidan, B.A.; Pietrucha-Urbanik, K.; Negm, A.M.; Armanuos, A.M. Modelling Approach for Assessment of Groundwater Potential of the Moghra Aquifer, Egypt, for Extensive Rural Development. *Water* **2024**, *16*, 1562. https://doi.org/10.3390/w16111562.

water

MDPI

*Article*

# Delineation of Groundwater Potential Zones Using Geospatial Techniques. Case Study: Roman City and the Surrounding Area in the Northeastern Region, Romania

**Petrut-Liviu Bogdan [1,2,*], Valentin Nedeff [1], Mirela Panainte-Lehadus [1], Dana Chitimuș [1], Narcis Barsan [1] and Florin Marian Nedeff [1]**

[1] Faculty of Engineering, Vasile Alecsandri University of Bacau, 157 Calea Marasesti, 600115 Bacau, Romania; vnedeff@ub.ro (V.N.); mirelap@ub.ro (M.P.-L.); dana.chitimus@ub.ro (D.C.); narcis.barsan@ub.ro (N.B.); florin_nedeff@ub.ro (F.M.N.)

[2] National Administration "Romanian Waters"—Siret River Basin Administration, Cuza Voda No. 1, 600274 Bacau, Romania

* Correspondence: bogdan.p.liviu@gmail.com

**Abstract:** Effective groundwater management is crucial under the current climatic conditions, addressing both qualitative and quantitative aspects. An important step in delineating groundwater potential zones involves remote sensing (RS) data and geographic information systems (GISs), facilitating resource assessment, and the implementation of suitable field data management. This study introduces the delineation of potential groundwater zones using seven layers and the Multi-Criteria Decision Analysis (MCDA) method. Satty's Analytic Hierarchy Process (AHP) was employed to rank the seven selected parameters, contributing to the advancement of groundwater research and resource assessment. All seven thematic layers (Rainfall, Geology, Land Use/Land Cover, Drainage Density, Elevation, Slope, and Soil) were prepared and analyzed to delineate groundwater potential zones. The resulting groundwater potential zone map was categorized into four classes, Very Good, Good, Moderate, and Poor, covering areas of 81.53 $km^2$ (45.1%), 56.36 $km^2$ (31.2%), 19.54 $km^2$ (10.8%), and 23.17 $km^2$ (12.8%) of the total area, respectively. The accuracy of the output was validated by comparing it with information on groundwater prospects in the area, and the overall accuracy of the method was approximately 72%. High-yield boreholes were drilled and concentrated in the Very Good groundwater potential zones, while low-yield ones were developed in the Poor areas.

**Keywords:** groundwater potential; multi-criteria decision-maker; AHP; GIS techniques; remote sensing

## 1. Introduction

Climate change plays a significant role in and has important implications for groundwater resources and the hydrological cycle [1–4]. In recent decades, due to a growing global population, the demand for freshwater has increased substantially, with the necessity of water for irrigation and industrialization leading to more frequent instances of water scarcity [5–8].

Groundwater is one of the most important resources, stored in various geological formations, showing variations in quantity and quality, as well as different parameters, including temperature, chemistry, flow direction, and interactions with surface water systems [9–13]. Preserving and conserving groundwater resources is a pressing issue of the 21st century, with water management posing a significant challenge for governments worldwide [11,14,15].

The study area is located in the Siret River Basin, and the primary source of supply comes from groundwater, specifically from the ROSI03 aquifer, a phreatic groundwater body, utilized in varying proportions: 67.7% for meeting the population's demands, 2.3% for agriculture, and 30% for industrial sector [16]. The examined region is positioned upstream from the confluence of two major rivers, namely the Moldova River and the Siret River, situated in the Siret River watershed, and is primarily characterized as an agricultural zone.

The conventional methodologies employed for the identification, delineation, and mapping of groundwater potential zones primarily rely on on-site surveys utilizing geophysical, geological, and hydrogeological tools, but these methods are typically characterized by high costs and time-intensive processes [4,14,17]. The introduction of geospatial tools in the field of geosciences is an efficient approach, particularly when it comes to groundwater exploration before resorting to conventional exploration and exploitation techniques [18]. Remote sensing technologies (RI) and the Geographic Information System (GIS) provide a rapid and cost-effective method for estimating underground resources [19–21]. In the context of delineating groundwater potential zones and mapping them, this study employs a combination of GIS technology and the Analytical Hierarchy Process (AHP). AHP, developed by Thomas Saaty in 1980, proves to be an effective method for simplifying complex decision-making through direct comparisons to a series of pairwise comparisons and for summarizing result integration [22]. Choosing AHP for this study was a decision based on the advantages of the method, particularly its ability to utilize multi-criteria decision-making for integrating diverse thematic layers, which is essential for delineating groundwater potential zones. Although other methods such as Machine Learning, Fuzzy Logic, and Artificial Neural Networks exist, AHP was selected for its simplicity, ease of application, and validation in various studies [14,23,24]. Furthermore, AHP is a suitable technique for ensuring result consistency, thereby minimizing subjective influences in the decision-making process [25]. The GIS was used to perform weighted overlay analysis on the layer maps, considering the priorities determined by the AHP, resulting in the identification of potential groundwater zones [26–28].

Various studies worldwide have focused on identifying potential groundwater zones utilizing GIS and mathematical models. For instance, Shreeya Baghel et al. [14] in their study on groundwater potential in the catchment of the Mahandi Basin, D. C. Jhariya et al. [10] in their paper assessing groundwater potential zones in Raipur city, India, Aysha Akter et al. [21] with the study predicting groundwater potential zones using GIS techniques, and Mustafa et al. [29] in their study of the Demirci district (Turkey) highlighted the effective integration of various factors, including geology, geomorphology, rainfall, land use/land cover (LULC), drainage density, slope, groundwater level depth, and soil texture. These studies employed the weighted index analytical hierarchy process (AHP) method to develop groundwater potential models, contributing to a clear understanding of groundwater potential. These findings affirm the method's utility across varied geographical contexts and suggest that the results obtained are similar to those presented in this study.

Taking all aspects into consideration, the primary goal of this study, which combines AHP with RS and GIS, is to outline, recognize, and map groundwater potential zones to facilitate the sustainable management of water resources in the region.

## 2. Materials and Methods

The present study was conducted in the northeast of Romania on the east part of Neamț county. The region of interest (ROI) described in this paper is situated in the south extremity of Fălticeni Plateau, to the confluence of two major rivers, Moldova on the west boundary and Siret on the east boundary [30]. The study area has a total surface of 180.6 square km, shown in Figure 1. The coordinate system used in this study is Stereo 70(EPSG31700), specific to Romania and and for all maps used in this article. For better international recognition, the equivalent coordinates in WGS 84(EPSG4326) are provided as follows: the study area is located between latitudes 46°54′28.22″ N and 47°05′30.26″ N, and longitudes 26°46′43.81″ E and 26°58′15.50″ E.

The area comprises the city of Roman, with the highest economic and industrial development, and the surrounding northern municipalities, which, in general, are predominantly characterized by agricultural activities [31]. The average elevation of the study area varies from 175 to 260 m above the mean sea level.

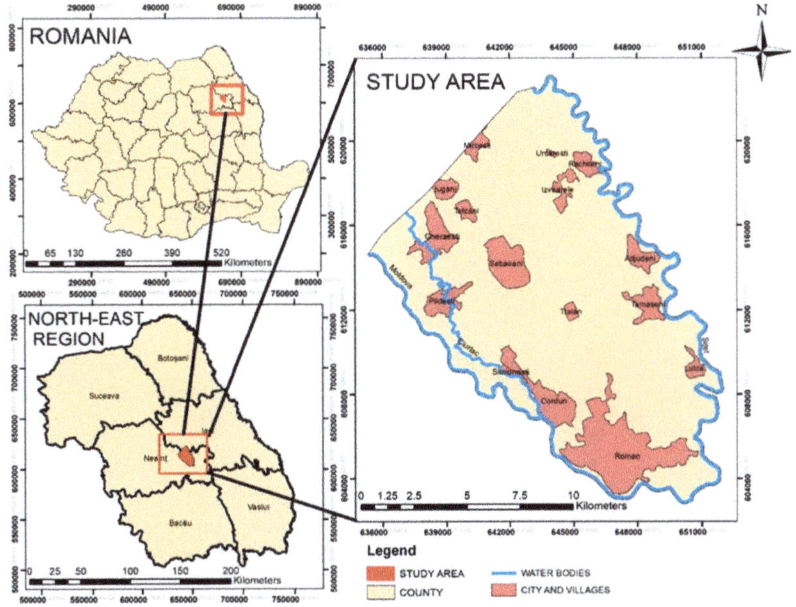


**Figure 1.** Study area.

## 2.1. Data Acquisition

In the process of identifying potential groundwater zones within the region of interest (ROI), a variety of geospatial techniques were employed. The methodology involves digital image processing, digital elevation model (DEM) evaluation, and field studies including hydrogeological data.

To delineate the groundwater potential zone (GPZ) for the study area, seven parameters such as geology, rainfall, slope, drainage density, Land Use and Cover (LULC), soil and elevation, as presented in Table 1, were analyzed. One of the datasets required for the study is represented by the digital elevation model (DEM) which has been acquired from the United States Geological Survey through the Shuttle Radar Topography Mission (SRTM) with a resolution of 30 m (cell size = 30 × 30 m/pixel). The DEM was used to delineate the boundaries of the ROI, drainage density and the slope using different techniques and spatial tools in ArcGis 10.8 software. Precipitation data were acquired from pluviometric gauging stations belonging to the Siret River Basin Administration Bacău. The data were processed to create rainfall maps with ArcGIS software using the IDW interpolation tool. The geological data were processed by digitizing the Geological map of Romania, L-25-IX, 13 Piatra Neamț at a scale of 1:200,000, and the soil data were prepared by digitizing the Soil Map of Romania at a scale of 1:200,000. The Land Use and Land Cover (LULC) map was generated from LANDSAT 8 OLI (Operational Land Imager) data from 2020 with a 30 m resolution using supervised image classification in ArcGIS10.8.

The entire workflow process is described in Figure 2 according to the data shown in Table 1. This representation summarizes the steps involved in delineating groundwater potential zones, integrating various data. This visualization aids in effectively communicating the methodology employed and facilitates a deeper understanding of the groundwater potential assessment process.

**Table 1.** Principal database used in groundwater potential study.

| S. No | Data Used | Parameters | Scale/Resolution | Source | Application |
|---|---|---|---|---|---|
| 1 | Elevation Data | Drainage Density, Slope, Location of the study area | 30 m | [32] | Boundary Area, Calculate GWPZ |
| 2 | Meteorological data | Rainfall map | 30 m | [16,33] | Calculate GWPZ |
| 3 | Soil Map | Soil Texture, Soil type | 1:200,000 | [34] | Calculate GWPZ |
| 4 | Geological Map | Geology type | 1:200,000 | [35] | Calculate GWPZ |
| 5 | LANDSAT 8 OLI | Land Use and Cover map (LULC) | 30 m | [31] | Calculate GWPZ |
| 6 | Hydrogeological maps | Hydrogeological data (groundwater table, groundwater flow rate) | 1:200,000 | [33] | Result Validation |

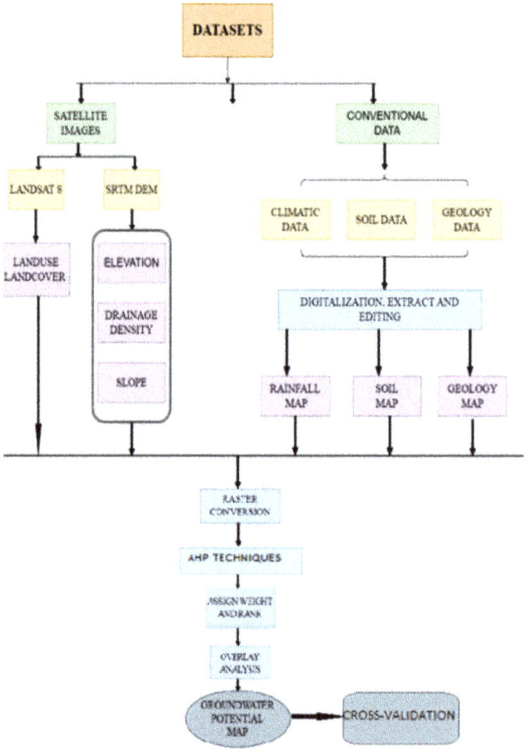


**Figure 2.** Flowchart.

*2.2. Multi-Criteria Decision Analysis Using AHP*

Analytical Hierarchical Process (AHP) is one of the most used and well-known GIS method for delineating groundwater potential zones by integrating all thematic layers. In this study, we used seven different layers with sub-criteria that have a different weight in controlling groundwater flow and storage in the area. These factors are weighted according to their local impact on the presence of groundwater and expert judgement/opinion. In the specialized literature, there are studies that establish similar weights for factors, depending on their importance in different hydrogeological contexts [5,10,14].

The Analytical Hierarchy Process (AHP), first introduced by L. Saaty, is a technique for multicriteria decision-making [36]. It utilizes a pairwise comparison matrix to determine scale ratios between different criteria. According to Saaty's scale, presented in Tables 2 and 3, in pairwise comparisons, a value of 1 is assigned to two thematic layers when they are considered to have equal importance for groundwater potential, while a value of 9 is given to represent extreme importance in such pairs [22]. According to each layer importance, we created a pairwise comparison matrix for seven variables, presented in Table 4. Various weights were allocated to thematic layers, reflecting their significance in terms of water retention capacity. The total weight (TW) was obtained by the total value of the scale weight divided by the cumulative number of parameters for all thematic layers [22]. The Normalized Pairwise Comparison Matrix (NPCM) was constructed by calculating ratios derived from the division of the total value and each individual judgment value within the same column. Following that, the total of these ratio values within each row was divided by the count of parameters in the pairwise comparison matrix [37].

**Table 2.** Analytical Hierarchy Process scale [22,36].

| Intensity of Importance | Definition | Explanation |
|---|---|---|
| 1 | Equal importance | Two elements contribute equally to the goal |
| 3 | Moderate importance | Experience and judgment slightly favor one element over another |
| 5 | Strong Importance | Experience and judgment strongly favor one element over another |
| 7 | Very strong importance | One element is favored very strongly over another, and its dominance is demonstrated in practice |
| 9 | Extreme importance | The evidence favoring one element over another is of the highest possible order of affirmation |
| 2,4,6,8 used to express intermediate values preference in weights | | |

The Consistent Ratio (CR) is used to validate the accuracy of the weights derived from the normalized pairwise comparison matrix. It compares the ratio between the Consistency Index (CI) and a Random Consistency Index (RI) that depends on the number of criteria used [25]:

$$CR = CI/RI \tag{1}$$

where CI is the Consistency Index; and RI is the Random Consistency Index (values obtained from Saaty's standard (Table 3)).

$$CI = (\lambda_{max} - n)/(n - 1) \tag{2}$$

where n is the number of criteria (layers); and $\lambda_{max}$ is the average value of the Consistency Index (Eigen value).

**Table 3.** Saaty's ratio index for different values of n [37].

| N | 1 | 2 | 3 | 4 | 5 | 6 | 7 |
|---|---|---|---|---|---|---|---|
| RI | 0 | 0 | 0.58 | 0.89 | 1.12 | 1.24 | 1.32 |

According to Saaty, if CR is equal to 0.1 or less, it is acceptable to continue the analysis, but if the consistency value is greater than 0.1, it is necessary for the judgment to be revised [36].

$$\text{Maximum Eigen value } (\lambda_{max}) = 7.37$$

$$n = 7$$

$$\text{Consistency Index CI} = (\lambda_{max} - n)/(n - 1) = 0.06$$

$$\text{Random Index} = 1.32$$

$$\text{Consistency Ratio (CR)} = CI/RI = 0.047$$

**Table 4.** Pairwise comparation matrix for all thematic layers.

| Variable | Rainfall | Geology | LULC | Drainage Density | Elevation | Slope | Soil | Normalized Weight |
|---|---|---|---|---|---|---|---|---|
| Rainfall | 1 | 2 | 4 | 3 | 2 | 3 | 3 | 0.28 |
| Geology | 1/2 | 1 | 4 | 4 | 1 | 3 | 3 | 0.21 |
| LULC | 1/4 | 1/4 | 1 | 1/3 | 1/4 | 1/3 | 1 | 0.05 |
| Drainage Density | 1/3 | 1/4 | 3 | 1 | 1/3 | 1/2 | 3 | 0.09 |
| Elevation | 1/2 | 1 | 4 | 3 | 1 | 2 | 3 | 0.19 |
| Slope | 1/3 | 1/3 | 3 | 2 | 1/2 | 1 | 3 | 0.12 |
| Soil | 1/3 | 1/3 | 1 | 1/3 | 1/3 | 1/3 | 1 | 0.06 |

### 2.3. *Weighted Overlay Analysis (WOA)*

Following the transformation to a cell size of 30 × 30 m, the thematic layers were reclassified using the obtained weighting values. Weighted Overlay Analysis (WOA) is a method that empowers users to address intricate spatial site suitability factors by integrating diverse input parameters, leading to the delineation of groundwater potential zones (GZPs). In the Weighted Overlay Analysis (WOA) tool, all reclassified raster maps are overlaid and appropriate weights are assigned. The generated raster layer was categorized into four groups of groundwater potential zones (GPZs) using consistent weight ranges in accordance with recommendations from the United Nations' Food and Agricultural Organization [38].

$$S = \sum_{i=1}^{n} (wi.xi), \tag{3}$$

where S is the total GPZ score; wi denotes the weight of the GPZ criteria; xi indicates the sub-criteria score of i GPZ criteria; and n represents the total number of GPZ criteria.

Each sub-criteria in each thematic layer received scores ranging from 1 to 5, taking into consideration favorable conditions and their importance in identifying groundwater zones. The least suitable sub-parameters were assigned a minimum score of 1, and the highest suitable sub-parameters were assigned a maximum score of 5, while sub-parameters of moderate suitability were given intermediate values for groundwater potential zone (GPZ) identification, as presented in Table 5.

**Table 5.** Percentage influence and scale value of individual themes for the overlay analysis.

| Parameter | Influence (%) | Sub-Criteria | Feature Weight |
|---|---|---|---|
| RAINFALL | 28 | 650.73–667.57 mm | 2 |
| | | 667.58–681.04 mm | 3 |
| | | 694.52–710.14 mm | 3 |
| | | 681.05–694.51 mm | 3 |
| | | 710.14–728.8 mm | 3 |
| GEOLOGY | 21 | Holocene gravel and sand deposits | 4 |
| | | Pleistocene diluvial and proluvial deposits | 1 |
| | | Holocene recent alluvium | 3 |

**Table 5.** *Cont.*

| Parameter | Influence (%) | Sub-Criteria | Feature Weight |
|---|---|---|---|
| **LULC** | 5 | Water bodies | 3 |
| | | High vegetation | 3 |
| | | Low vegetation | 3 |
| | | Agriculture land | 3 |
| | | Built-up area | 1 |
| **DRAINAGE DENSITY** | 9 | 0–0.31 km/km$^2$ | 4 |
| | | 0.32–0.88 km/km$^2$ | 3 |
| | | 0.89–1.4 km/km$^2$ | 3 |
| | | 1.5–2.0 km/km$^2$ | 2 |
| | | 2.1–3.3 km/km$^2$ | 1 |
| **ELEVATION** | 19 | 175.5–189.9 m | 4 |
| | | 190.0–201.2 m | 3 |
| | | 201.3–216.8 m | 2 |
| | | 216.9–233.5 m | 2 |
| | | 233.6–260.4 m | 1 |
| **SLOPE** | 12 | 0.0–1.3% | 5 |
| | | 1.4–3.8% | 4 |
| | | 3.9–8.6% | 3 |
| | | 8.7–16.0% | 2 |
| | | 16.1–53% | 1 |
| **SOIL** | 6 | Clay and silt loam | 2 |
| | | Soil with variate texture | 2 |
| | | Sandy loam | 3 |
| | | Marshes and pounds | 4 |
| | | Clayey | 1 |

Verification and validation of the obtained GWPZ map were conducted by comparing the groundwater yield data from eleven boreholes provided by the Siret River Basin Administration in Bacău.

## 3. Results and Discussion

The following sections explore the outcomes and discussion of this study. This part involves the creation of thematic layers, delineation of groundwater zones, and the evaluation of map accuracy through validation. This is accomplished by comparing actual yield descriptions with the expected yield class derived from the predictive map.

### *3.1. Rainfall*

One of the main factors influencing groundwater recharge in the study area is represented by precipitation. The intensity and duration of rainfall play an important role in infiltration dynamics. High-intensity, short-duration rain events tend to limit infiltration, favoring surface runoff, whereas low-intensity, long-duration rainfall supports greater infiltration, which can be fundamental for groundwater recharge and sustainable water resource management. The spatial distribution of the multiannual rainfall map was classified into five categories, which include (650.73–667.57) mm/year, (667.58–681.04) mm/year,

(681.05–694.51) mm/year, (694.52–710.14) mm/year, and (710.14–728.8) mm/year, as presented in Figure 3a. It is noticeable that the precipitation amounts gradually decrease near the study area's boundary.

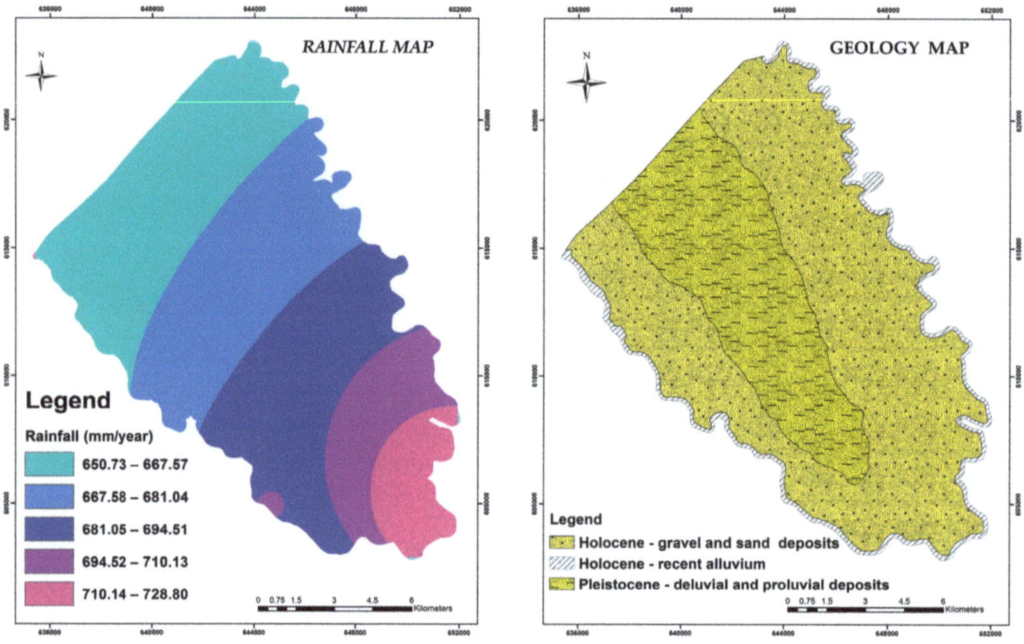


(a) (b)

**Figure 3.** (**a**) Spatial distribution of precipitation; (**b**) geology map of the interest area.

### *3.2. Geology*

Geology plays a critical role in determining the characteristics and distribution of groundwater, influencing the porosity and permeability of groundwater potential in an area, and is used in various analyzed studies where geological maps are integrated into GIS. A comparison between the geological layer used in this study and those in existing literature shows significant similarities in methodology and application [10,14]. A substantial area within the study region, presented in Figure 3b, is marked by Holocene gravel and sand deposits, while higher elevation zones display Pleistocene diluvial and proluvial deposits. According to the lithological characteristics, a higher feature weight was assigned to Holocene deposits, while a lower weight was assigned to Pleistocene deposits.

### *3.3. Land Use Land Cover (LULC)*

Land Use and Land Cover (LULC) data are important in providing essential information regarding infiltration, soil moisture, groundwater, and surface water, and these offer perspectives on groundwater requirements. The study area's land classification was categorized into five classes, which included water bodies, high vegetation, low vegetation, agricultural land, and built-up areas presented in Figure 4a. The predominant category in the study area is represented by agricultural land, accounting for 60.2% of the total area, with a concentration in the central part of the region. Built-up areas are notably prevalent in the southern part, particularly around the Roman city and its adjacent counties. The land use and land cover were validated using Google satellite images and ground-truthing data, with a Kappa coefficient of 0.71, and the overall classification accuracy was found to be 76%, as previously reported in other studies [31]. Based on various criteria such as water transmission and storage capacity, the "Built-Up Area" category received a low feature weight score of 1, while the remaining classes were assigned higher feature weights in general.

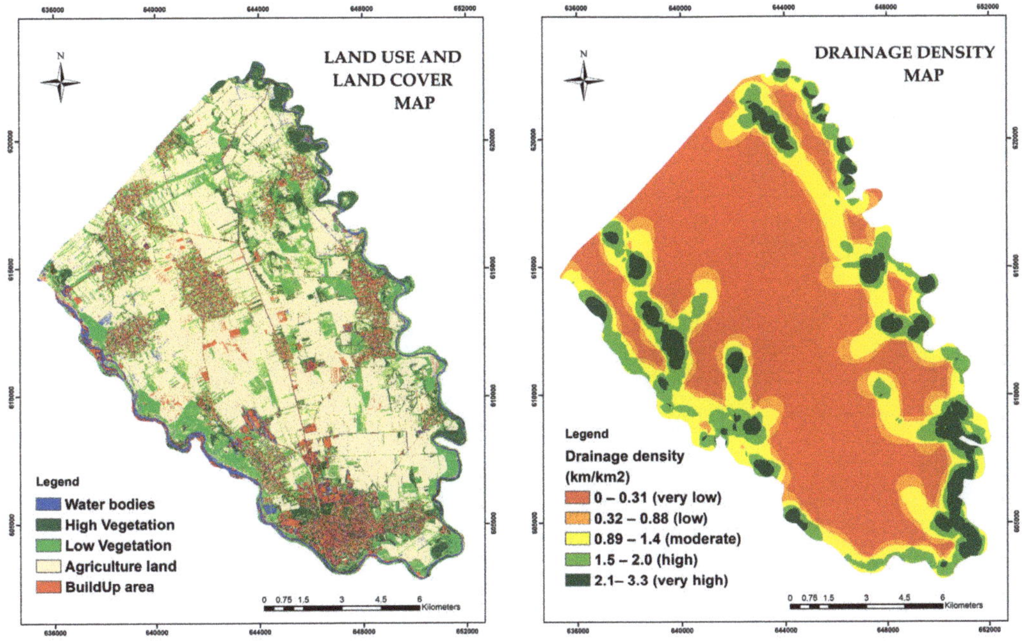


(**a**) (**b**)

**Figure 4.** (**a**) Land use and Land Cover map; (**b**) drainage density map.

### *3.4. Drainage Density*

Drainage density is obtained by dividing the total length of all the rivers in a drainage basin by the total area of the drainage basin [39]. It is a fundamental indicator of hydrological landscapes, and it serves as a valuable tool for analyzing landforms in the context of groundwater potential [10]. It plays a crucial role in determining both infiltration rates and the characteristics of the underlying lithology [40]. In this context, a region with low drainage density is more favorable for high groundwater potential compared to a region with high drainage density. This is why, in this research, a higher weight was assigned to areas with low drainage density. The drainage density was reclassified and sorted into the following categories presented in Figure 4b: Very low (0–0.31 km/km$^2$), Low (0.32–0.88 km/km$^2$), Moderate (0.89–1.4 km/km$^2$), High (1.5–2.0 km/km$^2$), and Very high (2.1–3.3 km/km$^2$).

### *3.5. Elevation*

The elevation information collected by the Shuttle Radar Topography Mission (SRTM) was used to create visual representations or maps of the terrain using ArcGis software. To verify the accuracy of the SRTM data with a 30 m resolution, the digital model (DEM) was compared with ground control measurements taken in 206 points using a GNSS RTK system. The maximum error observed was 0.7 m at point P198, while the minimum error recorded was −1.3 m at point P68, resulting in a median error of −0.19 m. This error can be considered acceptable for this type of digital terrain model, as illustrated in Figures 5 and 6.

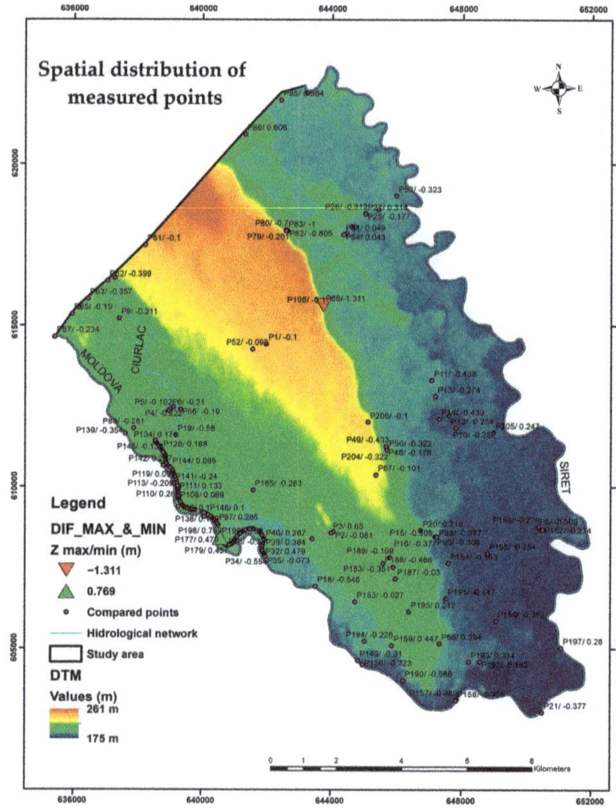


**Figure 5.** Spatial distribution of compared points.

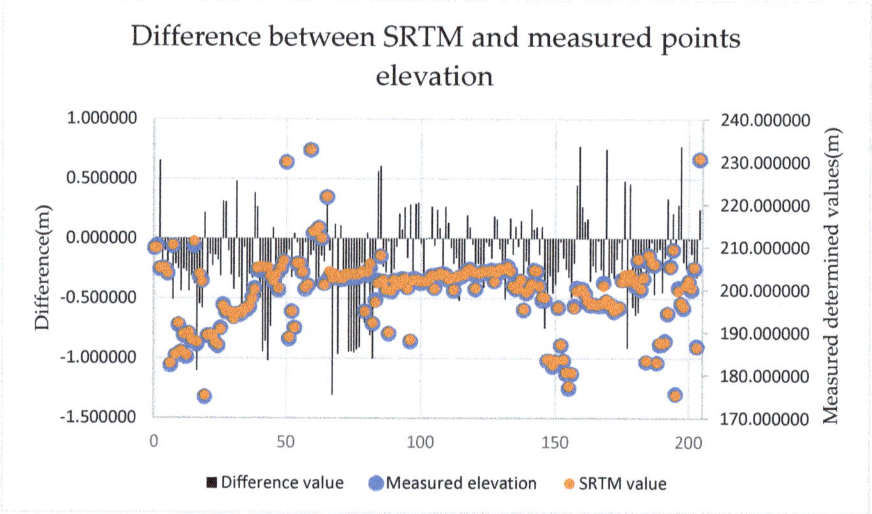


**Figure 6.** Difference between SRTM elevation and measured point elevation.

In the study area, the elevations range from a minimum of 175 m to a maximum of 261 m. In this case, low rankings were assigned to areas with higher elevations, while high rankings were designated for those with lower elevations. This ranking takes into account various hydrological characteristics, such as infiltration rates, run-off patterns, and slope steepness. The elevation was reclassified into five categories, ranging from low to high, as illustrated in Figure 7a. Different feature weights were assigned to these categories: a weight of 3 for the lower elevation zones, which range from 175.5 m to 201.2 m; a weight of 2 for the moderate elevation zones, spanning from 201.3 m to 216.8 m; and a weight of 1 for the highest elevation zone, covering the range from 233.6 m to 260.4 m.

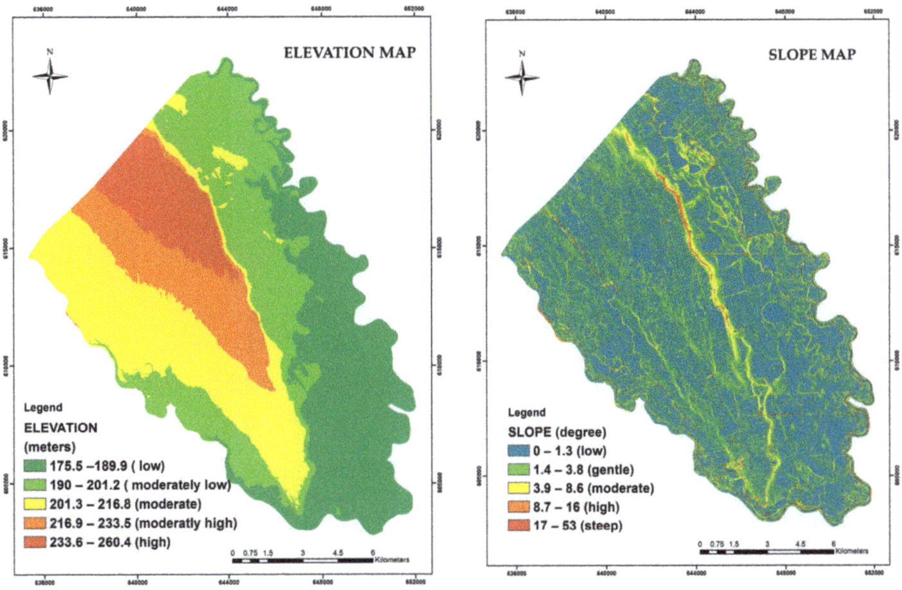


(a) (b)

**Figure 7.** (**a**) Elevation map; (**b**) slope map.

*3.6. Slope*

Slope, a crucial determinant of runoff and infiltration rates, is influenced by the spacing of contour lines. In general, closely spaced contours in vector representation indicate steeper slopes, while in raster representation, each cell provides a slope value where lower values suggest a flatter terrain and higher values denote steeper slopes. The slope values were reclassified and grouped into five categories, like in Figure 7b: Low (0–1.3), Gentle (1.4–3.8), Moderate (3.9–8.6), High (8.7–16.0), and Steep (17–53). Higher slopes result in reduced recharge as precipitation water quickly flows down steep slopes during rainfall events. Therefore, a lower weight (1) was assigned to steep slopes, while a higher weight (3) was allocated for low-to-gentle slopes.

*3.7. Soil*

Soil serves as a crucial parameter for identifying potential groundwater occurrence zones. In the study area, five soil types are present, including sandy loam, soil with variable texture, clay and silt loam, clay soil, and marshes and ponds (as shown in Figure 8). The soil's texture plays a vital role in controlling surface runoff and the infiltration of rainwater. Sandy soil has a low runoff rate and high groundwater potential, whereas clay and silty loam soil have a high runoff rate and very low groundwater storage capacity.

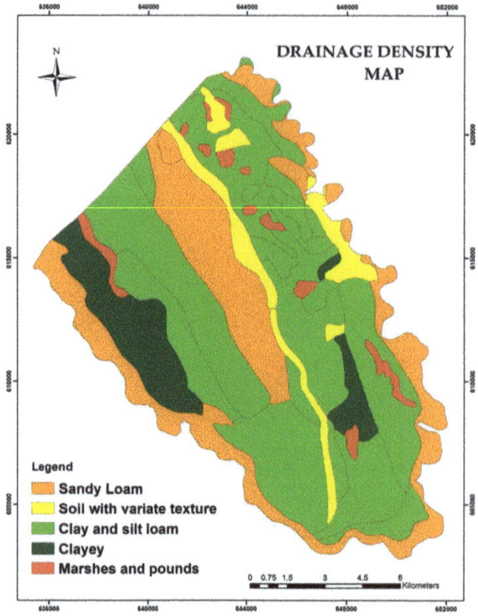


**Figure 8.** Soil map.

*3.8. Ground Water Potential Zones and Validation*

Ground Water Potential Zones

The result of the groundwater potential map in the study area indicated four distinct zones, representing areas with Very Good, Good, Moderate, and Poor groundwater potential, as shown in Figure 9 and Table 6.

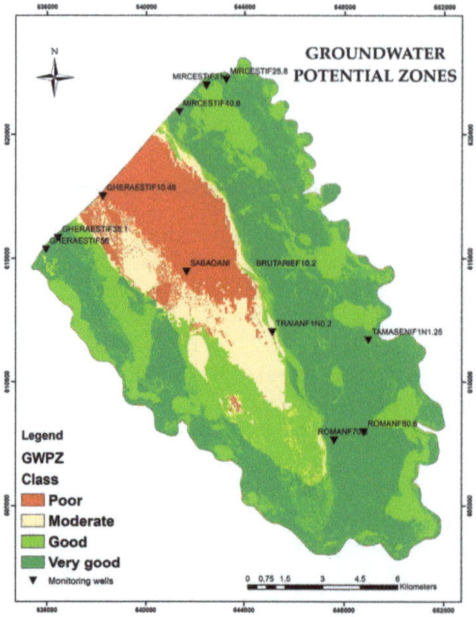


**Figure 9.** Groundwater potential map of the study area.

**Table 6.** Groundwater potential zones area.

| Class | Area (sq. km.) | Area (%) |
|---|---|---|
| Poor | 23.17 | 12.8 |
| Moderate | 19.54 | 10.8 |
| Good | 56.36 | 31.2 |
| Very Good | 81.53 | 45.1 |

Regions with Very Good and Good groundwater potential are generally located in low-altitude areas, such as along rivers, and on their terraces, and in regions with low slopes. Precipitation is a significant factor in the classification of these areas, but the values are relatively close between classes due to the relatively small size of the study area. Geology plays a crucial role, with areas characterized by lithology with high permeability and water storage capacity, such as sandy deposits, sand, or recent alluvial deposits primarily on terraces and riverbeds, indicating a higher potential for groundwater resources.

The accuracy assessment of the forecasting model is crucial to avoid errors and enhance the decision-making process in environmental studies. The potential zones derived from the integration of various techniques, including Remote Sensing (RS), Geographic Information Systems (GISs), and Multi-Criteria Decision Analysis (MCDA), were cross-validated with data from eleven boreholes in the study area. The spatial distribution of the 11 boreholes can be classified as plausible, considering that they are spread across the entire area, in zones with different geology, exposed soils with various characteristics, and in areas with significant elevation differences. The comparison between the predicted and actual yield data from the pumping test, as outlined in Table 7, highlights the accuracy of the prediction, standing at 72.7%. This validation confirms the reliability and precision of the methodology utilized in this study.

The accuracy verification of the predicted values is determined as follows:
Total number of boreholes = 11.
Number of boreholes with matching values between actual and expected yield = 8.
Number of boreholes with differing values between actual and expected yield = 3.
Prediction accuracy = $(8/1) \times 100 = 72.7\%$.

**Table 7.** Validation of groundwater potential zones area.

| S. No. | Borehole | Longitude | Latitude | Top Aquifer | Aquifer Bed | Yield from Drilled Borehole (L/s) | Actual Yield Description | Expected Yield Class from the Prediction Map | Agreement Between Expected and Actual Yields |
|---|---|---|---|---|---|---|---|---|---|
| 1 | Roman F7 | 647,590.50 | 607,626.57 | 4.30 | 6.30 | 1.71 | Good | Good–Very Good | Agree |
| 2 | Roman F8 | 648,789.46 | 607,947.26 | 3.60 | 6.30 | 0.60 | Moderate | Good–Very Good | Disagree |
| 3 | Tamaseni F1N | 648,965.90 | 611,672.02 | 5.30 | 8.90 | 1.25 | Good | Good–Very Good | Agree |
| 4 | Traian F1N | 645,126.58 | 611,998.20 | 23.40 | 27.80 | 0.20 | Poor | Poor–Moderate | Disagree |
| 5 | Gheraesti F5 | 635,949.01 | 615,347.45 | 1.40 | 6.50 | 6.00 | Very Good | Very Good | Agree |
| 6 | Gheraesti F3 | 636,440.61 | 615,816.37 | 4.00 | 7.90 | 5.10 | Very Good | Very Good | Agree |
| 7 | Gheraesti F1 | 638,219.22 | 617,495.75 | 9.80 | 14.80 | 0.48 | Poor | Poor | Agree |
| 8 | Mircesti F4 | 641,338.54 | 620,925.06 | 11.20 | 12.80 | 0.60 | Moderate | Good–Very Good | Disagree |
| 9 | Mircesti F3 | 642,450.36 | 621,983.51 | 5.70 | 7.50 | 1.30 | Good | Good–Very Good | Agree |
| 10 | Mircesti F2 | 643,245.01 | 622,203.72 | 6.50 | 9.80 | 5.80 | Very Good | Good–Very Good | Agree |
| 11 | Sabaoani B1 | 641,640.80 | 614,460.89 | 21.80 | 22.60 | 0.20 | Poor | Poor | Agree |

## 4. Conclusions

The present study aims to delineate groundwater potential zones in the eastern part of Neamț county, where groundwater plays a crucial role, particularly for domestic use. Additionally, groundwater is utilized in the industrial sector, with the city of Roman being a focal point due to its significant economic and industrial development. The study area is predominantly characterized by agricultural land, accounting for 60.2% of the total region, with a concentration in the central part. This sector relies on both river and groundwater, making efficient water management essential. The groundwater potential map classifies areas into Very Good (45.1%), Good (31.2%), Moderate (10.8%), and Poor (12.8%) potential zones. The demarcation of these zones is notably influenced by factors such as precipitation, geology, and elevation. The validation of this model was conducted using observation boreholes, resulting in an accuracy assessment of 72.7%.

The obtained result is comparable with similar studies in the literature that used the same techniques. For example, a study conducted in the Ambo area of the Blue Nile Basin in Ethiopia achieved an accuracy of 69.5% [23], and another study on landslides in the Upper Wabe–Shebele River Basin of Southeastern Ethiopia reported a higher accuracy of 81% [24].

Even though the hydrogeological conditions, climate, hydrology, and geographical location differ, by applying the same techniques, which take into account the weight of various factors, it is observed that the obtained results show good accuracy. This strengthens the case for using these techniques in the delineation of groundwater potential zones.

Identifying areas with groundwater potential is critical for the economic development of the region. This research represents an important step in making decisions regarding the selection of sites for water consumers, as well as in the sustainable use of resources. Additionally, it is essential for regional development planning to understand how to manage resources and reduce the risks associated with their depletion.

Therefore, it can be concluded that these results are valuable for planning new projects and expanding existing ones that involve groundwater utilization in the study area, making significant contributions to the sustainable planning and development of water resources in the region.

**Author Contributions:** Conceptualization, P.-L.B.; Methodology, P.-L.B.; Formal Analysis, P.-L.B.; Investigation, P.-L.B.; Writing—Original Draft, P.-L.B.; Writing—Review and Editing, N.B.; Visualization, M.P.-L., D.C. and F.M.N.; Supervision, V.N. All authors have read and agreed to the published version of the manuscript.

**Funding:** This research received no external funding.

**Data Availability Statement:** The datasets used and/or analyzed during the current study are available from the corresponding author upon reasonable request.

**Conflicts of Interest:** The authors declare no conflicts of interest.

## References

1. Taylor, R.G.; Scanlon, B.; Döll, P.; Rodell, M.; van Beek, R.; Wada, Y.; Longuevergne, L.; Leblanc, M.; Famiglietti, J.S.; Edmunds, M.; et al. Ground water and climate change. *Nat. Clim. Chang.* **2013**, *3*, 322–329. [CrossRef]
2. Fitts, C.R. *Groundwater Science*; Elsevier: Amsterdam, The Netherlands, 2002.
3. Croitoru, A.-E.; Minea, I. The impact of climate changes on rivers discharge in Eastern Romania. *Theor. Appl. Climatol.* **2015**, *120*, 563–573. [CrossRef]
4. Senanayake, I.P.; Dissanayake, D.M.D.O.K.; Mayadunna, B.B.; Weerasekera, W.L. An approach to delineate groundwater recharge potential sites in Ambalantota, Sri Lanka using GIS techniques. *Geosci. Front.* **2016**, *7*, 115–124. [CrossRef]
5. Abijith, D.; Saravanan, S.; Singh, L.; Jennifer, J.J.; Saranya, T.; Parthasarathy, K.S.S. GIS-based multi-criteria analysis for identification of potential groundwater recharge zones—A case study from Ponnaniyaru watershed, Tamil Nadu, India. *HydroResearch* **2020**, *3*, 1–14. [CrossRef]
6. Bond, N.R.; Burrows, R.M.; Kennard, M.J.; Bunn, S.E. Chapter 6—Water Scarcity as a Driver of Multiple Stressor Effects. In *Multiple Stressors in River Ecosystems*; Sabater, S., Elosegi, A., Ludwig, R., Eds.; Elsevier: Amsterdam, The Netherlands, 2019; pp. 111–129.

7. Brown, A.; Matlock, M.D. A review of water scarcity indices and methodologies. *White Pap.* **2011**, *106*, 19.
8. Irimia, O.; Tomozei, C.; Valentin, N.; Lehadus, M.P. Experimental results regarding the groundwater quality in rural zone in Romania. *Int. Multidiscip. Sci. GeoConf. SGEM* **2017**, *17*, 463–470.
9. Nasir, M.J.; Khan, S.; Zahid, H.; Khan, A. Delineation of groundwater potential zones using GIS and multi influence factor (MIF) techniques: A study of district Swat, Khyber Pakhtunkhwa, Pakistan. *Environ. Earth Sci.* **2018**, *77*, 367. [CrossRef]
10. Jhariya, D.C.; Khan, R.; Mondal, K.C.; Kumar, T.K.I.; Singh, V.K. Assessment of groundwater potential zone using GIS-based multi-influencing factor (MIF), multi-criteria decision analysis (MCDA) and electrical resistivity survey techniques in Raipur city, Chhattisgarh, India. *AQUA—Water Infrastruct. Ecosyst. Soc.* **2021**, *70*, 375–400. [CrossRef]
11. Mace, R. *The Importance of Groundwater Sustainability*; Springer International Publishing: Cham, Switzerland, 2023; pp. 1–20.
12. Ionuț, M.; Iosub, M.; Boicu, D. Groundwater Resources from Eastern Romania under Human and Climatic Pressure. *Sustainability* **2020**, *12*, 10341. [CrossRef]
13. Cochiorca, A.; Nedeff, V.; Bârsan, N.; Sandu, I.; Emilian, M.; Chiţimus, D.; Irimia, O. Surface Water and Groundwater Quality Evaluation in a Mining Area. *Rev. Chim.* **2020**, *71*, 124–131. [CrossRef]
14. Baghel, S.; Tripathi, M.P.; Khalkho, D.; Al-Ansari, N.; Kumar, A.; Elbeltagi, A. Delineation of suitable sites for groundwater recharge based on groundwater potential with RS, GIS, and AHP approach for Mand catchment of Mahanadi Basin. *Sci. Rep.* **2023**, *13*, 9860. [CrossRef] [PubMed]
15. Venkatesan, G.; Pitchaikani, S.; Saravanan, S. Assessment of Groundwater Vulnerability Using GIS and DRASTIC for Upper Palar River Basin, Tamil Nadu. *J. Geol. Soc. India* **2019**, *94*, 387–394. [CrossRef]
16. Romanian Water National Administration. *River Basin Management Plan for Hydrographic Basin SIRET 2023-2027*; Romanian Water National Administration Siret River Basin; Romanian Water National Administration: Bucharest, Romania, 2022.
17. Singh, P.; Thakur, J.K.; Kumar, S. Delineating groundwater potential zones in a hard-rock terrain using geospatial tool. *Hydrol. Sci. J.* **2013**, *58*, 213–223. [CrossRef]
18. Nampak, H.; Pradhan, B.; Abd Manap, M. Application of GIS based data driven evidential belief function model to predict groundwater potential zonation. *J. Hydrol.* **2014**, *513*, 283–300. [CrossRef]
19. Hinton, J. GIS and remote sensing integration for environmental applications. *Int. J. Geogr. Inf. Syst.* **1996**, *10*, 877–890. [CrossRef]
20. Jaiswal, R.; Mukherjee, S.; Krishnamurthy, J.; Saxena, R. Role of remote sensing and GIS techniques for generation of groundwater prospect zones towards rural development--an approach. *Int. J. Remote Sens.* **2003**, *24*, 993–1008. [CrossRef]
21. Akter, A.; Uddin, A.M.H.; Wahid, K.B.; Ahmed, S. Predicting groundwater recharge potential zones using geospatial technique. *Sustain. Water Resour. Manag.* **2020**, *6*, 24. [CrossRef]
22. Saaty, T.L. How to make a decision: The analytic hierarchy process. *Eur. J. Oper. Res.* **1990**, *48*, 9–26. [CrossRef]
23. Haile, M.; Yadeta, J.; Gambe, L. Delineation of groundwater potential zones using geo-spatial and AHP techniques: Ambo area, Blue Nile Basin, Ethiopia. *Sustain. Water Resour. Manag.* **2022**, *8*, 36. [CrossRef]
24. Sheko, S.; Moges, A.; Ulsido, M. Assessing Groundwater Potential Zones and Their Implications for Landslides in the Upper Wabe-Shebele River Basin of Southeastern Ethiopia. *J. Eng.* **2024**, *2024*, 7044576. [CrossRef]
25. Saaty, T.L. Decision making with the analytic hierarchy process. *Int. J. Serv. Sci.* **2008**, *1*, 83–98. [CrossRef]
26. Madrucci, V.; Taioli, F.; de Araújo, C.C. Groundwater favorability map using GIS multicriteria data analysis on crystalline terrain, Sao Paulo State, Brazil. *J. Hydrol.* **2008**, *357*, 153–173. [CrossRef]
27. Achu, A.; Thomas, J.; Reghunath, R. Multi-criteria decision analysis for delineation of groundwater potential zones in a tropical river basin using remote sensing, GIS and analytical hierarchy process (AHP). *Groundw. Sustain. Dev.* **2020**, *10*, 100365. [CrossRef]
28. Ghosh, D.; Mandal, M.; Karmakar, M.; Banerjee, M.; Mandal, D. Application of geospatial technology for delineating groundwater potential zones in the Gandheswari watershed, West Bengal. *Sustain. Water Resour. Manag.* **2020**, *6*, 14. [CrossRef]
29. Topuz, M.; Deniz, M. Application of GIS and AHP for land use suitability analysis: Case of Demirci district (Turkey). *Humanit. Soc. Sci. Commun.* **2023**, *10*, 115. [CrossRef]
30. Posea, G. *Geomorfologia României*; Editura Fundaţiei România de Mâine: Bucharest, Romania, 2002; pp. 27–40.
31. Bogdan, P.-L.; Nedeff, V.; Barsan, N.; Panainte-Lehaduș, M.; Chitimus, D.; Nedeff, F.; Tomozei, C. Applications of the GIS and Remote Sensing methods in the Land use and cover maps. Case study: Roman city and the surroundings of the Northern side area. In Proceedings of the 2022 8th International Conference on Energy Efficiency and Agricultural Engineering (EE&AE), Ruse, Bulgaria, 30 June–2 July 2022; pp. 1–5. [CrossRef]
32. Jet Propulsion Laboratory, NASA. *NASA Shuttle Radar Topography Mission Global 1 Arc Second*; Jet Propulsion Laboratory, NASA: La Cañada Flintridge, CA, USA, 2013. [CrossRef]
33. Romanian Water Administration—Siret River Basin Administration. *Basin Forecast Service, Hydrology, Hydrogeology—Data Archive*; Romanian Water Administration: Bucharest, Romanian, 2022.
34. Florea, N.; Bălăceanu, V.; Munteanu, I.; Asvadurov, H.; Conea, A.; Oancea, C.; Cernescu, N.; Popovăţ, M. *The Soil Map of Romania at the Scale 1: 200,000*; Legenda Generală. (The Soil Map of Romania at the scale 1: 200,000. General Legend); National Institute for Research and Development in Pedology, Agrochemistry and Environmental Protection—ICPA Bucharest: Bucharest, Romania, 1994; p. 1.
35. Alexandrescu, G.; Joja, T.; Mirauta, E. *Geological map of Romania (1:200,000), Sheet Piatra Neamt*; Harta Geologica 1:200,000; HG-13; Romania Institutul Geologic: Bucharest, Romania, 1968.
36. Saaty, T.L. A scaling method for priorities in hierarchical structures. *J. Math. Psychol.* **1977**, *15*, 234–281. [CrossRef]

37. Saaty, T.L. *The Analytic Hierarchy Process*; McGrawhill Juc.: New York, NY, USA, 1980.
38. FAO. The State of Food and Agriculture 2019. In *Moving Forward on Food Loss and Waste Reduction*; United Nation's Food and Agricultural Organization: Rome, Italy, 2019. [CrossRef]
39. Horton, R.E. Erosional development of streams and their drainage basins; hydrophysical approach to quantitative morphology. *Geol. Soc. Am. Bull.* **1945**, *56*, 275–370. [CrossRef]
40. Roy, S.; Hazra, S.; Chanda, A.; Das, S. Assessment of groundwater potential zones using multi-criteria decision-making technique: A micro-level case study from red and lateritic zone (RLZ) of West Bengal, India. *Sustain. Water Resour. Manag.* **2020**, *6*, 4. [CrossRef]

*water*

MDPI

*Article*

# Advancing the Understanding of Complex Piezometric Information: A Methodological Approach Integrating Long-Term Piezometry, Surface Nuclear Magnetic Resonance, and Fracture Analysis Using Insights from the "Calcaires du Barrois" Series, France

**Mathieu Bertrand [1,2,*], Catherine Bertrand [1], Naomi Mazzilli [3], Sylvain Gigleux [2], Sophie Denimal [1], Rémi Valois [3], Lise-Marie Girod [1,2], Guillaume Cinkus [3], Valentine Busquet [1,2] and Konstantinos Chalikakis [3,*]**

[1] UMR6249 ChronoEnvironnement (CNRS-UFC), University of Franche-Comté, 25000 Besançon, France; catherine.bertrand@univ-fcomte.fr (C.B.); sophie.denimal@univ-fcomte.fr (S.D.); lise-marie.girod@mines-paristech.fr (L.-M.G.); vambusquet@orange.fr (V.B.)

[2] Agence Nationale pour la Gestion des Déchets Radioactifs (Andra), RD960, BP9, 55290 Bure, France; sylvain.gigleux@andra.fr

[3] UMR 1114 EMMAH (AU-INRAE), Avignon Université, 84029 Avignon, France; naomi.mazzilli@univ-avignon.fr (N.M.); remi.valois@univ-avignon.fr (R.V.); guillaume.cinkus@univ-avignon.fr (G.C.)

* Correspondence: mathieu.bertrand@univ-fcomte.fr (M.B.); konstantinos.chalikakis@univ-avignon.fr (K.C.)

**Abstract:** This study aims to analyze spatio-temporal piezometric data and integrate them with geological, geotechnical, and geophysical data to enhance their interpretation. The research focuses on a site located in the Meuse and Haute-Marne departments of France, which has been under investigation since 1994 as part of the surface facilities characterization for the Meuse-Haute-Marne underground laboratory and the CIGEO (Centre Industriel de stockage Géologique) Andra project. These investigations span different spatial and temporal scales. We observed the contribution of water masses associated with external forcing and identified two types of aquifer responses: a rapid response to rainfall events when fractures are well-connected, and minor reactivity at the matrix level. Additionally, we demonstrated that the matrix compartment can be finely characterized through a combined interpretation of piezometric response analysis, fracture analysis, and surface nuclear magnetic resonance (SNMR) soundings. The methodology developed in this project offers an improved understanding of karst piezometry and/or unsaturated zone extension, which is essential for comprehend ding flow dynamics and better constraining the functioning of karst aquifers. Furthermore, this site serves as an ideal workshop for studying flow in fractured media, providing valuable insights into hydrodynamic behavior in complex subsurface environments.

**Keywords:** hydrogeology; karst; piezometric analysis; fracturing; SNMR

## 1. Introduction

Karst aquifers generally have significant water resources and reserves, which are still relatively under-utilized and in some regions of the world represent the essential, if not the only, resource [1–3]. As a result, the search for new water resources (water for food, water shortage support, agriculture...) leads to an increasing interest in karst systems, either to optimize seasonal exploitation (active management) or to establish systems to protect quality (protection perimeters and protection works) or quantity (sizing of works in urbanized areas) [1]. Additionally, studying these systems is essential during geotechnical or civil engineering investigations [4].

If karst aquifers are first to be studied using a descriptive and comparative approach [5], then piezometric records are fundamental data for characterizing these systems. They are measured in the field, either at the source or by direct measurement of the water column in the borehole [6]. It is an essential measure for understanding the behavior of an aquifer [7,8], for characterizing and evaluating its functioning and its hydrodynamic parameters (storage coefficient and hydraulic conductivity) [9–12], and also for detecting interactions between different structures during its operation [13]. These parameters can be extracted by analyzing signals acquired under active stress (usual hydraulic tests) or passive stress (tidal effect, [14]). These data are essential for the exploitation of water resources [15,16] as well as for civil engineering investigations [17,18].

However, in complex systems such as karst aquifers, the structure and the functioning of aquifers are difficult to understand due to the high heterogeneity degree and the different types of porosity that are involved in groundwater circulation. Indeed, the piezometric levels measured in boreholes are influenced by different types of water volumes coming from the matrix, fractures, or conduits [19–21]. Boreholes that are only a few meters apart can have piezometric levels that differ significantly, in part due to different dynamics between matrix and fractures and/or karst conduits. These differences in piezometric levels may also be due to varying degrees of saturation within the aquifer. Water circulation in this type of environment is thus constrained by hydrodynamic heterogeneities, which induce preferential circulation and, therefore, non-uniform water (and pressure) transfers [22,23], with a continuum between very fast, non-dispersive flows (piston or conduit flows) and very slow, highly dispersive flows (diffuse flows) [24–26]. Borehole logs, wall imaging techniques, and core analysis provide high-resolution fracturing information. Combined with hydraulic tests, they provide relevant information on the hydrodynamic functioning of the aquifer, such as fracturing, karstification degree, and connectivity. Nonetheless, the obtained information only reflects the well spatial scale, so there is a high uncertainty far from the wells. In these subsurface geological environments, flows are generally not linked to a particular structure or homogeneous reservoir but are distributed at all scales [27]. This complexity poses the problem of detecting, measuring (and being representative of), and monitoring these flows in 4D [19,28].

Non-intrusive techniques such as geophysical methods can complement and spatialize the geological and hydrogeological knowledge gained by analyzing the water levels in boreholes [29]. These methods are based on analyzing the propagation of physical variables (e.g., electromagnetic field, acoustic waves). This propagation is influenced by various factors, including the presence or absence of water in the environment being studied [30]. Among these methods, surface nuclear magnetic resonance (SNMR) provides a signal directly linked to the presence of groundwater. The initial amplitude of the SNMR signal is directly related to the presence of mobile water in the subsurface, while its relaxation time can be related to pore size [31,32]. This method thus provides direct information on groundwater hydrodynamic properties in karst environments [33].

When industrial activities are spread out above or within fractured and/or karstified aquifers, it is important to develop a comprehensive hydrodynamic functioning model to ensure the durability of the resources and minimize related risks. As an example, the French National Agency for the Management of Radioactive Waste (Andra) aims to securely dispose of high and low-intermediate level long-lived radioactive waste at the CIGEO (Centre Industriel de stockage Géologique) project (Meuse, France). The work support zone, and the ramp zone facilities, will be used for digging and building the underground structures located at a depth of approximately 500 m and for receiving and inspecting waste packages and preparing them for transfer to the underground facilities via a ramp these zones measuring around 550 hectares lies on the "Calcaires du Barrois" series. The basal part of the "Calcaires du Barrois" is the dominant formation at the outcrop, consisting of a complex hydrosystem with two horizons of differing permeability. The upper horizon is fractured, resulting from the decompression and cryoturbation phenomena of the last deglaciations, while the underlying horizon exhibits diffuse karstification. However, this karstification is not concentrated enough to produce large-scale dissolution features such as caves or metric drains. To properly size surface facilities and reservoir access shafts, a better understanding of the flow dynamics within this formation is essential. Additionally, the hydrogeological characterization of the "Calcaires du Barrois" is a key part of the project, aimed at establishing an effective environmental monitoring network.

The hydrodynamic behavior of the "Calcaires du Barrois" is directly linked to the fracturing of the environment. Numerous geological and geotechnical investigations have been conducted over hundreds of boreholes at this site over the years, providing data on fracturing, and petrophysics. Piezometry and hydrodynamics investigation has been conducted according to dozens of piezometers. Nineteen boreholes were re-used as piezometers. These were selected from an area of 13 $km^2$ as they have provided good quality data series for more than 8 years. Because this density of investigation is not often available for studying karst systems [29], we designed this project to use these facilities to study the hydrogeologic complexities of this "fractured-rock karst site".

We decided to bring together the data collected at the study site and evaluate the data as a whole, complemented by SNMR. This investigation is directly related to the presence of mobile water in the subsurface and provides direct information on the hydrodynamic properties of mobile groundwater in karst environments. It is particularly suited to the study area. We thus combined (1) the analysis of long piezometric records with (2) full fracture analysis from borehole logs and (3) geophysical measurements to gain further insight into the complex "Calcaires du Barrois" system. This approach is used to shed more light on hydrodynamic responses and their origins in the complex "Calcaires du Barrois" system. We also provide methodological insights that are easily transposable to other studies aiming at water body characterization in karst environments.

## 2. Context

### 2.1. Study Site

The Meuse/Haute-Marne region is located on the south-eastern edge of the Paris Basin, between the valleys of the Marne and the Meuse rivers (France, Figure 1A) and, more precisely, in the outcrop zone of the Upper Jurassic formations ("Calcaires du Barrois" series), at the boundary of the presence of Cretaceous terrains. The "Calcaires du Barrois" series is a succession of limestone strata (Sublithographic limestone, Dommartin limestone, "Oolithe de Bure" formation, and Decayed limestones) and a marl layer that is approximately 10 m thick ("Pierre Chaline" formation), separating the Sublithographic limestone and the Dommartin limestone. The Kimmeridgian marls form the bedrock on which the sedimentary sequence rests (Figure 1A,B,D). This series dips northwest towards the center of the Paris Basin with an average dip of about 1 degree [34]. This shallow dip is associated with a plateau morphology: the "Calcaires du Barrois" series plateau. This plateau is cut by relatively narrow, incised valleys. It is bordered to the east by the Ormançon River and to the west by the Saulx River. Between these two rivers flow the Bureau and Orge rivers, tributaries of the Saulx.

The study site covers 13 $km^2$ and is located at the head of the Orge catchment. It is bounded by the Orge River to the west and the Bureau River to the east (Figure 1C). These two rivers are characterized by dry periods from April to October. The Bureau joins the Orge near Bure (Figure 1A). In this sector, only the Sublithographic limestone is exposed. In detail, this limestone is composed of five levels, which are distinguished by variations in facies (multi-centimetric limestone-mudstone alternation at the bottom of the series) and thickness (up to 15 m for level N5) (Figure 1B).

Between 2002 and 2023, rainfall was measured at the CMHM (Centre Meuse-Haute-Marne) site with a weather station (hourly and daily). The average annual rainfall (over 21 years) recorded at the CMHM was 946 mm. The annual precipitation at CMHM ranged between 750 mm (in 2004) and 1200 mm (in 2007).

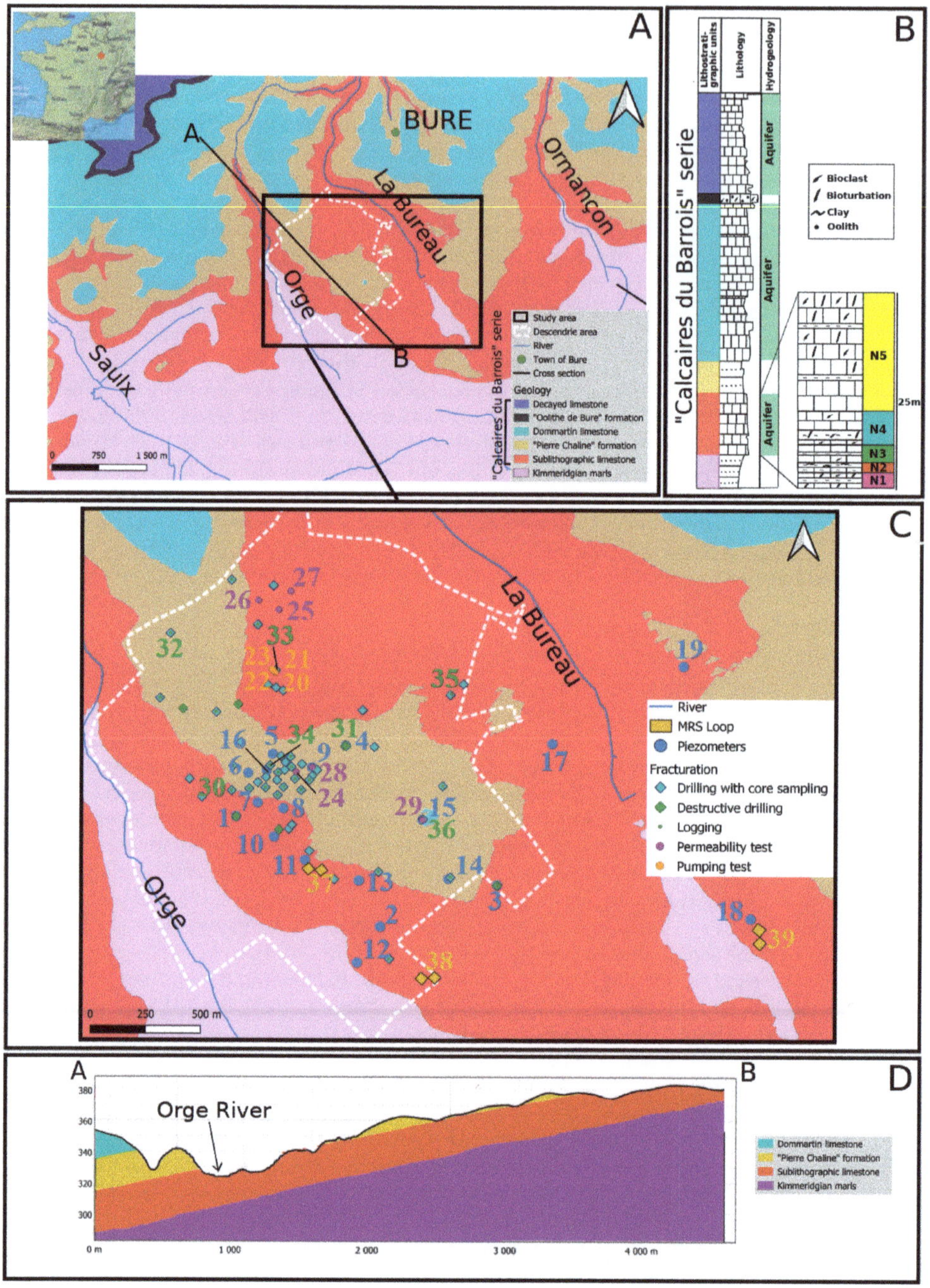


**Figure 1.** (**A**)—Geological map of the study area. The white dashes represent the location of the future waste reception, control, and preparation area of the CIGEO site area called Descendrie area. (**B**)—Lithostratigraphic succession of the study area and location of the aquifers in this succession. (**C**)—Focus on the study area, showing the location of piezometers, geological boreholes, and SNMR soundings. The numbers correspond to the piezometer names in Appendix A. (**D**)—North-West (**A**)/South-East (**B**) cross-section with vertical exaggeration (10×).

### 2.2. Geological and Hydrogeological Characteristics of the "Calcaires du Barrois" Series

The local hydrogeological system of the "Calcaires du Barrois" series consists of three more-or-less well-connected aquifer levels where the series is complete (Figure 1B). These are, from the bottom to the top [34]:

- The Sublithographic limestones (25 m thick) are lithographic micritic limestones with rare bioclastic beds and rare marly interbeds of decimetric thickness.
- The Dommartin limestones (63 m thick) consist of about 25 m of micritic limestones with bioclastic beds and marl interbeds at the base, an interval of about 25 m of micritic limestones with very rare bioclasts and bioclastic packstones at the top of this interval, and about 13 m of micritic limestones with very little bioclasticity at the top of the Dommartin limestones.
- The Decayed limestones (36 m—thick), which correspond to micritic limestones with little bioclasticity.

These levels are separated by aquitards such as the "Pierre Chaline" formation, a predominantly marly level, and the "Oolithe de Bure" formation, formed by a massive bed of beige limestone with oolites and a sparitic matrix. Depending on the degree of fracturing, these aquitards can be relatively permeable. The Kimmeridgian marl with exogyres that underlie the "Calcaires du Barrois" series forms an impermeable layer. The upper and lower limestone layers contained in these marls are also aquifers to a lesser extent. The aquifers in this series supply numerous springs with low flow rates [34].

The water table in the Sublithographic limestones is unconfined in the upstream part of the catchment but becomes captive north of the town of Bure under the "Pierre Chaline" formation. The water table in the Dommartin limestones is semi-captive due to the limited extent of the Oolithe de Bure. Over the extent of the study area, the water table in the Decayed limestones is unconfined.

Karstification is well developed within the "Calcaires du Barrois" series. In addition, typical karst features can be observed at the surface (sinkholes, wells). Speleological explorations to the north-west of the area [35] have highlighted the extensive development of karstification in the "Calcaires du Barrois" series. In the study area, numerous losses and resurgences of the Orge and Ormançon have been observed where these rivers cross the "Pierre Chaline" marls [36].

## 3. Data and Methods

The proposed methodology is based on spatial analysis of piezometric and fracture records, coupled with imaging and monitoring of water content using the SNMR method.

### 3.1. Hydrogeological Investigations: Piezometry Analyses

The extensive instrumentation at the site provides a considerable amount of borehole data. Our first objective was to extract from these records statistical or correlative indicators of the hydrodynamic behavior of the structures, and then to use these indicators to establish groups of similar hydrodynamic functioning.

Seven piezometric time series indicators were defined in this project:

1. The variance, which is measured via time series variation coefficient, an indicator of the dispersion of the distribution of piezometric levels.
2. The maximum amplitude (difference between the "Calcaires du Barrois" series base and the water level extremum piezometric levels). The base of the "Calcaires du Barrois" series is used as a reference to compare the height of piezometric variations measured in the boreholes.
3. The slopes of piezometric rises and recession. The positive and negative slopes of the piezometric curves provide information on the rate of piezometric rise and the rate at which wells are drying up. These indicators are determined by studying the derivative of piezometric variations as a function of time. These derivatives are calculated at all points of a hydrograph using the finite difference method. The indicators provide

information on the hydrodynamic parameters of the system. They therefore provide an insight into the structure of the system, such as the relative fracture density in the vicinity of the studied piezometers.

4. System inertia or memory effect. This parameter quantifies the time taken for an event to affect the chronicle or, in other words, the time $n$ taken for an event occurring at time $t$ to not affect the event recorded at time $t + n$. The inertia of the system is expressed by the autocorrelation function [37]. This function is a measure of the degree of dependence on events in the same chronicle by assessing their repetitive nature. A significant memory effect reflects a certain degree of flow regulation by the system and, therefore, the importance of a capacitive character [38,39]. The memory effect is an indicator of the structure of the system and fracturing.
5. Response time and cross-correlation coefficient. These indicators examine the strength of the relationship between the input signal, rainfall, and what can be considered as an approximation of the output signal, piezometry. It is assumed that there is a causal relationship between these two variables. The cross-correlation function corresponds to the transfer function between the two signals and provides information about the structure of the system.

These indicators were grouped by hydrodynamic behavior using principal component analysis and hierarchical ascending classification (HAC).

This approach was applied to 19 piezometers spread over the Descenderie area and over two time periods (Appendix B). Four of the 19 piezometers have a continuous series of over 25 years (piezometers 16, 17, 18, and 19) and the other piezometers have a continuous series since 2016. Datasets from recordings on the 19 piezometers between November 2016 and July 2021 were selected for use in this project because they provide the longest records to harmonize the processing, and they have been homogenized at hourly time steps. Before statistical processing, an analysis of the raw data was performed to quantify the missing data for all records (water samples or well tests). Over the 8 years of recording, the missing data represent 6.5% of the total data collected from the 19 piezometers.

Data on piezometric variations (mNGF) have been continuously recorded at quarter-hourly or hourly intervals by pressure sensors (Diver, Ven Essen Instrument) (Appendix B). Ranges are 10 m for pressure with an accuracy of 0.5 cm $H_2O$, and between −20 °C and 80 °C with an accuracy of +/−0.1 for temperature.

### 3.2. *Geological Investigations: Fracturing Analysis*

The fracturing and diffuse karstification degrees of the "Calcaires du Barrois" series have been described using borehole logging coupled with core analysis when possible. One of the objectives was to estimate the effect of lithology on the density and vertical extent of fracturing and karstification along the marl–limestone sequence where the flows seem to occur.

We used the following logs:

1. Optical Televiewer borehole wall imaging (sensor OPTV), which provides a complete, continuous scan of the borehole and, thus, provides a developed, oriented image of the borehole walls that can be used for subsequent image processing allowing measure dips and identify and orient planes of stratification, fracturing, and schistosity, (shown in polar plots). The characteristics of the Optical Televiewer are 360° camera 2 m long and 50 mm in diameter with 1 mm resolution. The quality of the images obtained depends mainly on the acquisition conditions in the borehole: the turbidity of the water, the diameter of the borehole, and the stability of the ground.
2. Natural radioactivity logging or "Gamma Ray" (GR) measures the natural gamma radioactivity of rocks. It measures the content of naturally radioactive elements in the rocks and thus clarifies and correlates the lithological cross-section obtained by sampling the ground during drilling. In the study area, it can be used to distinguish between marl intervals (high values) and limestone intervals (low values). Correlations between boreholes are based on this measurement.

We used 54 boreholes that had been drilled in the study area; they traversed both the Sublithographic limestones and the "Pierre Chaline" formation, depending on their location (Figure 1C). Forty-eight (48) of the 54 boreholes were cored and investigated using two types of logging: optical and gamma-ray. Thirty were logged optically and by gamma-ray; one was optically logged and 17 were investigated by gamma-ray logging. Of the 6 boreholes drilled destructively, only the gamma-ray logs could be used. Drilling and logging were conducted in 2015 and 2016. In most cases, wall imaging in boreholes is not relevant below the water level due to issues with turbidity.

### 3.3. Geotechnics Investigations: Hydraulic Tests

Two types of hydraulic tests were carried out: (i) Lefranc tests and (ii) pumping tests, carried out after the groundwater table had stabilized.

#### 3.3.1. Lefranc Tests

The Lefranc water test is an in situ test used to determine the value of the local permeability coefficient (KL), which may differ significantly from the bulk permeability coefficient. The Lefranc water test applies to all fine or gravelly soils below the water table, where the assumed permeability coefficient is greater than approximately $10^{-6}$ m/s. It can be conducted in exploratory boreholes as work progresses.

Lefranc tests are conducted either under constant load (water is injected into the borehole at a constant rate until the level in the borehole stabilizes) or under variable load (a volume of water is injected into the borehole and the drop in level is monitored as a function of time). These tests are carried out in 1 m increments as the borehole progresses.

One month after drilling completion, two constant flow injection tests were conducted throughout the borehole ({28}, {29}; Figure 1C). For {28}, the tests were carried out by injecting water at a constant rate of 0.02 l/s for 2 h and then, after a recovery time of 21 h, an injection of 0.03 L/s for 7 h. The protocol used for {29} was more or less the same in terms of injection time and recovery time. However, the injected flow rates are doubled, i.e., 0.04 L/s and then 0.06 L/s. Pressure measurements in the borehole with a DIVER pressure probe were continued for 450 h after the end of the tests. The different flow rates used for these tests were chosen by the companies conducting the geotechnical investigations.

Results of the constant load tests were based on the expression of [40], which expresses the permeability coefficient as a function of flow rate, a factor that depends only on the borehole shape and its position with respect to the aquifer boundaries, the water column height, and the borehole diameter. Only measurements recorded after stabilization were used to determine the permeability coefficient.

Twenty-six hydraulic tests (Lefranc tests) were conducted [41] by a company specialized in geotechnical studies during the destructive drilling of 4 boreholes (Figure 1C). A total of 16 slug tests were performed by injecting water at a constant rate of 1 L/s for 1 h into three boreholes {25}, {26} and {27}. Tests that did not produce significant measurements (very low infiltration) were continued for one hour with a variable load. This consisted of 10 tests, including 5 in borehole {24}, 1 in borehole {25}, 1 in borehole {27}, and 3 in borehole {26}.

#### 3.3.2. Pumping Tests

The constant flow test, which is the most commonly used pumping test, consists of lowering the piezometric surface of the water table by pumping and measuring changes in the level of this surface, as a function of time, as well as the flow rate pumped. This allows the permeability coefficient of the tested layer and the storage factor to be determined.

The pumping was carried out in borehole {20} (source borehole) and the changes in the piezometric surface over time were monitored using 3 piezometers (observation boreholes) located 3 m {21}, 8 m {22}, and 20 m {23} from the borehole (Figure 1C).

The pump test was carried out in December 2015 according to the NF P94-130 standard [42]. The lowering of the groundwater level was measured using pressure sensors in the 3 control piezometers. The pumping test was stopped after 1 h to prevent the pump from draining. The piezometric level was measured in the source borehole being pumped and in the 3 observation piezometers during a recovery period of 21.5 h after pumping was stopped. The interpretations of the pumping test are derived from the analytical solution developed by Moench [43] for the source borehole and from the modified Theis [44] solution for one of the observation piezometers. The analytical solution developed by Moench [43] takes into account the incomplete borehole effect and the parietal effect in a homogeneous and anisotropic-free aquifer with a drip. The analytical solution developed by Theis [44] was originally developed for use in confined aquifers, but can also be used for unconfined aquifers by correcting the drawdown data [45].

The permeability coefficient k obtained from the pumping test is the horizontal permeability coefficient of the layer. It is a global or "large" coefficient, representative of the average hydraulic behavior of the volume of the layer under test. The storage factor or S coefficient is dimensionless. It gives the change in the volume of water contained in a vertical prism with a horizontal section of unit area for a unit change in the hydraulic load.

### 3.4. Hydrogeophysics Investigations: Surface Nuclear Magnetic Resonance (SNMR)

We used SNMR to analyze the spatial and temporal variation of water content to estimate the location of water volumes passing through these fractured environments [30,32]. The measuring device was NUMISPoly equipment from IRIS Instruments, with a coincident transmitting/receiving eight-square loop configuration composed of two squares with 40 m sides.

SNMR signal parameters (initial amplitude E0 and observed transverse relaxation time T2*) are derived from the SNMR signal envelope assuming a mono exponential decay [33]. The SNMR water content (mobilizable water content) was derived from E0. The effective porosity in which the water is stored has been was derived from the apparent transverse relaxation time T2* [32]. The hydraulic parameters of the aquifers are then estimated qualitatively: The water content gives an image of the drainage porosity of the reservoir, and the T2* constant provides information on the characteristics of water-filled pores [46]. Measuring T2* is easier and faster than measuring T1 because the T1 assessment requires the use of a two-pulse protocol with a variable delay between pulses [47]. In the absence of magnetic heterogeneities, the T2* estimate is considered reliable [47], so we used T2* in our survey.

Several procedures have been used to improve the quality of SNMR data in this study [48]. To compensate for the fact that electromagnetic (EM) noise has a large influence on the measurement, an EM noise measurement campaign was conducted prior to drilling to select the areas with the least EM noise. An eight-square loop shape was favored to further reduce the EM noise effect on the soundings. The excitation pulse was repeated and the recording sequences were averaged to reduce the influence of random noise.

SNMR signal filtering is indispensable to improve the signal-to-noise ratio and to allow the estimation of SNMR parameters. The filtering procedure was chosen based on tests carried out using the raw noise data on a synthetic signal and performed using NumPro $12 \times 2$ from the SAMOVAR suite. A notch filter of the type global Narrow was used to attenuate the disturbance field due to the 50 Hz harmonics of the power lines present near the sounding [49]. The maximum noise level for the use of a stack was set at 500 nV, and the maximum average noise level was set to 50 nV. Damage to the signal caused by filtering was calculated using the RMSE (Root Mean Square Error). The RMSE compares the square root of differences between predicted and observed values.

SNMR data quality was estimated based on the average signal-to-noise ratio (S/N) [50] derived from stacked and filtered measurements [51]. The higher the S/N, the higher the reliability of the SNMR survey. We consider SNMR soundings to be of acceptable quality when the S/N is $\geq 2$.

A total of 19 SNMR soundings were implemented in 3 acquisition zones, under different hydraulic regimes. Acquisition zones covered the entire sector with varying thicknesses of Sublithographic limestone (3 m, 12 m, and 25 m). In each zone, a single low-water campaign (summer 2021) and a high-water campaign (spring 2022) were conducted to characterize the seasonal water level variations. To better assess SNMR water content dynamics, additional measurements were performed at site 38 prior and after rainfall events during both low water (7 soundings over 38 days from 23 June 2021 to 30 July 2021) and high water (9 soundings over 8 days from 5 April 2022 to 12 April 2022).

Only soundings with an average number of recording sequences (stacks) greater than 200 and a signal-to-noise ratio greater than 2.5 are presented in this paper (Table 1).

**Table 1.** Table of SNMR surveys and their main characteristics (name, site, date, average number of stacks, signal-to-noise (S/N) ratio, maximum depth with good resolution, rainfall in the 7 days preceding the sounding (Rainfall D-7), and the hydraulic regime.

| Sounding Name | Site | Date | Average Number of Stacks | Ratio S/N | Maximum Depth Resolution (m) | Rainfall D-7 (mm) | Hydraulic Regime High-Water (HW) or Low-Water (LW) |
|---|---|---|---|---|---|---|---|
| {37} | {37} | 1 March 2022 | 299 | 2.51 | 24.41 | 7.8 | HW |
| {38}_1 | {38} | 23 June 2021 | 274 | 2.67 | 25.77 | 10.4 | LW |
| {38}_2 | {38} | 28 June 2021 | 288 | 2.7 | 25.77 | 21.0 | LW |
| {38}_3 | {38} | 11 April 2022 | 367 | 9.18 | 21.6 | 48.4 | HW |
| {38}_4 | {38} | 12 April 2022 | 243 | 6.63 | 21.6 | 48.4 | HW |
| {39} | {39} | 3 March 2022 | 289 | 7.78 | 24.41 | 4.4 | HW |

A smooth inversion was performed with SVD (singular value decomposition) type regularization. We chose a relatively low regularization (15 for water content and 20 for T2*) based on a preconception of the geology of the soundings. A sharp contrast in hydrodynamic properties is indeed expected between the upper part, the limestone part and the marl, and the lower part of the sounding. The uncertainty in the water content is reflected in the 95% confidence interval of the SAMOVAR plots.

## 4. Results

### *4.1. Piezometric Data Processing (2016–2021)*

Table 2 shows the values of the piezometric indicators determined for each borehole, together with the percentage of missing data for that borehole. These results will be used as the basis for the multi-criteria analysis that will characterize the comparative phase of this study.

**Table 2.** Table of values of piezometric indicators for each structure.

| Name | 1-Variance | 2-Amplitude (m) | 3-Flood Rise Gradient | 3-Recovery Gradient | 4-Inertia (Days) | 5-Response Time (Hours) | 5-Rainfall-Piezometry Correlation | Missing Data (%) |
|---|---|---|---|---|---|---|---|---|
| {1} | 1.71 | 8.7 | $1.72 \times 10^{-6}$ | $-1.25 \times 10^{-7}$ | 69 | 25 | 0.14 | 1.59 |
| {2} | 1.14 | 8.88 | $1.13 \times 10^{-6}$ | $-1.24 \times 10^{-7}$ | 61 | 16 | 0.24 | 3.23 |
| {3} | 1.52 | 7.02 | $6.61 \times 10^{-6}$ | $-1.26 \times 10^{-7}$ | 62 | 37 | 0.18 | 7.05 |
| {4} | 1.7 | 15.93 | $7.88 \times 10^{-7}$ | $-1.91 \times 10^{-7}$ | 115 | 30 | 0.03 | 1.17 |
| {5} | 0.65 | 8.82 | $3.31 \times 10^{-7}$ | $-9.18 \times 10^{-8}$ | 40 | 18 | 0.23 | 0.02 |
| {6} | 0.82 | 6.71 | $5.57 \times 10^{-7}$ | $-1.23 \times 10^{-7}$ | 56 | 199 | 0.09 | 0.35 |
| {7} | 2.08 | 18.87 | $1.84 \times 10^{-6}$ | $-2.98 \times 10^{-7}$ | 74 | 16 | 0.17 | 3.77 |
| {8} | 0.45 | 8.11 | $9.26 \times 10^{-8}$ | $-3.98 \times 10^{-9}$ | 2 | 15 | 0.24 | 0 |
| {9} | 0.92 | 10.13 | $3.41 \times 10^{-7}$ | $-1.10 \times 10^{-7}$ | 54 | 18 | 0.2 | 1.59 |
| {10} | 0.88 | 12.89 | $1.49 \times 10^{-7}$ | $-3.35 \times 10^{-8}$ | 30 | 17 | 0.21 | 0 |

**Table 2.** *Cont.*

| Name | 1-Variance | 2-Amplitude (m) | 3-Flood Rise Gradient | 3-Recovery Gradient | 4-Inertia (Days) | 5-Response Time (Hours) | 5-Rainfall-Piezometry Correlation | Missing Data (%) |
|---|---|---|---|---|---|---|---|---|
| {11} | 0.23 | 3.71 | $9.99 \times 10^{-8}$ | $-6.81 \times 10^{-9}$ | 8 | 4 | 0.36 | 0 |
| {12} | 0.4 | 5.11 | $7.88 \times 10^{-8}$ | $-1.32 \times 10^{-9}$ | 7 | 13 | 0.21 | 0.02 |
| {13} | 1.4 | 10.86 | $2.31 \times 10^{-6}$ | $-1.04 \times 10^{-7}$ | 74 | 16 | 0.18 | 0.28 |
| {14} | 0.64 | 4.83 | $4.93 \times 10^{-9}$ | $-5.02 \times 10^{-10}$ | 42 | 20 | 0.23 | 0.06 |
| {15} | 0.55 | 2.44 | $3.95 \times 10^{-7}$ | $-3.12 \times 10^{-8}$ | 68 | 176 | 0.1 | 0 |
| {16} | 0.82 | 10.91 | $6.49 \times 10^{-7}$ | $-1.26 \times 10^{-8}$ | 6 | 13 | 0.29 | 0.76 |
| {17} | 0.99 | 5.28 | $1.43 \times 10^{-6}$ | $-1.31 \times 10^{-7}$ | 65 | 100 | 0.16 | 2.28 |
| {18} | 3.2 | 11.66 | $3.79 \times 10^{-6}$ | $-3.63 \times 10^{-7}$ | 71 | 68 | 0.11 | 70.61 |
| {19} | 0.99 | 5.16 | $4.87 \times 10^{-6}$ | $-5.66 \times 10^{-8}$ | 60 | 39 | 0.19 | 29.38 |

#### 4.1.1. Monocriteria Statistical Analysis

##### Strict Signal Indicators

For the 19 piezometers studied during the period 2016–2021, the coefficient of variation ranges from 0.23 to 3.2. This indicator is difficult to interpret in isolation from the other indicators. However, it is relevant to describe the variability of the signal. The maximum amplitude between the lowest and highest levels in each series is highly variable. It ranges from 2.4 to 18.9 m at piezometers {15} and {7}, which are about 750 m apart. The average maximum amplitude is 8.73 m with a dispersion of values of 4.2. These indicators reflect the heterogeneity of piezometric behavior in the structures.

##### Hydrodynamic Parameter Indicators

Piezometer {14} has the lowest rates of piezometric rise and recovery with values of $4.93 \times 10^{-9}$ and $-5.02 \times 10^{-10}$. The highest velocities are observed for piezometer {3}, a piezometric rise with a value of $6.6 \times 10^{-6}$, and structure {18}, for recovery with a value of $-3.63 \times 10^{-7}$. The slope values give a relative indication of the rate at which changes occur once a response begins after a rainfall event. The variation in these values supports the hypothesis of a heterogeneous environment with quite different behavior from one structure to another. The variation gives an idea of the hydrodynamic characteristics of the environment around the structures.

##### System Structure and Fracture Indicators

The inertia values derived from interpretation of the autocorrelation functions (Table 2) again highlight a divergence in the memory effect of the system in the vicinity of the investigated structures. The values range from 2 to 115 days, 400 m apart. In general, the interpretation of the autocorrelation curves shows that the memory effect can be quite different between one structure and another, despite the short distance between them. This is the case for holes {16} and {6}, 80 m apart, with memory effects of 6 and 69 days, respectively. These results are considered to be statistically robust.

The phase shift values between input and output signals, derived from the interpretation of the rain–piezometry cross-correlation curves, reflect variable response times ranging from 4 h {11} to 8.5 days {6} (Table 2).

The associated correlation coefficients are also variable, ranging from 0.03 {4} to 0.36 {11}. In absolute terms, these "rainfall–piezometric" correlation values appear to be quite low, reflecting the inertia of the system.

#### 4.1.2. Multi-Criteria Statistical Analysis

Multi-criteria analyses have the advantage of being able to characterize how the indicators work together and to determine which indicators are likely to discriminate between the piezometric signals from different boreholes (Principal Component Analyses (PCA)). Hierarchical Ascending Classifications classify the piezometers based on the coordinates of the points in the PCA.

The PCA was performed on all seven indicators for the 19 piezometers located in the study area. On the F1F2 axis, the percentage of information explained was 75.48%.

The distribution of the indicators in the variable space shows that all the indicators are significant on the F1F2 axes (due to their proximity to the circle of radius 1), except the slope of the flood rise (Figure 2C).

The amplitude of the variations correlates with the variance and the slopes of the piezometric rises in the series and is anti-correlated with the recovery slopes (Figure 2C). The inertia is anti-correlated with the rainfall–piezometric parameter. Response time shows no correlation with other indicators.

According to the distribution of structures in the space of individuals (Figure 2B,D), points {6} and {15} are strongly expressed on the F2 axis and seem to be discriminated by their response time.

Points {4}, {7}, and {18}, which are strongly expressed on the F1 axis, can be explained by their dispersion and variation amplitude values, which vary inversely to recovery slopes, which are also strongly expressed in the variation of these structures.

Points {8}, {11}, {12}, {14}, and {16} are expressed on the F1 axis and are strongly explained by their "rainfall–piezometry" correlation value.

On the other hand, the remaining points {1}, {2}, {3}, {5}, {9}, {10}, {13}, {17}, and {19} form a median group that is moderately expressed on both axes (Figure 2B,D).

A HAC was then used to discriminate between groups of piezometers based on the coordinates of the points in the PCA. Five groups stood out on the dissimilarity dendrogram. The HAC was performed for a significance threshold of 0.08, symbolized by a solid line on the dendrogram. At the end of this analysis, five groups stand out (Figure 2A).

The first group (A1, pink) discriminates between points {4}, {7} and {18}. This can be explained by the F1 axis, which is mainly responsible for the "rainfall–piezometry" correlation, amplitude, and dispersion. The representative piezometer in this group is {18}. The group is representative of highly transmissive behavior.

A second group (A2, red), which essentially shows "rainfall–piezometry" correlation values expressed primarily on the F1 axis, consists of points {8}, {11}, {12}, {14}, and {16}. The representative piezometer in this group is {16}. This group is related to the fracturing of the environment and therefore has a transmissive behavior.

A third group (A3, green), representing the response time indicator, is strongly expressed on the F2 axis and consists of points {6} and {15}. The representative piezometer in this group is {15}, which is typical of a capacitive environment.

A fourth group (A4, purple), the median, differs from the first by the best flood gradients, an indicator that is not very representative on the F1F2 axes and consists of points {3} and {19}. The most representative piezometer in this group is {19}. This group presents both transmissive and capacitive behavior.

A fifth and final group (A5, black), median, which has no axis and is therefore a factor of influence, consists of points {1}, {2}, {5}, {9}, {10}, {13}, and {17}. The representative piezometer in this group is {17}. This group has an above transmissive and capacitive behavior.

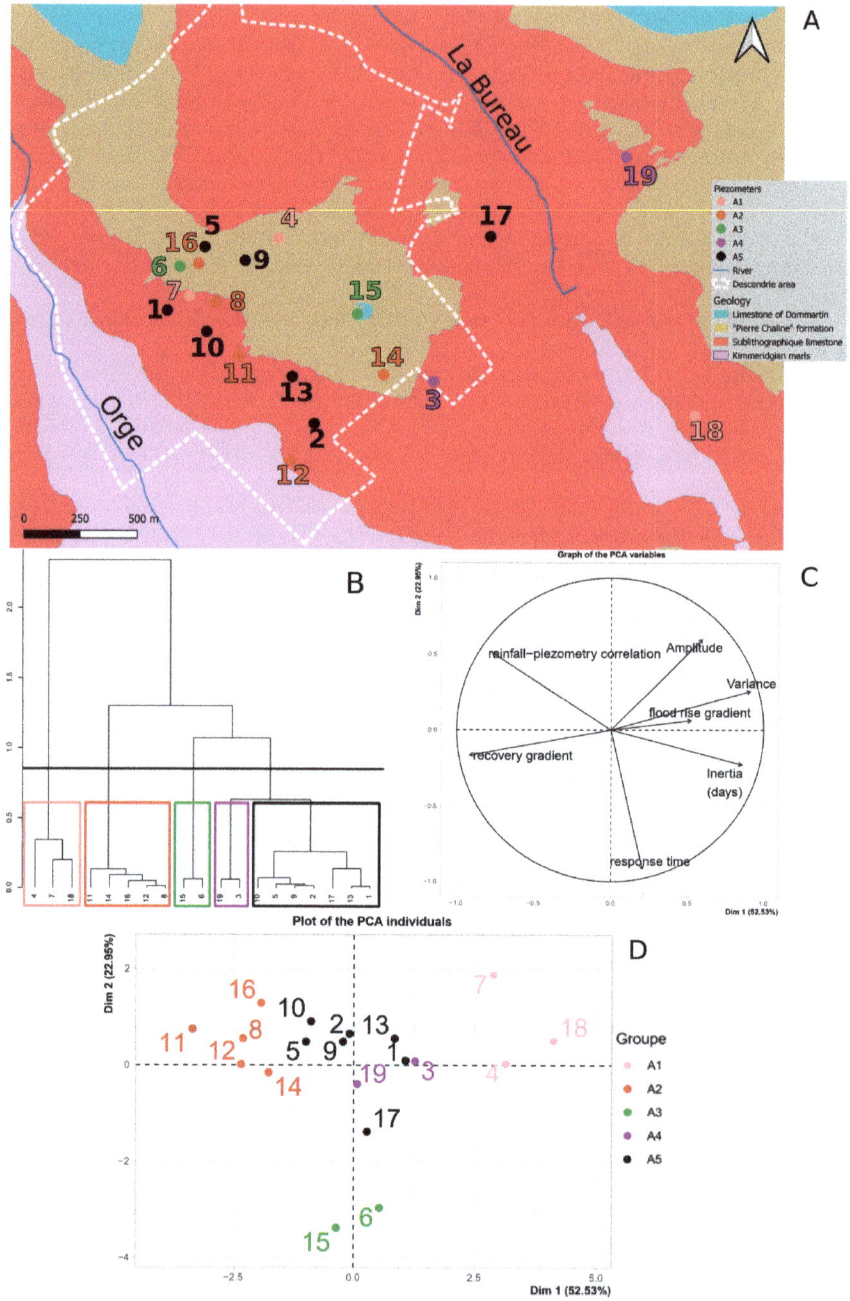


**Figure 2.** Result of multi-criteria statistical analysis. (**A**)—Mapping of piezometer positions according to the groups resulting from the hierarchical classification. The colors are associated with the groups determined by the hierarchical classification. (**B**)—Hierarchical classification of piezometers according to the selected criteria described in methodology. (**C**)—PCA variable graph. (**D**)—PCA individuals' graph.

*4.2. Processing Geological and Geotechnical Data*

4.2.1. Fracturing Data

Gamma Ray logs are used as a reference for all geological and fracture studies in the study area and are used to correlate Sublithographic limestone levels from one borehole to another. Areal fracturing has led to the development of a karst network that is visible on optical logs, as shown in Figure 3C,E,F).

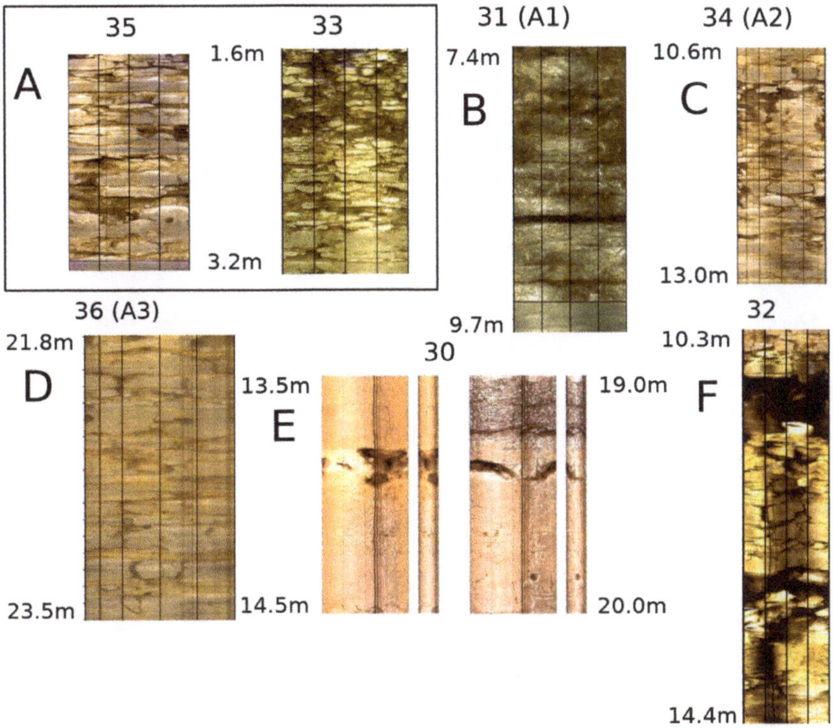


**Figure 3.** (**A**)—Weathered zone. (**B**)—Fracturing in a piezometer of group A1 (poorly developed fracturing). (**C**)—Fracturing in a borehole near a piezometer of group A2 {16}, with highly developed fracturing. (**D**)—Fracturing in a borehole near a Group A3 piezometer {29}, with vertical and horizontal fracturing visible. (**E**)—Dissolution marks in a borehole distant from the piezometers and therefore not able to be connected to a functioning group. (**F**)—Highly developed and partly karstified fracturing (presence of conduit) in a borehole that cannot be connected to a piezometric functioning group.

Below and sometimes at some distance from the altered surface zone (Figure 3A), centimeter-scale horizontal drains are most commonly found in the limestone intervals of the most marly levels, following the bedding planes (Figure 3B,D). These drains, which show clear traces of dissolution, are often found in the limestone interbeds of N4 and N2 (Figure 1B) or in the upper part of the Exogyres marls (Figure 3B,E,F).

Fracture orientation detected by wall imagery from 25 boreholes (3 of which are to the north of the study area) is summarized in Figure 4. A primary direction centered on N150°E is visible, with a scatter of measurements around this value. In detail (Figure 4), perpendicular directions dominate in some of the boreholes.

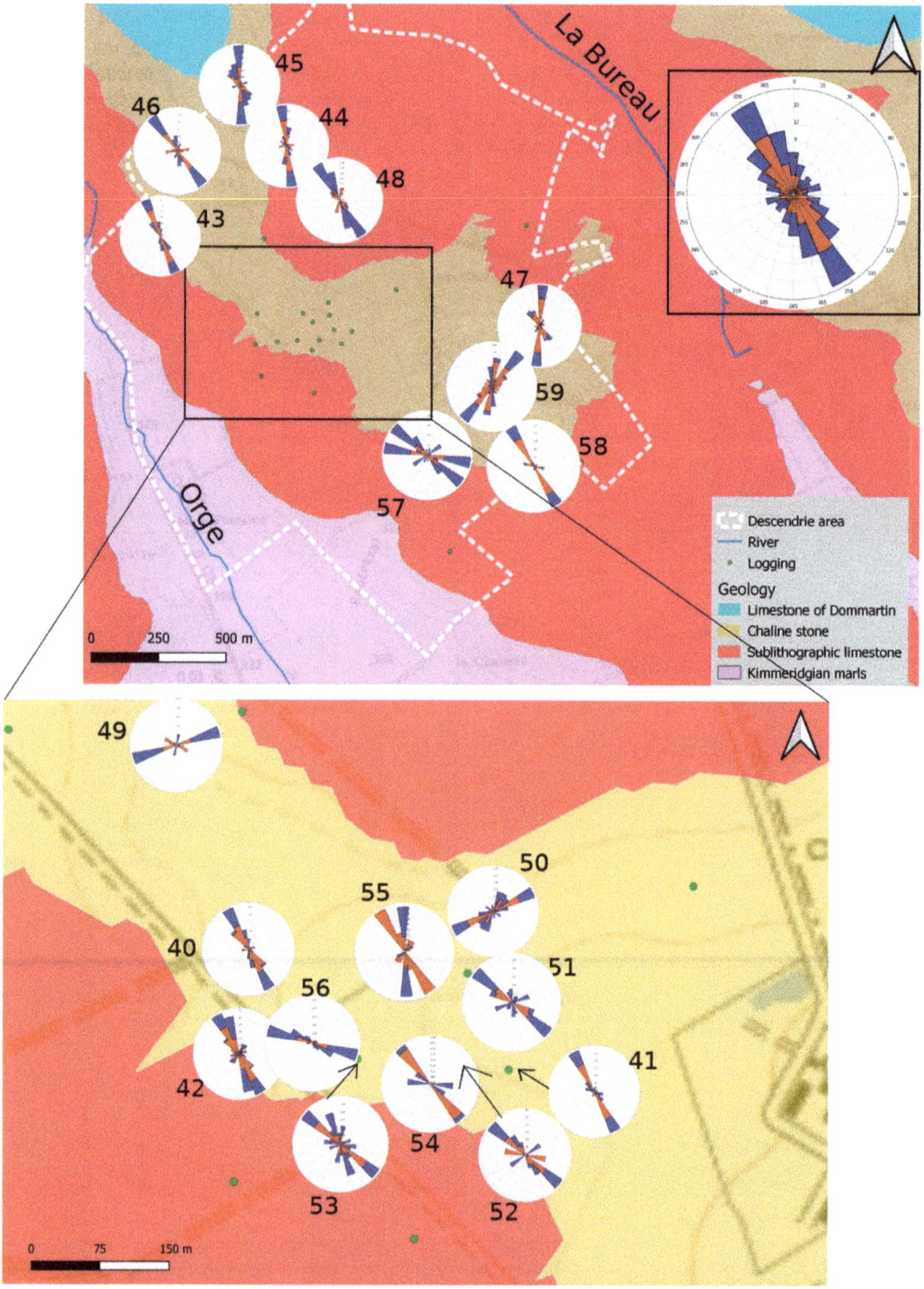


**Figure 4.** Map of the distribution of oriented fracturing based on optical logs, with differentiation between open fracturing (in red) and closed fracturing (in blue). The synthesis of all rose diagrams is represented by the large rose.

The fractures identified by the methods used are local and difficult to match from one borehole to another.

4.2.2. Hydraulics Tests

Lefranc Tests: Permeability on the Scale of the Borehole

Results of the tests conducted on boreholes {24}, {25}, {26}, and {27} are shown in Figure 5.

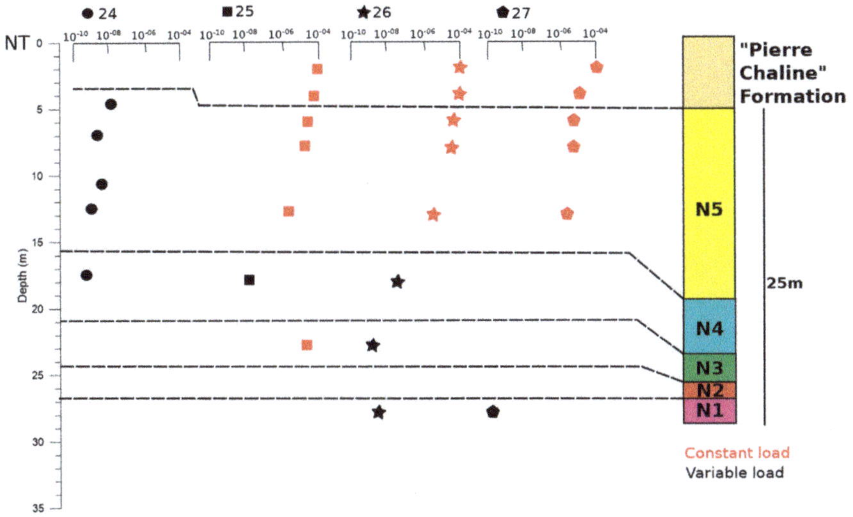


**Figure 5.** Permeability tests (Lefranc) at various normalized depths in relation to the natural terrain (NT). Legend colors correspond to those shown in Figure 1B.

Stratigraphically, the tests were conducted in the "Pierre Chaline" formation to a depth of 4.5 m, in the upper part of the Sublithographic limestone (N5) at depths from 4.5 m to 13.5 m, in the middle part of the Sublithographic limestone (N4) at depths from 17 to 18.5 m, and in the lower part of the Sublithographic limestone (N3 to N1) at depths from 22.5 m to 30 m. Kimmeridgian marls were also encountered in a single borehole {28}.

Permeability results from constant load tests range from $10^{-4}$ to $10^{-6}$ m/s, with permeability decreasing with depth. Permeability results from variable load tests range from $10^{-8}$ to $10^{-10}$ m/s, regardless of the formation and test depth.

The formations are permeable to semi-permeable ($10^{-5}$ to $10^{-6}$ m/s) down to a depth of 17 m (N5). Then, in the middle to lower part of the Sublithographic limestone (N4 to N1), they reach values of $10^{-8}$ to $10^{-10}$ m/s. A more permeable zone at $10^{-5}$ m/s in the lower part of the limestone (N3) was encountered in {25}.

The permeability differences measured reflect the general anisotropy of the various lithologies of the Sublithographic limestones. However, they also reflect anisotropy on a larger scale. For example, borehole {24} crosses the upper and middle parts of the Sublithographic limestones (N5 and N4), which have low permeabilities in the order of $10^{-8}$ to $10^{-10}$ m/s, whereas it is located 800 m from the other boreholes, which have higher permeability ranges (Figure 5). The borehole {24} is also very close to Group A2 and A5 piezometers, which have more-or-less transmissive behavior.

The results for the N5 formation remain consistent among nearby boreholes {25}, {26} and {27}. For the N3 formation, the variation in values at the same depth is significant, with differences of $10^{-4}$ m/s between two nearby boreholes ({25} and {26}). The N3 formation contains many more marly layers than the N5 formation. This explains the much greater permeability variation within this formation and also between the two formations.

Pumping Tests: Permeability at Aquifer Scale

Analyses of the pumping test and the disturbance in the only responsive observation piezometer give permeability values between $10^{-8}$ and $10^{-7}$ m/s. The difference between these values can be explained by the fact that different volumes were studied due to the heterogeneity of the medium being excited by the tests.

The storage coefficient for both pumped and interference boreholes obtained from these tests is approximately $10^{-3}$ m/s.

### *4.3. SNMR Data*

#### 4.3.1. SNMR Signal

The lowest initial amplitude (E0) over all soundings is between 5 and 10 nV. Initial amplitudes associated with the highest pulse moments range between 1 and 5 nV. These very low initial amplitude values do, however, allow some interpretation, since the signal-to-noise ratio of the soundings (Table 1) is always greater than 2.5.

The SNMR signals are provided in Figure 6. Uncertainty bounds are derived by considering minimum and maximum values from all SNMR signal parameter sets (signal amplitude e and T2* relaxation time) that yield a 10% increase in the root mean squared error (RMSE) fitting error as compared to the optimal RMSE, as proposed by [51].

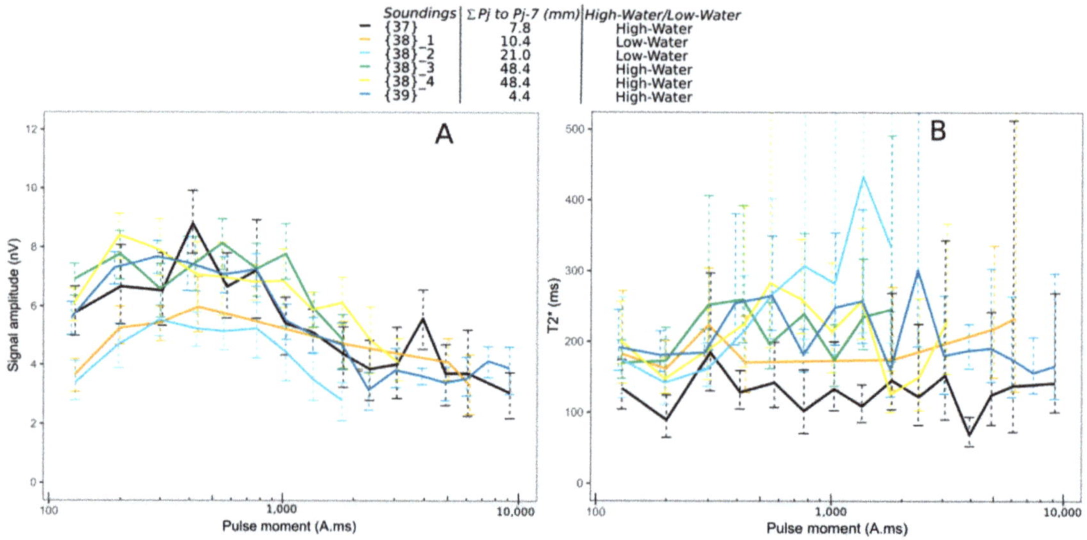


**Figure 6.** SNMR signal parameters as a function of the energizing pulse, for all soundings (Σ Pj to Pj − 7 = sum of rainfall over the 7 days preceding the sounding). (**A**)—Initial signal amplitude. (**B**)—Apparent transverse relaxation time T2*.

Two groups of responses stand out. The first group, consisting of soundings {38}_1 and {38}_2, has initial signal amplitudes approximately 3 nV lower at low pulse moments than the group consisting of soundings {38}_3, {38}_4, {37} and {39}, which suggests lower water content in the upper part of the sounding. High uncertainties are associated with T2* estimates. Sounding {37} stands out with relatively lower T2*, which may be associated with small porosity dimensions. These original data (Figure 6) are unaffected by any inversion artefacts. It is therefore interesting to compare these trends with the inverted results (Figure 7).

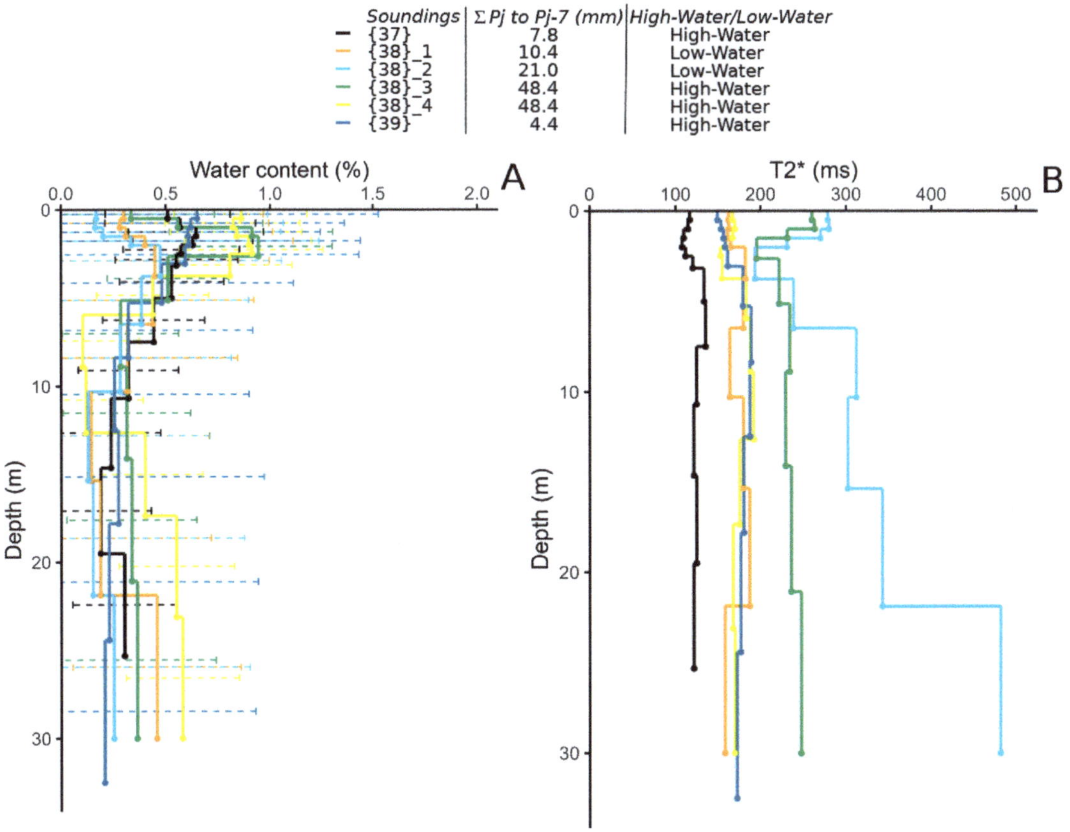


**Figure 7.** SNMR inversion for all sites. (**A**)—SNMR water content. (**B**)—Apparent transverse relaxation time T2*.

4.3.2. Inversion Results

The inversion results are displayed in Figure 7 with a 95% confidence interval on SNMR water content derived from the SVD inversion [52]. This approach provides an estimate of the uncertainty bounds of the water content in each layer considered separately. It accounts for the fact the buried water layers that have equal water volumes can, to some extent, yield equivalent SNMR responses. The results show very low to low water contents, which do not exceed 0.9% (1.5% considering the maximum inversion uncertainties), and that have an average value of 0.42%, with highest water content being reached in the first 0-5m (Figure 7A). These results indicate and quantify the low water storage capacity of the Sublithographic limestone matrix. In addition, the trends seen in Figure 6 are confirmed with these inverted data, since the same trends are present (two response groups and a very limited quantity of water content).

The T2* estimates are consistent with a filtered, non-inverted SNMR signal, with significantly lower values associated with Sounding {37} (see Section 4.3.1). High values associated with Sounding {38}_2 are not significant as they are associated with high signal uncertainties.

Soundings performed under different hydrologic conditions exhibit contrasting behaviors in the upper 0–5 m. Soundings {38}_1 and {38}_2 were performed in low water levels with significant antecedent precipitations. They have have lowest water content (0.3% in average). Soundings {37} and {39} were performed in high water levels with little antecedent precipitations. The have intermediate water content (0.55% in average). Soundings {38}_3 and {38}_4 were performed in high water levels with important antecedent precipitations. The have the highest water content (0.8% in average). SNMR inversions thus evidence both a primarily seasonal time-scale variation on the limestone water content, and a significant but secondary short (daily) time-scale variation.

The uncertainties provided in Figure 7 show the minimum and maximum water content that may be associated with each water layer, taken separately. Maximum water content for a specific water layer is associated with minimum or at least reduced water content in other layers. Such representation of uncertainty is not adequate to assess the minimum and maximum water content over the whole sounding. As a complement to Figure 7, we thus provide uncertainty bounds for sounding {38} for equivalent models that yield the minimum and the maximum water volumes over the investigated depth [53] (Figure 8). The average water content is between 0.57–0.61% and 0.51–0.63% for soundings {38}_4 and {38}_3, respectively, while it is only between 0.19–0.25% and 0.33–0.35% for soundings {38}_2 and {38}_1, respectively.

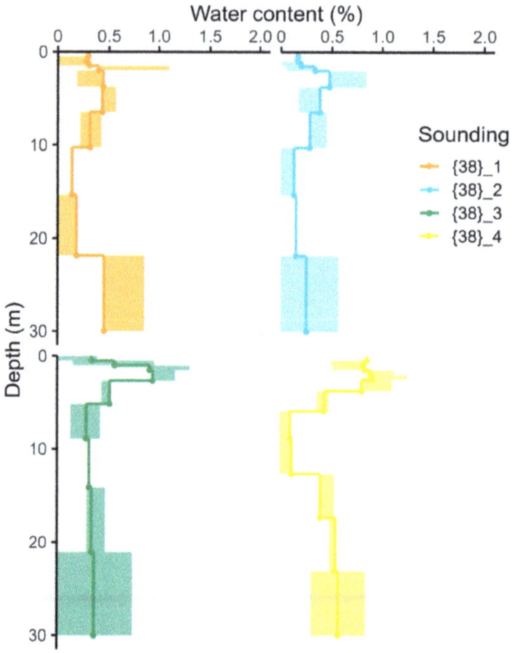


**Figure 8.** SNMR inversion for soundings performed at location {38}. Temporal water content variability between high ({38}_3 and {38}_4) and low ({38}_1 and {38}_2) water levels can be significant.

## 5. Discussion

Large borehole networks are often installed with geotechnical objectives in mind, and the data collected as part of these studies are often not linked to hydrogeological studies as such. Although these boreholes can be reused as piezometers, they were not originally designed as such. In many cases, these boreholes were dug throughout the structure and are not cased. This greatly complicates the interpretation of water level variations in response to internal or external forcing. Our study site benefits from such a network of discrete boreholes set up for geotechnical purposes. In the following discussion, we examine the strengths and weaknesses of this network, as well as possible ways of overcoming them.

### 5.1. Piezometric Response and Fracture Network Connectivity

Statistical analysis of these chronicles requires a large dataset and, depending on the indicators used, makes it possible to classify the different piezometric variations according to their response to rainfall. This descriptive analysis made it possible to characterize five behaviors, grouping 19 piezometers and identifying transmissive (groups A1 and A2), an inertial (group A3), and a group with mixed behaviors (groups A4 and A5) (Figure 9).

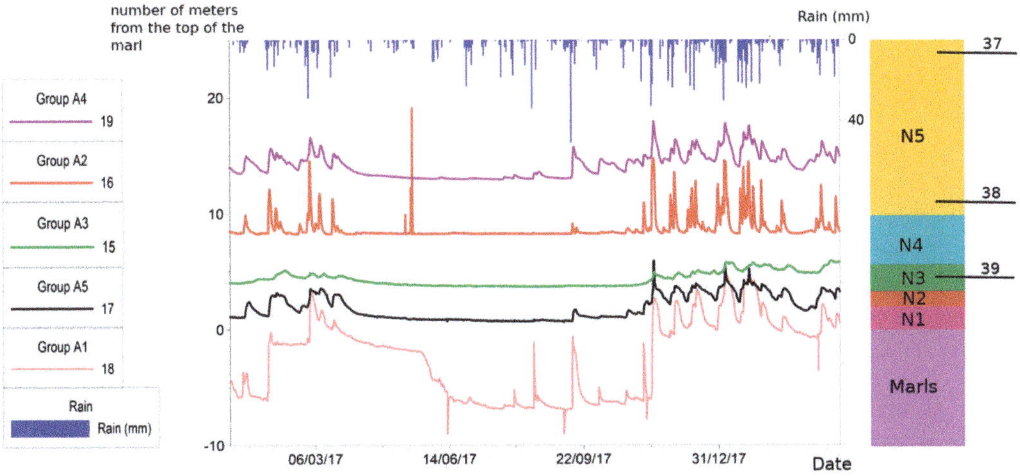


**Figure 9.** Classification of hydrodynamic behavior of water levels in piezometers in response to rainfall events.

Considered alone, piezometric response does allow some relevant insights into the hydrodynamics behavior. This is the case for Groups A4 and A5, where the analysis of the piezometric response provides information on the more-or-less inertial functioning of the boreholes, and suggests the presence of fractures in the process of development or a poorly connected fracture network. It is also the case for Group A1, for which neither geological drillings allowing fracture analysis, nor geophysical logs, are available near the piezometers. According to the descriptive analysis, Group A1 behaves like a permeable medium with a rapid response time to rainfall events but a relatively long memory effect with a slow decrease in water level. We associated such behavior with a moderate degree of fracturing (Figure 3B), with the rapid water level rise in the piezometer being simply due to the filling of the piezometer and not significative of the hydrodynamic response of the water table in the piezometer area. The SNMR results confirm this interpretation, as the significantly lower T2* associated with sounding {37} hints at a significative matrix contribution to the effective porosity.

Detailed fracture analysis may evidence discrete features that are not relevant for understanding hydrodynamic responses. For example, group A3 is associated with matrix functioning, despite the presence of open fractures near one piezometer in this group (Figure 3D). The orientation of these open fractures is perpendicular to the main flow, which may explain the inertial behavior of this piezometer.

In other cases, detailed fracture analysis using geological boreholes and SNMR can avoid erroneous water level dynamics interpretation. In the example provided in Figure 10, optical imaging allows to identify a karst conduit that is responsible for the regulation of the groundwater level. SNMR sounding {39}, located close to piezometer {16} (Figure 10), did not record evidence of significative water content below this fracture. We thus formulate the hypothesis that the water level below this base level does not contribute to the flow and may not be considered as representative from the water table.

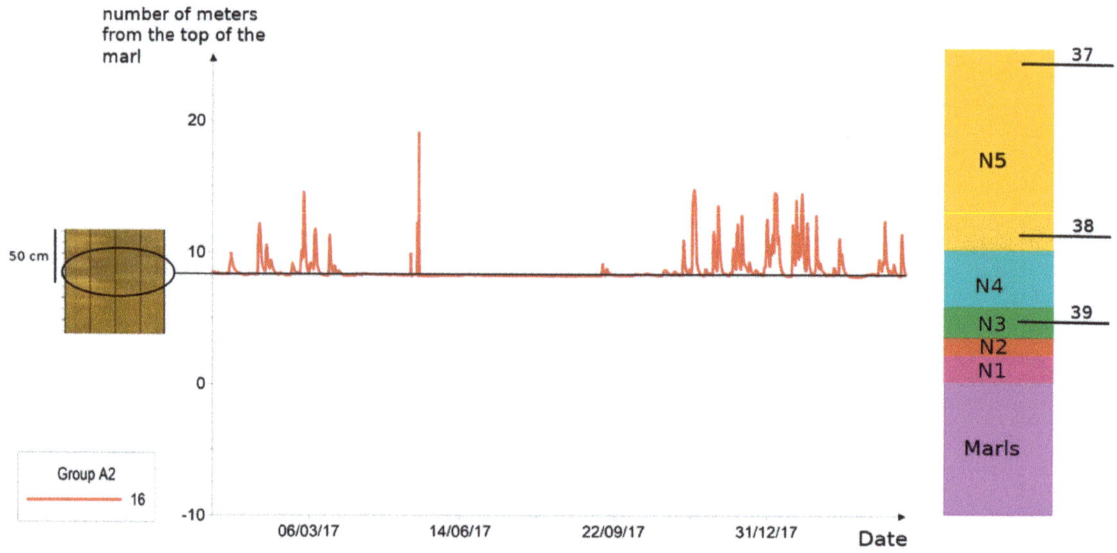


**Figure 10.** Link between conduit and piezometric level: the karst conduit maintains a base level in the piezometer. Only water levels above this base level contribute to the flow.

### 5.2. Hydraulic Tests and Local Matrix Characterization

In karstified environments, permeabilities can vary by several orders of magnitude depending on the scale of measurement (laboratory, borehole, and catchment) [27]. For each of these permeability ranges, different measurement methods are used. Hydraulic tests carried out in this study are mostly sensitive to the near borehole media.

Permeability measurements carried out indicate higher permeability at the surface of the Sublithographic limestones and lower permeability at depth in these limestones, in the marls, and in the marlier limestone layers. These values highlight a semi-permeable behavior for limestones to be impermeable when marls predominate, whereas the piezometric response tends to highlight the reactivity of the system.

In our study, the duration of the Lefranc tests is too short and the pumping test flow rates are too low to attract the more distant water contained within the carbonate matrix or connected fractures, so they mainly provide local matrix characterization.

### 5.3. SNMR and Water Content Quantification

Information on water mobilization in response to forcing obtained from piezometric responses, fracturing analysis, and hydraulic tests remains incomplete as they do not quantify the water saturation of the medium. SNMR measurements provide access to the water content and its variation over a wider access radius than those investigated by borehole tests or piezometric variations.

In our study, very low (3 and 8 nV) SNMR signals associated with high signal-to-noise ratio allow us to quantify very low (<0.6%) water content with significative variability.

We relate this variability to both a primarily seasonal time-scale variation of the limestone water content and a significant but secondary short (daily) time-scale variation. This result reinforces the interpretation of the piezometric behavior of matrix-like and mixed matrix/transmissive groups, with mostly seasonal matrix dynamics and daily-to-weekly fissures and fracture dynamics.

## 6. Conclusions

This study utilized a variety of methods to better understand the functioning of karst groundwater systems. With karst outcrops covering 10-15% of the world's continental surface [21], understanding these hydrosystems is crucial for addressing the challenges associated with the exploitation of groundwater resources. It is estimated that around 25% of the world's population relies on groundwater extracted from karst areas for drinking water [52]. Additionally, studying these systems is essential during geotechnical or civil engineering investigations [4].

Karst hydrosystems are complex due to their triple porosity (matrix, fractures, and conduits) and inherent heterogeneity. Both the fracture properties and the hydrodynamic properties of the carbonate matrix determine the functioning of karst aquifers. Piezometric information is often the basis for understanding flow organization. Many studies rely on discrete piezometric data to characterize these complex environments where flows are distributed according to permeability constraints.

Our study benefits from a large network of boreholes initially implemented for geotechnical purposes. Despite the density of this observation network, the collection of water level variations is not easily linked to the hydrogeological properties of the aquifer. While piezometric responses alone provide some insights into hydrodynamic behavior, detailed fracture analysis can reveal discrete features that are not always relevant for understanding these responses. However, in some cases, combining detailed fracture analysis from geological boreholes with SNMR is necessary to avoid erroneous interpretations of water level dynamics.

High-density and high-quality data in karst environments are rarely available due to the complexity and heterogeneity of these systems. Comprehensive datasets are crucial for accurate analysis and modeling, yet they are challenging to obtain. By leveraging the extensive and high-quality data from this unique site, we can gain valuable insights into the hydrogeologic complexities of fractured-rock karst systems, providing a strong basis for effectively understanding and managing these environments

The combination of geological, hydrogeological, and geophysical measurements in this project highlights the contribution of water masses with two types of aquifer responses: a rapid response at hourly-to-daily timescales to rainfall events in well-connected fractures, and a slower response at seasonal timescales in the matrix. The characterization of the dynamics and the quantification of the water content in the matrix and fissured/fractured compartments were enabled by the combined interpretation of piezometric response analyses and SNMR data.

Thus, coupling various separately collected measurements provides a key to understanding the aquifer as a whole. Future work could include longer pumping tests or modelling fluctuations of base groundwater levels to better assess matrix flows.

**Author Contributions:** Conceptualization, M.B., C.B. and N.M.; methodology, M.B., L.-M.G., V.B., N.M., R.V., K.C., C.B., S.D., S.G. and G.C.; software, M.B., N.M. and G.C.; validation, N.M., R.V. and K.C.; formal analysis, M.B.; investigation, M.B., L.-M.G. and V.B.; resources, S.G.; data curation, M.B., L.-M.G. and V.B.; writing—original draft preparation, M.B., C.B. and N.M.; writing—review and editing, C.B., N.M., K.C., R.V., S.D. and S.G.; Supervision, C.B., N.M., K.C. and S.G.; project administration, C.B., K.C. and S.G.; funding acquisition, S.G., C.B., K.C., N.M. and S.D. All authors have read and agreed to the published version of the manuscript.

**Funding:** This research was funded by ANDRA PhD grant number 20080158.

**Data Availability Statement:** Data are contained within the article.

**Acknowledgments:** We would like to thank the reviewers for their work and constructive comments and suggestions. We would also like to express our sincere thanks to Patricia Bobeck for her proof-reading and for her contribution to the English version of the manuscript.

**Conflicts of Interest:** The authors declare no conflict of interest.

## Appendix A

**Table A1.** Links between number, name, and information.

| Number | Name | Information | Number | Name | Information |
|---|---|---|---|---|---|
| 1 | CIG1028 | Piezometry | 31 | CIG1031 | Borehole imaging |
| 2 | CIG1029 | Piezometry | 32 | CIG1080 | Borehole imaging |
| 3 | CIG1030 | Piezometry | 33 | CIG1088 | Borehole imaging |
| 4 | CIG1031 | Piezometry | 34 | CIG1120 | Borehole imaging |
| 5 | CIG1227 | Piezometry | 35 | CIG1122 | Borehole imaging |
| 6 | CIG1228 | Piezometry | 36 | CIG1254 | Borehole imaging |
| 7 | CIG1229 | Piezometry | 37 | EST3BIS | SNMR |
| 8 | CIG1230 | Piezometry | 38 | EST14 | SNMR |
| 9 | CIG1237 | Piezometry | 39 | EST18 | SNMR |
| 10 | CIG1245 | Piezometry | 40 | CIG1005 | Fracturing |
| 11 | CIG1246 | Piezometry | 41 | CIG1009 | Fracturing |
| 12 | CIG1247 | Piezometry | 42 | CIG1011 | Fracturing |
| 13 | CIG1248 | Piezometry | 43 | CIG1077 | Fracturing |
| 14 | CIG1249 | Piezometry | 44 | CIG1078 | Fracturing |
| 15 | CIG1250 | Piezometry | 45 | CIG1079 | Fracturing |
| 16 | EST1012 | Piezometry | 46 | CIG1080 | Fracturing |
| 17 | EST1021 | Piezometry | 47 | CIG1086 | Fracturing |
| 18 | EST1037 | Piezometry | 48 | CIG1088 | Fracturing |
| 19 | EST5071 | Piezometry | 49 | CIG1090 | Fracturing |
| 20 | CIG1023 | Pumping tests | 50 | CIG1108 | Fracturing |
| 21 | CIG1024 | Pumping tests | 51 | CIG1114 | Fracturing |
| 22 | CIG1025 | Pumping tests | 52 | CIG1116 | Fracturing |
| 23 | CIG1026 | Pumping tests | 53 | CIG1117 | Fracturing |
| 24 | CIG1137 | Lefranc tests | 54 | CIG1118 | Fracturing |
| 25 | CIG1138 | Lefranc tests | 55 | CIG1120 | Fracturing |
| 26 | CIG1139 | Lefranc tests | 56 | CIG1121 | Fracturing |
| 27 | CIG1140 | Lefranc tests | 57 | CIG1252 | Fracturing |
| 28 | CIG1237 | Lefranc tests | 58 | CIG1253 | Fracturing |
| 29 | CIG1250 | Lefranc tests | 59 | CIG1254 | Fracturing |
| 30 | CIG1011 | Borehole imaging | | | |

## Appendix B

**Table A2.** Characteristics of the piezometers used in this study. Type and data-acquisition start dates.

| Name | Depth (m) | Borehole Screen | Geological Level | Data | Data Acquisition Start On |
|---|---|---|---|---|---|
| {16} | 25.5 | Ø 112/122.5 mm + gravel packing | Sublithographic limestone (N1) | Pressure<br>Temperature | 13 April 1996<br>30 June 1999 |
| {17} | 30 | Ø 113/125 mm opening 1 mm + gravel packing | Kimmeridgian marls | Pressure<br>Temperature<br>Sampling | 4 April 2000<br>6 July 2000<br>6 July 2000 |
| {18} | 20.2 | Ø 113/125 mm opening 1 mm + gravel packing | Kimmeridgian marls | Pressure<br>Temperature<br>Sampling | 5 April 2000 |
| {19} | 39.06 | Ø 119.8/125 mm | Kimmeridgian marls | Pressure<br>Temperature | 20 December 1995<br>10 July 1999 |
| {1} | 20 | gravel packing | Kimmeridgian marls | Pressure | 25 January 2016 |
| {2} | 14 | gravel packing | Kimmeridgian marls | Pressure | 17 February 2016 |
| {3} | 28 | gravel packing | Kimmeridgian marls | Pressure | 25 February 2016 |
| {4} | 30 | gravel packing | Kimmeridgian marls | Pressure | 27 January 2016 |
| {5} | 30.3 | gravel packing | Kimmeridgian marls | Pressure<br>Temperature | 31 August 2016 |
| {6} | 30.3 | gravel packing | Kimmeridgian marls | Pressure<br>Temperature | 31 August 2016 |
| {7} | 30.3 | gravel packing | Kimmeridgian marls | Pressure<br>Temperature | 22 August 2016 |
| {8} | 32 | gravel packing | Kimmeridgian marls | Pressure<br>Temperature | 31 August 2016 |
| {9} | 30.3 | gravel packing | Kimmeridgian marls | Pressure<br>Temperature | 31 August 2016 |
| {10} | 20.3 | gravel packing | Kimmeridgian marls | Pressure<br>Temperature | 18 August 2016 |
| {11} | 20.3 | gravel packing | Sublithographic limestone (N1) | Pressure<br>Temperature | 16 June 2016 |
| {12} | 15.4 | gravel packing | Kimmeridgian marls | Pressure<br>Temperature | 29 September 2016 |
| {13} | 30.35 | gravel packing | Kimmeridgian marls | Pressure<br>Temperature | 22 August 2016 |
| {14} | 15.3 | gravel packing | Sublithographic limestone (N5) | Pressure<br>Temperature | 24 November 2016 |
| {15} | 30.3 | gravel packing | Sublithographic limestone (N4) | Pressure<br>Temperature | 3 August 2016 |

## References

1. Stevanović, Z. Global Distribution and Use of Water from Karst Aquifers. *Geol. Soc. Lond. Spec. Publ.* **2018**, *466*, 217–236. [CrossRef]
2. Ravbar, N.; Šebela, S. The Effectiveness of Protection Policies and Legislative Framework with Special Regard to Karst Landscapes: Insights from Slovenia. *Environ. Sci. Policy* **2015**, *51*, 106–116. [CrossRef]
3. Hartmann, A.; Goldscheider, N.; Wagener, T.; Lange, J.; Weiler, M. Karst Water Resources in a Changing World: Review of Hydrological Modeling Approaches: Karst Water Resources Prediction. *Rev. Geophys.* **2014**, *52*, 218–242. [CrossRef]
4. Chen, C.-T.; Hu, J.-C.; Lu, C.-Y.; Lee, J.-C.; Chan, Y.-C. Thirty-Year Land Elevation Change from Subsidence to Uplift Following the Termination of Groundwater Pumping and Its Geological Implications in the Metropolitan Taipei Basin, Northern Taiwan. *Eng. Geol.* **2007**, *95*, 30–47. [CrossRef]

5. Mangin, A. Pour une meilleure connaissance des systèmes hydrologiques à partir des analyses corrélatoire et spectrale (The use of autocorrelation and spectral analyses to obtain a better understanding of hydrological systems). *J. Hydrol.* **1984**, *67*, 25–43. [CrossRef]
6. Nistor, M.M.; Rahardjo, H.; Satyanaga, A.; Hao, K.Z.; Xiaosheng, Q.; Sham, A.W.L. Investigation of Groundwater Table Distribution Using Borehole Piezometer Data Interpolation: Case Study of Singapore. *Eng. Geol.* **2020**, *271*, 105590. [CrossRef]
7. De Luca, D.A.; Lasagna, M.; Debernardi, L. Hydrogeology of the Western Po Plain (Piedmont, NW Italy). *J. Maps* **2020**, *16*, 265–273. [CrossRef]
8. Fiorillo, F. Tank-Reservoir Drainage as a Simulation of the Recession Limb of Karst Spring Hydrographs. *Hydrogeol. J* **2011**, *19*, 1009–1019. [CrossRef]
9. Bennett, G.; Van Camp, M.; Shemsanga, C.; Kervyn, M.; Walraevens, K. Assessment of Spatial and Temporal Variability of Groundwater Level in the Aquifer System on the Flanks of Mount Meru, Northern Tanzania. *J. Hydrol. Reg. Stud.* **2022**, *44*, 101212. [CrossRef]
10. Giese, M.; Reimann, T.; Bailly-Comte, V.; Maréchal, J.-C.; Sauter, M.; Geyer, T. Turbulent and Laminar Flow in Karst Conduits Under Unsteady Flow Conditions: Interpretation of Pumping Tests by Discrete Conduit-Continuum Modeling. *Water Resour. Res.* **2018**, *54*, 1918–1933. [CrossRef]
11. Praamsma, T.; Novakowski, K.; Kyser, K.; Hall, K. Using Stable Isotopes and Hydraulic Head Data to Investigate Groundwater Recharge and Discharge in a Fractured Rock Aquifer. *J. Hydrol.* **2009**, *366*, 35–45. [CrossRef]
12. Maréchal, J.-C.; Ladouche, B.; Dörfliger, N.; Lachassagne, P. Interpretation of Pumping Tests in a Mixed Flow Karst System: Pumping Tests in Karst System. *Water Resour. Res.* **2008**, *44*, W05401. [CrossRef]
13. Banks, E.W.; Simmons, C.T.; Love, A.J.; Shand, P. Assessing Spatial and Temporal Connectivity between Surface Water and Groundwater in a Regional Catchment: Implications for Regional Scale Water Quantity and Quality. *J. Hydrol.* **2011**, *404*, 30–49. [CrossRef]
14. Valois, R.; Rau, G.C.; Vouillamoz, J.-M.; Derode, B. Estimating Hydraulic Properties of the Shallow Subsurface Using the Groundwater Response to Earth and Atmospheric Tides: A Comparison With Pumping Tests. *Water Resour. Res.* **2022**, *58*, e2021WR031666. [CrossRef]
15. Bon, A.F.; Aoudou Doua, S.; Banakeng, L.A.; Narke, C.; Chouto, S.; Mbouombouo Ndam, A. Contribution of a Geostatistical Model of Electrical Conductivity in the Assessment of the Water Pollution Index of the Quaternary Aquifer of the Lake Chad Basin (Kousseri-Cameroon). *Arab. J. Geosci.* **2020**, *13*, 170. [CrossRef]
16. Dewandel, B.; Aunay, B.; Maréchal, J.C.; Roques, C.; Bour, O.; Mougin, B.; Aquilina, L. Analytical Solutions for Analysing Pumping Tests in a Sub-Vertical and Anisotropic Fault Zone Draining Shallow Aquifers. *J. Hydrol.* **2014**, *509*, 115–131. [CrossRef]
17. Castellazzi, P.; Arroyo-Domínguez, N.; Martel, R.; Calderhead, A.I.; Normand, J.C.L.; Gárfias, J.; Rivera, A. Land Subsidence in Major Cities of Central Mexico: Interpreting InSAR-Derived Land Subsidence Mapping with Hydrogeological Data. *Int. J. Appl. Earth Obs. Geoinf.* **2016**, *47*, 102–111. [CrossRef]
18. Haimson, B.C.; Doe, T.W. State of Stress, Permeability, and Fractures in the Precambrian Granite of Northern Illinois. *J. Geophys. Res.* **1983**, *88*, 7355–7371. [CrossRef]
19. Dausse, A.; Leonardi, V.; Jourde, H. Hydraulic Characterization and Identification of Flow-Bearing Structures Based on Multi-Scale Investigations Applied to the Lez Karst Aquifer. *J. Hydrol. Reg. Stud.* **2019**, *26*, 100627. [CrossRef]
20. Goldscheider, N.; Drew, D. *Methods in Karst Hydrogeology*; International Contributions to Hydrogeology; Taylor & Francis: London, UK, 2007; ISBN 978-0-415-42873-6.
21. Kiraly, L. Karstification and Groundwater Flow. *Speleogenesis Evol. Karst Aquifers* **2003**, *1*, 1–26.
22. Ford, D.C.; Williams, P.W. *Karst Hydrogeology and Geomorphology*; Chapman and Hall: London, UK, 2007. [CrossRef]
23. Bakalowicz, M. Karst Groundwater: A Challenge for New Resources. *Hydrogeol. J.* **2005**, *13*, 148–160. [CrossRef]
24. Bakalowicz, M. Le milieu karstique: Études et perspectives, identification et caractérisation de la ressource. In Proceedings of the CFH—Colloque Hydrogéologie et Karst au Travers des Travaux de Michel Lepiller 17 Mai 2008, Orléans, France, 16–17 May 2008.
25. Atkinson, T.C. Diffuse Flow and Conduit Flow in Limestone Terrain in the Mendip Hills, Somerset (Great Britain). *J. Hydrol.* **1977**, *35*, 93–110. [CrossRef]
26. Shuster, E.T.; White, W.B. Seasonal Fluctuations in the Chemistry of Lime-Stone Springs: A Possible Means for Characterizing Carbonate Aquifers. *J. Hydrol.* **1971**, *14*, 93–128. [CrossRef]
27. Clauser, C. Permeability of Crystalline Rocks. *Eos* **1992**, *73*, 233–238. [CrossRef]
28. Jeannin, P.-Y. Structure et Comportement Hydraulique des Aquifères Karstiques (Structure and Hydraulic Behavior of Karst Aquifers). Ph.D. Thesis, Université de Neuchâtel, Neuchâtel, Switzerland, 1996.
29. Delbart, C.; Valdés, D.; Barbecot, F.; Tognelli, A.; Couchoux, L. Spatial Organization of the Impulse Response in a Karst Aquifer. *J. Hydrol.* **2016**, *537*, 18–26. [CrossRef]
30. Chalikakis, K.; Plagnes, V.; Guerin, R.; Valois, R.; Bosch, F.P. Contribution of Geophysical Methods to Karst-System Exploration: An Overview. *Hydrogeol. J.* **2011**, *19*, 1169–1180. [CrossRef]
31. Chalikakis, K.; Nielsen, M.R.; Legchenko, A. MRS Applicability for a Study of Glacial Sedimentary Aquifers in Central Jutland, Denmark. *J. Appl. Geophys.* **2008**, *12*, 176–187. [CrossRef]
32. Legchenko, A.; Valla, P. A Review of the Basic Principles for Proton Magnetic Resonance Sounding Measurements. *J. Appl. Geophys.* **2002**, *50*, 3–19. [CrossRef]

33. Mazzilli, N.; Chalikakis, K.; Carrière, S.D. Surface Nuclear Magnetic Resonance Monitoring Reveals Karst Unsaturated Zone Recharge Dynamics during a Rain Event. *Water* **2020**, *12*, 3183. [CrossRef]
34. Scholz, E. *Synthèse Hydrogéologique des Calcaires du Barrois;* ANDRA: Bure, France, 2011; p. 97.
35. Jaillet, S. Un Karst de Bas Plateau: Le Barrois, Structure, Fonctionnement, Evolution (A Low Plateau Karst: The Barrois, Structure, Functioning, Evolution). Ph.D. Thesis, Université de Bordeaux, Bordeaux, France, 2000.
36. Decloux, J.-P. *Hydrologie et Hydrogeologie des Aquiferes de Surface sur les Bassins de la Saulx et de L'ornain—Rapport de Tâche 1 (RT1);* ANDRA: Bure, France, 2018; p. 47.
37. Mangin, A. Contribution à L'étude Hydrodynamique des Aquifères Karstiques. Ph.D. Thesis, Université de Dijon, Dijon, France, 1975.
38. Cholet, C. Fonctionnement Hydrogéologique et Processus de Transport dans les Aquifères Karstiques du Massif du Jura. Ph.D. Thesis, Université Bourgogne Franche-Comté, Besançon, France, 2017.
39. Delbart, C. Variabilité Spatio-Temporelle du Fonctionnement d'un Aquifère Karstique du Dogger: Suivis Hydrodynamiques et Géochimiques Multifréquences; Traitement du Signal des Réponses Physiques et Géochimiques. Ph.D. Thesis, Université Paris-Sud—Paris XI, Paris, France, 2013.
40. Hvorslev, J. *Time Lag and Soil Permeabilty in Ground;* Waterways Experiment Station: Vicksburg, MS, USA, 1951; Volume 50.
41. *ISO 22282-2;* Reconnaissance et Essais Géotechniques—Essais Géohydrauliques—Partie 2: Essai de Perméabilité à L'eau Dans un Forage en Tube Ouvert. ISO: Geneva, Switzerland, 2013.
42. *NF P94-130 Standard;* Soils: Investigation and Testing—Pumping Test. AFNOR: Paris, France, 2002.
43. Moench, A.F. Flow to a Well of Finite Diameter in a Homogeneous, Anisotropic Water Table Aquifer. *Water Resour. Res.* **1997**, *33*, 1397–1407. [CrossRef]
44. Theis, C.V. The Relation between the Lowering of the Piezometric Surface and the Rate and Duration of Discharge of a Well Using Ground. *Eos Trans. AGU* **1935**, *16*, 519–524. [CrossRef]
45. Kruseman, G.P.; De Ridder, N.A. *Analysis and Evaluation of Pumping Test Data;* International Institute for Land Reclamation and Improvement: Wageningen, The Netherlands, 1991; Volume 47.
46. Vouillamoz, J.-M. La Caractérisation des Aquifères par une Méthode Non Invasive: Les Sondages par Resonance Magnetique Protonique. Ph.D. Thesis, Université de Paris XI, Paris, France, 2003.
47. Legchenko, A.; Baltassat, J.-M.; Beauce, A.; Bernard, J. Nuclear Magnetic Resonance as a Geophysical Tool for Hydrogeologists. *J. Appl. Geophys.* **2002**, *50*, 21–46. [CrossRef]
48. Legchenko, A.; Valla, P. Processing of Surface Proton Magnetic Resonance Signals Using Non-Linear Fitting. *J. Appl. Geophys.* **1998**, *39*, 77–83. [CrossRef]
49. Legchenko, A.; Valla, P. Removal of Power-Line Harmonics from Proton Magnetic Resonance Measurements. *J. Appl. Geophys.* **2003**, *53*, 103–120. [CrossRef]
50. Legchenko, A. MRS Measurements and Inversion in Presence of EM Noise. *Bol. Geol. Min.* **2007**, *118*, 489–508.
51. Mazzilli, N.; Boucher, M.; Chalikakis, K.; Legchenko, A.; Jourde, H.; Champollion, C. Contribution of Magnetic Resonance Soundings for Characterizing Water Storage in the Unsaturated Zone of Karst Aquifers. *Geophysics* **2016**, *81*, WB49–WB61. [CrossRef]
52. Legchenko, A. *Magnetic Resonance Imaging for Groundwater;* Wiley-ISTE: Hoboken, NJ, USA, 2013; ISBN 978-1-84821-568-9.
53. Legchenko, A.; Comte, J.-C.; Ofterdinger, U.; Vouillamoz, J.-M.; Lawson, F.M.A.; Walsh, J. Joint Use of Singular Value Decomposition and Monte-Carlo Simulation for Estimating Uncertainty in Surface NMR Inversion. *J. Appl. Geophys.* **2017**, *144*, 28–36. [CrossRef]

water

MDPI

Article

# Modelling Approach for Assessment of Groundwater Potential of the Moghra Aquifer, Egypt, for Extensive Rural Development

**Ahmed Shalby [1], Bakenaz A. Zeidan [1], Katarzyna Pietrucha-Urbanik [2,*], Abdelazim M. Negm [3] and Asaad M. Armanuos [1]**

1 Irrigation and Hydraulics Engineering Department, Faculty of Engineering, Tanta University, Tanta 31733, Egypt; ahmed.shalby@f-eng.tanta.edu.eg (A.S.); b.zeidan@f-eng.tanta.edu.eg (B.A.Z.); asaad.matter@f-eng.tanta.edu.eg (A.M.A.)
2 Department of Water Supply and Sewerage Systems, Faculty of Civil, Environmental Engineering and Architecture, Rzeszow University of Technology, Al. Powstańców Warszawy 6, 35-959 Rzeszow, Poland
3 Department of Water and Water Structures Engineering, Faculty of Engineering, Zagazig University, Zagazig 44519, Egypt; amnegm85@yahoo.com
* Correspondence: kpiet@prz.edu.pl

**Abstract:** Groundwater-dependent cultivation is imperative to meet the ever-increasing food demands in Egypt. To explore the Moghra aquifer's potential, where a large-scale rural community is being established, a finite element groundwater flow (i.e., FEFLOW®) model was invoked. The developed model was calibrated against the observed water levels. GRACE-based groundwater storage was incorporated into the tuning procedure of the developed model. Eight abstraction rates from 1000 wells, changing from 800 to 1500 $m^3$/day/well, were simulated for a 100-year test period. The maximum resulting drawdown values, respectively, ranged from 59 to 112 m equating to about 20–40% of the aquifer's saturated thickness. The implications of the climate change from gradual sea level rise and an increase in crop consumptive water use were investigated. Extending seawater invasion into the aquifer caused a slight increase in the piezometric levels within a narrow strip along the seaside. Applying a chronologically increasing withdrawal rate to meet the projected increment in crop water requirements raised the maximum resulting drawdown by about 7.5%. The sustainable exploitation regime was defined as a time-increasing withdrawal rate adequate to reclaim 85,715 acres (34,688 ha). The recommended development scheme is compatible with the withdrawal rationing rule, aiming to maintain that the resulting drawdown does not exceed one meter a year.

**Keywords:** FEFLOW; agriculture; sustainable management; aquifer depletion; over-abstraction

## 1. Introduction

Globally, water and food shortage problems have aggravated during the last few decades. Intensifying groundwater withdrawal has taken place to satisfy freshwater demands for domestic purposes [1]. The malpractice of uncontrolled pumping is further exacerbated to meet increasing agricultural water requirements [2]. Even fossil (non-renewable) aquifers are highly exploited, particularly in semiarid and arid countries [3]. Overdraft resulting in aquifer depletion is an emerging issue owing to the development of such non-renewable aquifers [4]. Thus, it is widely urged to rationalize pumping and judiciously use non-renewable aquifers [5]. However, achieving sustainable management of such resources is a key challenge, particularly in regions plagued with water scarcity. Egypt is within an arid climate region featuring extinct precipitation, high temperatures, and elevated humidity, which leads to an increased rate of evapotranspiration [6,7]. Due to climate change (CC) and the damming construction at the Nile River upstream, Egypt would be unable to draw as much as its water quota [8,9]. Accordingly, Egypt shall witness an imminent challenge for the agriculture sector to cater to the demands of the exponentially growing human population [10]. Rural development is therefore crucial in

Egypt to meet the growing food needs [11,12]. Accordingly, the government has initiated an ambitious rural development strategy through a mega groundwater-dependent cultivation project [13].

A mega reclamation project is now being implemented in Egypt under the framework of moving towards horizontal expansion in barren lands distanced from the Nile River [14]. Thus, desert fossil aquifers have become the sole source of freshwater [11,15]. Currently, groundwater withdrawal is well underway in the newly cultivated areas, mainly for irrigation purposes [16]. The pumping of groundwater is projected to increase year in and year out [17]. A concentrated number of wells that are not spatially distributed over wide areas, are being established [18]. Mismanagement or lack of pumping management, with its consequent overconsumption of the underground reserve, restricts groundwater availability. It, therefore, endangers the exploited aquifer through depletion and water quality degradation. Additionally, the severe decline in groundwater level might result in adverse consequences, including land subsidence, diminished surface vegetation, etc. [19]. Deficient replenishment of arid-region aquifer systems poses additional intricacy in their management. Attainment of a sustainable yield from an aquifer is, therefore, a major constraint for groundwater-beholden agribusiness [20]. Thus, a management approach should be set to maintain the aquifer storage in terms of quantity and quality aspects.

Groundwater flow models are increasingly recognized as a supportive utility for a sustainable groundwater management approach [21]. Groundwater models provide an approximate solution for the partial differential equation that visualizes water flow through a porous medium [22]. Modelling is, therefore, an essential tool for assessing an aquifer's potentiality and exemplifying its stages under pumping stresses [23]. Several numerical simulation codes have been programmed based on finite difference and finite element approaches [24]. MODFLOW is the most commonly used finite difference-based code used in the field of groundwater modelling. The Finite Element subsurface FLOW (FEFLOW®) is a relatively new 3D groundwater simulation code [25]. However, it has become one of the simulators widely utilized for groundwater flow and transport problems. FEFLOW has a substantial geometrizing skill over alternative finite-difference models (e.g., MODFLOW) [26]. Many studies have employed the FEFLOW for assessing trade-offs among potential water pumping policies and accounting for the drivers that have resulted in aquifer depletion and deteriorating water quality [27–30].

The recently drilled wells into the Moghra aquifer (Egypt) may be a clear-cut cause for water table drawdown and quality degradation. Thus, groundwater managers urged defining the sustainable management scheme of the aquifer. Herein, a hydrogeological investigation of the subsurface Moghra aquifer system is provided through the robust development of a 3D conceptual numerical model. The groundwater flow mechanism of the Moghra aquifer and its behaviour under stressful pumping conditions were thoroughly simulated utilizing the FEFLOW® numerical model. The main goal is to provide an exploitation approach that sustains groundwater in the target development area. CC, the resulting SLR, and increases in crop water requirements were considered. The developed hydrologic model could provide additive support for decision-makers to maintain the sustainability of ongoing agribusiness activities.

## 2. Study Area

The availability of a huge amount of stored water, convenient climatic conditions, and favourable soil texture render the Moghra desert one of the most promising reclamation plots in Egypt [14,15]. Accordingly, the Moghra desert was incorporated into a mega-rural development project. Previously, the study area of the Moghra desert was almost barren, with no land cover feature that may hinder such proposed large-scale cultivation activities [14]. It is located within the Marmarican Homoclinal plateau, which covers the uninhabited desert region between the Mediterranean Sea and the Qattara depression. It is in an arid climate region featuring long, hot summers and short, warm, rainless winters. It occupies a non-perennial area of 150,000 acres (60,702 ha), extending between latitude

28°11′ and 31°38′ N and between longitude 28°14′ and 31°22′ E, as shown in Figure 1. It includes the proposed cultivation swaths suggested within the national strategic plan, where the fossil groundwater reserves would be the main water source. The wells field is located 75 km away from the Mediterranean shoreline and away from the coastal area that is highly susceptible to seawater intrusion [31]. It includes two conspicuous features, namely, the Qattara depression to the west and the Moghra swamp (locally called Moghra Lake), which is located at 38 m below the mean sea level (MSL), as shown in Figure 1. Recently, the region has been subject to significant land use changes since 2015, when intense agribusiness activities of the national rural development project were initiated. The cultivation project is well underway in the area. As deduced by the analysis of Gravity Recovery and Climate Experiment (GRACE) satellite mission-derived groundwater storage (GWS), the GWS within the development area is currently overexploited, a fact that bears adverse impacts on the whole aquifer extent [32]. The Moghra aquifer is the main water-bearing formation, covering the majority of the study area [33]. The Moghra aquifer occupies a surface area of approximately 5 million hectares (ha), stretching between the Qattara depression to the west and the Nile River to the east and between the El-Fayoum depression to the south and the Mediterranean Sea to the north [34].

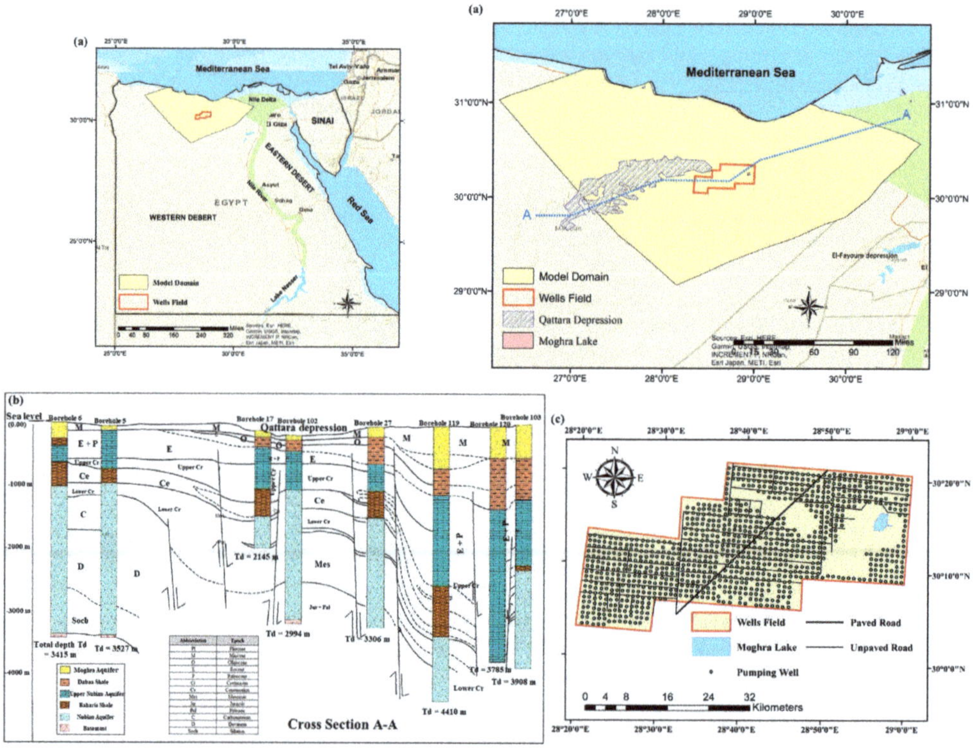


**Figure 1.** (**a**) Location and main features of the study area, (**b**) geological cross-section of the Moghra aquifer (adapted with permission from Ref. [35]. 2024, Elsevier), and (**c**) spatial distribution of the production wells within the development area.

The Moghra Formation is a well-identified geological unit composed of sand and sandstone that belongs to the Lower Miocene epoch [36]. The deposition is intercalated by clay and siltstone that increase in the northward direction [37]. The depth of the aquifer ranges from 0 to 500 m, underlain by Oligocene Dabaa shale and Eocene Baharia shale, as shown in Figure 1b [34]. The aquifer is almost phreatic, but it seems confined within the narrow strip adjacent to the Nile delta aquifer [38]. Paleo (fossil) groundwater constitutes the main bulk of the aquifer storage [17]. However, the aquifer may be recharged by lateral seepage from the adjacent Nile Delta aquifer (Pleistocene formation) and the Nile River in the east [39]. The groundwater level varies from about −50 m at the Qattara depression in the west to −10 m in the eastern portions close to the Nile Delta. The main groundwater flow is from the north and northeast towards the Qattara depression in the southwest [36].

An area of 142,716 ha (about 352,659 acres) to the east of the Qattara depression was dedicated as a wells field within the reclamation project, as shown in Figure 1a. Installing the production wells was initiated by the end of 2015 with an ambitious plan to cultivate 135,000 acres (about 54,633 ha) [39]. Pumping and replacement experiments conducted on some existing pumping and test wells within the development area indicated high potentiality in terms of groundwater storage and productivity [37]. Moderate conductivity in a range between 5 and 25 m/d along with high storativity in a range between 0.1 and 0.001 were reported [40]. The transmissivity values range between 550 and 3500 $m^2/d$ [31].

## 3. Conceptual Model Development

The conceptual model is the digital 3D hydro-geologic representation of the physical groundwater system that is organized in a computerized format. FEFLOW (version 7.5) numerical software is adopted for model compilation and development. FEFLOW is favourable for aquifers with complex geological structures [41]. FEFLOW facilitates local mesh refinements along the model domain borders; hence, the boundary conditions (BCs) are precisely defined. Further, the ability of mesh refinements within the area of interest (e.g., the wells fields) helps to avoid the creation of excessive element numbers [42]. Herein, the conceptual model assumes that the aquifer system comprises an unconfined, one-layered formation. A hydraulic connection with the underlying Nubian Sandstone Aquifer (NSA) through an aquitard was suggested [17]. However, the vertical leakage remains poorly defined. The exchange is hereby negligible due to a lack of precise data to quantify water exchange by the interlayer vertical flow.

### 3.1. Digital Hydro-Geologic Framework Model

To describe the modelled system geometry, ArcGIS was used to prepare the input files in a format accepted by FEFLOW. ArcGIS is vital to ensure all input parameters are properly projected into Zone 35 N of the Universal Transverse Mercator (UTM) coordinate system. Figure 1 shows the map that denotes the computational domain and embedded stress area (i.e., the wells field). The modelled system covers a total area of about 6 M ha, and it is 445 km in length and 224 km wide at its longest and widest parts, respectively, as shown in Figure 1. The location of the installed and proposed pumping wells was retrieved from the Egyptian Countryside Development Company (ECDC's website at http://elreefelmasry.com; accessed on 1 May 2023). During the first stage of the reclamation project, from 2016 to 2019, 450 wells were drilled [17]. As a development expansion, 550 wells are proposed to be added [43]. Worth mentioning is that the wells field was found within the most favourable areas for such a cultivation project according to environmental and economic criteria [14].

### 3.2. Hydro-Geologic Boundary Conditions

The accurate denotation of hydrogeologic conditions along the outer boundaries of the conceptual model area is essential. The BCs control fluid flows into or out of the modelled system during steady-state and time-dependent (transient state) simulations [44]. All boundaries were selected far enough from the effect of pumping wells to be regarded as constant head boundaries, as shown in Figure 2. Along the outer border of the modelled area, the first kind (Dirichlet) BC that specifies a time-constant or time-varying head value is applied. The northern and northeastern BCs were chosen to be natural boundaries (i.e., water bodies). The Mediterranean Sea completely controls the head along the northern boundary. The water level of 0.0 m was initially assigned as the head value for northern

BC. A wide main canal, locally called the "El-Nubariya" canal, appears on the right side of the studied area and extends for 127 km in length. It starts as a tributary of the "El-Rayah El-Beheiry" at the "Pollin" barrage area and cuts through the sandy soil, reaching the Mediterranean Sea. Its water levels are controlled by a few hydraulic structures (e.g., regulators) and temporally change according to the releasing discharges, with an average water level of 9.03 m at its mouth [45]. Accordingly, it acts as an influent stream and is considered a source for groundwater recharge. Dirichlet BC of head linearly change from 0.0 to 9.03 m was assumed along the line striking at almost N-E direction, as shown in Figure 2. Along other borders, no natural BC or fault (water divide) exists close to the boundary. Instead, head contour lines that represent the aquifer's water levels before initializing the development project were used [33]. The head values at each intersection of the model borders with the contour lines were assigned. Then, the interactive linear 1D interpolation tool was used to interpolate hydraulic head values along the BCs, as shown in Figure 2. Pumping wells were set as nodal BCs (i.e., the fourth kind of BC) within the development area, based on the wells map that depicts the original locations of pumping wells. Several pumping rates were assigned through a linked attribute table containing the pumping rate, radius, and screen's top and bottom elevation of each well. The well's diameter (outer/inner) was 420/280 mm (16.5/11″). The penetration depth reaches a maximum of 140 m into the Moghra geological formation. The screen depth was an average of 30 m.

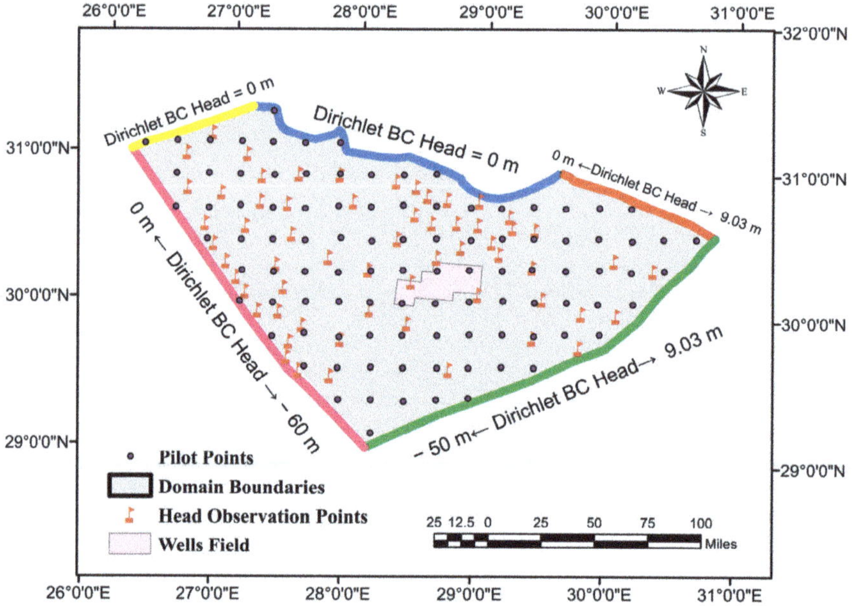


**Figure 2.** Applied boundary conditions for the modelled domain and distribution of observation heads and pilot points to define the aquifer's anisotropic conductivities.

Episodic recharge that may result from infrequent, low-intensity precipitation events in such arid climates was neglected [46,47]. Additionally, the water table throughout the aquifer extent is below the extinction depth of 0.5 m to include evaporation [48]. Thus, no evaporation of groundwater from such bare, sandy soil occurs.

### *3.3. Domain Discretization*

Obtaining a proper spatial discretization of the model domain into well-shaped prisms that have no distorted elements reduces simulation difficulties and the risk of instabilities [42,49]. So much attention was paid to discretizing the finite element mesh of the computational model domain. The available element types allow the selection of either quadrilateral or triangular elements, where the fluid flow is calculated over each element node [41]. The triangulation algorithm was selected to discretize the model domain into 2D-triangle meshes. This mesh generation algorithm is favourable for such a complex model domain, allowing a minimum angle to be determined for all accommodated complex configurations of finite elements [29]. Local mesh refinements with a maximum element size along the domain margins and within the area of production wells were carried out. Mesh smoothing was typically applied to generate better mesh quality. "The denser the mesh, the better the numerical accuracy, and the higher the computational effort" [41]. Satisfactory element numbers and geometrical criteria optimized with computational time have been obtained after many trials. Figure 3 depicts the super-element mesh that represents the basic structure of the model domain. The mesh area averaged 4.56 km$^2$, but with element concentration and refinements along the domain borders and throughout the wells field, as shown in Figure 3. The mesh configuration led to 13,064 elements and 7004 nodes on each slice (surface). Then, this mesh has been expanded in depth according to the lithology information, resulting in six-node finite elements.

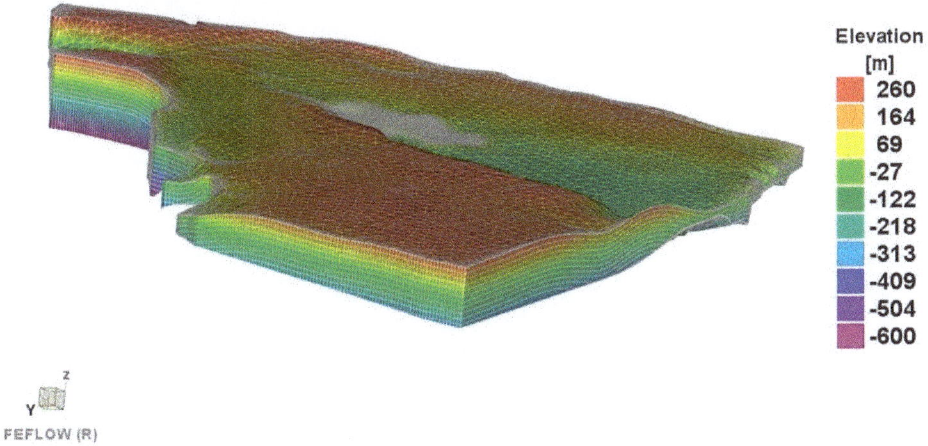


**Figure 3.** Configured unstructured meshes along with a 3D view of the constructed groundwater flow model (depth is scaled up by 30 times).

### 3.4. Subsurface Stratigraphy

The conceptualized hydrogeological system has a simplistic physiographic feature. It is vertically composed of one main Miocene formation resting on Oligocene basaltic sheets that create thick impervious base strata, as shown in Figure 1b [50]. The topography of the model domain was retrieved using the Digital Elevation Model (DEM) processed from the Shuttle Radar Topography Mission (SRTM) data. The ground surface elevation within the model domain ranges from −111 m below MSL to 269 m above MSL, as shown in Figure 3. Two data sources were employed to locate the depth down to the impervious shale base and construct the geological stratification of the modelled area: (1) fifty-four groundwater exploration wells were excavated by the Egyptian Ministry of Water Resources and Irrigation (MWRI) to locate groundwater supplies in the Western Desert of Egypt [51], and (2) forty-two deep boreholes were drilled by the national oil companies to explore for fossil fuels in the Western Desert [52]. The subsurface lithologic elevation information of the wells was utilized in combination with surface geologic data for the expansion of the modelled grid into three dimensions (3D). Akima's bivariate data regionalization method, along with the linear interpolation technique, was applied to assigning the elevation to a group of adjacent nodes. The aquifer's total depth ranges from 302 to 824 m. The aquifer was then discretized into a few sublayers of varied depths with an average of 50 m. A 3D view of the constructed model domain, composed of 12 slices and 11 layers, is shown in Figure 3. A total of 143,704 elements and 84,048 nodes, which cover an area of about $5.9 \times 10^4$ km$^2$ and a volume of $2.4 \times 10^4$ km$^3$, were created. The flexible, unstructured discretization and local grid refinements appropriately fulfil the irregular geometry and topography of the model domain without nesting. Figure A1 in Appendix A depicts the super-element mesh that represents the basic structure of the model domain.

### 3.5. Hydrogeological and Pumping Data

Following the 3D mesh expansion, the properties of the depositing formation and pumping conditions were set up. The development progress in the Moghra region has been divided into three successive stages. The ECDC is technically and administratively supervising the project [43]. The first stage was initiated in early 2016 by drilling 450 pumping wells; each well is about 1 km away from the others. According to the wells' pumping test analysis, the clear separation distance between the adjacent wells is sufficient to prevent pumping interference. The pre-development static water levels within the installed wells field ranged from −35 to −49 m, and the depth down to the water level was about 55 m. A drip irrigation system, as a high-efficiency distribution technique, is adopted, and the average water duty is 14 m$^3$/acre/day (≈34.6 m$^3$/hectare/day). The developed lands in the first stage were distributed to small investors and young people [43]. Further, 550 wells are being established during the following phases, as shown in Figure 1c. The groundwater salinity (as measured by TDS measurements) reaches up to 8000 mg/L, rating it as brackish water. Thus, only salt-tolerant crops such as canola, barley, quinoa, olive, and jojoba are cultivated in the project area. Through a field survey report with the investors and farmers in the study area, a close picture of overall groundwater pumping was obtained. The pumping pattern for each well was difficult to retrieve due to irregular monitoring frequencies of pumped water. The pumping hydrograph shows a seasonal pattern that peaks in summers while troughs occur in winters. Solar energy, as a renewable energy resource, is used to pump water. The sunshine limits the well's daily operation time to eight hours and restricts the maximum pumping rate to 120 m$^3$/h.

The hydraulic parameters were compiled from diverse literature, field surveys, and recently executed pumping and recovery analysis tests. A wide range of hydrogeological parameters for the extended Moghra aquifer were reported in the literature [13]. In general, the values of hydraulic parameters are higher in the central portion of the model domain, where the wells field is. The horizontal hydraulic conductivity ($K_x$) of the entire Moghra aquifer ranges between 0.2 and 100 m/d. The transmissivity values vary from 760 m$^2$/d to 7600 m$^2$/d. The vertical conductivity ($K_z$) was found to be 10% of the horizontal conductivity ($K_x$). In such an unconfined aquifer, the specific yield "$S_y$" is used to represent the aquifer's storativity (S). Ten pumping tests that have been conducted in the Moghra aquifer demonstrated a relatively narrow range of specific storage ($S_s$) values varying from $5 \times 10^{-4}$ to $1 \times 10^{-4}$ m$^{-1}$ with the majority ranging around $13 \times 10^{-5}$ m$^{-1}$ [37]. While the specific yield ($S_y$) varied from 0.005 to 0.13 [37]. The average storativity is about 0.15. The spatial distribution and the actual values of the initially assigned parameters were reviewed during the calibration process.

## 4. Numerical Model Set-Up

A saturated stratum, where Darcy's Law is applicable, was configured. By default, the whole model domain was initially kept as unconstrained heads. The upper slice was defined as the free surface of a movable slice. The simulation was initiated in a steady state and then run under transient conditions.

### 4.1. Steady-State Simulation

Accurate simulation of a flow problem through an aquifer requires achieving a reliable steady state. Up until late 2015, before initializing the rural development project, the

Moghra aquifer was almost untapped. Water levels within the aquifer system were near equilibrium, and the effect of wide-distributed private tube wells was negligible.

4.1.1. Model Calibration

The calibration is imperative to ensure a reliable simulation result by adjusting the input hydrogeological parameters [53]. The steady-state calibration was executed as a fully implicit simulation in which the pre-development water heads (by the year 2015) were used as a target condition. The system is regarded as steady when the overall balance for the entire domain is achieved. Due to the importance of the conductivity and the model's high sensitivity to its alteration, anisotropic conductivity was adopted, in which the conductivity is calibrated along the X, Y, and Z axes. Given their swiftness and ease, non-linear parameter estimation techniques (PEST) are now commonly used for the proper identification of adjustment parameters. FePEST® is the graphical user interface for the development of the FEFLOW model with the PEST utilities. In FePEST, two spatial characterization techniques (i.e., pilot points and constant zones) are used to represent heterogeneities of hydraulic properties within the modelled hydrogeological basin [26]. The pilot points approach was applied in this study as it provides an improved heterogeneity representation of the system. A uniform distribution of 101 pilot points was generated throughout the model domain, as shown in Figure 2. Exaggeration in the number of pilot points would require extensive modelling times and could result in over-parameterization and large variations in the parameter values over small distances [26]. Herein, the Moghra model was calibrated by adjusting conductivity values such that simulated groundwater tables matched the observed hydraulic heads. $K_x$ was bounded between 0.1 and 200 m/day (minimum and maximum limits), according to prior information given in the literature. $K_y$ and $K_z$ values were tied to changes in $K_x$ by applying an initial relationship: $K_x = K_y = 10\ K_z$. The calibration fit was assumed when a divergence of less than 1 m (about 10% of the difference between the maximum and minimum water levels) was achieved between the computed and observed heads.

The model's results were assessed by comparing simulated heads against observed heads at 60 points scattered throughout the model domain, as shown in Figure 2. The observed heads ranged from 0 to −80 m. The model demonstrated a normalized root mean square error (NRMSE) of 1.8% (1.6 m) and a correlation factor (r) of 0.9973. Figure 4 shows a scatter plot depicting the comparison conducted between the monitored water levels and the simulated results. The value of the obtained root mean square error (RMSE) was 1.94 m. The conductivities were largely consistent with values reported in the literature, varying between a maximum of $2.4 \times 10^{-3}$ m/s and a minimum of $1.2 \times 10^{-6}$ m/s. Model simulations thereby replicated flow patterns in terms of hydraulic gradients and flow directions. The estimated hydraulic head throughout the model domain under the steady-state condition is shown in Figure 5. Using the particle tracking technique, streamlines of regional groundwater flow were visualized, as shown in Figure A2 in Appendix A. The flow's main direction is from the north and northeast towards the south and then radially south-westward.

4.1.2. Model Sensitivity

It is essential to analyse the uncertainty and sensitivity related to the input hydrologic parameters. The sensitivity analysis identifies uncertainties in input parameters for the model. Additionally, this analysis addresses an overall understanding of the modelled system and enhances the calibration procedures [54]. In this context, FePEST, through the "SENSAN" utility, enables estimating sensitivities for a set of input parameters. The model's sensitivity was examined by multiplying calibrated conductivities throughout the model's interior nodes by factors of 3.0, 2.5, 2.0, 1.5, 1.2, 0.8, 0.4, and 0.2.

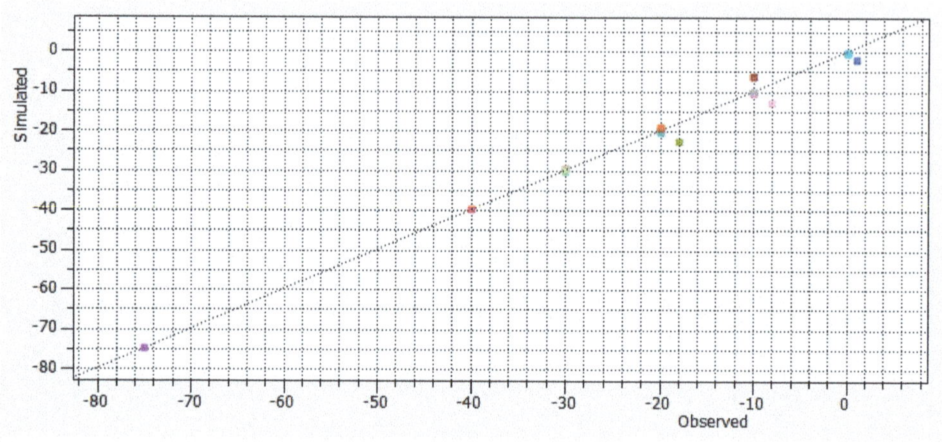


**Figure 4.** Comparison between observed and simulated water levels for steady-state calibration.

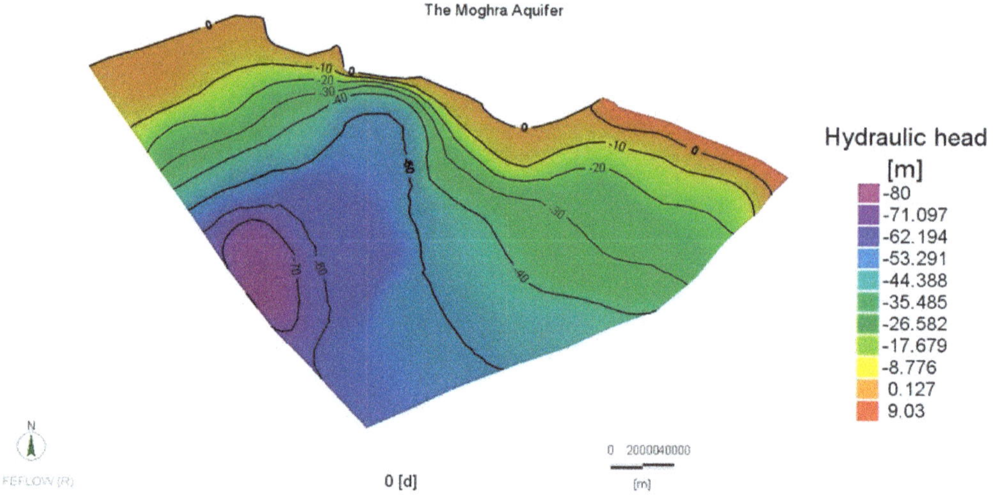


**Figure 5.** Steady state simulated hydraulic heads for the year 2015 over the modelled domain.

Table 1 summarizes the sensitivity analysis to changes in K of the steady-state model. The model is reasonably sensitive to alterations in hydraulic conductivity, referring to its importance in model adjustment. A change of 20% in conductivity values resulted in an average sensitivity value of 0.22. Applying a constant change factor for both longitudinal and transverse conductivities throughout the model domain maintained the general groundwater flow direction. But a high-water budget imbalance was indicated when applying higher and lower conductivity values.

**Table 1.** Percentage of groundwater head variations versus the changes in hydraulic conductivity.

| Change Factor of Conductivity | % Change of Hydraulic Head |
|---|---|
| 0.2 | 0.16 |
| 0.4 | 0.26 |
| 0.6 | 0.16 |
| 0.8 | 0.30 |
| 1.2 | 0.26 |
| 1.5 | 0.20 |
| 2.0 | 0.16 |
| 2.5 | 0.22 |
| 3.0 | 0.20 |

### 4.2. Time-Dependent Simulation

#### 4.2.1. Transient Calibration Using Pumping Test Data

Adjusting the input properties under transient state conditions was performed by employing data from continuous pumping experiments conducted on 39 recently installed production wells [37]. The pumping analysis test is transient data commonly utilized for calibration of the groundwater flow model in cases of insufficient temporal variation in pumping records and water level [55]. The storativity was adjusted to fit the observed drawdown in response to pumping. As shown in Figure 6, the computed drawdown matches well with those observed during the pumping tests.

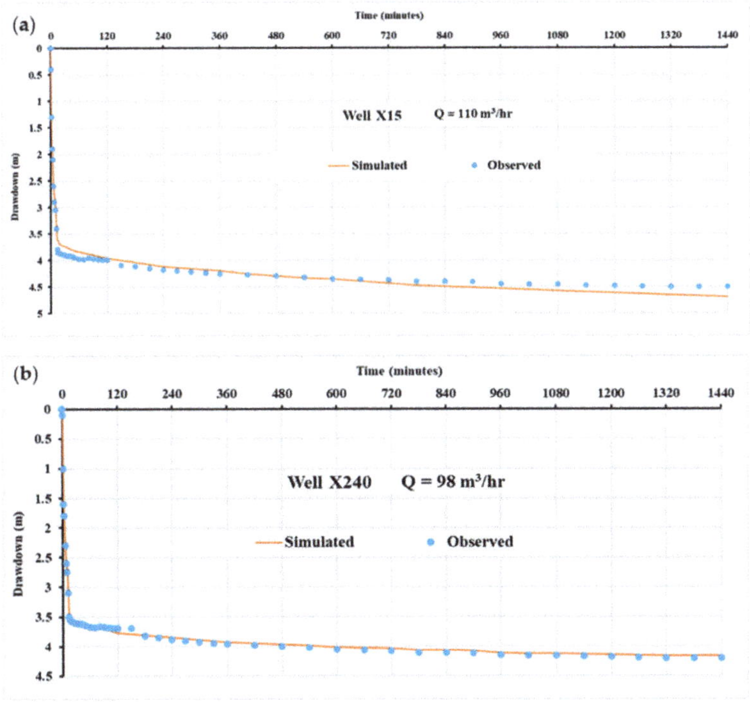


**Figure 6.** Transient calibration using (**a**) 24 h pumping test data of Well X15 @ (longitude 28.48° and latitude 30.12°) and (**b**) 48 h pumping test data of Well X136 @ (longitude 28.56° and latitude 30.27°).

#### 4.2.2. Transient Verification Using GRACE-Retrieved Data

Motivated by the observed capacity of the GRACE mission for tracking changes in GWS, this study explores the use of GRACE-derived GWS changes for the transient verification of the developed model [56]. GRACE-based GWS alterations were incorporated in the tuning procedure of groundwater flow models to overcome insufficient in situ measurements [57,58]. GRACE was also employed to estimate groundwater storage coefficients, including the storativity or specific yield ($S_y$), as appropriate [59,60]. The time series that represents the change in terrestrial water storage (ΔTWS) was extracted from the recently released gridded GRACE dataset provided by the three processing centres (i.e., GRACE's science data centres) [32]. This dataset consists of a temporal monthly TWS anomaly, a mass deviation from the baseline, which is expressed in terms of equivalent water height (EWH) and posted on a 1° grid resolution (PODAAC website at https://grace.jpl.nasa.gov/data/get-data/monthly-mass-grids-land; accessed on 1 March 2023). Within such a semiarid area with no perennial surface waters, groundwater, and soil moisture (SM) account for all inferred alterations in the TWS. Therefore, ΔGWS was estimated by subtracting the SM's contribution from GRACE-derived TWS [61]. The SM was estimated based on the outputs of the land surface model (GLDAS-NOAH) [62]. During the verification period (2016–2021), the water table had undergone an average decline of 0.96 m within the wells field. Minor alterations in groundwater tables were observed in the rest of the model domain.

The temporal pattern of GRACE-inferred GWS has been incorporated as an additional regularization dataset for verification of the developed model. The FEFLOW software allows the control of the simulation time via an integration scheme [41], while the simulation results can be depicted according to the recorded output time steps. An automatic time control scheme using the Adams-Bashforth/backward trapezoid (AB/TR) predictor-corrector scheme was adopted [30]. Using the "Period Budget" utility, the developed model can capture the inflows and outflows quantity during a simulation period, and thereby the corresponding storage loss/gain is estimated. The simulated annual mean storage anomaly from 2016 to 2021 was obtained as 16.27 $Mm^3$ (about 0.276 cm EWH), which is close to the GRACE-derived GWS anomaly (0.26 cm/year), as shown in Figure 7.

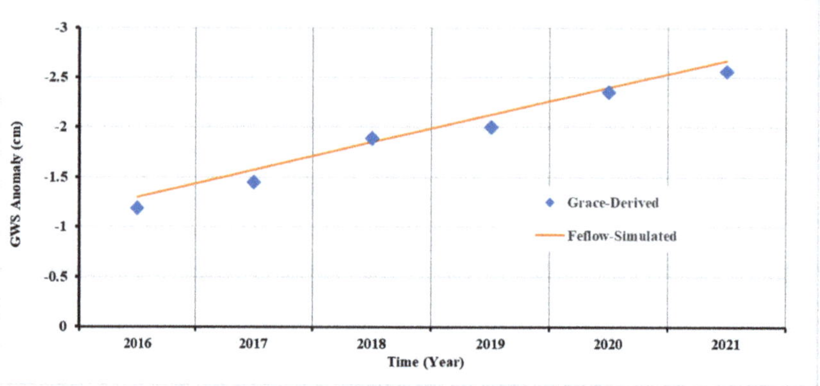


**Figure 7.** Comparison of GRACE-derived and simulated GWS changes from 2016 to 2021 over the model domain.

*4.3. Management Schemes*

The calibrated model was run afterwards for a 100-year test period as a predictive simulation. Estimated water levels during the steady-state simulation were regarded as the initial conditions for the transient simulation. The BCs and the material properties (e.g., hydraulic conductivity and storativity) were kept unchanged. The added input to the model was the pumping wells. To investigate the CC implications, time-varying boundary-condition data were applied.

The adjusted model was employed to mimic the aquifer's behaviour under variant exploitation strategies. The traditional safe yield concept, meant to mean discharge equals recharge, does not apply to such a non-renewable aquifer [63]. The pumping-induced loss in aquifers' storage is not offset, and the resulting drawdown is inevitable. However, a rationing restriction is to be set to ensure the long-term water supply and, thereby, the sustainability of the dependent cultivation activities [37]. For the Moghra aquifer, the abstraction-regulating rule states the drawdown after 100 years should not exceed a third of the saturation depth. Hereby, the maximum allowable drawdown rate is one meter each year.

Sustainable management should consider CC and the resulting SLR. The Moghra model is employed to investigate the combined implications of SLR, and the water duty increment due to global warming. The Representative Concentration Pathways (RCPs) are a suite of scenarios addressed in the Fifth Assessment Report (AR5) that was released by the Intergovernmental Panel on Climate Change (IPCC) to project future climatic conditions [64]. RCP 85 is the CC worst-case scenario that involves a global average temperature increase of about 4.3 °C by this century's end [65]. The corresponding average SLR is projected in a range of 53 to 98 cm [66]. However, due to the distinct features of the Mediterranean Sea as a semienclosed basin connected with the ocean through the narrow Gibraltar Strait, its local level change will not exceed 0.5 m [67,68].

Crop consumptive use is often referenced to the evapotranspiration (ET) that can be estimated using different theoretical approaches (e.g., the Blaney–Criddle formula). Global warming causes an increase in crop water requirements that differ regionally. Regional climate models (RCMs) are therefore employed to project climate conditions on a regional scale and thus provide accurate estimates over general circulation models (GCMs) [7]. The outputs of "MPI-ESM-LR/RCA4, GCM/RCM" ensemble simulations, provided through the CORDEX project (CORDEX website at https://esg-dn1.nsc.liu.se/projects/esgf-liu/; accessed on 1 January 2024) were used to project the increase in potential evapotranspiration (PET) over the wells field [69]. Figure 8 shows the projected percentage change in PET during the upcoming 100 years. The average annual increase in PET is 0.0058 mm, which implies an annual increase in water duty by 0.019 $m^3$/acre/day (≈0.047 $m^3$/hectare/day).

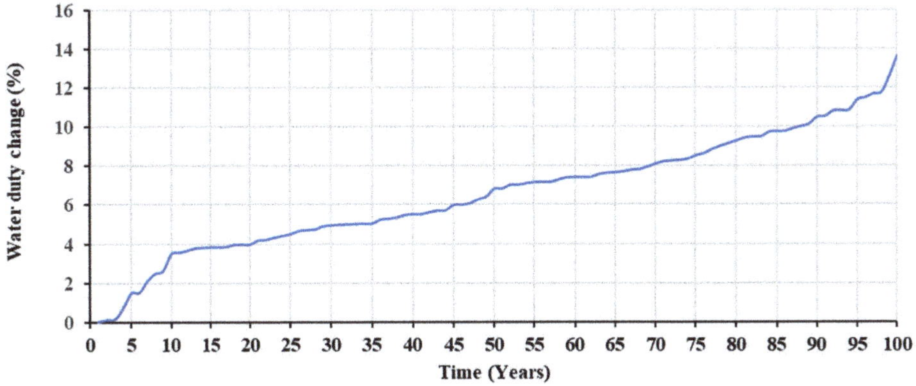


**Figure 8.** Percentage increment in crop water consumptive use under the climate change worst-case scenario.

## 5. Results and Discussions

### *5.1. Pumping Scenarios*

Invoking the calibrated model, a set of alternative pumping policies was designed and reliably tested to identify the optimal water resources management scheme. Eight pumping regimes ranging from 0.8 to 1.5 $Mm^3$/day were simulated. The continuous pumping has resulted in decreasing groundwater levels throughout the model domain. The maximum drawdown was observed within the wells field. Table 2 summarizes the resulting drawdown breakthrough under the applied pumping policies during a 100-year test period.

**Table 2.** Maximum drawdown (m) for different pumping rates along a 100-year test period.

| Scenario | Total Pumping "Mm³/Day" | Cultivated Area "Acres" * | Time (Years) | | | | | | | | | |
|---|---|---|---|---|---|---|---|---|---|---|---|---|
| | | | 10 | 20 | 30 | 40 | 50 | 60 | 70 | 80 | 90 | 100 |
| 1 | 0.80 | 57,143 | 8.72 | 15.75 | 22.34 | 28.55 | 34.29 | 39.66 | 44.72 | 49.54 | 54.15 | 58.59 |
| 2 | 0.90 | 64,286 | 9.81 | 17.72 | 25.15 | 32.15 | 38.62 | 44.68 | 50.40 | 55.84 | 61.06 | 66.09 |
| 3 | 1.00 | 71,429 | 10.90 | 19.70 | 27.96 | 35.75 | 42.96 | 49.71 | 56.09 | 62.17 | 68.00 | 73.62 |
| 4 | 1.10 | 78,571 | 11.99 | 21.67 | 30.77 | 39.63 | 47.32 | 54.76 | 61.81 | 68.35 | 74.98 | 81.20 |
| 5 | 1.20 | 85,714 | 13.08 | 23.65 | 33.59 | 42.98 | 51.68 | 59.83 | 67.55 | 74.91 | 81.99 | 88.82 |
| **6** | **1.30** | **92,857** | **14.17** | **25.63** | **36.41** | **46.6** | **56.05** | **64.91** | **73.31** | **81.33** | **89.04** | **96.49** |
| 7 | 1.40 | 100,000 | 15.26 | 27.61 | 39.23 | 50.23 | 60.43 | 70.00 | 79.09 | 87.77 | 96.12 | 104.2 |
| 8 | 1.50 | 107,143 | 16.36 | 29.59 | 42.06 | 53.86 | 64.82 | 75.12 | 84.89 | 94.24 | 103.24 | 111.96 |

Notes: * 1 acre = 0.404686 hectares = 0.00404686 $km^2$. Bold font refers to recommended pumping scenario.

According to the obtained results, it is concluded that the optimum management pathway is to pump a total water quantity of 1.3 $Mm^3$/day through 1000 wells (i.e., Pumping Scheme No. 6). Considering a daily water duty of 14 $m^3$/acre (34.6 $m^3$/ha), the cultivated areas would reach up to 92,857 acres (37,578 ha). The resulting drawdown is up to about 96.49 m, which is compatible with the MWRI's regulations stating that the drawdown threshold should not exceed 1.0 m in a year, as shown in Table 2. Figure 9 shows the simulated hydraulic head and drawdown under this pumping regime after 100 years. Appendix B showcases detailed information on the recommended pumping scheme to be adopted in the Moghra region. Figure A3 depicts the maximum drawdown resulting through a 100-year test period from different investigated pumping rates, where the optimum one is marked. Figure A4 depicts the rate budget analysis of the groundwater system under this pumping scheme during the simulation period. The net storage was estimated as the difference between captured and released storage, which showed an average storage loss of 1.31 $Mm^3$/day. Figure A5 depicts the resulting drawdown every ten years on a cross-sectional view (Section A-A, Figure 1b) of the model domain.

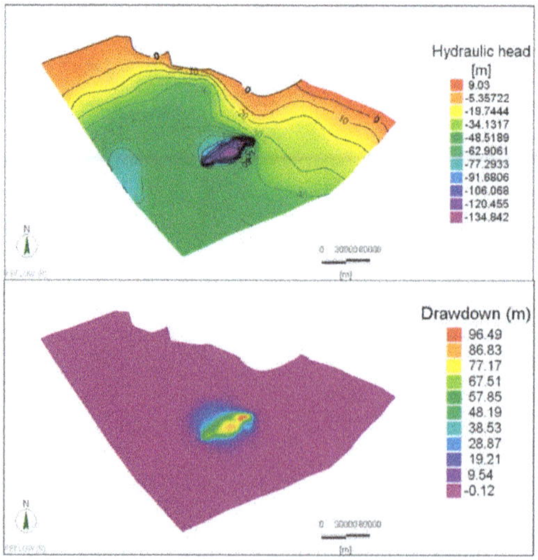


**Figure 9.** Simulated hydraulic heads (**Up**) and drawdowns (**Down**) for the recommended pumping scheme (1.3 $Mm^3$/day) after 100 years.

### *5.2. Climate Change Impacts*

As mentioned, the maximum pumping rate is 1.3 $Mm^3$/day through 1000 wells, which is sufficient to irrigate about 92,850 acres (about 37,575 ha) with an average water duty of 14 $m^3$/acres/day, as shown in Table 2. However, CC would impose significant pressure on the aquifer to meet the anticipated increase in crop consumptive water use. To avoid the loss of a portion of the reclaimed areas in the future, a time-increasing abstraction rate should be applied. Accordingly, the developed model was utilized to investigate the combined effect of SLR and the increase in crop-consumptive water use. A time series representing a linear rise by 0.5 m in sea level during the upcoming 100 years was assigned for the northern BC along the Mediterranean Sea. Additionally, the daily pumping rate to maintain the cultivated area was defined as a time series considering the increase in PET under the worst-case CC scenario (i.e., RCP 85), as shown in Figure 8.

Figure 10 shows the resulting heads and drawdown after a 100-year simulation period under a constant pumping rate of 1300 $m^3$/day/well, considering a gradual SLR of up to 0.5 m after 100 years. The water levels within a narrow strip along the seaside were marginally raised, but they were not altered over the rest of the model domain. The low effect of SLR on water levels is attributed to the vast area of the model domain. Accordingly, the resulting drawdown within the wells field was almost unaffected, as shown in Figure 10. On the contrary, the resulting drawdown from assigning a time-increased withdrawal rate was 103.7 m, as shown in Figure 10, compared to 96.5 m when applying a fixed pumping rate of 1300 $m^3$/day/well. The simulation affirmed the exceeding of this management scenario under CC implications for the maximum rationalizing limit of drawdown. The combined effect of SLR and pumping increases in all management scenarios on the modelled domain is shown in Table 3. The exploitation scheme No. 5, which involves cultivating 85,714 acres, may be more consistent with the drawdown rationalizing limit. The resulting drawdown by the end of the simulation period under the combined effect of global warming and the resulting SLR is shown in Figure 11. The resulting drawdown is 95.4 m by the simulation period's end. This value is compatible with the MWRI's regulations, which state that the maximum annual drawdown should not exceed 1.0 m. Appendix C showcases more information about the implications of a variant pumping scheme considering climate change impacts. Figure A6 depicts the maximum drawdown resulting through a 100-year test period from different investigated pumping rates, where the optimum one is marked. Figure A7 depicts the average rate budget analysis of the groundwater system over 100 years of applying time-increasing abstraction rates under the combined effects of climate change. The net storage was estimated as the difference between captured and released storage, which showed an average storage loss of 1.362 $Mm^3$/day.

**Table 3.** Maximum drawdown (m) along a 100-year test period under a chronologically increasing abstraction rate corresponding to climate change implications on crop water requirements.

| Scenario | Cultivated Area "Acres" * | Time (Years) | | | | | | | | | |
|---|---|---|---|---|---|---|---|---|---|---|---|
| | | 10 | 20 | 30 | 40 | 50 | 60 | 70 | 80 | 90 | 100 |
| 1 | 57,143 | 8.85 | 16.19 | 23.09 | 29.67 | 35.81 | 41.64 | 47.19 | 52.53 | 57.74 | 62.89 |
| 2 | 64,286 | 9.96 | 18.22 | 25.99 | 33.41 | 40.34 | 46.92 | 53.18 | 59.22 | 65.12 | 70.94 |
| 3 | 71,429 | 11.07 | 20.25 | 28.90 | 37.16 | 44.87 | 52.21 | 59.20 | 65.95 | 72.53 | 79.05 |
| 4 | 78,571 | 12.17 | 22.28 | 31.81 | 40.91 | 49.42 | 57.52 | 65.23 | 72.70 | 79.98 | 87.20 |
| **5** | **85,714** | **13.28** | **24.31** | **34.73** | **44.67** | **53.98** | **62.84** | **71.30** | **79.48** | **87.48** | **95.40** |
| 6 | 92,857 | 14.39 | 26.35 | 37.64 | 48.43 | 58.55 | 68.18 | 77.39 | 86.3 | 95.01 | 103.65 |
| 7 | 100,000 | 15.5 | 28.38 | 40.56 | 52.2 | 63.13 | 73.54 | 83.5 | 93.14 | 102.59 | 111.96 |
| 8 | 107,143 | 16.61 | 30.42 | 43.49 | 55.98 | 67.72 | 78.92 | 89.63 | 100.02 | 110.21 | 120.32 |

Notes: * 1 acre = 0.404686 hectares = 0.00404686 $km^2$. Bold font refers to recommended pumping scenario.

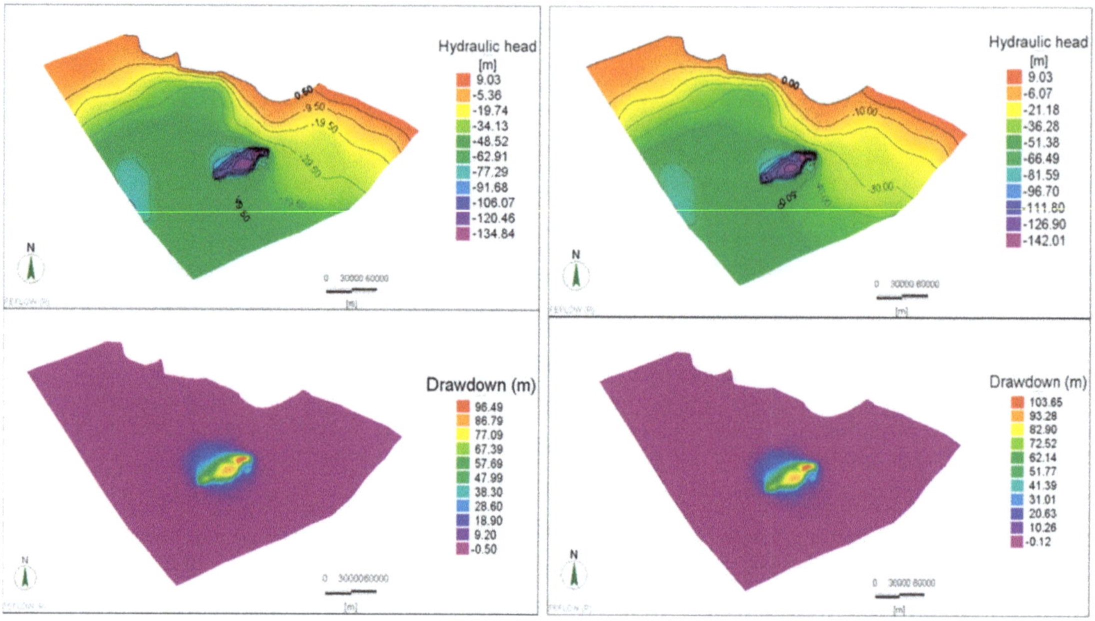


**Figure 10.** Simulated hydraulic heads (**Up**) and drawdown (**Down**) after 100 years under a constant pumping rate of 1.3 $m^3$/day/well considering (**Right**) a sea level rise of 0.5 m and (**Left**) an increase in crop water requirements.

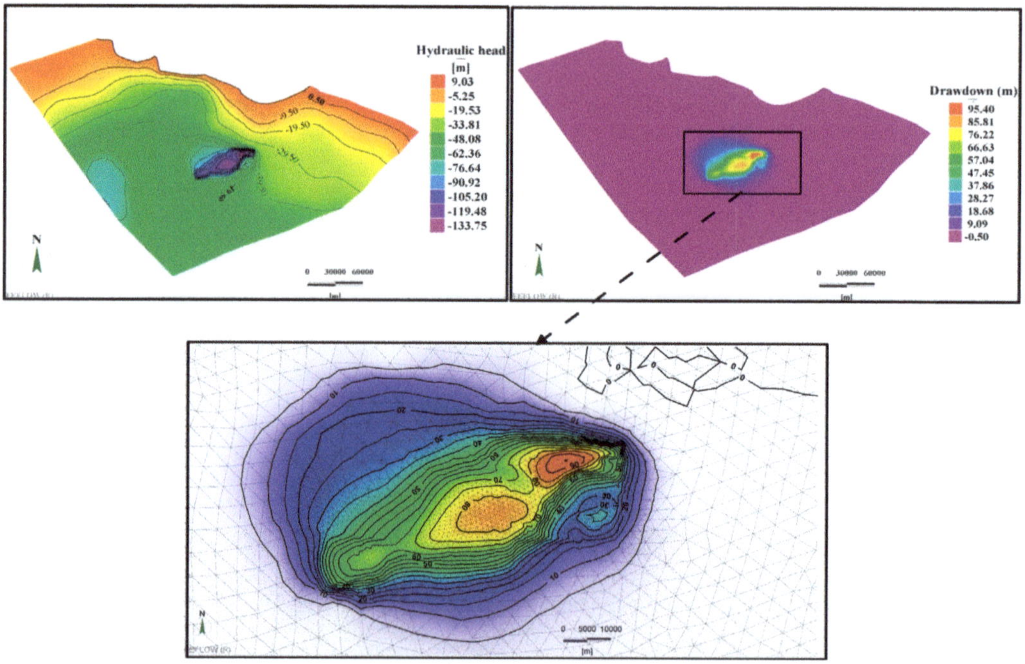


**Figure 11.** Simulated hydraulic heads (**Left**) and drawdowns (**Right**) after 100 years of applying the recommended development regime under the combined effects of climate change.

## 6. Conclusions

To cope with the rising food shortage issue in Egypt, there is a crucial need to expand cultivation swaths outside of the overpopulated Nile Delta and Valley. Thus, a groundwater-dependent reclamation project was advocated to reclaim the barren desert areas. The allocated aquifers for the proposed cultivation activities are almost non-renewable systems. Nevertheless, there is still substantial room for exploitation of this water resource, provided its pumping is regulated by principles of wisdom and rationality. Herein, a groundwater flow model was generated using the FEFLOW software to explore the Moghra aquifer's potential for extensive rural development. A simulation of reality is performed by representing the aquifer's boundary conditions, hydrogeological parameters, and groundwater extraction rates. The developed model was calibrated against the observed water levels in the archive and continuous pumping experiments. Simulated losses in groundwater storage (GWS) from 2016 to 2021 were compatible with the GRACE-derived GWS anomaly. Subsequently, the calibrated model was utilized to produce the groundwater equipotential surface and hydraulic head gradient under the influence of changing abstraction regimes. The persistently high abstraction rates play a significant role in lowering groundwater heads within the development area. To provide an integrated management approach, this study has also investigated the effects of global warming. SLR would increase the extent of saltwater encroachment into the aquifer, causing a slight increase in the piezometric levels within the coastal plain. It, therefore, has no effect on the heads within the well field. The projected increase in crop water requirement was estimated for climate change's worst-case scenario (i.e., RCP 85). The corresponding temporal-increasing abstraction rate to irrigate the cultivation area without any reduction was assigned. The maximum pumping regime that coincides with the regulation rule that was set by the MWRI for the next 100 years has been concluded. There is a potential for sustainable exploitation of this water resource to irrigate a total area of 85,715 acres (34,687 ha). The resulting drawdown is expected to reach 95.4 m.

This study addressed the first modelling of the Moghra aquifer using a finite-element-based code. The developed model can be invoked by decision-makers to beneficially use its outcomes to define sustainable management approaches. It is suggested to upgrade the model to evaluate the degree to which the production wells would intrude saltwater into the aquifer system.

**Author Contributions:** Conceptualization, A.S., B.A.Z. and A.M.A.; methodology, A.S., A.M.N. and A.M.A.; software, A.S. and A.M.A.; validation, A.S. and A.M.A.; formal analysis, A.M.A. and K.P.-U.; investigation, A.S., A.M.A. and K.P.-U.; resources, K.P.-U. and A.M.A.; data curation, A.S.; writing—original draft preparation, A.S.; writing—review and editing, A.S., A.M.A. and K.P.-U.; visualization, A.S. and A.M.A.; supervision, B.A.Z. and A.M.A. All authors have read and agreed to the published version of the manuscript.

**Funding:** Financial support received from the Science, Technology & Innovation Funding Authority STDF-Egypt (Grant number: 46278) has facilitated this collaborative research as a part of a project titled "Optimal Exploitation Strategies for Sustainable Utilization of Fossil Groundwater Reserves in Egypt".

**Data Availability Statement:** Research data are available from the corresponding author on request.

**Acknowledgments:** The authors would like to thank the STDF-Egypt for the financial support received under the framework of a project (ID#46278). The authors are beholden to the Egyptian Countryside Development Company ECDC's officials for providing in situ observation, reports, surveys, and hosting during the field campaign. Thanks must go to the DHI group for providing technical support for the FEFLOW software. All authors acknowledge the reviewers' time and efforts to revise and improve the manuscript.

**Conflicts of Interest:** The authors state no conflicts of interest that could influence the study reported in this paper. The funders had no role in the design of this study; in the collection, analyses, or interpretation of data; in the writing of the manuscript; or in the decision to publish the results.

## Appendix A. Model Development and Calibration

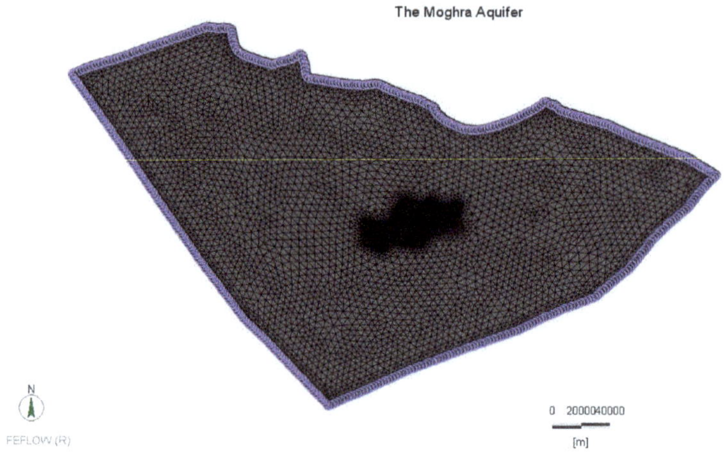


**Figure A1.** Configured mesh and boundary conditions of the modelled domain.

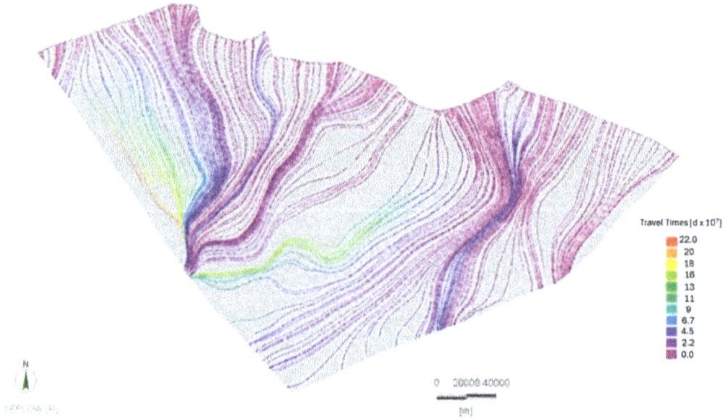


**Figure A2.** Visualization of paths and travel times of virtual particles released at the domain's boundaries.

## Appendix B. Optimum Pumping Scheme without Considering Climate Change Impact

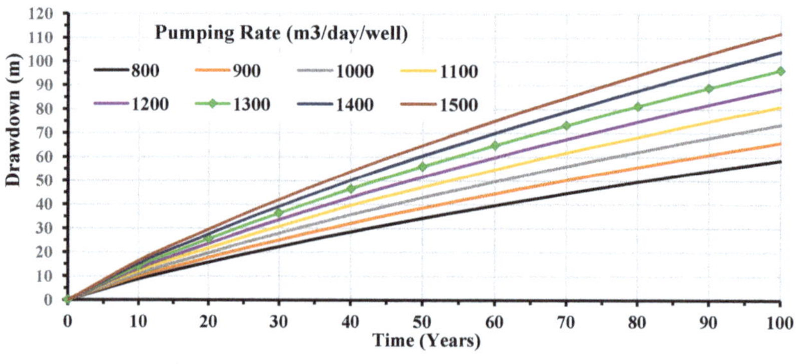


**Figure A3.** Predicted maximum drawdown resulting from different pumping rates through a 100-year test period.

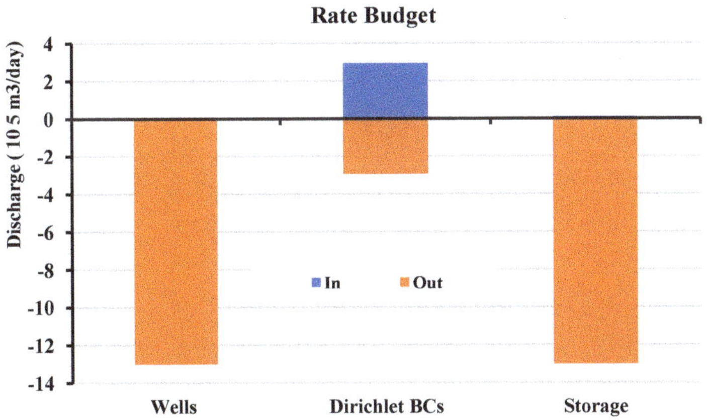


**Figure A4.** Components of the groundwater budget analysis (+) for inflow and (−) for outflow for the recommended pumping scheme (1.3 $Mm^3/day$) after 100 years.

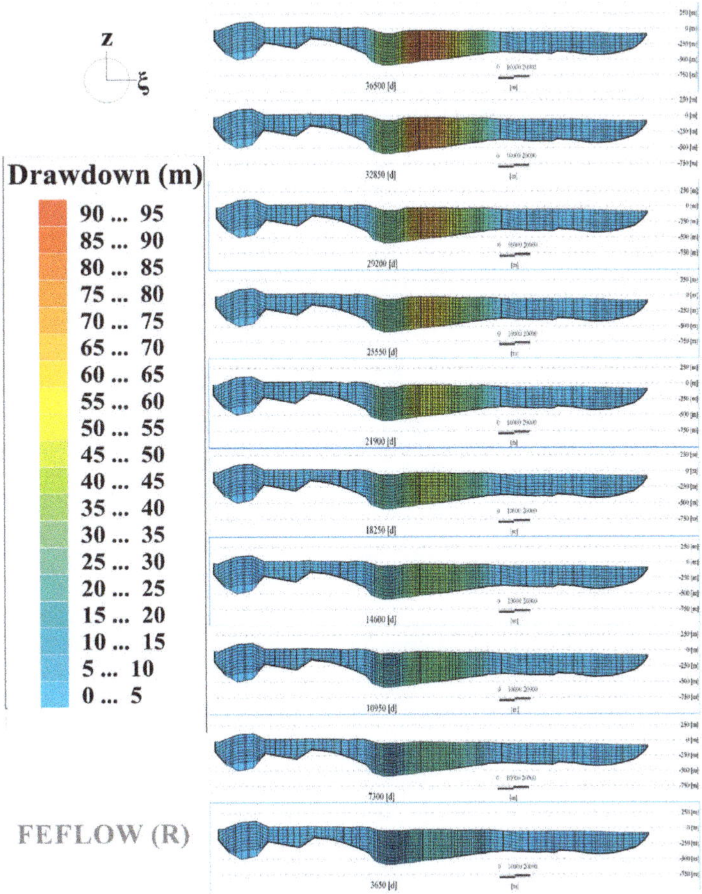


**Figure A5.** Cross-section (A-A) view showing the resulting drawdown from the recommended pumping scheme (1.3 $Mm^3/day$) (depth is scaled up by 30 times).

## Appendix C. Optimum Pumping Scheme Considering Climate Change Impact

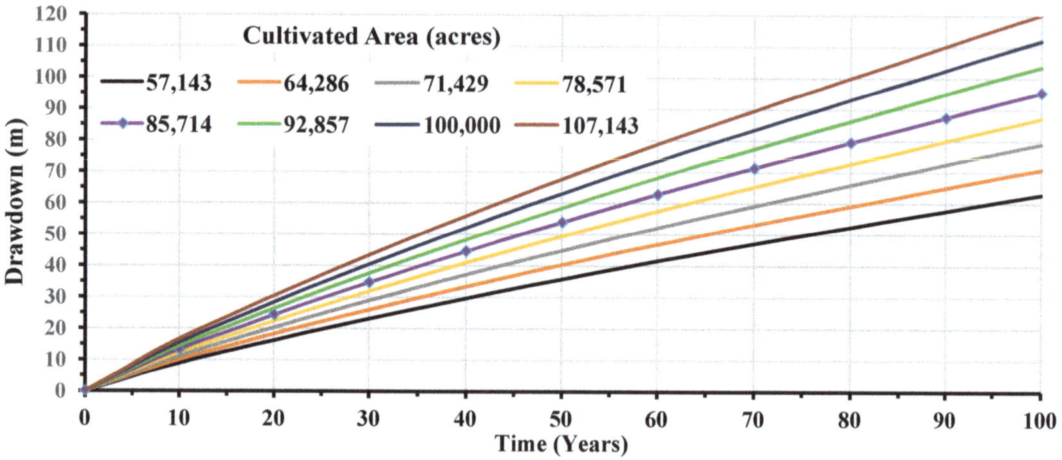


**Figure A6.** Maximum drawdown resulting from time-increasing abstraction rates through a 100-year test period.

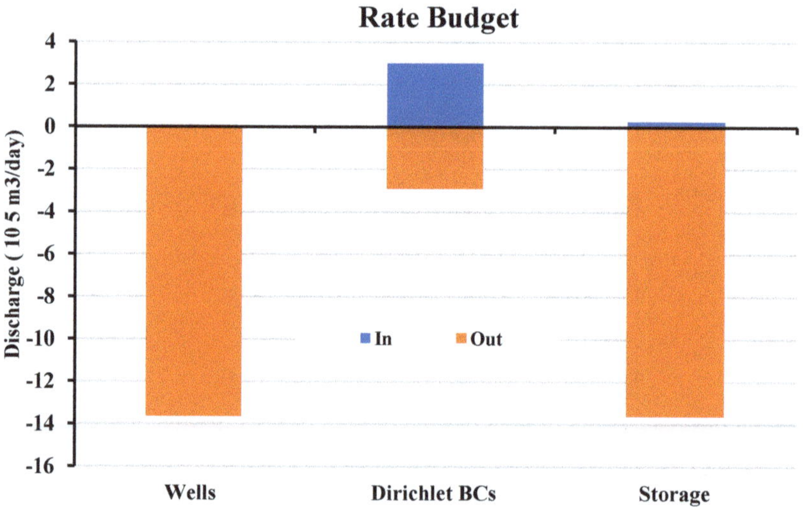


**Figure A7.** Components of the groundwater budget analysis (+) for inflow and (−) for outflow after 100 years of applying the recommended development regime (a growing pumping rate initiated by 1.3 $m^3$/day/well) under the combined effects of climate change.

## References

1. Frappart, F.; Ramillien, G. Monitoring groundwater storage changes using the Gravity Recovery and Climate Experiment (GRACE) satellite mission: A review. *Remote Sens.* **2018**, *10*, 829. [CrossRef]
2. Closas, A.; Rap, E. Solar-based groundwater pumping for irrigation: Sustainability, policies, and limitations. *Energy Policy* **2017**, *104*, 33–37. [CrossRef]
3. Rubio-Aliaga, Á.; Sánchez-Lozano, J.M.; García-Cascales, M.S.; Benhamou, M.; Molina-García, A. GIS based solar resource analysis for irrigation purposes: Rural areas comparison under groundwater scarcity conditions. *Sol. Energy Mater. Sol. Cells* **2016**, *156*, 128–139. [CrossRef]
4. Zektser, S.; Loáiciga, H.A.; Wolf, J.T. Environmental impacts of groundwater overdraft: Selected case studies in the southwestern United States. *Environ. Geol.* **2005**, *47*, 396–404. [CrossRef]

5. Khater, A.R.; Water, N. Intensive groundwater use in Middle East and North Africa. In *Intensive Use of Groundwater: Challenges and Opportunities*; CRC Press: Boca Raton, FL, USA, 2019; pp. 355–386.
6. Gado, T.A.; El-Hagrsy, R.M.; Rashwan, I.M.H. Spatial and temporal rainfall changes in Egypt. *Environ. Sci. Pollut. Res.* **2019**, *26*, 28228–28242. [CrossRef]
7. Gado, T.A.; Mohameden, M.B.; Rashwan, I.M.H. Bias correction of regional climate model simulations for the impact assessment of the climate change in Egypt. *Environ. Sci. Pollut. Res.* **2022**, *29*, 20200–20220. [CrossRef]
8. Farrag, A.A. The hydraulic and hydrochemical impacts of the Nile system on the groundwater in upper Egypt. *Assuit Univ. Bull. Environ. Resour* **2005**, *8*, 87–102.
9. Aziz, S.A.; Zeleňáková, M.; Mésároš, P.; Purcz, P.; Abd-Elhamid, H. Assessing the Potential Impacts of the Grand Ethiopian Renaissance Dam on Water Resources and Soil Salinity in the Nile Delta, Egypt. *Sustainability* **2019**, *11*, 7050. [CrossRef]
10. Shalaby, M.Y.; Al-Zahrani, K.H.; Baig, M.B.; Straquadine, G.S.; Aldosari, F. Threats and challenges to sustainable agriculture and rural development in Egypt: Implications for agricultural extension. *J. Anim. Plant Sci.* **2011**, *21*, 581–588.
11. Aly, M.M.; Elhamid, A.M.I.A.; Abu-Bakr, H.A.-A.; Shalby, A.; Fayad, S.A.K. Integrated Management and Environmental Impact Assessment of Sustainable Groundwater-Dependent Development in Toshka District, Egypt. *Water* **2023**, *15*, 2183. [CrossRef]
12. Aly, M.M.; Sakr, S.A.; Fayad, S.A.K. Evaluation of the impact of Lake Nasser on the groundwater system in Toshka under future development scenarios, Western Desert, Egypt. *Arab. J. Geosci.* **2019**, *12*, 553. [CrossRef]
13. El-Agha, D.E.; Molle, F.; Metwally, M.I.; Emara, S.R.; Shalby, A.; Armanuos, A.M.; Negm, A.; Gado, T.A. Review: Toward sustainable management of groundwater in the deserts of Egypt. *Hydrogeol. J.* **2023**. [CrossRef]
14. Shalby, A.; Emara, S.R.; Metwally, M.I.; Negm, A.M.; Gado, T.A.; Armanuos, A.M. Geospatial model for allocating favorable plots for groundwater-dependent cultivation activities in Egypt. *Egypt. J. Remote Sens. Sp. Sci.* **2023**, *26*, 777–788. [CrossRef]
15. Sayed, E.; Riad, P.; Elbeih, S.; Hagras, M.; Hassan, A.A. Multi criteria analysis for groundwater management using solar energy in Moghra Oasis, Egypt. *Egypt. J. Remote Sens. Sp. Sci.* **2019**, *22*, 227–235. [CrossRef]
16. Elmenshawy, M.R.; Shalaby, S.M.; Elshinnawy, A.I.; Gado, T.A. Groundwater desalination for agricultural purposes using reverse osmosis and nanofiltration technologies: Case study wadi El-natrun, Egypt. *Process Saf. Environ. Prot.* **2023**, *180*, 669–685. [CrossRef]
17. Gomaa, S.M.; Hassan, T.M.; Helal, E. Assessment of seawater intrusion under different pumping scenarios in Moghra aquifer, Egypt. *Sci. Total Environ.* **2021**, *781*, 146710. [CrossRef]
18. Elmansy, Y.; Mohamed, M.S.; Hassan, A.A.; Riad, P. Sustainable Agricultural Groundwater Management for New Reclaimed Areas in Farafra Oasis, Western Desert of Egypt. *Int. J. Eng. Adv. Technol.* **2020**, *9*, 118–127. [CrossRef]
19. Singh, A. Groundwater resources management through the applications of simulation modeling: A review. *Sci. Total Environ.* **2014**, *499*, 414–423. [CrossRef]
20. Alley, W.M.; Leake, S.A. The Journey from Safe Yield to Sustainability. *Ground Water* **2004**, *42*, 12–16. [CrossRef]
21. Janipella, R.; Pujari, P.R. Review on Groundwater Flow and Solute Transport Modelling in India: Recent Advances and Future Directions. *J. Geol. Soc. India* **2022**, *98*, 278–284. [CrossRef]
22. Dibaj, M.; Javadi, A.A.; Akrami, M.; Ke, K.Y.; Farmani, R.; Tan, Y.C.; Chen, A.S. Modelling seawater intrusion in the Pingtung coastal aquifer in Taiwan, under the influence of sea-level rise and changing abstraction regime. *Hydrogeol. J.* **2020**, *28*, 2085–2103. [CrossRef]
23. Zhou, Y.; Li, W. A review of regional groundwater flow modeling. *Geosci. Front.* **2011**, *2*, 205–214. [CrossRef]
24. Strack, O.D.L. Finite differences and finite elements. In *Analytical Groundwater Mechanics*; Cambridge University Press: Cambridge, UK, 2017; Volume 7, pp. 403–422. ISBN 9781316563144.
25. Koukidou, I.; Panagopoulos, A. Application of Feflow for the Simulation of Groundwater Flow at the Tirnavos (Central Greece). *Bull. Geol. Soc. Greece* **2010**, *43*, 1747–1757. [CrossRef]
26. Arenas, M.C.; Pescador, J.P.; Garzón, L.D.D.; Saavedra, E.Y.; Obando, P.F.A. Hydrogeological modeling in tropical regions via feflow. *Earth Sci. Res. J.* **2020**, *24*, 283–293. [CrossRef]
27. Dufour, R.; Aguirre, C.; Sanchez, M.; Maqueda, A.; Zwinger, J.; Renz, A.; Cho, J.; Evans, D. *Pit Dewatering Optimisation of a 3D Feflow Unstructured Groundwater Model at Geologically Complex Antamina Mine Site in Peru*; Australian Centre for Geomechanics: Perth, Australia, 2020; pp. 1329–1348. [CrossRef]
28. Normi Idris, A.; Zaharin Aris, A.; Sheikhy Narany, T.; Praveena Mangala, S.; Suratman, S.; Tawnie, I.; Khairul Nizar Shamsuddin, M.; Sefie, A. Simulation of Saltwater Intrusion in Coastal Aquifer of Kg. Salang, Tioman Island, Pahang, Malaysia. *MATEC Web Conf.* **2017**, *103*, 4024. [CrossRef]
29. Chu, J.; Liu, N.; Miao, Y.; Lin, X.; Tan, H.; Liu, F. Numerical simulation and pollution prediction of karst groundwater in water-conducting faults distribution area. *IOP Conf. Ser. Earth Environ. Sci.* **2020**, *585*, 012097. [CrossRef]
30. Sefelnasr, A.; Sherif, M. Impacts of Seawater Rise on Seawater Intrusion in the Nile Delta Aquifer, Egypt. *Groundwater* **2014**, *52*, 264–276. [CrossRef] [PubMed]
31. Emara, S.R.; Armanuos, A.M.; Shalby, A. Groundwater for Sustainable Development Appraisal seawater intrusion vulnerability for the Moghra coastal aquifer, Egypt—Application of the GALDIT index, sensitivity analysis, and hydro-chemical indicators. *Groundw. Sustain. Dev.* **2024**, *25*, 101166. [CrossRef]
32. Shalby, A.; Emara, S.R.; Metwally, M.I.; Armanuos, A.M.; El-Agha, D.E.; Negm, A.M.; Gado, T.A. Satellite-based Estimates of Groundwater Depletion over Egypt. *Environ. Monit. Assess.* **2023**, *105*, 594. [CrossRef]

33. Abdel Mogith, S.; Ibrahim, S.; Hafiez, R. Groundwater Potentials and Characteristics of El-Moghra Aquifer in the Vicinity of Qattara Depression. *Egypt. J. Desert Res.* **2013**, *63*, 1–20. [CrossRef]
34. Ahmed, M.; Chen, Y.; Khalil, M.M. Isotopic composition of groundwater resources in arid environments. *J. Hydrol.* **2022**, *609*, 127773. [CrossRef]
35. Sayed, E.; Riad, P.; Elbeih, S.F.; Hassan, A.A.; Hagras, M. Sustainable groundwater management in arid regions considering climate change impacts in Moghra region, Egypt. *Groundw. Sustain. Dev.* **2020**, *11*, 100385. [CrossRef]
36. Yousf, A.F.; El Fakharany, M.A.; Abu Risha, U.A.; Afifi, M.M.; Al Sayyad, M.A. Contributions to the geology of the New Valley area, Western Desert, Egypt. *J. Basic Environ. Sci.* **2018**, *5*, 1–19.
37. el-Aziz Abu-Bakr, H.A.; Abdelmoniem, M. Groundwater potentiality delineation in Moghra, Egypt. *Ain Shams Eng. J.* **2023**, 102337. [CrossRef]
38. Morad, N.A.; Masoud, M.H.; Abdel Moghith, S.M. Hydrologic factors controlling groundwater salinity in northwestern coastal zone, Egypt. *J. Earth Syst. Sci.* **2014**, *123*, 1567–1578. [CrossRef]
39. Selim, T.; Moghazy, N.H.; Elasbah, R.; Elkiki, M.; Eltarabily, M.G. Sustainable agricultural development under different climate change scenarios for El Moghra region, Western Desert of Egypt. *Environ. Dev. Sustain.* **2024**, *26*, 14957–14979. [CrossRef]
40. El Osta, M.; Masoud, M.; Alqarawy, A.; Badran, O. Utilizing of aquifer hydraulic parameters to assess the groundwater sustainability in the new reclamation area of Moghra Oasis: Western Desert—Egypt. *Appl. Water Sci.* **2023**, *13*, 238. [CrossRef]
41. Diersch, H.J.G. *FEFLOW: Finite Element Modeling of Flow, Mass and Heat Transport in Porous and Fractured Media*; Springer: Berlin/Heidelberg, Germany, 2014; ISBN 9783642387395.
42. Zheng, Y.; Chen, S.; Qin, H.; Jiao, J.J. Modeling the spatial and seasonal variations of groundwater head in an urbanized area under low impact development. *Water* **2018**, *10*, 803. [CrossRef]
43. ECDC "Egyptian Countryside Development Company" EGDC "Egyptian Countryside Development Company". Available online: http://elreefelmasry.com (accessed on 1 May 2023).
44. Sarwar, A.; Eggers, H. Development of a conjunctive use model to evaluate alternative management options for surface and groundwater resources. *Hydrogeol. J.* **2006**, *14*, 1676–1687. [CrossRef]
45. Sobeih, M.M.; El-Arabi, N.E.; Helal, E.E.D.Y.; Awad, B.S. Management of water resources to control groundwater levels in the southern area of the western Nile delta, Egypt. *Water Sci.* **2017**, *31*, 137–150. [CrossRef]
46. Gado, T.A.; Shalaby, B.A.; Guo, Y.; Rashwan, I.M.H. Assessment of Satellite-Based Precipitation Estimates over Egypt. *J. Hydrol. Eng.* **2024**, *29*, 05023027. [CrossRef]
47. Thomas, B.F.; Behrangi, A.; Famiglietti, J.S. Precipitation intensity effects on groundwater recharge in the southwestern United States. *Water* **2016**, *8*, 90. [CrossRef]
48. Shah, N.; Nachabe, M.; Ross, M. Extinction depth and evapotranspiration from ground water under selected land covers. *Ground Water* **2007**, *45*, 329–338. [CrossRef]
49. Campoli, D.; Crestaz, E.; Fehervari, M.; Patata, L.; Scipioni, P. 3D density-dependent groundwater modeling in assessing optimisation of a pumping strategy within a chemical plant. *Water Resour. Manag. III* **2005**, *80*, 131–141.
50. Wassef, R.; Schüttrumpf, H. Impact of sea-level rise on groundwater salinity at the development area western delta, Egypt. *Groundw. Sustain. Dev.* **2016**, *2–3*, 85–103. [CrossRef]
51. Abdel Azeem, H.; El-Didy, S.; Eizeldin, M.; Helmi, A. Hydrogeological Evaluation of Massive Salty Lagoons, (Case Study: Qattara Depression). *Al-Azhar Univ. Civ. Eng. Res. Mag.* **2017**, *39*, 277–291.
52. El Shazly, E.M.; Abdel Hady, M.A.; El Ghawaby, M.A.; Khawasik, S.M.; El Shazly, M.M. Landsat satellite imagery of the Qattara Depression area, Egypt. *Acta Astronaut.* **1978**, *5*, 947–957. [CrossRef]
53. Moharram, S.H.; Gad, M.I.; Saafan, T.A.; Allah, S.K. Optimal groundwater management using Genetic Algorithm in El-Farafra Oasis, Western Desert, Egypt. *Water Resour. Manag.* **2012**, *26*, 927–948. [CrossRef]
54. Al-Muqdadi, S.W.H.; Abo, R.; Khattab, M.O.; Abdulhussein, F.M. Groundwater flow-modeling and sensitivity analysis in a hyper arid region. *Water* **2020**, *12*, 2131. [CrossRef]
55. Sultan, M.; Yan, E.; Sturchio, N.; Wagdy, A.; Abdel Gelil, K.; Becker, R.; Manocha, N.; Milewski, A. Natural discharge: A key to sustainable utilization of fossil groundwater. *J. Hydrol.* **2007**, *335*, 25–36. [CrossRef]
56. Haddad, O.B.; Tabari, M.M.R.; Fallah-Mehdipour, E.; Mariño, M.A. Groundwater Model Calibration by Meta-Heuristic Algorithms. *Water Resour. Manag.* **2013**, *27*, 2515–2529. [CrossRef]
57. Hu, L.; Jiao, J.J. Calibration of a large-scale groundwater flow model using GRACE data: A case study in the Qaidam Basin, China. *Hydrogeol. J.* **2015**, *23*, 1305–1317. [CrossRef]
58. Sun, A.Y.; Green, R.; Swenson, S.; Rodell, M. Toward calibration of regional groundwater models using GRACE data. *J. Hydrol.* **2012**, *422–423*, 1–9. [CrossRef]
59. Lo, M.H.; Famiglietti, J.S.; Yeh, P.J.F.; Syed, T.H. Improving parameter estimation and water table depth simulation in a land surface model using GRACE water storage and estimated base flow data. *Water Resour. Res.* **2010**, *46*. [CrossRef]
60. Sun, A.Y.; Green, R.; Rodell, M.; Swenson, S. Inferring aquifer storage parameters using satellite and in situ measurements: Estimation under uncertainty. *Geophys. Res. Lett.* **2010**, *37*, 1–5. [CrossRef]
61. Richey, A.S.; Thomas, B.F.; Lo, M.H.; Reager, J.T.; Famiglietti, J.S.; Voss, K.; Swenson, S.; Rodell, M. Quantifying renewable groundwater stress with GRACE. *Water Resour. Res.* **2015**, *51*, 5217–5237. [CrossRef] [PubMed]

62. Rodell, M.; Chen, J.; Kato, H.; Famiglietti, J.S.; Nigro, J.; Wilson, C.R. Estimating groundwater storage changes in the Mississippi River basin (USA) using GRACE. *Hydrogeol. J.* **2007**, *15*, 159–166. [CrossRef]
63. Bierkens, M.F.P.; Wada, Y. Non-renewable groundwater use and groundwater depletion: A review. *Environ. Res. Lett.* **2019**, *14*, 43. [CrossRef]
64. Shalby, A.; Elshemy, M.; Zeidan, B.A. Assessment of climate change impacts on water quality parameters of Lake Burullus, Egypt. *Environ. Sci. Pollut. Res.* **2019**, *27*, 32157–32178. [CrossRef] [PubMed]
65. Moss, R.; Babiker, M.; Brinkman, S.; Calvo, E.; Carter, T.; Edmonds, J.; Elgizouli, I.; Emori, S.; Erda, L.; Hibbard, K.; et al. Towards New Scenarios for Analysis of Emissions, Climate Change, Impacts and Response Strategies. In Proceedings of the IPCC Expert Meeting Report, Noordwijkerhout, The Netherlands, 19–21 September 2008.
66. Gregory, J. *Projections of Sea Level Rise*; Climate Research: Cambridge, UK; New York, NY, USA, 2014.
67. Shaltout, M.; Tonbol, K.; Omstedt, A. Sea-level change and projected future flooding along the Egyptian Mediterranean coast. *Oceanologia* **2015**, *57*, 293–307. [CrossRef]
68. Shalby, A.; Elshemy, M.; Zeidan, B.A. Modeling of climate change impacts on Lake Burullus, coastal lagoon (Egypt). *Int. J. Sediment Res.* **2020**. [CrossRef]
69. Alkhawaga, A.; Zeidan, B.; Elshemy, M. Climate change impacts on water security elements of Kafr El-Sheikh governorate, Egypt. *Agric. Water Manag.* **2022**, *259*, 107217. [CrossRef]

water

MDPI

Article

# Identification of Microplastics Using μ-Raman Spectroscopy in Surface and Groundwater Bodies of SE Attica, Greece

**Maria Perraki [1,*], Vasilios Skliros [1], Petros Mecaj [2,3], Eleni Vasileiou [1], Christos Salmas [2], Ioannis Papanikolaou [2] and Georgios Stamatis [2]**

1 School of Mining and Metallurgical Engineering, Division of Geo-Sciences, National Technical University of Athens, 15773 Zografou, Greece; sklirosbill@gmail.com (V.S.); elvas@metal.ntua.gr (E.V.)
2 Mineralogy-Geology Laboratory, Department of Natural Resources and Agricultural Engineering, Agricultural University of Athens, 11855 Athens, Greece; petros.mecaj@univ-lyon1.fr (P.M.); c.salmas@aua.gr (C.S.); i.pap@aua.gr (I.P.); stamatis@aua.gr (G.S.)
3 Université Claude Bernard Lyon 1, LEHNA UMR 5023, CNRS, ENTPE, F-69622 Villeurbanne, France
* Correspondence: maria@metal.ntua.gr

**Abstract:** Sixteen surface (5) and groundwater (11) samples were collected from the south-eastern part of Attica, Greece, and analysed for physico-chemical parameters and microplastics (MPs) by optical microscopy and Raman microspectroscopy (RS). A total of 3399 particles were optically identified in all sixteen samples, ranging from only 16 particles/L in a sample from a deeper borehole to 513 particles/L in a sample from a shallow water well. They were then visually classified into eight categories based on their color, texture, size, reflectivity, shape, and general morphological properties. Raman microspectroscopy was performed on the particles on the filters and revealed four different types of MPs, namely polyethylene (PE, 35%), polypropylene (PP, 30%), polystyrene (PS, 10%), and polyethylene terephthalate (PET, 25%). The samples from the shallow phreatic aquifer contained more MPs than the samples from the deeper borehole, which contained fewer MPs and categories. This is to be expected, since the phreatic aquifer (a) is generally more contaminated, as it is close to human activities that generate MPs and its infiltration depth is only a few metres, which means that many microplastics can infiltrate at such shallow depths, and (b) it is exposed to the atmosphere, so they can also be suspended in the air. On the other hand, it is interesting to note that MPs, especially PET and PE, were detected in the borehole sample, suggesting that MPs can migrate to greater depths through water infiltration. Chemical analyses of the groundwater samples revealed high values of E.C., $Ca^{2+}$, $Mg^{2+}$, $Cl^{-}$, and $Na^{+}$, which indicate that seawater intrusion is taking place in the coastal aquifer system of the Erasinos basin. The increased concentrations of $NO_3^{-}$ and $PO_4^{3-}$ also indicate the impact of agricultural activities.

**Keywords:** polyethylene; polypropylene; polystyrene; polyethylene terephthalate microplastics; groundwater; Attica; Raman spectroscopy

## 1. Introduction

Environmental geochemistry is an important application of Raman spectroscopy (RS) in the geosciences. From inorganic to organic and biological contaminants in soil, water and air, RS is an important tool for their characterization. In the last decade, RS has been successfully used to identify microplastics (MPs) [1,2]. "Microplastics" were defined as plastic particles smaller than 5 mm by the National Oceanic and Atmospheric Administration (NOAA) International Research Workshop on the occurrence, effects, and fate of microplastic marine debris (2009, [3]). Since then, MPs have been systematically detected in soils, sediments, and in marine and aquatic environments around the world, raising scientific interest and public health concerns [4]. The main types of polymers that can be detected in the environment are polyethylene (PE), polypropylene (PP), and polystyrene (PS). Microplastics are either primary MPs, which include production pellet nurdles and microbeads, or secondary MPs, which are formed by the mechanical and chemical decomposition of larger plastic waste and fibers from synthetic textiles [5]. Although MPs are mostly colorless, colored MPs are also frequently found. This is due to the increasing trend of adding color pigments to plastics to obtain more attractive plastic products or to absorb part of the ultraviolet light as a light stabilizer to prevent or delay photodegradation and thus extend the life of plastic products [6].

From the Mediterranean [7] to the Pacific Ocean [8], Arctic Sea ice [9], and freshwater lakes [10], MPs are reported in an increasing number of studies worldwide.

The identification of MPs is of great significance and follows the process of separation, extraction, or concentration. The physicochemical properties of MPs, such as size, shape, and composition, are crucial for studying their pathways in soil, marine, and aquatic environments. Microscopic methods such as optical microscopy and scanning electron microscopy mainly provide information on the size, color, and shape of the MPs, whereas the vibrational spectroscopic analytical methods, Fourier Transform Infrared Spectroscopy (FT-IR) and RS, are employed to reveal the chemical composition of the MPs [11]. More advanced analytical methods such as Flow Cytometry (FCM), Dynamic Light Scattering (DLS), hyperspectral imaging (HIS), and Atomic Force Microscopy (AFM) have also been applied to the study of MPs [12–14] and others have yet to be adopted.

Capable of detecting MPs with particle sizes as small as 1 μm [15], RS spectroscopy combined with an RS microscope (μ-RS) offers rapid, generally non-destructive analysis and is becoming a conventional tool in MPs identification, as evidenced by the rapidly growing number of related publications. Raman imaging has also been applied in identifying MPs on filters [16].

Identifying the type of MPs found in soils and waters provides information on the origin of MPs, as well as the weathering/degradation processes that follow the original deposition/release. This may further contribute to monitoring and assessing the different types of MPs in the environment, setting up future prevention measures.

Even though MPs are considered a significant factor in groundwater pollution, a limited number of studies have linked them to other quality parameters such as electric conductivity (EC), pH, temperature (T), dissolved oxygen (DO), oxidation–reduction potential (ORP), total dissolved solids (TDS), turbidity, and salinity. Measuring and assessing these physical parameters in situ could improve the determination of the sources and behavior of MPs as a contaminant in aquifer systems [17]. However, the existing knowledge on how these parameters affect water or interact with MPs is still very limited. It is assumed that microplastics adsorb and transfer organic pollutants and ionic heavy metal pollutants (e.g., Cu, Pb, Hg, Cd, and Cr) in the aquatic environment [18]. The adsorption process of MPs is also controlled by various anions and cations [19]. It is believed that the physical parameters and chemical composition of groundwater are important factors influencing the occurrence of MPs in groundwater [17]. In the coming years, the big challenge for researchers will be the following: how do MPs interact with the chemical composition of groundwater and other contaminants?

There are only a limited number of published studies on the occurrence of MPs in Greek marine and aquatic ecosystems, mostly focussing on seawater samples and sediments. The analytical methods of SEM and ATR-FTIR have been mainly used to study the degradation of plastic materials in different marine environments in Greece [20], plastic pellets and fragments of polyethylene (PE) and expanded polystyrene (EPS) from two beaches of Salamina Island [21], and PE and PP as ingested microplastics in *Boops Boops* fish in the Southern Ionian Sea and Saronikos Gulf [22]. In addition, the temporal and spatial distribution of plastic pellets and fragments on nine sandy beaches along the coast of Northern Crete was investigated using Gas Chromatography—Ion Trap Mass Spectrometry (GC-ITMS) [23]. However, the possible presence of MPs in groundwater from wells and boreholes in Greece has yet to be explored, limiting our knowledge to the coastal and marine environment.

In this work, we conducted a thorough investigation of MPs detected in the surface water and groundwater of the Erasinos River basin in the southeastern part of Attica, Greece, using RS, after filtering and detailed visual presorting. In addition, chemical analyses of water samples were also carried out, in order to investigate the relationship between the chemical status of the groundwater and the occurrence of MPs in the Erasinos basin. The study area was selected because it is characterized by certain anthropogenic activities and land uses (Eleftherios Venizelos International Airport, with more than 25 million passengers annually, large shopping malls). In the Erasinos basin, there are also a variety of potential sources of MPs entering the groundwater system, such as surface runoff, agricultural activities (agrochemical products), atmospheric deposition, and wastewater, which can contribute to MP contamination. The aim was to determine the morphological characteristics and type(s) of MPs present, the impact of degradation, and the possible dependence on the depth of the aquifer. This will further contribute to build up a database of MPs present in Greek ecosystems, including underground aquifers of different types and depths.

## 2. Materials and Methods

### 2.1. Study Area

The study area of the Erasinos basin is located in the south-eastern part of Attica Prefecture, covering an area of approximately 204 km$^2$. The basin is bordered to the east by the coastline of Rafina–Artemis–Porto Rafti. The Erasinos River flows into the southern Gulf of Evoikos. The climate of the region is characterized by low precipitation and high summer temperatures. The average annual precipitation is estimated at 406 mm per year.

The Erasinos basin is part of the Attico-Cycladic Massif. The dominant geological formations in the basin are as follows [24,25] (Figure 1): (a) the crystalline basement of Paleozoic–Upper Cretaceous age (schists and carbonate rocks, mainly limestones and dolomite marbles), and (b) the clastic depositions of Neogene and Quaternary age (marl limestones, marls, clays, sandstones, conglomerates, and other unconsolidated materials).

A knowledge of the hydrogeological conditions of the basin is a crucial factor for the migration of MPs. In general, the groundwater flows towards the sea and discharges into Vavrona bay, while in some parts of the basin, it flows into the Erasinos River [26]. The groundwater flow is locally determined mainly by impermeable schists.

The main aquifer systems in the area are the following [27]:

- A karstic aquifer with a high capacity (discharge rate over 100 m$^3$/h) as it is also fractured. Along the coasts of Porto Rafti and Artemis, scattered underwater sources of this karst water have been identified, confirming the hydraulic connection with seawater.
- The alternation of permeable and impermeable layers favors the development of unconfined and confined aquifers in the Neogene deposits. The capacity of these systems is low (discharge rates up to 5 m$^3$/h). The shallow aquifer is overexploited by extensive pumping to meet irrigation and industrial needs in the region.
- A phreatic aquifer has developed in Quaternary deposits with low capacity (discharge rates of 15 up to 35 m$^3$/h). The phreatic aquifer is exploited by many wells, resulting in seawater intrusion in the coastal parts of the aquifer system. This fact is also reflected in the groundwater quality, as it is characterized as brackish.
- Fractured aquifers are mainly developed in the Rafina area, due to secondary porosity, created by metamorphic and tectonic processes and the weathering of impermeable formations.

The main recharge of the aquifer system, investigated here, originates from the following: (a) direct infiltration of precipitation and (b) infiltration of surface runoff, rivers, and streams, and (c) lateral recharge from other aquifer systems that are hydraulically connected.

Land uses are mainly as follows: (i) agriculture, (ii) industry, (iii) pastures, (iv) transportation (Eleftherios Venizelos International Airport), (v) residential, and (vi) commercial.

The area is intensively used for agriculture, resulting in aquifer degradation, with significant concentrations of inorganic pollutants such as $NO_3^-$, $NO_2^-$, and $PO_4^{3-}$. In addition, high concentrations of sodium ($Na^+$) and chlorine ($Cl^-$) are also traced due to seawater intrusion from excessive pumping [26,27]. The area is also exposed to significant pollution from plastic materials due to intensive agriculture and anthropogenic processes, such as the Eleftherios Venizelos International Airport and the neighbouring city of Koropi.

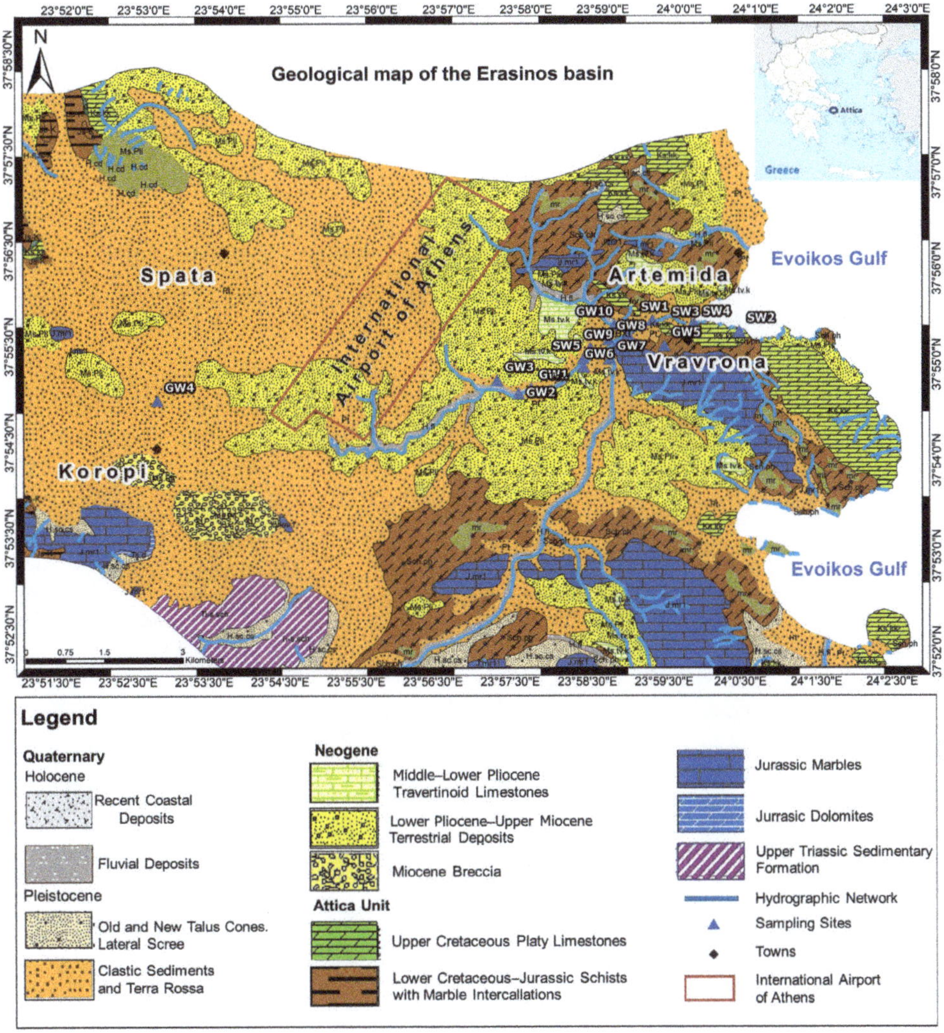


**Figure 1.** Geological map of the Erasinos basin, with the locations of the surface and groundwater samples studied here (modified from the 1:50,000 geological map published by the Institute of Geology and Mineral Exploration of Greece [28]).

## *2.2. Sampling*

Sixteen (16) samples were collected in August 2020 from surface waters, groundwater wells, and a borehole from the catchment area of the Erasinos River in the area of Vravrona and Koropi; five (5) from surface waters (SW1, SW2, SW3, SW4, SW5), one (1) of which

(SW2) was from the river exit into the sea, ten (10) from wells (GW1, GW2, GW3, GW5, GW6, GW7, GW8, GW9, GW10, GW11), and one (1) from a water borehole (GW4) (Figure 2). All groundwater samples, except GW4, were collected from the phreatic water table (up to 8 m), and the surface water samples SW1, SW3, SW4, and SW5 were collected from the Erasinos River estuary. Borehole sample GW4 was taken from a deeper aquifer (70 m) within the Neogene sediments. In all cases, the water was used exclusively for irrigation purposes. The sampling sites are presented in Table 1. The water samples were collected in two 500 mL glass bottles for microplastic analysis and 1000 mL polyethylene bottles for chemical analysis. In each case, the bottles were pre-cleaned, acid-washed, and thoroughly rinsed first with distilled water and then with Milli-Q deionized water [29,30]. Before sampling, the polyethylene bottles were rinsed three times with the water to be sampled. In addition, each sample was stored immediately after sampling without being treated—apart from transport—prior to laboratory analysis [29]. The physical parameters temperature (T), pH, Eh, dissolved oxygen (DO), and electrical conductivity (EC) were measured in situ immediately after sampling, using the YSI Professional Digital Sampling System (ProDSS).

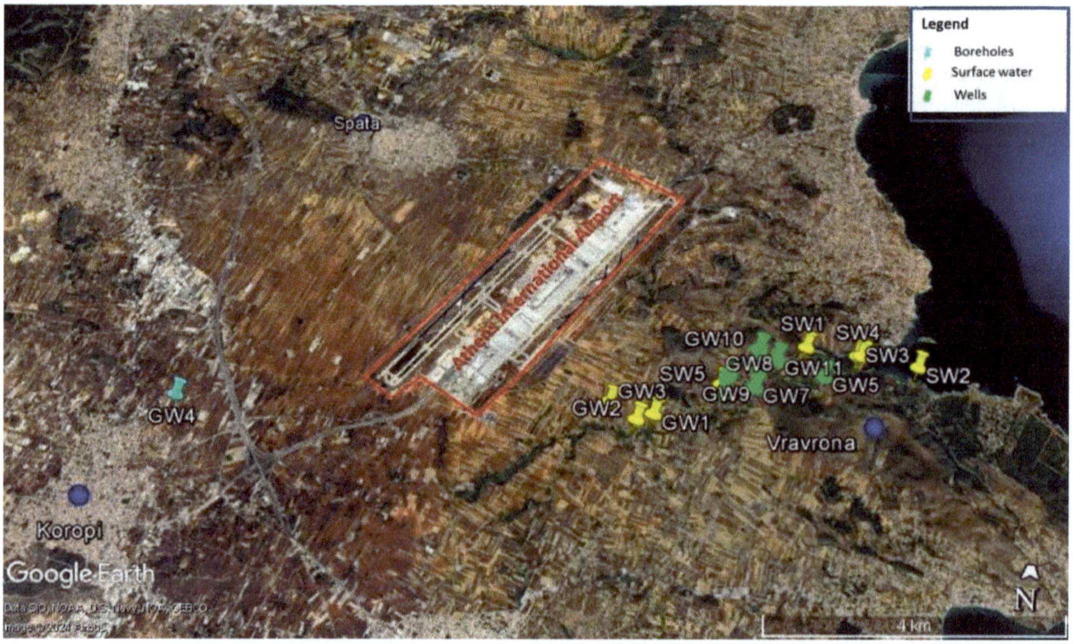


**Figure 2.** Sampling sites in the Erasinos basin (Google Earth).

Twelve selected water samples (SW1, GW5, SW2, SW3, SW4, SW5, GW6, GW7, GW8, GW9, GW10, and GW11) were analyzed to determine major cations (i.e., $Ca^{2+}$, $Mg^{2+}$, $Na^{+}$, and $K^{+}$) and anions (i.e., $Cl^{-}$, $SO_4^{2-}$, $HCO_3^{-}$, $NO_3^{-}$, and $PO_4^{3-}$). The saturation indices for calcite and dolomite were calculated, depending on the site and the prevailing hydrogeological conditions. The analyses of major ions were determined by atomic absorption spectrometry ($Ca^{2+}$, $Mg^{2+}$, $Na^{+}$, and $K^{+}$), by spectrophotometry ($NO_3^{-}$, $NO_2^{-}$, $NH_4^{+}$, and $PO_4^{3-}$), by titrimetry ($Cl^{-r}$ and $HCO_3^{-}$), and by turbidimetry ($SO_4^{2-}$). The total dissolved solids (TDS) content was calculated directly from the sum of the major ions. All QC protocols were applied to the blank samples on site and in the laboratory, as well as to the duplicate analyses of the samples. The analytical accuracy of the ion measurement was determined by calculating the charge balance error in each sample and was less than ±10%.

**Table 1.** Sampling sites.

| Sample ID | X Longitude | Y Latitude | Elevation (m.a.s.l.) * | Notes |
|---|---|---|---|---|
| GW1 | 23°57′57.77″ E | 37°54′53.56″ N | 25 m | Well—phreatic water table |
| GW2 | 23°57′43.44″ E | 37°54′50.45″ N | 33 m | Well—phreatic water table |
| GW3 | 23°57′27.13″ E | 37°54′58.95″ N | 41 m | Well—phreatic water table |
| GW4 | 23°52′59.76″ E | 37°54′53.20″ N | 103 m | Borehole |
| SW1 | 23°59′29.80″ E | 37°55′33.56″ N | 10 m | Surface water |
| GW5 | 23°59′38.50″ E | 37°55′17.50″ N | 9.5 m | Well—phreatic water table |
| SW2 | 24°0′37.90″ E | 37°55′24.90″ N | 0 m | Surface water—river mouth to the sea |
| SW3 | 24°0′0.85″ E | 37°55′29.64″ N | 8.5 m | Surface water |
| SW4 | 24°0′3.50″ E | 37°55′29.85″ N | 0 m | Surface water |
| SW5 | 23°58′33.4″ E | 37°55′7.40″ N | 19.5 m | Surface water |
| GW6 | 23°58′32.55″ E | 37°55′5.16″ N | 19.8 m | Well—phreatic water table |
| GW7 | 23°58′55.15″ E | 37°55′10.10″ N | 15.2 m | Well—phreatic water table |
| GW8 | 23°58′55.11″ E | 37°55′22.59″ N | 12 m | Well—phreatic water table |
| GW9 | 23°58′40.60″ E | 37°55′17.65″ N | 14 m | Well—phreatic water table |
| GW10 | 23°58′59.85″ E | 37°55′31.39″ N | 13.5 m | Well—phreatic water table |
| GW11 | 23°59′10.14″ E | 37°55′25.90″ N | 5.2 m | Well—phreatic water table |

Note: * m.a.s.l.: meters above sea level.

### 2.3. Geochemical Modeling

The geochemical software PHREEQC version 3.1.2 was used to calculate the saturation indices (SIs) of the water samples. The mineral SIs for calcite, dolomite, and quartz were employed to define the mineral dissolution and precipitation processes in the water samples of the Erasinos basin.

A positive SI indicates that the mineral is oversaturated or supersaturated with respect to the solution [31]; thus, the mineral could precipitate. Conversely, a negative SI indicates that the solution is undersaturated with respect to the selected mineral, suggesting that the mineral is dissolved in the groundwater to reach equilibrium.

### 2.4. Sample Extraction

Each sample was filtered using a vacuum pump (−100 mbar pressure) to extract the MPs. Filtration was performed on GN–6 Metricel mixed cellulose ester membranes (47 mm diameter, Pore Size 0.45 μm, 152 μm thickness, gamma irradiated, gridded) at a constant flow rate. The samples were then dried at room temperature and the filters were covered with Petri glass dishes to prevent possible external contamination and the loss of particles. Samples were analyzed microscopically using a Dino X-light digital microscope with 200× optics (AM413MT, Dino-Lite, Almere, The Netherlands). The lowest resolution limit per particle size was set at 6μm. Bright field was used to visualize the samples, and color images were extracted to examine the results. The image analysis software DINOCAPTURE 2.0 was used. The size of the particles was estimated based on the Feret min diameter.

A protocol was applied to avoid contamination within the laboratory during the analysis. All work surfaces were disinfected with double-filtered 70% ethanol before handling the samples, and cotton gowns and nitrile gloves were used throughout the process. Samples were processed in a laminar flow cabinet (switched off) and plastic-free laboratory equipment was used whenever possible [30]. Considering the literature on microplastic sample preparation, three blanks (3) were used in each major analytical step [29]. All the blanks were open 50 mm Petri dishes [32] placed as follows: (1) in the room where the samples were stored for 24 h to dry, (2) under the laminar flow cabinet where the samples were processed while the process was running, and (3) next to the samples that were analyzed under the microscope at the time of analysis. The blanks were subjected to microscopic analysis, during which 1–5 fibers were removed. Blank #3 contained 5 fibers, which can be explained by the fact that the microscopy room and sample inspection as an action are associated with increased human movement [30]. All fibers that resulted from blanks were categorized but not included in our results.

Subsequently, a non-destructive Raman study was performed on 194 suspect particles of each of these MP categories directly on the filters, without further treatment, employing a Renishaw inVia Reflex micro-Raman (Renishaw, Gloucestershire, UK), at the School of Mining and Metallurgical Engineering of the Technical University of Athens, Greece. The spectra were excited at room temperature with the 785 nm (near-infrared) excitation wavelength of a diode laser, in the spectral range 100 $cm^{-1}$–4000 $cm^{-1}$. The light was scattered by a diffraction grating with 1800 lines/mm. The laser beam was focused on the samples by means of a $\times$100 short working distance objective under a Leica microscope. The laser spot on the surface had a diameter of ~1 μm and a power of 4 mW. Wavenumber calibration was checked periodically by measuring the LO phonon mode at 521 $cm^{-1}$ of an internal silicon reference standard. An exposure time of 20 s and 2 accumulation cycles for each spectrum were applied. Attention should be paid to the laser power used to excite the Raman spectra of the MPs, as some of them are very temperature-sensitive, and a laser beam with high radiant energy can lead to their degradation during Raman analysis. The spectra were processed with the software WIRE 3.4. The peaks were assigned by comparing the Raman peaks with reference values from the literature.

## 3. Results

Table 2 shows the chemical analyses and the saturation indices of all water samples. All groundwater samples, with the exception of GW4, originate from the developed shallow unconfined aquifer system in Quaternary sediments. The water depth of this aquifer system is less than 8 m for all water wells. The borehole from which sample GW4 was taken also exploits this aquifer at a depth of 70 m.

**Table 2.** Analyzed physical and chemical parameters of the investigated water samples in the Erasinos basin.

| Sample ID | pH | TDS | E.C. | Temp | $Na^+$ | $K^+$ | $Ca^{2+}$ | $Mg^{2+}$ | $Cl^-$ | $HCO_3^-$ | $NO_3^-$ | Si | $SO_4^{2-}$ | $PO_4^{3-}$ |
|---|---|---|---|---|---|---|---|---|---|---|---|---|---|---|
| | | mg/L | μS/cm | °C | mg/L | mg/L | mg/L | mg/L | mg/L | mg/L | mg/L | mg/L | mg/L | mg/L |
| SW1 | 7.4 | 1126 | 1732 | 21 | 99.3 | 4 | 136 | 39.4 | 212.8 | 366 | 31.7 | 10.7 | 120 | 0.034 |
| GW5 | 7.4 | 1807 | 2780 | 22.6 | 346 | 3.4 | 208.8 | 74.2 | 531.9 | 317.2 | 72.2 | 6 | 470 | 0.14 |
| SW2 | 8.1 | 35,750 | 55,000 | 20 | 10,300 | 420 | 435.2 | 1393.6 | 12,198.6 | 170.8 | 11 | 0.8 | 2640 | 0.055 |
| SW3 | 7.8 | 1251 | 1925 | 22 | 206.6 | 5 | 147.2 | 81.4 | 425.5 | 396.5 | 31.2 | 10.3 | 162.5 | 0.145 |
| SW4 | 7.8 | 11,440 | 17,600 | 21 | 208 | 70 | 227.2 | 506.6 | 6241.1 | 549 | 22 | 8.1 | 195.5 | 0.13 |
| SW5 | 7.7 | 965 | 1485 | 24.8 | 93.3 | 3.2 | 116.8 | 43.7 | 248.2 | 329.4 | 36.1 | 9.8 | 43.8 | 0.069 |
| GW6 | 7.4 | 1043 | 1605 | 22.5 | 95.1 | 4 | 142.4 | 47.7 | 283.7 | 347.7 | 40.5 | 10.5 | 115 | 0.11 |
| GW7 | 7.3 | 1040 | 1600 | 22 | 101 | 3.8 | 123.2 | 69.1 | 319.2 | 427 | 16.7 | 8.9 | 72.5 | 0.07 |
| GW8 | 7.9 | 917 | 1410 | 21.5 | 86.6 | 6.4 | 120 | 39.3 | 248.2 | 384.3 | 12.3 | 9.5 | 17.5 | 0.935 |
| GW9 | 7.1 | 1339 | 2060 | 21.2 | 114 | 3.6 | 174.4 | 57.5 | 372.3 | 366 | 61.2 | 13.1 | 125 | 0.14 |
| GW10 | 6.9 | 1411 | 2170 | 24 | 122 | 2.2 | 209.6 | 53.2 | 407.8 | 396.5 | 48 | 9.9 | 87.5 | 0.07 |
| GW11 | 7.5 | 897 | 1380 | 25 | 91.6 | 4.2 | 123.2 | 66.7 | 319.2 | 372.1 | 36.5 | 8 | 50 | 0.155 |

As shown in Table 3, the pH values range from 6.9 to 8.1; the water samples are characterized as neutral to alkaline, and the electrical conductivity (E.C.) values range from 1380 to 55,000 μS/cm. The measurements of major ions show strong fluctuations in the Erasinos basin. The concentration of $Ca^{2+}$ ranges from 116.8 to 435.2 mg/L, of $Mg^{2+}$ from 39.3 to 1393.6 mg/L, of $K^+$ from 2.2 to 420 mg/L, of $Na^+$ from 91.6 to 10,300 mg/L, of $Cl^-$ from 212.8 to 12,198 mg/L, and of $NO^{3-}$ from 11 to 72.2 mg/L. All saturation indices are oversaturated except for the sample SW2, in which the SI for quartz is undersaturated.

**Table 3.** Calculated saturation indices of the investigated water samples in the Erasinos basin.

| Sample ID | SI (Calcite) | SI (Dolomite) | SI (Quartz) |
|---|---|---|---|
| SW1 | 0.45 | 0.66 | 0.62 |
| GW5 | 0.49 | 0.84 | 0.35 |
| SW2 | 0.71 | 2.25 | −0.42 |
| SW3 | 0.87 | 1.8 | 0.59 |
| SW4 | 0.94 | 2.55 | 0.52 |
| SW5 | 0.7 | 1.32 | 0.52 |
| GW6 | 0.46 | 0.77 | 0.59 |
| GW7 | 0.39 | 0.85 | 0.53 |
| GW8 | 0.93 | 1.68 | 0.56 |
| GW9 | 0.19 | 0.2 | 0.71 |
| GW10 | 0.13 | 0.01 | 0.55 |
| GW11 | 0.6 | 1.28 | 0.44 |

Table 4 displays the number of particles identified in the sixteen samples. A total of 3399 particles were identified in all sixteen samples. The mean value is 212.5 particles per liter of water. The highest value (513 particles/L) was found in an uncapped well (GW5), while the lowest value (16 particles/L) was found in the borehole sample (GW4), indicating that potential airborne microplastic pollution should be considered and quantified in future work. Specifically, 2008 MPs particles were detected in the 11 groundwater samples, with a mean value of 182.5 MPs/L. On the other hand, the surface water samples showed a higher microplastic accumulation, with a mean value of 278.2 MPs/L. The closer we get to the banks of the Erasinos River (SW4), the more microplastics were detected, indicating that pollution is accumulating throughout the catchment area. The predominant visual category was F, with a total number of more than 2500 particles. The microplastic particles in the form of fibers and fragments were predominantly blue in color.

**Table 4.** Number of MPs particles/L per category identified in the investigated water samples in the Erasinos basin.

| Category/Samples No of MPs Particles/L | A | B | C | D | E | F | G | H | Total |
|---|---|---|---|---|---|---|---|---|---|
| SW1 (MPs particles/L) | 26 | 7 | 2 | 4 | 2 | 89 | 0 | 40 | 170 |
| GW5 (MPs particles/L) | 11 | 4 | 3 | 10 | 1 | 458 | 0 | 26 | 513 |
| SW2 (MPs particles/L) | 16 | 6 | 7 | 7 | 3 | 347 | 1 | 16 | 403 |
| SW3 (MPs particles/L) | 14 | 2 | 2 | 7 | 0 | 166 | 1 | 15 | 207 |
| SW4 (MPs particles/L) | 27 | 7 | 7 | 44 | 12 | 303 | 13 | 72 | 485 |
| SW5 (MPs particles/L) | 3 | 0 | 3 | 8 | 1 | 95 | 0 | 16 | 126 |
| GW6 (MPs particles/L) | 17 | 5 | 4 | 16 | 0 | 220 | 0 | 18 | 280 |
| GW7 (MPs particles/L) | 8 | 2 | 2 | 7 | 0 | 29 | 1 | 8 | 57 |
| GW8 (MPs particles/L) | 12 | 1 | 1 | 5 | 3 | 229 | 0 | 13 | 264 |
| GW9 (MPs particles/L) | 13 | 0 | 5 | 8 | 20 | 362 | 2 | 21 | 431 |
| GW10 (MPs particles/L) | 9 | 2 | 2 | 8 | 3 | 77 | 0 | 11 | 112 |
| GW11 (MPs particles/L) | 11 | 1 | 1 | 5 | 2 | 128 | 0 | 9 | 157 |
| GW1 (MPs particles/L) | 17 | 1 | 0 | 2 | 3 | 31 | 4 | 9 | 67 |
| GW2 (MPs particles/L) | 18 | 0 | 3 | 6 | 2 | 36 | 0 | 5 | 70 |
| GW3 (MPs particles/L) | 1 | 0 | 0 | 2 | 0 | 37 | 0 | 1 | 41 |
| GW4 (MPs particles/L) | 4 | 0 | 0 | 0 | 0 | 5 | 6 | 1 | 16 |
| Total per category | 207 | 38 | 42 | 139 | 52 | 2612 | 28 | 281 | 3399 |

## 3.1. Visual Presorting

Microplastics were classified into eight categories based on their color, texture, size, reflectivity, shape, and general morphological features (Figure 3). These categories are described below:

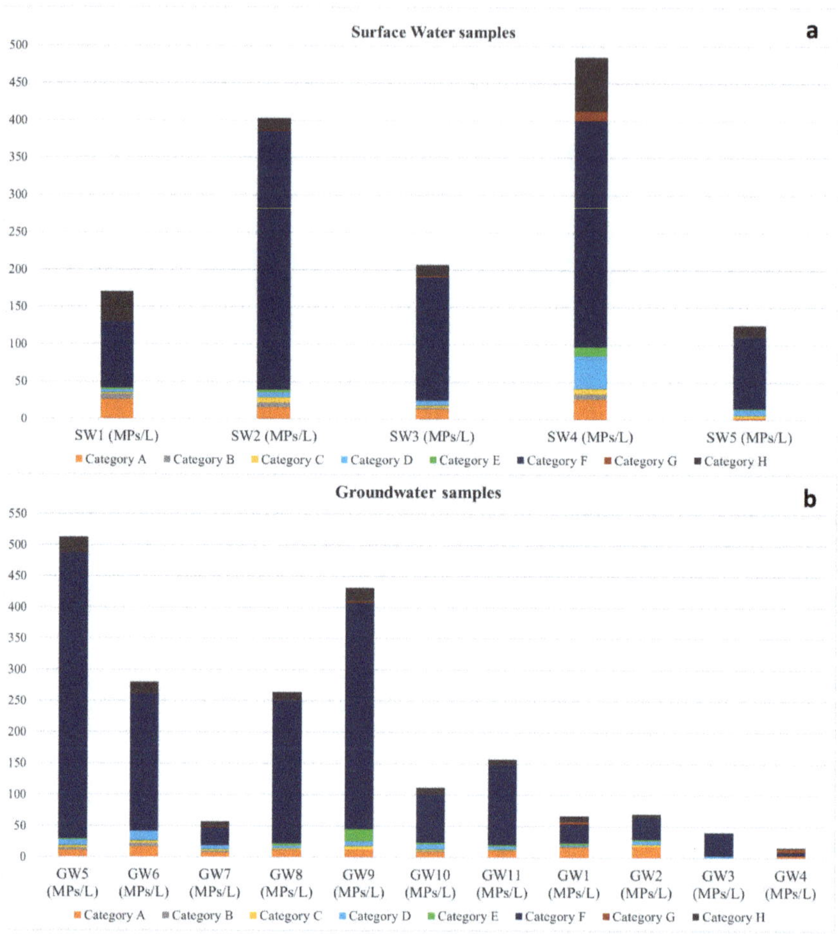


**Figure 3.** (**a**) MPs/L/Surface water, (**b**) MPs/L/Groundwater.

Category A: Blue–colored fibers with a length of >150 μm, characterized mostly by uniform coloring throughout their length and a fairly stable thickness (minor or no thickness differentiations along strike). Some of them may have scissor-tailed tips.

Category B: Light blue–/azure fibers. All other characteristics are similar to category A.

Category C: Reddish–colored fibers. All other characteristics are similar to category A.

Category D: Blue–colored particles. Their color ranges from light blue, turquoise, and cyan to dark blue. Their size >30 μm and their shape is characterized as angular, irregular, and in a few cases as needle-shaped. Their color is uniform throughout their length, except for some needle-shaped particles where local discoloration can be observed.

Category E: Red-to-orange particles and granules. Their color is uniform throughout their length, their size >25 μm and their shape ranges from irregular, angular to rounded.

Category F: White–/whitish-gray particles and granules. Their coloring ranges from pale yellow, beige, whitish, and white to gray and their size >20 μm. Their shape is characterized as irregular, angular up to medium/half-rounded. Gray-colored particles tend to be transparent.

Category G: Highly reflective particles. They are highly reflective in a vertical light source, but if the light source forms an angle with the particles, then they are either low- or non-reflective. Their size >40 μm, and their shape is characterized as irregular and angular.

Category H: Blue–colored fibers, but different from category A. Their color varies from light blue, blue, dark blue, up to black. Their size is mostly >140 μm. Their morphological features and texture differ significantly from category A fibers, since they are characterized by heterogeneity in shape, texture, and thickness.

*3.2. Raman Spectroscopy*

Considering that most of the samples contained particles of each visual category, four randomly selected samples (three well samples and one borehole sample) covering most of the Erasinos catchment were subjected to μ-Raman analysis.

The 194 microplastics from each sample and category were analyzed using RS; RS analysis showed that each visual category was indicative of microplastics. In total, four different types of MPs were identified in the four samples based on their characteristic Raman peaks, namely, polyethylene (PE), polypropylene (PP), polystyrene (PS), and polyethylene terephthalate (PET) (Table 5, Figures 4–7). In general, the peaks in the spectral range 2780–2980 $cm^{-1}$ of the Raman spectra of MPs correspond to the stretching vibrations of $CH/CH_2/CH_3$ groups; in the spectral range 1580–1640 $cm^{-1}$, to the aromatic bending vibrations, and in the spectral range 709–759 $cm^{-1}$, to the symmetric stretching vibrations of $CF_2$ [15,33]. In total, four different types of MPs polymers can be identified by their Raman peaks in these three spectral regions. In detail:

Polyethylene (PE) was identified in three samples (GW1, GW2, GW4), mainly by the two intense peaks at ~2850 and 2885 $cm^{-1}$ corresponding to $CH_2$ stretching vibrations. Moreover, the antisymmetric and symmetric stretching vibrations of C–C at ~1062 $cm^{-1}$ and 1129 $cm^{-1}$, respectively, and the bending vibrations of $CH_2$ at 1296 and 1440 $cm^{-1}$, were clearly visible in the Raman spectra of PE. The weak peak at ~2725 $cm^{-1}$ is an overtone of wavenumbers in the range 1400–1495 $cm^{-1}$ (–$CH_2^-$ bonds). In the spectral region below 1000 $cm^{-1}$, stretching and bending vibrations of C–C are expected, possibly together with $CH_2$ twisting and rocking vibrations of methylene units in a Gauche relation. A typical spectrum of PE is shown in Figure 4a. In Figure 4b,c, in addition to the bands of PE (at ~1062, 1129, 1296, 1417, 1440, 1460, 2725, 2850, and 2885 $cm^{-1}$), bands appear at ~483, 678, 746, 955, 1145, and 1530 $cm^{-1}$. These bands are attributed to the blue pigment copper phthalocyanine (Pigment Blue 15, IRUG collection #ROD00126 [34]), which is often used to color plastics and is consistent with the blue color of these MPs. Polyethylene was identified in samples GW1, GW2, and GW4 (40% of the total Raman analyses) mainly in category A, followed by D, B, and E.

Isotactic polypropylene (iPP) was found in three samples (GW1, GW2, and GW3, 35%). It is characterized by intense peaks at 396 $cm^{-1}$ attributed to $CH_2$ wagging and CH bending, at ~809 $cm^{-1}$ corresponding to the rocking vibrations of $CH_2$ and stretching vibrations of C–C and C–$CH_3$, at 841 $cm^{-1}$ due to the rocking of $CH_2$ and $CH_3$ and the stretching of C–C and C–$CH_3$, at 973 $cm^{-1}$ assigned to $CH_3$ rocking and C–C stretching, at 1153 $cm^{-1}$ resulting by stretching vibrations of C–C and C–$CH_3$, bending vibrations of CH, and rocking vibrations of CH, at 1330 $cm^{-1}$ attributed to bending vibrations of CH, at 1360 $cm^{-1}$ due to the symmetric bending of $CH_3$ and CH bending, and at 1460 $cm^{-1}$ due to the asymmetric bending of $CH_3$ and $CH_2$ bending (Figure 5b,c). The peaks of $CH_2$ stretching vibrations between 2800 and 3000 $cm^{-1}$ were relatively weak in some spectra, and a broad and strong band was also observed in the range of 2100–2200 $cm^{-1}$. In Figure 5b,c, additional peaks at ~682, 750, and 1532 $cm^{-1}$ are assigned to the pigment copper phthalocyanine (Pigment Blue 15, IRUG collection #ROD00126 [34]); this is consistent with the blue color of these two microplastic particles.

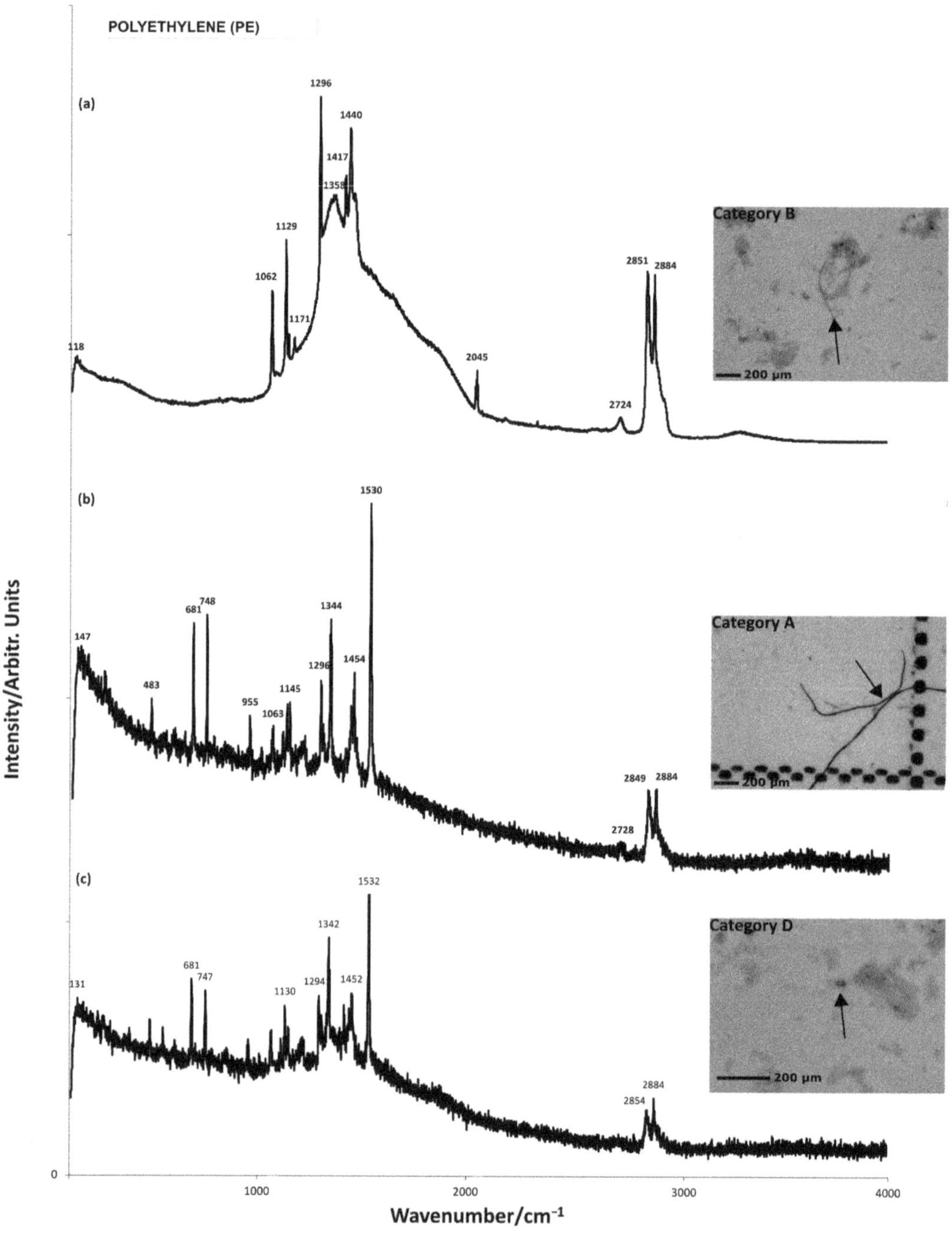


**Figure 4.** (**a**–**c**) Raman spectra of polyethylene-type MPs, found in the studied samples. The black arrow on the inset shows the analyzed MP. In (**b**,**c**) additional peaks at ~682, 750, and 1532 $cm^{-1}$ are assigned to the pigment copper phthalocyanine (Pigment Blue 15, IRUG collection #ROD00126 [34]).

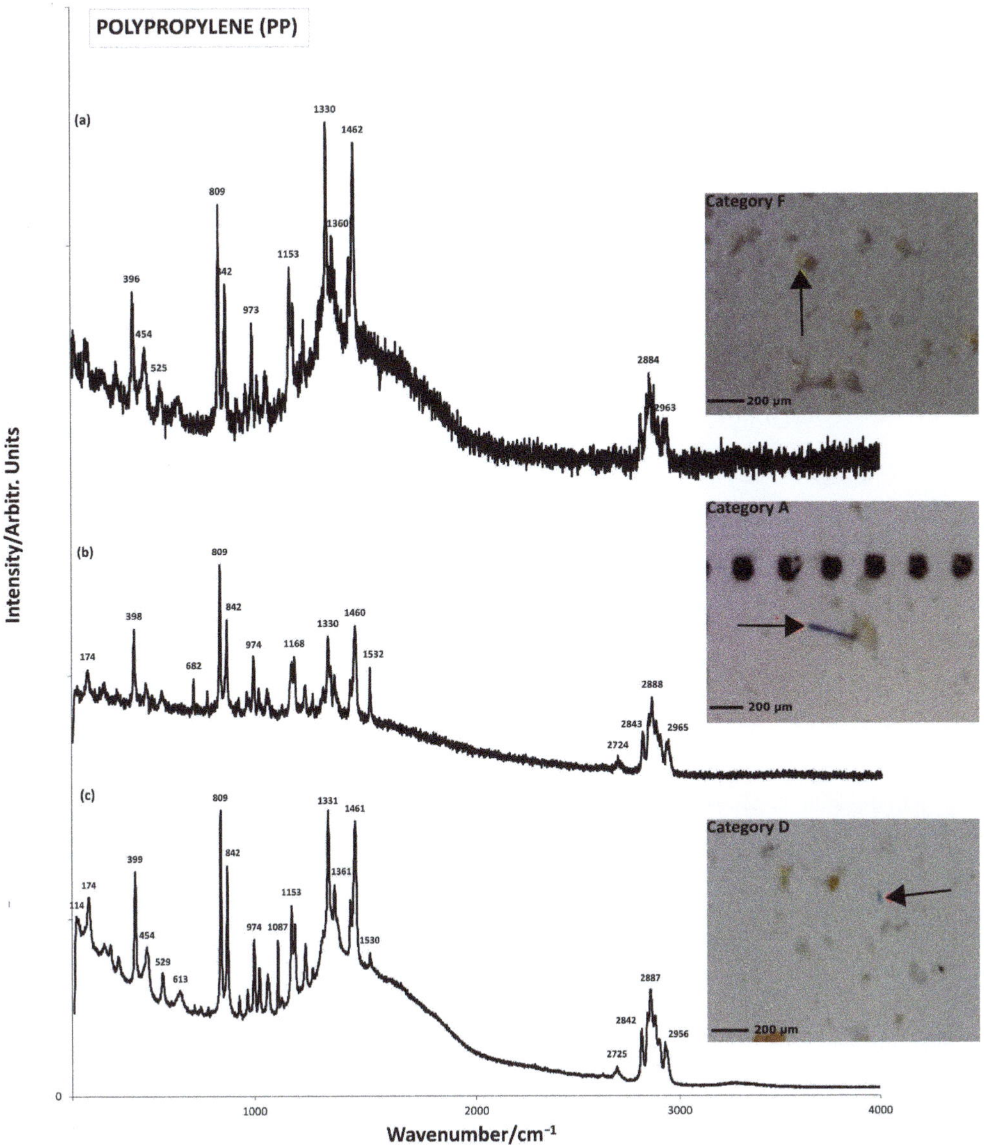


**Figure 5.** (**a–c**) Raman spectra of polypropylene-type MPs, found in the studied samples. The black arrow on the inset shows the analyzed MP.

Polyethylene terephthalate (PET) was identified in all samples studied (25%), mainly in category F, followed by D and G. The peak at approximately 1100 $cm^{-1}$ corresponds to C–O and C–C stretching or C–O–C bending. A shoulder at about 1120 $cm^{-1}$ indicates an amorphous phase. The peak at ~1615 $cm^{-1}$ is due to the C-C ring of phenyl, and the peak at ~1730 $cm^{-1}$ is assigned to the stretching vibrations of C=O. Between 2800 and 3000 $cm^{-1}$, the peaks of $CH_2$ stretching vibrations arise (Figure 6). The peak at 817 $cm^{-1}$ in Figure 6b is attributed to C−H out-of-plane bending vibrations in para-xylene (p-xylene) [35,36].

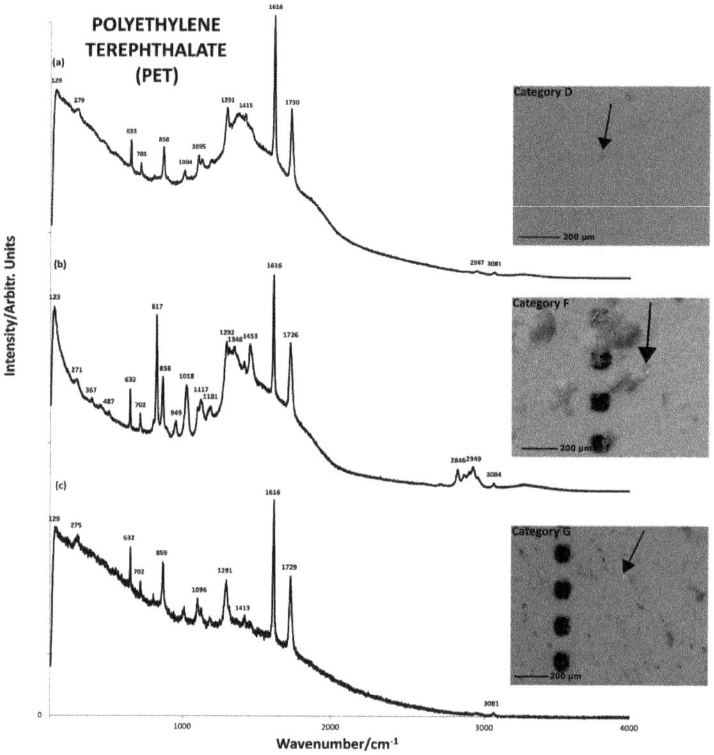


**Figure 6.** (**a–c**) Raman spectra of polyethylene terephthalate-type MPs, found in the studied samples. The black arrow on the inset shows the analyzed MP.

Polystyrene (PS), identified only in sample GW1 (10%), provided intense Raman peaks at 1002 $cm^{-1}$, corresponding to C–C symmetric stretching vibrations (Figure 7). Other characteristic PS bands were observed at ~621, 797, 1033, 1157, 1450, and 1604 $cm^{-1}$, corresponding to the $\nu_{6b}$ radial ring-stretching mode, $\nu_1$ symmetric ring-stretching mode, $\nu_{18a}$ tangential C–H bending mode, $\nu_{15}$ mode, $\nu_{19b}$ stretching or $CH_2$ bending vibrations, and $\nu_{12}$ C–C stretching mode, respectively [37,38]. Overlapping bands at 3060 $cm^{-1}$ appear due to the C–H bonds stretching on the benzene ring.

Finally, some blue fibers analyzed yielded the Raman spectrum of the pigment indigo blue (250, 545, 599, 1310, 1228, 1575, and 1584 $cm^{-1}$ [39]. Due to the high background of these spectra, it was not possible to identify the substrate of the pigment.

Raman spectroscopy on microparticles in the groundwater samples investigated also revealed the presence of inorganic mineral phases, mainly carbonates (calcite, dolomite) and quartz.

By combining the results of virtual sorting and Raman microspectroscopy, we found the following:

Polyethylene (PE) was found mainly in the form of dark and light blue fibers >150 µm long (visual categories A and B). A limited number of blue angular, irregular, and in a few cases acicular particles with a size of >30 µm (visual category D), and very rare red-to-orange particles and granules with a size of >25 µm, were also identified as PE.

Polypropylene (PP) was found mainly in the form of blue fibers >150 µm long (visual category A). Angular, irregular particles of a blue (visual category D) and whitish-gray (visual category E) color sized >25 µm were also identified as PP.

Polyethylene terephthalate (PET) was found mainly as irregular, angular-to-medium/semi-rounded particles and granules of a color ranging from pale yellow, beige, whitish, and white to gray and of a size of >20 μm (visual category F). A few angular, irregular particles of a blue color (visual category D) and a limited number of highly reflective angular, irregular particles in vertical reflected light (visual category G) were also identified as PET.

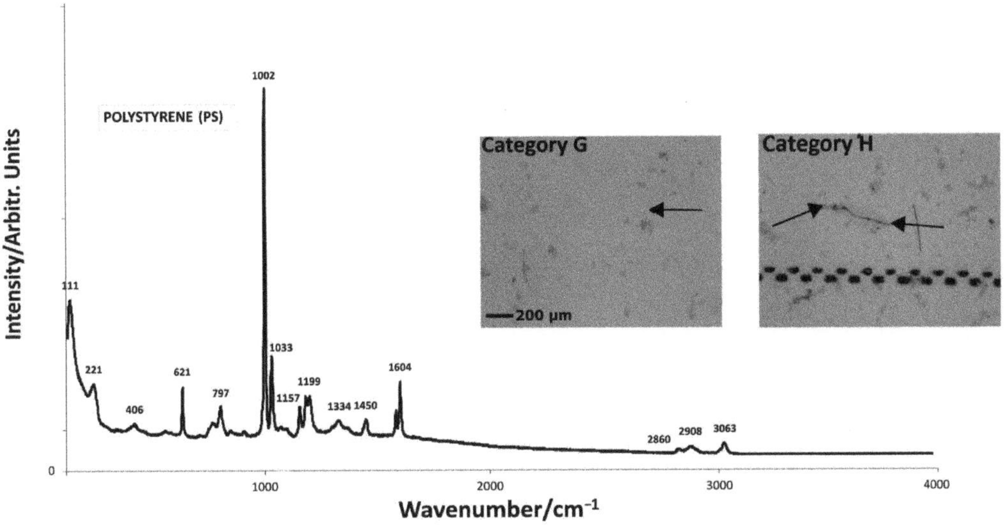


**Figure 7.** Raman spectra of polystyrene MPs, found in the studied samples. The black arrow on the inset shows the analyzed MP.

A very limited number of PS MPs were identified as highly reflective irregular, angular particles (visual category G) and blue–colored fibers characterized by a high heterogeneity in shape, texture, and thickness (visual category H).

**Table 5.** Identified polymer types in the studied samples based on their Raman peaks.

| Polymer Types | Polymer Formula | Raman Peaks (cm$^{-1}$) | Sample | Visual Categories |
|---|---|---|---|---|
| PE | 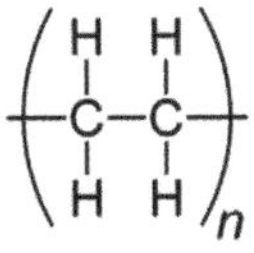<br>$(C_2H_4)_n$ | $1062_s$ ($\nu_{as}$C–C) [40–42]<br>~$1129_s$ ($\nu_s$C–C) [40–42]<br>$1296_s$ ($\tau CH_2$) [40–42]<br>~$1417_m$ ($\omega CH_2$) [42]<br>~$1440_s$ ($\delta CH_2$) [40,41]<br>$1512_w$ ($\delta CH_2$) [40,41]<br>~2850 ($\nu_s CH_2$) [43,44]<br>~$2884_s$ ($\nu_{as}CH_2$) [43,44] | GW1, GW2, GW4 | A,B,D,E |
| PP | 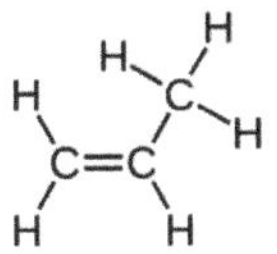<br>$(C_3H_6)_n$ | $396_s$ ($\omega$CH2,$\delta$CH) [45,46]<br>$809_{vs}$ ($\rho CH_2$, νCC, νC–$CH_3$ [42,45,46]<br>$841_s$ ($\rho CH_2$, ν$CC_b$, νC–$CH_3$, $\rho CH_3$) [45,46]<br>$973_m$ ($\rho CH_3$ + νC–C chain) [45,46]<br>$1153_s$ (ν$CC_b$, νC–$CH_3$, δCH, $\rho CH_3$) [45,46]<br>~$1330_s$ (δCH) [45,46]<br>~$1460_s$ ($\delta CH_3$) [45,46]<br>~$2842_m$(ν$CH_2$) [43,44]<br>$2860_m$(ν$CH_2$) [43,44]<br>~$2885_s$ ($\nu_s CH_3$) [43,44]<br>2954-$2963_m$ ($\nu_a CH_3$) [43,44] | GW1, GW2, GW3 | A,C,D,F |

**Table 5.** *Cont.*

| Polymer Types | Polymer Formula | Raman Peaks ($cm^{-1}$) | Sample | Visual Categories |
|---|---|---|---|---|
| PS | 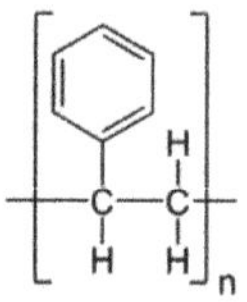<br>$(C_8H_8)_n$ | 621s (νC–C) [37,38]<br>797m ($\nu_s$C–C) [37,38]<br>$1002_s$ (νC–C) [45]<br>1033 (δC–H) [37]<br>$1157_w$ (ν15 mode) [37]<br>$1450_w$ [ν19b or δ(CH2)] [37]<br>$1604_m$ ($\nu_s$C–C) [37]<br>$2860_w$ ($\nu_s CH_2$) [42]<br>$2908_w$ ($\nu_{as} CH_2$) [42]<br>$3063_w$ (νC–H) [42] | GW1 | G,H |
| PET | 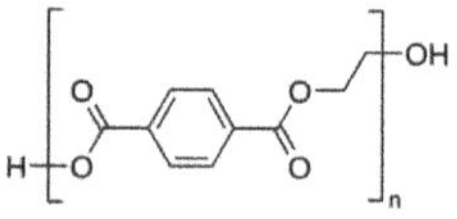<br>$(C_{10}H_8O_4)_n$ | $860_s$ ($\rho CH_2$)<br>$1097_m$ (νC–O, νC–C, δC–O–C) [47]<br>1612-$1617_s$ (C–C ring of phenyl) [43,47]<br>1726-$1730_s$ (νC=O) [47]<br>$2850_w$,m ($\nu CH_2$) [43,47]<br>$2883_w$,m ($\nu CH_2$) [43,47]<br>$3081_w$ (C–H of phenyl) [43,47] | GW1, GW2, GW3, GW4 | D,F,G |

## 4. Discussion

In view of the above, visual presorting, while a very time-consuming method, does not provide an unambiguous identification of the type of MPs in a sample; the latter is only possible with the use of a microspectroscopy method such as RS. However, visual presorting is faster in quantifying the total number of MPs in a sample. Future challenges in this field include the use of Raman hyperspectral mapping on filters aiming not only the identification but also the quantification of MPs, and/or the in situ identification of MPs with portable Raman microspectrometers.

Polyethylene and PP were the most abundant MPs in the samples studied. This is consistent with the fact that these two plastic polymers are the most commonly produced in the world and are therefore frequently found in relevant studies, especially in the marine environment (e.g., [48]).

Although one might attribute the abundance of blue MPs in the studied samples to possibly biased visual presorting, this could be explained by the faster photoaging of blue plastics compared to red or yellow plastics. As stated by Zhao et al. [6] in the visible light region, red or yellow pigments with longer wavelengths may absorb short-wave light with higher energy, while blue pigments absorb long-wavelength light with lower energy. The corresponding light energy transmitted to the blue plastic is consequently higher than that transmitted to red or yellow plastics, which makes the photoaging more intense. Because blue plastics cannot absorb UV light effectively, they age faster when exposed to sunlight, and higher levels of bluish microplastics are often found in the environment, especially in the smallest sizes [49].

Copper phthalocyanine (Pigment Blue 15), a common blue pigment used for dyeing plastics (e.g., [50]), was identified in both PP and PE MPs. In general, the MPs in the form of blue-colored fibers mainly correspond to either PP or PE and, less frequently, PS. Overall, RS confirmed that all identified particles were indeed MPs. Even though MPs are visually similar and belong to the same optical categories, RS has shown that they can differ in nature. Nevertheless, there is a trend that separates PP and PE, in one visual group (A, B, D), from PS and PET, which are part of another visual group (G, H, E). This trend needs further research.

Samples SW2 and SW4 are the only samples that have MPs from all eight categories, while GW4 has the least number of MPs and fewer categories. This was to be expected, as samples SW2 and SW4 were collected from surface water which is (a) generally more contaminated due to its proximity to human activities that generate MPs and (b) exposed to the atmosphere, so that MPs can also be enriched by air. Furthermore, it is very important to mention that they were collected at the estuary of Erasinos (SW4) and at the sea, where the river flows in and has accumulated MPs from the entire catchment area. This is also justified by the fact that more MPs were detected in samples SW1, GW5, SW2, SW3, SW4, SW5, GW6, GW7, GW8, GW9, GW10, and GW11, collected downstream, than in samples GW1, GW2, GW3, and GW4, which were collected further upstream of the catchment. Comparing the results in Figure 3, it can be assumed that surface water is generally more susceptible to MP contamination than groundwater, as on average, more MP particles/L of water samples were detected. On the other hand, it is interesting to note that MPs, and in particular PET and PE, were detected in the sample from the water borehole, suggesting that MPs can migrate to greater depths through water infiltration. Another alternative scenario for tracing MPs at greater water depths would be the possibility of waste injection through other boreholes.

The role of inorganic mineral phases such as carbonates and quartz in connection with microplastics needs further investigation in the future. Column experiments have shown that the presence of iron and manganese oxides, as well as calcite, not only changes the transport behavior of plastics, but can also shield plastics from UV light (e.g., [51,52]). In this context, it was investigated whether the groundwater tends to dissolve or precipitate the minerals calcite, dolomite, and quartz in particular. The positive SI value of these minerals indicates that the water is oversaturated in relation to the solution, which leads to the precipitation of these minerals. These mineral phases were also identified by Raman analyses, which confirmed their precipitation. Saturation-induced precipitation leads to the formation of small crystals on the surface of MPs.

A major challenge for researchers is the mechanism of how microplastics migrate through the vadose zone into the saturated zone and how they contaminate groundwater. The hydrogeological conditions of a region appear to play a key role in the transport of microplastics via different pathways. The most important hydrogeological factors that have recently been suggested to influence the distribution and migration of microplastics in groundwater are as follows [53–55]: (a) the permeability and porosity of an aquifer, which favor the size of microplastics that can migrate into groundwater, (b) the recharge of the aquifer by surface runoff, (c) the structure of the vadose zone, (d) the seasonal periods of recharge, (e) the hydraulic conductivity of an aquifer, (f) the water–soil interaction, (g) the groundwater velocity, (h) the lateral hydraulic connection between aquifers or the intrusion of seawater, and (i) the water depth. Another important factor for the migration of microplastics into groundwater is the type of sources, i.e, diffuse linear sources (torrents, rivers, streams) or point sources (spills, etc.), as well as the land cover of an area [56]. However, the extent to which these individual factors influence the migration of MPs and their interactions is an open field for future research.

The physico-chemical properties of the analyzed water samples (surface and groundwater) indicate the decisive role of anthropogenic activities in the area. According to Wang et al. [57], the pH of water could theoretically influence the sorption of ionic chemicals by MPs. The high values of E.C., $Ca^{2+}$, $Mg^{2+}$, $Cl^-$, and $Na^+$ indicate the intrusion of seawater into the coastal area of the Erasinos basin due to the overexploitation of aquifers for irrigation. Agricultural activities are also a potential source of microplastics in surface and groundwater, mainly due to plastic bottles and the plastic in greenhouses. Agricultural activities in the study area were confirmed by the high $NO_3^-$ and $PO_4^{3-}$ concentrations depicted in Table 2. In addition, major anions and cations are abundant in the groundwater environment [19]. It is hypothesized that the increased concentrations of $Ca^{2+}$ forms increase the number of adsorption sites and enhance the adsorption of microplastics in the media. In contrast, it is assumed that the influence of increased concentrations of $HCO_3^-$ and $SO_4^{2-}$, in the water solution reduces the adsorption of MPs on the surface of the aquifers [19]. Monitoring the chemical properties of groundwater, in conjunction with the quantification and identification of MPs, will provide sufficient results when considering the correlation between chemical parameters and the presence of MPs. The latter may facilitate tracing the influence of different sources of MPs among the samples. This study can contribute to future databases that may correlate the presence of certain MPs with specific sources and environments supported by the geochemical analyses.

In the Erasinos basin, groundwater recharge from seepage of the Erasinos River and other surface runoff (torrents, rivers, streams) could potentially enhance the migration of MPs to the saturated zone. As seawater infiltrates into the nearshore part of the basin, there is a hydraulic connection between the aquifer system and the sea, which could represent an additional source of MPs.

This study is the first systematic work on MPs in groundwater in Attica, Greece, and one of the few worldwide [58–60] that have recently investigated the presence and transport of microplastics in groundwater. The results indicate that the investigation of microplastics in other aquifers needs to be expanded in order to monitor the extent of contamination and trace the sources where possible. As groundwater is in many cases not only used for irrigation but also as drinking water, it is of great importance and relevant for public health to investigate the occurrence of microplastics more systematically.

The concentrations of MPs are not stable and standardized in a region and vary due to the geomorphological characteristics, hydrogeological conditions, atmospheric conditions, and land use of each area, since they define the MPs sources. For example, the amount of MPs in shallow aquifers close to industrial and agricultural activities can be as high as 6832 MP particles/L [61]. In contrast, an average of 6.4 particles/L [53] was observed in karstic aquifers with low anthropogenic activities. In our study area, the high amount of microplastics (3399 particles/ per 16 Liters) is related to the intensive anthropogenic activities (agricultural, industrial, and urban) in this area, combined with the shallow aquifer.

## 5. Conclusions

In this work, we have investigated and documented for the first time the occurrence of MPs in groundwater in shallow and deep aquifers in Attica, Greece. Based on our results, the following conclusions can be drawn:

- The characterization of MPs by Raman microspectroscopy revealed the presence of the following:
  a. PE and PP in the form of fibers and fragments;
  b. PET as angular irregular particles;
  c. PS as reflective irregular angular particles and blue-colored fibers.
- Microplastics are more prevalent in shallow aquifers, as expected, reaching up to 513 MPs/L. In the deeper aquifers (water table ~70 m), the abundance of MPs is lower (up to 16 MPs/L); however, this is no less significant, as it may indicate that MPs are able to migrate to greater depths through water infiltration.
- The types of MPs appear to vary due to different land use and the vulnerability of groundwater systems. A high amount of microplastics (3399 particles/ per 16 Liters) was found where intensive anthropogenic activities (agricultural, industrial, and urban) take place.
- The mechanism of MPs admission into the deep (water table ~70 m) aquifer needs further investigation. Seawater intrusion, which is reflected in the qualitative characteristics of groundwater, could potentially be a source of microplastics. The extent at which seawater MP contamination interacts with other potential sources such as agricultural activities requires further in-depth geochemical investigations, to document the correlation between MPs and key chemical parameters.

This study can contribute to future databases that, supported by geochemical analyses, will link the presence of MPs to specific sources and environments.

**Author Contributions:** Conceptualization, M.P., E.V. and I.P.; methodology, M.P., E.V. and I.P.; resources: P.M., E.V., C.S., I.P. and G.S.; investigation, all; data curation, P.M., V.S. and C.S.; writing—original draft preparation, M.P. and V.S.; writing—review and editing, all; supervision, M.P., E.V. and I.P. All authors have read and agreed to the published version of the manuscript.

**Funding:** This research received no external funding.

**Data Availability Statement:** Data are contained within the article.

**Conflicts of Interest:** The authors declare no conflicts of interest.

## References

1. Cowger, W.; Gray, A.; Christiansen, S.H.; DeFrond, H.; Deshpande, A.D.; Hemabessiere, L.; Lee, E.; Mill, L.; Munno, K.; Ossmann, B.E.; et al. Critical Review of Processing and Classification Techniques for Images and Spectra in Microplastic Research. *Appl. Spectrosc.* **2020**, *74*, 989–1010. [CrossRef] [PubMed]
2. Silva, A.B.; Bastos, A.S.; Justino, C.I.L.; da Costa, J.P.; Duarte, A.C.; Rocha-Santos, T.A.P. Microplastics in the Environment: Challenges in Analytical Chemistry—A Review. *Anal. Chim. Acta* **2018**, *1017*, 1–19. [CrossRef] [PubMed]
3. Arthur, C.; Baker, J.E.; Bamford, H.A. NOAA Technical Memorandum NOS-OR&R-30. In Proceedings of the International Research Workshop on the Occurrence, Effects, and Fate of Microplastic Marine Debris, University of Washington Tacoma, Tacoma, WA, USA, 9–11 September 2008.
4. Ziani, K.; Ioniță-Mîndrican, C.-B.; Mititelu, M.; Neacșu, S.M.; Negrei, C.; Moroșan, E.; Drăgănescu, D.; Preda, O.-T. Microplastics: A Real Global Threat for Environment and Food Safety: A State of the Art Review. *Nutrients* **2023**, *15*, 617. [CrossRef]
5. An, L.; Liu, Q.; Deng, Y.; Wu, W.; Gao, Y.; Ling, W. Sources of Microplastic in the Environment. In *Microplastics in Terrestrial Environments: Emerging Contaminants and Major Challenges*; He, D., Luo, Y., Eds.; The Handbook of Environmental Chemistry; Springer International Publishing: Cham, Switzerland, 2020; pp. 143–159. [CrossRef]
6. Zhao, X.; Wang, J.; Yee Leung, K.M.; Wu, F. Color: An Important but Overlooked Factor for Plastic Photoaging and Microplastic Formation. *Environ. Sci. Technol.* **2022**, *56*, 9161–9163. [CrossRef]
7. Suaria, G.; Avio, C.G.; Mineo, A.; Lattin, G.L.; Magaldi, M.G.; Belmonte, G.; Moore, C.J.; Regoli, F.; Aliani, S. The Mediterranean Plastic Soup: Synthetic Polymers in Mediterranean Surface Waters. *Sci. Rep.* **2016**, *6*, 37551. [CrossRef]
8. Isobe, A.; Iwasaki, S.; Uchida, K.; Tokai, T. Abundance of Non-Conservative Microplastics in the Upper Ocean from 1957 to 2066. *Nat. Commun.* **2019**, *10*, 417. [CrossRef]
9. Peeken, I.; Primpke, S.; Beyer, B.; Gütermann, J.; Katlein, C.; Krumpen, T.; Bergmann, M.; Hehemann, L.; Gerdts, G. Arctic Sea Ice Is an Important Temporal Sink and Means of Transport for Microplastic. *Nat. Commun.* **2018**, *9*, 1505. [CrossRef]
10. González-Pleiter, M.; Velázquez, D.; Edo, C.; Carretero, O.; Gago, J.; Baron-Sola, Á.; Hernández, L.E.; Yousef, I.; Quesada, A.; Leganés, F.; et al. Fibers Spreading Worldwide: Microplastics and Other Anthropogenic Litter in an Arctic Freshwater Lake. *Sci. Total Environ.* **2020**, *722*, 137904. [CrossRef]
11. Nguyen, N.B.; Kim, M.-K.; Le, Q.T.; Ngo, D.N.; Zoh, K.-D.; Joo, S.-W. Spectroscopic Analysis of Microplastic Contaminants in an Urban Wastewater Treatment Plant from Seoul, South Korea. *Chemosphere* **2021**, *263*, 127812. [CrossRef]
12. Hidalgo-Ruz, V.; Gutow, L.; Thompson, R.C.; Thiel, M. Microplastics in the Marine Environment: A Review of the Methods Used for Identification and Quantification. *Environ. Sci. Technol.* **2012**, *46*, 3060–3075. [CrossRef] [PubMed]
13. Shan, J.; Zhao, J.; Liu, L.; Zhang, Y.; Wang, X.; Wu, F. A Novel Way to Rapidly Monitor Microplastics in Soil by Hyperspectral Imaging Technology and Chemometrics. *Environ. Pollut.* **2018**, *238*, 121–129. [CrossRef] [PubMed]
14. Fu, W.; Min, J.; Jiang, W.; Li, Y.; Zhang, W. Separation, Characterization and Identification of Microplastics and Nanoplastics in the Environment. *Sci. Total Environ.* **2020**, *721*, 137561. [CrossRef] [PubMed]
15. Käppler, A.; Fischer, D.; Oberbeckmann, S.; Schernewski, G.; Labrenz, M.; Eichhorn, K.-J.; Voit, B. Analysis of Environmental Microplastics by Vibrational Microspectroscopy: FTIR, Raman or Both? *Anal. Bioanal. Chem.* **2016**, *408*, 8377–8391. [CrossRef]
16. Käppler, A.; Windrich, F.; Löder, M.G.J.; Malanin, M.; Fischer, D.; Labrenz, M.; Eichhorn, K.-J.; Voit, B. Identification of Microplastics by FTIR and Raman Microscopy: A Novel Silicon Filter Substrate Opens the Important Spectral Range below 1300 $Cm^{-1}$ for FTIR Transmission Measurements. *Anal. Bioanal. Chem.* **2015**, *407*, 6791–6801. [CrossRef]
17. Lee, J.-Y.; Cha, J.; Ha, K.; Viaroli, S. Microplastic Pollution in Groundwater: A Systematic Review. *Environ. Pollut. Bioavailab.* **2024**, *36*, 2299545. [CrossRef]
18. Ma, H.; Chao, L.; Wan, H.; Zhu, Q. Microplastic Pollution in Water Systems: Characteristics and Control Methods. *Diversity* **2024**, *16*, 70. [CrossRef]
19. Li, S.; Yang, M.; Wang, H.; Jiang, Y. Adsorption of Microplastics on Aquifer Media: Effects of the Action Time, Initial Concentration, Ionic Strength, Ionic Types and Dissolved Organic Matter. *Environ. Pollut.* **2022**, *308*, 119482. [CrossRef] [PubMed]
20. Ioakeimidis, C.; Fotopoulou, K.N.; Karapanagioti, H.K.; Geraga, M.; Zeri, C.; Papathanassiou, E.; Galgani, F.; Papatheodorou, G. The Degradation Potential of PET Bottles in the Marine Environment: An ATR-FTIR Based Approach. *Sci. Rep.* **2016**, *6*, 23501. [CrossRef]
21. Tziourrou, P.; Megalovasilis, P.; Tsounia, M.; Karapanagioti, H.K. Characteristics of Microplastics on Two Beaches Affected by Different Land Uses in Salamina Island in Saronikos Gulf, East Mediterranean. *Mar. Pollut. Bull.* **2019**, *149*, 110531. [CrossRef]
22. Tsangaris, C.; Digka, N.; Valente, T.; Aguilar, A.; Borrell, A.; de Lucia, G.A.; Gambaiani, D.; Garcia-Garin, O.; Kaberi, H.; Martin, J.; et al. Using *Boops Boops* (Osteichthyes) to Assess Microplastic Ingestion in the Mediterranean Sea. *Mar. Pollut. Bull.* **2020**, *158*, 111397. [CrossRef]

23. Karkanorachaki, K.; Kiparissis, S.; Kalogerakis, G.C.; Yiantzi, E.; Psillakis, E.; Kalogerakis, N. Plastic Pellets, Meso- and Microplastics on the Coastline of Northern Crete: Distribution and Organic Pollution. *Mar. Pollut. Bull.* **2018**, *133*, 578–589. [CrossRef] [PubMed]
24. Jacobshagen, V. Geologie von Griechenland. Available online: https://www.schweizerbart.de/publications/detail/isbn/9783443110192/Geologie_von_Griechenland (accessed on 27 January 2024).
25. Katsikatsos, G. *Geology of Greece*; University of Patras: Patras, Greece, 1992.
26. Champidi, P.; Stamatis, G.; Zagana, E. Groundwater Quality Assessment and Geogenic and Anthropogenic Effect Estimation in Erasinos Basin (E. Attica). *Eur. Water* **2011**, *33*, 11–27.
27. Stamatis, G.; Lambrakis, N.; Alexakis, D.; Zagana, E. Groundwater Quality in Mesogea Basin in Eastern Attica (Greece). *Hydrol. Process.* **2006**, *20*, 2803–2818. [CrossRef]
28. Latsoudas, X. *Geological Map "Koropi-Plaka", Scale 1:50.000*; Institute of Geology and Mineral Exploration: Athens, Greece, 2003.
29. Koelmans, A.A.; Mohamed Nor, N.H.; Hermsen, E.; Kooi, M.; Mintenig, S.M.; De France, J. Microplastics in Freshwaters and Drinking Water: Critical Review and Assessment of Data Quality. *Water Res.* **2019**, *155*, 410–422. [CrossRef]
30. Kutralam-Muniasamy, G.; Shruti, V.C.; Pérez-Guevara, F.; Roy, P.D.; Elizalde-Martínez, I. Common Laboratory Reagents: Are They a Double-Edged Sword in Microplastics Research? *Sci. Total Environ.* **2023**, *875*, 162610. [CrossRef] [PubMed]
31. Merkel, B.; Friedrich, B. *Groundwater Geochemistry: A Practical Guide to Modeling of Natural and Contaminated Aquatic Systems*; Nordstorm, D.K., Ed.; Springer: Berlin/Heidelberg, Germany, 2005. [CrossRef]
32. Shruti, V.C.; Kutralam-Muniasamy, G. Blanks and Bias in Microplastic Research: Implications for Future Quality Assurance. *Trends Environ. Anal. Chem.* **2023**, *38*, e00203. [CrossRef]
33. Jin, N.; Song, Y.; Ma, R.; Li, J.; Li, G.; Zhang, D. Characterization and Identification of Microplastics Using Raman Spectroscopy Coupled with Multivariate Analysis. *Anal. Chim. Acta* **2022**, *1197*, 339519. [CrossRef]
34. Scherrer, N.C.; Stefan, Z.; Francoise, D.; Annette, F.; Renate, K. Synthetic Organic Pigments of the 20th and 21st Century Relevant to Artist's Paints: Raman Spectra Reference Collection. *Spectrochim. Acta Part A Mol. Biomol. Spectrosc.* **2009**, *73*, 505–524. [CrossRef]
35. Vedad, J.; Reilly, L.; Desamero, R.Z.B.; Mojica, E.-R.E. Quantitative Analysis of Xylene Mixtures Using a Handheld Raman Spectrometer. In *Raman Spectroscopy in the Undergraduate Curriculum*; ACS Symposium Series; American Chemical Society: Washington, DC, USA, 2018; Volume 1305, pp. 129–152. [CrossRef]
36. Lapa, H.M.; Martins, L.M.D.R.S. P-Xylene Oxidation to Terephthalic Acid: New Trends. *Molecules* **2023**, *28*, 1922. [CrossRef]
37. Zhou, X.-X.; Liu, R.; Hao, L.-T.; Liu, J.-F. Identification of Polystyrene Nanoplastics Using Surface Enhanced Raman Spectroscopy. *Talanta* **2021**, *221*, 121552. [CrossRef]
38. Kellar, E.J.C.; Galiotis, C.; Andrews, E.H. Raman Vibrational Studies of Syndiotactic Polystyrene. 1. Assignments in a Conformational/Crystallinity Sensitive Spectral Region. *Macromolecules* **1996**, *29*, 3515–3520. [CrossRef]
39. Caggiani, M.C.; Cosentino, A.; Mangone, A. Pigments Checker Version 3.0, a Handy Set for Conservation Scientists: A Free Online Raman Spectra Database. *Microchem. J.* **2016**, *129*, 123–132. [CrossRef]
40. Gall, M.J.; Hendra, P.J.; Peacock, C.J.; Cudby, M.E.A.; Willis, H.A. Laser-Raman Spectrum of Polyethylene: Part 1. Structure and Analysis of the Polymer. *Polymer* **1972**, *13*, 104–108. [CrossRef]
41. Fischer, J.; Wallner, G.M.; Pieber, A. Spectroscopical Investigation of Ski Base Materials. *Macromol. Symp.* **2008**, *265*, 28–36. [CrossRef]
42. Nava, V.; Frezzotti, M.L.; Leoni, B. Raman Spectroscopy for the Analysis of Microplastics in Aquatic Systems. *Appl. Spectrosc.* **2021**, *75*, 1341–1357. [CrossRef] [PubMed]
43. Dong, M.; Zhang, Q.; Xing, X.; Chen, W.; She, Z.; Luo, Z. Raman Spectra and Surface Changes of Microplastics Weathered under Natural Environments. *Sci. Total Environ.* **2020**, *739*, 139990. [CrossRef] [PubMed]
44. Kniggendorf, A.-K.; Wetzel, C.; Roth, B. Microplastics Detection in Streaming Tap Water with Raman Spectroscopy. *Sensors* **2019**, *19*, 1839. [CrossRef] [PubMed]
45. De Baez, M.A.; Hendra, P.J.; Judkins, M. The Raman spectra of oriented isotactic polypropylene. *Spectrochim. Acta Part A Mol. Biomol. Spectrosc.* **1995**, *51*, 2117–2124. [CrossRef]
46. Andreassen, E. Infrared and Raman Spectroscopy of Polypropylene. In *Polypropylene*; Karger-Kocsis, J., Ed.; Polymer Science and Technology Series; Springer: Dordrecht, The Netherlands, 1999; Volume 2, pp. 320–328. [CrossRef]
47. Ciera, L.; Beladjal, L.; Almeras, X.; Gheysens, T.; Van Landuyt, L.; Mertens, J.; Nierstrasz, V.; Van Langenhove, L. Morphological and Material Properties of Polyethyleneterephthalate (PET) Fibres with Spores Incorporated. *Fibres Text. East. Eur.* **2014**, *22*, 29–36.
48. Prata, J.C.; da Costa, J.P.; Duarte, A.C.; Rocha-Santos, T. Methods for Sampling and Detection of Microplastics in Water and Sediment: A Critical Review. *TrAC Trends Anal. Chem.* **2019**, *110*, 150–159. [CrossRef]
49. Martí, E.; Martin, C.; Galli, M.; Echevarría, F.; Duarte, C.M.; Cózar, A. The Colors of the Ocean Plastics. *Environ. Sci. Technol.* **2020**, *54*, 6594–6601. [CrossRef]
50. Gregory, P. Industrial Applications of Phthalocyanines. *J. Porphyr. Phthalocyanines* **2000**, *4*, 432–437. [CrossRef]
51. Leiser, R.; Jongsma, R.; Bakenhus, I.; Möckel, R.; Philipp, B.; Neu, T.R.; Wendt-Potthoff, K. Interaction of Cyanobacteria with Calcium Facilitates the Sedimentation of Microplastics in a Eutrophic Reservoir. *Water Res.* **2021**, *189*, 116582. [CrossRef] [PubMed]

52. Dodhia, M.S.; Rogers, K.L.; Fernández-Juárez, V.; Carreres-Calabuig, J.A.; Löscher, C.R.; Tisserand, A.A.; Keulen, N.; Riemann, L.; Shashoua, Y.; Posth, N.R. Microbe-Mineral Interactions in the Plastisphere: Coastal Biogeochemistry and Consequences for Degradation of Plastics. *Front. Mar. Sci.* **2023**, *10*, 1134815. [CrossRef]
53. Panno, S.V.; Kelly, W.R.; Scott, J.; Zheng, W.; McNeish, R.E.; Holm, N.; Hoellein, T.J.; Baranski, E.L. Microplastic Contamination in Karst Groundwater Systems. *Groundwater* **2019**, *57*, 189–196. [CrossRef] [PubMed]
54. Severini, E.; Ducci, L.; Sutti, A.; Robottom, S.; Sutti, S.; Celico, F. River–Groundwater Interaction and Recharge Effects on Microplastics Contamination of Groundwater in Confined Alluvial Aquifers. *Water* **2022**, *14*, 1913. [CrossRef]
55. Gong, X.; Tian, L.; Wang, P.; Wang, Z.; Zeng, L.; Hu, J. Microplastic Pollution in the Groundwater under a Bedrock Island in the South China Sea. *Environ. Res.* **2023**, *239*, 117277. [CrossRef] [PubMed]
56. Ye, S.; Pei, D. Relationships between Microplastic Pollution and Land Use in the Chongqing Section of the Yangtze River. *Front. Ecol. Evol.* **2023**, *11*, 1202562. [CrossRef]
57. Wang, F.; Shih, K.M.; Li, X.Y. The Partition Behavior of Perfluorooctanesulfonate (PFOS) and Perfluorooctanesulfonamide (FOSA) on Microplastics. *Chemosphere* **2015**, *119*, 841–847. [CrossRef]
58. Shockley, D.J.; Lapworth, D.J. Microplastics in Groundwater: A Literature Review. Available online: https://nora.nerc.ac.uk/id/eprint/532669/ (accessed on 27 January 2024).
59. Alvarado-Zambrano, D.; Rivera-Hernández, J.R.; Green-Ruiz, C. First Insight into Microplastic Groundwater Pollution in Latin America: The Case of a Coastal Aquifer in Northwest Mexico. *Environ. Sci. Pollut. Res.* **2023**, *30*, 73600–73611. [CrossRef]
60. Sangkham, S.; Aminul Islam, M.d.; Adhikari, S.; Kumar, R.; Sharma, P.; Sakunkoo, P.; Bhattacharya, P.; Tiwari, A. Evidence of Microplastics in Groundwater: A Growing Risk for Human Health. *Groundw. Sustain. Dev.* **2023**, *23*, 100981. [CrossRef]
61. Mu, H.; Wang, Y.; Zhang, H.; Guo, F.; Li, A.; Zhang, S.; Liu, S.; Liu, T. High Abundance of Microplastics in Groundwater in Jiaodong Peninsula, China. *Sci. Total Environ.* **2022**, *839*, 156318. [CrossRef] [PubMed]

*water*

MDPI

*Article*

# Recharge Assessment in Greek Karst Systems: Methodological Considerations and Implications

**Eleni-Anna Nanou *, Konstantinos Perdikaris, Panagiotis Tserolas and Eleni Zagana**

Laboratory of Hydrogeology, Department of Geology, Faculty of Natural Sciences, University of Patras, Rion, 26504 Patras, Greece; zagana@upatras.gr (E.Z.)
* Correspondence: e.nanou@ac.upatras.gr

**Abstract:** The recharge of karst aquifers is significant for the effective management of groundwater resources, and its estimation should be tailored to accommodate their specific hydrogeological characteristics. This study provides a two-step methodological approach for the determination of karst aquifer recharge. Initially, mean annual recharge rates were estimated in the karst system of Ziria (Southern Greece) utilizing the APLIS and modified APLIS methods in order to decipher which was the most suitable version for recharge assessments. The results indicated similar mean recharge rate values at 42.7% and 41.4%, respectively, but significant differences in the spatial distribution. The modified methodology emerged as a more accurate and realistic approach, mainly due to the incorporation of permeability assessments. The final phase of the methodological approach involved the application of modified APLIS in two additional karst hydrosystems, Planitero and Xiromero, while a quantitative cross-comparison of the recharge rates was obtained for a deeper understanding of the factors controlling the groundwater recharge process. In Ziria, recharge rates exhibit a relatively uniform distribution throughout the area, with a median value of 46.7%. Conversely, in Planitero, High recharge rates (60–80%) occupy 56.8% of the surface, while in Xiromero, Moderate recharge rates (40–60%) dominate, representing 53.4% of the land coverage. These variations underscore the spatial heterogeneity of recharge within the karst systems, highlighting the importance of considering local geological and hydrological conditions in its assessments. The methodological approach of this study is flexible and can be adapted to different karst sites for the determination of recharge regimes, contributing to the alleviation of the groundwater depletion issue.

**Keywords:** karst; groundwater recharge; geographic information systems; APLIS method

## 1. Introduction

Groundwater resources play an immense role in today's society [1,2], since the universal demand for water supply is crucially increasing. The achievement of sustainable groundwater management is a pressing and often addressed issue due to its critical role in meeting basic human requirements [3]. The escalating need for freshwater, driven by global population growth and rapid urbanization, can cause groundwater reservoirs to face over-extraction and depletion, compromising their sustainability. Simultaneously, the issue of groundwater contamination has become extensive, with pollutants from agricultural runoff, industrial discharges, and improper waste disposal infiltrating aquifers.

Besides the ramifications that can arise from human activities, climatic causes may burden aquifers further, with dire impacts on their resources. In an era dominated by the pervasive effects of climate change, Earth's hydrological cycle faces unprecedented challenges with shifts in temperature regimes, alterations in precipitation patterns, and intensified extreme weather events. These climate-induced changes can cause direct and indirect repercussions on global reserves [4], disrupt traditional recharge mechanisms, and intensify water scarcity. The increasing occurrence of droughts may lead to changes in the water balance, while rising global temperatures contribute to heightened evaporation rates, further amplifying water scarcity in vulnerable regions. The challenges facing groundwater resources have reached a critical juncture, potentially influencing their global distribution in the future. This underscores the imperative to address numerous water-related issues in a sustainable manner. Groundwater depletion constitutes a universal concern due to the unbalanced abstractions in comparison to the natural recharge. Obviously, the study of groundwater recharge is quite challenging, especially in karst aquifers.

At the forefront of hydrological environments, karst aquifers provide high-quality groundwater that is utilized either for drinking water supply or irrigation purposes. Climate-induced shifts can cause severe repercussions, leading to both a degradation in the availability and a compromise in the quality of water resources [5–8]. At the same time, with distinct characteristics and processes, karst aquifers are able to provide substantial amounts of groundwater, often resulting in their overexploitation [9,10], in order to meet the water demands of the constantly increasing population.

In Greece, there are numerous carbonate massifs distinguished by a diverse array of geological, geomorphological, and hydrogeological features, leading to a substantial variation in the degree of karstification across the country. Over the last decade, increasing research efforts have been directed towards the characterization, protection, and sustainable management of native karst aquifers, highlighting their significant value as water sources in Greece [11]. Up to 80% of the domestic karst systems are defined by good groundwater quality and quantity conditions, while qualitative issues are recorded mainly in coastal aquifers on islands, represented by a percentage of 5% [12].

The viable management of groundwater use requires the aversion of resource depletion. Recharge is one of the most significant parameters to be assessed in order to sustain the overall effective function of an exploited aquifer. Its approximation is essential to evaluating the maximum volume of water obtainable and to avoid overutilization [13]. Stable isotopes [14–17], lump and spatial models (e.g., Karstmod, VarKarst, Modflow-CFP), as well as index-based methods (e.g., APLIS, KARSTLOP), are widely used in order to determine the recharge process in karst environments [11,18–24]. Due to the heterogeneity of porosity and permeability that transcends karst aquifers [25,26] and significantly affects many of their qualities, the quantification of recharge constitutes a challenging task. The APLIS method was developed by Andreo et al. [27] and later modified by Marin [28] for the estimation of the recharge rate, expressed as a percentage of precipitation. Its parametric methodology is based on a geographic information system (GIS) and was developed specifically for carbonate aquifers under Mediterranean climatic conditions. APLIS has been implemented as well as compared to several other methods by numerous researchers [29–32], while many modifications have been made since to accommodate varying circumstances [33–35].

In this research, we applied a two-step-based methodological approach. Initially, the efficiency of APLIS and its modified version were compared in the karst system of Ziria in Southern Greece. The quantitative comparison of the two methods showed that modified APLIS is more suitable for Greek karst systems. The next step was the application of modified APLIS at two additional karst sites, Planitero and Xiromero, located in Southern and Western Greece, respectively. A comprehensive examination of the method's performance was provided across distinct hydrogeological settings. The outcomes from all three study sites underwent statistical analysis in order to decipher the hydrogeological factors that influence the recharge process. To the best of our knowledge, this methodological approach was initially applied within this manuscript. Considering that groundwater recharge constitutes the most important parameter in the fight against groundwater depletion, this study contributes to the quantification and deeper understanding of the infiltration regime in karst hydrosystems.

## 2. Materials and Methods

### 2.1. Background Information and Geological Overview of the Study Sites

In the frame of this research, the recharge rate was estimated at three Greek karst sites. Similarities and differences are both present between the study areas in regards to the geological-topographical setting and the precipitation regime. From east to west, the aforementioned sites are: (a) Ziria, (b) Planitero, and (c) Xiromero.

The karst groundwater system of Ziria (EL0200220) was delineated in the framework of River Basin Management Plans [36] that were drafted according to the requirements of the 2000/60/EC Directive [37]. It is located in Northern Peloponnese, Southern Greece, approximately 45 km west of the city of Corinth, and covers an area of roughly 200 km$^2$. The aquifers of the system are formed within a large mountainous area of an almost circular extension known as Mount Killini or Ziria (2368 m.a.s.l.). A large karstic polje with an orientation of almost E-W is situated in the southern part of the study area, within which Stymphalia Lake was developed. The topography is rough, with often steep slopes, since the mountain occupies a considerable extent of the hydrological system. Ziria is primarily composed of Upper Triassic–Cretaceous and undifferentiated limestones and dolomites (Tripolis Zone) along with Cretaceous limestones (Olonos–Pindos nappe) that were thrust over the latter (Figure 1) [38,39]. The metamorphic bedrock of the area consists of quartzites, phyllites, and schists that mainly surface in the SW and NW regions. A substantial part of the karst system is covered by Quaternary deposits, including the polje, which, according to geoelectrical investigations [40], is mostly filled by impermeable clays.

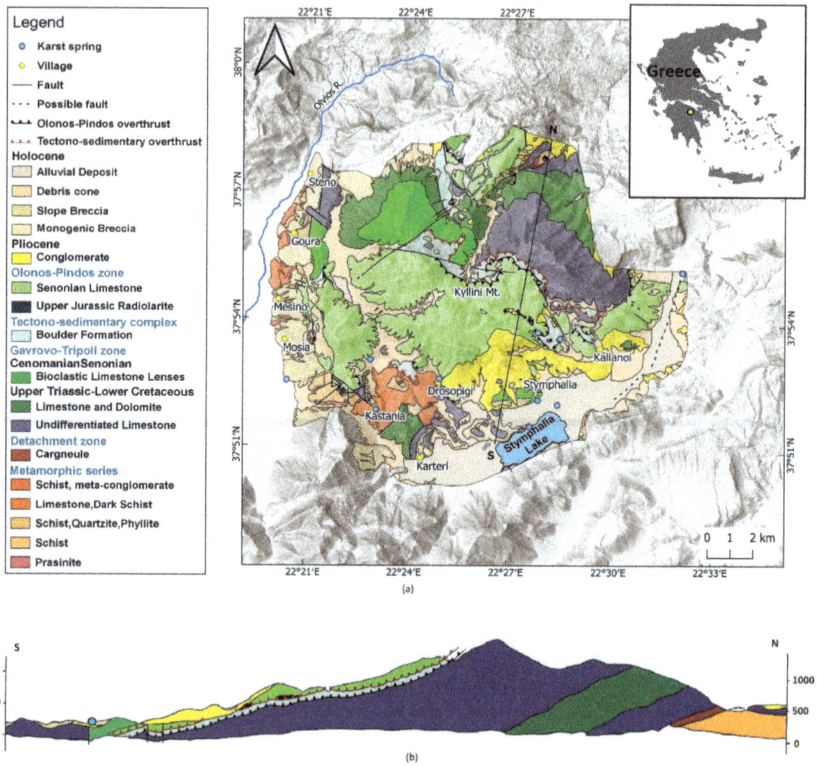


**Figure 1.** (**a**) Geological setting of the karst groundwater system of Ziria (modified by [38,39]), (**b**) Geological cross section (modified by [38]).

The karstification degree is prominent with the presence of many karstic landforms, such as dolines, swallow holes, and caves, while the recharge of the aquifers occurs in both diffuse and concentrated forms. The front of Stymphalia springs discharges one of the largest aquifers in the area and provides the drinking water supply for the cities of Corinth and Kiato. The groundwater of the system is also heavily used for irrigation purposes since large sections of the polje and the adjacent areas are being cultivated.

The aquifers of Planitero manifest within the Aroania (Chelmos) Mountains, which are located approximately 20 km west of Mount Ziria, so, due to their close proximity, the geological setting is very similar. This site constitutes a part of the much larger Ladon system (EL0100030), severed in order to include the recharge area of Planitero springs. It occupies 112 km$^2$ of land, and it is mainly composed of karstified carbonate rocks (Tripolis and Pindos zones) (Figure 2A) [38,41,42]. A considerable segment of the site is covered by Quaternary deposits, while there are outcrops of flysch and metamorphic bedrock. The aquifers discharge from a front of springs that supplement the Aroanios River. The presence of multiple faults and overthrusts makes the tectonic stress of the region indisputable. The resemblance between Ziria and Planitero is also evident in the steep terrain, with abrupt altitudinal inclinations and high elevations over 2300 m.a.s.l.

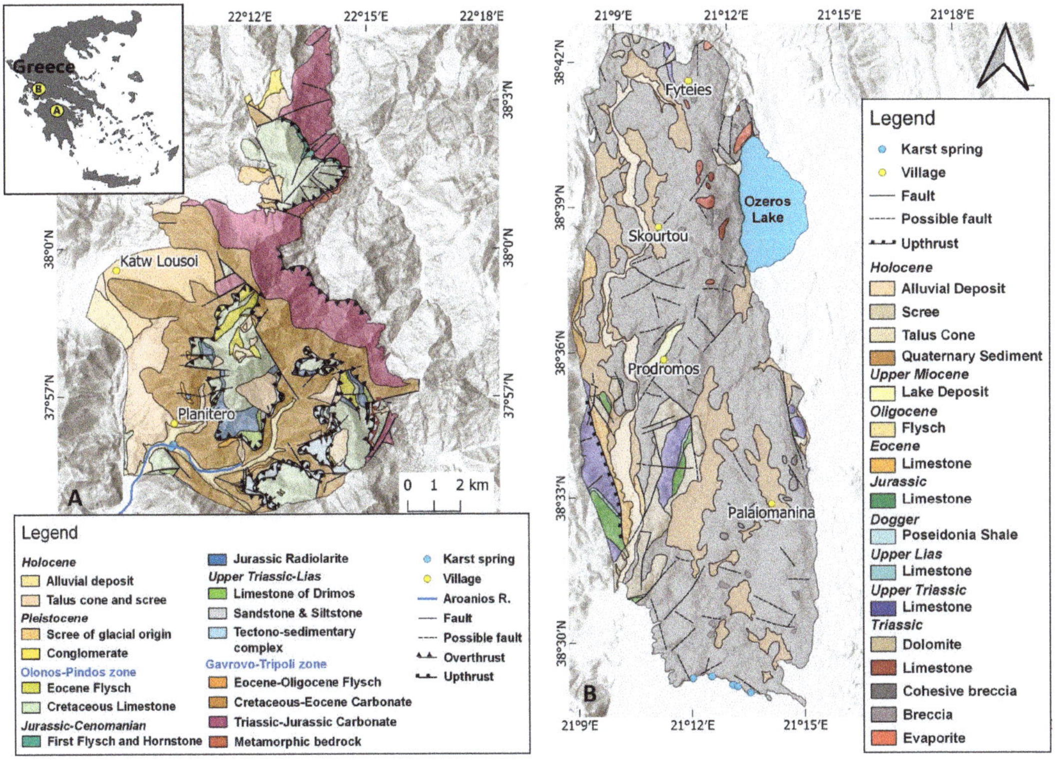


**Figure 2.** Geological setting of (**A**) Planitero (modified by [38,41,42]) and (**B**) Xiromero (modified by [43,44]).

Xiromero (approximately 170 $km^2$) is located in Western Greece and presents many dissimilarities compared to the other study sites. Like Planitero, it is also a delimited section of the Katouna–Lesini system (EL0400050), due to the emersion of evaporites in the northern part of the area, near Fyteies, functioning as a hydrological barrier. The aquifers are formed within the karstified Triassic breccia that prevail in the area and were developed by the fragmentation of adjacent Triassic limestones and dolomites (Figure 2B). They discharge from the Lambra–Agios Dimitrios front of springs, providing groundwater to accommodate the needs of adjacent villages. Quaternary deposits consisting mainly of clays and sandy materials, along with small outcrops of various limestones, complete the geological setting of the area [43,44]. Compared to the other study sites, the topography is much gentler, with altitudes not exceeding 510 m.a.s.l. Due to its location, Xiromero is also defined by greater rainfall amounts, since the precipitation pattern in Greece shows a general latitudinal decrease from west to east [45].

### 2.2. Application of the APLIS Method

The initial APLIS method [27] was applied in the karst system of Ziria, followed by the implementation of its modified version [28] in order to obtain a cross-comparison between the two adaptations. Modified APLIS was subsequently applied in the Planitero system, while in Xiromero the method was updated from Zagana et al. [29], with karst landforms emerging from remote sensing techniques derived from a more recent study [46].

In order to express recharge in mm/year, the percentage (%) maps resulting from modified APLIS were multiplied with the spatial distribution of precipitation in each study site. Precipitation data corresponding to time series spanning 30 years were collected from meteorological stations in the vicinity of the sites. The methodology employed for the spatial analysis considered the variation in precipitation according to elevation.

#### 2.2.1. Assessing Recharge with APLIS and Modified APLIS in the Karst System of Ziria

APLIS estimates the mean annual recharge rate of carbonate aquifers. The assessed recharge originates solely from precipitation, and it is expressed as its percentage. The methodology behind APLIS also enables the acquisition of maps with the spatial distribution of the estimated recharge rates. To achieve that, the method proposes the combination of individual intrinsic variables that can potentially influence recharge in karst aquifers, such as Altitude (A), Slope (P), Lithology (L), Infiltration landforms (I), and Soil type (S) [27]. The APLIS method was later modified by Marín [28], who introduced refinements such as the reclassification of slopes, adjustments to infiltration scorings, and the incorporation of a correction factor (Fh) that is associated with the permeability of the formation outcropping on the surface. To apply the method, the prior variables were developed as information layers within a Geographic Information System (GIS). Scoring values, ranging from 1 (minimum) to 10 (maximum), were then assigned according to the influence on recharge. All the map layers that were developed for the application of APLIS and modified APLIS in Ziria are depicted in Figures 3 and 4.

The altitude map (A) was derived from a Digital Elevation Model (DEM) of the country, which in turn originated the slope map (P) in percent rise as required by the method. The elevation in the study area ranges from 600 to 2368 m.a.s.l., and the altitude was classified in 6 sequences of 300 m intervals. The slopes span from 0%, mainly in the polje of Stymphalia, to over 100% on the steep inclinations of Mount Ziria. The scoring of the first two variables was attributed so that the higher the altitude, the greater the recharge due to the increase in precipitation, and the greater the slope, the lower the recharge into the aquifer [27] (Table 1). In both scenarios, the altitude map and its corresponding values remain consistent. However, the classification and scoring of the slopes differ in APLIS and its modified version, with the latter presenting an additional class.

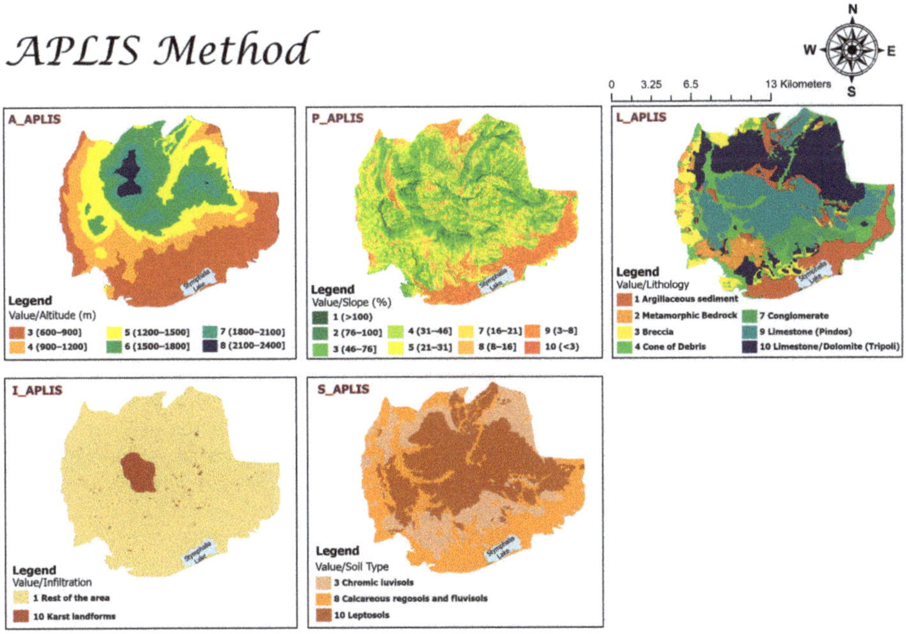


**Figure 3.** Maps of all the variables developed for the application of the APLIS method.

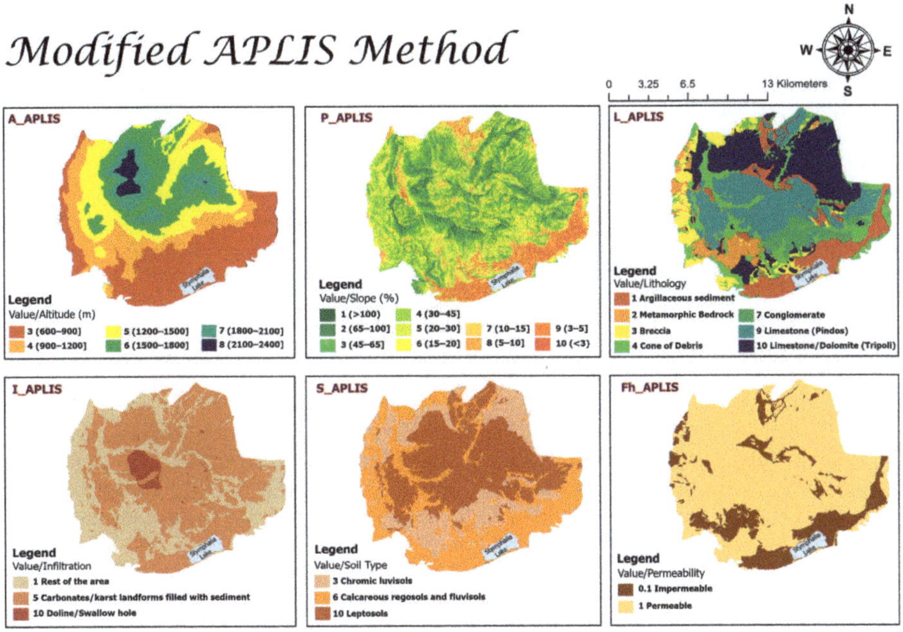


**Figure 4.** Maps of all the variables developed for the application of modified APLIS.

**Table 1.** Scoring of the Altitude (A) and Slope (P) variables for the karst system of Ziria.

| Scoring | 1 | 2 | 3 | 4 | 5 | 6 | 7 | 8 | 9 | 10 |
|---|---|---|---|---|---|---|---|---|---|---|
| Altitude (m) | - | - | (600–900] | (900–1200] | (1200–1500] | (1500–1800] | (1800–2100] | (2100–2400] | - | - |
| Slope (%) | | | | | | | | | | |
| *APLIS* | >100 | (76–100] | (46–76] | (31–46] | (21–31] | - | (16–21] | (8–16] | (3–8] | <3 |
| *Modified APLIS* | >100 | (65–100] | (45–65] | (30–45] | (20–30] | (15–20] | (10–15] | (5–10] | (3–5] | <3 |

Considering the lithological aspect (L), there are seven major units dominating the site that all received scoring based on imputes from previous studies [47,48] and field observations. All the carbonates received high scores due to their karstification degree and tectonic stress. The limestones of Tripolis are considered to be more karstified than the Cretaceous limestones of Pindos because the latter often intervene with thin layers of radiolarite [48], therefore they were attributed the values 10 and 9, respectively. The Quaternary formations exhibit fluctuations due to changes in grain size and composition, along with the presence of argillaceous materials. The alluvial deposits in the polje, along with the outcrop at the north-eastern part of the area, received lower values than the rest. Finally, the conglomerates were attributed a significantly higher score than what is proposed by APLIS since they are considered to be karstic [49].

The areas of preferential infiltration (I) were located with remote sensing techniques, using aerial photographs and stereo pairs in Erdas Imagine 2010 software that enabled the three-dimensional representation of the system. Predominantly, these landforms were observed on the limestones of the Mountain, where the dominance of bare rocks and the absence of vegetation facilitated the clear delineation of such topographic depressions. On the Cretaceous limestones of Pindos, fields of dolines can be visually distinguished through satellite images and photographs, with no further assistance from remote sensing software. These landforms were collectively grouped rather than individually digitized, occupying, in such a way, a significant section of the limestones. Regarding this variable, APLIS provides only two rankings: a score of 10, which was attributed to the preferential infiltration landforms, and a score of 1, which was assigned to the rest of the system. Karst landforms that are filled with Quaternary sediments were assigned the lowest value (1), due to the method's limited scoring options. In modified APLIS, an additional intermediate value of 5 was introduced. It was received by the polje and other karstic features that are covered by various materials, as well as attributed to all the carbonate outcrops, in order to be distinguished from the non-carbonate formations.

Due to the lack of specific soil data, the evaluation of the soil types (S) was based on the combination of information provided by different Soil maps [50–52]. A substantial section of the Mountain is characterized by exposed rock, devoid of soil and vegetation. Leptosols, assigned the highest value of 10, were designated as the soil type for limestone areas exhibiting minimal to no vegetation. Chromic luvisols were attributed to formations with denser vegetation, while calcareous regosols and fluvisols were predominantly assigned to Quaternary sediments. There are minimum scoring differences between the two methods.

For the estimation of the correction factor in modified APLIS, permeable formations such as carbonates, conglomerates, and most of the Quaternary deposits were identified with aquifer potential and were assigned the correction coefficient (Fh) 1. Impermeable formations like the metamorphic series and argillaceous sediments were assigned the value 0.1.

The reliability of the used data for the application of the methods is provided in the Supplementary Materials.

The scorings that were attributed to the rest of the variables for the application of modified APLIS are displayed in Table 2. For the estimation of the annual recharge rate and the creation of maps with its spatial distribution in both scenarios, the expressions of the original (1) and modified method (2) were utilized in ArcGIS 10.8 software:

$$R = (A + P + 3 \times L + 2 \times I + S)/0.9 \tag{1}$$

$$R = [(A + P + 3 \times L + 2 \times I + S)/0.9] \times Fh. \tag{2}$$

**Table 2.** Scoring values of the variables L, I, and S for the application of modified APLIS in Ziria.

| Lithology (L) | *Scoring* | Infiltration Landforms (I) | *Scoring* |
|---|---|---|---|
| Argillaceous sediment | 1 | Rest of the area | 1 |
| Metamorphic bedrock | 2 | Carbonates/karstic landforms filled with sediment | 5 |
| Breccia | 3 | Doline/Swallow hole | 10 |
| Cone of Debris | 4 | **Soil Type (S)** | ***Scoring*** |
| Conglomerate | 7 | Chromic luvisols | 3 |
| Cretaceous Limestone (Pindos) | 9 | Calcareous regosols and fluvisols | 6 |
| Limestone/Dolomite (Tripoli) | 10 | Leptosols | 10 |

2.2.2. Application of Modified APLIS to Planitero and Xiromero

The neighboring Planitero system presents many similarities to Ziria, and the values that were assigned to each variable are depicted in Table 3, while the scoring of the first two parameters (A and P) corresponds to Table 1.

**Table 3.** Assigned values to the variables of the modified APLIS for Planitero.

| Lithology (L) | Scoring | Infiltration Landforms (I) | Scoring |
|---|---|---|---|
| Flysch/Argillaceous sediment | 1 | Rest of the area | 1 |
| Metamorphic bedrock | 2 | Carbonates/karstic landforms filled with sediment | 5 |
| Quaternary deposits | 3 | Doline | 10 |
| Talus Cone/Scree | 4 | **Soil Type (S)** | **Scoring** |
| | | Chromic luvisols | 3 |
| Carbonates (Pindos Zone) | 9 | Calcareous regosols and fluvisols | 6 |
| Carbonates (Tripolis Zone) | 10 | Leptosols | 10 |

The application of modified APLIS in Xiromero is thoroughly described in [29], while Table 4 presents the updated values that were attributed to each variable.

**Table 4.** Assigned values to the variables of modified APLIS for Xiromero—modified from [29].

| Lithology (L) | Scoring | Infiltration Landforms (I) | Scoring |
|---|---|---|---|
| Quaternary Sediment | 1 | Rest of the area | 1 |
| Evaporites/non karstified limestones | 2 | Karstified Limestone/Breccia | 5 |
| Talus Cone/Scree | 3 | Doline | 10 |
| Alluvial Deposit | 4 | **Soil Type (S)** | **Scoring** |
| | | Fluvisols | 6 |
| Dolomite | 7 | Calcareous Lithosols | 6 |
| Triassic Limestone/Breccia | 9 | Leptosols | 10 |

## 3. Results and Discussion

### 3.1. Comparison of the Results That Each Version of the APLIS Method Provided in the Karst System of Ziria

The spatial distribution of recharge rates in Ziria reveals significant disparities when assessed by the original and modified methodologies. Considering the first case (Figure 5a), the recharge rates in the area are predominantly Low (20–40%) and Moderate (40–60%), while those identified as High (60–80%) and Very High (>80%) are in correlation with preferential infiltration zones or the presence of flattened slopes in carbonates. The annual recharge rate estimated by APLIS ranges from 16.6% to 85.5%, with a mean annual value of 42.7%.

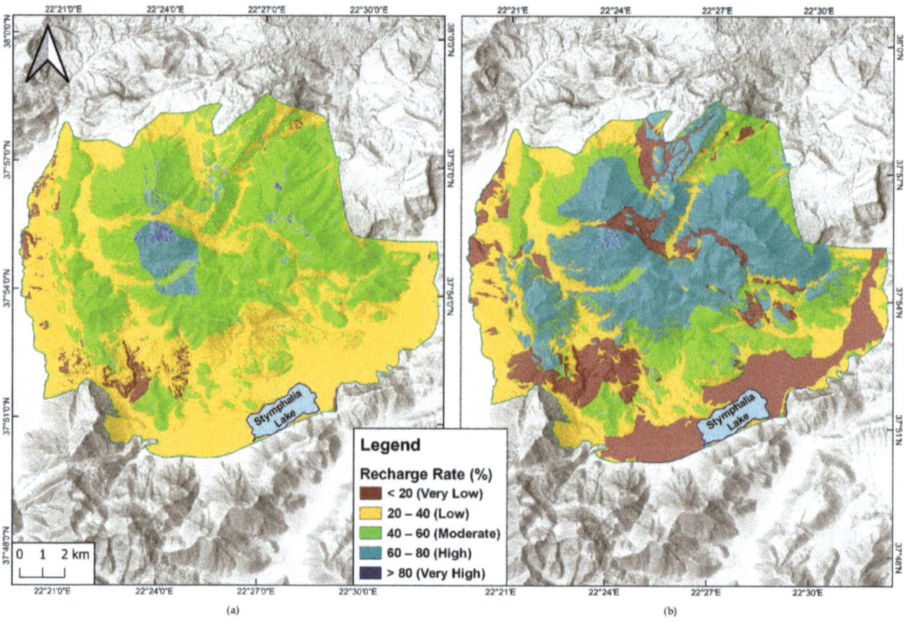


**Figure 5.** Spatial distribution of the annual recharge rate (%) in the karst system of Ziria utilizing (**a**) the APLIS method and (**b**) modified APLIS.

In modified APLIS (Figure 5b), the spatial distribution of recharge has been significantly impacted by the introduction of the correction factor (Fh). Notably, the recharge rates in a substantial portion of the area, encompassing the polje and other outcrops of low-permeability materials, have been categorized as Very Low (<20%). Furthermore, there is a noteworthy shift in the recharge classification for carbonate outcrops in Mount Ziria, elevating them from the previous Moderate class (40–60%) to High recharge rates (60–80%) due to the additional value (5) in the preferential infiltration parameter that was attributed to all the carbonates. A similar upgrading is observed for conglomerates, transitioning from Low (20–40%) to Moderate (40–60%) recharge rates. Despite the significant divergence in spatial distribution compared to APLIS, the simultaneous expansion of formations with Very Low (<20%) and High (60–80%) recharge rates has resulted in a marginal alteration of the mean annual recharge rate, which has slightly decreased to 41.4%. With the modified version of APLIS, the recharge rate in the study area ranges from 1.6% to 85.5%.

The spatial distribution of the recharge rate resulting from both methods seems to be in unison for a significant part of the system classified as Low and Moderate, mainly referring to some of the Quaternary deposits and sections of the carbonates.

In a hydrological balance study carried out in the area, the infiltration coefficient was estimated at 51% for the carbonates of Mount Ziria, 35% for the conglomerates, 15% for the Quaternary deposits, and 2% for the tectono-sedimentary complex [53]. According to Röckel and Hötzl [49], on the other hand, the infiltration was roughly estimated at 50% for limestones and conglomerates since both of them are considered karstified, 20% for Quaternary gravels, and 4% for clays and marls. It is also suggested that, especially for the carbonate outcrops on Mount Ziria, 60% of infiltration can be expected because of the dominance of bare karst on vast areas.

The inputs of these prior studies align more closely with the outcomes derived from modified APLIS, where, first and foremost, due to the correction factor (Fh), the areas of very low permeability and recharge below 8% can be recognized, contrary to the original methodology. In the alluvial deposits of the polje, along with the tectono-sedimentary complex, the recharge rates span between 1.6% and 3.5%. The percentages that are essentially identified in the conglomerates are in the range of 44–52%, while for the carbonates of Mount Ziria, they are 51–70%. In contrast, APLIS indicates different recharge rates for the same formations: approximately 16–28% for the alluvial deposits in the polje, 24–33% for the tectono-sedimentary complex, 35–42% for the conglomerates, and 44–64% for the limestones and dolomites. The highest percentage of recharge in both methods is 85%, but the range of 70% to 85% primarily refers to specific karstic landforms.

Data distribution and statistics contributed to the comparison between the two outcomes (Figure 6). In APLIS, the median recharge value is 38.9%, with 25% of the data falling below the threshold of 31.1% (representing the first quartile), while 75% of the data falls below the rate of 53.3% (corresponding to the third quartile). In modified APLIS, the median and first quartile values are 46.7% and 27.8%, respectively, while the third quartile stands at 62.2% (Figure 6a). The distribution of data within each recharge class appears to be more evenly distributed in the modified methodology. Conversely, in the original APLIS method, there is a noticeable deviation between the classes (Figure 6b). Low (20–40%) and Moderate (4060%) recharge rates dominate the area, collectively covering nearly 94% of the karstic system.

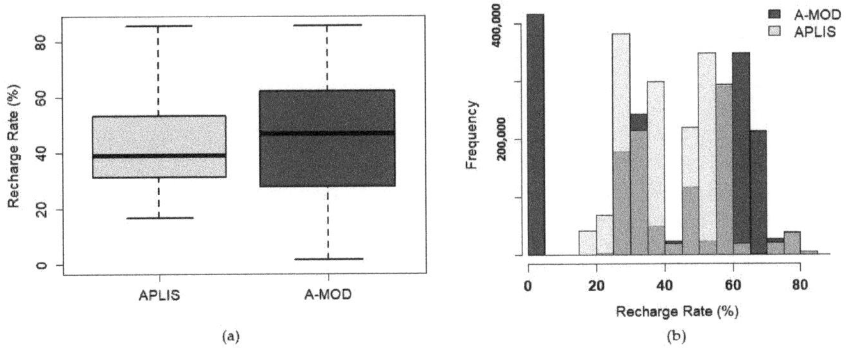


**Figure 6.** (**a**) Data visualization and comparison with boxplots; (**b**) Histogram with data distribution and range of values (A-MOD represents modified APLISand the light green color portrays areas of overlap).

In the modified methodology, High recharge rates (60–80%) appear to be the most prevalent, occupying 63 km$^2$ of carbonate sediments. There is a substantial difference between the methods regarding the Very Low (<20%) recharge class, where the variation in the two data volumes translates to 2.1% (APLIS) and 21% (modified) with respect to area coverage. In both methods, Very High (>80%) recharge rates are the least widespread, taking up an area of 0.5 and 0.25 km$^2$, respectively. Since the scoring of each variable corresponding to the karstic landforms of the mountain that primarily exhibit such a recharge class is exactly the same, the differentiation of the occupied surface between the methods is most likely due to the diverse classification of the slope factor (P).

To conclude, the modifications made to the APLIS method have significantly enhanced the accuracy of recharge rate estimations, leading many researchers to adopt the adapted methodology for their assessments. The attribution of additional values and the reclassification of specific variables have proven to be pivotal in shaping the final outcome. Modified APLIS distinguishes the infiltration landforms in three categories (1, 5, and 10), in contrast to the initial method, which has only two categories (1 and 10). Consequently, the category with a score of 5 is more representative for many formations of the studied hydrosystem compared to the initial score of 1. The greater extent of the lowest recharge rates is due to the correction factor (Fh) that was introduced in the modified methodology and empowers users to provide a more precise permeability assessment for the geological formations. Inevitably, in the final map, extreme values of recharge rates (e.g., <20% and 60–80%) cover a greater area. Consequently, the mean annual value of the recharge rate estimated by the modified APLIS is lower than the initial method. The permeability factor has notably addressed the limitation of the original method, which could not estimate recharge rates below 8%. In non-permeable formations, APLIS estimated recharge rates between 16% and 33%, as it relied solely on the minimum attributed values of the lithological parameter. High elevations, flattened slopes, and thin soil coverage, however, can increase the recharge rate of an impermeable formation that receives high scores in all three categories.

Other researchers have also compared the two adaptations of the method. Espinoza et al. [34] applied APLIS and its modified version to a fractured karst aquifer in Peru, estimating the mean annual recharge rates at 48% and 24%, respectively, and emphasizing the key role of the correction factor in this differentiation. In another study, AHP, APLIS, and modified APLIS were implemented to evaluate the infiltration potential of the Roein Esfarayen Basin in Iran [54], with modified APLIS demonstrating the highest correlation coefficient (0.85) among the three methods, underscoring its efficacy in capturing and describing the complexities of recharge dynamics. Overall, the methodology behind modified APLIS consistently produces a more coherent and realistic portrayal of the recharge rate.

### *3.2. Statistical and Spatial Comparison of the Annual Recharge Rates in All Three Karst Sites Utilizing Modified APLIS*

The estimation of the annual recharge rate was conducted using modified APLIS in both Planitero and Xiromero, resulting in maps illustrating its spatial distribution (as shown in Figure 7a,b).

In Planitero, recharge rates exhibit a broad range, spanning from 1.3% to 86.6%, with a mean value of 48.6%. Notably, the site is primarily composed of carbonate formations, which accounts for the prevalence of High recharge rates (60–80%) covering 62.7 $km^2$ of the total area (Figure 8a). A very small percentage, approximately 0.007%, is classified as Very High (>80%), mainly associated with infiltration landforms. Very Low (<20%) and Moderate (40–60%) recharge rates are nearly equally represented, encompassing 14.3 $km^2$ and 13.4 $km^2$ of the system, respectively.

The annual recharge rate in Xiromero ranges from 1.4% to 76.6%, with a mean annual value of 46.9%. It stands as the only site among the three that does not exhibit recharge rates that exceed 80% (Figure 7b). This differentiation can be attributed to the significantly lower elevations in the area, along with the absence of the highest lithological scoring (10). The dominant recharge class in Xiromero is Moderate (40–60%), occupying more than half of the entire surface (Figure 8a), and it is primarily associated with karstified breccia and limestone outcrops. High recharge rates (60–80%) correspond to 21.9% of the system, with a higher prevalence of this recharge class observed in both Ziria and Planitero.

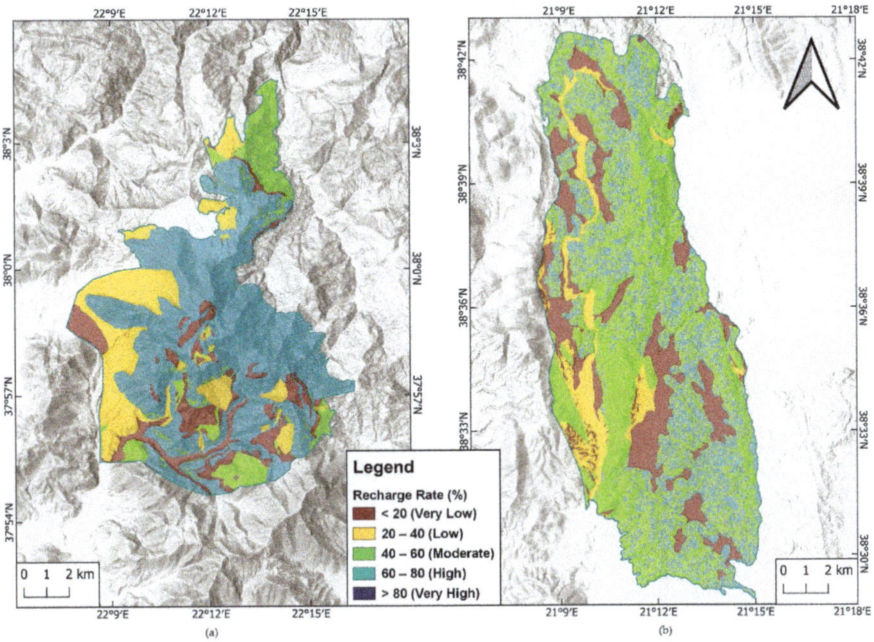


**Figure 7.** Annual recharge rate estimated with modified APLIS (**a**) in Planitero, Northern Peloponnese, and (**b**) in Xiromero, Western Greece.

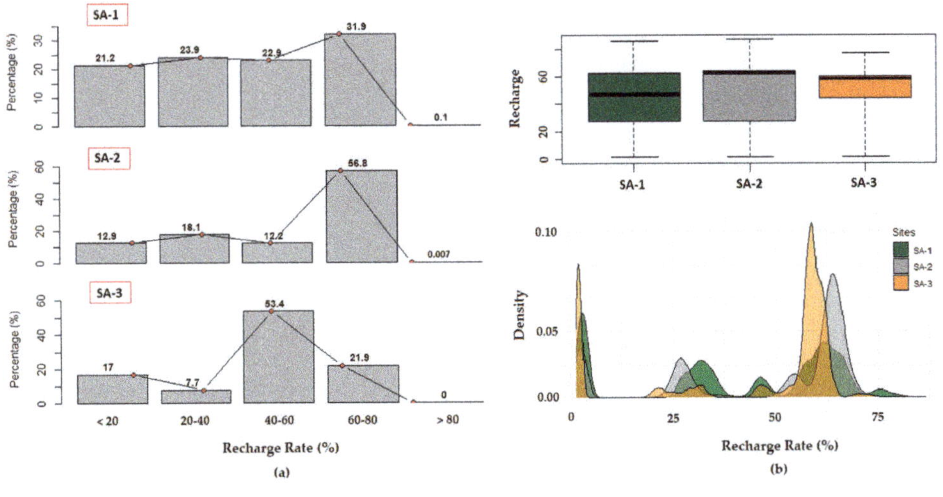


**Figure 8.** (**a**) Surface coverage of each recharge class in Ziria (SA-1), Planitero (SA-2), and Xiromero (SA-3), and (**b**) Boxplots illustrating the distribution of the annual recharge rate data in SA-1, SA-2, and SA-3 (**top**) and a density plot (**bottom**) showing its allocation between the sites.

A comparative analysis of all three sites (Figure 8a,b) reveals noteworthy distinctions in the annual recharge rates. In Ziria, apart from the Very High Class (>80%), recharge rates are much more evenly distributed throughout the area (median value 46.7%), since it is the only site that is not dominated by a specific geological formation. While Ziria and Planitero share several similarities, the range of recharge values in the latter is broader. The median and third quartile values are 62.2% and 64.4%, respectively, suggesting a tendency towards higher recharge values (Figure 8b). In contrast, the range of recharge rates in Xiromero is comparatively narrower, with the median value being 57.8% and the third quartile being 60%. These variations underscore the unique hydrogeological characteristics and geological formations of each system, which influence the distribution of recharge rates.

The quantification of recharge in mm/year was accomplished by multiplying the spatial distribution of the assessed recharge rates (%) with precipitation (mm/year). Due to the close proximity of Ziria and Planitero, the same equation was utilized in order to convert contours into isohyets lines. In Ziria, recharge ranges from 13.7 to 1423.4 mm/year with a mean annual value of 498 mm/year, while in Planitero, due to the dominance of limestones and dolomites, the mean annual recharge is 618 mm/year. Xiromero does not exhibit Very High recharge rates (>80%), and the elevations in the area are considerably lower than in the other systems. However, due to the different precipitation regime in Western Greece, recharge in Xiromero spans from 13 to 852.5 mm/year, and the mean annual value is 439 mm/year.

Several methods and models have been developed over the years for the determination of aquifer recharge. The methodology behind APLIS provides reasonable results with an easy GIS-based application without requiring data that is difficult to acquire or not available, unlike many other approaches. It also enables the determination of the spatial distribution of recharge, something that separates it from many other methods. Its methodology has been widely used, so if it is necessary to have the spatial distribution of mean annual recharge rates, APLIS constitutes the most suitable method.

### 3.3. *Implications/Limitations and Future Works*

The estimation of recharge is a very intricate subject for karst environments due to their hydrogeological uniqueness. APLIS has yielded valuable insights, but it is essential to acknowledge the implications of this approach. The accuracy of the method heavily depends on the quality of input data and the attribution of scoring values, while it is not difficult to generate unreliable results. One major challenge arose from the assignment of values to the Soil types (S) within the study sites. The problem stems from the reliance on global and continental-scale soil maps that often provide generic and generalized data. This can introduce a level of uncertainty into the method, as local variations in soil characteristics may not be adequately represented. Consequently, the accuracy and reliability of the recharge assessment may be compromised, necessitating caution in interpreting the results and highlighting the need for higher-resolution local soil data for more precise assessments.

Another important consideration pertains to the variables Lithology and Preferential infiltration, which are of paramount importance in the expression of recharge assessment. The values assigned in Lithology in particular entail a degree of subjectivity, as they are left to the discretion of the user. This introduces a degree of user-dependent variability into the method, leading to variations in the outcomes based on individual judgment. It is crucial to communicate this inherent subjectivity, emphasizing that the accuracy of the recharge assessment is inherently linked to the quality and appropriateness of the values selected.

While the APLIS method is effective in mapping the spatial distribution of annual recharge within a given area, it exhibits a limitation in its capacity to estimate its temporal variability. This limitation stems from the inherent nature of the method, which predominantly focuses on spatial aspects, neglecting the intricate temporal dynamics of recharge processes and providing averaged annual values of recharge rates. Recharge in karst aquifers is not a constant phenomenon but can vary significantly over time, influenced by factors such as seasonal variations in precipitation, land use changes, and climate fluctuations. Kirn et al. [55] extended the spatial dimension of the method to include the aspect of time variance and concluded that APLIS tends to overestimate recharge in dry years and underestimate it in wet years.

APLIS is a promising method for assessing the mean annual recharge that has been applied to several karst aquifers under Mediterranean climatic conditions, yielding satisfactory results. Given the intricate nature of karst systems and the challenges associated with verifying recharge assessments, the adoption of a multi-method approach is essential. In the future, further research will be conducted in order to strengthen the premise of this ongoing project. A comparison of APLIS' results with lump simulation models is strongly suggested. The implementation of advanced modeling at the study sites will provide a more holistic view of recharge, aid in error reduction, and enhance the reliability of findings associated with APLIS as an individual technique. Sensitivity analyses and robustness testing will be conducted in order to assess the potential impact of varying the parameters on the method's results, and tectonic lineaments will be introduced as possible infiltration landforms. Discharge measurements and hydrograph analysis from the front of springs in the karst system of Ziria will contribute to the verification of the results. The determination of recharge rates on a hydrogeological basin scale is valuable for the understanding of karst aquifer functions. Nevertheless, we strongly suggest comparing the results with the hydrological component of infiltration at the hydrological basin scale [56]. This approach could better determine underground runoff, which is critically important in the groundwater extraction process for the sustainable water supply of the community. Future research will incorporate hydro-chemical data, high-frequency discharge measurements, and isotopic data from groundwater and rain.

## 4. Conclusions

The two-step methodological approach that was applied in this study revealed that the geological structure influences the recharge rates of a karst aquifer. The conclusions are summarized below:

- The discretization of formation permeability increases the reliability of recharge rate determinations.
- Modified APLIS, compared to the initial method, provided extended zones with Very Low (<20%) and High (60–80%) recharge rates.
- The modified APLIS method estimated a relatively lower recharge rate (41.4%) compared to the initial method (42.7%) in Ziria, mainly due to the imported parameter of permeability.
- The cross-comparison of the results between the different karst systems revealed that the coverage of karst aquifers determines the recharge rate.
- The highest mean recharge rate was estimated in the Planitero karst system (48.6%) due to the absence of coverage of the karstified rocks.

This study revealed that modified APLIS constitutes a flexible method for the estimation of karst aquifer recharge rates in the Mediterranean region. The quantification and spatial distribution of mean annual recharge rates can increase the accuracy of water balance estimations, which is a fundamental step for sustainable groundwater resource management.

**Supplementary Materials:** The following supporting information can be downloaded at: https://www.mdpi.com/article/10.3390/w16040568/s1, Reliability of used data.

**Author Contributions:** Conceptualization, E.-A.N. and E.Z.; data curation and processing, E.-A.N., P.T. and K.P.; software, E.-A.N., P.T. and K.P.; writing-original draft preparation, E.-A.N.; writing–review and editing, E.-A.N. and E.Z.; supervision, E.Z.; funding acquisition, E.-A.N. All authors have read and agreed to the published version of the manuscript.

**Funding:** Part of the research work was supported by the Hellenic Foundation for Research and Innovation (HFRI) under the HFRI PhD Fellowship grant (Fellowship Number: 684).

**Data Availability Statement:** The presented data are available on request from the corresponding author.

**Acknowledgments:** A section of this study has been presented at the 12th International Hydrogeological Conference held in Nicosia, Cyprus, in 2022.

**Conflicts of Interest:** The authors declare no conflict of interest.

## References

1. Wada, Y.; Van Beek, L.P.H.; Van Kempen, C.M.; Reckman, J.W.T.M.; Vasak, S.; Bierkens, M.F.P. Global depletion of groundwater resources. *Geophys. Res. Lett.* **2010**, *37*, L20402. [CrossRef]
2. Scanlon, B.R.; Fakhreddine, S.; Rateb, A.; de Graaf, I.; Famiglietti, J.; Gleeson, T.; Grafton, R.Q.; Jobbagy, E.; Kebede, S.; Kolusu, S.R.; et al. Global water resources and the role of groundwater in a resilient water future. *Nat. Rev. Earth Environ.* **2023**, *4*, 87–101. [CrossRef]
3. Mishra, B.K.; Kumar, P.; Saraswat, C.; Chakraborty, S.; Gautam, A. Water Security in a Changing Environment: Concept, Challenges and Solutions. *Water* **2021**, *13*, 490. [CrossRef]
4. Taylor, R.G.; Scanlon, B.; Döll, P.; Rodell, M.; Van Beek, R.; Wada, Y.; Longuevergne, L.; Leblanc, M.; Famiglietti, J.S.; Edmunds, M.; et al. Ground water and climate change. *Nat. Clim. Chang.* **2012**, *3*, 322–329. [CrossRef]
5. Jia, Z.; Zang, H.; Zheng, X.; Xu, Y. Climate Change and Its Influence on the Karst Groundwater Recharge in the Jinci Spring Region, Northern China. *Water* **2017**, *9*, 267. [CrossRef]
6. Lukač Reberski, J.; Rubinić, J.; Terzić, J.; Radišić, M. Climate Change Impacts on Groundwater Resources in the Coastal Karstic Adriatic Area: A Case Study from the Dinaric Karst. *Nat. Resour. Res.* **2020**, *29*, 1975–1988. [CrossRef]
7. Klaas, D.K.S.Y.; Imteaz, M.A.; Sudiayem, I.; Klaas, E.M.E.; Klaas, E.C.M. Assessing climate changes impacts on tropical karst catchment: Implications on groundwater resource sustainability and management strategies. *J. Hydrol.* **2020**, *582*, 124426. [CrossRef]
8. Sivelle, V.; Jourde, H.; Bittner, D.; Mazzilli, N.; Tramblay, Y. Assessment of the relative impacts of climate changes and anthropogenic forcing on spring discharge of a Mediterranean karst system. *J. Hydrol.* **2021**, *598*, 126396. [CrossRef]
9. Taheri, K.; Taheri, M.; Parise, M. Impact of intensive groundwater exploitation on an unprotected covered karst aquifer: A case study in Kermanshah Province, western Iran. *Environ. Earth Sci.* **2016**, *75*, 1221. [CrossRef]
10. Zhang, Z.; Xu, Y.; Zhang, Y.; Guo, L.; Wang, Z.; Zheng, Q. Impact of groundwater overexploitation on karst aquifer and delineation of the critical zones: Case study of Jinci spring in Shanxi, China. *Carbonates Evaporites* **2022**, *37*, 68. [CrossRef]
11. Kazakis, N.; Chalikakis, K.; Mazzilli, N.; Ollivier, C.; Manakos, A.; Voudouris, K. Management and research strategies of karst aquifers in Greece: Literature overview and exemplification based on hydrodynamic modelling and vulnerability assessment of a strategic karst aquifer. *Sci. Total Environ.* **2018**, *643*, 592–609. [CrossRef]
12. Voudouris, K.S. Status and codification of karst aquifer systems in Greece. *Bull. Geol. Soc. Greece* **2021**, *57*, 23–51. [CrossRef]
13. Seiler, K.P.; Gat, J.R. *Groundwater Recharge from Run-Off, Infiltration and Percolation*; Springer: Dordrecht, The Netherlands, 2007.
14. Sappa, G.; Vitale, S.; Ferranti, F. Identifying Karst Aquifer Recharge Areas using Environmental Isotopes: A Case Study in Central Italy. *Geosciences* **2018**, *8*, 351. [CrossRef]
15. Dar, F.A.; Jeelani, G.; Perrin, J.; Ahmed, S. Groundwater recharge in semi-arid karst context using chloride and stable water isotopes. *Groundw. Sustain. Dev.* **2021**, *14*, 100634. [CrossRef]
16. Iacurto, S.; Grelle, G.; De Filippi, F.M.; Sappa, G. Karst Recharge Areas Identified by Combined Application of Isotopes and Hydrogeological Budget. *Water* **2021**, *13*, 1965. [CrossRef]
17. Petitta, M.; Banzato, F.; Lorenzi, V.; Matani, E.; Sbarbati, C. Determining recharge distribution in fractured carbonate aquifers in central Italy using environmental isotopes: Snowpack cover as an indicator for future availability of groundwater resources. *Hydrogeol. J.* **2022**, *30*, 1619–1636. [CrossRef]
18. Radulovic, M.; Stevanovic, Z.; Radulovic, M. A new approach in assessing recharge of highly karstified terrains-Montenegro case studies. *Environ. Earth Sci.* **2012**, *65*, 2221–22230. [CrossRef]
19. Hartmann, A.; Barbera, J.A.; Lange, J.; Andreo, B.; Weiler, M. Progress in the hydrologic simulation of time variant recharge areas of karst systems—Exemplified at a karst spring in Southern Spain. *Adv. Water Resour.* **2013**, *54*, 149–160. [CrossRef]
20. Malard, A.; Sinreich, M.; Jeannin, P.-Y. A novel approach for estimating karst groundwater recharge in mountainous regions and its application in Switzerland. *Hydrol. Process.* **2016**, *30*, 2153–2166. [CrossRef]
21. Hosseini, S.M.; Ataie-Ashtiani, B.; Simmons, C.T. Spring hydrograph simulation of karstic aquifers: Impacts of variable recharge area, intermediate storage and memory effects. *J. Hydrol.* **2017**, *552*, 225–240. [CrossRef]
22. Jukić, D.; Denić-Jukić, V.; Lozić, A. An alternative method for groundwater recharge estimation in karst. *J. Hydrol.* **2021**, *600*, 126671. [CrossRef]
23. Messerschmid, C.; Aliewi, A. Spatial distribution of groundwater recharge, based on regionalised soil moisture models in Wadi Natuf karst aquifers, Palestine. *Hydrol. Earth Syst. Sci.* **2022**, *26*, 1043–1061. [CrossRef]
24. Berthelin, R.; Olarinoye, T.; Rinderer, M.; Mudarra, M.; Demand, D.; Scheller, M.; Hartmann, A. Estimating karst groundwater recharge from soil moisture observations—A new method tested at the Swabian Alb, southwest Germany. *Hydrol. Earth Syst. Sci.* **2023**, *27*, 385–400. [CrossRef]
25. White, W.B. Conceptual models for karstic aquifers. In *Karst Modeling*; Palmer, A.N., Palmer, M.V., Sasowsky, I.D., Eds.; The Karst Waters Institute: Charles Town, WV, USA, 1999; pp. 11–16.
26. Bakalowicz, M. Karst groundwater: A challenge for a new resources. *Hydrogeol. J.* **2005**, *13*, 148–160. [CrossRef]
27. Andreo, B.; Vías, J.; Durán, J.J.; Jiménez, P.; López-Geta, J.A.; Carrasco, F. Methodology for groundwater recharge assessment in carbonate aquifers: Application to pilot sites in southern Spain. *Hydrogeol. J.* **2008**, *16*, 911–925. [CrossRef]

28. Marín, A.I. Los Sistemas de Información Geográfica Aplicados a la Evaluación de Recursos Hídricos y a la Vulnerabilidad a la Contaminación de Acuíferos Carbonatados: Caso de la Alta Cadena (Provincia de Málaga). Ph.D. Thesis, University of Malaga, Malaga, Spain, 2009.
29. Zagana, E.; Tserolas, P.; Floros, G.; Katsanou, K.; Andreo, B. First outcomes from groundwater recharge estimation in evaporate aquifer in Greece with the use of APLIS method. In *Advances in the Research of Aquatic Environment*; Lambrakis, N., Stournaras, G., Katsanou, K., Eds.; Springer: Berlin/Heidelberg, Germany, 2011; Volume 2, pp. 89–96.
30. Martos-Rosillo, S.; Rodriguez-Rodriguez, M.; Pedrera, A.; Cruz- SanJulian, J.J.; Rubio, J.C. Groundwater recharge in semi-arid carbonate aquifers under intensive use: The Estepa Range aquifers (Seville, southern Spain). *Environ. Earth Sci.* **2013**, *70*, 2453–2468. [CrossRef]
31. Hartmann, A.; Mudarra, M.; Andreo, B.; Marín, A.; Wagener, T.; Lange, J. Modeling spatiotemporal impacts of hydroclimatic extremes on groundwater recharge at a Mediterranean karst aquifer. *Water Resour. Res.* **2014**, *50*, 6507–6521. [CrossRef]
32. Mudarra, M.; Hartmann, A.; Andreo, B. Combining Experimental Methods and Modeling to Quantify the Complex Recharge Behavior of Karst Aquifers. *Water Resour. Res.* **2019**, *55*, 1384–1404. [CrossRef]
33. Farfán, H.; Corvea, J.L.; De Bustamante, I. Sensitivity Analysis of ALPIS Method to Compute Spatial Variability of Karst Aquifers Recharge at the national Park of Viñales (Cuba). In *Advances in Research in Karst Media*; Andreo, B., Carrasco, F., Durán, J.J., LaMoreaux, J.W., Eds.; Springer: Berlin/Heidelberg, Germany, 2010; pp. 19–24.
34. Espinoza, K.; Marina, M.; Fortuna, J.H.; Altamirano, F. Comparison of the APLIS and Modified-APLIS methods to estimate the recharge in fractured karst aquifer, Amazonas, Peru. In *Hydrogeological and Environmental Investigations in Karst Systems*; Andreo, B., Carrasco, F., Durán, J.J., Jiménez, P., LaMoreaux, J.W., Eds.; Springer: Berlin/Heidelberg, Germany, 2015; Volume 1, pp. 83–90.
35. Alem, H.; Soudejani, A.E.; Farmanieh, S.N. Ground water recharge assessment in the karst aquifers of North Khorasan, Iran using APLIS method. *Acta Carsologica* **2017**, *46*, 283–294. [CrossRef]
36. Special Secretariat for Water. *Management Plans of Northern Peloponnese River Basin District*; Ministry of Environment and Energy: Athens, Greece, April 2013.
37. WFD. Directive 2000/60/EC of the European Parliament and of the Council of 23 October 2000 establishing a framework for Community action in the field of water policy. *Off. J. Eur. Communities* **2000**, *L 327*, 1–72.
38. De Wever, P.; Dercourt, J.; Fleury, J.J. *Kandhila Sheet. 1:50.000*; Institute of Geology and Mineral Exploration: Athens, Greece, 1982.
39. Tataris, A.; Maragoudakis, N.; Kounis, G.; Christodoulou, G.; Tsaila-Monopoli, S. *Nemea Sheet. 1:50.000*; Institute for Geology and Subsurface Research: Athens, Greece, 1970.
40. Hötzl, H.; Merkler, G.; Reichert, B.; Röckel, T.; Watzel, R. Interpretation of geophysical measurements in the poljes of Alea-Scotini and Stymfalia. In *Karst Hydrogeology of the Central and Eastern Peloponnesus (Greece), Proceedings of the 5th International Symposium on Underground Water Tracing, Athens, Greece, 22–27 September 1986*; Steirische Beiträge zur Hydrogeologie; Morfis, A., Zojer, H., Eds.; Springer: Cham, Switzerland, 1986; Volumes 37/38, pp. 44–54.
41. Meilliez, F.; Flament, J.M.; De Wever, P.; Dercourt, J.; Fleury, J.J. *Dhafni Sheet. 1:50.000*; Institute of Geology and Mineral Exploration: Athens, Greece, 1978.
42. Psonis, K.; Skourtsi-Koroneou, V.; Carras, N.; Varti-Mataragha, M. *Aeghion Sheet. 1:50.000*; Institute of Geology and Mineral Exploration: Athens, Greece, 2013.
43. Psonis, K.; Tsapralis, V.; Skourtsi-Koroneou, V. *Astakos Sheet. 1:50.000*; Institute of Geology and Mineral Exploration: Athens, Greece, 1986.
44. Kouris, C.; Tsaila-Monopolis, S. *Echinades Sheet. 1:50.000*; Institute of Geology and Mineral Exploration: Athens, Greece, 1989.
45. Gofa, F.; Mamara, A.; Anadranistakis, M.; Flocas, H. Developing gridded climate data sets of precipitation for Greece based on homogenized time series. *Climate* **2019**, *7*, 68. [CrossRef]
46. Kakavas, M.; Nikolakopoulos, K.G.; Kyriou, A.; Zagana, H. Assessment of freely available DSMs for automatic karst feature detection. *Arab. J. Geosci.* **2018**, *11*, 388. [CrossRef]
47. Morfis, A.; Zojer, H. (Eds.) *Karst Hydrogeology of the Central and Eastern Peloponnesus (Greece), Proceedings of the 5th International Symposium on Underground Water Tracing, Athens, Greece, 22–27 September 1986*; Steirische Beiträge zur Hydrogeologie; Springer: Cham, Switzerland, 1986; Volumes 37/38, pp. 1–301.
48. Zervogiannis, G. *Final Hydrogeological Study for the Reformation of the Stymphalian Springs in the Prefecture of Corinth*; Ministry of Agriculture: Athens, Greece, 1991. (In Greek)
49. Röckel, T.; Hötzl, H. Polje of Stymfalia. In *Karst Hydrogeology of the Central and Eastern Peloponnesus (Greece), Proceedings of the 5th International Symposium on Underground Water Tracing, Athens, Greece, 22–27 September 1986*; Steirische Beiträge zur Hydrogeologie; Morfis, A., Zojer, H., Eds.; Springer: Cham, Switzerland, 1986; Volumes 37/38, pp. 127–152.
50. Katakouzinos, D.S. *Soil Map of Greece. 1:1.000.000*; Ministry of Agriculture: Athens, Greece, 1967.
51. FAO; UNESCO. *Soil Map of the World. 1:5.000.000*; UNESCO: Paris, France, 1978.
52. European Soil Bureau Network. *Soil Atlas of Europe*; European Commission: Luxembourg, 2005; p. 128.
53. Voudouris, K.; Panagopoulos, A.; Markantonis, K. Hydrological balance and possibilities of sustainable water resources management in Stimfalia basin (prefecture of Korinthos). In Proceedings of the 6th Pan-Hellenic Geographical Conference of the Hellenic Geographical Society, Thessaloniki, Greece, 3–6 October 2002; Volume 2, pp. 327–334.
54. Rad, M.M.; Mokhtari, L.G.; Bahrami, S.; Asadi, M.Z. Assessment of infiltration potential using the AHP, APLIS and modified APLIS models (Case study: Esfarayen Basin). *Quant. Geomorphol. Res.* **2018**, *7*, 117–139.

55. Kirn, L.; Mudarra, M.; Marín, A.; Andreo, B.; Hartmann, A. Improved assessment of groundwater recharge in a Mediterranean karst region: Andalusia, Spain. In *EuroKarst 2016, Neuchâtel*; Renard, P., Bertrand, C., Eds.; Springer: Cham, Switzerland, 2017; pp. 117–125.
56. Malov, A.I. Groundwater resources of the White Sea Basin (NW Russia). In *The Arctic: Current Issues and Challenges*; Malov, A.I., Pokrovsky, O.S., Kirpotin, S.N., Eds.; Nova Science Publishers: New York, NY, USA, 2020; pp. 239–266.

*water*

MDPI

*Article*

# Nature-Based Solutions for the Restoration of Groundwater Level and Groundwater-Dependent Ecosystems in a Typical Inland Region in China

**Min Liu [1,2], Zhenlong Nie [1,2,*], Le Cao [1,2,*], Lifang Wang [1,2] and Huixiong Lu [3]**

1 Institute of Hydrogeology and Environmental Geology, Chinese Academy of Geological Sciences, 268 Zhonghua North Street, Shijiazhuang 050061, China; agnes0505@163.com (M.L.); wanglifang@mail.cgs.gov.cn (L.W.)
2 Key Laboratory of Groundwater Sciences and Engineering, Ministry of Natural Resources, 268 Zhonghua North Street, Shijiazhuang 050061, China
3 Airborne Survey and Remote Sensing Center of Nuclear Industry, Shijiazhuang 050002, China; 13933164939@163.com
* Correspondence: nzlngj@163.com (Z.N.); caole@mail.cgs.gov.cn (L.C.)

**Abstract:** The rapid drop in groundwater level and the degradation of groundwater-dependent ecosystems (GDEs) has led to serious imbalances in the development of the socio-ecological system (SES) in Minqin Oasis, northwest China. In this study, the interactions between the elements of the SES were analyzed, and the effects of the implementation of nature-based solutions (NbS) and the resilience of the SES were assessed using the methods of linear trend analysis, correlation, and fuzzy comprehensive evaluation. The results showed that: (i) the most important contributing factor to the increase in groundwater depth (GLD) and the degradation of GDEs was the continuous groundwater extraction for the irrigation of farmland; and (ii) the natural vegetation began to be degraded when the GLD surpassed 5 m; (iii) the initial goal of NbS was achieved, the increase in GLD has been curbed, and the vegetation and the wetland area of Qingtu lake have begun to be restored. The SES resilience index increased by a factor of 1.82 from 2009 to 2017; however, it was still below 0.6, and thus the NbS should be continued in the future. This study provides a reference for the sustainable development of the SES in other similar areas under various environmental challenges.

**Keywords:** groundwater depth; vegetation ecology; wetland ecology; resilience of socio-ecological system; Minqin Oasis

## 1. Introduction

The crises currently facing the world are constantly escalating, and the interactions between humans and the environment continue to deteriorate, suggesting that humans and other life on Earth may be approaching a catastrophic tipping point [1]. Global issues such as climate change, food shortages, groundwater depletion, and loss of biodiversity pose significant challenges to social and ecological systems on which humans rely for survival. In response, human beings urgently need innovative solutions to these problems, and to reconsider the relationship between humans and nature. It is time to shift from the disorderly utilization of natural resources to respecting nature, cooperating with nature, and maximizing the power of nature. Thus, the concepts and methods of Nature-based Solutions (NbS) are emerging [2–8].

NbS is a new concept that has arisen during the past two decades, and there are currently multiple definitions and interpretations of the term [9,10]. In the fifth session of the United Nations Environment Assembly of the United Nations Environment Programme, NbS was defined as "actions to protect, conserve, restore, sustainably use and manage natural or modified terrestrial, freshwater, coastal and marine ecosystems, which address social, economic, and environmental challenges effectively and adaptively, while simultaneously providing human well-being, ecosystem services and resilience and biodiversity benefits" [11]. Many such NbS initiatives, including ecological restoration and protection, holistic management, groundwater recharge, and blue-green infrastructure practices, involve multiple methods of cooperating with nature to achieve social benefits. These methods are being implemented globally, especially in rural communities and among people in low- and middle-income countries [12,13]. Although NbS may provide more productive and cost-effective remedies than conventional techniques, monitoring and assessment methods for implementing these solutions are not documented across multiple scales and resources; how NbS practices can be transformed into the direct application of management activities remains an open question.

The Shiyang River Basin (SRB) is a typical inland arid basin in northwest China. Minqin Oasis, located downstream of the SRB, was once one of the seventeen poverty-stricken counties in Gansu Province that accounted for 40% of the "Silk Road Economic Belt" in China. The area has a dry climate with scarce water resources. Groundwater is the main water source for maintaining local socio-economic development, agriculture, and the oasis environment. Since 1970, continuous groundwater exploitation in the SRB has led to the transformation of the downstream Minqin Oasis into a pure pumping-well irrigation area, and there has been a rapid decline in the groundwater level [14,15]. By 2007, the groundwater depth (GLD) in Minqin Oasis was generally between 20 and 40 m, far lower than the ecologically suitable GLD for surface vegetation [16,17]. Thus, the vegetation ecology in the oasis–desert ecotone (ODE) was severely degraded. Sandstorms were frequent, and the wetland area of the terminal lake (Qingtu Lake) was significantly reduced. Due to the intrusion of sandstorms from the northwest Badain Jaran Desert, there has been extensive sand intrusion and human retreat in the oasis. Faced with the serious imbalance of the socio-ecological system (SES) in Minqin Oasis, the Chinese government proposed a series of NbS, called the Comprehensive Administration Project of the SRB (CASRB). This program aimed to restore the groundwater level (GL) and vegetation of the Minqin Oasis in the lower SRB, and in the meantime lift the county out of poverty by increasing the per-capita income of the residents, ultimately achieving sustainable development of the SES. With regard to the effect of the CASRB, the recovery of the GL [16,18] and the ecological functions of the groundwater [17,19] have been investigated by some researchers. However, there are few reports on the evaluation of this NbS from the perspective of the sustainable development of regional SES.

Therefore, the objectives of this study were: (1) to analyze the changes in hydro-climatic elements, land use and land cover, and the normalized difference vegetation index (NDVI) in the oasis–desert ecotone (ODE) during the past several decades; (2) to investigate the close interactions among the elements of the SES both in agroecological area and natural ecological area; (3) to reveal the effects of the CASRB on the restoration of the groundwater and groundwater-dependent ecosystems (GDEs) by primary indicators; and (4) to assess the changes in the resilience of the SES in the post-CASRB period by using the fuzzy comprehensive evaluation method based on the above interaction analysis and the selected indicators. This study can enrich the theory and practice of NbS in combination of irrigation oasis and natural oasis in arid regions, and provide decision-making references for the sustainable development, planning, and management of the SES in the SRB under various environmental challenges.

## 2. Materials and Methods

### 2.1. Study Area

The SRB is located in the central part of Gansu Province and the eastern edge of the Hexi Corridor; it forms one of the three arid inland basins in China. The upper reaches of the SRB are in the Qilian Mountain, and the middle reaches comprise the Wuwei sub-basin, with Liangzhou County as the capital. The main study area of Minqin Oasis is in the downstream area of the SRB (Figure 1). In its natural state, runoff flows into Minqin after being consumed in the middle sub-basin, then flows through the Caiqi section located upstream of Hongyashan Reservoir. Minqin Oasis is located between longitude 101°49′41″ E–104°12′10″ E and latitude 38°3′45″ N–39°27′37″ N [20]. Due to being sandwiched between the northwest Badain Jaran Desert and the southeast Tengger Desert, the ecology of Minqin Oasis is a vital natural barrier that prevents the two deserts from converging. It has a typical continental arid desert climate, with an average annual temperature of 7.8 °C, an average annual precipitation of 116.5 mm, and an average annual evaporation of 2308 mm during 1971–2010 [21]. The irrigation areas in Minqin Oasis mainly consist of the Dam, Quanshan, and Lake irrigation areas (Figure 1). In 2017, the GLD in Minqin Oasis was generally between 20 and 30 m in the irrigation area and was generally less than 10 m in the Qingtu Lake area [17]. The natural vegetation in the oasis–desert ecotone (ODE) of Minqin Oasis is largely temperate desert (arid desert) vegetation dominated by arid and super-arid shrubs and semi-shrubs. The main shrub species include *Tamarix* spp., *Nitraria* spp., *Kalidium foliatum*, *Haloxylon ammodendron*, *Artemisia arenaria* and *Reaumuria soongorica*. Arbors include *Populus gansuensis*, *Populus euphratica*, *Angustifolia elaeagnus* and *Salix matsudana*. Herbs include *Achnatherum splendens*, *Phragmites australis*, *Bassia dasyphylla*, *Sophora alopecuroides*, and *Agriophyllum squarrosum* [16,17,22,23].

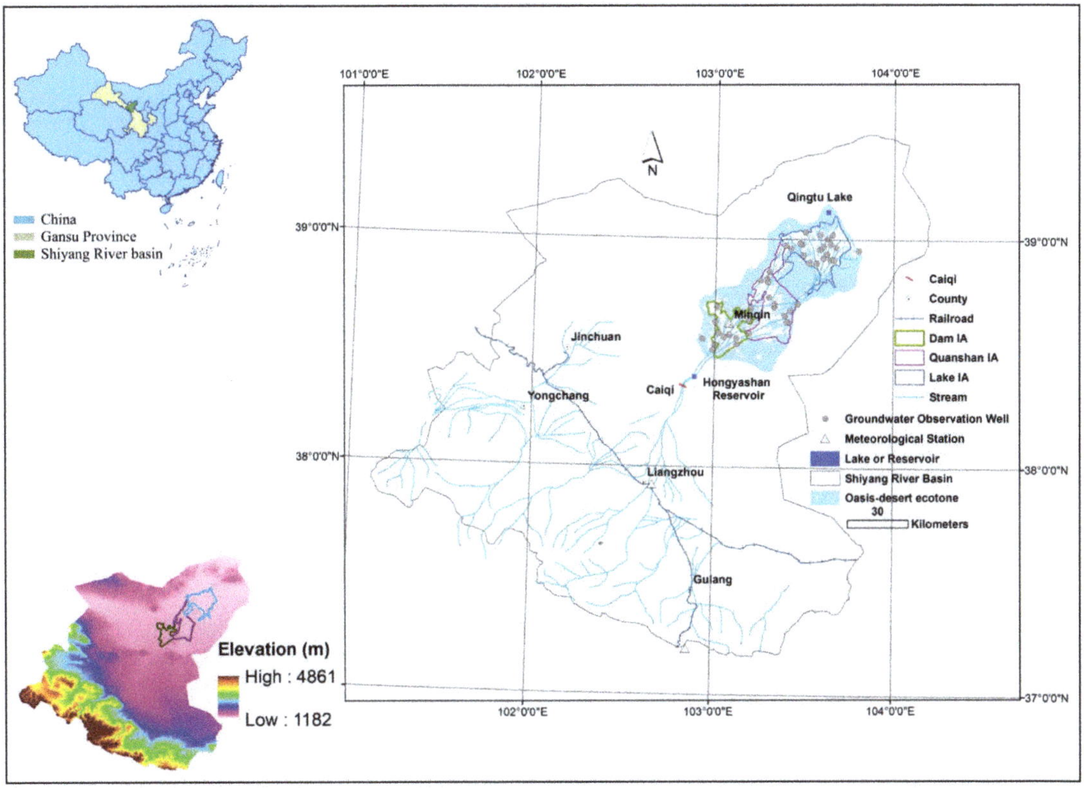


**Figure 1.** Study area (IA: irrigation area).

## *2.2. Data*

In this study, the meteorological data were collected from the National Meteorological Administration, and the runoff and water supply data were collected from the Wuwei Water Affairs Authority. The GLD data were collected from the Institute of Hydrogeological Engineering Geology, Gansu Provincial Bureau of Geology and Mineral Exploration and

Development. There were 46 observation wells for phreatic aquifers. The wells were mostly distributed in irrigation areas (Figure 1). The time series for these data were monthly during 1980–2017. The land-use and land-cover data of 1985, 2008, and 2017 were derived from the thematic mapper (TM), enhanced thematic mapper (ETM), and operational land imager (OLI) data obtained by the US Landsat satellite-5, the US Landsat satellite-7, and the US Landsat satellite-8. The annual mean normalized difference vegetation index (NDVI) for 2000–2017 were also obtained from the Landsat satellites. The area of the Qingtu Lake wetland for 2009–2017 was obtained from both the Landsat satellites and the China Gaofen-1 satellite and Gaofen-2 satellite due to that the high resolution of Gaofen satellites has great advantage in the interpretation of small areas. The image map data were derived by atmospheric correction, geometric correction, image fusion, image stitching, and band calculation [24]. The species richness data and corresponding GLD were from the natural vegetation quadrat survey, in which the species richness is defined as the number of species within a quadrat. The per-capita income data of Minqin County during 2009–2017 were obtained from the Statistical Yearbook of Wuwei City.

### *2.3. Methods*

#### 2.3.1. Mathematical Statistics Method

The mathematical statistics method studies the regularity of the inter-relationships between the sample and the population, as well as the regularity of the inter-relationships among various related factors. A linear trend analysis and a correlation analysis was applied in this study.

1. Linear trend analysis

Linear trend analysis is a highly general and therefore very flexible data-analytic system that can be used for studying specific quantitative relationships between variables [25]. It is normally used to establish mathematical models that reflect the specific quantitative relationships between geographic elements, namely regression models [26]. The form of the relationship is not constrained; it may be univariate or multivariate. In this study, a univariate linear regression model was used to examine the variation rates of the air temperature, precipitation, runoff, the GLD of each observation well, and the wetland area of Qingtu lake. The least squares estimation was used for the parameter estimation [26].

2. Correlation analysis

Correlation analysis is a statistical analysis method that investigates the relationship between two or more random variables of equal status. The general meaning of the concept is that knowledge with regard to one of the variables carries information about the other variable [27]. It is the process of describing the degree of closeness between objective things and expressing it with appropriate statistical indicators [26]. A correlation analysis was used to analyze the relationships between the elements of the SES in this study.

#### 2.3.2. GLD—Natural Vegetation Quadrat Joint Survey

The GLD—natural vegetation ecology quadrat joint survey was adapted from previously published guidelines [28] while considering the actual distribution of the groundwater-dependent vegetation communities in the SRB. Therefore, this survey added GLD measurements near each vegetation quadrat. If no groundwater observation well was available near the vegetation quadrat, the investigators would dig a simple groundwater measuring well on site, and GLD was measured once the groundwater level had stabilized. The survey was conducted in August and September 2018. The procedure used in the joint survey is described in Liu et al. [17].

#### 2.3.3. Fuzzy Comprehensive Evaluation Method

The fuzzy comprehensive evaluation method was used to assess the resilience of the SES during the post-CASRB period of Minqin Oasis. The theory was introduced as follows [26,29].

For two finite sets, $U = \{U_1, U_2, \ldots\ldots, U_m\}$ and $V = \{V_1, V_2, \ldots\ldots, V_m\}$, where $U$ represents a set of indicators based on which comprehensive evaluation is performed and $V$ is a set of judgments, a fuzzy comprehensive evaluation is a fuzzy transformation process as follows:

$$B = A \circ R \quad (1)$$

In Equation (1), $A$ is a fuzzy subset of $U$, and the evaluation result $B$ is a fuzzy subset of $V$. $A = \{a_1, a_2, \ldots\ldots, a_m\}, 0 \leq b_j \leq 1$, where $a_i$ is the degree of membership of $U$ to $A$. $a_i$ represents the relative importance (weight) of the indicator $U_i$ among all of the evaluation indicators, and also represents the ability of $U_i$ to determine the grade, satisfying $\sum_{i=1}^{m} a_i = 1$. $b_j$ is the degree of membership of grade $V_j$ to $B$, representing a comprehensive evaluation result.

The fuzzy transformation matrix is:

$$R = \begin{bmatrix} r_{11}, & r_{12}, & \cdots & r_{1n} \\ r_{21}, & r_{22}, & \cdots & r_{2n} \\ \cdots & \cdots & & \cdots \\ r_{m1}, & r_{m2}, & \cdots & r_{mn} \end{bmatrix}, \quad (2)$$

where $r_{ij}$ is the degree of membership of grade $V_j$ for $U_i$, and thus, the $i$th row $R_i = (r_{i1}, r_{i2}, \ldots\ldots, r_{in})$ in the matrix $R$ represents the evaluation results of $U_i$.

In this study, the indicators used for comprehensive evaluation were determined based on the factors influencing the sustainable development of SES in Minqin. The evaluation procedure involved the following steps: (1) selection of the resilience indicators of the SES, (2) determination of grading values and the corresponding rating scores, (3) determination of the weight for each resilience indicator, and (4) composite evaluation index calculation of the resilience of SES.

## 3. Results

### 3.1. Changes in Elements of Climate, Water Resources, and Ecology

#### 3.1.1. Variations in Natural Hydro-Climatic Elements

Annual air temperature in Minqin showed a significant upward trend under global warming. The rate of increase was 0.4 °C/10a during 1956–2017 (Figure 2a). Temporally, the air temperature increased slowly during 1956–1986, then rose significantly after 1987, and the rising trend reached a new level in 1998. Annual precipitation showed a nonsignificant increase rate of 2.8 mm/10a (Figure 2b). Temporally, precipitation decreased in the 1960s, then increased in the 1970s, and then decreased again from the 1980s, finally increasing since 1994. Pan evaporation (E601) showed a nonsignificant increase rate of 9.8 mm/10a (Figure 2c). The runoff at the mountain pass in the upper SRB showed no clear trend. However, the runoff had a high flow rate before 1970, then entered a normal flow period during the next two decades before experiencing a low flow period in the 1990s. The runoff has increased since 2003 (Figure 2d). This was in accordance with the conclusion that the largest tributary of the Shiyang River entered a wet period in 2003 by Song et al. [30]. The natural runoff at the Caiqi section would eventually flow into the Minqin sub-basin after being consumed in the middle SRB. There was a significant decreasing trend with the rate of 81 million cubic meters of water per decade (Figure 2d).

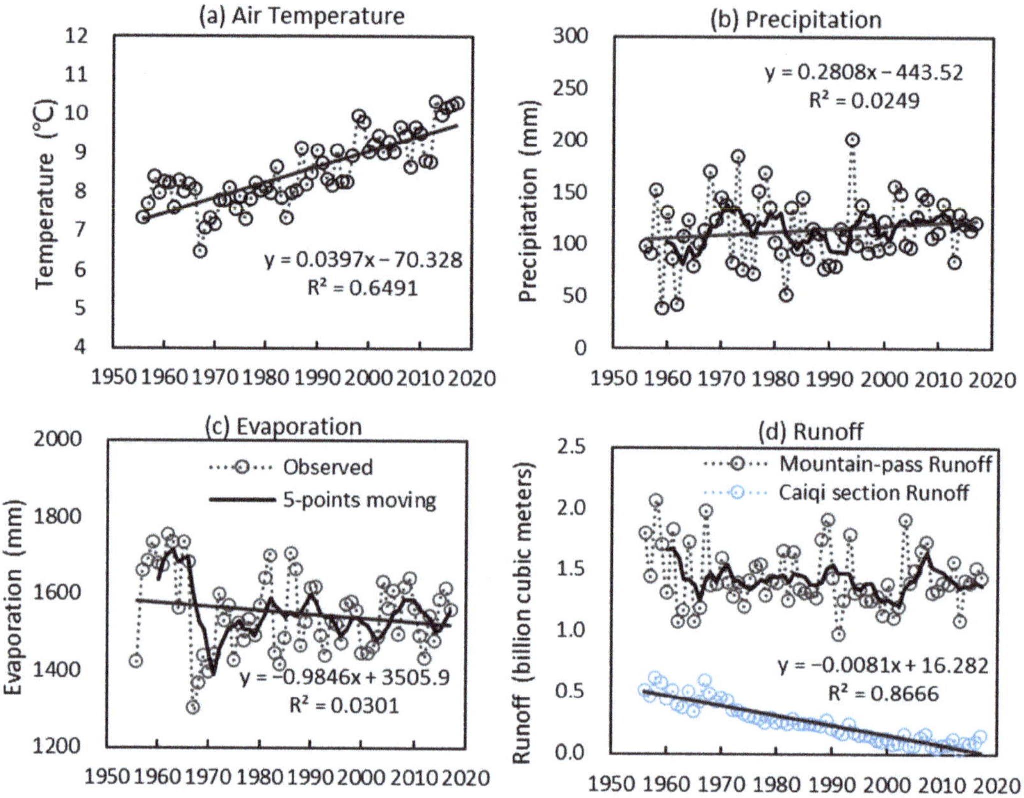


**Figure 2.** Variations in climatic elements in Minqin and runoff at the mountain-pass. (The dotted line represented 'Observed'; the grey solid line represented the linear trend; the black solid line represented 5-points moving).

### 3.1.2. Changes in Groundwater Depth

GLD increased dramatically during 1980–2017 (Figure 3). The trend of the GLD at all of the observation wells was significant at the 0.05 level. The increase rates of the GLD of the wells were mostly between 4 m/10a and 6 m/10a, and the rates were greater in the center of the irrigation areas, while becoming less in the lake irrigation area and the eastern peripheral areas of Minqin Oasis (Figure 3a). On the irrigation area scale, the increase rate was larger in the Dam and Quanshan irrigation areas, while being relatively smaller in the lake irrigation area (Figure 3b). Temporally, the change in the GLD experienced two periods (Figure 3b); i.e., the rapid increasing period before the implementation of the CASRB, and a stable or recovery trend from then on, was due to the initial effect of the implementation of CASRB.

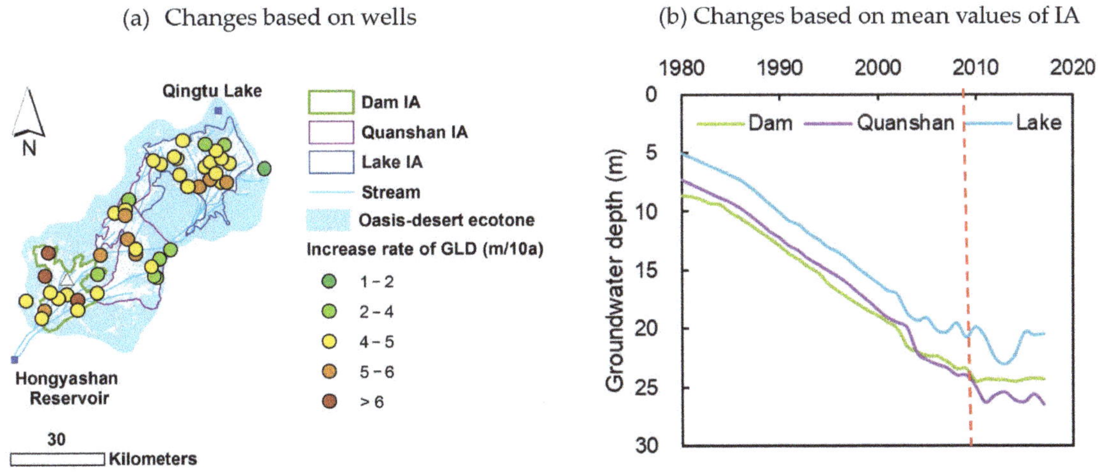


**Figure 3.** Changes in the groundwater depth in the irrigation areas of Minqin Oasis during 1980–2017 (GLD: groundwater depth; IA: irrigation area; the red dotted line represented the turning point year).

3.1.3. Changes in Land Use and Land Cover (LULC)

The area of the natural oasis and the natural wetland showed decreasing trends from 1985 to 2017. The area of the natural oasis in 2017 was just 78% of that in 1985, and the area of the natural wetland in 2017 was just 18.5% of that in 2017 (Figure 4). Temporally, the decreasing trend became milder after 2008. In the meantime, the area of the artificial oasis, including the farmland and the artificial wetland, increased dramatically until 2008, when the trend leveled off (Figure 4). The farmland expanded by a factor of 1.25 from 1985 to 2008; after which the area was stable from 2008 to the present. The area of the artificial wetland was 5.8 times that in 1985. Furthermore, the area of planted forest (including arbors, shrubs, and grasses) showed an initial decreasing trend before 2008, and then an increasing trend from 2008 (Figure 4). There was also a turning-point year in the land-use type at around 2008, the year after the implementation of the CASRB. The construction land has always shown an increasing trend (Figure 4).

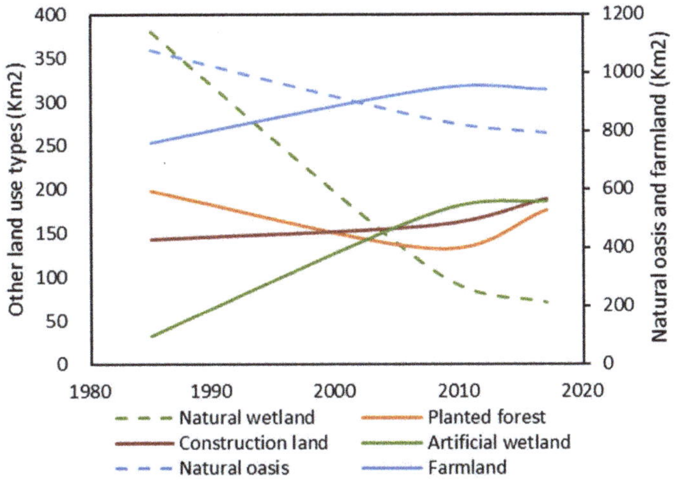


**Figure 4.** Changes in land-use types in Minqin Oasis (other land use types represent planted forest, construction land, artificial wetland, natural wetland, and farmland).

3.1.4. Changes in NDVI in Oasis–Desert Ecotone

The annual regional mean value of NDVI was between 0.07 and 0.10 during 2000–2011 (Figure 5). This indicated that there was almost no vegetation growth in large areas of the oasis–desert ecotone, and most of the area experienced desertification. From 2011 to 2017, the annual mean NDVI of natural vegetation increased from 0.09 to 0.21, reflecting an improvement in vegetation growth. This was due to the artificial afforestation implemented by the CASRB to prevent the desertification [16] and the restoration of the groundwater ecological function in the areas with a shallow groundwater depth [17].

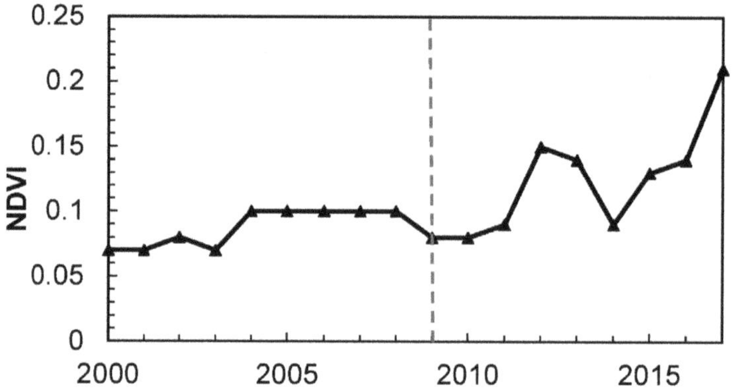


**Figure 5.** Annual regional mean NDVI value in the Minqin Oasis–desert ecotone [16] (the red dotted line represented the turning point year).

*3.2. Interactions among the Elements of Social and Ecological Systems*

3.2.1. Farmland–Groundwater Abstraction–Groundwater Depth–Natural Oasis

With the dry climate and the shortage of surface water, the SRB is a region whose socio-economic development, especially regarding agricultural production, was primarily supported by groundwater irrigation. Therefore, the interactions between the land-use type and the groundwater extraction in Minqin Oasis were analyzed based on the data before 2008, when the CASRB had not been fully implemented. The results showed that the area of farmland had a high positive correlation with the groundwater extraction; when the area of the farmland increased, the groundwater abstraction increased (Figure 6b). This finding indicated that groundwater was extracted primarily for agricultural irrigation. Furthermore, the GLD had a strong correlation with the accumulated groundwater extracted during the period of overexploitation. Every increase of 10 million cubic meters of water extracted from the aquifer would lead to a 3 cm decline in the GL (Figure 6d). Eventually, the GL declined beyond the level where most of the surface vegetation could reach, thereby leading to a shrinkage of the natural oasis (Figure 6c). Every increase of 1000 square kilometers of farmland would lead to a decrease of 957 square kilometers of natural oasis (Figure 6a). This indicated that the expansion of the irrigated farmland was the dominant factor to the significant decline in GL and the degradation of the groundwater-dependent vegetation in Minqin Oasis.

The accumulated groundwater extraction instead of annual groundwater extraction was adopted when analyzing the interactions with the GLD. The reasons were as follows. When the groundwater is over-extracted for a long period, this has a direct and accumulated impact on the GL due to the low rate of the aquifer recharge. In other words, the gap between the extraction and the input continuously increased, and thus the GL declined.

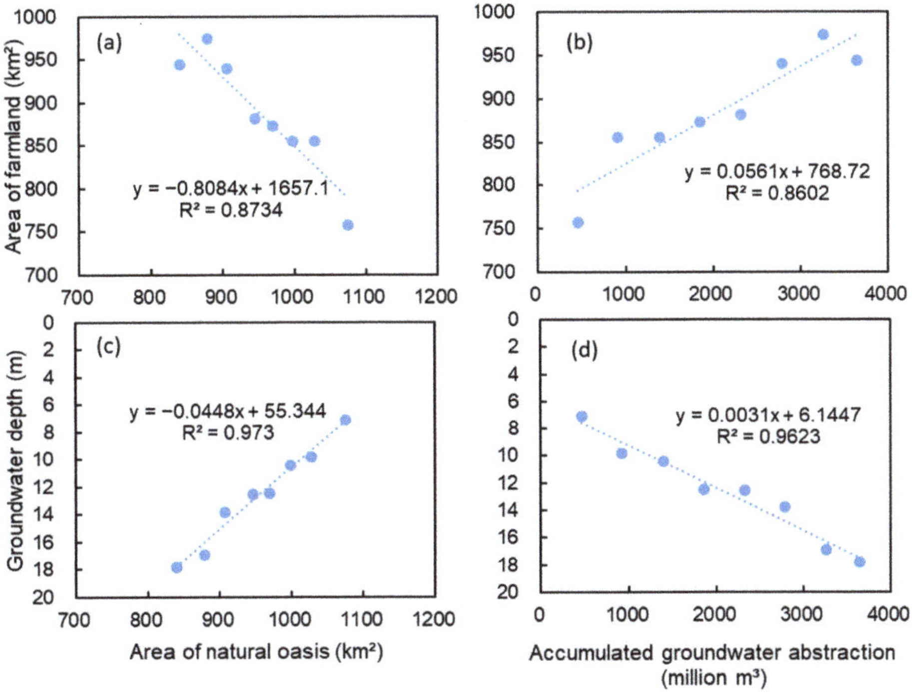


**Figure 6.** Interactions among the farmland–groundwater abstraction–groundwater depth–natural oasis. ((**a**): represented the correlation between areas of farmland and natural oasis; (**b**): represented that of farmland area and the groundwater abstraction; (**c**): represented that of groundwater depth and area of natural oasis; (**d**): represented that of groundwater depth and groundwater abstraction).

3.2.2. The Relationship of Groundwater to the Main Vegetation Indicators

An ecologically suitable GLD was estimated by analyzing the relationship between GLD with the NDVI of the natural vegetation in the ODE during 2000–2017. The results showed that in 94% of the years considered, the GLD corresponding to the maximum value of NDVI ranged from 2.2 to 3.7 m, and the mean GLD corresponding to the peak value of NDVI was 3.0 m (Figure 7a). After the GLD had increased more than 5.0 m, the value of NDVI was generally less than 0.1 (except for several years the vegetation was planted since 2012), and there was no significant correlation between the GLD and NDVI. When the GLD was less than 5 m, the two indicators were significantly negatively correlated, with a correlation coefficient (r) reaching 0.85 (Figure 7c) [17]. This was in agreement with the results of Cao et al. [16]. Furthermore, an ecological survey of vegetation quadrats and groundwater was conducted, and the results indicated that the species richness reached a maximum at a GLD of about 3 m. That is to say, with the GLD increased from 1 m to 3 m, species richness increased, peaking at approximately 3 m at 10 species; and then slowly become less as the GLD increased greater than 3 m (Figure 7b) [17]. Due to the sparse vegetation in the northwest desert area, vegetation development was only considered when the NDVI value of vegetation is greater than a certain value, such as in Ejina Banner Oasis, where the index was bounded by 0.08 [31]. Therefore, when the GLD exceeded 10 m, there was almost no vegetation, and the transient vegetation growth was influenced more by individual rainfall events and shallow soil moisture [16,17]. In addition, the close relationship between the GLD and vegetation was only observed for soil salinity below 5 g/kg. Our research work demonstrated that soil salinity may also be a restrictive factor for the growth and development of vegetation. The vegetation coverage began to decrease when the soil salinity was greater than 5 g/kg, and when this reached 20 g/kg, the vegetation nearly disappeared and desertification occurred [32].

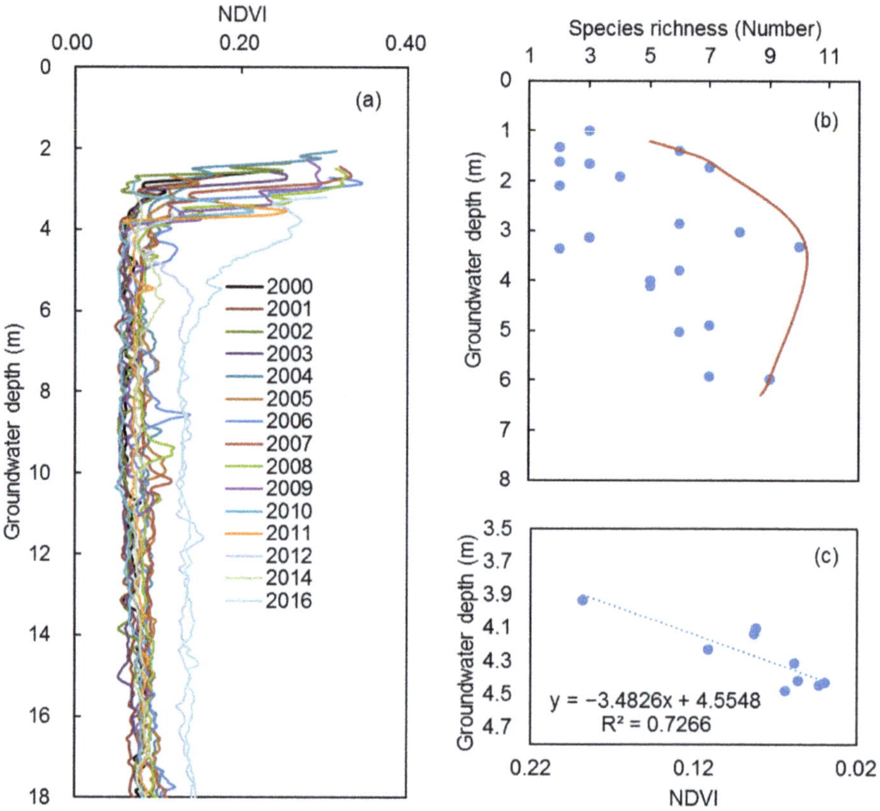


**Figure 7.** The relationship of groundwater depth with NDVI and species richness in the Minqin Oasis–desert ecotone [17]. ((**a**): the annual relationship between groundwater depth and NDVI based on grid; (**b**): the relationship between groundwater depth and species richness based on survey of vegetation quadrats; (**c**): the relationship between annual regional mean groundwater depth and NDVI in Qingtu Lake area).

All in all, under arid climate conditions and without irrigation, there can be no agriculture. The rapid increases in the population, farmland area, and intense agricultural irrigation activities have led to the excessive use of groundwater resources. These disorganized ways of utilizing water and land resources were the direct cause of the significant increase in GLD and the main reason for the degradation and shrinkage of natural vegetation in the ODE. This seriously restricted the sustainable development of SES in SRB. The ODE plays an important role in the stability of the oasis and the control of desertification, despite it being a key ecological sensitivity zone. Therefore, it is extremely important to protect and restore the ecosystem in ODE.

## 4. Discussion

### *4.1. The Effects of the Nature-Based Solutions*

Based on the concept of NbS, the local government adopted a series of measures to protect and restore the groundwater resources and the GDEs in Minqin Oasis. This effort was called the CASRB and was launched in 2007. The concrete measures primarily consisted of increasing the surface water supply through water transfer projects, decreasing groundwater mining by locking the pumping wells, and formation of water user associations, planting natural vegetation instead of crops, enhancing industrial water-saving technologies, and the adjustment of the agricultural planting structure by planting low-water-consuming crops [18,33]. These measures, combining the recovery of GLD with the afforestation project, aim to gradually restore the natural ecological function of the groundwater resources and eventually promote the resilience of the SES.

1. The proportions of surface water and groundwater to total water supply amounts

Water transfer projects have greatly increased the amount of surface water entering Minqin Oasis. The effort primarily involved two water transfer projects [18,34]. The first was the Jingdian Project Phase II that diverted water from the Yellow River starting in 2001; the second was diverting water from the Xiying River—a tributary of the Shiyang River—that began in 2006. With the additional water supply from the two water diversion projects, the amount of water diverted continued to increase, and the cross-sectional flow at Caiqi station showed a significant increasing trend [34]. In 2019, the cross-sectional flow reached 401 million $m^3$, almost the same as that in the 1970s (Figure 2b), of which 208 million $m^3$ was the amount of diverted water.

In addition to the measures for reducing groundwater extraction, the consumption of groundwater for agriculture continued to decrease. The proportion of the water supply used for agriculture gradually became less than 90% [34]. The groundwater exploitation has gradually decreased and stabilized. Since 2009, the groundwater extraction has been constantly below 110 million $m^3$. Analysis on the statistical data from the Wuwei Water Authority illustrated that the water supply composition of Minqin has been transformed from primarily emphasizing on groundwater before 2009 to emphasizing on both surface water and groundwater (Figure 8).

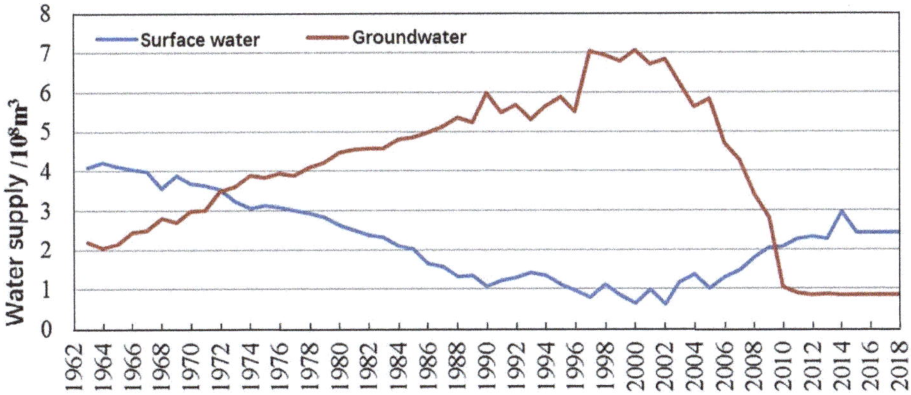


**Figure 8.** Changes in proportions of water supply amounts in Minqin Oasis.

2. Groundwater level recovery

By increasing the surface water supply and reducing groundwater exploitation, as well as implementing other water-saving policies and technologies, the GL has shown a recovery trend since 2009 compared to the previous period (Figures 3 and 9). The increasing rate of the GLD became less, from mostly between 4 m/10a and 6 m/10a (Figure 3a) to less than 4 m/10a (Figure 9), especially in the lake irrigation area, where the GLD of parts of the observation wells began to show a negative change trend. The CASRB thus has had a remarkable effect on the prevention of increased GLD.

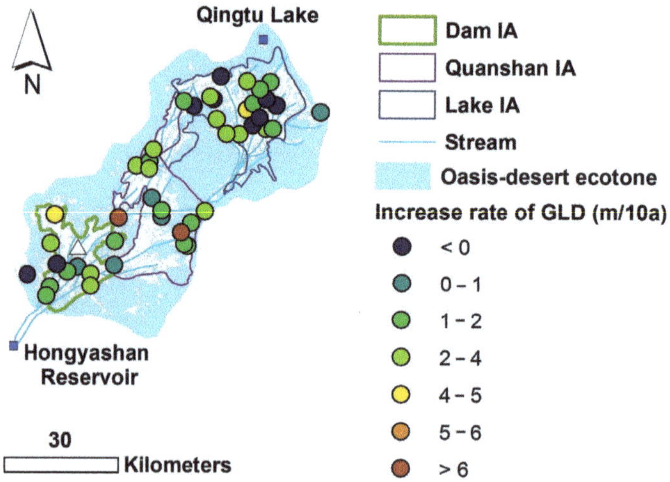


**Figure 9.** Change rate of the groundwater depth during 2009–2017.

3. Wetland area of Qingtu Lake

In addition to carrying out the above solutions, 280 million cubic meters of water was directly added to the Qingtu Lake from the Hongyashan Reservoir from 2010 to 2019 [34]; this greatly increased the area of Qingtu Lake (Figure 10) and the surrounding vegetation. The increasing rate of the Qingtu Lake was about 2.48 $km^2$ per year (Figure 10). The addition also played a role in the rise of the GL around Qingtu Lake in the downstream area [19,35].

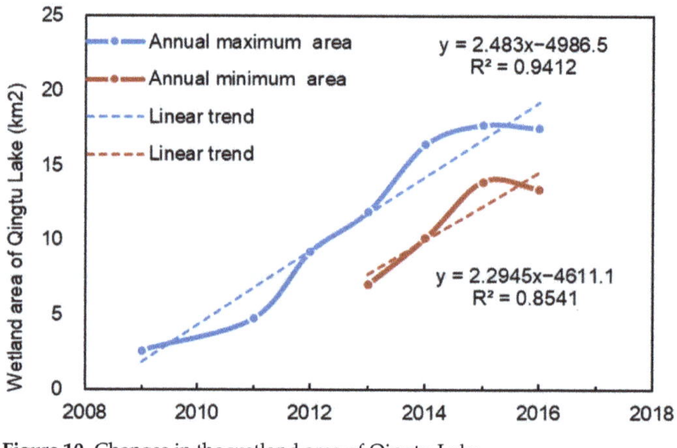


**Figure 10.** Changes in the wetland area of Qingtu Lake.

4. Vegetation ecology in the oasis–desert ecotone

From 2009 to 2017, the annual mean NDVI of natural vegetation gradually increased from 0.09 to 0.21 [34]. This demonstrated the effects of the implementation of the CRSRB on the improvement in vegetation ecology of the ODE. Currently, the areas with increasing NDVI are distributed in the eastern and southeastern peripheral areas of Minqin Oasis, the western part of Minqin Oasis, and the areas around Qingtu Lake. The increasing rate of the annual mean NDVI in these areas reached 0.1–0.3/10a [16]. Combined with the GLD of these areas and the correlation between the GLD and NDVI, we also found that the improvement of the vegetation in the western part of Minqin Oasis where the GLD was still about 20–30 m, was largely attributed to the afforestation project; the improvement of the vegetation in the lake irrigation area and the peripheral areas of the oasis were attributed to both the recovery of the GLD and the afforestation project. Under these interventions, the ecological function of the groundwater was gradually restored.

5. Lifting residents out of poverty

Based on the NbS of the CASRB, and in accordance with the requirements of high efficiency, new modes, new varieties, and new technologies, the Minqin County government continued to adjust the agricultural structure and resolutely prohibited the planting of high-water-consuming crops such as onions and corn. The county has vigorously developed efficient water-saving agriculture, agricultural practices, green organic agriculture, and ecological circular agriculture as well as the creation of a variety of characteristic vegetable- and fruit-based products. The agricultural water production has increased from 2.66 yuan/m$^3$ in 2005 to 21.7 yuan/m$^3$ in 2020, and the per capita disposable income of farmers in the rural area has increased from 3869 yuan in 2007 to 15,466 yuan in 2020. These measures have successfully eliminated the farmers' poverty and achieved solutions for the sustainable utilization of water resources, ecological protection, and increased income for farmers.

### 4.2. Comprehensive Evaluation of the Resilience of the Socio-Ecological System

The composite evaluation index of the SES resilience during 2009–2017 was calculated by the fuzzy comprehensive assessment method. The steps were as follows.

1. Selection of the resilience indicators of the SES

There are various resilience indicators of the SES. Since the main issues regarding the sustainable development of the SES in Minqin were the drop in GL, the degradation of GDEs, and the poverty of the residents, the NDVI, the wetland area of the Qingtu, the per-capita income, the GLD, and its change rate were selected as the indicators to assess the resilience of the SES (Table 1). The units of the indicators are shown in Table 1.

**Table 1.** Rating criteria of the resilience indicators of socio-ecological system.

| Indicators | V1 | V2 | V3 |
|---|---|---|---|
| NDVI | 0.50 | 0.15 | 0.08 |
| Area of the Qingtu wetland (km$^2$) | 17.73 | 11.46 | 2.63 |
| Per-capita income (Yuan/person) | 31,128 | 19,110 | 9160 |
| GLD (m) | 3.0 | 5.0 | 10.0 |
| Change rate of GLD (%) | −5.0 | 0.0 | 5.0 |
| Rating score | 0.95 | 0.50 | 0.05 |

2. Determination of grading values and the corresponding rating score

According to the theory of fuzzy mathematics, the values of the above-selected indicators were divided into three levels of rating score, i.e., V1, V2, and V3. Among these, V1 indicates that the resilience of the SES is relatively strong; V3 indicates that the resilience of the SES is close to the minimum value. V2 is between V1 and V3, indicating a medium level of resilience. To quantitatively reflect the resilience of the SES, V1, V2, and V3 were scored between 0 and 1 (Table 1). The composite index (CI) of the SES was calculated using the Equation (1), in which a is the CI; b is the weight of each indicator, and $\alpha$ is the rating score. The thresholds of indicators corresponding to V1, V2, and V3 were determined by the following method. That of NDVI was determined based on the analysis of changes detailed above and by previous studies on the desert vegetation [17,31,32]. The thresholds of the wetland area were determined by the maximum, mean, and minimum values during the evaluation period. The thresholds of per-capita income were determined according to the maximum, mean, and minimum values for Wuwei City. The thresholds of GLD were determined according to the interactions between the GLD and vegetation, as in

Liu et al. [17]. The thresholds for the change rate were determined by the increasing or decreasing trends of the GLD due to the fact that a decreasing trend is favorable for the restoration of groundwater and its dependent ecology.

$$\sum_{j=1}^{3} b_j^k \cdot \alpha_j / \sum_{j=1}^{3} b_j^k \tag{3}$$

3. Determination of the weight for each resilience indicator

The indicators were classified into three categories, i.e., the ecology, the socio-economy, and the groundwater resources (Table 2). Combining the principles of analytic hierarchy processes and expert scoring methods, the relative weights of NDVI, wetland area, per-capita income, GLD, and the change rate of GLD were finally adjusted to 0.198, 0.132, 0.340, 0.198, and 0.132.

**Table 2.** Weights of the indicators.

| Classification | Weights | Indicators | Weights |
|---|---|---|---|
| Ecology | 0.33 | NDVI<br>Area of the Qingtu wetland ($km^2$) | 0.6<br>0.4 |
| Socio-economy | 0.33 | Per-capita income (yuan/person) | 1.0 |
| Groundwater | 0.34 | GLD (m)<br>Relative rate of GLD (%) | 0.6<br>0.4 |

4. Composite evaluation index of the resilience of SES

Changes in the composite index of the SES (*Vc*) during 2009–2017 were shown in Figure 11. The results illustrated that *Vc* in most of the years was below 0.6, indicating a relatively low resilience of the SES. However, the *Vc* showed an increasing trend, especially increasing significantly from 2011 to 2017, the second stage of the CASRB. The value in 2017 was 1.73 times that in 2011. In the meantime, *V*3 showed a decreasing tendency, while *V*1 had an increasing tendency. This illustrated the dramatic effects of the CASRB. Temporally, the increase in the CI was initially attributed more to the increase in *V*2, and in turn to more to *V*1. This process suggested that the resilience of SES was gradually stronger after the implementation of the CASRB. Among the three rating levels, *V*2 still accounted for most of the current state, being almost twice that of *V*1. Therefore, the resilience of the SES in Minqin Oasis could still be enhanced.

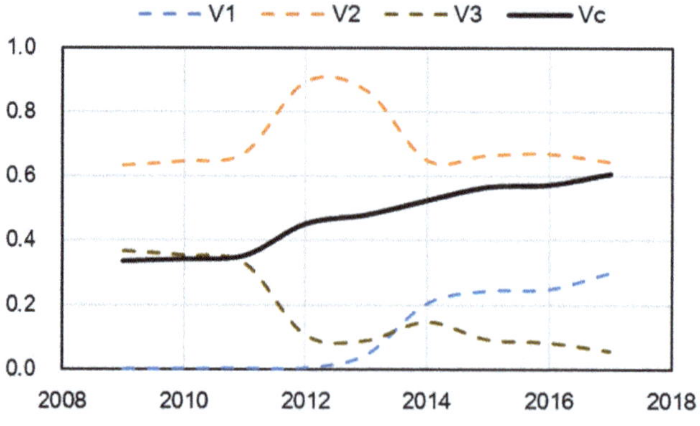


**Figure 11.** Changes in the composite index of the resilience of the socio-ecological system.

### 4.3. Recommendations and Measures

Based on the above analysis, we conclude that the implementation of the CASRB has had a positive effect on the resilience of the SES of Minqin Oasis. The GLD reached a relatively balanced status, the ecological environment has been on a path of restoration, and residents of Minqin County have been comprehensively lifted out of poverty. However, the overall GLD in the Minqin Oasis is still very large, especially in the western part of Minqin, where the current GLD is 20–40 m. The artificially planted *Haloxylon ammodendron* grows poorly in dry years and is still insufficient to withstand the intrusion of the Badain Jilin Desert. Therefore, it is necessary to continue to strengthen this comprehensive governance. Furthermore, the authors recommended that in the later stages of the CASRB, different measures need to be taken for different regions within the oasis.

(1) The ODE is a natural vegetation protection region. The critical GLD of soil salinization (2 m) and the critical vegetation-degradation value of groundwater ecological function (5 m) are recommended as GLD thresholds [17,34]. In the near future (i.e., by the year 2035), the main goal is to restore the ecological functioning of groundwater and prohibit its overexploitation. In the long term (after 2035), attention should be paid to controlling the GLD less than 2 m to avoid salt desertification. Artificial afforestation should be suited to local conditions, and shrub and grass planting projects should conform to the local climate and environment, selecting drought- and salt-tolerant plant species.

(2) In agroecosystem regions, groundwater should be extracted rationally. Multiple factors such as the amount of groundwater recharge, the soil salinization threshold of the GLD, and the influence of groundwater mining on the surrounding natural ecology should be comprehensively considered. The GLD values of 2 m and 10 m are recommended as the minimum and maximum thresholds, and the amount of groundwater extraction could be determined according to these GLD thresholds. Therefore, the near future goal of the agroecosystem region is food security. The local government should reasonably determine the scale of agricultural land and irrigate the farmland by the joint regulation of surface water and groundwater. In the long term, attention should be paid to controlling the GLD greater than 2 m to avoid causing secondary salinization of the soil and affecting the quality and yield of crop.

(3) In the urban ecological security guarantee region, the safety guarantee of underground buildings is the primary task. The local government should reasonably determine the maximum GLD threshold and appropriately extract groundwater by holistically coordinating regional water level changes. The authors recommend the minimum and maximum thresholds as 10 m and 15–20 m, respectively, due to the conclusion that a high GL would lead to the instability of underground buildings.

## 5. Conclusions

(1) The natural runoff flowing into Minqin Oasis decreased significantly at a rate of 81 million cubic meters per decade. Combined with the expansion of irrigated farmland in Minqin Oasis, GLD increased dramatically, and the natural vegetation ecology of the ODE was seriously degraded. There was a strong correlation between each couple factor among the four factors on the causal chain; i.e., expansion of farmland–increase in groundwater abstraction amount–increase in GLD–degradation of natural ecosystem.

(2) Since the implementation of the NbS in 2007, the dropping trend of GL has been curbed, the natural vegetation of the ODE has been partially restored, the wetland area of Qingtu Lake has increased, and people's income has also increased. The SES resilience index has increased by a factor of 1.82 from 2009 to 2017, but it was still below 0.6, and the GLD has not reached the ecologically suitable depth.

(3) To continually improve the resilience of the SES of Minqin Oasis, the idea of harmonious coexistence between humans and nature should continue to be adhered to in the later comprehensive treatment of the SRB, and the implementation of the CASRB

should focus on comprehensive management, systematic management, and zoning management to promote the integrated protection and restoration of mountains, rivers, forests, fields, lakes, grass, and sand.

**Author Contributions:** Conceptualization, Z.N. and M.L.; methodology, M.L., Z.N. and L.C.; software, M.L., L.W. and H.L.; validation, L.W. and L.C.; formal analysis, M.L.; investigation, M.L., L.C. and L.W.; resources, M.L., L.W., L.C. and H.L.; data curation, L.W. and L.C.; writing—original draft preparation, M.L.; writing—review and editing, Z.N. and L.C.; supervision, Z.N.; project administration, Z.N.; funding acquisition, M.L. and Z.N. All authors have read and agreed to the published version of the manuscript.

**Funding:** This research was funded by the National Natural Science Foundation of China [No. 41902262]; the Innovation Capability Enhancement Plan of Hebei Province [No. 225A4204D], and the Geological Survey Project of China [No. DD20230432].

**Data Availability Statement:** The data presented in this study are available on request from the corresponding author.

**Acknowledgments:** We thank Wang Jinzhe, Wang Qian, Cui Haohao, et al. in our team for their help in the joint investigation of groundwater and ecological quadrat. We also thank the three anonymous reviewers for spending time on reviewing our manuscript and providing us with a list of constructive comments.

**Conflicts of Interest:** The authors declare no conflict of interest.

## References

1. Intergovernmental Panel on Climate Change. *Global Warming of 1.5 °C*; Cambridge University Press: Cambridge, UK; New York, NY, USA, 2018.
2. World Bank. *Biodiversity, Climate Change, and Adaptation: Nature-Based Solutions from the World Bank Portfolio*; World Bank: Washington, DC, USA, 2008.
3. Cohen-Shacham, E.; Walters, G.; Maginnis, S.; Janzen, C. *Nature-Based Solutions to Address Global Societal Challenges*; IUCN: Gland, Switzerland, 2016.
4. Cohen-Shacham, E.; Andrade, A.; Dalton, J.; Dudley, N.; Jones, M.; Kumar, C.; Maginnis, S.; Maynard, S.; Nelson, C.; Renaud, F.; et al. Core principles for successfully implementing and upscaling Nature-based Solutions. *Environ. Sci. Policy* **2019**, *98*, 20–29. [CrossRef]
5. United Nations Development Programme. *Human Development and Planetary Health: The Role of Nature-Based Solutions*; United Nations: San Francisco, CA, USA, 2019. Available online: https://hdr.undp.org/content/human-development-and-planetary-healththe-role-nature-based-solutions (accessed on 16 September 2023).
6. International Union for Conservation of Nature. *Guidance for Using the IUCN Global Standard for Nature-Based Solutions: A User-friendly Framework for the Verification, Design and Scaling up of Nature-Based Solutions*, 1st ed.; International Union for Conservation of Nature: Gland, Switzerland, 2020.
7. European Environment Agency; Castellari, S.; Zandersen, M.; Davis, M.; Veerkamp, C.; Förster, J.; Marttunen, M.; Mysiak, J.; Vandewalle, M.; Medri, S.; et al. *Nature-Based Solutions in Europe Policy, Knowledge and Practice For Climate Change Adaptation and Disaster Risk Reduction*; Publications Office of the European Union, LU: Luxembourg, 2021.
8. Pérez-Cirera, V.; Cornelius, S.; Zapata, J. *Powering Nature: Creating the Conditions to Enable Nature-Based Solutions*; WWF: Gland, Switzerland, 2021. Available online: https://lp.panda.org/powering-nature-report (accessed on 16 September 2023).
9. European Union. Towards an EU Research and Innovation Policy Agenda for Nature-based Solutions & Renaturing Cities. 2015. Available online: https://publications.europa.eu/en/publication-detail/-/publication/fb117980-d5aa-46df-8edc-af367cddc202 (accessed on 16 September 2023).
10. International Union for Conservation of Nature. Call for Wetland Decade under the UN Decade on Ecosystem Restoration (2021–2030). 2019. Available online: https://www.iucn.org/news/water/201903/call-wetland-decade-under-un-decade-ecosystem-restoration-2021--2030 (accessed on 16 September 2023).
11. United Nations. Nature-based Solutions for Supporting Sustainable Development, Resolution 5 of United Nations Environment Assembly Adopted on 2 March 2022, 2022 Retrieved on 5 June 2022. Available online: https://wedocs.unep.org/bitstream/handle/20.500.11822/39864/NATURE-BASED%20SOLUTIONS%20FOR%20SUPPORTING%20SUSTAINABLE%20DEVELOPMENT.%20English.pdf?sequence=1&isAllowed=y (accessed on 16 September 2023).
12. Mehta, D.; Pandey, R.; Gupta, A.K.; Juhola, S. Nature-based Solutions in Hindu Kush Himalayas: IUCN Global Standard Based Synthesis. *Ecol. Indic.* **2023**, *154*, 110875. [CrossRef]
13. Yu, R.; Mu, Q. Implementation progress of NbS in China: A Global Comparative Review. *NbS* **2023**, *4*, 100075.

14. Fan, X. The Characteristics of Groundwater Resources and the Environmental Problems Caused by Development of Water Resource, the Arid Area in the Northwest of China. *Hydrol. Eng. Geol.* **1990**, *1*, 3–7, 12–16. (In Chinese)
15. Ding, H.; Wang, G.; Huang, X. Runoff Reduction into Hongyashan Reservoir and Analysis on Water Resources Crisis of Minqin Oasis. *J. Des. Res.* **2003**, *23*, 84–89.
16. Cao, L.; Nie, Z.; Liu, M.; Wang, L.; Wang, J.; Wang, Q. Changes in Natural Vegetation Growth and Groundwater Depth and Their Relationship in the Minqin Oasis in the Shiyang River Basin. *Hydro. Eng. Geol.* **2020**, *47*, 25–33. (In Chinese)
17. Liu, M.; Nie, Z.; Cao, L.; Wang, L.; Lu, H.; Wang, Z.; Zhu, P. Comprehensive Evaluation on the Ecological Function of Groundwater in the Shiyang River Watershed. *J. Ground. Sci. Eng.* **2021**, *9*, 326–340. [CrossRef]
18. Hao, Y.; Xie, Y.; Ma, J.; Zhang, W. The Critical Role of Local Policy Effects in Arid Watershed Groundwater Resources Sustainability: A Case Study in the Minqin Oasis, China. *Sci. Total Environ.* **2017**, *601–602*, 1084–1096. [CrossRef]
19. Huang, F.; Chunyu, X.; Zhang, D.; Chen, X.; Ochoa, C.G. A Framework to Assess the Impact of Ecological Water Conveyance on Groundwater-Dependent Terrestrial Ecosystems in Arid Inland River Basins. *Sci. Total Environ.* **2020**, *709*, 136155. [CrossRef]
20. Zhang, Z.; Yang, X.; Li, H.; Li, W.; Yan, H.; Shi, F. Application of A Novel Hybrid Method for Spatiotemporal Data Imputation: A Case Study of the Minqin County Groundwater Level. *J. Hydrol.* **2017**, *553*, 384–397. [CrossRef]
21. Wang, X.; Chen, X.; Ding, Q.; Zhao, X.; Wang, X.; Ma, Z.; Lian, J. Vegetation and Soil Environmental Factor Characteristics, and Their Relationship at Different Desertification Stages: A Case Study in the Minqin Desert-Oasis Ecotone. *Acta Ecolo Sinica* **2018**, *38*, 1569–1580, (In Chinese with English abstract).
22. Zhang, Y.; Chang, Z. Study on the Dynamic Vegetation Degeneration in the Minqin Desert Area. *Arid Zone Res.* **2008**, *25*, 16–21, (In Chinese with English abstract). [CrossRef]
23. Liu, H.; Liu, S.; Li, Y.; Yang, Z.; Li, Y.; Zhang, Y.; Guo, C.; Liu, S. Response of Riparian Vegetation to the Change of Groundwater Level at Middle and Lower Reaches of the Shiyang River. *Arid Zone Res.* **2012**, *29*, 335–341, (In Chinese with English abstract).
24. Lu, H.; Nie, Z.; Liu, M.; Feng, B.; Cheng, X.; Wang, J.; Wang, Q.; Cui, H.; Fan, F. Research on Land Cover Changes in Shiyang River Basin in Recent 50 Years Based on RS and GIS. *Geogr. Res.* **2020**, *29*, 8, (In Chinese with English abstract).
25. Hess, A.; Iyer, H.; Malm, W. Linear Trend Analysis: A Comparison of Methods. *Atmos. Environ.* **2001**, *35*, 5211–5222. [CrossRef]
26. Xu, J. *Mathematical Methods in Modern Geography*, 2nd ed.; Higher Education Press: Beijing, China, 2002; pp. 305–335.
27. Cohen, P.; Cohen, P.; West, S.G.; Aiken, L.S. *Applied Multiple Regression/Correlation Analysis for the Behavioral Sciences*, 2nd ed.; Psychology Press: London, UK, 1983. [CrossRef]
28. Fang, J.; Wang, X.; Shen, Z.; Tang, Z.; He, J.; Yu, D.; Jiang, Y. Methods and Protocols for Plant Community Inventory. *Biodi. Sci.* **2009**, *17*, 533–548. (In Chinese)
29. Lou, S. *Fuzzy Mathematics*; Science Press: Beijing, China, 1983.
30. Song, S.; Nie, Z.; Geng, X.; Shen, X.; Wang, Z.; Zhu, P. Response of Runoff to Climate Change in the Area of Runoff Yield in Upstream Shiyang River Basin, Northwest China: A Case Study of the Xiying River. *J. Groundw. Sci. Eng.* **2023**, *11*, 89–96. [CrossRef]
31. Jin, X. Quantitative Relationship Between the Desert Vegetation and Groundwater Depth in Ejina Oasis, the Heihe River Basin. *Earth Sci. Front.* **2010**, *17*, 181–186, (In Chinese with English abstract).
32. Cao, L.; Nie, Z.; Liu, M.; Wang, L.; Wang, Q. The Ecological Relationship of Groundwater–Soil–Vegetation in the Oasis–Desert Transition Zone of the Shiyang River Basin. *Water* **2021**, *13*, 1642. [CrossRef]
33. Chen, L.; Feng, Q. Geostatistical Analysis of Temporal and Spatial Variations in Groundwater Levels and Quality in the Minqin Oasis, Northwest China. *Environ. Earth Sci.* **2013**, *70*, 1367–1378. [CrossRef]
34. Nie, Z.; Liu, M.; Liu, P. *Acceptance Report of the Research and Demonstration of Rational Exploitation and Utilization of Groundwater Resources and the Protection of Its Ecological Function in the Shiyang River Basin*; Ministry of Science and Technology of the People's Republic of China: Beijing, China, 2021.
35. Gu, J.; Guo, P.; Huang, G. Achieving the Objective of Ecological Planning for Arid Inland River Basin Under Uncertainty Based on Ecological Risk Assessment. *Stoch. Environ. Res. Risk Assess.* **2016**, *30*, 1485–1501. [CrossRef]

water

MDPI

*Article*

# A Participatory Approach to Exploring Nexus Challenges: A Case Study on the Pinios River Basin, Greece

**Dimitrios Malamataris [1], Anna Chatzi [1], Konstantinos Babakos [1], Vassilios Pisinaras [1], Evangelos Hatzigiannakis [1], Barbara A. Willaarts [2], Manuel Bea [2], Alessandro Pagano [3] and Andreas Panagopoulos [1,*]**

1 Soil and Water Resources Institute, Hellenic Agricultural Organization "DEMETER", 57400 Thessaloniki, Greece; d.malamataris@swri.gr (D.M.); a.chatzi@swri.gr (A.C.); k.babakos@swri.gr (K.B.); v.pisinaras@swri.gr (V.P.); e.hatzigiannakis@swri.gr (E.H.)

2 ICATALIST S.L., Las Rozas de Madrid, 28232 Madrid, Spain; bwillaarts@icatalist.eu (B.A.W.); mbea@icatalist.eu (M.B.)

3 Consiglio Nazionale delle Ricerche, Istituto di Ricerca Sulle Acque (IRSA-CNR), 70132 Bari, Italy; alessandro.pagano@ba.irsa.cnr.it

* Correspondence: a.panagopoulos@swri.gr; Tel.: +30-2310798790 (ext. 205)

**Abstract:** The conservation of natural resources is indispensable for supporting sustainable development. Water–Energy–Food–Ecosystem (WEFE) nexus management is widely acknowledged as the utmost priority of our time. Considering the specificities of the different nexus sectors, conflicting views of sustainability may arise from different stakeholders. This paper presents and analyses an effective stakeholders' participation strategy aiming at identifying WEFE challenges in the Pinios River Basin (PRB), the most agriculturally productive and natural resource-stressed basin of Greece. The presented methodology was developed in the context of a Learning and Action Alliance (LAA) scheme adopted by the H2020 REXUS project, intended to support resilient nexus system management with the engagement of stakeholders. The proposed comprehensive methodology comprises multiple phases, including "framing", "mapping", "involvement", "co-production", and "sharing" phases. The involvement phase is the most interactive one, including the organisation and outputs of efficient crowdsourcing theme sessions organised in the framework of the first REXUS PRB stakeholders' workshop. This paper illustrates and analyses stakeholders' perceptions regarding nexus status in the PRB and reveals the most critical challenges in the pilot basin, along with their interdependencies and correlations. The determination of nexus challenges insights by stakeholders could significantly contribute to redefining policies so that they align with sustainable development aims. This methodology is proposed to form the baseline strategy in stakeholders' engagement for future nexus management studies.

**Keywords:** WEFE nexus; Pinios River Basin; stakeholder engagement; sustainability; Thessaly; Mediterranean area; natural resource management; crowdsourcing; expert knowledge

## 1. Introduction

### *1.1. Natural Resource Security and Nexus*

Human activities have currently reached a threshold that could damage the Earth's systems [1]. Worldwide, the agricultural sector accounts for about 70% of freshwater abstractions, the energy sector accounts for about 60% of total greenhouse gas emissions [2], and land use changes have affected 32% of the globe's land area since 1960 [3–5]. The global population is projected to grow to 8.5 and 9.7 billion in 2030 and 2050, respectively, according to the latest projections [2]; thus, we are expected to need 55% more water, 80% more energy, and 60% more food by 2050 [6,7].

Resolving natural resource scarcity, particularly in arid and semi-arid regions, including the Mediterranean region, constitutes a critical challenge for supporting human lives and improving well-being in the context of sustainable development [8–10]. These regions face common natural pressures, such as increasing temperature, decreasing precipitation, and spatiotemporal imbalances between natural resource supply and demand [11], apart from the management challenges related to stakeholders' involvement in the decision and policy making processes [12,13]. Primarily, water, energy, food/agriculture, and ecosystems/climate are the most precious and essential sectors for sustaining healthy ecosystem functioning to facilitate improved human well-being, stable economic development, and strengthened social cohesion [14–18]. Improved natural resource management constitutes a prerequisite for achieving sustainability in accordance with the Sustainable Development Goals [19]. In this direction, a coordinated and integrated management of resources across sectors, supporting the cost-effective, efficient, and sustainable resources use, is being increasingly promoted [20–22].

Integrated approaches emphasising effective governance [23,24] and the promotion of innovative solutions, such as Nature-based Solutions (NbS) [25], have recently emerged as crucial factors to guarantee natural resource security. The characteristics of governance have also shifted from a state-centric to a decentralised and more transparent approach characterised by the broad participation of stakeholders [26,27]. Increasing attention is being given to cross-sectoral stakeholder collaboration, since impacts of sectoral decisions and policies could easily affect other domains and may result in resource shortages and conflicts among users [28–31].

In 2011, the concept of the Water–Energy–Food (WEF) nexus was conceived and put forward for the first time; water is the core element of this nexus system [32]. In the same year, the Global Risks Report [4] recognised the complex relationships among WEF nexus sectors, characterising their risks as one of the three most important global risks. The WEF link relationship was then further explained by the Asian Development Bank [33] and the Food and Agriculture Organization of the United Nations [5]. Since then, nexus approaches have become widely recognised by academic communities [34–42]. The essence of the nexus is the interconnections between different sectors, and the nexus is inextricably connected to the concept of resource security. Achieving the UN's Sustainable Development Goals (SDGs), including those regarding food (SDG 2), water (SDG 6), and energy (SDG 7), will require substantial, if not transformative, efforts across many sectors and landscapes. Nexus thinking advocates for water, energy, and food systems to be viewed collectively and holistically in order to achieve water, energy, and food (WEF) security. The nexus concept has been also broadened to include other factors, such as ecosystems, land, climate, or emissions [39].

### *1.2. Stakeholders' Engagement in Nexus Studies*

An effective natural resource management of complex nexus systems requires the participation of stakeholders [43–46]. Despite the challenges and potential conflicts related to stakeholder involvement, consultation among stakeholders is fundamental in order to guarantee that existing interlinkages regarding the management of natural resources are highlighted and in order for a line to be drawn between the use and protection of certain limited natural resources [47,48]. Stakeholder participation can be understood as a means through which authorities of different levels and sectors, including end users, local managers, policy and decision makers, academics, organised citizens, and every single user, try to contribute to improving the understanding and management of natural resources [13,49].

Successful stakeholder participation is centred around incorporating, amongst other things, a broad spectrum of heterogeneous backgrounds and knowledge levels of participants, as well as possibly conflicting interests regarding resource allocation to create a "common sense" consensus [50–52]. There is evidence that transparent and democratic stakeholder engagement may result in more well-founded decisions [53], potentially leading to less social resistance, easier implementation [54], and higher efficiency [55]. Stakeholder engagement forms and tools can vary, ranging from questionnaires to forums and other means of participation [56]. However, whatever the form of engagement is, a participatory approach that involves both qualitative and quantitative methods is thought to be the most appropriate for addressing uncertainties and enhancing the understanding of nexus systems [39,57].

A number of studies involving engaging stakeholders in nexus systems analysis have been carried out. Howarth and Monasterolo investigated the interactions between water, energy, and food in the United Kingdom through the organisation of five themed workshops, recognising the added value of an interdisciplinary approach focused on facilitating a constructive dialogue and more efficient decision making [47,58]. Johnson and Karlberg proposed a participatory scenario building process involving stakeholders with different levels of knowledge and experience and priorities to address water, energy, and food nexus challenges in Ethiopia and Rwanda [59]. Hoolohan et al. collected insights from stakeholders through interviews, case studies, and workshops to examine the value of stakeholder engagement in terms of addressing water, energy, and food nexus challenges [60]. Brethaut et al. analysed results obtained from interviews and workshops to better understand water, food, and energy nexus conflicts in the WWF Conservation Mekong Flooded Forest Landscape in Cambodia, Asia [61]. Kropf et al. conducted semi-structured interviews and a workshop to elicit sectoral stakeholders' perceptions on water, ecosystems, food, and climate nexus sectors' relationships in the Seewinkel agricultural region, Austria [62]. Through stakeholder workshops, Karutz et al. integrated local stakeholder perspectives into a coherent framework for food, water, and energy analysis in Bhima basin, India [63]. Vinca et al. [64] adopted a co-production approach through a series of workshops to identify nexus challenges in the Indus Basin and stakeholders' perspectives on these challenges, and they used this information to co-develop a number of quantitative future scenarios. De Vito et al. [65] used System Dynamics tools to facilitate a dialogue among stakeholders, particularly focusing on water resource management for agriculture, trying to understand the impact of policies and decisions across multiple nexus sectors.

### *1.3. Literature Gaps and Objective Focus*

One of the major obstacles to achieving WEFE nexus sustainability is involving multiple stakeholders with various interests into all of the resource planning and implementation processes [66]. Nonetheless, most of the publications in the literature on stakeholders' engagement in the context of nexus sustainability frameworks focus either on nexus challenges and identifying their interrelations or on the determination of measures. Focusing on the Pinios River Basin (PRB), a highly stressed basin in Greece, the current study investigates and presents an effective stakeholder engagement strategy that can be adapted for any other nexus system. For this study, the perceptions and interpretations of stakeholders in terms of the main PRB WEFE challenges and their interconnections were investigated. Particular attention was given to (i) the collaborative identification of the main challenges for the area in terms of the conservation of natural resources, (ii) and the identification and analysis of the main cross-sectoral interconnections and dependencies related to resource use (and security), (iii) and the collaborative identification of potential sustainability pathways for the study area.

The proposed stakeholder engagement strategy aims to bridge the gap between science and implementation, facilitating the active involvement of stakeholders on the whole roadmap towards achieving nexus sustainability, i.e., from the identification of nexus challenges related to the security of natural resources to the identification and selection of potential measures. To the authors' knowledge, the design and development of a holistic and integrated WEFE nexus system regarding the PRB and shifting from "Nexus Thinking to Nexus Doing" within the context of the climate crisis and based on different stakeholders' viewpoints has not been described in the literature previously.

The remainder of this article is structured as follows: Section 2 presents the WEFE characteristics of the Pinios River Basin and the strategy developed for effective stakeholder participation. Section 3 is divided into five sub-sections dedicated to illustrating and analysing the stakeholder engagement strategy, including framing, mapping, involvement, co-production, and sharing phases. Emphasis is placed on the presentation of the first REXUS PRB stakeholder workshop organisation and the challenges identified in the study area and the interrelations between them.

## 2. Materials and Methods

### 2.1. Study Area

The Pinios River Basin (Figure 1) is situated in Thessaly, Central Greece, covering an area of ca 11,000 $km^2$. It is characterised by rugged terrain, predominantly along the northern and western–southwestern margins of the basin, with two main plains gently sloping to the southeast and a considerable topographic elevation range spanning from 0 to 2810 m a.s.l. The hydrologic basin is mainly drained by the Pinios River, which has a length of 213 km, constituting the third longest river in Greece. The Pinios River course starts at the northwestern part of the Thessaly plain, crosses a large part of Central Greece, and discharges into the Aegean Sea, forming a NATURA convention-protected delta area. The major tributaries of the Pinios River are the Malakasiotiko, Portaikos, Pamisos, Enippeas, Ion, Lithaios, Neochoritis, and Titarisios rivers, which all drain extensive and heterogenous areas. Agricultural areas predominate, comprising 50.90% of the total area in the PRB, followed by forest and seminatural areas (45.32%), artificial surfaces (2.80%), water bodies (0.82%), and wetlands (0.15%). In-depth information on PRB water balance and detailed analyses of the hydrodynamic mechanisms in the region can be provided and supported, respectively, by the Pinios Hydrologic Observatory (PHO), which was established in the PRB in 2015, belonging to the International Long-Term Ecological Research Network, which aims to promote sustainable ecosystems and natural resource management [67]. However, several studies have been elaborated utilising hydrologic and climatic monitoring data series of the past 5 decades maintained by several public bodies.

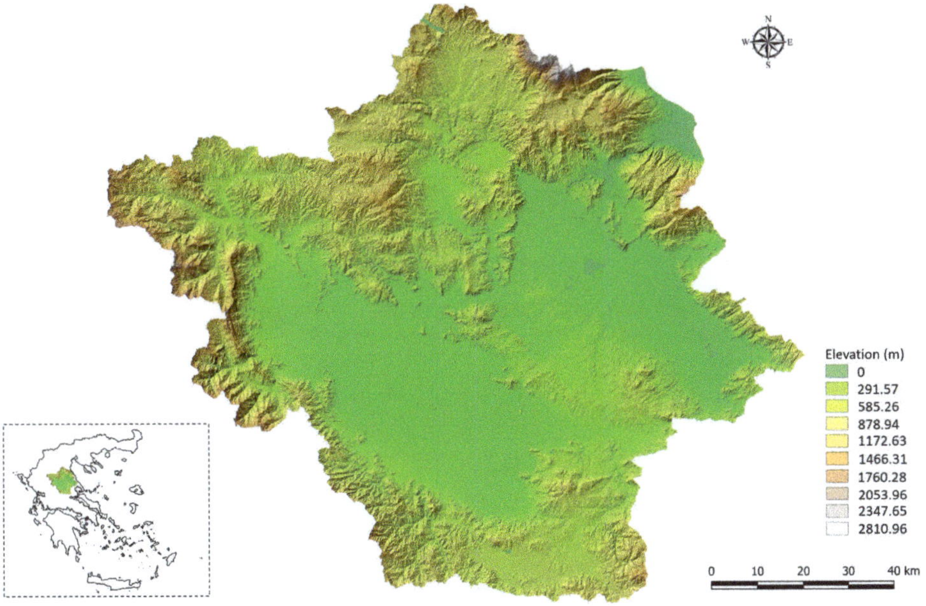


**Figure 1.** Pinios River Basin pilot area—DEM generated by the Hellenic Cadastre S.A at 5 m spatial resolution.

The PRB presents highly diversified climatic, geological, and hydrogeological conditions [68], supporting developed economic activities that, in turn, create several pressures, thus shaping a regional nexus scheme that is rather complex to comprehend, analyse, and manage. Climatic conditions vary considerably eastwards, following the dominant topographic discretisation in the eastern central and western parts. In particular, at the western and central parts, the continental climate type is dominant, while at the eastern part of the basin, the typical Mediterranean climate type is apparent. According to Pisinaras et al. [68], long-term average precipitation between 1971 and 2000 was 770 mm/year, ranging between 500 mm/year in the eastern Thessaly plain around the city of Larissa and 1800 mm/year in the western boundary of the basin. Most of the PRB's area is covered by permeable geological formations (44%), followed by impermeable formations (40%), and karstified marbles (16%). The aquifer systems of the plains are rich in the thick alluvial deposits that fill the eastern and western basins and are characterised by high groundwater potential that largely covers the regional water demands. The karstic marbles mostly crop out at the margins of the plains and give rise to high-potential aquifers that play a key role in the hydrodynamic evolution of the alluvial aquifers [69] (Figure 2, in GGRS87 coordinate system).

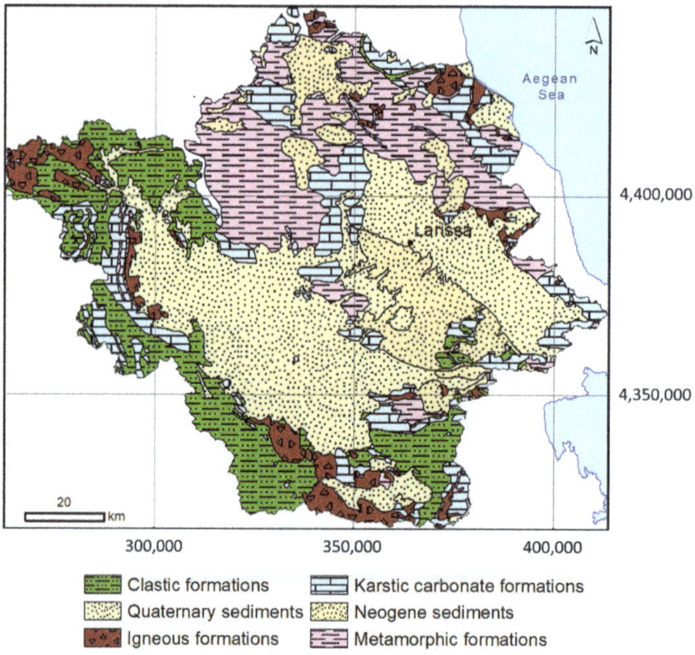


**Figure 2.** Hydrolithology of the Pinios River Basin.

The Pinios River Basin is considered one of the most characteristic examples of non-sustainable WEFE resource management in Greece, with contradicting opinions regarding the appropriate solutions amongst those involved in the relevant sectors, often leading to many disputes and confrontation, as well as maintaining the long-standing tradition of stakeholder inactivity and disengagement. Intensive agricultural activities have led to the over-exploitation of natural resources and increased environmental and energy pressures, while a notable water deficit for more than the last four decades has been established [70,71]. The identification of the main challenges in the PRB by stakeholders varies significantly, since they present different views and priorities according to differing criteria, including their area(s) of interest, main economic activity, and even social perception, adding to the complexity of the situation. As such, the process of identifying the main PRB challenges becomes a fairly complicated process.

The PRB Water sector: The PRB covers almost 85% of the spatial extent of the Thessaly water district (EL08). In the framework of the 1st Revision of River Basin Management Plans of Thessaly Water District (implemented in accordance with the European Union Water Framework Directive), 64 rivers, 2 natural lakes, 1 artificial lake, 2 coastal water systems, and 40 groundwater systems have been identified in the PRB (Figure 3a, in WGS84 coordinate system). Regarding water allocation, agricultural use constitutes the major reason for water consumption in the PRB, accounting for about 93% of total water consumption (1292 $hm^3$), followed by domestic (5.5%), livestock (1%), and industrial (0.5%) uses [69]. The systematic over-exploitation of the area's groundwater resources, as well as its surface water resources, was initiated in the PRB in the late 1980s. The highly intensified agricultural activity, in conjunction with irrational irrigation practices, has resulted in a remarkable need for an increase in irrigation water, and attempts to address this need have mainly involved the over-exploitation of groundwater resources. The exploitation of non-renewable groundwater resources, along with nitrate contamination, constitute the most critical threats to groundwater sustainability [72]. The concentration of nitrates in the regional aquifers is expected to increase in the future, mainly due to the decrease in groundwater recharge and increase in water abstraction for several uses [73]. In the past decade, signs of reductions in documented groundwater heads, leading to declining trends and small-scale reversal trends, have been noticed as a result of the rationalisation of the practiced irrigation, the reduction in irrigated land, and the shift to less water-demanding crops [74]. The persistent groundwater deficit underscores the urgent need to develop collective irrigation networks to reduce the excessive reliance on private wells and promote sustainable groundwater resource management [75].

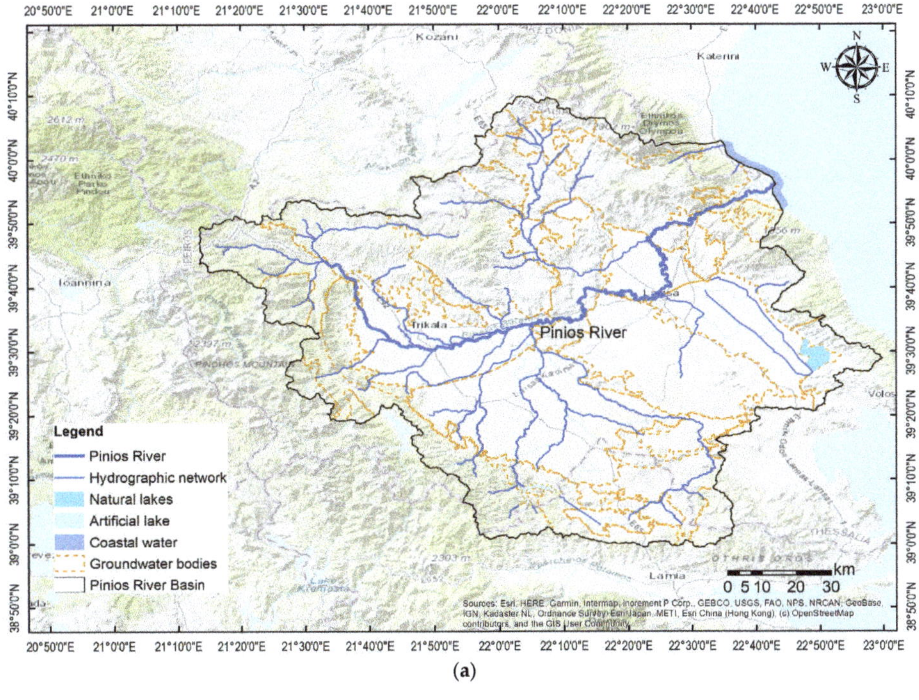


(a)

**Figure 3.** *Cont.*

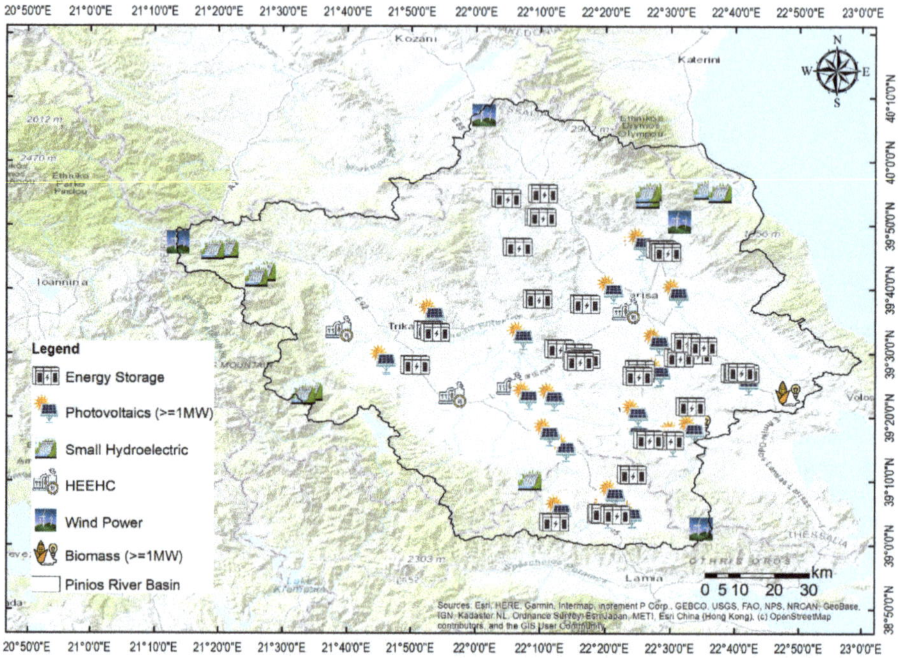


(b)

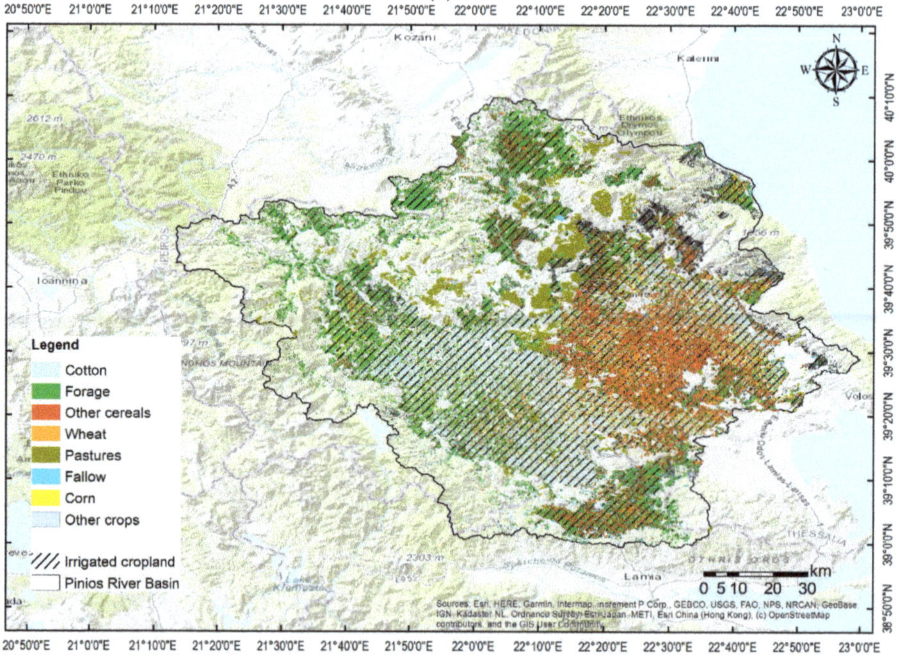


(c)

**Figure 3.** *Cont.*

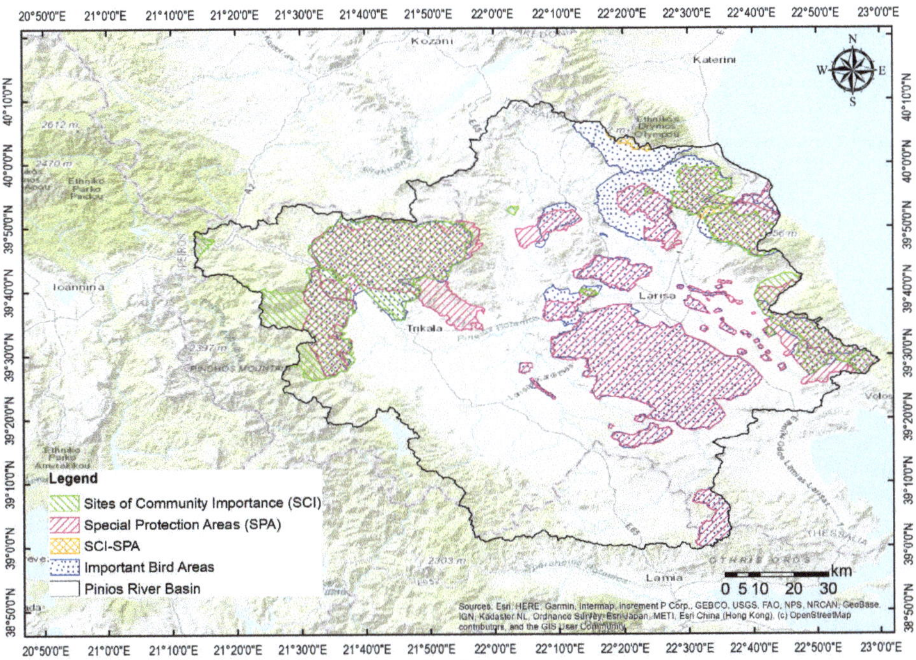


(d)

**Figure 3.** (**a**) Water sector in the Pinios River Basin. (**b**) Energy sector in the Pinios River Basin. (**c**) Food/Agriculture sector in the Pinios River Basin. (**d**) Ecosystems/Climate sector in the Pinios River Basin.

The PRB Energy sector: The water-intensive agricultural sector in the PRB has led to an increase in energy consumption for irrigation purposes, especially considering that most needs are covered by groundwater at declining levels, i.e., greater abstraction depths. This, in turn, results in a significant increase in agricultural production costs, magnified in recent years by the energy crisis. To this end, a wide expansion of installed Renewable Energy Sources (RES) is thought to be imperative for Greece, among other countries, in view of meeting the European Union RES expansion targets in the framework of the post-lignite era developments. However, the availability of suitable sites, land use competitiveness (mainly with agriculture), and the limited capacity of the national power grid pose additional constraining factors in the complex PRB nexus framework. Based on the latest available data provided by the Geospatial Map of the Hellenic Regulatory Authority for Energy [76], the currently installed and licensed RES units that operate in the PRB include 36 energy storage units, 33 photovoltaic stations (each with a power capacity of over 1 MW), 15 small hydroelectric stations, 5 high-efficiency electricity–heat cogeneration (HEEHC) stations, 4 wind power stations, and 3 biomass stations (each with a power capacity of more than 1 MW) (Figure 3b, in WGS84 coordinate system).

The PRB Food/Agriculture sector: The PRB constitutes the most productive and intensively cultivated agricultural basin in Greece [77]. The plains cover almost 45% of the PRB's spatial extent and are divided through a low-lying hill split into two parts: (a) a western part, including the cities of Trikala and Karditsa, and (b) an eastern part, including the city of Larisa. Cotton is the dominant crop in the PRB (19.81%), followed by forage (17.29%), other cereals (15.80%), wheat (13.52%), pastures (7.72%), fallow (4.63%), corn (4.20%), nuts (3.04%), energy crops (2.00%), olives (1.77%), legumes (1.70%), seeds for sowing (1.08%), other forest crops (1.01%), and other crops (6.43%), according to data from

2021 from the records of the Payment and Control Agency for Guidance and Guarantee Community Aid (Figure 3c, in WGS84 coordinate system). Almost 55% of the cropland area is under irrigation, of which almost 35–40% is dedicated to collective irrigation networks, while the rest is dedicated to private irrigation networks [78].

The PRB Ecosystems/Climate sector: The PRB is considered one of the most important hydro-ecosystems in Central Greece, with 34% of its total area, i.e., 3762 $km^2$, included in the EU Natura2000 protected areas and Important Bird Areas. In particular, in the PRB, there are 10 Sites of Community Importance (SCI), which amount to a total area of 1516 $km^2$, 13 Special Protection Areas (SPA), covering a total area of 2947 $km^2$, 2 SCI-SPA areas, covering a total are of 288 $km^2$, and 11 Important Bird Areas (IBA), spanning a total area of 3170 $km^2$ (Figure 3d, in WGS84 coordinate system).

### 2.2. Stakeholders' Participation

Based on the presented analysis of the basin, the PRB is characterised by critical issues that need to be resolved in a coordinated way to ensure that it has a long-lasting future. Complex nexus interdependencies need to be analysed in-depth in order to address the challenges it faces in a comprehensive and efficient manner. To this end, ideally, a broad spectrum of stakeholders covering all 4 identified key sectors of the WEFE nexus, all levels of hierarchy, and the entire productive web of the basin needs to be recruited into an actively participating group that collaborate to unravel the challenges of the basin and shape appropriate measures for their alleviation.

The stakeholders' participatory activities developed in the PRB have followed backbone guidelines proposed in the REXUS project. These guidelines provide overall strategy clustering as core (common in all pilots) and additional (pilot-specific) activities for efficient stakeholder engagement from the perspectives of both conceptual and practical implementation. This means that the developed strategy, from its onset, is meant to be replicable and flexible enough to be customised to different contexts, even beyond the scope of the European Union's goals.

Stakeholders' sampling and approaching processes are dictated by the resources and time available, although the inclusion of multiple different views and opinions is important. Participatory workshops, forums, surveys, or other methods are useful for capturing stakeholders' perceptions, depending also on the level of desired detail concerning the collected information.

Background information on the stakeholders could be helpful if external factors are thought to be influencing their perceptions and priorities, limiting discussions and negotiations among stakeholders, and, in turn, inhibiting them from establishing a common ground of understanding.

## 3. Results and Discussion

The proposed stakeholder engagement strategy comprises both individual and group activities. The use of individual activities (in the form of semi-structured interviews) primarily allowed us to directly take into account individual viewpoints regarding the 'sectoral' perspective each stakeholder could provide. This analysis included a preliminary investigation of the main needs and challenges (related to resources security) each stakeholder typically faced (including, e.g., conflicting uses of resources and the impacts of climate change on resource availability and state). Participants were initially offered the option to express themselves freely in private in their own environment, without the pressure of time and/or potential challenges in the form of refusal, opposition, or provocation by other members of the group, all of which could have been limiting factors. Subsequently, the analysts worked on developing a 'summary' of the main issues that emerged from the interviews, asking either the stakeholders or the leaders of the study for specific clarifications on potential inconsistencies or conflicting views. Lastly, a summary of the outcomes of the individual interviews was presented and discussed—without any reference to individual stakeholders—in group exercises during a stakeholder workshop.

During the workshop, some results of modelling activities were also presented in order to supplement the discussion with scientific evidence. The support of facilitators during the workshop was crucial to building a consensus while discussing potentially contrasting viewpoints and keeping track of the differences in perspectives.

Throughout the whole stakeholder engagement process, key principles have to be followed to optimise the results. In particular, creating a friendly environment allows everyone to become invested and join in, limiting the possibility of anyone feeling excluded from the participatory process. Transparency in every decision taken and methodological step followed has to be ensured, thus avoiding the feeling of lobbying in decision making. The viewpoints and perceptions of participants have to be respected and valued, disregarding the level of appeal they may have to wider society. The time devoted by the participants and their privacy have to be respected to encourage free and goal-oriented discussions. The adoption of a neutral position among all of the facilitators also supports the freedom of the discussions. In addition, iterative processes that allow for the building of knowledge and progressive shaping of actions have to be followed to develop sound and acceptable pathways to safety. Regarding all the Learning and Action Alliances (LAAs) actions, the "co-" prefix constitutes the key synthetic in terms of working with the stakeholders, for the stakeholders, and their working and living ecosystem. The needs of the stakeholders have to be identified and addressed to the best possible degree, as these may relate to expressing their viewpoints, their complaints, their ideas, and their proposals, in a context shaped to ensure an amicable, equitable, and respectful environment. In light of this, the "mere" data provision process has to be distinguished from stakeholder engagement, and in no case should stakeholders get the impression that they have been approached or recruited to simply act as data providers. Furthermore, patience and time availability always have to be considered, since stakeholders need time to warm and open up and feel like they are in a non-hostile, inviting environment wherein their voice is respected and taken into consideration beyond formalities, thus building a sphere of trust and a sense of belonging among the project's scientific team and the stakeholders. Stakeholders' specificities have to be respected, but boundaries must be set, and stakeholders should not order others around or impose their ideas on the group, regardless of their level of expertise of influence.

The developed strategy comprises five main steps, consisting of framing, mapping, involvement, co-production, and sharing processes (Figure 4).

The following sub-sections contain more in-depth discussions on the proposed approach and the key points that should be considered when applying the methodology.

### 3.1. Framing

Initial Study of Challenges: Firstly, the key challenges to be addressed were identified. A baseline investigation of the study area, which helped to identify the key problems faced across the PRB, was carried out through conducting research in the study area and utilising the research team's expertise and key consultancy skills. The identification of the PRB's challenges not only enabled the research team to pinpoint the key sectors to focus efforts on but also allowed for an initial characterisation of the basin and the identification of sub-basins for more specific discussions. The preliminary identification of challenges was also supported by the information provided by stakeholders in the round of individual interviews. Indeed, part of the questions focused on the selection of their main needs related to the use of natural resources and potential barriers to fulfilling these needs (including biophysical and governance aspects). Subsequently, the key approaches and methodological frameworks to be tested were proposed. This process was accompanied by the definition of a wish list of data requirements. Key sectoral challenges, cross-sectoral interrelations, data requirements and availability, potential levels of intervention, and the spatial distribution of issues were preliminarily determined. This step also involved the preliminary selection of which stakeholders should be invited to participate in the LAA.

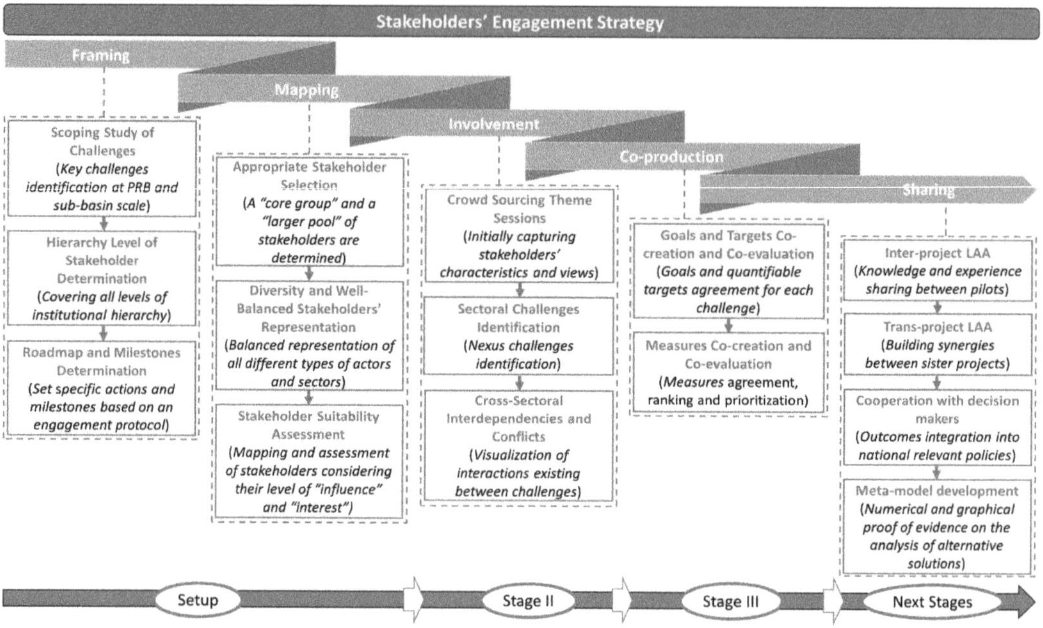


**Figure 4.** Stakeholder engagement strategy as part of the PRB LAA development process.

Determination of the Hierarchy Levels of Stakeholders: Identifying the hierarchy levels that need to be covered and the institutional profiles that need to be included in the pool of stakeholders greatly depends upon the size of the study area, the nature of the identified issues, the envisaged problems' resolution plans, the need to perform in depth analyses at small scales, and the need to consider potential institutional changes, awareness raising, and capacity building initiatives. Evidently, the more comprehensive and wide the pool of key stakeholders is, the more representative it is, lowering the possibility of omitting/missing criticalities in the construction of a holistic management plan. On the other hand, the wider the pool of stakeholders, the greater the likelihood of conflicts arising and the consultation concluding without a consensus being reached. In the case of the PRB, in view of the multitude of complex issues that need to be tackled, an expanded stakeholders' panel was set up. Therefore, the adopted approach aimed to cover all levels of institutional hierarchy, such as end users, decision makers, strategic planning/policy makers, non-governmental organisations, experts, research institutions, and private companies, to the best possible degree, thus potentially facilitating interaction throughout the entire pyramid (i.e., from the bottom to the top and from the top to the bottom).

Roadmap and Milestone Construction: A roadmap and milestones for the stakeholder engagement strategy were then constructed. Starting from when the stakeholders were initially approached, a specific engagement protocol was drawn up and followed to ensure the highest possible engagement level and a low incidence of stakeholders declining to participate. This protocol was based on systematic and frequent follow-up contact on a personalised level, starting with phone calls and also involving regular e-mail communication, followed, in due time, by individual in-person interviews and small group meetings. Achieving high success rates in recruitment necessitates making use of existing personal contacts, gaining new personal contacts, and developing relationships in a continuous effort to build and maintain/confirm trust. Stakeholders' individual characteristics and profiles, along with the profiles of the body they represent, were studied prior to approaching them and collaborating with them.

The stakeholder engagement strategy has been strategically shaped to address the demands of the project whilst taking into account stakeholders' characteristics, the key features of the study area, the institutional capacity and structure surrounding the project, and the level of knowledge among the stakeholders about the nexus approach and nexus concepts. Based on the tasks to be accomplished by the project and the focus developed for the study, a roadmap mapping the whole trajectory of the project and specific actions and milestones was established in collaboration with the involved partners. Hence, a series of meetings, workshops, and desk work sessions have been defined to be elaborated with specific content and target and a precise timeline to reach the required result. In every step of the process, it has to be clearly understood that successful stakeholder engagement is an elaborate and long investment that needs to be developed on the basis of mutual understanding and respect.

### *3.2. Mapping*

Appropriate Stakeholder Selection: An initial list of potential stakeholders was drafted, based largely on the work experience of the REXUS research team regarding the region of Thessaly, and augmented by the personal relations that were developed among them. In parallel, a wide search for additional stakeholders was performed; this search for stakeholders was carried out through using the snowball technique, partially ensuring their high level of interest, and an online search for potential stakeholders who represent significant actors in the study area; however, the a priori assurance of their willingness to actively participate in the project was not part of this search. An initial list of more than 150 stakeholders was compiled.

Starting with this list, specific criteria were used to reduce the final number of stakeholders; these criteria were as follows: representativeness of all sectors, coverage of all hierarchy levels, participation of key actors, wide geographical coverage, presence of critical policy and decision making as well as implementation bodies, representation of all key socio-economic activities, pluralism in stated viewpoints (including the academic literature, consultancy, production, society, technocrats, etc.). Subsequently, the list of preselected stakeholders comprised approximately 50 people, all of which were invited to participate in the REXUS project. This number was assumed to be realistic in helping the research team achieve effective and representative results. Invitations were extended on a personalised level through telephone calls, e-mails, and in-person visits when requested to communicate the targets of the project and explain the role of the stakeholders and their engagement in the active participatory approach towards developing holistic management solutions. Approximately 35 people responded positively to the invitation, forming the initial group of stakeholders for the project. This number was increased to 44 following the suggestions offered by the participating stakeholders, as some nominated their contemporaries to ensure a higher degree of representativeness.

Orientation criteria and hints for the selection of appropriate stakeholders were employed, resulting in the construction of a list of two distinct groups of stakeholders. The "core group" consisted of stakeholders who mainly focused on most of the feedback in the co-mapping of the challenges, the co-identification of the constraints and criticalities, and the co-design and redefinition of a potential set of solutions in the holistic management strategy. These are the people that presented the highest interest, influential capacity, and decision making power while possessing the deepest knowledge of the study area. The "larger pool" incorporated the aforementioned group but also included a large number of stakeholders that essentially formed a conclusive list of existing representatives of all sectors and levels with variable degrees of knowledge and scholarly education, involvement, abilities to influence and/or motivate, and decision capacities, but every one still had a vast knowledge of the study area and/or the sectoral issues.

Both groups were dynamically altered as the project progressed to adjust to its needs and ensure alignment with the responses of the stakeholders. To this end, the opinions and suggestions of the stakeholders were seriously considered to inform their inclusion in either group. Overall, the key criteria considered for stakeholders' inclusion in the core group included "representativeness", "coverage", "performance", "experience", "influence", "collaboration", and "satisfaction/preference". Adequate representativeness of the nexus sectors and societal groups at regional and local scales under all the identified institutional levels has to be ensured. Both a holistic coverage of sectors and topics that ensures no critical challenge, impacted system, or activity is left unconsidered and a trans-sectoral coverage enabling improved and more efficient designs of acceptable and applicable "working" holistic solutions have to be achieved. Also, the entire spectrum of performance (from poor to excellent) has to be covered with regard to resource management at the regional level, as long as stakeholders directly or indirectly become involved in problem solving for the study area. The adoption of best practices by stakeholders and their responses to troublesome, inadequate, or mismanaged resources and occurrences in their respective sectors, along with everything in between, are of great interest. Obviously, well-performing stakeholders were prioritised, as they are normally aware of issues, needs, and problems and capable of actively assisting in shaping best practices for holistic management with the aim of increasing resilience and enhancing sectoral security. However, it is strongly believed that considering the viewpoints and analyses of poorly performing stakeholders may help considerably in mapping problems and perhaps even more so in understanding the reasons behind their poor performance. In the end, these are the stakeholders that need to participate in the co-decision process on the development of solutions that merit social consent, inclusivity, justice, and acceptability in order to ensure full-scale implementation. In addition, having a wealth of collective experiences to share and diffuse amongst the core group of stakeholders is of the utmost importance to achieve a deep and conclusive understanding of the current conditions of the study area and improve the likelihood of the designed solutions being applied successfully. The potential influence of stakeholders should be considered to improve the feasibility of the co-designed solutions and ensure they are efficiently communicated, as increasing the probability their actual implementation is of great interest, as is the ability of stakeholders to collaborate with other stakeholders and scientific teams, ensuring a healthy, fruitful, and overall amicable collaboration with the highest possible degree of efficiency. Moreover, sometimes, the appointed representative of the stakeholders' board of directors does not necessarily choose with the most appropriate or obvious choice amongst the available options, but ensuring that a consensus is met is critical in finding and, more importantly, implementing solutions.

The group of engaged "core stakeholders" comprised 44 members and was synthesised as presented synoptically in Figure 5 and detailed in Appendix A (Table A1). A unique code was provided to each different stakeholder based on their main related nexus sector (i.e., the prefixes W, EN, F, and EC refer to the water, energy, food, and ecosystem sectors, respectively). Overall, the water, energy, food, and ecosystem sectors were represented by 11, 3, 23, and 7 stakeholders, respectively (Figure 5a). In addition, stakeholders were categorised according to their type; i.e., there were 15 decision makers and policy-makers (POL), 11 representatives of end users and individual end users (USER), 2 citizens and NGOs (CIT), 3 individual experts (EXP), 3 representatives of private companies (COM), and 10 research and academia representatives (RES) (Figure 5b). Levels of influence and interest were also assigned to each stakeholder, as presented in Table A1.

Diversity and Well-Balanced Stakeholders Representation: Diversity and a well-balanced representation of all WEFE nexus sectors and institutional/governance levels and types constitutes a prerequisite for a nexus approach and therefore has to be ensured. The higher number of stakeholders representing the food and water nexus components is attributed to the dominance of the agricultural and water-related sectors in the PRB; however, these sectors were not treated favourably compared to the other sectors since the engaged stakeholders were asked to express their views on all of the nexus components. This is illustrated in Figure 5, where it is clear to see that the water and food sectors outnumber the others, especially the energy sector. Still, as energy is strongly interrelated to the water and food sectors, especially due to the financial aspects escalating from contemporary critical issues, the energy sector was indirectly but responsibly covered in equal terms by the group of stakeholders. Interestingly enough, the assessments and categorisations made by our research team for each stakeholder with regard to the key sector each one represents best on the basis of their profiles shifted towards the water sector, as proved by the self-determination exercise practiced in the very first meeting that was held, as explained in Section 3.3.

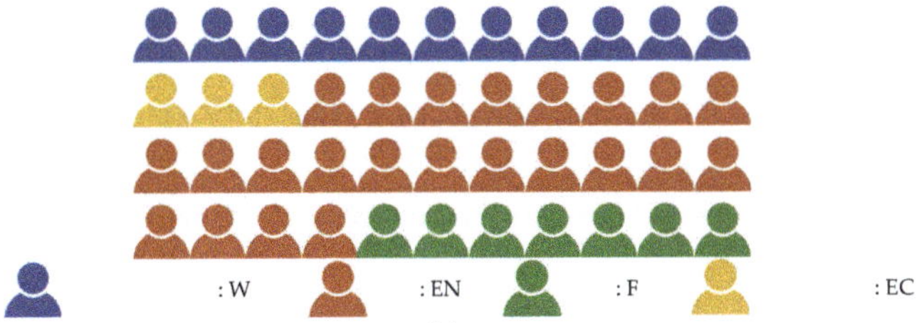


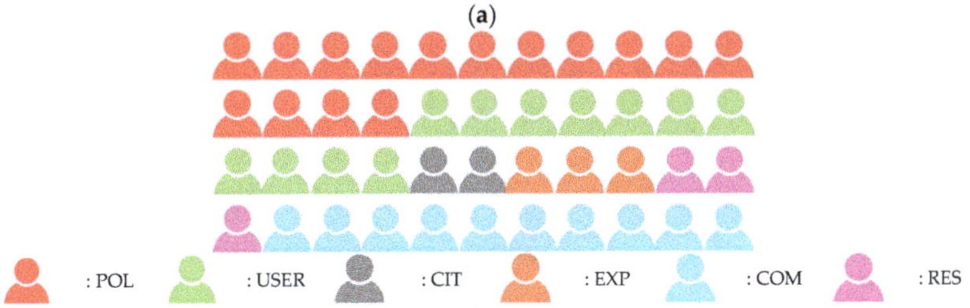


**Figure 5.** Characteristics of engaged stakeholders based on (**a**) their related nexus sector and (**b**) their type.

Stakeholders' Suitability Assessment: Once the list of selected stakeholders was created, stakeholders were characterised and mapped in terms of their influence and interest criteria. The idea of developing a stakeholder power grid to visualise the influence and interest of stakeholders was conceived by Mendelow in 1991 [79]. Stakeholders differ in how much influence they have and in how much they are interested in a particular topic. The term "influence" refers to the capacity/power of stakeholders to solve or exacerbate a problem, while the term "interest" refers to the importance of a problem in the eyes of stakeholders regarding the level of relevance it has to their business. Our research team's approach involved the stakeholders being categorised into four classes: "promoters", "defenders", "latents", and "apathetics" [80]. Promoters are those who usually have both great interest in an issue and the power to help resolve it successfully, defenders are stakeholders who have a vested interest and can voice their support in the community but may have little actual power to influence the issue in any way, latents are those who have no particular interest or involvement in an issue but have the power to influence it greatly if they become interested, and apathetics include stakeholders who have little interest, little power, and may not even be aware of an issue. The stakeholders were mapped on a four-quadrant plot based on their levels of influence and interest, and this plot is depicted in Figure 6.

The distribution of stakeholders in the four quadrants reflects the perception of the analysts. However, in our case, to facilitate LAA activities through the active engagement of the different groups of stakeholders, the selection procedure involved assessing each individual's level of influence and interest. Hence, members of the group that were assigned to the bottom two quadrants (apathetics and latents) were filtered out and not considered further. All the engaged stakeholders were therefore distributed in the upper two quadrants of the mapping diagram, meaning that they were characterised as promoters and defenders and that they were all interested in the nexus issues. The filtering out of the latents and apathetics in the developed PRB LAA activities was carried out to avoid the negative effect they may have on the implementation of a project and its outcomes. About a third of engaged stakeholders (15) were characterised as promoters (those with high levels of influence and interest), with the remaining 29 being classed as "defenders". Interestingly, the distribution between the two categories is rather balanced for three of the four sectors too. As already stated, the mapping described herein is the result of filtering the initially compiled stakeholders' pool to ensure a group of stakeholders with the potential to leverage decision making on nexus management throughout the whole project. However, stakeholder characterisation and mapping can dynamically change as a result of several factors, including potential erroneous assessments by the scientific team members, changes in the prioritisation and obligations of the stakeholders, failing to the attract attention of the stakeholders to the scope of the project/process, a lack of genuine interest in the process among participants, and disproportionate requirements for investing time in the process.

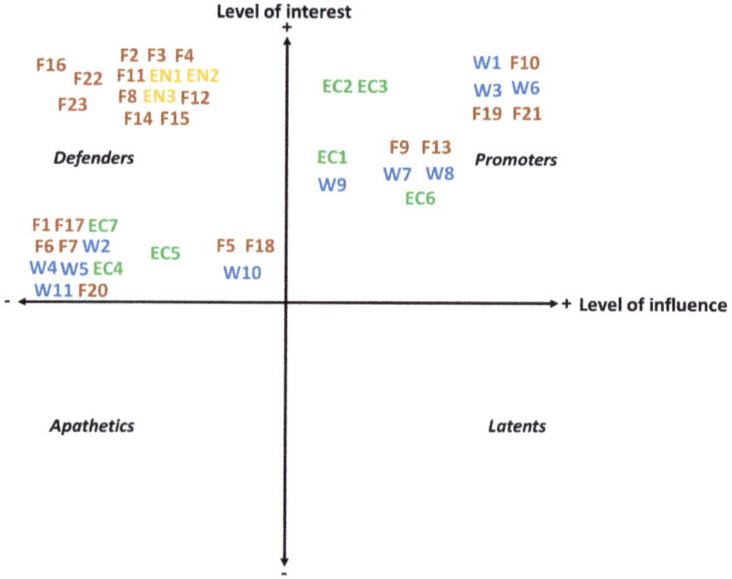


**Figure 6.** Selected stakeholders' classification based on influence and interest criteria.

### 3.3. Involvement

This phase is central for the stakeholder engagement, as it centres around enabling the active participation of relevant stakeholders through a series of participatory activities, details of which are provided in the following sub-sections. This phase is centred on the 'diagnosis' of PRB conditions and is carried out to both select relevant sectoral challenges for the area and to better understand cross-sectoral interdependencies.

Crowd Sourcing Theme Sessions: The next step consisted of the preparation and realisation of the first meeting of the REXUS PRB stakeholders, which took place online on 30 November 2021. The meeting consisted of three crowd sourcing theme sessions to initially capture stakeholders' characteristics and views and two virtual roundtable discussions to involve all the stakeholders in a common virtual discussion.

In particular, 15 online questions were included in the framework of the three crowd sourcing theme sessions of the REXUS PRB Learning and Action Alliance kick-off meeting, allowing for immediate feedback on the overall status of the stakeholders' perceptions. The questions, along with an analysis of the stakeholders' responses, are presented synoptically below and in-depth in Appendix B. Each session involved asking five questions to stoke the interest of the audience regarding the nexus system issues and register their viewpoints on the fly. The Mentimeter online platform was employed for this purpose.

In fact, the first session acted as an ice breaking session for the stakeholders, elucidating their linkages to the PRB, perceptions of the nexus approach, and their expectations for the REXUS project. Almost 80% of the participated stakeholders have ties to the water and food/agriculture sectors, revealing the agricultural nature of the PRB. The vast majority of them are familiar with the PRB nexus system, and they could decisively contribute to the analysis of cross-sectoral challenges, while half of them are mainly ministers and academics who could influence the decision making process. A high level of nexus principle understanding was also revealed by the fact that most stakeholders acknowledged the need for each WEFE sector to be approached equally, while this perspective did not seem to be clear enough among stakeholders related to the food/agriculture sector, highlighting the complexity of the agricultural sector in the PRB. The main reasons for stakeholders' involvement were related to their intentions to express their interest in and concerns about the challenges associated with the PRB, along with their desire to co-design a sustainable future.

The second session was developed to capture stakeholders' viewpoints on the PRB's conditions. The management framework of the PRB nexus system was characterised as very complex or complex among the vast majority of stakeholders, exposing the limited effectiveness of already implemented measures. The water sector of the PRB faces the most crucial challenges, a fact that was recognised by stakeholders from all the different nexus sectors. In addition, the water and food/agriculture sectors appeared to have the highest importance and were the most thoroughly investigated PRB sectors.

The third session was conducted to capture the perception of stakeholders on the implemented management practices. The efficiency of the already implemented measures was characterised as moderate, with a selection of these measures mainly being affected by the influential capacity of some stakeholders. The water and food/agriculture sectors present the strongest interaction and are either already being significantly affected or expected to be affected by the climate crisis.

The Identification of Sectoral Challenges: In the framework of the first REXUS meeting of the PRB stakeholders, a roundtable discussion was organised and conducted to select the most critical nexus challenges. As mentioned before, a series of individual semi-structured interviews were conducted, either in-person or virtually, with each different stakeholder, and this resulted in an extensive list of challenges that should be addressed to ensure the sustainability of the study area (Figure 7). As highlighted in the figure below, the focus was firstly on the nexus sectors in isolation. Interestingly enough, governance issues were identified as central issues in all sectors. The process was not influenced by any direct or indirect indications on the challenges to consider. As a result, the identification of the most important ones was based on the stakeholders' own critical judgment only through a specific session organised during the stakeholders' workshop.

A set of critical challenges was thus identified for each sector, in addition to the major issue of governance. Governance is a multidimensional concept that refers to the political, social, economic, and administrative systems in place that influence the development and management of resources such as water, land, or energy. In the water sector, several frameworks have been developed to diagnose water governance and support the development of action planning [81–84]. Despite the different focuses of governance assessments and target groups, diagnosis focuses on analysing formal structures (i.e., the existence, adequacy, and coherence of the policy and regulatory framework and the roles and responsibilities of the different institutions managing the resources and financing) and how well processes work in practice (i.e., level of coordination, cooperation, implementation, and performance). As the Organisation for Economic Co-operation and Development [85] describes, it is about exploring the effectiveness and efficiency of formal and procedural aspects, as well as trust and engagement (Figure 8).

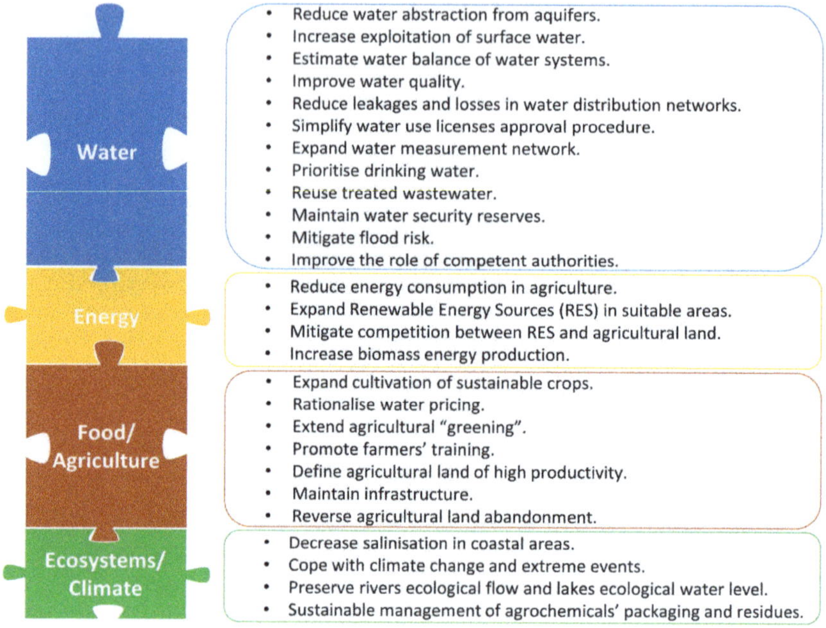


**Figure 7.** Critical nexus challenges with respect to the Pinios River Basin.

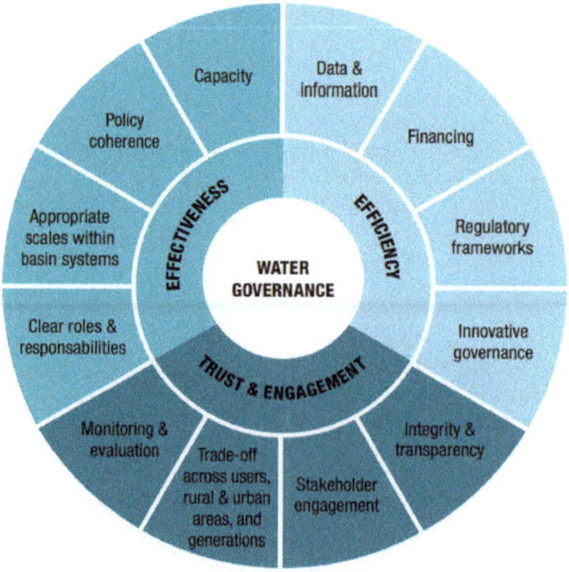


**Figure 8.** OECD principles regarding water governance [85].

Effective administration and governance in the PRB is of paramount importance and directly or indirectly reflected in all four sectors as a challenge. Its importance lies in the fact that none of the measures defined and proposed for implementation in order to address the identified challenges may be successful and meaningfully impact the system unless the governance structure is well developed and supplemented by efficient administrative units.

Complex legislation framework and scattered responsibilities among competent authorities were identified as the most critical barriers in the PRB, hindering effective administration and governance.

The water-associated challenges primarily relate to ensuring sufficiency in terms of fair spatiotemporal distribution to ensure the coverage of all socio-economic activities without disregarding quality, especially in specific parts of the basin and with respect to agriculture-related pollutants. In parallel, as part of the management practices proposed to address the water-related challenges, flood risk mitigation, resilience to shortages due to droughts, and the utilisation of treated domestic effluents are high on the agenda. The food-related challenges pertain to the development of viable and sustainable agriculture, towards which production cost rationalisation and greening are thought to be of paramount importance, as are the productive utilisation of agricultural land and the training of farmers. The rehabilitation and preservation of ecosystems and their functions are the key challenges relating to the environment and, along with the sustainable management of agrochemical' residues and the preservation of ecological river flows and minimum lake levels. High use efficiency lies at the essence of the identified energy-related challenges; the energy sector's challenges include the expansion of renewable and biomass energy production to resolve issues such as the antagonistic use of high-productivity land and reducing energy consumption in the agricultural sector.

The PRB challenges were also critically analysed through reviewing previous studies. Groundwater overexploitation and water quality deterioration are triggered mainly by non-sustainable agricultural management practices which have been carried out since the mid-1980s [67]. Mean annual precipitation in the PRB is projected to decrease, while temperature and evaporation are projected to increase until 2100, resulting in a significant decrease in water availability, which is strongly related to the projected declining groundwater recharge trend and increasing trend in relation to irrigation water demand. Ecological flow is difficult to preserve in some rivers in the PRB, presenting low to zero discharge during the summer months in recent years [78], while the total runoff in the PRB is projected to decrease further from 22% to 66% until 2080 compared to the 1980–2000 period [14,71]. Flood occurrences increased between 1990–2010 and 1979–1989 in the PRB, with most flood events being recorded in the southern part of the PRB and most of the damage occurring in rural areas [86]. The PRB has also experienced severe, extreme, and persistent droughts that have affected large areas since the late 1970s and became more evident and frequent during the last three decades, especially over the past 5–7 years.

Cross-Sectoral Interdependencies and Conflicts: During one stakeholders' meeting, a second roundtable discussion was conducted to explore the transectoral relations between the identified challenges in order to facilitate a holistic approach. The identified interrelations were then finalised, and further details on them were obtained from personal interviews with stakeholders. The final interrelations were drawn up and visualised (Figure 9). The colour selected for each connecting line refers to the nexus sector of the related driving challenge (i.e., the blue, yellow, brown, and green lines refer to water-, energy-, food/agriculture- and ecosystems/climate-related challenges, respectively).

Cross-sectoral interdependencies provide a comprehensive understanding of the bonds between nexus sectors, helping to identify how and where exactly each sector interacts and depends on another. Unless these dependencies are fully mapped, understood, and taken into account, any attempt to design and successfully implement management measures will inevitably be of limited efficiency and impact. Successful holistic approaches involve a global overview and examination of the sectors that form the nexus of a system and require a common understanding of the identified issues, challenges, and interdependencies in order to reach a wide consensus on ways to improve progress towards sustainable management.

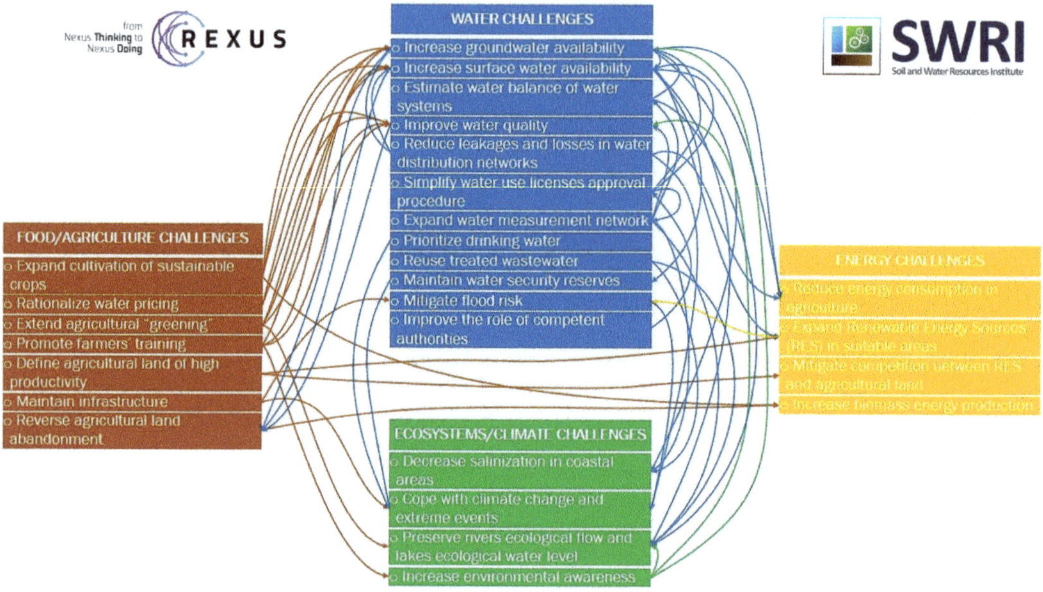


**Figure 9.** Interrelations of critical challenges in the Pinios River Basin.

The canvas illustrated in Figure 9, which summarises the feedback from participants, clearly shows the complexity of the studied nexus and the tight bonds amongst its sectors. Especially between the water and food/agriculture sectors, multiple and multi-level dependencies are denoted for the PRB, clearly suggesting that managing an element of one sector would inevitably influence the other. Interestingly enough, a very strong dependency was identified between the former two sectors and the ecosystems/climate sector, which mainly reflects concerns surrounding the current management conditions and perceived sustainability status. Evidently, the stakeholders acknowledge that there are less interdependencies for the energy sector. This can be attributed to the limited knowledge on the local relevance of the sector, considering the challenges that are faced worldwide.

### 3.4. Co-Production

The Co-creation and Co-evaluation of Goals and Targets: The challenges that emerged from the involvement phase helped in setting the goals for achieving nexus sustainability. Clear goal setting could help policy and decision makers to improve their performance in terms of sustainable development [87]. In addition, quantifiable targets have to be determined to measure the satisfaction level of defined goals and thus promote WEFE security.

To promote their applicability and effectiveness, targets should be in compliance with international and national WEFE legislative regimes. The Sustainable Development Goals (SDGs) included in the 2030 Agenda for Sustainable Development [19] were established in compliance with the rights and commitments of the United Nations, and they are considered political goals, not legal rules. The SDGs and associated targets serve as inspirational statements that constitute crucial parts of international legislative instruments and aim to trigger action that will yield results in critical areas for the well-being of humans and the planet by 2030. The SDGs attempt to orchestrate the wide variety and diversity of targets already embedded in various international agreements.

Regarding the PRB, proposed goals and targets should be in accordance with European and national policies. European Union (EU) legislation pervades the national policies of all member states while regulating most of aspects of the WEFE nexus. In terms of the water sector, the Water Framework Directive (2000/60/EC) sets out rules to prevent water status deterioration and achieve a good status for all deteriorated surface and groundwater bodies [88]. Regarding the energy sector, the recent Revision of the Renewable Energy Directive (COM/2022/230) proposes to further increase the target of renewable energy sources in the EU to 45% by 2030 [89]. In the field of agriculture and rural development, the Common Agricultural Policy (CAP) 2023–27, implemented on 1 January 2023, aims to ensure stable food supply, ensure farmers' incomes, and keep rural areas vibrant within the context of environmental sustainability [90]. In terms of EU climate action, initiatives include engaging all parts of society in climate-benefitting actions, further reducing emissions by 2030, and making Europe the world's first climate-neutral continent by 2050 [91].

The proposed goals and targets may interact among each other, mirroring the interrelations and critical trade-offs among the WEFE challenges. In particular, some goals and targets seem to be independent of each other, while others may reinforce or impose constraints on others due to the competition for scarce natural resources. However, effective stakeholder engagement and involving stakeholders in common discussions could mitigate potential policy conflicts, resulting in the development of a coherent set of goals and targets and a complete consensus, ultimately shifting goals and targets from being centred around achieving acceptable trade-offs to achieving win–win scenarios.

The Co-creation and Co-evaluation of Measures: Defined goals and targets call for both sustained and sustainable development through the compilation and implementation of specific measures. Policy makers and decision makers need to elaborate on how to integrate goals and targets into management policies to address nexus challenges. A wide range of sustainable measures could be implemented, including supply-side and demand-side options, across all nexus sectors. Supply measures are mainly related to infrastructure and treatment technologies, while demand management measures include resource efficiency promotion, changes in resource allocation mechanisms, and improvements in resource productivity and availability. However, each measure presents specific strengths and weaknesses which have to be properly considered. Efforts should be made to analyse, rank, and prioritise measures using as wide a range of objective criteria as possible, covering technical, social, and other characteristics. The creation of effective measures may secure reliable supplies of natural resources to meet future demands.

There is an increasing awareness of Nature-based Solutions (NbS), which have the potential to address both climate mitigation and adaptation challenges at a quite low cost while promoting human and nature benefits. NbS promote the sustainable use and management of natural resources while tackling economic and environmental challenges and sustaining human well-being. In addition, NbS emerge as an integrated approach that mitigate trade-offs and promotes synergy among the SDGs. The importance of promoting NbS is even greater considering that some SDGs, including those centred around stabilisation and adaptation to climate change (SDG13) and biodiversity protection (SDGs 14 and 15), are unlikely to be met by 2030 [92]. The importance of NbS was recognised by their endorsement in the Climate Change and Land Report of the Intergovernmental Panel on Climate Change (IPCC) [93] and the fact that they were included in a COP cover policy for the first time during the 27th Conference of the Parties to the United Nations Framework Convention on Climate Change (COP27).

### 3.5. Sharing

Inter- and trans-project approaches can significantly advance nexus management towards resilient WEFE systems and drive sustainability transitions.

Inter-project LAA: Horizontal knowledge is expected to be enhanced through the development of an inter-project LAA centred around knowledge sharing between pilot areas of the project. Dialogue on policy integration needs and gears across pilot areas is promoted. In addition, cross-fertilisation amongst the project's pilots is being practiced to improve the developed strategies for nexus management, rectifying what has been found to not yield results and strengthening the efficient parts of the existing strategies.

Trans-project LAA: The produced experience will be further disseminated across other projects under the nexus scheme through the development of a trans-project LAA. This way, alignment with similar projects and building synergy among them will be achieved. The role of the trans-project LAA can prove crucial in a much broader context since integrating the lessons learnt from engagement strategies in areas with different particularities provide give a significant boost to align agendas and policies in a trans-sectoral perspective.

Cooperation with Decision Makers: Experience and information sharing may be practiced among influential decision making bodies and authorities, again aiming to improve the decisions to be made and perfect the outcomes of the nexus management project. To this end, within the framework of the REXUS project, the research team has initiated a twofold external collaboration (a) through creating a focal group of the Water Resources Management Plan (WRMP) of the Thessaly district that PRB belongs to and (b) establishing a focal group with the official consultancy for the implementation of the WRMPs on a nationwide level. Focal group (a) consists of high-ranking specialist executives of the Ministry of the Environment and Energy and specialist consultants of the consortium responsible for the second revision of the WRMP at the PRB. The objective of this collaboration is to strengthen and promote the holistic approach towards the effective and sustainable management of the PRB by exchanging viewpoints and aligning information and data that enhances pluralism and integration. The ultimate aim of this collaboration is to establish a revised plan that has a higher chance of being successfully implemented and yielding results. Focal group (b) consists of highly experienced executive technocrats that are responsible for guiding and consulting the ministerial and regional services towards aligned efforts for the efficient implementation of the WRMPs at national level through the consultancy work carried out by several consortia and the regional authorities' scientific staff. They share the vision of holistic nexus management, and the target of this collaboration is to disseminate the virtues and needs for holistic nexus management nationwide through the systematic, active engagement of stakeholders whilst offering help in efforts towards training scientists involved in this activity.

Meta-Model Development: A meta-modelling approach can be employed to visualise and more easily present the main information and outcomes of a project or task. A meta-model transforms produced implicit knowledge into straightforward results, supporting the analysis of information and thus the accumulation of knowledge. The meta-model's layout and component selection could be based on users' opinions, including the indicators, scenarios, and measures that a meta-model could present and the type of visualisations that a meta-model could offer. Specific answers to these generic questions mainly depend on the target audience; thus, meta-model co-development constitutes an inextricably linked component of the whole stakeholder engagement process. An efficient meta-model design could allow users to easily gain insights into the complex nexus systems, supporting decision making processes within a holistic point of view. Its development for the PRB is largely shaped on the needs and requirements of the stakeholders, who we envisage to be amongst the key users of it.

## 4. Conclusions

The main objective of the present work was to present an effective strategy for incorporating different stakeholders' views and priorities through their active engagement in group discussions to define the critical challenges of WEFE nexus components in the Pinios River Basin. Defining critical challenges and their interconnections in WEFE-stressed areas could help in redefining policies and seeking solutions within an integrated sustainable development context. The findings of the present work are summarized below:

- The identification of key challenges, along with their interconnections and interdependencies, is crucial to supporting policy makers in enhancing the resilience of the study area.

- Despite the scientific team's extensive experience regarding the study area, the responses of the stakeholders offered new viewpoints and challenges that had not been realised in previous activities.
- The use of a holistic approach as a basis for the engagement process revealed impressive interdependencies amongst the key nexus sectors. More importantly, governance emerged as a hotspot challenge by itself, having been indicated in several ways and shapes as such in all four examined sectors.
- The involvement of a broad spectrum of stakeholders not only improves the representativeness of the identified challenges but also renders the participatory process more transparent. As a consequence, the identified solutions are expected to have wider social acceptance and an increased chance of being successfully and efficiently implemented.

In conclusion, the proposed methodology shows a clear and detailed path of actions that need to be taken to support stakeholder engagement in nexus studies. In the methodology, particular attention is given to the process of stakeholder selection and analysis, as well as to their involvement both in individual activities (interviews) and group exercises. The purpose of this is to have information related to each sector in isolation (e.g., sectoral challenges) but also, even more importantly, on cross-sectoral interdependencies and potential conflicts. The whole approach comprises multiple phases and steps and is flexible enough to be adapted to the specificities of other case studies.

Future research could include the analysis of changes in stakeholders' viewpoints after their most intensive engagement in the actions of the REXUS project, also in view of assessing the effectiveness of the engagement roadmap.

**Author Contributions:** Conceptualisation, A.P. (Andreas Panagopoulos) and V.P.; methodology, A.P. (Andreas Panagopoulos), V.P., E.H., B.A.W., M.B. and A.P. (Alessandro Pagano); resources, D.M., A.C. and K.B.; data curation, D.M.; writing—original draft preparation, D.M., A.P. (Andreas Panagopoulos) and V.P.; writing—review and editing, A.P. (Andreas Panagopoulos), V.P., B.A.W., and A.P. (Alessandro Pagano); visualisation, D.M.; supervision, A.P. (Andreas Panagopoulos); project administration, A.P. (Andreas Panagopoulos) and V.P.; funding acquisition, A.P. (Andreas Panagopoulos) and V.P. All authors have read and agreed to the published version of the manuscript.

**Funding:** This research article was funded by the European Union's Horizon 2020 research and innovation program under grant agreement No. 101003632, project REXUS.

**Data Availability Statement:** Data is contained within the article.

**Acknowledgments:** Then: authors would like to thank the Pinios stakeholders for participating in the stakeholders' meeting.

**Conflicts of Interest:** The authors declare no conflict of interest.

## Appendix A

**Table A1.** Characteristics of the participating stakeholders, as assessed by the research team.

| Code | Stakeholder's (Body) Description | Type | Level of Influence | Level of Interest |
|---|---|---|---|---|
| F1 | Legal advisor, collective irrigation organisations | EXP | - - - | + |
| F2 | Irrigation Organisation A | USER | - - | + + + |
| F3 | Irrigation Organisation B | USER | - - | + + + |
| W1 | National-level authority, water services planning and management | POL | + + + | + + + |
| EC1 | International NGO related to environment protection | CIT | + | + + |
| F4 | Agricultural interprofessional organisation, Irrigation Organisation C | USER | - - | + + + |
| EC2 | Local environmental protection authority | CIT | + | + + + |

**Table A1.** *Cont.*

| Code | Stakeholder's (Body) Description | Type | Level of Influence | Level of Interest |
|---|---|---|---|---|
| F5 | Former representative of relevant chamber and currently involved in several actions for the improvement of water resource management in the basin | EXP | - | + |
| F6 | Professor, expert in crop production and rural environment | RES | - - - | + |
| F7 | Professor, expert in irrigation and water resource management | RES | - - - | + |
| W2 | Professor, expert in water resource management and representative of international organisation for sustainable development | RES | - - - | + |
| EC3 | Professor, expert in ecology and ecosystems | POL | + | + + + |
| F8 | Farmers' Cooperative of Thessaly | USER | - - | + + + |
| F9 | Regional authority related to agricultural development | POL | + + | + + |
| EC4 | Former member of NGOs related to environmental protection | EXP | - - - | + |
| EN1 | Wind farm development and installation company | COM | - - | + + + |
| F10 | National-level authority, cultivation systems and crop production | POL | + + + | + + + |
| F11 | Agricultural Association A | USER | - - | + + + |
| F12 | Agricultural Association B | USER | - - | + + + |
| W3 | National-level authority, water environment protection and management | POL | + + + | + + + |
| F13 | Regional-level authority related to agricultural economy and veterinary services | POL | + + | + + |
| W4 | Professor, expert in water resource management | RES | - - - | + |
| F14 | Irrigation Organisation D | USER | - - | + + + |
| F15 | National-scale authority related to irrigation water management | USER | - - | + + + |
| EN2 | National-scale corporation related to electricity supply | COM | - - | + + + |
| W5 | Professor specialising in water resource management | RES | - - - | + |
| F16 | Food production and distribution company | COM | - - - | + + + |
| W6 | National-level authority, water services costing and pricing | POL | + + + | + + + |
| F17 | Research organisation related to crop quality control, classification, and standardisation | RES | - - - | + |
| F18 | Regional division of chamber related to geotechnical professionals | POL | - | + |
| W7 | Regional authority responsible for water resource management | POL | + + | + + |
| F19 | National authority related to land reclamation and mechanisation | POL | + + + | + + + |
| W8 | Regional authority related to hydro-economy | POL | + + | + + |
| W9 | Consortium of private companies involved in the development of water resource management studies in the basin | POL | + | + + |
| EC5 | Research organisation related to groundwater exploitation | RES | - - | + |
| EN3 | Operator of hydroelectric plant | USER | - - | + + + |
| F20 | Organisation related to agricultural education and training | RES | - - - | + |
| F21 | National authority related to land reclamation, soil, and water | POL | + + + | + + + |
| F22 | Young farmer | USER | - - - | + + + |
| EC6 | Regional authority related to environment and spatial planning | POL | + + | + + |
| W10 | Regional department of chamber related to geotechnical professionals | POL | - | + |
| EC7 | National research organisation related to biodiversity and ecosystems | RES | - - - | + |
| W11 | Professor, expert in water resource management and environment protection | RES | - - - | + |
| F23 | Agricultural Cooperative | USER | - - - | + + + |

## Appendix B

*Appendix B.1 Session A: Ice Breaking*

Q1: Which of the following sectors are you mostly associated with? (select 1 out of 4 predefined answers).

A1: More than half (54%) of the stakeholders who participated in the first meeting of the PRB pilot area self-identified as having ties to the water sector, and 25% had ties to the food/agriculture sector; a fact that reveals the importance of these sectors in the PRB. Representatives of the ecosystems/climate and energy sectors accounted for almost 21% of the participants.

Q2: What is your relation with the Pinios River Basin (work-residence-knowledge of PRB's nexus system)? (select 1 out of 4 predefined answers).

A2: The vast majority (90%) of the stakeholders who participated declared that they were familiar with the PRB nexus system, and therefore, they could play a significant role in the framework of the identification of management problems and solutions. Half (50%) of the participants did not reside or work in the basin, and these participants were mainly high-level stakeholders, including ministers and academics. The highest proportion of the stakeholders who were not familiar with the PRB nexus system were the stakeholders related to the energy sector (33%); a fact that is attributed to both the early development stage and the multivariate complexity of this specific sector in the study area.

Q3: How can problems of the Pinios River Basin be resolved to achieve resilience and sustainability? (select 1 out of 3 predefined answers).

A3: Most of the stakeholders (85%) acknowledged the need for equally approaching WEFE sectors to achieve resilience and sustainability; a fact that reveals a high level of understanding of the necessity for a holistic approach in resource management. Only a small proportion of stakeholders (4%) were pessimistic and believed that the problems in the PRB cannot be resolved. More than half (57%) of the stakeholders related to the food/agriculture sector appeared not to have a clear view regarding the pathway to achieve sustainability in the PRB. This is mainly attributed to the complexity of the agricultural sector in the study area, the viability of which presents probably the highest interdependency to the other sectors, and this sector has already suffered the impacts of economic and energy crises and climate change.

Q4: What do you expect to gain from being involved as a stakeholder in the REXUS project? (select maximum 3 out of 4 predefined answers).

A4: Almost 40% of the stakeholders were involved in a project centred around co-developing solutions in the context of a holistic management framework; a fact that demonstrates their high levels of motivation and strong desire for a sustainable future for the PRB, along with a strong belief and a demand for participating in shaping their future by directly affecting the sustainability of the basin or contributing their expertise to shape the future, even if they do not reside in the PRB.

Q5: What do you expect to gain from your participation in the current meeting? (select maximum 2 out of 4 predefined answers).

A5: None of the involved stakeholders considered their participation in the first PRB stakeholders' meeting as a chore. In particular, more than 40% of the stakeholders participated in the meeting in order to become more informed about the ways that they can be actively involved in the tasks and activities of the project. Almost 58% of the stakeholders participated in the meeting to express their interest/concerns regarding the PRB's challenges and also to become informed about the other stakeholders' perceptions.

The answers of stakeholders to Q1-Q5 are presented in Figure A1, where questions 1–5 correspond to responses a–e, respectively.

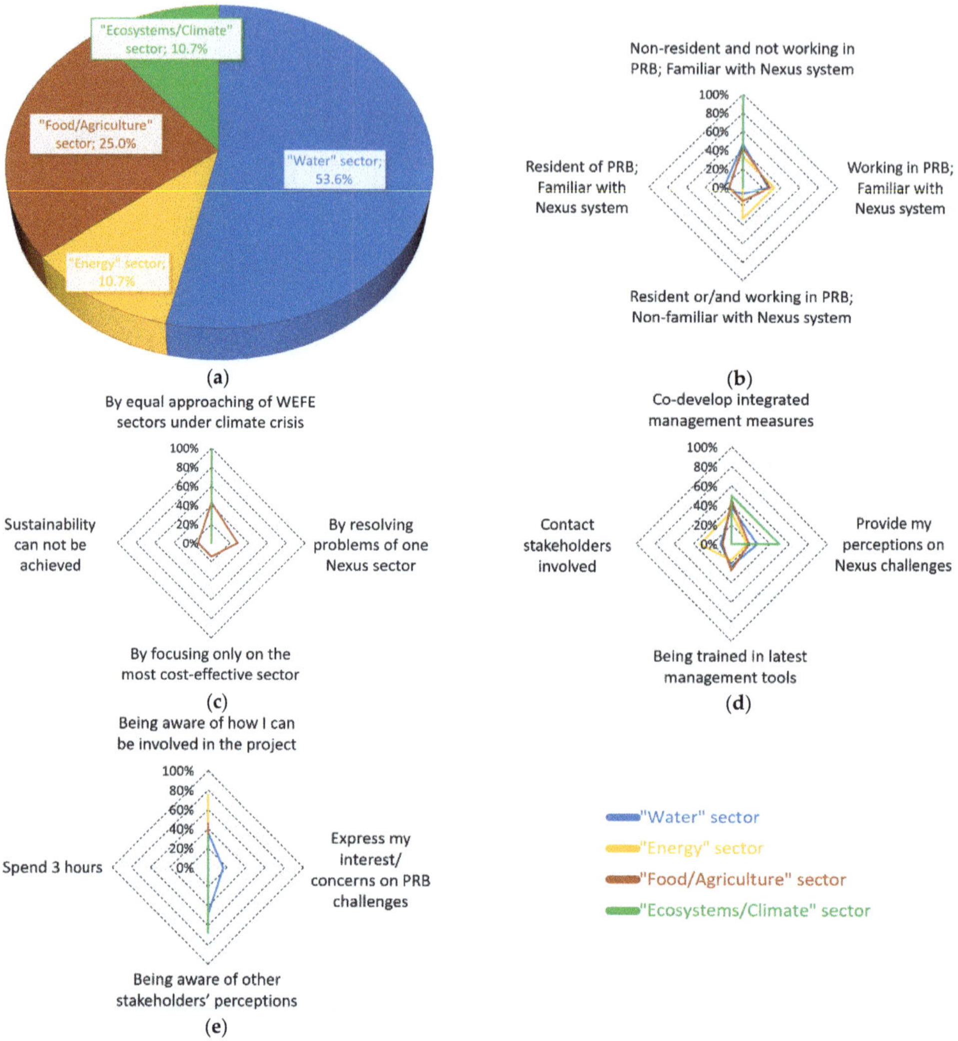


**Figure A1.** Responses of Pinios River Basin stakeholders to (**a**) Q1, (**b**) Q2, (**c**) Q3, (**d**), Q4 and (**e**) Q5 of the first crowd sourcing theme session.

*Appendix B.2 Session B: Pinios River Basin Current Conditions*

Q6: How can the management framework of the nexus system of Pinios River Basin be characterised? (select 1 out of 4 predefined answers).

A6: The vast majority of the stakeholders (90%) position the management framework of the PRB nexus system as a very complex spectrum, while only 10% of the stakeholders characterise it as moderately complex or simple. This fact reveals the limited effectiveness of previously implemented policies and poses higher expectations on the ongoing project. Only a small percentage (14%) of the food/agriculture-related stakeholders characterised the PRB's management framework as a simple one, believing that the promotion of new water transfer projects can simply address the relevant challenges.

Q7: Which sector of the nexus system faces the most crucial challenges? (select maximum 2 out of 4 predefined answers).

A7: Almost half (49%) of the participants consider the water sector as the PRB nexus sector that faces the most crucial challenges, followed by the ecosystems/climate (30%), food/agriculture (18%), and energy (3%) sectors. It should be highlighted that the water-related challenges are recognised as the most crucial ones by the stakeholders of all of the nexus sectors. The energy-related challenges are mostly recognised by the stakeholders related to the food/agriculture sector and are associated with the increased required energy amount and cost and thus agricultural production cost. Although the recognition of the high importance of the energy sector, the water, ecosystems/climate, and food/agriculture sectors seem to face most crucial challenges in the agricultural areas of the PRB in terms of the promotion of its socio-economic stability and development. The impulse response of the stakeholders may have been stimulated by the research teams' perception on the current challenges of the PRB, as expressed in a presentation that preceded the crowd sourcing meeting.

Q8: Which sector of the nexus system presents the highest complexity? (select maximum 2 out of 4 predefined answers).

A8: The water and ecosystems/climate sectors were suggested to present the highest level of complexity, according to the beliefs of 87% of the participating stakeholders.

Q9: How are the nexus sectors ranked according to their importance? (1-low to 4-high).

A9: The water sector (3.6 out of 4) was perceived as the most important sector in the PRB, followed by the food/agriculture (2.8 out of 4), ecosystems/climate (2.3 out of 4), and energy (1.5 out of 4) sectors. The water sector's high importance was unanimously recognised by stakeholders related to all PRB nexus sectors.

Q10: Which sector has already been investigated more thoroughly? (select maximum 2 out of 5 predefined answers).

A10: The food/agriculture sector (58%) was identified as the sector that has been the most extensively investigated, according to the perceptions of the participants, followed by the water (38%) and energy (4%) sectors, while none of the participants recognised the ecosystems/climate sector as the most studied one. This fact reveals that previous efforts and policies have mainly focused on the food/agriculture and water sectors, thus reflecting their importance for the PRB. Apparently, this approach denotes the "traditional" tendency to disregard, at least to an extent, the roles of the ecosystems/climate and energy sectors in sustaining the nexus system of a basin, despite the fact that large portions of the PRB's area have been included in one or more lists of environmental protection areas.

*Appendix B.3 Session C: Management Practices*

Q11: How do you characterise the efficiency of the management framework related to the challenges of your sector? (select 1 out of 5 predefined answers).

A11: Most (63%) of the stakeholders characterised the management framework efficiency of the sector they are involved in as moderate, followed by those who believe that the management context is low (25%), high (1%), or very low (1%). The moderate characterisation was found to be the most popular answer among the stakeholders from all the different nexus sectors. This fact reflects the efficiency of the previously implemented management policies in the PRB, which were deemed to have moderate impacts.

Q12: Which pairs of nexus sectors present the strongest interaction? (select maximum 2 out of 6 predefined answers).

A12: Half (50%) of the participants believe that the water–food/agriculture pair present the highest inter-dependency in the PRB, followed by the water–ecosystems/climate (30%), water–energy (7%), energy–ecosystems/climate (7%), food–ecosystems/climate (3%), and energy–food/agriculture (3%) pairs. It should be highlighted that the top three pairs with the strongest interaction in the PRB include the water nexus component, which denotes the importance attributed to the sector by all the different groups of stakeholders.

Q13: To what extent do the existing management practices consider more than one sectors? (select 1 out of 5 predefined answers).

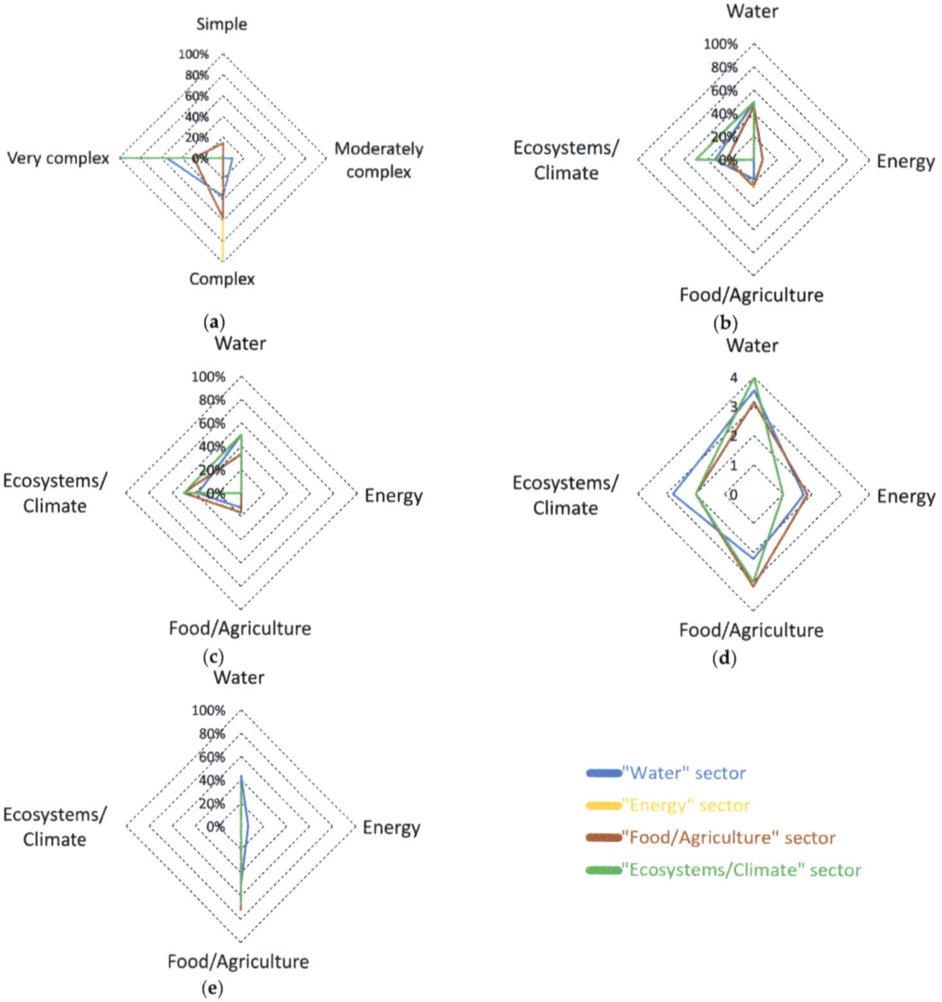


**Figure A2.** Responses of Pinios River Basin stakeholders to (**a**) Q6, (**b**) Q7, (**c**) Q8, (**d**) Q9, and (**e**) Q10 of the second crowd sourcing theme session.

A13: The vast majority (88%) of the stakeholders believe that the existing practices consider the other nexus sectors in the PRB to a moderate or low degree, followed by those who believe that other sectors are considered to a very low (6%) or high (6%) degree.

Q14: Which is/are the main basis/-es of management practices' selection? (select maximum 2 out of 5 predefined answers).

A14: The influential capacity of some stakeholders was recognised as the major factor affecting the selection of management practices by almost half of the stakeholders (44%), followed by the active stakeholder engagement (22%), non-active stakeholder engagement (15%), acceptance by local society (11%), and scientific evidence and sustainable development (8%), according to the perceptions of the participating stakeholders. The influential capacity of some stakeholders was found to be the most popular answer by the participants across all nexus sectors.

Q15: How is your sector affected or will be affected due to climate change? (select 1 out of 5 predefined answers).

A15: The vast majority (93%) of stakeholders believed their associated sector is already or will be highly or immensely affected by the climate crisis. Only 7% of the stakeholders believed that their sector is or will be moderately affected by climate change, while none of them assumed that climate change will not affect their related sector at all or minimally. It is also worth mentioning that most of the stakeholders related to the water and food/agriculture sectors believed that climate change is already having or expected to have a very high impact on their sectors. This perception is probably influenced by the intense and prolonged drought periods and the extreme precipitation events that recently occurred in the PRB, including the tropical-like cyclone named Ianos, which impacted the wider area in September 2020. The Ianos cyclone was the most intense medicane ever recorded in the Mediterranean region [94], causing casualties, severe floods, extensive infrastructure destruction, and heavy agricultural damage.

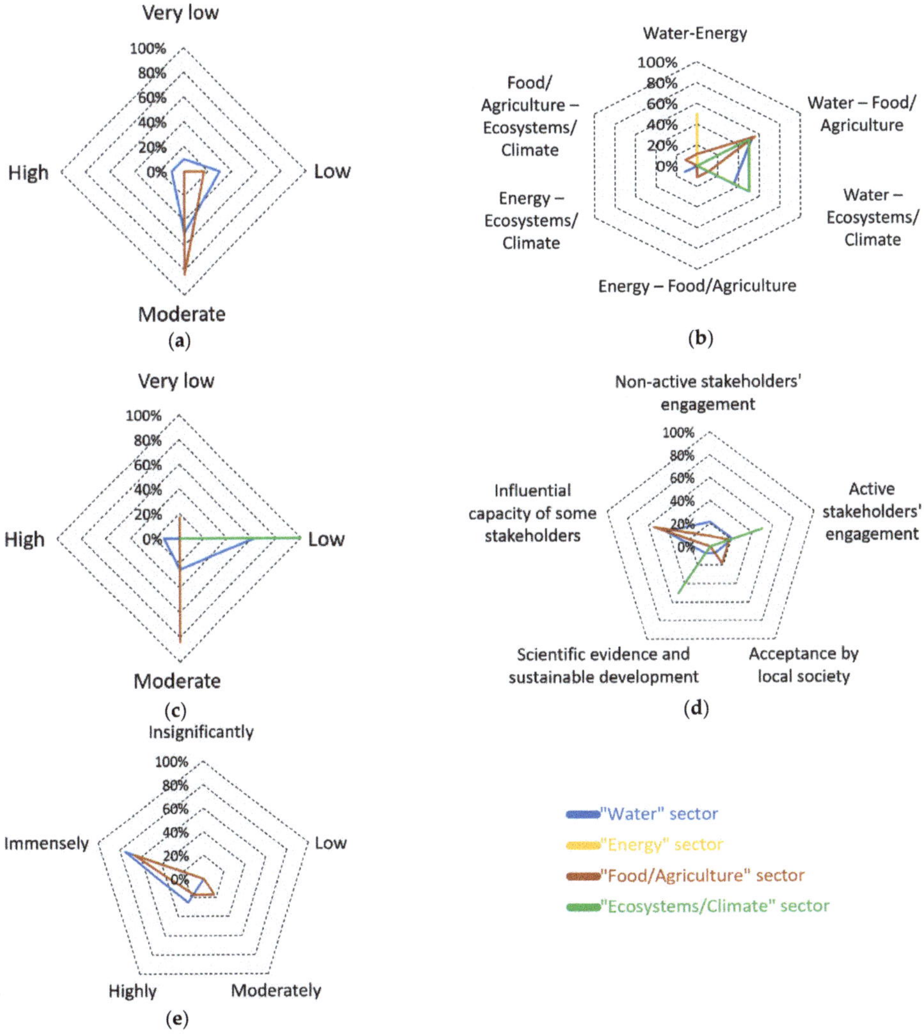


**Figure A3.** Responses of Pinios River Basin stakeholders to (**a**) Q11, (**b**) Q12, (**c**) Q13, (**d**) Q14, and (**e**) Q15 of the third crowd sourcing theme session.

## References


1. Persson, L.; Carney Almroth, B.M.; Collins, C.D.; Cornell, S.; De Wit, C.A.; Diamond, M.L.; Fantke, P.; Hassellöv, M.; Macleod, M.; Ryberg, M.W.; et al. Outside the Safe Operating Space of the Planetary Boundary for Novel Entities. *Environ. Sci. Technol.* **2022**, *56*, 1510–1521. [CrossRef] [PubMed]
2. United Nations (UN) Sustainable Development Goals. Available online: https://www.un.org/sustainabledevelopment/ (accessed on 16 September 2022).
3. Winkler, K.; Fuchs, R.; Rounsevell, M.; Herold, M. Global Land Use Changes Are Four Times Greater than Previously Estimated. *Nat. Commun.* **2021**, *12*, 2501. [CrossRef] [PubMed]
4. World Economic Forum. *Global Risks: An Initiative of the Risk Response Network*; World Economic Forum: Cologny, Switzerland, 2011; ISBN 9295044479.
5. Food and Agriculture Organization of the United Nations (FAO). *The Water-Energy-Food Nexus A New Approach in Support of Food Security and Sustainable Agriculture*; FAO: Rome, Italy, 2014.
6. International Renewable Energy Agency (IRENA). *Renewable Energy in the Water, Energy and Food Nexus*; International Renewable Energy Agency: Abu Dhabi, United Arab Emirates, 2015.
7. Organisation for Economic Co-operation and Development (OECD); Food and Agriculture Organization of the United Nations (FAO). *OECD-FAO Agricultural Outlook 2021–2030*; OECD Publishing: Paris, France, 2021.
8. Pisinaras, V.; Paraskevas, C.; Panagopoulos, A. Investigating the Effects of Agricultural Water Management in a Mediterranean Coastal Aquifer under Current and Projected Climate Conditions. *Water* **2021**, *13*, 108. [CrossRef]
9. Sánchez, J.M.; Simón, L.; González-Piqueras, J.; Montoya, F.; López-Urrea, R. Monitoring Crop Evapotranspiration and Transpiration/Evaporation Partitioning in a Drip-Irrigated Young Almond Orchard Applying a Two-Source Surface Energy Balance Model. *Water* **2021**, *13*, 2073. [CrossRef]
10. Arampatzis, G.; Hatzigiannakis, E.; Pisinaras, V.; Kourgialas, N.; Psarras, G.; Kinigopoulou, V.; Panagopoulos, A.; Koubouris, G. Soil Water Content and Olive Tree Yield Responses to Soil Management, Irrigation, and Precipitation in a Hilly Mediterranean Area. *J. Water Clim. Chang.* **2018**, *9*, 672–678. [CrossRef]
11. Willaarts, B.A.; Lechón, Y.; Mayor, B.; de la Rúa, C.; Garrido, A. Cross-Sectoral Implications of the Implementation of Irrigation Water Use Efficiency Policies in Spain: A Nexus Footprint Approach. *Ecol. Indic.* **2020**, *109*, 105795. [CrossRef]
12. Chitsaz, N.; Azarnivand, A. Water Scarcity Management in Arid Regions Based on an Extended Multiple Criteria Technique. *Water Resour. Manag.* **2017**, *31*, 233–250. [CrossRef]
13. Villada-Canela, M.; Muñoz-Pizza, D.M.; García-Searcy, V.; Camacho-López, R.; Daesslé, L.W.; Mendoza-Espinosa, L. Public Participation for Integrated Groundwater Management: The Case of Maneadero Valley, Baja California, Mexico. *Water* **2021**, *13*, 2326. [CrossRef]
14. Panagopoulos, A.; Arampatzis, G.; Tziritis, E.; Pisinaras, V.; Herrmann, F.; Kunkel, R.; Wendland, F. Assessment of Climate Change Impact in the Hydrological Regime of River Pinios Basin, Central Greece. *Desalin. Water Treat.* **2016**, *57*, 2256–2267. [CrossRef]
15. Osann, A.; Nino, P.; Zajicek, A.; Evangelou, E. Evaluation of Environmental and Economic Sustainability of Different Nutrient Management Strategies. In Proceedings of the 6th Congress on Soil and Water Resources, Izmir, Turkey, 12–14 November 2019.
16. Malamataris, D.; Kolokytha, E.; Loukas, A. Integrated Hydrological Modelling of Surface Water and Groundwater under Climate Change: The Case of the Mygdonia Basin in Greece. *J. Water Clim. Chang.* **2020**, *11*, 1429–1454. [CrossRef]
17. Li, J.; Ma, J.; Yu, L.; Zuo, Q. Analysis and Regulation of the Harmonious Relationship among Water, Energy, and Food in Nine Provinces along the Yellow River. *Water* **2022**, *14*, 1042. [CrossRef]
18. Bazilian, M.; Rogner, H.; Howells, M.; Hermann, S.; Arent, D.; Gielen, D.; Steduto, P.; Mueller, A.; Komor, P.; Tol, R.S.J.; et al. Considering the Energy, Water and Food Nexus: Towards an Integrated Modelling Approach. *Energy Policy* **2011**, *39*, 7896–7906. [CrossRef]
19. United Nations (UN). *Transforming Our World: The 2030 Agenda for Sustainable Development*; United Nations (UN): New York, NY, USA, 2015.
20. Winters, Z.S.; Crisman, T.L.; Dumke, D.T. Sustainability of the Water-Energy-Food Nexus in Caribbean Small Island Developing States. *Water* **2022**, *14*, 322. [CrossRef]
21. Fjäder, C. Developing Partnerships for Building Resilience. In *Handbook of Disaster Risk Reduction for Resilience*; Eslamian, S., Eslamian, F., Eds.; Springer: Cham, Switzerland, 2021; pp. 261–278.
22. Global Commission on Adaptation. Adapt Now: A Global Call for Leadership on Climate Resilience. Available online: https://gca.org/wp-content/uploads/2019/09/GlobalCommission_Report_FINAL.pdf (accessed on 5 September 2022).
23. Jiménez, A.; Saikia, P.; Giné, R.; Avello, P.; Leten, J.; Lymer, B.L.; Schneider, K.; Ward, R. Unpacking Water Governance: A Framework for Practitioners. *Water* **2020**, *12*, 827. [CrossRef]
24. Loukas, A.; Garrote, L. Feature Papers of Water Resources Management, Policy and Governance. *Water* **2022**, *14*, 2191. [CrossRef]
25. Penning, E. Using Nature-Based Solutions to Address Water Security Changes. *Hydrolink* **2022**, *2*, 45–47.
26. Graham, J.; Amos, B.; Plumptre, T. *Principles for Good Governance in the 21st Century. Policy Brief No. 15*; Institute on Governance: Ottawa, ON, Canada, 2003.
27. Blühdorn, I.; Deflorian, M. The Collaborative Management of Sustained Unsustainability: On the Performance of Participatory Forms of Environmental Governance. *Sustainability* **2019**, *11*, 1189. [CrossRef]

28. Villholth, K.G.; Sharma, B.R. Creating Synergy Between Groundwater Research and Management in South and South East Asia. In Proceedings of the IWMI-ITP-NIH International Workshop, Roorkee, India, 8–9 February 2005.
29. Kolokytha, E.; Malamataris, D. Integrated Water Management Approach for Adaptation to Climate Change in Highly Water Stressed Basins. *Water Resour. Manag.* **2020**, *34*, 1173–1197. [CrossRef]
30. Aldaya, M.M.; Sesma-Martín, D.; Schyns, J.F. Advances and Challenges in the Water Footprint Assessment Research Field: Towards a More Integrated Understanding of the Water-Energy-Food-Land Nexus in a Changing Climate. *Water* **2022**, *14*, 1488. [CrossRef]
31. Liu, J.; Bao, Z.; Wang, G.; Zhou, X.; Liu, L. The Spatial and Temporal Assessment of the Water-Land Nexus in a Changing Environment: The Huang-Huai-Hai River Basin (China). *Water* **2022**, *14*, 1905. [CrossRef]
32. Hoff, H. Understanding the Nexus. In Proceedings of the Water, Energy and Food Security Nexus Solutions for the Green Economy, Stockholm, Sweden, 16–18 November 2011.
33. Asian Development Bank. *Thinking about Water Differently: Managing the Water–Food–Energy Nexus*; Asian Development Bank: Metro Manila, Philippines, 2013.
34. Hamidov, A.; Helming, K. Sustainability Considerations in Water–Energy–Food Nexus Research in Irrigated Agriculture. *Sustainability* **2020**, *12*, 6274. [CrossRef]
35. Higgins, C.W.; Najm, M.A. An Organizing Principle for the Water-Energy-Food Nexus. *Sustainability* **2020**, *12*, 8135. [CrossRef]
36. Purwanto, A.; Sušnik, J.; Suryadi, F.X.; de Fraiture, C. Water-Energy-Food Nexus: Critical Review, Practical Applications, and Prospects for Future Research. *Sustainability* **2021**, *13*, 1919. [CrossRef]
37. Taghdisian, A.; Bukkens, S.G.F.; Giampietro, M. A Societal Metabolism Approach to Effectively Analyze the Water-Energy-Food Nexus in an Agricultural Transboundary River Basin. *Sustainability* **2022**, *14*, 9110. [CrossRef]
38. Wang, J.; Ju, K.; Wei, X. Where Will Water-Energy-Food Research Go Next? —*Visualisation Review and Prospect. Sustainability* **2022**, *14*, 7751. [CrossRef]
39. González-Rosell, A.; Blanco, M.; Arfa, I. Integrating Stakeholder Views and System Dynamics to Assess the Water–Energy–Food Nexus in Andalusia. *Water* **2020**, *12*, 3172. [CrossRef]
40. Kurian, M. The Water-Energy-Food Nexus: Trade-Offs, Thresholds and Transdisciplinary Approaches to Sustainable Development. *Environ. Sci. Policy* **2017**, *68*, 97–106. [CrossRef]
41. Liu, J.; Hull, V.; Godfray, H.C.J.; Tilman, D.; Gleick, P.; Hoff, H.; Pahl-Wostl, C.; Xu, Z.; Chung, M.G.; Sun, J.; et al. Nexus Approaches to Global Sustainable Development. *Nat. Sustain.* **2018**, *1*, 466–476. [CrossRef]
42. Albrecht, T.R.; Crootof, A.; Scott, C.A. The Water-Energy-Food Nexus: A Systematic Review of Methods for Nexus Assessment. *Environ. Res. Lett.* **2018**, *13*, 043002. [CrossRef]
43. van Buuren, A.; van Meerkerk, I.; Tortajada, C. Understanding Emergent Participation Practices in Water Governance. *Int. J. Water Resour. Dev.* **2019**, *35*, 367–382. [CrossRef]
44. Goodspeed, R.; Babbitt, C.; Briones, A.L.G.; Pfleiderer, E.; Lizundia, C.; Seifert, C.M. Learning to Manage Common Resources: Stakeholders Playing a Serious Game See Increased Interdependence in Groundwater Basin Management. *Water* **2020**, *12*, 1966. [CrossRef]
45. King, H.; Guehlstorf, N. Assessing the Variances of Citizen Stakeholder Justice in Watershed Governance: Is Action More Effective with Reactive or Proactive Administrations. *Water* **2022**, *14*, 878. [CrossRef]
46. Yuan, M.H.; Lo, S.L. Principles of Food-Energy-Water Nexus Governance. *Renew. Sust. Energ. Rev.* **2022**, *155*, 111937. [CrossRef]
47. Howarth, C.; Monasterolo, I. Understanding Barriers to Decision Making in the UK Energy-Food-Water Nexus: The Added Value of Interdisciplinary Approaches. *Environ. Sci. Policy* **2016**, *61*, 53–60. [CrossRef]
48. Vogler, D.; Macey, S.; Sigouin, A. *Stakeholder Analysis in Environmental and Conservation Planning*; Network of Conservation Educators and Practitioners, Center for Biodiversity and Conservation: New York, NY, USA, 2017; Volume 7, pp. 5–16.
49. Geaves, L.H.; Penning-Rowsell, E.C. Flood Risk Management as a Public or a Private Good, and the Implications for Stakeholder Engagement. *Environ. Sci. Policy* **2016**, *55*, 281–291. [CrossRef]
50. Ferraro, F.; Beunza, D. Creating Common Ground: A Communicative Action Model of Dialogue in Shareholder Engagement. *Organ. Sci.* **2018**, *29*, 1187–1207. [CrossRef]
51. Kapetas, L.; Kazakis, N.; Voudouris, K.; McNicholl, D. Water Allocation and Governance in Multi-Stakeholder Environments: Insight from Axios Delta, Greece. *Sci. Total Environ.* **2019**, *695*, 133831. [CrossRef]
52. Pagano, A.; Pluchinotta, I.; Pengal, P.; Cokan, B.; Giordano, R. Engaging Stakeholders in the Assessment of NBS Effectiveness in Flood Risk Reduction: A Participatory System Dynamics Model for Benefits and Co-Benefits Evaluation. *Sci. Total Environ.* **2019**, *690*, 543–555. [CrossRef]
53. Elster, J. *Deliberative Democracy*; Cambridge University Press: Cambridge, UK, 1998.
54. Bjerregaard, R. Getting Europe's Waters Cleaner: Getting the Citizens Involved. *Water Policy* **1998**, *1*, 73–80. [CrossRef]
55. Giatti, L.L.; Urbinatti, A.M.; De Carvalho, C.M.; Bedran-Martins, A.M.; De Oliveira Santos, I.P.; Honda, S.O.; Fracalanza, A.P.; Jacobi, P.R. Nexus of Exclusion and Challenges for Sustainability and Health in an Urban Periphery in Brazil. *Cad. Saude Publica* **2019**, *35*, 1–12. [CrossRef]
56. Endo, A.; Tsurita, I.; Burnett, K.; Orencio, P.M. A Review of the Current State of Research on the Water, Energy, and Food Nexus. *J. Hydrol. Reg. Stud.* **2017**, *11*, 20–30. [CrossRef]

57. Röckmann, C.; Ulrich, C.; Dreyer, M.; Bell, E.; Borodzicz, E.; Haapasaari, P.; Hauge, K.H.; Howell, D.; Mäntyniemi, S.; Miller, D.; et al. The Added Value of Participatory Modelling in Fisheries Management—What Has Been Learnt? *Mar. Policy* **2012**, *36*, 1072–1085. [CrossRef]
58. Howarth, C.; Monasterolo, I. Opportunities for Knowledge Co-Production across the Energy-Food-Water Nexus: Making Interdisciplinary Approaches Work for Better Climate Decision Making. *Environ. Sci. Policy* **2017**, *75*, 103–110. [CrossRef]
59. Johnson, O.W.; Karlberg, L. Co-Exploring the Water-Energy-Food Nexus: Facilitating Dialogue through Participatory Scenario Building. *Front. Environ. Sci.* **2017**, *5*, 24. [CrossRef]
60. Hoolohan, C.; Larkin, A.; McLachlan, C.; Falconer, R.; Soutar, I.; Suckling, J.; Varga, L.; Haltas, I.; Druckman, A.; Lumbroso, D.; et al. Engaging Stakeholders in Research to Address Water–Energy–Food (WEF) Nexus Challenges. *Sustain. Sci.* **2018**, *13*, 1415–1426. [CrossRef] [PubMed]
61. Bréthaut, C.; Gallagher, L.; Dalton, J.; Allouche, J. Power Dynamics and Integration in the Water-Energy-Food Nexus: Learning Lessons for Transdisciplinary Research in Cambodia. *Environ. Sci. Policy* **2019**, *94*, 153–162. [CrossRef]
62. Kropf, B.; Schmid, E.; Mitter, H. Multi-Step Cognitive Mapping of Perceived Nexus Relationships in the Seewinkel Region in Austria. *Environ. Sci. Policy* **2021**, *124*, 604–615. [CrossRef]
63. Karutz, R.; Omann, I.; Gorelick, S.M.; Klassert, C.J.A.; Zozmann, H.; Zhu, Y.; Kabisch, S.; Kindler, A.; Figueroa, A.J.; Wang, A.; et al. Capturing Stakeholders' Challenges of the Food–Water–Energy Nexus—A Participatory Approach for Pune and the Bhima Basin, India. *Sustainability* **2022**, *14*, 5323. [CrossRef]
64. Vinca, A.; Parkinson, S.; Riahi, K.; Byers, E.; Siddiqi, A.; Muhammad, A.; Ilyas, A.; Yogeswaran, N.; Willaarts, B.; Magnuszewski, P.; et al. Transboundary Cooperation a Potential Route to Sustainable Development in the Indus Basin. *Nat. Sustain.* **2021**, *4*, 331–339. [CrossRef]
65. de Vito, R.; Pagano, A.; Portoghese, I.; Giordano, R.; Vurro, M.; Fratino, U. Integrated Approach for Supporting Sustainable Water Resources Management of Irrigation Based on the WEFN Framework. *Water Resour. Manag.* **2019**, *33*, 1281–1295. [CrossRef]
66. Cansino-Loeza, B.; Ponce-Ortega, J.M. Sustainable Assessment of Water-Energy-Food Nexus at Regional Level through a Multi-Stakeholder Optimization Approach. *J. Clean. Prod.* **2021**, *290*, 125194. [CrossRef]
67. Pisinaras, V.; Panagopoulos, A.; Herrmann, F.; Bogena, H.R.; Doulgeris, C.; Ilias, A.; Tziritis, E.; Wendland, F. Hydrologic and Geochemical Research at Pinios Hydrologic Observatory: Initial Results. *Vadose Zone J.* **2018**, *17*, 180102. [CrossRef]
68. Pisinaras, V.; Herrmann, F.; Panagopoulos, A.; Tziritis, E.; McNamara, I.; Wendland, F. Fully Distributed Water Balance Modelling in Large Agricultural Areas—The Pinios River Basin (Greece) Case Study. *Sustainability* **2023**, *15*, 4343. [CrossRef]
69. Arampatzis, G.; Panagopoulos, A.; Pisinaras, V.; Tziritis, E.; Wendland, F. Identifying Potential Effects of Climate Change on the Development of Water Resources in Pinios River Basin, Central Greece. *Appl. Water Sci.* **2018**, *8*, 51. [CrossRef]
70. Vasiliades, L.; Loukas, A.; Liberis, N. A Water Balance Derived Drought Index for Pinios River Basin, Greece. *Water Resour. Manag.* **2011**, *25*, 1087–1101. [CrossRef]
71. Panagopoulos, A.; Herrmann, F.; Pisinaras, V.; Wendland, F. Impact of Climate Change on Irrigation Need and Groundwater Resources in Pinios Basin. *Proceedings* **2018**, *2*, 659. [CrossRef]
72. Sidiropoulos, P.; Mylopoulos, N.; Vasiliades, L.; Loukas, A. Stochastic Nitrate Simulation under Hydraulic Conductivity Uncertainty of an Agricultural Basin Aquifer at Eastern Thessaly, Greece. *Environ. Sci. Pollut. Res.* **2021**, *28*, 65700–65715. [CrossRef]
73. Sidiropoulos, P.; Tziatzios, G.; Vasiliades, L.; Mylopoulos, N.; Loukas, A. Groundwater Nitrate Contamination Integrated Modeling for Climate and Water Resources Scenarios: The Case of Lake Karla over-Exploited Aquifer. *Water* **2019**, *11*, 1201. [CrossRef]
74. Vrouhakis, I. Hydrogeological Research Using Modern Management Tools in the Framework of Qualitative and Quantitative Monitoring of the Groundwater System of Tirnavos Sub-Basin (Thessaly). Ph.D. Thesis, Agricultural University of Athens, Athens, Greece, 2022.
75. Tziatzios, G.; Sidiropoulos, P.; Vasiliades, L.; Lyra, A.; Mylopoulos, N.; Loukas, A. The Use of the Pilot Points Method on Groundwater Modelling for a Degraded Aquifer with Limited Field Data: The Case of Lake Karla Aquifer. *Water Supply* **2021**, *21*, 2633–2645. [CrossRef]
76. Hellenic Regulatory Authority for Energy. Geospatial Map for Energy Units & Requests. Available online: https://geo.rae.gr/ (accessed on 10 September 2022).
77. Loukas, A. Surface Water Quantity and Quality Assessment in Pinios River, Thessaly, Greece. *Desalination* **2010**, *250*, 266–273. [CrossRef]
78. Hatzigiannakis, E.; Filintas, A.; Ilias, A.; Panagopoulos, A.; Arampatzis, G.; Hatzispiroglou, I. Hydrological and Rating Curve Modelling of Pinios River Water Flows in Central Greece, for Environmental and Agricultural Water Resources Management. *Desalin. Water Treat.* **2015**, *57*, 11639–11659. [CrossRef]
79. Mendelow, A. Stakeholder Mapping. In Proceedings of the 2nd International Conference on Information Systems, Cambridge, MA, USA, 16–18 December 1991.
80. Winter, S. Organizational Perspective on Leadership Strategies for the Success of Cross-Disciplinary Science Teams. In *Strategies for Team Science Success. Handbook of Evidence-Based Principles for Cross-Disciplinary Science and Practical Lessons Learned from Health Researchers*; Hall, K.L., Vogel, A.L., Croyle, R.T., Eds.; Springer Nature: Cham, Switzerland, 2019; pp. 329–345.

81. Hermann-Friede, J.; Kropac, M.; Achermann, S.; Heeb, J.; Feuerstein, L.; Integrity Management Toolbox for Water Service Providers. Manual for Facilitators. Available online: https://sswm.info/sites/default/files/reference_attachments/WSP-Toolbox-Manual_WEB2.pdf (accessed on 25 August 2022).
82. Anon Global Water Partnership. GWP Integrated Water Resources Management Toolbox. Available online: https://www.gwp.org/en/learn/iwrm-toolbox/About_IWRM_ToolBox (accessed on 11 October 2023).
83. OECD. *Implementing the OECD Principles on Water Governance: Indicator Framework and Evolving Practices;* OECD Publishing: Paris, France, 2018.
84. Stein, U.; Bueb, B.; Knieper, C.; Tröltzsch, J.; Vidaurre, R.; Favero, F. The Diagnostic Water Governance Tool–Supporting Cross-Sectoral Cooperation and Coordination in Water Resources Management. *Environ. Sci. Policy* **2023**, *140*, 111–121. [CrossRef]
85. OECD. OECD Principles on Water Governance. Available online: www.oecd.org/governance/oecd-principles-on-water-governance.htm (accessed on 5 October 2023).
86. Bathrellos, G.D.; Skilodimou, H.D.; Soukis, K.; Koskeridou, E. Temporal and Spatial Analysis of Flood Occurrences in the Drainage Basin of Pinios River (Thessaly, Central Greece). *Land* **2018**, *7*, 106. [CrossRef]
87. Kim, R.E. The Nexus between International Law and the Sustainable Development Goals. *Rev. Eur. Comp. Int. Environ. Law.* **2016**, *25*, 15–26. [CrossRef]
88. European Parliament and Council. *Directive 2000/60/EC on Establishing a Framework for Community Action in the Field of Water Policy;* European Parliament and Council: Strasbourg, France, 2000; pp. L 327/1–L 327/72.
89. European Commission. *REPowerEU Plan: Communication from the Commission to the European Parliament, the European Council, the Council, the European Economic and Social Committee and the Committee of the Regions;* COM(2022) 230 Final; European Commission: Brussels, Belgium, 2022.
90. European Commission. The New Common Agricultural Policy: 2023–2027. Available online: https://agriculture.ec.europa.eu/common-agricultural-policy/cap-overview/cap-2023-27_en (accessed on 22 May 2023).
91. European Commission Climate Action. Available online: https://climate.ec.europa.eu/index_en (accessed on 11 June 2023).
92. Seddon, N.; Chausson, A.; Berry, P.; Girardin, C.A.J.; Smith, A.; Turner, B. Understanding the Value and Limits of Nature-Based Solutions to Climate Change and Other Global Challenges. *Philos. Trans. R. Soc. B Biol. Sci.* **2020**, *375*, 20190120. [CrossRef]
93. IPCC. *Climate Change and Land: An IPCC Special Report on Climate Change, Desertification, Land Degradation, Sustainable Land Management, Food Security, and Greenhouse Gas Fluxes in Terrestrial Ecosystems;* Shukla, P.R., Skea, J., Calvo Buendia, E., Masson-Delmotte, V., Portner, H.O., Roberts, D.C., Zhai, P., Slade, R., Connors, S., van Diemen, R., et al., Eds.; IPCC: Geneva, Switzerland, 2019.
94. Lagouvardos, K.; Karagiannidis, A.; Dafis, S.; Kalimeris, A.; Kotroni, V. Ianos—A Hurricane in the Mediterranean. *Bull. Am. Meteorol. Soc.* **2021**, *103*, E1621–E1636. [CrossRef]

*water*

MDPI

*Article*

# Hydrogeological Hazards in Open Pit Coal Mines–Investigating Triggering Mechanisms by Validating the European Ground Motion Service Product with Ground Truth Data

**Ploutarchos Tzampoglou [1,2,*] and Constantinos Loupasakis [1]**

1 Laboratory of Engineering Geology and Hydrogeology, Department of Geological Sciences, School of Mining and Metallurgical Engineering, National Technical University of Athens, 15780 Athens, Greece; cloupasakis@metal.ntua.gr
2 Department of Civil & Environmental Engineering, University of Cyprus, Nicosia 1678, Cyprus
* Correspondence: tzampoglou.ploutarchos@ucy.ac.cy

**Abstract:** This research focuses on the investigation of hydrogeological hazards in open pit coal mines. The study area is the Amyntaio sub-basin area, located in West Macedonia prefecture, Greece. A major part of the SE of this area is occupied by the Amyntaio open pit coal mine. In recent decades, the Amyntaio basin's aquifer has been overexploited both by dewatering wells of the open pit coal mine and irrigation wells, triggering extensive land subsidence in an area that extends 3 km around the mine. Additionally, one of the biggest mining landslides worldwide occurred on the South-West slopes of the open pit on 10 June 2017. The current study investigates the land subsidence phenomenon and the landslide, highlighting the influence and the interaction of their causal factors which were strongly affected by the groundwater management. To estimate ground surface movement, Earth Observation data from the European Ground Motion Service, of the Copernicus European Union's Earth observation program, were used for the period 1 January 2016–31 December 2020. The geologic, geotechnical and hydrogeologic data coming from the extensive ground truth survey have been incorporated with the Earth Observation data, highlighting the opposing mechanisms of the interacting geohazards.

**Keywords:** mining landslide; land subsidence; amyntaio basin; mining geohazards; ground water management; Copernicus; land monitoring

## 1. Introduction

Anthropogenic activities, such as mining, significantly affect the natural environment [1], causing various types of hazards. The most common geohazards are land subsidence and landslides; these catastrophic events can occur during mining as well as in the post-mining period (in cases where appropriate mitigation measures are not taken) and can be triggered by a range of hydrologic, geologic, and tectonic factors. Specifically, 132 different cases of slope failures have been documented worldwide since 1960 [2], causing hundreds of deaths and massive economic damage. Key examples include the Bingham Canyon Mine landslide in Utah, USA, in which two landslides displaced around 145 million tons of geomaterial on 10 Arpil 2013 [3–6], and the Jade Mine landslide in the northern state of Kachin, Myanmar, in which over 170 people were killed [7]. In contrast, land subsidence slowly affects extensive areas around mines, causing mainly significant economic damage [8–12].

To date, many studies have been conducted to investigate the failure mechanisms triggering natural hazards such as landslides [13], land subsidence [14,15], floods [16,17], and seasonal ground movement due to swelling formations [18] using state of the art Earth Observation technologies. The use of Synthetic Aperture Radar Interferometry satellite data (InSAR) combined with the geotechnical research helps to obtain new insights into the failure mechanisms of the geohazards. Furthermore, it can contribute significantly to estimating the main causal factors that play an active role in the occurrence of the geohazard. Thus, it can give the opportunity to take all the necessary measures for the mitigation of the catastrophic events.

The study area in this work is the Amyntaio sub-basin, which is located in the West Macedonia prefecture of Greece. A major part of the southeast of this area is occupied by the Amyntaio open pit coal mine. A 220 m deep coal mine that operated between 1989 and 2020, with 179 million tons of lignite production and 1595 million $m^3$ of total excavations.

There are four lakes and ten villages within the Amyntaio basin, and the basin's agricultural area exceeds 200 $km^2$. In recent decades, an extensive groundwater level drop has been recorded due to the overpumping of the aquifer, both for protecting the open pit coal mine's slopes (i.e., draining wells) and providing for the irrigation needs of the area's agricultural activity. Specifically, the groundwater level drawdown between May 1992 and May 2015 reaches 70 m near the mine, extending northwest to the village of Valtonera and west to Anargiroi village [19–23]. The drop in the groundwater level, combined with the area's general geotectonic setting, triggered extensive land subsidence in an area extending up to 2 km around the mine, causing damage to villages and infrastructure since 2002. Additionally, one of the world's largest mining landslides took place on the South-West slopes of the open pit mine on 10 June 2017. Based on previous studies, the landslide mass involved in this event was 80 million $m^3$ [24]. A timely prediction of this event allowed precautionary measures (such as stopping work inside the mine) to be taken and, therefore, loss of human life was avoided.

Aiming to further study the mechanisms of the catastrophic events that took place at the perimeter of the Amyntaio open pit coal mine, this study employs InSAR data from the European Ground Motion Service (EGMS) product covering the period 2016–2021. The EGMS provides reliable information regarding natural and anthropogenic ground motion over the Copernicus Participating States and across national borders, with millimeter accuracy [25–29].

The motivation of this study is to combine the ground truth datasets acquired for the study of the mining geohazards, namely land subsidence and landslide, with the EGMS InSAR data set, aiming to cross-verify their outcomes. The main objectives of the study are to (a) estimate the spatial distribution of the vertical deformation due to land subsidence and landslide by evaluating the EGMS InSAR datasets; (b) correlate the above data with the data obtain thought the geotechnical research highlighting opposite driving failures mechanisms and main causal factors and (c) estimate the interaction between these two catastrophic events.

The comparison of the ground truth survey data (that have been acquired for more than 15 years since 2005) with the INSAR datasets, provides an excellent case study to demonstrate the reliability and value of the EGMS product. The understanding of the failure mechanism of land subsidence and landslide as well the estimation of their interaction can help local and regional authorities to understand the arising risks. This knowledge will enable them to take appropriate mitigation measures to ensure the future safe landplaning development.

## 2. Study Area

### *2.1. Geologic, and Geotechnical Setting*

The study area forms part of the Kozani–Amyntaio–Ptolemaida–Florina basin and belongs to the Pelagonian geotectonic zone. The area consists of a Paleozoic and Mesozoic crystalline schist bedrock covered by Neogene and Quaternary deposits. Based on previous geotechnical studies [23,30–32], some useful insights can be gained relating to the geotechnical behavior of the formations developed in the study area. In detail, the study area's geological formations are as follows (Figure 1).

Recent alluvial deposits: including lacustrine and peat deposits. They obtain high to extremely high compressibility index (Cc) values of up to 0.187 and 1.707, respectively.

Despite their high compressibility index, these units occur superficially and, therefore, are unlikely to be significantly affected by changes in groundwater levels; thus, they likely do not play a major role in driving land subsidence due to overexploitation of the aquifer. However, these units exhibit land subsidence driven by oxidation as a result of their high organic content levels.

Anargiroi Formation: Quaternary formation comprising of silty–clayey sands and red clays with coarse-grained intercalations. This unit has high strength values, consistent with its soil grain distribution; its Cc value, which reaches 0.133 for the clay intercalations, is likely due to the presence of small amounts of swelling clay minerals. This formation occurs in hilly areas of the study region's lowlands, at heights above the aquifers' phreatic head. Thus, it is not directly affected by groundwater head variations and does not contribute to or interact with land subsidence phenomena.

Perdikas formation: Quaternary formation divided into sandy silty clay (65–75%), organic silty clay (10–20%), and coarse-grained (10–20%) layers with some organic silty–clayey intercalations in the upper part. This unit mainly occurs under the recent deposits across the majority of the basin. In terms of its geotechnical data, the Perdikas formation shows high Cc values due to the significant montmorillonite content, which reach 0.16 for the silty–clayey sand, 0.27 for the sandy silty clay, and 0.99 for the organic silty clay horizons. Therefore, this formation, as a result of its composition (around 65–75% clayey horizons) and its geotechnical setting, is prone to vertical deformation in areas where fluctuation of the aquifer occurs.

Proastio Formation: lower Quaternary formation consisting of sandy coarse-grained horizons with red silty–clayey sands, clays, and organic layers [33]. It can be separated into two distinct parts; the upper part consists of more than 80% coarse-grained material and the lower part contains over 30% silty clay layers. The average Cc values are 0.16 for the silty–clayey sands, 0.18 for the organic layers, and 0.16 for the sandy silty clay layers. Due to the minor contribution of the fine-grained layers in the stratigraphy of the Proastio formation, their impact on the overall land subsidence is considered insignificant.

The Neogene formations underlie the Proastio formation. They are separated into two series: (a) the lignite-bearing Ptolemaida formation (Amyntaio and Ptolemaida lignite deposits), which consists mainly of gray silty clays and clayey marls [34,35], and (b) the lignite-bearing Komnina formation (Vegora and Komnina lignite deposits), which consists of sand and clay layers with intercalations of xylite [35] and sandstones. These formations are over-consolidated and impermeable, occurring beneath the base of the aquifer. Thus, they are not considered prone to consolidation, even if the active stresses change. However, it should be stressed that throughout this formation, bedding-parallel shear zones with reduced shear strength values occur [36]. These zones are highly likely to contribute to the formation of failure planes along the working slopes. Therefore, although this formation is not susceptible to land subsidence, it is susceptible to landslide failures.

Alpine–pre-Alpine formations: these formations are divided into: (a) Paleozoic metamorphic rocks (gneiss and schists) and (b) Triassic–Lower Jurassic carbonate rocks (crystalline limestones and marbles) [37,38]. They mainly occur in the west of the basin.

The faults in the Amyntaio sub-basin can be separated into two groups [39–42]. The first, trending NW–SE, affects only the Alpine–pre-Alpine formations, and the second, trending NE–SW, extends up to the Quaternary deposits. Three active tectonic structures with NE-SW directions [43] occur in the Amyntaio basin (Figure 1): (a) the Petron–Xino Nero–Aetos fault (Fpxa) in the northwest; (b) the Anargiroi fault (Fang) in the south, and (c) the Vegoritida fault (Fveg) on the east to northeast side of the basin.

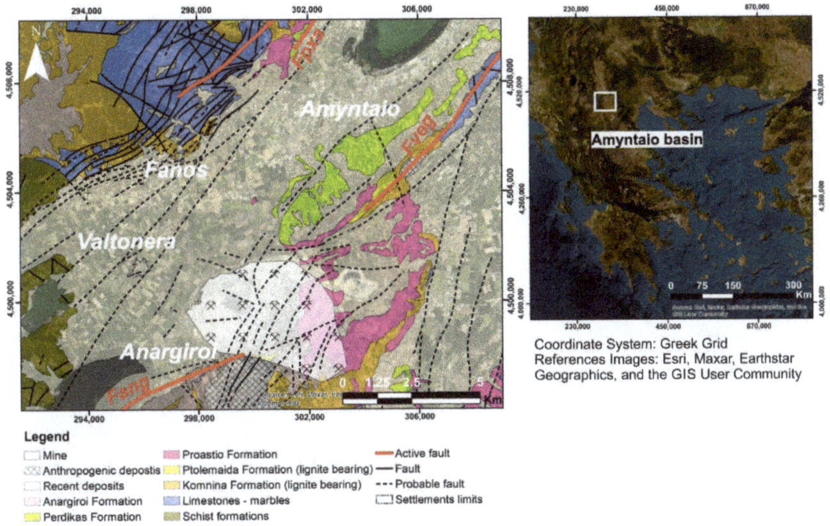


**Figure 1.** Geological map of the Amyntaio basin. The three active tectonic lines identified in the Amyntaio basin are clearly indicated.

## 2.2. *Hydrogeologic Setting*

Two aquifer systems can be identified in the Amyntaio sub-basin. The first one constitutes a karstic aquifer system within the Alpine–pre-Alpine bedrock formations, while the second one is developed in the Quaternary formations.

The piezometric head of the karstic aquifer system exceeds 100 m above the open pit bottom floor. Given that high piezometric loads were transferred on the slopes of the open pit mine in addition to the existence of bedding-parallel shear zones, negative slope stability conditions are expected [24]. Note that this aquifer does not affect the land subsidence phenomenon as it is not subjected to any kind of exploitation along its entire plane due to its depth, considering that it is completely isolated underneath the Neogene formations. The shallow semiconfined aquifer is recharged by both karstic water overflow, along the boundaries of the basin, and by filtration of surface water; its total depth does not exceed 120 m [44].

During recent decades, a significant level drop in this aquifer has been identified as a result of systematic overpumping both for protecting the open pit coal mine's slopes (i.e., draining wells) and providing for the irrigation needs of the area's agricultural activity. Considering the results of extensive field investigation measurement campaigns [20,45,46], some key conclusions concerning changes in the piezometric level are as follows (Figure 2A–E): according to the results of groundwater level measurements that were carried out at the end of the dry season (October 2014 and October 2015), the groundwater flow appears to occur toward the open pit mine, forming an extensive depression cone (Figure 2B,D). In addition, local dynamic depression cones of minor importance were observed in areas of agricultural activity. A similar groundwater flow trend was identified in data that were obtained from water level measurement campaigns during May 2015 and May 2016 (i.e., the end of the wet period). Therefore, currently the open pit along with the surrounding draining wells operate as an oversized (4 km wide) well, continuously draining a big part of the Amyntaio basin.

The evolution of the groundwater flow over time can be easily estimated by comparing equal-drawdown contour lines between May 1992–May 2015 [23] and May 1992–May 2016 [47]. Note that the data from May 1992 relate to the period before the operation of the open pit coal mine [48]. As shown in Figure 3A,B, the drawdown reaches 70 m near the working slopes of the mine and decreases with increasing distance away from the mine.

The depression cone extends to the west reaching to the villages of Valtonera and Anargiroi. On the contrary, as indicated in Figure 2A, before the operation of the mine, only a gentle groundwater flow from the west towards the east of the basin can be identified. It is clear that during the last three decades the operation of the mine and, secondarily, the increase of the agricultural activities radically changed the groundwater dynamics of the basin.

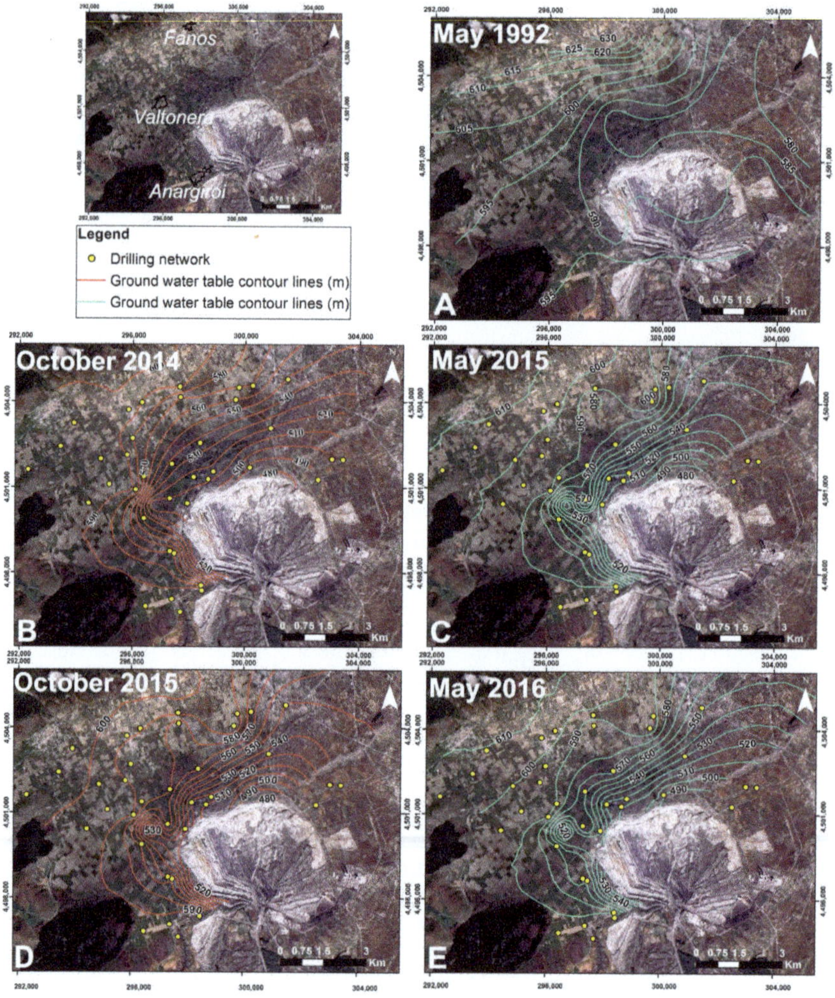


**Figure 2.** Ground water table contour lines based on the measurements carried out on: (**A**) May 1992 [48], (**B**) October 2014; (**C**) May 2015, (**D**) October 2016, and (**E**) May 2016.

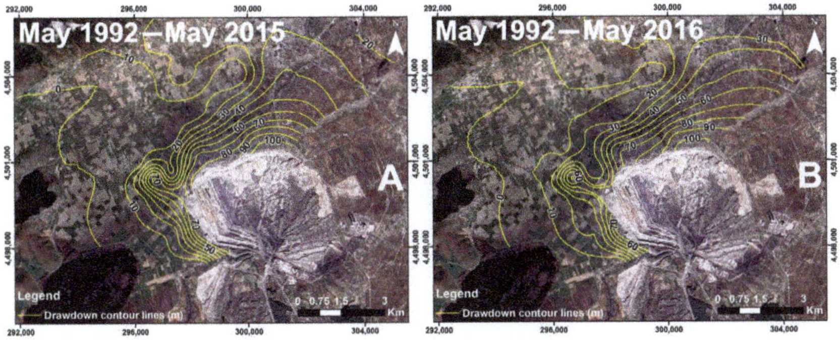


**Figure 3.** Equal drawdown contour lines between: (**A**) May 1992–May 2015 and (**B**) May 1992–May 2016.

## 3. Methodology for the Determination of the Catastrophic Events Driving Mechanism

Initially, all geological [34], tectonic [28,34,35,37,43], geotechnical [32] and hydrogeological [42] data, from previous studies, were collected and reevaluated during the 15 years long study, conducted by the authors of the current paper. New data sets have been added to the existing knowledge, including: (a) engineering geological mapping of the study area, (b) geotechnical laboratory tests' results from over 60 samples; (b) four ground water level measurements' campaigns (October 2014, May 2015, October 2015, May 2016) on a newly established network of over 40 wells and (c) detailed surface rapture mapping and infrastructure damage recording. All data sources taken into account are presented in the Supplementary Material (Table S1).

Aiming to validate the ground truth datasets as well as to identify the spatial distribution of deformation occurring in the villages surrounding the mine, data from the EGMS product were cross-validated with the above-described ground truth data sets. The EGMS is based on the multi-temporal interferometric analysis of Sentinel-1 radar images at full resolution. This technique allows for identifying reliable measurement points for which ground motion velocity values and time series of deformation are extracted [47]. The EGMS Ortho product provides purely vertical and purely east-west displacements over time with 100 m × 100 m output resolution cells. The data available for the studied site range from 5 January 2016 to 31 December 2021 [26], which includes the period of the landslide event that occurred on 10 June 2017.

The cross-validation of the Earth Observation data with the ground truth data sets provided a clear view on the driving mechanism of the land subsidence and landslide phenomena, highlighting their opposing triggering factors.

### 3.1. The Land Subsidence Mechanism

In 2001, Anargiroi village was affected by land subsidence. Until 2006, vertical deformation was recorded across the entire study area. To investigate the spatial distribution of surface ruptures in the Amyntaio basin, observations from previous studies were initially compiled [23,49–52]. Then, six field work campaigns were carried out, in which 15.5 km of surface ruptures were recorded (Figure 4). Most of these ruptures are oriented NW–SE and NE–SW, following the trends of the area's main fault lines. Thus, the tectonic faults acted as a pre-existing structural fabric that has controlled the orientation of later ruptures.

Three villages have been significantly affected by surface deformation (Anargiroi (Figure S1), Valtonera (Figure S2), and Fanos (Figure S3)) [23]. All these settlements share some common characteristics, namely (a) they are located in areas affected by groundwater level drawdown, (b) they are crossed by faults, and (c) they are underlain by the Perdikas formation.

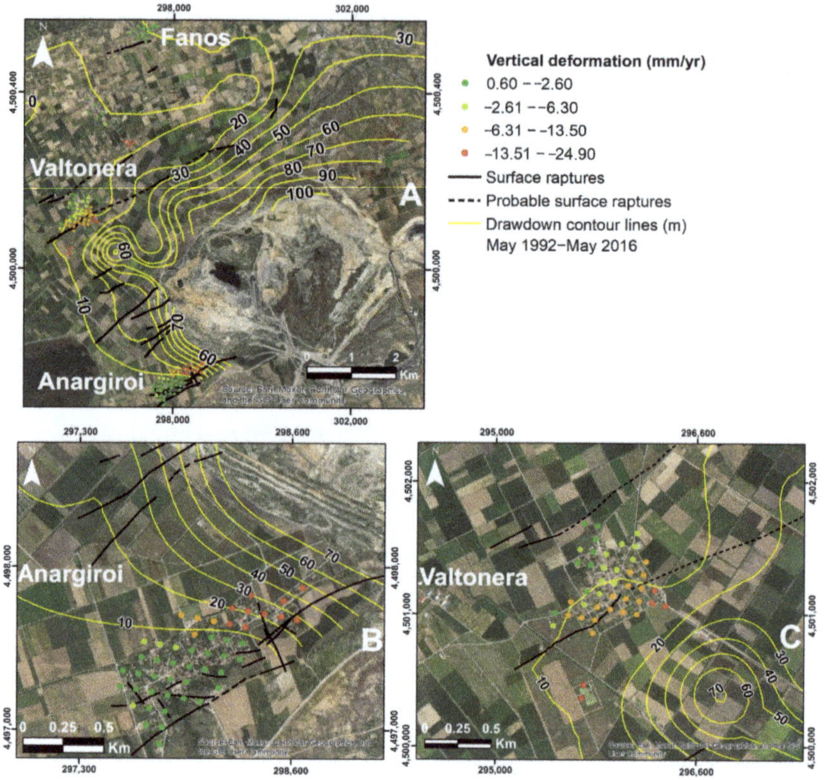


**Figure 4.** (**A**) Vertical deformations (mm/year) between 5 January 2016 and 31 December 2020 according to the InSAR data provided by Copernicus Land Monitoring Service [26] along with the recorded surfaces ruptures and the drawdown contour lines. (**B**,**C**) scale up view of the Anargiroi and Valtonera villages.

The vertical displacements of the surface ruptures appear to be proportional to the groundwater table drop and the distance from the mine. Specifically, the largest offsets (0.5 m to 1 m) are observed in areas near the mine where the groundwater withdrawal reaches 70 m. The differential displacements decrease with increasing distance from the mine and do not occur in areas where no water table level drop is recorded.

The correlation of the persistent scatterer (PS) point data obtained from the EGMS datasets (for the period between 5 January 2016 and 31 December 2020) with the recorded surface ruptures and drawdown contour lines for the period between May 1992 and May 2016 can be shown in Figure 4. Although the datasets do not refer to the exact same period, their correlation is applicable as the general hydrogeological conditions remain practically unchanged since 1992. The open pit keeps on operating as the main draining source while no extra irrigation wells have been drilled at the wider area. Evaluating the EGMS datasets, it is clear that the largest vertical displacement amplitudes, exceeding −20 mm/year, are observed to the northwest of Anargiroi village; the vertical deformation then decreases southeast of the village. This observation correlates closely with the recorded surface ruptures and drawdown contour lines, highlighting that land subsidence is still occurring in the study area.

Similar conclusions can be drawn from the PS points at Valtonera village. The observed water level drop in this area of around 10 to 20 m can trigger high-amplitude vertical displacements.

### 3.2. The Landslide Mechanism

On 10 June 2017, the largest recorded Greek mining landslide took place on the mining slopes of the Amyntaio open pit coal mine (Figure 5). According to estimates, the collapse mass reached 80 million $m^3$, extending over an area of 2 km in width and 2.2 km in length, while the foot of the landslide coincided with the lower point of the open pit mine at a depth of 210 m.

**Figure 5.** The spatial extent of the 10 June 2017 landslide at the Amyntaio mine, as shown by satellite imagery. Images obtained by Google Earth. (**A–C**) show oblique views of the landslide area. (**A**) one of the bucket wheel excavators buried inside the sliding mass; (**B**) panoramic view of the landslide scarp in front of the Anargyroi village; and (**C**) panoramic view of the landslide head looking towards East West.

The first signs of this failure were recorded some months before it occurred. In detail, eight months before the failure (November 2016), archaeological excavations, which were being carried out on the plain above the northwest side of the mine, were interrupted due to the occurrence of rapidly formed surface raptures caused by the increasing vertical deformation. In addition, five months before the failure (February 2017), deformation along the surface ruptures in the wider area of Anargiroi village began to increase. The timely recognition and prediction of the phenomenon allowed the evacuation of the open pit coal mine eight days before the failure, which contributed to avoiding the loss of lives.

Furthermore, the financial impacts of this landslide, which affected infrastructure both inside the mine and in the nearby area, were extremely large. Specifically, four bucket wheel excavators (weighing 500 to 1500 tons each), many small excavators, and several kilometers of conveyor were significantly damaged or buried inside the mine (Figure 5A). Additionally, 25 million tons of lignite were buried under the sliding mass, leading to further losses. Outside the open pit coal mine, the landslide caused enormous damage to the village of Anargiroi (Figure S4). In particular, the northwest part of Anargiroi village was severely damaged and permanently evacuated, the wastewater treatment plant collapsed into the pit (Figure 5B,C), and hundreds of meters of road network and electricity and water supply infrastructure were destroyed (Figure 6A–E). Furthermore, an archaeological settlement of lacustrine Neolithic age, which was located at the western perimeter of the mine, disappeared as it collapsed into the open cast pit.

**Figure 6.** Spatial distribution of surface ruptures identified behind the crown of the 10 June 2017 landslide. (**A**,**B**) surface ruptures within the basin between Anargiroi and Valtonera village (the height of the scarp is approximately 1 m), (**C**,**D**) pre-existing surface ruptures caused by the land subsidence phenomenon that were intensified by the landslide event in the NW part of Anargiroi village, and (**E**) damage along the Anargyroi village road network.

To constrain the potential failure mechanism for this event, the three main triggering factors, tectonic activity, groundwater pore pressure, and slope geometry, were investigated. In terms of tectonic activity, according to the National Observatory of Greece's earthquake database [53], no earthquakes of magnitude 2.5 or above were recorded in the area surrounding the open pit in the five-month period preceding the landslides. Therefore, the area's tectonic activity is deemed to have not played a major active role in activating this particular landslide or in the associated pre-failure vertical deformation. Nonetheless, it should be stressed that the faults crossing the mine affected the formations occupying the working slopes of the mine, creating zones with reduced shear strength. In particular, the Vegoritida fault ($F_{Veg}$) defines the western boundary of the landslide while the Anargiri fault ($F_{Ang}$) crosses the crown and eastern boundary of the landslide. In addition, secondary parallel faults cross the site, affecting the geometry of the landslide mass. Thus, the tectonic structure of the study area can be characterized as a preparatory factor, but not a triggering causal factor, of the landslide on 10 June 2017.

In regard to the groundwater pore pressure, two aquifer systems occur in the study area. The aquifer system developed in the Quaternary formations has been systematically overexploited during recent decades, aiming to protect the mining slopes from failure. The volume of water pumped out from the aquifers, via the mine protection wells, was reduced over time as the subsurface consolidation resulted in compaction of the aquifers and a gradual reduction of their hydraulic conductivity. In detail, the protection wells around the mine extracted 13 × $10^6$ $m^3$/year of water in 2002, 9 × $10^6$ $m^3$/year in 2009, and 3 × $10^6$ $m^3$/year in 2016 [51]. Furthermore, the pumping activities directly adjacent to Anargiroi village, according to PPC [51], were essentially stopped in 2016, aiming to eliminate the deformation caused by overpumping at the nearby village. These two factors led to a steady increase in the water table between Anargiroi village and the mine, thereby significantly degrading the mechanical properties of the Quaternary deposits. This increase became visible after the occurrence of the landslide when water from the aquifer began gushing from the main scarp of the landslides (Figure 5B). In contrast, this did not occur along the northwestern parts of the main scarp where pumping activities were active at least since the slope failure (Figure 5C). Additionally, the karstic aquifer system created negative slope stability conditions by transferring piezometric load to the lower Neogene formations at the bottom of the mine.

The geometry of the working slopes was uneven and steep, negatively impacting their stability. The evaluation of all available satellite images of the mine illustrates that the two upper benches were excavated first in order to unload the slope. Subsequently, the excavation of the lower benches of the coal seams was prioritized to meet production needs. As a result, the excavation of the middle benches, occupied by the upper Neogene and lower Quaternary formations, was systematically neglected. Accordingly, the geometry of the working slopes was uneven, as the upper parts were more gently sloped and the lower parts steeper. This configuration, with a steeper base of the working slopes, raises significant issues concerning the overall safety factor.

Overall, the stratigraphy of the Neogene formations, including shear zones, the tectonic structure of the plain with faults crossing the mine, and the hydrogeologic setting with two aquifer systems interacting with the mine are considered to be the main preparatory factors for the studied landslide event. Respectively, the main triggering factors of this failure were: (a) the increase in the water table in the Quaternary formation aquifer due to selective reduction of pumping and (b) the steep geometry of the working slopes.

## 4. Results and Discussion

Based on the vertical and east-west displacement time series from PS points at the villages of Anargiroi, Valtonera, and Fanos, various insights can be gained into the spatial distribution of land subsidence and the impact of the landslide.

In regard to the Anargiroi village, the maximum vertical displacements occur in areas closer to the mine (Figures 4 and S1). This area is underlain by the highly compressible Perdikas Formation; in addition, the maximum groundwater dropdown values of the mine's depression cone occur in this area. As anticipated, the vertical displacement decreases (from NE to SW) in proportion to the groundwater level drop (Figures 4 and S1). In contrast, at the southwestern end of the village, only slight seasonal groundwater fluctuations appear to affect the displacements (A7).

Considering the vertical displacement time series, the village appears to be affected by a steadily decreasing deformation trend (i.e., decreasing land subsidence) during the period January 2016 to December 2020. This observation is directly linked with changes in pumping activities, which have practically stopped since 2016 (Figure 7), allowing partial recovery of the unconfined aquifer.

The abrupt fluctuations of PS points in both the vertical and east/west displacements that can be observed in the period around 10 June 2017, especially in the NE part of the village (i.e., A1, A2, A3, and A6), highlights the significant impact of the landslide. Furthermore, as clearly shown from the east–west horizontal displacement time series (Figure 7), the majority of the PSs around the NE part of the village display a steady eastward displacement trend, towards the open pit, after the landslide. The westward displacement trend identified at two PS points (A2 and A3) can be attributed to rotational movements of the structures that operated as reflectors for those two PSs. Thus, the present trends suggest that the studied site is slowly moving toward the abandoned 220 m deep open pit located only 180 m from the first houses of the village. Physical evidence of the ongoing activity of the landslide can be obtained by the continuous deformations recorded at the damaged buildings located close to the open pit (Figure S7).

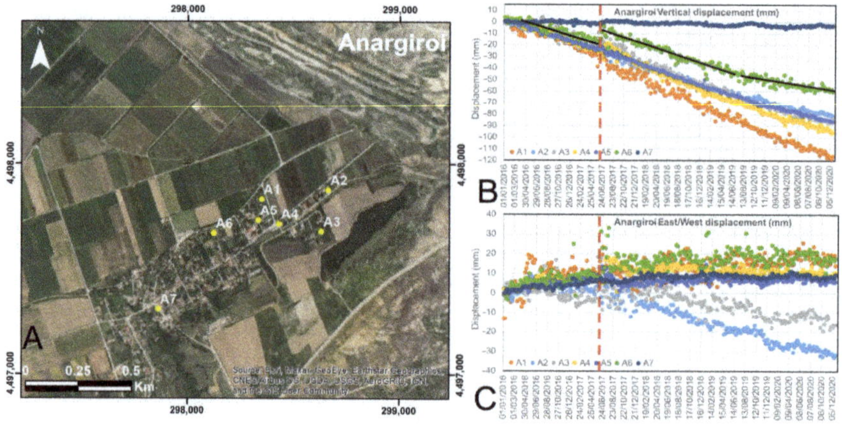


**Figure 7.** Spatial distribution of examined PS points (**A**) along with their displacement time series (**B**,**C**) at the village of Anargiroi. The dashed red line indicates the date of the mining landslide event. The trend lines added to the time series of point A6 indicate the variations of the land subsiding trend due to the draining activities before and after the occurrence of the landslide, as well as after the 2016 cessation of pumping from the wells located between the village and the open pit. Point A6 was selected as it clearly presents the disturbance caused by the landslide as well as the slight reduction of the deformations after the end of the end of the draining activities.

Consistent with the vertical displacement distribution identified at Anargiroi village, the deformation at Valtonera village increases toward the east, i.e., in areas closer to the mine (Figure 4). Similar to Anargiroi village, this area is also underlain by the highly compressible Perdikas formation and the depression cone of the mine is present. The vertical displacements clearly decreased in the western part of the village with increasing distance from the mine.

The vertical displacement time series in the village of Valtonera shows a steadily decreasing trend with pronounced annual fluctuations. This observation is highly consistent with the water table time series at Valtonera village, which shows a steady downward trend with seasonal fluctuations. The observed trend of a continuous groundwater level drop can be attributed to the draining of the aquifer toward the mines, whereas the seasonal fluctuation likely results from exploitation of the aquifers for irrigation purposes followed by a partial rebound of the groundwater head during the winter [47]. As shown in Figure 8, the vertical displacements each year exhibit a decreasing trend until October (end of the dry period) followed by an increasing trend until May (end of the rainy period). Note that as this village was unaffected by the landslide (Figure S5), no abrupt scatter in the vertical and east–west displacement time series can be identified during the landslide period (Figure 8).

Considering the east–west horizontal displacement time series (Figure 7), most of the PSs around the village indicate a steady eastward displacement trend toward the open pit. These movements cannot be attributed to any landslide activity due to the great distance of the village from the mine (1800 m) relative to the height of the slopes (220 m). Thus, this deformation can only be attributed to ongoing land subsidence-driven deformation. It should be noted that the village is crossed by the Vegoritida fault (Fveg) dipping to the SE and that the fault acted as a pre-existing structural fabric that controls the orientation of the later ruptures generating slight horizontal movements.

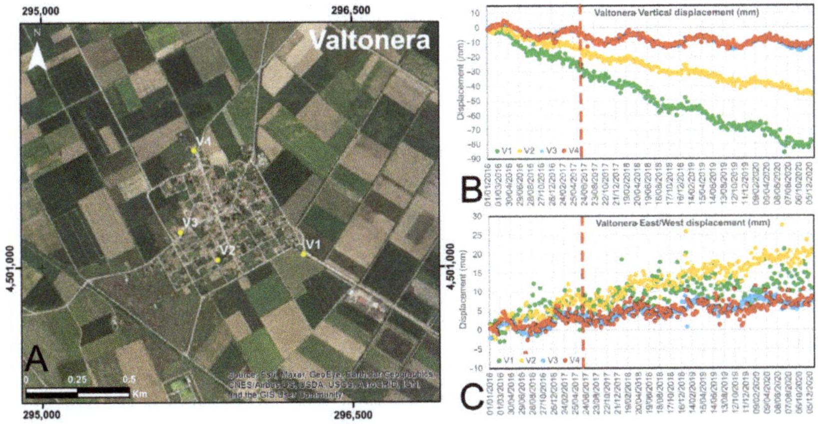


**Figure 8.** Spatial distribution of examined PS points (**A**) along with their displacement time series (**B**,**C**) at the village of Valtonera. The dashed line indicates the date of the mining landslide event.

The vertical displacements identified at the Fanos village are significantly less intense than those at the Valtonera and Anargiroi villages; nonetheless, this village also suffers from the occurrence of surface ruptures and differential deformation. These deformation types can be attributed to the pronounced stiffness and compressibility differences of the area's underlying formations, as the northern part of the village overlies Neogene rocks, and the western part overlies Quaternary formations.

Considering the vertical displacement time series (Figure 9), although annual fluctuations can be observed, there is no evidence of a gradually evolving long-term land subsidence trend. This implies that groundwater exploitation of the aquifer for irrigation needs is the main factor triggering the deformation that more strongly affects the softer and more compressible Quaternary formations. As shown in the deformation time series, no disturbance is recorded that can be attributed to the 10 June 2017 landslide (Figures 9 and S6); in addition, there are no clear trends observed in the east–west displacement values.

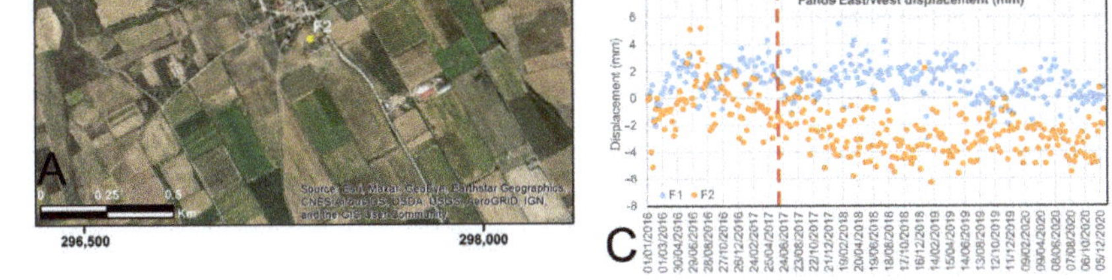


**Figure 9.** Spatial distribution of examined PS points (**A**) along with their displacement time series (**B**,**C**) at the village of Fanos. The dashed line indicates the date of the mining landslide event. The points were selected from both sides of the fault crossing the village, pointing to the displacements taking place on the stiff Neogene, point F1, and the soft Quaternary, point F2, formations.

## 5. Conclusions

In conclusion, the combination of ground truth data and information from EGMS datasets provides important insights into the mechanisms and interactions of geocatastrophic events taking place at the Amyntaio coal mine.

In particular, Anargiroi village is clearly affected by aquifer overexploitation via the mine's protection wells. The seasonal fluctuations in vertical deformation identified at Valtonera and Fanos villages, due to the operation of irrigation wells, are not visible in the vertical displacement time series acquired for Anargiroi village. Accordingly, farming activity is interpreted as not contributing significantly to land subsidence in the vicinity of Anargiroi village. This conclusion was largely expected as most of the farms in the village are irrigated from irrigation canals rather than from wells. In addition, the coincidence between the observed decreasing subsidence trends and the termination of pumping activities in 2016 between the open cast mine and the village also strongly supports the interpretation that the subsidence phenomenon is mainly caused by dewatering activities performed at the perimeter of the open cast mine. Regarding the effect of the 10 June 2017, landslide, the InSAR data set demonstrates that the failure affected and continues to affect at least the NE part of the village; in addition, slow horizontal movements toward the open pit are still taking place.

In terms of the deformation occurring at Valtonera village, the 10 June 2017, landslide did not affect the settlement due to the large distance between the village and the open cast mine. Subsidence in this area can be attributed to overpumping caused by both the mine's dewatering activity and farming activity. The mining activity appears to be responsible for the ongoing steady subsidence trend affecting the village, causing substantial damage. In contrast, farming activities in this area appear to cause gentle fluctuating deformations of a minor scale compared to those caused by mining activities.

The deformations at Fanos village are affected only by farming activity, as evidenced by the lack of any observed cross-seasonal or long-term deformation trend and the exclusive occurrence of seasonal deformation in this area. In addition, as expected, the fluctuating deformations appear to be more intense in areas of soft Quaternary formations rather than those of the stiff Neogene.

As a final conclusion, it should be stressed that EGMS datasets are a powerful resource for deciphering the different causal factors of deformations affecting the study area for more than two decades, highlighting the important role of the groundwater management. Because these datasets are provided by an independent and reliable international source, their conclusions are highly robust. Overall, InSAR techniques represent a highly important approach for the study of geohazards.

**Supplementary Materials:** The following supporting information can be downloaded at: https://www.mdpi.com/article/10.3390/w15081474/s1, Figure S1: The impact of land subsidence at the Anarigoi village. The photos were capture in May 2016 (before the landslide event that occurred on 10 June 2017); Figure S2: The impact of land subsidence at the Valtonera village. The photos were capture in May 2016 (before the landslide event that occurred on 10 June 2017); Figure S3: The impact of land subsidence at the Fanos village. The photos were capture in May 2016 (before the landslide event that occurred on 10 June 2017); Figure S4: The impact of landslide event in areas previously affected by land subsidence, at the Anarigoi village. The expansion of the surface raptures and the intensification of the damages on the constructions is clearly precented; Figure S5: Images taken at the Valtonera village before and after the occurrence of the major landslide event, that took place on 10 June 2017. It is clear that, as expected due to the large distance of the village from the open pit, the landslide did not cause any further damages to the already damaged, by the land subsidence, constructions of the village; Figure S6: Images taken at the Fanos village before and after the occurrence of the major landslide event, that took place on 10 June 2017. It is clear that, as expected due to the large distance of the village from the open pit, the landslide did not cause any further damages to the already damaged, by the land subsidence, constructions of the village; Figure S7: Images taken at the Anargyroi village after the occurrence of the major landslide event, that took place on 10 June 2017. The deformations taking place due to the continuous landslide and land subsidence movements since the occurrence of the landslide are clearly presented; Table S1: Ground truth dataset.

**Author Contributions:** Methodology, P.T. and C.L.; Validation, P.T. and C.L.; Formal analysis, P.T.; Investigation, P.T.; Writing—original draft, P.T.; Writing—review & editing, C.L.; Supervision, C.L. All authors have read and agreed to the published version of the manuscript.

**Funding:** This research received no external funding.

**Data Availability Statement:** The datasets used and/or analyzed during the current study are available from the authors upon request.

**Acknowledgments:** The authors would like to acknowledge the European Land Monitoring Service of the European Environment Agency for freely providing the INSAR data.

**Conflicts of Interest:** The authors declare no conflict of interest.

## References


1. Passariello, B.; Giuliano, V.; Quaresima, S.; Barbaro, M.; Caroli, S.; Forte, G.; Carelli, G.; Iavicoli, I. Evaluation of the environmental contamination at an abandoned mining site. *Microchem. J.* **2002**, *73*, 245–250. [CrossRef]
2. WISE_Uranium_Project. Chronology of Major Tailings Dam Failures. Available online: https://www.wise-uranium.org/mdaf.html (accessed on 30 August 2022).
3. Williams, C.; Ross, B.; Zebker, M.; Leighton, J.; Gaida, M.; Morkeh, J.; Robotham, M. Assessment of the available historic RADARSAT-2 synthetic aperture radar data prior to the manefay slide at the bingham canyon mine using modern InSAR techniques. *Rock Mech. Rock Eng.* **2021**, *54*, 3469–3489. [CrossRef]
4. Moore, J.R.; Pankow, K.L.; Ford, S.R.; Koper, K.D.; Hale, J.M.; Aaron, J.; Larsen, C.F. Dynamics of the Bingham Canyon rock avalanches (Utah, USA) resolved from topographic, seismic, and infrasound data. *J. Geophys. Res. Earth Surf.* **2017**, *122*, 615–640. [CrossRef]
5. Hibert, C.; Ekström, G.; Stark, C.P. Dynamics of the Bingham Canyon Mine landslides from seismic signal analysis. *Geophys. Res. Lett.* **2014**, *41*, 4535–4541. [CrossRef]
6. Pankow, K.L.; Moore, J.R.; Hale, J.M.; Koper, K.D.; Kubacki, T.; Whidden, K.M.; McCarter, M.K. Massive landslide at Utah copper mine generates wealth of geophysical data. *GSA Today* **2014**, *24*, 4–9. [CrossRef]
7. Lin, Y.N.; Park, E.; Wang, Y.; Quek, Y.P.; Lim, J.; Alcantara, E.; Loc, H.H. The 2020 Hpakant Jade Mine Disaster, Myanmar: A multi-sensor investigation for slope failure. *ISPRS J. Photogramm. Remote Sens.* **2021**, *177*, 291–305. [CrossRef]
8. Sun, Y.; Zhang, X.; Mao, W.; Xu, L. Mechanism and stability evaluation of goaf ground subsidence in the third mining area in Gong Changling District, China. *Arab. J. Geosci.* **2015**, *8*, 639–646. [CrossRef]
9. Carnec, C.; Delacourt, C. Three years of mining subsidence monitored by SAR interferometry, near Gardanne, France. *J. Appl. Geophys.* **2000**, *43*, 43–54. [CrossRef]
10. Wolkersdorfer, C.; Thiem, G. Ground water withdrawal and land subsidence in northeastern Saxony (Germany). *Mine Water Environ.* **1999**, *18*, 81–92. [CrossRef]
11. Tang, W.; Motagh, M.; Zhan, W. Monitoring active open-pit mine stability in the Rhenish coalfields of Germany using a coherence-based SBAS method. *Int. J. Appl. Earth Obs. Geoinf.* **2020**, *93*, 102217. [CrossRef]
12. Mohammady, M.; Pourghasemi, H.R.; Amiri, M. Assessment of land subsidence susceptibility in Semnan plain (Iran): A comparison of support vector machine and weights of evidence data mining algorithms. *Nat. Hazards* **2019**, *99*, 951–971. [CrossRef]
13. Hong, H.; Tsangaratos, P.; Ilia, I.; Loupasakis, C.; Wang, Y. Introducing a novel multi-layer perceptron network based on stochastic gradient descent optimized by a meta-heuristic algorithm for landslide susceptibility mapping. *Sci. Total Environ.* **2020**, *742*, 140549. [CrossRef]
14. Kaitantzian, A.; Loupasakis, C.; Tzampoglou, P.; Parcharidis, I. Ground Subsidence Triggered by the Overexploitation of Aquifers Affecting Urban Sites: The Case of Athens Coastal Zone along Faliro Bay (Greece). *Geofluids* **2020**, *2020*, 8896907. [CrossRef]
15. Svigkas, N.; Loupasakis, C.; Papoutsis, I.; Kontoes, C.; Alatza, S.; Tzampoglou, P.; Tolomei, C.; Spachos, T. InSAR Campaign Reveals Ongoing Displacement Trends at High Impact Sites of Thessaloniki and Chalkidiki, Greece. *Remote Sens.* **2020**, *12*, 2396. [CrossRef]
16. Nhu, V.-H.; Thi Ngo, P.-T.; Pham, T.D.; Dou, J.; Song, X.; Hoang, N.-D.; Tran, D.A.; Cao, D.P.; Aydilek, İ.B.; Amiri, M. A new hybrid firefly–PSO optimized random subspace tree intelligence for torrential rainfall-induced flash flood susceptible mapping. *Remote Sens.* **2020**, *12*, 2688. [CrossRef]
17. Ilia, I.; Tsangaratos, P.; Tzampoglou, P.; Chen, W.; Hong, H. Flash flood susceptibility mapping using stacking ensemble machine learning models. *Geocarto Int.* **2022**, *37*, 1–27. [CrossRef]
18. Tzampoglou, P.; Loukidis, D.; Koulermou, N. Seasonal Ground Movement Due to Swelling/Shrinkage of Nicosia Marl. *Remote Sens.* **2022**, *14*, 1440. [CrossRef]

19. Loupasakis, C. *Study of the Geotechnical Conditions of the Amintaio Coalmine Slopes Close to the Anargiri Village, Aetos Municipality, Florina Prefecture, Greece*; IGME: Athens, Greece, 2006; p. 48.
20. Tzampoglou, P.; Loupasakis, C. Updated ground water piezometry data of the Amyntaio sub-basin and their effect to the manifestation of the land subsidence phenomena. In Proceedings of the 11th International Hydorgeological Congress of Greece, Athens, Greece, 4–6 October 2017.
21. Loupasakis, C.; Angelitsa, V.; Rozos, D.; Spanou, N. Mining geohazards—Land subsidence caused by the dewatering of opencast coal mines: The case study of the Amyntaio coal mine, Florina, Greece. *Nat. Hazards* **2014**, *70*, 675–691. [CrossRef]
22. Tzampoglou, P.; Loupasakis, C. Mining geohazards susceptibility and risk mapping: The case of the Amyntaio open-pit coal mine, West Macedonia, Greece. *Environ. Earth Sci.* **2017**, *76*, 542. [CrossRef]
23. Tzampoglou, P.; Loupasakis, C. Evaluating geological and geotechnical data for the study of land subsidence phenomena at the perimeter of the Amyntaio coalmine, Greece. *Int. J. Min. Sci. Technol.* **2018**, *28*, 601–612. [CrossRef]
24. Loupasakis, C. Contradictive mining–Induced geocatastrophic events at open pit coal mines: The case of Amintaio coal mine, West Macedonia, Greece. *Arab. J. Geosci.* **2020**, *13*, 1–12. [CrossRef]
25. European Ground Motion Service. End-to-end Implementation and Operation of the European Ground Motion Service (EGMS). In *Product User Manual*; European Environment Agency (EEA): Copenhagen, Denmark, 2022.
26. European Union; European Environment Agency (EEA). Copernicus Land Monitoring Service. 2022. Available online: https://egms.land.copernicus.eu/ (accessed on 21 September 2022).
27. Crosetto, M.; Solari, L.; Mróz, M.; Balasis-Levinsen, J.; Casagli, N.; Frei, M.; Oyen, A.; Moldestad, D.A.; Bateson, L.; Guerrieri, L. The evolution of wide-area DInSAR: From regional and national services to the European Ground Motion Service. *Remote Sens.* **2020**, *12*, 2043. [CrossRef]
28. Crosetto, M.; Solari, L.; Balasis-Levinsen, J.; Bateson, L.; Casagli, N.; Frei, M.; Oyen, A.; Moldestad, D.; Mróz, M. Deformation monitoring at european scale: The copernicus ground motion service. *Int. Arch. Photogramm. Remote Sens. Spat. Inf. Sci.* **2021**, *XLIII-B3-2*, 141–146. [CrossRef]
29. Costantini, M.; Minati, F.; Trillo, F.; Ferretti, A.; Novali, F.; Passera, E.; Dehls, J.; Larsen, Y.; Marinkovic, P.; Eineder, M. European Ground Motion Service (EGMS). In Proceedings of the 2021 IEEE International Geoscience and Remote Sensing Symposium IGARSS, Brussels, Belgium, 11–16 July 2021; pp. 3293–3296.
30. Doukissa, K. *Geotechnical Study of the Bridge in the Central Trench of Aanargiroi Area*; Municipality of Aetos: Thessaloniki, Greece, 2010; p. 24.
31. Dhmaras, K.; Georgiadhs, M. *Lifting and Waterproofing Study of the Existing Embankment of Lake Cheimaditida Florina*; Geognosi Public Limited Company: Thessaloniki, Greece, 2002.
32. Tzampoglou, P. Mining geohazards in Greece. The case of the Amyntai open-pit coal mine. Ph.D. Thesis, National Technical University of Athens, Athens, Greece, 2017.
33. Koukouzas, C.; Kotis, T.; Ploumidis, M.; Metaxas, A. *Coal Exploration of Anargiri-Amynteon Area, Mineral Deposit Research*; Institution of Geology and Mineral Exploration: Athens, Greece, 1979; Volume 9.
34. Institution of Geology and Mineral Exploration. *Geological Map of Greece, Scale 1:50,000, Ptolemaida Sheet*; Institution of Geology and Mineral Exploration: Athens, Greece, 1997.
35. Metaxas, A.; Karageorgiou, D.; Varvarousis, G.; Kotis, T.; Ploumidis, M.; Papanikolaou, G. Geological evolution-stratigraphy of Florina, Ptolemaida, Kozani and Saradaporo graben. *Bull. Geol. Soc. Greece* **2007**, *40*, 161–172. [CrossRef]
36. Leonardos, M.; Terezopoulos, N. Rim slope failure mechanism in the Greek deep lignite mines–A case study. *Min. Technol.* **2003**, *112*, 197–204. [CrossRef]
37. Mountrakis, D. *Geology of Greece*; University Studio Press: Thessaloniki, Greece, 1985.
38. Spyropoulos, N. The geological structure of Pelagonian zone in the Mount of Askio D. Macedonia. Ph.D. Thesis, Aristotle University of Thessaloniki, Thessaloniki, Greece, 1992.
39. Pavlides, S.; Mountrakis, D. Neotectonics of the Florina-Vegoritis-Ptolemais Neogene Basin (NW Greece): An example of extensional tectonics of the greater Aegean area. *Ann. Géol. Pays Hell.* **1985**, *33*, 311–327.
40. Pavlides, S. Neotectonic evolution of the Florina-Vegoritis-Ptolemais basin (W. Macedonia, Greece). Ph.D. Thesis, Aristotle University of Thessaloniki, Thessaloniki, Greece, 1985.
41. Pavlides, S.; Mountrakis, D. Extensional tectonics of northwestern Macedonia, Greece, since the late Miocene. *J. Struct. Geol.* **1987**, *9*, 385–392. [CrossRef]
42. Koukouzas, C.; Kotis, T.; Ploumidis, M.; Metaxas, A.; Dimitriou, D. *Coal Exploration of Komninon-Ptolemaidas Area. Mineral Deposit Research*; Institution of Geology and Mineral Exploration: Athens, Greece, 1983; Volume 2.
43. Ganas, A.; Oikonomou, I.; Tsimi, A. NOAfaults: A digital database for active faults in Greece. *Bull. Geol. Soc. Greece* **2015**, *47*, 518–530. [CrossRef]
44. Stamos, A.; Giannoulopoulos, P. *Hydrogeological Report, Geotechnical Work to the Anargiri Village, Aetos Municipality, Florina Prefecture, Greece*; Institution of Geology and Mineral Exploration: Athens, Greece, 2010; p. 34.
45. Tzampoglou, P.; Loupasakis, C. New data regarding the ground water level changes at the Amyntaio basin-Florina Prefecture, Greece. *Bull. Geol. Soc. Greece* **2016**, *50*, 1006–1015. [CrossRef]
46. Tzampoglou, P.; Loupasakis, C. Land subsidence due to the overexploitation of the aquifer at the Valtonera village. *Bull. Geol. Soc. Greece* **2017**, *50*, 1006. [CrossRef]

47. Tzampoglou, P.; Loupasakis, C. Numerical simulation of the factors causing land subsidence due to overexploitation of the aquifer in the Amyntaio open coal mine, Greece. *HydroResearch* **2019**, *1*, 8–24. [CrossRef]
48. Dimitrakopoulos, D. Hydrogeological conditioning at the Amyndeon mine. Problems occuring during the exploitation and their mitigation actions. Ph.D. Thesis, NTUA, Athens, Greece, 2001.
49. Tsourlos, P. *Research for the Appearance of Surface Cracks in the Valtonera Village, Amyntaio Municipality, Florina Region*; School of Geology of the Aristotle University of Thessaloniki Aristotelian University of Thessaloniki: Thessaloniki, Greece, 2015; p. 17.
50. Soulios, G.; Tsapanos, T.; Voudouris, K.; Kaklis, T.; Mattas, C.; Sotiriadis, M. Ruptures on surface and buildings due to land subsidence in Anargyri village (Florina Prefecture, Macedonia). *Environ. Earth Sci.* **2011**, *5*, 505–512.
51. Dimitrakopoulos, D.; Koumantakis, I. Hydrodynamic regime of Amynteon basin. Influence of open lignite mines. In Proceedings of the 11th International Hydrogeological Congress of Greece, Athens, Greece, 4–6 October 2017; pp. 101–112.
52. Tzampoglou, P.; Loupasakis, C. Land subsidence susceptibility and hazard mapping: The case of Amyntaio Basin, Greece. In Proceedings of the Fifth International Conference on Remote Sensing and Geoinformation of the Environment (RSCy2017), Paphos, Cyprus, 20–23 March 2017; p. 104441L.
53. NOA. Earthquake Database Search. Available online: https://www.gein.noa.gr/en/services-products/database-search/ (accessed on 25 September 2022).

water

MDPI

*Article*

# Repurposing of a Closed Surface Coal Mine with Respect to Pit Lake Development

**Georgios Louloudis [1,*], Christos Roumpos [1,*], Emmanouil Louloudis [2], Eleni Mertiri [1] and Georgios Kasfikis [1]**

1 Public Power Corporation of Greece, Department of Mining Engineering and Closure Planning, 10432 Athens, Greece
2 Department of Statistics, Athens University of Economics and Business, 10434 Athens, Greece
* Correspondence: g.louloudis@dei.gr (G.L.); c.roumpos@dei.gr (C.R.); Tel.: +30-697-000-5090 (G.L.); +30-697-979-9291 (C.R.)

**Abstract:** In the coal phase-out era, achieving sustainable mine closure is significant and prioritizes targets for the mining industry. In this study, the already closed lignite mine of Kardia, North Greece, is investigated, where the mine void left is naturally filled with water. The viability of different repurposing land uses is evaluated, and the natural water level development inside the mine pit is investigated concerning its future uses. The potential for solar photovoltaic (PV) panels developed on mining land and its surrounding area is evaluated in combination with the application of pumped hydro storage (PHS) technology, utilizing pit lake water. Except for electricity system planning, other end-uses that offer multiple, mutually reinforcing and lasting benefits are investigated, such as recreation parks, terrestrial wildlife, aquaculture and agriculture. All repurposing scenarios are evaluated with regard to the spatiotemporal evolution of the lake, by generating forecasts of the dependent variables (rainfall and temperature) via linear (autoregressive integrated moving average) and non-linear (artificial neural network) models. The prediction of pit lake natural development redefines the new land use layout and the land repurposing decisions. This is essential for strategic planning, considering the Greek lignite mining industry's priority regarding transitioning from the current coal-based electricity to renewable energy sources (RES) technology.

**Keywords:** mine closure; autoregressive integrated moving average (ARIMA); artificial neural network (ANN); post-mining; land use; energy; sustainability; spatiotemporal development; economic transition

## 1. Introduction

In the region of the South Ptolemais lignite-bearing basin, in Macedonia, N.W.( northwest) Greece (Figure 1), the Public Power Corporation (PPC) of Greece had developed extensive mining activity in order to provide adequate lignite quantities for 3000 MW power plants. However, Greece's commitment to the European Commission's long-term strategic vision to achieve net-zero greenhouse-gas emissions [1,2] has provided motivations for sustainable economic growth and spurred the modernization of the economy. The PPC operates within the framework of corporate environmental responsibility [3] and has established an environmental policy that reflects its commitment to comply with all applicable national and international laws and support continuous environmental improvement.

Over the last decade, particular policy attention has been paid to the transition from coal energy to clean energy technologies to reduce the risks of climate change impacts and contribute to climate-resilient pathways [4] for sustainable development. The goal is to integrate renewable sources and energy efficiency while simultaneously providing the lowest possible cost. As a result, multi-level planning is critical to designing and implementing effective strategies, exploring repurposing options, and making decisions about a post-mining future and economic transitions, considering regional and policy contexts [5].

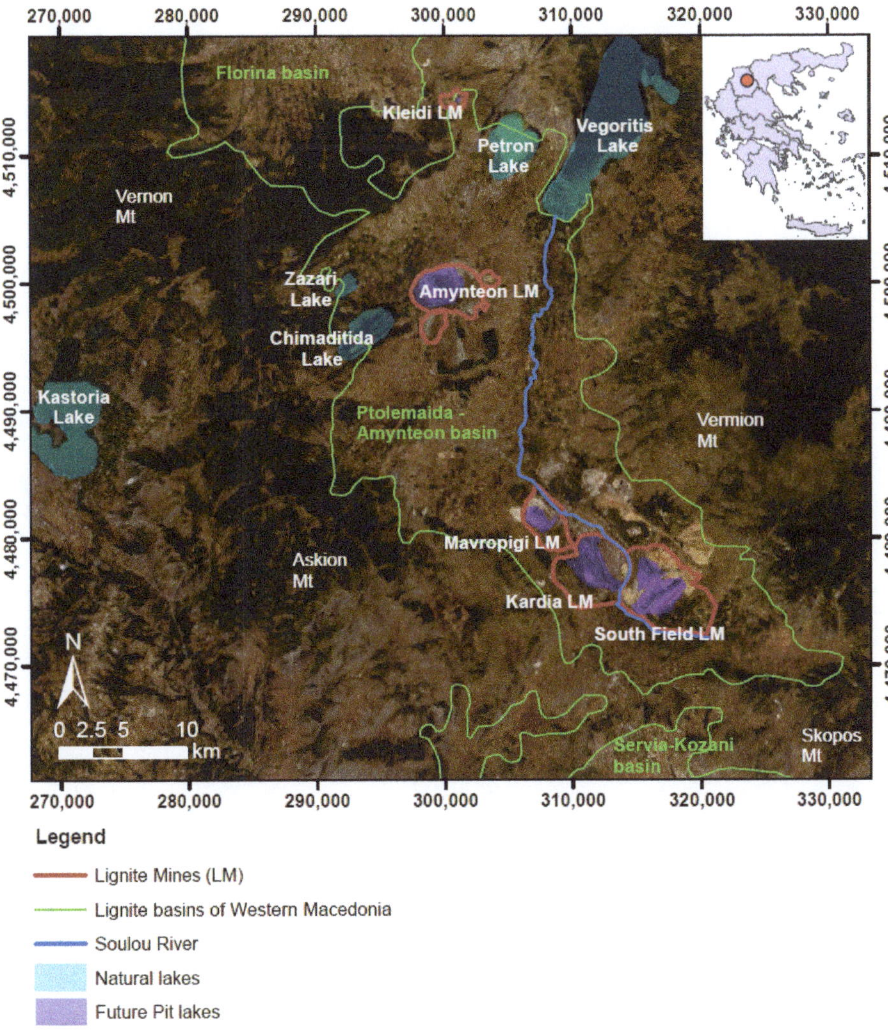


**Figure 1.** Research area and location of the Kardia Lignite Mine (LM), Greece. Basemap obtained from https://www.esri.com/en-us/home accessed on 30 March 2022 (Coordinate Reference System GGRS87/Greek Grid).

Recently, the PPC has investigated and evaluated alternative repurposing options to produce energy using renewable energy sources (RES) and has planned post-mining land uses to be sustainable [6] in the long term. In this context, the PPC is planning to transition from coal-based electricity to renewable energy. Specifically, the PPC's research is focused on the sustainable post-mining development of different RES plant types in the remaining post-mining landscapes and consequently evaluating whether implementing such technology could fulfill the electricity demand.

Regarding RES technology application, it is essential to consider the intermittent power outcome. In such a case, integrating energy storage technology with renewable energy production could be an effective solution. Therefore, the PPC uses a holistic approach to mine transition planning by examining energy storage technology, such as pumped hydro storage (PHS) technology.

A considerable volume of literature has been published on applying PHS technology in closed mines in combination with renewable energy sources, such as photovoltaic (PV) parks. According to Bódis et al. [7] the post-mining available landscape area for installing PV systems is abundant in the coal regions of the European Union. In addition, solar PV systems could replace the current coal-fired power plants, provided storage-flexible units with sufficient capacities are added. The first innovative project of a hybrid energy system, which combines PHS technology with multi-stage solar and wind renewable energy units in an abandoned mine, is the Kidston Clean Energy Hub Project [8] in Kidston, Far-North Queensland in Australia, which is expected to be in operation by 2024. According to Holcombe, pit voids in abandoned mines could become an asset underpinning RES development at mining sites and creatively reusing them [9]. Regarding economic feasibility, Wessel et al. reported that hybrid PHS–RES projects could be developed and operated at reasonably low costs [10].

Except for the production of electricity using RES, there is also a growing body of literature on other post-mining transitions and repurposing land-use scenarios [3,11,12], such as wildlife habitat, parks for recreation, eco-tourism and rehabilitation practices concerning pit lakes, i.e., development of fishing farms, recreational sport fisheries, angling, diving, hiking around the lake and even using pits for water storage and flood protection.

Therefore, water quality is an important parameter that could enable new beneficial end uses. Poor water quality could be significantly improved by rehabilitation practices, such as permanent flushing with river water, diverting seasonal floods into the pit lake [13,14], or even permanent river flow-through [15], which can mitigate the risk of eutrophication. Other rehabilitation practices have been reported in post-mining areas, such as appropriate landscaping of the pit lake banks with gentle sloping to emulate natural lakes if possible, filling the bottom of the open pit with alkaline substances such as re-excavated waste lime materials and application of riparian zone vegetation [14,16,17]. These rehabilitation practices could help the development of crucial aquatic biodiversity. Thus, a closure monitoring program is a prerequisite in post-mining areas to ensure long-term pit lake quality, since improving long-term environmental outcomes [18] could enhance pit lake ecosystem development and mitigate environmental damage.

Open pits include large areas that could also be used for waste disposal. Woodlawn is another innovative hybrid electric power project and an example of a circular economy, where the abandoned open pit was repurposed and is now converted into a bio-reactor landfill [12]. In this example, municipal wastes are disposed into the pit, and bio-gas power is produced, feeding the electricity grid. At the same time, other land repurposing is also applied in the surrounding area of the mine. It includes the development of RES technology using wind and solar farms to generate electricity, agricultural operations for farming practices, aquaculture and horticulture operations utilizing captured waste heat in hydroponic horticulture and to cultivate fish, and also the operation of a biological treatment facility to extract the organic content from waste and select recyclable materials or produce compost [19].

Another post-mining land-use transition is transforming mining landscapes into memorial pathways and places of remembrance, demonstrating historical artifacts to memorialize the local mining legacy [12]. In addition, these landmarks could become symbols of the region and thus help the development of tourism and recreation in the area [20].

This study aims to provide a case analysis to examine and evaluate land use repurposing scenarios in a mined-out pit of a closed mine. For this purpose, the mined-out pit of the already closed lignite mine of Kardia, northern Greece, was used as the case study. According to the planning studies for the closure of the lignite mines of western Macedonia [6], rehabilitation includes the development of lakes inside the open pits, photovoltaic (PV) parks developed in the surrounding area, areas used for agriculture or recreational activities and forest areas. In the Kardia mine, the construction of a PHS facility is proposed to work as an energy storage unit due to the major setback in the performance of PV panels, i.e., the difficulty in adjusting to power demands at all times [7,12,21,22]. More specifically, the pit lake is repurposed as a PHS plant. It is used as an energy storage unit to supplement variable renewable capacity and enhance the proposed photovoltaic panels' performance, increasing the PV park's capacity factor. The conceptual idea is to utilize the pit lake that will start to develop after the end of mining activities, ensure the cessation of dewatering measures (dewatering wells and pumping stations) and create another or use an existing upper reservoir at higher altitudes. Then, taking advantage of the head difference between the pit lake (lower reservoir) and upper reservoir, the whole system could be used as a storage battery, releasing power when needed to compensate for renewable energy fluctuations.

In this framework, the feasibility of applying PHS technology is examined concerning the prediction of the natural development lake inside the open pit. In addition, considering the lake's potential growth, the application of different land repurposing scenarios in the Kardia mine is evaluated. Finally, all potential repurposing land uses are evaluated regarding the energy transition potential in the specific region, long-term socio-economic viability and successful adaptation. The intent of this paper is to recommend strategies for improving post-mining land use transition and transforming abandoned surface lignite mines—including pit lakes—into post-mining assets for the communities. In addition, our intent is to contribute to further advancing successful pit lake end uses, as pit lakes could become assets to the clean energy sector.

## 2. Materials and Methods

### *2.1. Research Area—Morphology of the Basin*

The Ptolemais Lignite Field is located in Kozani Prefecture in north Greece, surrounded by the mountains Vermion, Askion and Skopos (Figure 1). The northern side of the basin is bounded by Komanos horst. The Public Power Corporation of Greece has exploited three lignite reserves in this area: Mavropigi, Kardia, and South Field. The Kardia deposit lies in the middle, with exploitable reserves of 273 million tons (since January 2020) that hosted a 1250 MW lignite-fired power plant for almost 40 years. Inside dumping was conducted mainly with spreaders (continuous surface mining) and, in some cases, with trucks and loaders (non-continuous surface mining). The internal waste dumps are estimated in the order of $\approx$1.0 billion $m^3$. Currently, the region's economy depends on agriculture and energy fuel reserves.

The deepest part of the Kardia mine excavation was +450 m a.s.l. when lignite exploitation activities terminated in 2021. By the end of 2022, backfilling of the bottom of the mine pit is planned to achieve slope stability conditions, and thus, the bathymetry of the Kardia excavation will alter, with the deepest part forming at approximately +523 m a.s.l. (Figure 2). The range of ground elevations in the pit is between $\approx$+523 m and +670 m a.s.l.

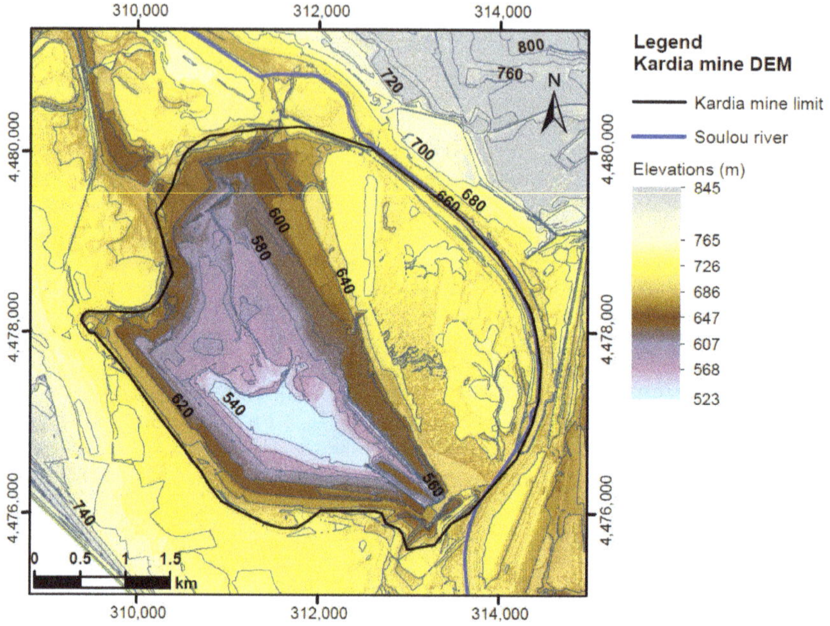


(a)

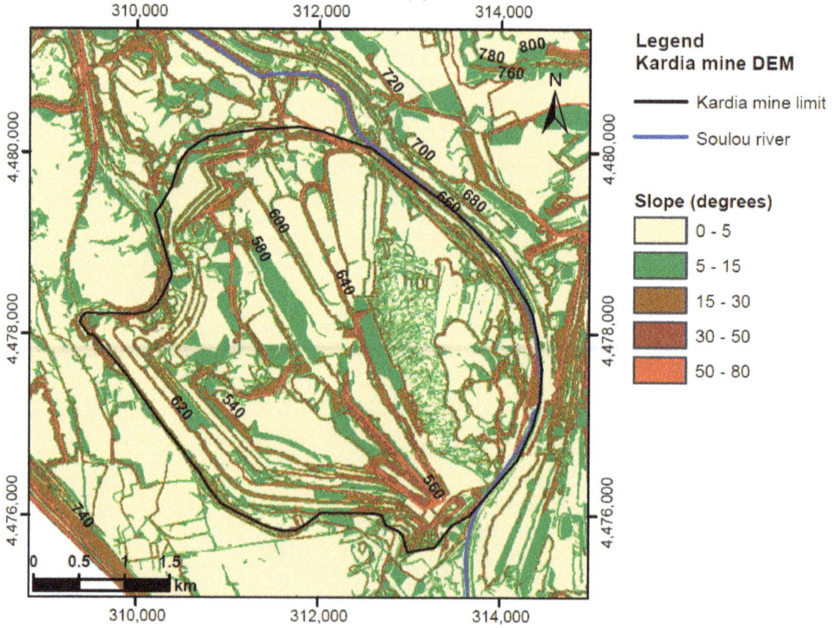


(b)

**Figure 2.** Kardia lignite mine geomorphology, Greece (Coordinate Reference System GGRS87/Greek Grid): (**a**) digital elevation model (DEM) of the open pit; (**b**) slope model (in degrees).

### 2.2. Hydrological and Hydrogeological Setting

The broader research area is part of larger lignite basins of western Macedonia, formed during the Pliocene. The region has a complex of natural lakes (Chimaditida, Zazari, Petron and Vegoritis) (Figure 1). Vegoritis Lake is located in the homonymous closed hydrological basin in western Macedonia, Greece, and is the final receiving body of the basin runoff, with a total area of 2000 $km^2$. Vegoritis Lake is also the surficial feature of a major karstic aquifer [23]. Therefore, any alteration in the surface or groundwater conditions within this area would affect the level of the lake. The primary water discharge into the lake is achieved via the Soulou River [24,25]. Therefore, the main priority of the PPC is not only the protection of the natural lakes but also the monitoring of the artificial lake evolution developed in the open voids of the closed lignite mines, as well as the evaluation of the chemical properties of lake water and of the new aquatic ecosystem that is formed.

Kardia open pit is an exhausted mine located on the western margins of the South Ptolemais lignite-bearing basin (Sarigiol basin) in the foothills of Askion Mountain. Quaternary sediments are met on the broader area, which consists mainly of intercalations of sands, clays, sandy marls and conglomerates. These formations overlie Paleozoic schists, ophiolites and granite formations. In addition, Mesozoic dolomitic limestones, volcanic sediments and flysch overlay the Paleozoic formations [6,26]. The lignite sequence underlies the Quaternary sediments and consists of impermeable Pliocene formations of marls and lignite. Lignite sequence formation is significant for the South Field Lignite Mine (LM) and Kardia LM operations due to the potential slope instabilities associated with sliding along a sub-horizontal lignite-to-clay/marl interface, mainly caused by the groundwater pressures [27].

Regarding the hydrolithological conditions in the area of the Kardia mine, two aquifers are generally recognized, the upper (Sarigiol aquifer) and the lower (Komanos aquifer). These aquifers are separated via the impermeable lignite sequence. The upper unconfined aquifer consists of high permeability layers (sands, conglomerates) with intercalations of low permeability layers (clayey interlayers, fluvial deposits). This unit consists of a unique hydraulically connected network that behaves as one aquifer [24].

The hydrogeological basin of Ptolemais South Field (Sarigiol basin) is surrounded by the Vermion (east side), Askion (west side) and Skopos (south side) mountains. The northern part of the basin is bounded by the Komanos tectonic horst. Most of the streams originate from the surrounding mountains, where the general water flow direction is towards the Soulou River. Part of the surface run-off is discharged into the Soulou River via the consolidated sediment channels formed on the ground surface, while part of the water is infiltrated inside the sediment formations. The stream network and the Soulou River are presented in Figure 3. The recharge area of the Sarigiol aquifer is mainly the Vermion and Askion foothills conglomerates and talus cones, whereas the Soulou River is the main outflow channel of the basin, which directs the water flow into Vegoritis Lake, north of the Kardia lignite mine (Figure 1).

Regarding the Sarigiol aquifer, the piezometric surface is conducted by considering the water levels measured in the boreholes. Based on the piezometry in the area of interest, the piezometric level is met at elevations between $\approx$+640 m and +660 m a.s.l. (Figure 3). The predominant direction of groundwater flow in the Sarigiol aquifer body south of the Kardia mine is generally NW to SE, while locally, next to the open pit Kardia mine location, the direction that prevails is S to N. It is noted that in the same area, the lignite sequence is met at elevations between $\approx$+350 m and +500 m a.s.l., while during excavations inside the Kardia open pit, the bottom of the lignite sequence was not revealed. Thus, in the conceptual model of Kardia pit lake formation, the simulation was oriented only in the aquifer above the lignite sequence, i.e., the Sarigiol aquifer.

In Figure 4, the methodological approach is presented briefly illustrating the modeled processes.

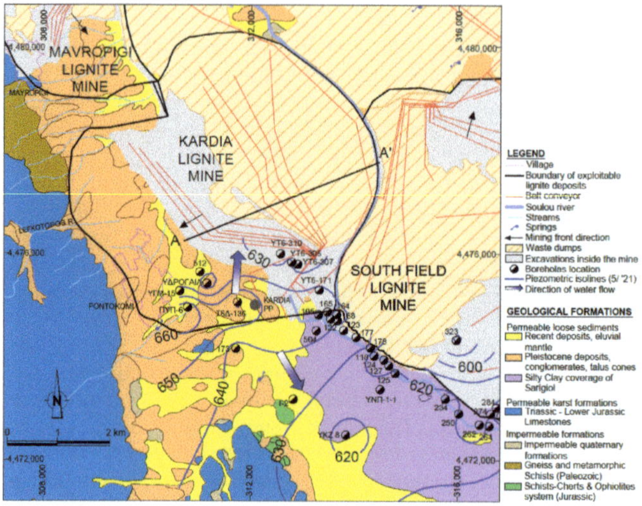


**Figure 3.** Geological and hydro-lithological map in the Ptolemais lignite field, northern Greece (Coordinate Reference System GGRS87/Greek Grid).

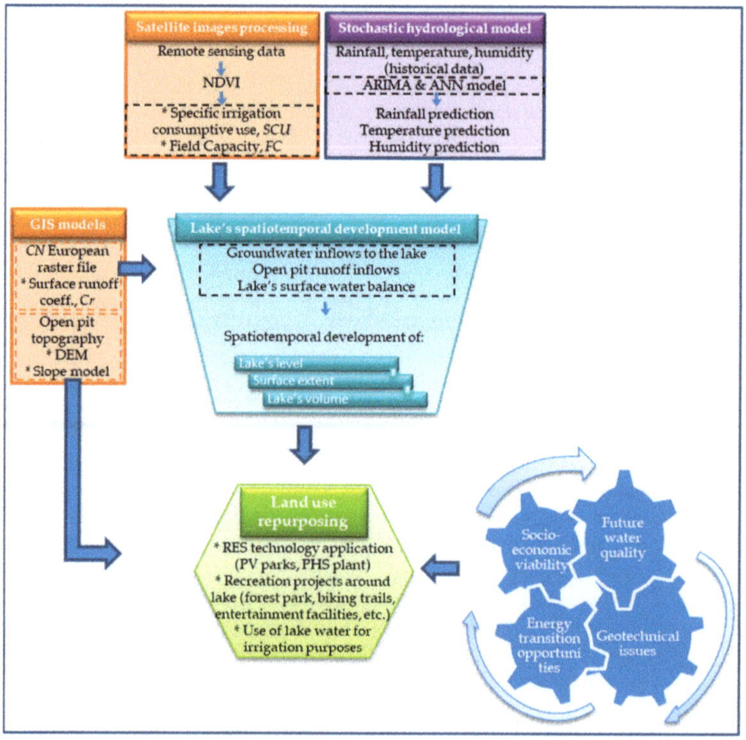


**Figure 4.** Flowchart of the methodological approach.

The renewable resource of the aquifer is the rainfall infiltration through the Quaternary sediments. Meteorological data (temperature, rainfall) were obtained from Kozani WMO No 16 632 weather station, recorded by the Hellenic National Meteorological Service, and spanning a period of 44 years, from 1970–1997 and 2006–2021 (Figure 5). This station is approximately at the same elevation as the Ptolemais lignite field, located 14 km south of the Kardia LM. Meteorological data were also obtained from Ptolemaida LG26 weather station, recorded by the National Observatory of Athens (NOA), and spanning a period of 15 years, from 2006–2021.

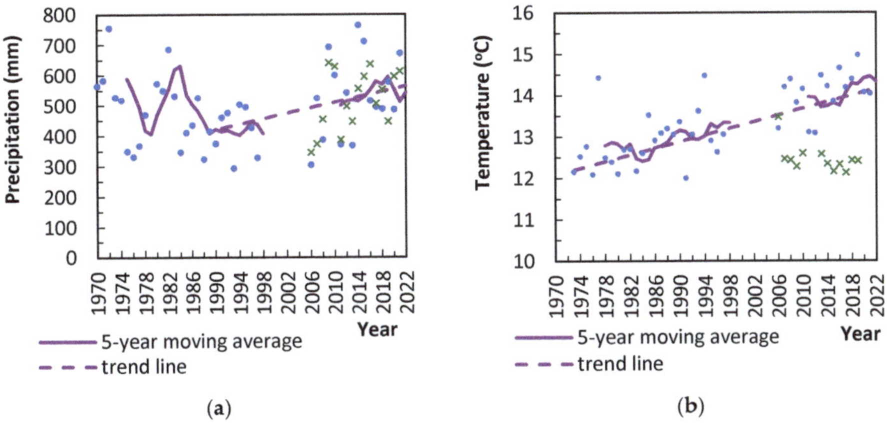


**Figure 5.** Mean annual: (**a**) rainfall and (**b**) temperature from the Kozani weather station (blue color) and Ptolemaida weather station (green color), northern Greece.

According to the meteorological stations' data, the study location in the Kardia lignite mine, in Ptolemais, had a Köppen climate classification of Cfa (moist with mild winters, wet all seasons with a hot summer), which is consistent with the updated 1-km resolution global maps of the Köppen–Geiger climate classification [28]. The average annual rainfalls at Ptolemais were 498.1 mm (±132.1 mm s.d.), and November was the wettest month with average monthly precipitation of 56.7 mm, while August was the driest month with an average monthly precipitation of 31.7 mm. The average annual temperature was 13.3 °C (±0.8 °C s.d.), with January being the coldest month, with an average temperature of 2.4 °C and the highest temperature of 6.1 °C. According to the meteorological data, there is a visible trend of long-term increase in temperature, presenting an average increase of approximately 2.2 °C over the last 50 years. Regarding precipitation, there seems to be an increasing trend during the period of the last 30 years.

Thornthwaite's water balance method [29,30] was used to predict the mean annual runoff for the unsaturated zone in the pit's catchment area and the Sarigiol hydrological basin. This method can provide mean annual estimates in close agreement with measured values [31]. The unsaturated zone's water balance during the time interval $(t, t+dt)$ is described through the moisture content $B$. The moisture content $B(t+dt)$ at the end of time $dt$ is described by the following equation [30]:

$$B(t+dt) = B(t) + (N - V - R - SL)\,dt \tag{1}$$

where all quantities are expressed to the unit area, and $B\ (t)$ is the water content stored inside the unsaturated zone at the time $t$ (mm), $N$ is the precipitation rate (mm/month), $V$ is the real evaporation (mm/month), $R$ is the surface runoff outflow (mm/month), $SL$ is the percolation rate (mm/month).

In the calculations, it was considered that deep percolation takes place when soil moisture reaches 70% of the field capacity ($FC$) [32], where $FC$ is the amount of water remaining when downward drainage has markedly decreased. Therefore, the following equations were applied for the different cases [30]:

$$B(t) + (N - V - R)dt \le 0.7\,FC \text{ then } (SL)dt = 0 \quad (2)$$

$$F \le B(t) + (N - V - R)dt \ge 0.7\,FC \quad \text{then } (SL)dt = 0.1\,(N - V - R)dt \quad (3)$$

$$B(t) + (N - V - R)dt \ge FC \qquad \text{then } (SL)dt = B(t) + (N - V - R)dt - FC \quad (4)$$

The real evaporation $V$ was estimated as a portion of Thornwaite potential evaporation $P$. It was assumed that real evaporation takes place when soil moisture reaches 70% of the field capacity, and thus, the following equations were applied for the different cases:

$$B(t + dt) \ge 0.7\,FC \text{ then } V = P \quad (5)$$

$$B(t + dt) \le 0.7\,FC \text{ then } V = P\left[0.188 + 2\,(B/FC) - 1.2\,(B/FC)^2\right] \quad (6)$$

The potential evaporation $P$ is derived by the following equation:

$$P = 1.6\,(10\,t_n/I)^a \quad (7)$$

where $t_n$ is the mean temperature of month $n$, $I$ is a temperature coefficient of monthly temperature coefficients $i_n$ summation ($I = \sum_{n=1}^{12} i_n$), where $i_n = (t_n/5)^{1.514}$ and a value can be determined as a polynominal function of the temperature coefficient $I$, where $a = 0.49239 + \left(1792 \times 10^{(-5)}\,I\right) - \left(771 \times 10^{(-2)}\,I^2\right) + \left(675 \times 10^{(-9)}\,I^3\right)$.

In addition, in the hydrological model it was assumed that real evaporation cannot exceed the available water derived from precipitation and soil moisture and thus, $(V)dt \le (N - R)dt + B(t)$.

Considering the climate conditions in Greece, for the specific area, it is expected that soil moisture content reaches $FC$ in April and thus, regarding the initial conditions of the unsaturated water balance at the time $t = 0$, then $B(0) = FC$ for April.

Sensitivity analysis performed towards field capacity $FC$ for the unsuturated zone water balance Thornwaite model revealed the following linear relation between $FC$ and percolation coefficient as $SL$ percentage (%) of rainfall, which is taken as a percentage of precipitation, presenting a coefficient of determination $R^2 = 0.964$ (Figure 6):

$$SL = 32.429 - 0.0903\,FC \quad (8)$$

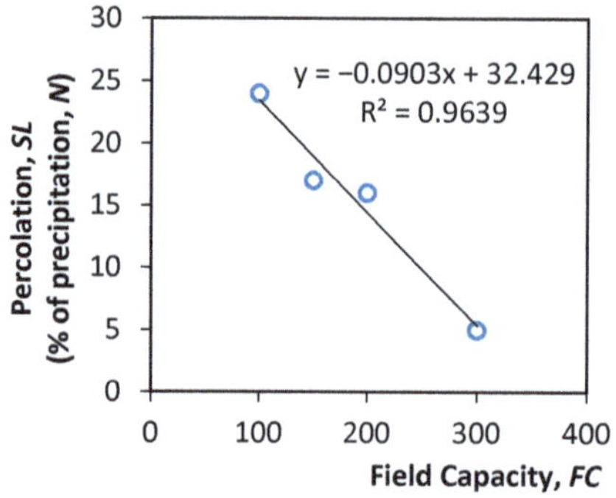


**Figure 6.** Sensitivity analysis of percolation rate $SL$, as a percentage of precipitation $N$, towards field capacity $FC$ at the wider area of Ptolemais lignite field, northern Greece.

This relation helped in transforming $FC$ values (deduced from $NDVI$ values) into percolation rate values $SL$.

According to the results (Table 1), for the hydrological cycle in the Sarigiol basin (2006–2021), the mean annual precipitation that the loose sediment basin receives is approximately 584 mm, while the real annual evaporation is estimated to be approximately 425 mm. Also, based on the 15-hydrological year annual water balance output, only 86 mm per year percolates down and recharges the unsaturated zone, which constitutes the long-term renewable recharge of the aquifer. Therefore, there is only one discharge from the Sarigiol basin, i.e., the surface runoff that occurs through the outflow of the Soulou River. In different historical time periods, either the Sarigiol aquifer discharges into the Soulou River or the Soulou River recharges the Sarigiol aquifer, depending on the groundwater level.

**Table 1.** Mean annual water balance for the unsaturated zone of the Sarigiol loose sediment basin.

| Hydrological Cycle in Sarigiol Basin (2006–2021) | mm/yr | $10^6$ $m^3$/yr | Precipitation (%) |
|---|---|---|---|
| Mean annual precipitation | 584 | 123 | 100 |
| Mean annual real evaporation | 425 | 90 | 73 |
| Mean annual percolation | 86 | 18 | 15 |
| Mean annual runoff | 72 | 15 | 12 |

## 2.3. Conceptual Model for the Pit Lake Long-Term Water Balance

In order to calculate the spatiotemporal evolution of the lake inside the Kardia open pit, a simplified conceptual hydrogeological model was developed (Figure 7), following the hydrogeological conceptual model of Louloudis et al. [33], in which the different origins of water inflows into the pit lake's body are presented. The inflows include (a) the groundwater inflows from the slope seepage face, (b) the surface runoff into the open pit, considering the open pit as a catchment area, and (c) water losses from the lake surface that occur due to direct evaporation processes which result in reducing the net recharge over the lake surface area (i.e., the hydrological processes of precipitation minus evaporative losses over the lake surface area).

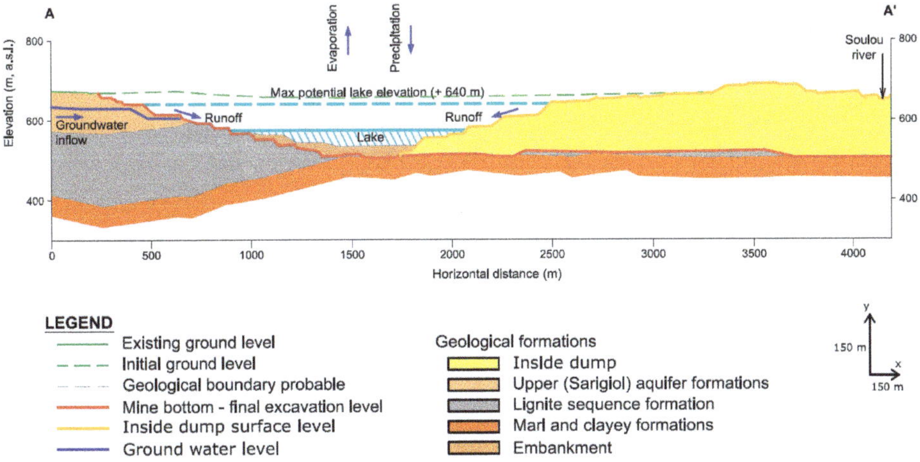


**Figure 7.** Hydrogeological conceptual model for Kardia pit lake, Greece (the location of the cross section A-A' is shown in Figure 3).

### 2.3.1. Groundwater Inflows

Since the dewatering measures adopted for mine protection will cease after the mine closure by the end of the year 2022, the inflowing groundwater will then begin to increase.

As a result, the open pit will fill with groundwater rebound, and the groundwater level will rise. The discharging water derives from the upper aquifer (i.e., above lignite stratification) and the confined aquifers inside the recently developed internal dumps.

For the initial conditions of the model in the year 2022 ($q_{(2022)}$) the groundwater inflow was calculated by Darcy's law [34] to be around $1.7 \times 10^6$ m$^3$. This quantity was cross-validated by the records of the water sump pumped quantities during the summer period in 2020, considered as baseflow of surficial hydraulic protection. Specifically, approximately $1.2 \times 10^6$ m$^3$ of groundwater inflow is pumped from the sumps, and $0.5 \times 10^6$ m$^3$ is pumped from the dewatering wells during the summer period. In addition, the seepage face of groundwater inflow into the excavation is also subjected to an evaporation procedure at mine's benches. For the Kardia mine, it is estimated that 70% of this quantity is evaporated; thus, only 30% of the groundwater inflow finally reaches and recharges at the pit lake's water body (i.e., around $0.6 \times 10^6$ m$^3$).

For each year $i$ of the prediction period, the groundwater inflow $q(i)$ was estimated with respect to the surface lake water level $h$ increase. More specifically, the groundwater inflow rate $Q$ per unit face length $A$ is calculated according to the following equation [35]:

$$q(i) = Q/A = K\left(H^2 - h^2\right)/2L \quad (9)$$

where $K$ is the hydraulic conductivity, $H$ is the aquifer's thickness, $L$ is the influence radius of the open pit and $h$ is the height of the lake water level calculated from the surface of the lignite sequence (considered as impermeable formation).

Hydraulic conductivity $K$ ranges from 1.03 to 25.92 m/day, according to pumping tests analysis [36], with an average value of 13.48 m/day. The upper aquifer's thickness $H$ is estimated in the order of 100 m [24,36] and thus, transmissivity $T$ presents an average value in the order of 50 m$^2$/day on the west side of Kardia excavation and a range of 200–1200 m$^2$/day in the south region, while the average hydraulic gradient towards the excavation is approximately 2%, as depicted in Figure 3.

To delineate the groundwater inflow rate in the future, groundwater attenuation is considered due to the excessive irrigation consumptive use in the area. In the methodology followed for estimating the aquifer's depletion, the groundwater recharge of the Sarigiol aquifer around Kardia excavation was considered and the specific consumptive use $SCU$ of irrigation water. The renewable recharge of the aquifer was estimated via Thornthwaite's water balance method (Table 1), while the $SCU$ was derived from the evapotranspiration $ET$.

For the calculation of the crop water needs $P_C$, i.e., the water amount needed to meet the water loss through evapotranspiration, the following general relationship between the reference crop and the crop actually grown was used:

$$P_C = Kc \times ETo \quad (10)$$

where $ETo$ is the reference crop evapotranspiration calculated by using the Blaney–Criddle method (mm/month) [37], while $Kc$ is the crop factor derived from the normalized difference vegetation index ($NDVI$) [38] which was estimated equal to $Kc = 0.58$:

$$ETo = 30\ p\ (0.46\ T_{mean} + 8) \quad (11)$$

$$Kc = 1.18\ NDVI + 0.04 \quad (12)$$

$$NDVI = (NIR - RED)/(NIR + RED) \quad (13)$$

where $T_{mean}$ is the mean daily temperature (°C), and $p$ is the mean daily percentage of annual daytime hours depending on the latitude of the area. In addition, satellite-based $NDVI$ is computed from the red and near-infrared bands of the satellite by processing a satellite image acquired in 2018, as presented in Figure 8.

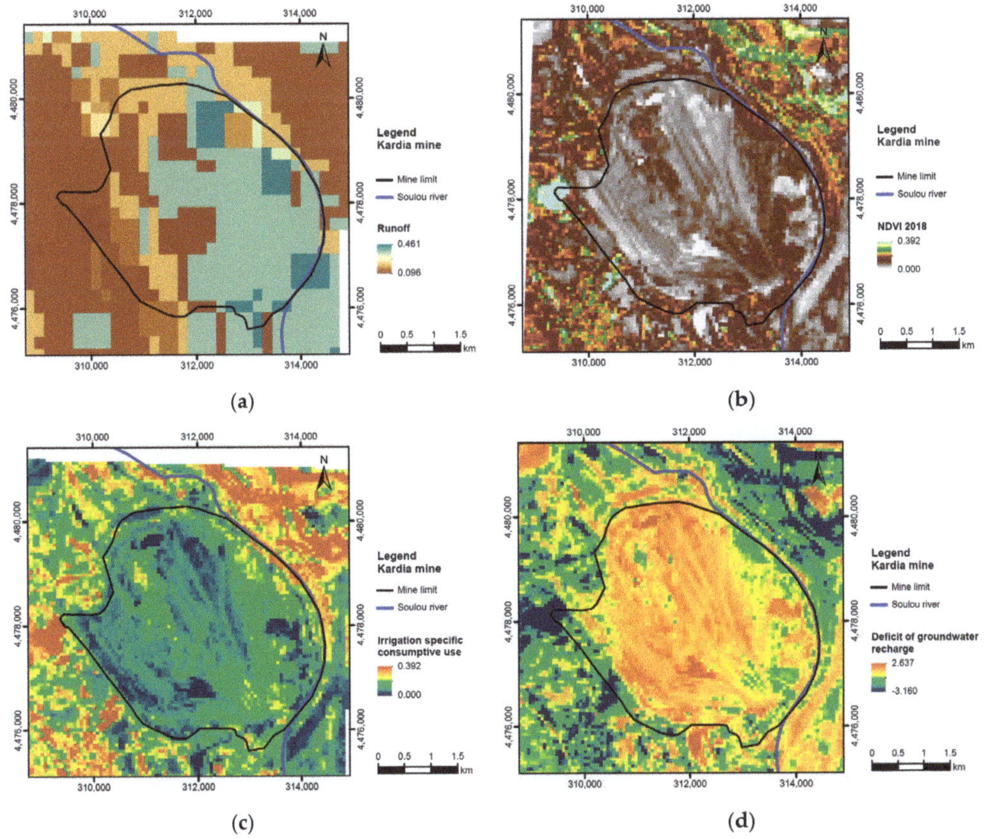


**Figure 8.** Spatial distribution of: (**a**) runoff coefficient by processing $CN$ European raster; (**b**) spectra indices of $NDVI$ using Landsat-8 (image acquired in July 2018); (**c**) irrigation-specific consumptive use $SCU$ ($m^3/m^2$); (**d**) annual drawdown in m, corresponding to the deficit of groundwater recharge due to excessive irrigation consumptive use of groundwater in Kardia lignite field, northern Greece (Coordinate Reference System GGRS87/Greek Grid).

The crop water need $P_C$ is supplied to the crops by a combination of irrigation and rainfall. Therefore, the irrigation water need ($IN$, mm of water column/month) is the difference between the crop water need $P_C$ and the effective rainfall $Ne$, i.e., the part of the rainfall effectively used by the plants:

$$IN = P_C - Ne \tag{14}$$

where $Ne$ is estimated based on an empirical relationship presented by the Greek Land Reclamation Service of the Ministry of Agriculture [39] for the irrigation period from May to October:

$$Ne = N - (12 + 0.125\ N) \tag{15}$$

where $N$ is the precipitation rate (mm/month).

Subsequently, the specific consumptive use $SCU$ ($m^3/m^2$) was derived from the irrigation water need $IN$ (mm/ month), using the methodology used by Louloudis and Stathopoulos [40]. A statistically significant correlation between irrigation water $SCU$ ($m^3/m^2$) and $NDVI$ is shown in the Figure 9, presenting a coefficient of determination $R^2 = 0.781$. The linear regression analysis was performed by using the electric power consumption of irrigation pumps in the area of the Sarigiol basin for 2018 and also by using a Landsat image acquired in July 2018 (Landsat-8: 15 July 2018). Based on the results, the following relationship between the irrigation water $SCU$ ($m^3/m^2$) and $NDVI$ was used for the calculation of the $SCU$ spatial distribution (Figure 8) for the Sarigiol basin area:

$$SCU = 1.539\ NDVI - 0.150 \tag{16}$$

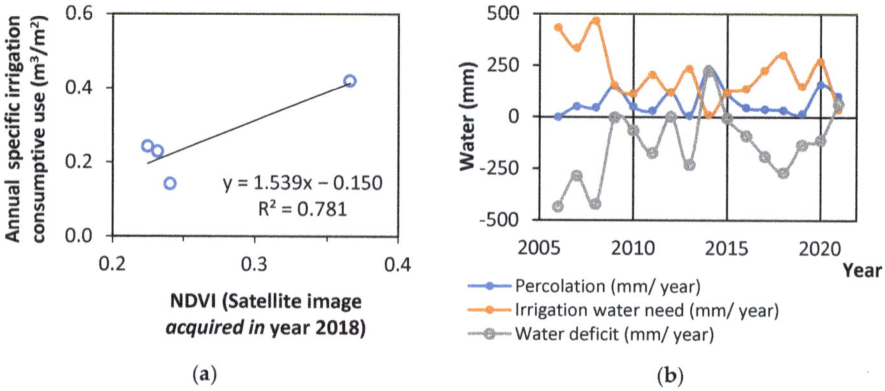


**Figure 9.** (**a**) Irrigation-specific consumptive use ($m^3/m^2$) versus $NDVI$ using Landsat-8 (image acquired in July 2018) and (**b**) Percolation, irrigation and groundwater deficit time series during the period 2006–2021 in Sarigiol basin, Greece.

In addition, the deficit of groundwater recharge due to excessive irrigation consumptive use of groundwater near Kardia lignite field is presented in Figure 9, and its spatial distribution for the Sarigiol basin area is shown in Figure 8. The aquifer's annual drawdown due to the irrigation for the period from July to September was calculated as a difference between annual percolation and irrigation water need, divided by the storage coefficient considered equal to 0.1.

### 2.3.2. Surface Runoff

Surface runoff is the most essential inflow to the lake's body. Surface runoff also appears to be the more sensitive parameter in the lake's filling model so there is a need for precision in surface runoff coefficients introduced to the model. In order to achieve a reliable estimation of the surface runoff coefficient, $Cr$, in the hydrologically ungauged Kardia area, $CN$ European raster file [41] was processed and the following equations were used [42,43]:

$$Cr = \left(P_{vol} - 0.2\left(\frac{25,400}{CN} - 254\right)\right)^2 / P_{vol}\left(P_{vol} + 0.8\left(\frac{25,400}{CN} - 254\right)\right) \tag{17}$$

$$P_{vol} = I_{TC}\ T_C \tag{18}$$

where $P_{vol}$ is the volume of rainfall in mm, $CN$ is the curve number, $I_{TC}$ is the mean rainfall intensity (mm/h) for the time of the concentration $T_C$ (h) which is described by Giandotti equation [44].

In Figure 8 spatial distribution of the runoff coefficient using $CN$ European raster file is presented. Based on the results, inside the active mine area, high values of the runoff coefficient are calculated, in the order of 30% to 40%, whereas outside the perimeter of the excavation, in the zone of the loose sediments between Kardia excavations and Askion Mountain the calculated values hardly exceed 15%. In the western area on the consolidated sediments of Askion Mountain, a runoff coefficient of 20% in magnitude is calculated, while northern of this area near Mavropigi village insignificant values in the order of 5%

are revealed. Also, Lefkotopos and Lagorema streams in the region of Skopos Mountain consolidate carbonate sediments with a considerable runoff coefficient of 10% up to 20% in magnitude.

2.3.3. Water Losses from the Lake's Surface

There are many expressions to evaluate evaporation from the lake's water surface. The well-known, and highly accurate, Penman–Monteith approach involves many variables (air temperature, solar radiation, relative humidity, wind speed) [45–47] which in some cases may not be available. In these cases, adjusted equations with fewer weather parameters are used. Eagleman proposed a simple correlation between these parameters [48]:

$$E_A = 0.035\, e_S\ (100 - RH)^{1/2} \tag{19}$$

where $E_A$ is the adjusted pan evaporation in inches per month (free surface evaporation), $e_S$ is the saturation vapour pressure [49] corresponding to the mean monthly temperature in millibars (mb), and $RH$ is the mean monthly percent relative humidity:

$$RH = (e_S\ / e) \times 100 \tag{20}$$

$$e_S\ = 6.11\ exp\ [17.269\ T/(273 + T - 36)] \tag{21}$$

$$e = 6.11\ exp\ [17.269\ T/(273 + T - 36)]\ (RH/100) \tag{22}$$

In the case of the Kardia mine, mean values are used, i.e., the mean monthly temperature (°C), the mean monthly rainfall (mm) and the mean monthly relative humidity (%).

2.3.4. Spatiotemporal Development of the Pit Lake

A crucial research question that arises from this study is whether the pit lake can be naturally filled in a reasonable time frame or is there a need for artificial recharge by flooding with river water. Thus, the spatiotemporal development of the water level inside the Kardia pit void was assessed by generating forecasts of the dependent variables (rainfall and temperature) via linear (autoregressive integrated moving average—ARIMA) and non-linear (artificial neural network—ANN) models, following the same methodology described by Louloudis et al., for the Amynteon mine pit lake [33].

*2.4. Energy Transition Strategies in the Post-Mining Era*

In the Kardia mine, the PPC energy transition strategy is based on the PV parks that will be developed in the surrounding area, inside and outside the mining exploitation area. Thus, energy storage units are required to supplement the variable renewable capacity. In this mine, the best conditions prevail for a PHS facility installation thanks to its morphology, high exploitation depth of approximately 150 m, and sizable pit lake volume. Therefore, a PHS facility could be used as a storage battery and would provide sufficient storage capacity to compensate for renewable energy fluctuations.

Analyzing the electricity demand is a prerequisite for optimal planning of a hybrid PHS system. Figure 10 presents a representative scheme for the hybrid pumped hydro storage (HPHS) plant operational schedule during daytime hours for the case study area of the Kardia mine. This scheme presents the integration of RES electricity production and the utilization of a hybrid plant. A linear programming optimization model using the simplex method [50] has been developed, where the objective function value addressed to the capacity factor [51] of the PV park is maximized. The main restriction is that the sum of the PV power produced, the energy released in PHS and the grid supplementary power should meet load demands. Another restriction concerns the water balance in the PHS system, where the quantity of the water pumped into the upper reservoir equals the quantity of the water released downstream into the lower reservoir.

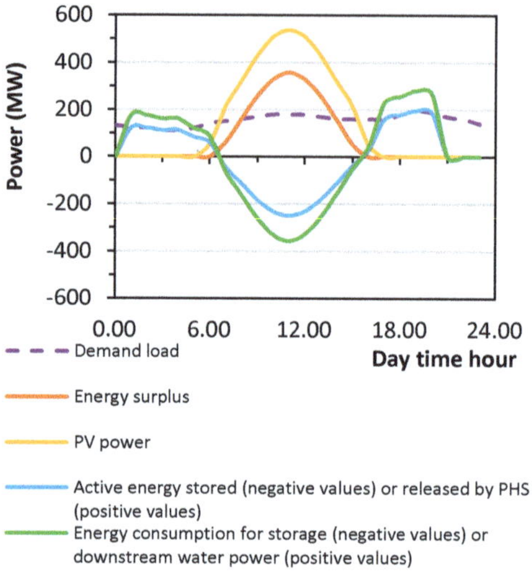


**Figure 10.** Representative scheme for the hybrid pumped hydro storage (HPHS) plant operational schedule during daytime hours and photovoltaic (PV) integration, considering the installation of a 270 MW PV plant in the Kardia mine.

In this idealized model, no legislative restrictions are taken into account regarding the deployment of energy storage by the PHS plant. According to the National Law provision (Law 3468, Government Gazette 129/A/27 June 2006) the total amount of electricity taken in from the network on an annual basis does not exceed 30 percent of the total amount of energy consumed for the filling of the storage system in that station. Thus only 30% of the energy produced by the PV plant is allowed to be stored using the PHS plant by utilizing the potential energy of water. Another assumption considered in the optimization problem is that the load demand distribution during daytime is set directly proportional to the electrical power distribution network flow [52]. In addition, the distribution system performance of the PV park during the daytime is set directly proportional to the distribution of the solar radiation production [53] during the daytime.

As depicted in Figure 10, the operation of the HPHS system dictates a pumping procedure between 7:00 and 16:00 and energy production in the rest of the 24 h cycle. The designed energy system for the PHS plant must be credible to cover the demand load of electricity during the whole day, providing efficient energy generation and a guaranteed rate. Consequently, when the energy system supplies more power than that required by the users (i.e., the total load produced is higher than the demand load), then this additional load is used to supply the PHS plant, and thus, some additional water can be stored in the upper reservoir of the plant. However, the "real" PHS storage load is lower due to energy losses.

For this purpose, representative synthetic time series, including load prediction data, must be generated during the design phase for the optimal planning of the HPHS system, aiming at the prediction of the potential energy storage. The potential energy storage $P_e$ in a PHS complex required to supplement the PV park can be estimated by calculating the pumping energy consumption via the following formula, which is an expansion of Bernoulli's equation [54,55]:

$$P_e = (Q\ H)/(102\ n) \tag{23}$$

where $P_e$ is the total power required (kW), $Q$ is the flow rate (lt/s), $H$ is the total head, i.e., the pressure head and elevation head (m), $n$ is the combined effect of pumping efficiency and derating, and 102 is a unit conversion factor value. Pumping efficiency and derating account for the operating efficiency of pumps, and efficiency losses between the energy required at the pump shaft and the total energy required [54].

The total power required to be stored in order to return it to the electric grid when needed can be obtained using two alternatives: (a) either by having a great hydraulic difference between the upper and lower reservoir, with a relatively small quantity of water per daily cycle uplift, (b) or by having a smaller hydraulic difference between the reservoirs, though with a larger quantity of pumped water. In addition, the head difference influences the efficiency of pumps and turbines [56]. The selection between the two options is mainly a financial decision since the magnitude of the project (mainly the number of units of pumps and generators, excavation volume of both reservoirs and area covered for PHS facilities to be mounted) is proportional to the required quantity of stored water. Thus, it affects the total cost. So, the option of a significant head difference could have the advantage of minimum fixed cost.

The required quantity of stored water and the head difference set the limits for the pit lake level variation $LLV$, which is a critical component in the design [10]. Based on this acknowledgment, a diagram that can help to select the appropriate altitude of the PHS upper reservoir water level can be the total power required $P_e$ per unit m of water level daily fluctuation $LLV$ in relation to the pit lake's level, as presented in the Section 3. This approach could help in the decision-making process and achieving a PHS plant's economic viability.

### 2.5. Economic Value

The prediction of the spatiotemporal evolution of the lake is crucial for achieving the financial efficiency of the HPHS project, by evaluating different levels of the lake's filling. The financial efficiency of each water level in the pit lake can be determined via performance indicators, i.e., net present value ($NPV$) and net cash recovery ($NCR$) of a cash flow series.

In the case of the Kardia pit lake, initial calculations of the economic feasibility of a PHS plant were performed with the help of the $NPV$ method, where the values of alternative scenarios were evaluated. The $NPV$ value indicates whether the overall lifetime costs of the project (benefits and costs of different scenarios), i.e., the investment, is less or more than the income generated. For example, for an investment project with a minimum design operational lifetime of up to 50 years, the following equation was used to estimate $NPV$ [10]:

$$NPV_{real} = -I_o + \sum_{t=1}^{50} \{(R_t - C_t)/q_r^t\} \tag{24}$$

where $NPV_{real}$ is the $NPV$ in real terms, considering also the inflation rate $r$ and the interest rate $i$, $C_t$; $R_t$ are costs and revenues at different points in time $t$, $I_o$ is the initial investment at time $t$ = 0 and $q_r^t$ is the discounting factor which reflects the value of money in time $t$:

$$q_r^t = \{(1+i)/(1+r)\}^t \tag{25}$$

The component $q_r^t$ in the equation above helps make the different cash flows in time comparable [10].

### 2.6. Geotechnical Issues to Be Considered in the Post-Mining Design

Several studies have sought to determine successful and unsuccessful cases of pit lakes in the post-closure mine phase. It has been documented that stability is improved as water levels increase inside the pit lake. Kavvadas et al. investigated the stability conditions that prevailed in decommissioning surface lignite mines in Greece [57], using the Amyntaion mine as a case study, where the pit is gradually filled with water. According to this report, it is shown that stability conditions are reinstated when the depth of water (critical depth) reaches approximately 15–35% of the final equilibrium condition, whereas, at higher lake levels, the safety factor is expected to increase significantly due to the beneficial effect of the lake water pressure acting on the slope. Table 2 presents preliminary predictions in slope stability for different scenarios of lake water development inside the open pit [57].

**Table 2.** Variation in slope stability conditions related to the pit lake development in decommissioning surface lignite mines in Greece, used for preliminary prediction.

| Lake Water Depth (Hw)/Height of the Slope (H) | FS/FSo* |
|---|---|
| 0 | 1.0 |
| 0.1 | <1.0 |
| 0.2 | 1.0 |
| 0.4 | 1.075 |
| 0.6 | 1.25 |
| 0.8 | 1.6 |

Note(s): * FS = Safety factor, FSo = initial FS when water depth Hw = 0. Source: [57].

In addition, in the early stages of the lake's development, more favorable slope stability conditions are expected to be achieved in only a few years. This result is associated with the fact that the surface area of the pit is smaller at deeper levels; thus, the depth of the water inside the pit increases rapidly in the early stages of the lake development.

Regarding the operation of the HPHS complex facility, in the daily cyclic process in the lower reservoir (i.e., the pit lake), excessive weathering and corrosion of the lake banks could appear, caused by the daily fluctuation of the head difference between the lower and the upper reservoir. Furthermore, the daily fluctuation induces an increase in water pressure at the pit lake banks; thus, slope instabilities could appear due to the fatigue effect. Consequently, adopting protection measures at the pit lake banks is a prerequisite, which could also affect the excavation volume and, as a result, the magnitude of the PHS complex facility.

## 3. Results and Discussion

### *3.1. Water Flux Trends and Forecast of Lake's Spatiotemporal Development*

Predictive models to forecast rainfall in the near future, as well as surface runoff, evaporation, percolation rates and consumptive irrigation rates are generated for the next decade, from which the water balance of the lake's surface is derived. The results from the implementation of the NNAR model (neural network autoregression) with no seasonal participation are presented in Figure 11.

According to the NNAR predictions for the next decade, the maximum expected groundwater attenuation of the Sarigiol aquifer is estimated at 10 m (i.e., approximately one meter per year on average). This drawdown is due to excessive irrigation water demands [40], which allows the expectation that the groundwater level will be successively lowered below the elevation +635 m during our conceptual model of the lake filling time interval. Time series forecasts of annual drawdown for the next two decades are computed. This groundwater level attenuation was incorporated into the mathematical model for predicting the lake's water balance, where a 20% reduction in groundwater influxes to the lake's water body was introduced.

In addition, the lake's surface balance is displayed in Figure 11, as a result of the difference between evaporation and precipitation over the lake's surface. The results showed an increasing trend of yearly water losses in the lake's surface water balance due to a gradual increase in the water surface, as evaporation also increases.

Figure 12 displays the lake's spatiotemporal development based on the model developed in this study and the results from implementing the NNAR model. To highlight the influence of the time variable datasets, the different water inflows into the pit lake's body have been plotted together and are associated with the time component. The results indicate that, the filling of the lake model is very sensitive to surface runoff, which is described by an average value of 0.3 for the area inside the open pit (Figure 8). There is a clear decreasing trend of annual surface runoff due to the gradual decrease in the watershed area.

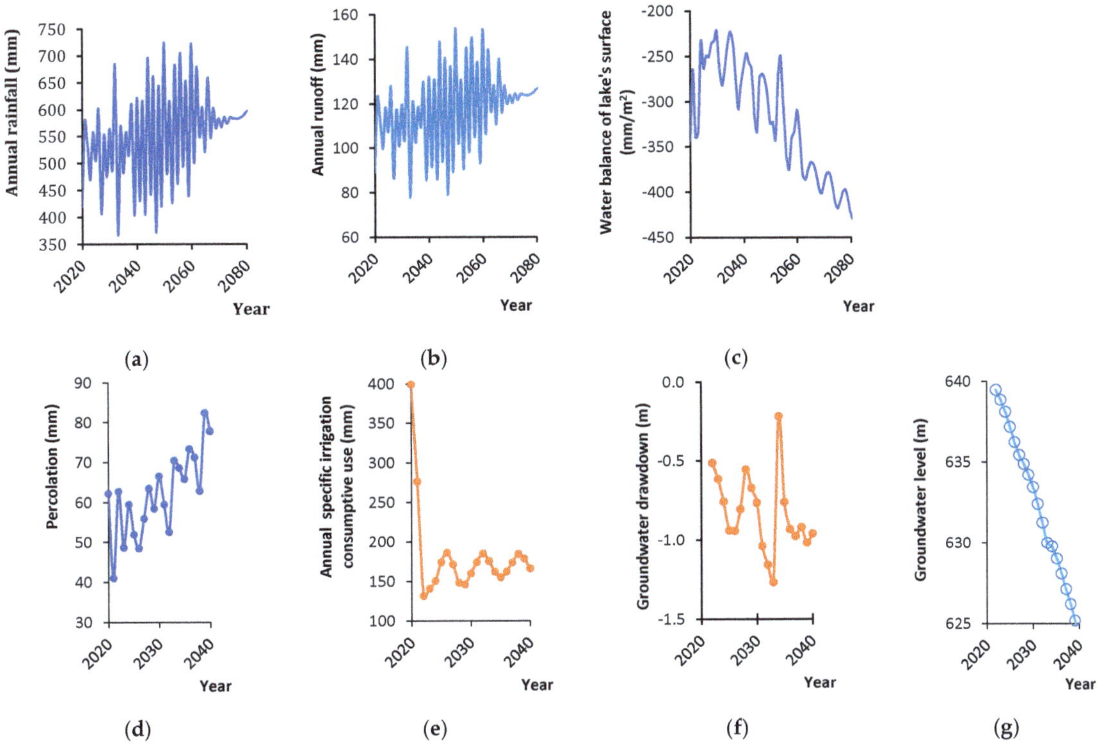


**Figure 11.** Artificial neural network prediction (NNAR) implemented for the Kardia lignite field area, northern Greece. NNAR results of the annual: (**a**) rainfall; (**b**) runoff; (**c**) water balance of lake's surface (precipitation minus evaporation on lake's surface); (**d**) percolation rate; (**e**) specific consumptive irrigation use; (**f**) drawdown (percolation minus groundwater attenuation as a result of irrigation consumption); (**g**) groundwater level in the Sarigiol aquifer (i.e., above the lignite sequence).

As the pit lake level and volume increase with time, the surface area of the pit decreases and thus, the surface runoff decreases. It is interesting, though, that, during the last phase of the lake's development, surface runoff still exceeds the lake's surface losses (water balance of lake's surface, i.e., precipitation minus evaporation on lake's surface). Despite the large water losses (in mm/m$^2$ of lake's surface) compared to surface runoff (in mm/m$^2$ of watershed surface) (Figure 11), the total annual surface runoff quantities exceed the water losses in the early stages of the lake's development (Figure 12).

The results of the lake's development displayed in Figure 12 indicate that the natural recharging of the lake is feasible. In the early stages, the lake expands rapidly, while in the last phase it does so very slowly. On the contrary, the lake's water body volume increases almost steadily. As a result, the maximum natural lake growth inside the pit can potentially cover almost 50% of the pit's total surface (Figure 2) in approximately 80 years, with a depth of approximately 117 m.

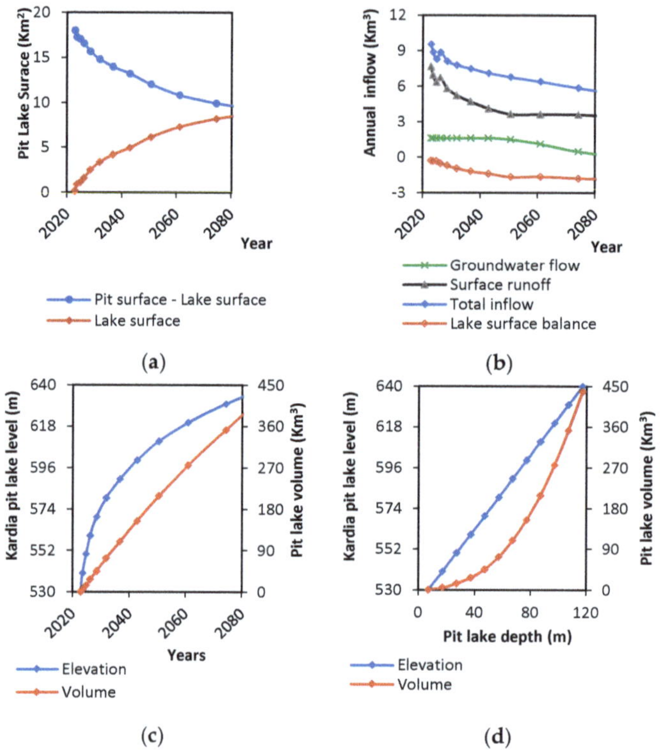


**Figure 12.** Forecast of the spatiotemporal development of the Kardia pit lake: (**a**) lake surface vs surface runoff decrease; (**b**) annual inflows and outflows; (**c**) the absolute elevations and gradual increase in stored water volume vs. time; (**d**) the absolute elevations and gradual increase in stored water volume vs. lake depth.

### *3.2. Evaluation of Potential Land Use Repurposing Scenarios*

For the already closed lignite mine of Kardia, a preliminary evaluation was conducted for potential land use repurposing scenarios, by evaluating the landscape formed after the lignite exploitation activities terminated. Generally, in abandoned open pits, the pit void will eventually be filled with water at a specific time period. Thus, the viability of any repurposing land use scenarios should be evaluated concerning the spatiotemporal lake level development and the water coverage area compared to the pit's total surface.

In evaluating different land use scenarios, the energy transition potential in the specific region is of primary importance. It is considered with respect to long-term socioeconomic viability in the area. At the same time, the transition should be in line with the holistic policy framework developed by Greece to further develop renewable technology in the electricity sector and eventually achieve net-zero greenhouse-gas emissions. Consequently, repurposing land uses includes RES technology application in the remaining post-mining landscapes. Moreover, "Smart Post-mining" has to be developed in concert with "Green and Climate Smart Mining" [58], in a holistic approach so that the climate-driven clean energy transition is managed responsibly and sustainably.

Based on topographical criteria, from the digital elevation model (DEM) and the slope model implemented (Figure 2), it can be seen that the area between elevations +650 m and +680 m could be characterized as relatively flat. The flatness of this area makes it ideal for installing PV parks. Consequently, lake growth should not exceed the elevation of +650 m. Also, based on preliminary predictions as described in Table 2, stability conditions are expected to be reinstated when the depth of lake water reaches approximately 22 m to 29 m, i.e., for pit lake elevations +545 m up to 552 m. Thus, adopting lower lake levels increases the risk of failure.

From the model developed in this study and the results from the implementation of the NNAR model, the maximum lake water level cannot exceed elevations +640 m ~ +650 m, as it is presented in Figure 12. This condition is due to hydraulic reasons, as this level is related to the groundwater level in the loose sediment aquifer overlying the lignite sequence. Considering the lake growth limitation, the flat area between elevations +650 m and +680 m could be examined for installing PV parks (Figure 13). This area is approximately 5 km$^2$, with an average inclination of approximately 3.5%.

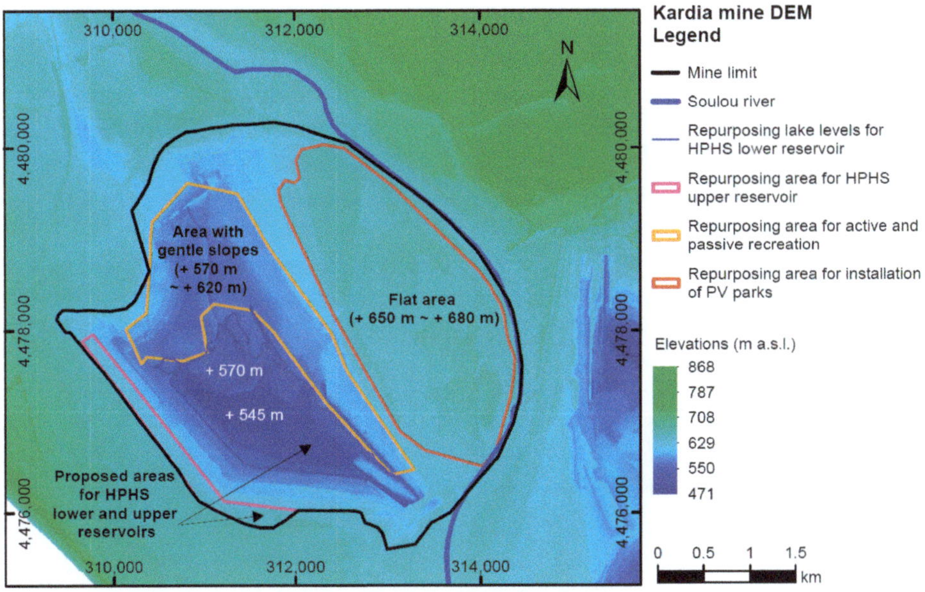


**Figure 13.** Land use repurposing inside the Kardia mine, Greece (Project Coordinate Reference System GGRS87/Greek Grid).

In addition, the area between elevations +570 m and +620 m on the lake's northeast side could be repurposed for active (such as swimming, diving, etc.) and passive recreation (such as biking trails, recreation forest park, road access network to the lake, entertainment facilities, etc.). This area appears to have gentle slopes presenting average inclinations in the order of 12~17% and covers approximately 3 km$^2$ of the surface.

An important issue emerging from these findings is that repurposing a PHS dam at the elevation of +545 m is feasible and can be achieved by naturally filling the pit lake for approximately three (3) years. Artificial recharging is also feasible if a fast filling is required to implement the PHS Project. More specifically, pumped water from Mavropigi and South field sumps could be diverted inside Kardia pit lake and, thus, fill the pit lake within a year.

Another alternative could be forming the lake water surface level at +570 m to implement the PHS and PV project. The findings from the implementation of the NNAR model suggest that this could be realized by naturally filling the pit lake in approximately seven (7) years. In the case of fast filling of the pit lake, this could be achieved in approximately three (3) years, i.e., four (4) years earlier.

The area on the southwest side of the lake is proposed for the upper reservoir installation of the PHS project. This area could be ideal for the location of the upper reservoir due to the proximity to both the lake (lower reservoir) and the Soulou River, where, after the PHS operational start, the required discharge of overflow is prospected to be rejected. This area covers approximately 0.5 $km^2$ of the surface, with an average inclination of approximately 3%.

Based on these results, our findings indicate that the level + 570 m could cover the needs of a hybrid PHS plant (HPHS) and PV park while providing enough space for recreation projects on the lake banks and simultaneously ensuring slope stability. The stabilizing effect of the lake water pressure ensures more favorable slope stability conditions at higher lake levels. In addition, the stability conditions are further improved in case the PHS upper reservoir is mounted outside the mine exploitation area, at a safe distance from the excavations' slopes.

However, water pumping is required in such cases to maintain the level of the pit lake at a certain elevation. Thus, the pit lake water quality is an important parameter that could enable new beneficial end uses, such as exploiting the overflow of the pit lake water for irrigation purposes. For example, Figure 14 shows that the pit lake water is expected to have good quality and thus could be suitable for irrigation purposes. Moreover, waters are marked by salinization risk (measured on the x-axis by conductivity) rather than alkalinization risk (measured on the y-axis by sodium).

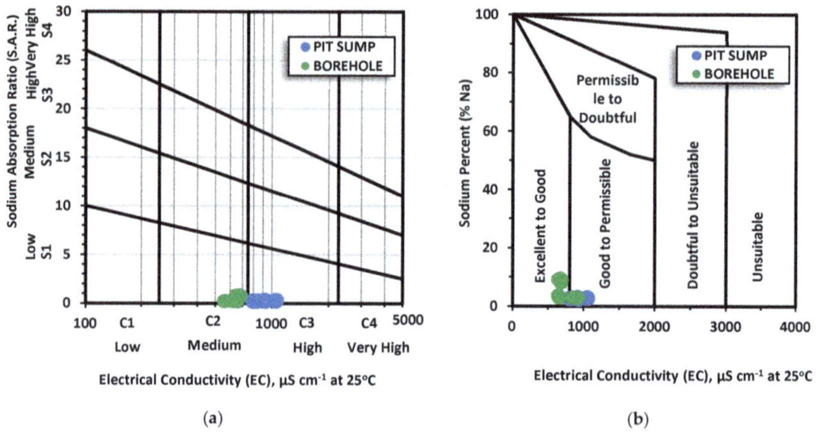


**Figure 14.** Classification of groundwater and the mine's internal sump water for irrigation purposes: (**a**) diagram of sodium hazard (sodium adsorption ration) and salinity hazard (electrical conductivity) [59,60]; (**b**) suitability of water for irrigation use, according to Wilcox diagram [61].

### *3.3. PHS Plant Efficiency Potential*

Figure 15 displays the potential energy storage $P_e$ per unit m of the daily water level fluctuation (MW/ m of water fluctuation) in relation to the pit lake's elevation and volume, considering two different scenarios: a) in the case of the upper reservoir mounted inside the mine at the elevation +640 m a.s.l. and b) in the case of the upper reservoir mounted outside the mine exploitation area (i.e., +670 m a.s.l.). The graph shows that in the case of the Kardia mine and for a specific pit water level at + 570 m a.s.l., the estimated potential energy storage $P_e$ is 40% higher in the case of a 30 m increase in the total head (*H*). In this example, lake level variation (*LLV*) at the pit lake was considered 1.0 m.

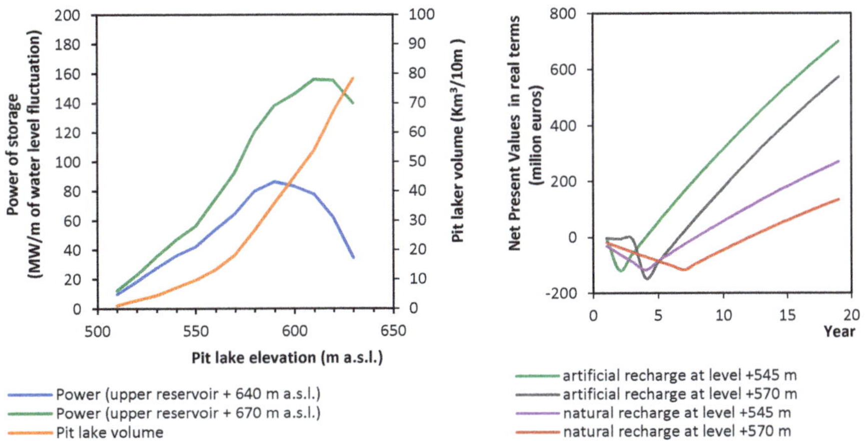


(a) (b)

**Figure 15.** (**a**) Potential power of storage provision (MW/ m of water fluctuation) in relation to pit lake volume and water level, considering 1.0 m for the pit lake level variation (*LLV*), in the case of the upper reservoir mounted inside the mine (+ 640 m a.s.l.) and outside the mine exploitation area (+ 670 m a.s.l.) and, (**b**) indicative net present value in real terms ($NPV_{real}$) in alternative scenarios of the pit lake's elevation and filling procedure, inside the Kardia mine, Greece.

In a cash flow of revenues and expenses, net present values in real terms are evaluated for the next 20 years, as presented in Figure 15. Specifically, in the case of natural recharging of the pit lake, it is evident that an increase in the net present value in the order of 34% is estimated for the pit lake elevation +545 m, compared to the scenario where the pit lake elevation is formed at +570 m. In addition, in the case of artificial recharging of the pit lake, an increase in the net present value in the order of 18% is estimated for the pit lake elevation +545 m, compared to the scenario where the pit lake elevation is formed at +570 m, due to the late recovery of investment capital. In this initial economic feasibility evaluation, constant values of inflation $r = 0.07$ and rate of interest $i = 0.10$ were considered, whereas the same value of the investment $I_0$ was considered in all scenarios.

## 4. Conclusions

In the abandoned Kardia lignite mine, the evaluation of land potential and resources is assessed regarding the lake water development inside the open pit area. Thus, the natural development of the lake is predicted, and different repurposing land uses are considered, concerning the energy transition potential in this specific region.

Initially, the spatiotemporal development of the lake in the final pit is calculated, and subsequently, land use repurposing is evaluated with regard to lake growth. Considering the energy transition strategies at this specific mine, it is concluded that on the east side of the lake, there is a relatively flat area of approximately 5 km$^2$, with an average inclination of approximately 3.5%. Therefore, this area is considered ideal for the installation of PV parks.

In addition, a hybrid PHS plant and PV park are examined to make provision for the energy surplus produced by the photovoltaic parks, aiming to meet electricity demand 24 h a day. Concentrating the solar power surplus with a PHS storage unit can provide flexible, renewable energy. This study has shown that repurposing a PHS dam at the elevation of + 545 m is feasible by naturally filling the pit lake for approximately three (3) years. Another scenario considering the elevation of +570 m could be achieved by recharging the pit within seven (7) years. Artificial recharging is also feasible in case fast filling is required due to the proximity of the Kardia lignite mine to two (2) other mines, the Mavropigi and

South field lignite mines. Therefore, a fast filling could be achieved in less than one (1) year for a lake elevation of +545 m and less than three (3) years for an elevation of +570 m.

From an engineering point of view, forming the lake at the level of +570 m is considered a better option for implementing the HPHS plant, as the stabilizing effect of the lake water pressure ensures more favorable slope stability conditions at higher lake levels. Moreover, the slope stability conditions are further improved in case the PHS upper reservoir is mounted outside the mine exploitation area, at a safe distance from the excavations' slopes.

Based on these results, our findings indicate that the surface lignite mine under investigation could cover both the needs of a hybrid PHS plant and PV park and, at the same time, provide enough space for activities (such as swimming, diving, etc.) and passive recreation (such as biking trails, recreation forest park, road access network to the lake, entertainment facilities, etc.) on the lake banks.

Forecasting the spatiotemporal development of a pit lake could help plan post-mining land repurposing, evaluate different scenarios, and find the optimum water level of the lake surface. The aim is to optimize land use repurposing in terms of economical design, efficiency and sustainability. In addition, the proximity of the Kardia lignite mine to the Mavropigi and South Field lignite mines is a crucial component to consider in the final design of the hybrid electric power project to make this project more economical, efficient and effective.

However, further studies need to be conducted to evaluate land repurposing, such as forecasting future water quality problems. In addition, different land repurposing projects are prospected to be developed under unfavorable geological settings of an abandoned open pit. This implies the need for detailed geotechnical (slope stability, bank erosion, corrosion, etc.) studies with the implementation of appropriate stability measures, in combination with a careful geotechnical monitoring application. To this end, continuous monitoring of ground movements is imperative. Therefore, this research question is an issue for further investigation and future model calibration when designing different repurposing land uses.

The whole system—pit lakes and alternative land repurposing scenarios—should be incorporated into a rational resources management program to optimize benefits to the land, water and electricity demands.

**Author Contributions:** Conceptualization, C.R. and G.L.; methodology, G.L., C.R., E.L. and E.M.; software, G.L. and E.M; validation, G.L., C.R., E.L., E.M. and G.K.; formal analysis, G.L., C.R., E.L, E.M. and G.K.; investigation, G.L., C.R., E.M. and G.K.; data curation, G.L., C.R., E.L., E.M. and G.K.; writing—original draft preparation, G.L. and E.L.; writing—review and editing, G.L., C.R. and E.M.; visualization, G.L., C.R., E.L., E.M. and G.K.; supervision G.L. and C.R.; project administration, C.R and G.L. All authors have read and agreed to the published version of the manuscript.

**Funding:** This research received no external funding.

**Data Availability Statement:** Data will be made available by the authors upon request.

**Conflicts of Interest:** The authors declare no conflict of interest.

## References

1. European Commission. *A Clean Planet for All a European Strategic Long-Term Vision for a Prosperous, Modern, Competitive and Climate Neutral Economy*; Document EUR-Lex-52018DC0773. COM(2018) 773 Final; Publications Office of the European Union: Brussels, Belgium, 2018. Available online: https://eur-lex.europa.eu/legal-content/EN/TXT/?uri=CELEX:52018DC0773 (accessed on 11 January 2021).
2. European Commission. *The European Green Deal*; Document EUR-Lex-52019DC0640. COM(2019) 640 Final; Publications Office of the European Union: Brussels, Belgium, 2019. Available online: https://eur-lex.europa.eu/legal-content/EN/TXT/PDF/?uri=CELEX:52019DC0640&from=EN (accessed on 11 January 2021).
3. Sokratidou, A.; Gkouvatsou, E.; Roumpos, C.; Perdiou, A.; Tsagkarakis, G.; Kaimaki, S. Innovative Approaches to Coal Surface Mine Sites Rehabilitation: A Case Study of Megalopolis Lignite Fields. In Proceedings of the 14th International Symposium of Continuous Surface Mining, Thesaloniki, Greece, 24 September 2018; Sideri, D., Paraskevis, N., Eds.; pp. 437–452.

4. Pachauri, R.K.; Mayer, L.; Intergovernmental Panel on Climate Change (Eds.) *Climate Change 2014: Synthesis Report*; IPCC: Geneva, Switzerland, 2015; ISBN 978-92-9169-143-2. Available online: https://climate-adapt.eea.europa.eu/countries-regions/countries/greece (accessed on 11 January 2021).
5. Bistline, J.E.T.; Blanford, G.J. Value of Technology in the U.S. Electric Power Sector: Impacts of Full Portfolios and Technological Change on the Costs of Meeting Decarbonization Goals. *Energy Econ.* **2020**, *86*, 104694. [CrossRef]
6. Pavloudakis, F.; Roumpos, C.; Karlopoulos, E.; Koukouzas, N. Sustainable Rehabilitation of Surface Coal Mining Areas: The Case of Greek Lignite Mines. *Energies* **2020**, *13*, 3995. [CrossRef]
7. Bódis, K.; Kougias, I.; Taylor, N.; Jäger-Waldau, A. Solar Photovoltaic Electricity Generation: A Lifeline for the European Coal Regions in Transition. *Sustainability* **2019**, *11*, 3703. [CrossRef]
8. Nikolić, Z.; Nikolić, D. Practical Example of Solar and Hydro Energy Hybrid System—The Need for a Reversible Power Plant. In Proceedings of the International Conference on Renewable Electrical Power Sources, Belgrade, Serbia, 2 October 2018; Volume 6, pp. 165–171.
9. Keenan, J.; Holcombe, S. Mining as a Temporary Land Use: A Global Stocktake of Post-Mining Transitions and Repurposing. *Extr. Ind. Soc.* **2021**, *8*, 100924. [CrossRef]
10. Wessel, M.; Madlener, R.; Hilgers, C. Economic Feasibility of Semi-Underground Pumped Storage Hydropower Plants in Open-Pit Mines. *Energies* **2020**, *13*, 4178. [CrossRef]
11. McCullough, C.; Schultze, M.; Vandenberg, J. Realizing Beneficial End Uses from Abandoned Pit Lakes. *Miner* **2020**, *10*, 133. [CrossRef]
12. Holcombe, S.; Keenan, J. *Mining as a Temporary Land Use Scoping Project: Transitions and Repurposing*; The Social Aspects of Mine Closure Research Consortium; Centre for Social Responsibility in Mining, Sustainable Minerals Institute: Brisbane, Australia, 2020. Available online: https://www.mineclosure.net/elibrary/mining-as-a-temporary-land-use-industry-led-transitions-and-repurposing (accessed on 13 January 2021).
13. Gammons, C.H.; Harris, L.N.; Castro, J.M.; Cott, P.A.; Hanna, B.W. Creating Lakes from Open Pit Mines: Processes and Considerations, Emphasis on Northern Environments. *Can. Tech. Rep. Fish. Aquat. Sci. 2826* **2009**, 106.
14. Schultze, M. The Filling and Remediation of Pit Lakes in Former Open Cast Lignite Mines. Ph.D. Dissertation, Tech Univ Braunschweig Fac Archit Constr Eng Environ Sci, Braunschweig, Germany, 2012.
15. McCullough, C.D. Rehabilitation of Lake Kepwari: A Previously Acidic Mine Lake in Western Australia. In Proceedings of the IMWA—Mine Water Management for Future Generations; Stanley, P., Wolkersdorfer, C., Wolkersdorfer, K., Eds.; Cardiff University: Cardiff, Wales, UK, 2021; pp. 346–351. Available online: https://www.imwa.info/imwaconferencesandcongresses/proceedings/325-proceedings-2021.html (accessed on 20 January 2022).
16. Lund, M.; van Etten, E.; Polifka, J.; Vasquez, M.Q.; Ramessur, R.; Yangzom, D.; Blanchette, M.L. The Importance of Catchments to Mine-Pit Lakes: Implications for Closure. *Mine Water Environ.* **2020**, *39*, 572–588. [CrossRef]
17. Lund, M.A.; Blanchette, M. Can Saline Pit Lakes Offer Biodiversity Values at Closure? In Proceedings of the IMWA—Mine Water Management for Future Generations; Stanley, P., Wolkersdorfer, C., Wolkersdorfer, K., Eds.; Cardiff University: Cardiff, Wales, UK, 2021; pp. 300–306. Available online: https://www.imwa.info/imwaconferencesandcongresses/proceedings/325-proceedings-2021.html (accessed on 20 January 2022).
18. Blanchette, M.L.; Lund, M.A. Pit Lakes Are a Global Legacy of Mining: An Integrated Approach to Achieving Sustainable Ecosystems and Value for Communities. *Curr. Opin. Environ. Sustain.* **2016**, *23*, 28–34. [CrossRef]
19. Snashall, D.; Cox, K. Woodlawn Advanced Energy Recovery Centre Scoping Report; Australia. Available online: https://majorprojects.planningportal.nsw.gov.au (accessed on 11 May 2021).
20. Pytel, S.; Sitek, S.; Chmielewska, M.; Zuzańska-Żyśko, E.; Runge, A.; Markiewicz-Patkowska, J. Transformation Directions of Brownfields: The Case of the Górnośląsko-Zagłębiowska Metropolis. *Sustainability* **2021**, *13*, 2075. [CrossRef]
21. Ma, T.; Yang, H.; Lu, L.; Peng, J. Technical Feasibility Study on a Standalone Hybrid Solar-Wind System with Pumped Hydro Storage for a Remote Island in Hong Kong. *Renew. Energy* **2014**, *69*, 7–15. [CrossRef]
22. Matos, C.R.; Carneiro, J.F.; Silva, P.P. Overview of Large-Scale Underground Energy Storage Technologies for Integration of Renewable Energies and Criteria for Reservoir Identification. *J. Energy Storage* **2019**, *21*, 241–258. [CrossRef]
23. Kavvadas, M.J.; Marinos, P.G. Prediction of Groundwater Table Lowering for Lignite Open-Cast Mining in a Karstic Terrain in Western Macedonia, Greece. *Q. J. Eng. Geol. Hydrogeol.* **1994**, *27*, S41–S55. [CrossRef]
24. Louloudis, G. Hydrogeological conditions of South Lignite Bearing Field of Ptolemais region. Confrontation of groundwater problems during the exploitation. Ph.D. Dissertation, Natl Tech University Athen Sch Min Metall Eng, Athens, Greece, 1991. (In Greek)
25. Dimitrakopoulos, D.; Grigorakou, E. Aquatic Balance of Vegoritis Lake, West Macedonia, Greece, Ralated to Mining Activity. In *Proceedings of the Mine Water 2004—International Mine Water Association Symposium 1*; Jarvis, A.P., Dudgeon, B.A., Younger, P.L., Eds.; University of Newcastle: Newcastle upon Tyne, UK, 2004; pp. 201–207. Available online: https://www.imwa.info/imwaconferencesandcongresses/proceedings/189-proceedings-2004.html (accessed on 20 January 2022).
26. Koukouzas, N.; Ward, C.R.; Li, Z. Mineralogy of Lignites and Associated Strata in the Mavropigi Field of the Ptolemais Basin, Northern Greece. *Int. J. Coal Geol.* **2010**, *81*, 182–190. [CrossRef]
27. Kavvadas, M.; Roumpos, C.; Schilizzi, P. Stability of Deep Excavation Slopes in Continuous Surface Lignite Mining Systems. *Geotech. Geol. Eng.* **2020**, *38*, 791–812. [CrossRef]

28. Beck, H.E.; Zimmermann, N.E.; McVicar, T.R.; Vergopolan, N.; Berg, A.; Wood, E.F. Present and Future Köppen-Geiger Climate Classification Maps at 1-Km Resolution. *Sci. Data* **2018**, *5*, 180214. [CrossRef] [PubMed]
29. Thornthwaite, C.W.; Mather, J.R. *The Water Balance*; Drexel Institute of Technology, Laboratory of Climatology, N.J. Publications in Climatology, 1955; Volume VIII, p. 104. Available online: https://www.jstor.org/stable/210739?origin=crossref (accessed on 20 January 2022).
30. Kinzelbach, W. *Groundwater Modelling: An Introduction with Sample Programs in BASIC*; Developments in water science; Elsevier: Amsterdam, The Netherlands; Oxford, UK; New York, NY, USA; Tokyo, Japan, 1986; ISBN 978-0-444-42582-9.
31. Calvo, J.C. An Evaluation of Thornthwaite's Water Balance Technique in Predicting Stream Runoff in Costa Rica. *Hydrol. Sci. J.* **1986**, *31*, 51–60. [CrossRef]
32. Blau, R.V.; Hoehn, P.; Hufschmied, P.; Werner, A. Ermittlung Der Grundwasserneubildung Aus Niederschlägen. *Gas Wasser Abwasser Schweiz. Ver. Des Gas- Und Wasserfaches* **1983**, *63*, 45–54.
33. Louloudis, G.; Louloudis, E.; Roumpos, C.; Mertiri, E.; Kasfikis, G.; Chatzopoulos, K. Forecasting Development of Mine Pit Lake Water Surface Levels Based on Time Series Analysis and Neural Networks. *Mine Water Environ.* **2021**, *41*, 458–474. [CrossRef]
34. Manning, J.C. *Applied Principles of Hydrology*, 3rd ed.; Prentice Hall: Hoboken, NJ, USA, 1997.
35. Dupuit, J. *Theoretical and Practical Studies on the Movement of Water in Open Channels and through Permeable Soils*; Dunod: Paris, Frence, 1863. (In French)
36. Krassakis, P.; Pyrgaki, K.; Gemeni, V.; Roumpos, C.; Louloudis, G.; Koukouzas, N. GIS-Based Subsurface Analysis and 3D Geological Modeling as a Tool for Combined Conventional Mining and In-Situ Coal Conversion: The Case of Kardia Lignite Mine, Western Greece. *Mining* **2022**, *2*, 297–314. [CrossRef]
37. Zhan, G.; Lin Shelp, M. *Modified Blaney-Criddle Method—An Empirical Approach to Estimate Potential Evaporation Using Air Temperature*; The University of Western Australia: Perth, Australia, 2009; pp. 461–471.
38. Tasumi, M.; Allen, R.G.; Trezza, R. Calibrating Satellite-Based Vegetation Indices to Estimate Evapotranspiration and Crop Coefficients. In *Proceedings of the USCID Water Management Conference*; Wichelns, D., Anderson, S.S., Eds.; U.S. Committee on Irrigation and Drainage: Denver, CO, USA, 2006; pp. 103–112.
39. Memmos, N. *Soil Survey Report of the Drained Giannitsa Lake*; Ministry of Agriculture, Sector of Land Reclamation Service (YEB, Acronym in Greek): Thessaloniki, Greece, 1973; *unpublished*. (In Greek)
40. Louloudis, G.; Stathopoulos, N. Irrigation Water Consumptive Use Changes in South Ptolemais Lignite Bearing Hydrogeological Basin (Greece) Using NDVI Remotely Sensed Data. In *Proceedings of the 15th Internat ional Congress of the Geological Society of Greece*; Kranis, H., Ed.; Bulletin of the Geological Society of Greece; Harokopio University of Athens: Athens, Greece, 2019; Volume 7, pp. 611–612. Available online: https://ejournals.epublishing.ekt.gr/index.php/geosociety/issue/view/1265 (accessed on 20 January 2022).
41. Panagos, P.; Van Liedekerke, M.; Jones, A.; Montanarella, L. European Soil Data Centre: Response to European Policy Support and Public Data Requirements. *Land Use Policy* **2012**, *29*, 329–338. [CrossRef]
42. McCuen, R.H.; Bondelid, T.R. Relation Between Curve Number and Runoff Coefficient. *J. Irrig. Drain. Div.* **1981**, *107*, 395–400. [CrossRef]
43. Kim, N.; Shin, M.-J. Estimation of Peak Flow in Ungauged Catchments Using the Relationship between Runoff Coefficient and Curve Number. *Water* **2018**, *10*, 1669. [CrossRef]
44. Giandotti, M. Previsione delle piene e delle magre dei corsi d'acqua. In *Memorie e Studi Idrografici*; Ministero dei LL.PP., Servizio Idrografico Italiano: Rome, Italy, 1934; Volume 8, pp. 107–117.
45. Penman, H.L. Natural Evaporation from Open Water, Bare Soil and Grass. *Proc. R. Soc. Lond. A* **1948**, *193*, 120–145. [CrossRef]
46. Sellers, W.D. *Physical Climatology*; The University of Chicago Press: Chicago, IL, USA, 1965.
47. Tetens, O. About some meteorological terms. *J. Geophys.* **1930**, *6*, 297–309. (In German)
48. Eagleman, J.R. Pan Evaporation, Potential and Actual Evapotranspiration. *J. Appl. Meteorol. Climatol.* **1967**, *6*, 482–488. [CrossRef]
49. Monteith, J.; Unsworth, M. *Principles of Environmental Physics: Plants, Animals, and the Atmosphere*; Academic Press: Warszawa, Poland, 2013; Volume 4, ISBN 978-0-12-386910-4.
50. Gass, S.I. Linear Programming. In *Encyclopedia of Statistical Sciences*; Kotz, S., Read, C.B., Balakrishnan, N., Vidakovic, B., Johnson, N.L., Eds.; John Wiley & Sons, Inc.: Hoboken, NJ, USA, 2006; p. ess1466.pub2. ISBN 978-0-471-66719-3.
51. Boretti, A.; Castelletto, S.; Al-Kouz, W.; Nayfeh, J. Capacity Factors of Solar Photovoltaic Energy Facilities in California, Annual Mean and Variability. *E3S Web Conf.* **2020**, *181*, 02004. [CrossRef]
52. Tyralis, H.; Karakatsanis, G.; Tzouka, K.; Mamassis, N. Data and Code for the Exploratory Data Analysis of the Electrical Energy Demand in the Time Domain in Greece. *Data Brief* **2017**, *13*, 700–702. [CrossRef]
53. Koudouris, G.; Dimitriadis, P.; Iliopoulou, T.; Mamassis, N.; Koutsoyiannis, D. Investigation on the Stochastic Nature of the Solar Radiation Process. *Energy Procedia* **2017**, *125*, 398–404. [CrossRef]
54. Faour, K. *Farm Budget Handbook, Southern NSW-Irrigated Winter Crops 2001*; NSW Agriculture, 2001. Available online: https://lib.k-state.edu/gracroplands/cite.php?CiteID=3758 (accessed on 20 January 2022).
55. Gemitzi, A.; Lakshmi, V. Estimating Groundwater Abstractions at the Aquifer Scale Using GRACE Observations. *Geosciences* **2018**, *8*, 419. [CrossRef]

56. Pujades, E.; Orban, P.; Bodeux, S.; Archambeau, P.; Erpicum, S.; Dassargues, A. Underground Pumped Storage Hydropower Plants Using Open Pit Mines: How Do Groundwater Exchanges Influence the Efficiency? *Appl. Energy* **2017**, *190*, 135–146. [CrossRef]
57. Kavvadas, M.; Roumpos, C.; Servou, A.; Paraskevis, N. Geotechnical Issues in Decommissioning Surface Lignite Mines—The Case of Amyntaion Mine in Greece. *Mining* **2022**, *2*, 278–296. [CrossRef]
58. Jiskani, I.M.; Cai, Q.; Zhou, W.; Ali Shah, S.A. Green and Climate-Smart Mining: A Framework to Analyze Open-Pit Mines for Cleaner Mineral Production. *Resour. Policy* **2021**, *71*, 102007. [CrossRef]
59. Richards, L.A.; Allison, L.E.; Brown, J.W.; Hayward, H.E.; Richards, L.A.; Bernstein, L.; Fireman, M.; Pearson, G.A.; Wilcox, L.V.; Bower, C.A.; et al. *Diagnosis and Improvement of Saline and Alkali Soils*; Agricultural Handbook No. 60; Salinity Laboratory, Soil and Water Conservation Research Branch, Agricultural Research Service, US Department of Agriculture: Washington, DC, USA, 1954; pp. 69–82.
60. Wilcox, L.V. *Classification and Use of Irrigation Waters*; Salinity Laboratory, US Department of Agriculture: Washington, DC, USA, 1955; Circular No. 969; pp. 1–19.
61. Wilcox, L.V. *The Quality of Water for Irrigation Use*; Salinity Laboratory, US Department of Agriculture: Washington, DC, USA, 1948; Technical Bulletin No. 962; pp. 1–40. [CrossRef]

water

MDPI

*Article*

# Spatiotemporal Analysis of Groundwater Storage Changes, Controlling Factors, and Management Options over the Transboundary Indus Basin

**Kashif Mehmood [1,2,*], Bernhard Tischbein [1], Martina Flörke [3] and Muhammad Usman [4]**

[1] Center for Development Research (ZEF), University of Bonn, Genscherallee 3, 53113 Bonn, Germany
[2] Department of Irrigation & Drainage, University of Agriculture, Faisalabad 38040, Pakistan
[3] Engineering Hydrology and Water Resources Management, Ruhr-Universität Bochum, 44801 Bochum, Germany
[4] Department of Geoecology, Institute of Geosciences and Geography, Martin Luther University, 06099 Halle, Germany
* Correspondence: kashif.mehmood@uni-bonn.de

**Abstract:** Intensive groundwater abstraction has augmented socio-economic development worldwide but threatens the sustainability of groundwater resources. Spatiotemporal analysis of groundwater storage changes is a prerequisite to sustainable water resource management over river basins. To estimate the groundwater storage changes/anomalies (GWCs) in the Indus River Basin (IRB), where observation wells are sparse, Gravity Recovery and Climate Experiment, the Global Land Data Assimilation System, and the WaterGAP Hydrological Model data were employed. The groundwater storage changes and controlling factors were investigated at three tier levels (TTLs), i.e., the basin, river reach, and region, to explore their implications on regional water resource management and provide management options at each level. Overall, the IRB groundwater declined from January 2003 to December 2016, with a relatively higher rate during 2003–2009 than during 2010–2016. Spatially, according to a reach-specific analysis, 24%, 14%, and 2% of the upper, middle, and lower reaches of the IRB, respectively, were indicated by a 'severe groundwater decline' over the entire period (i.e., 2003–2016). The GRACE-based GWCs were validated with in situ data of two heterogeneous regions, i.e., Kabul River Basin (KRB) and Lower Bari Doab Canal (LBDC). The analysis showed a correlation ($R^2$) of 0.77 for LBDC and 0.29 for KRB. This study's results reveal that climatic variations (increase in evapotranspiration); anthropogenic activities, i.e., pumping for irrigation; and water allocations in these regions mainly drive the groundwater storage changes across the Indus Basin.

**Keywords:** Kabul River Basin; groundwater dynamics; GRACE; WGHM; GLDAS; anthropogenic activities

## 1. Introduction

According to global estimates, approximately 43,000 km$^3$ of renewable freshwater resources are being fed to rivers, lakes, and aquifers annually, among which 70% is consumed by agriculture, 19% by industry, and 11% by households [1,2]. Presently, 38% of the global irrigated area (i.e., 307 million hectares) is contingent on groundwater [3]. In the Indus River Basin (IRB), where canal water supplies are unreliable and insufficient, groundwater is used independently or in conjunction with surface water. As irrigation in this region is sourced mainly from groundwater [4,5], the IRB aquifer is on the brink of an alarming water crisis due to large-scale groundwater extraction [6]. Therefore, the basin's sustainability requires proper water governance, management, and usage, especially given the complex, diversified hydrology and geopolitical nature of IRB. The key drivers of the basin's economy are agriculture and industry. Competition among the primary users (agriculture, industry, domestic, and the environment) is spiraling as agriculture intensifies, the population grows, and people migrate to the cities [7]. Over the past several years, the continuously varying climate conditions and expansion of agricultural lands have altered the crop irrigation requirements, placing new demands on surface and groundwater resources [8,9]. Consequently, the groundwater resources in IRB have been overexploited [10].

In the Indus Basin, the overall water supply from groundwater has increased by 60% since 1960 as fresh aquifer systems are continuously pumped to meet the growing domestic and agricultural water demands [6,11]. Therefore, the groundwater aquifer system must be carefully assessed through critical long-term analysis of the groundwater storage changes (GWCs). Such an assessment builds a baseline for sustainable water use and its management. Ground-based measurement could provide a best estimate of changes in groundwater storage. However, ground-based observations are expensive and limited on spatial and temporal scales, especially in IRB, so groundwaters cannot be thoroughly monitored at different levels (regions and reaches). In particular, the limited network of observation wells reduces the understanding of the water cycle [12,13]. Policymakers and researchers believe that because there is a lack of in situ observations, groundwater regulation and the adoption of area-specific management strategies are severely compromised, especially under the effects of climate change [14]. To address these issues, many researchers are observing groundwater storage variations in real time by remote sensing observations supported by land-hydrological models at global and regional scales [5,15,16]. For example, Cheema et al. [5] employed satellite-based evapotranspiration (ET) and precipitation (P) data in the process-based Soil and Water Assessment Tool and estimated the groundwater depletion in the irrigated Indus Basin. Since the launch of the Gravity Recovery and Climate Experiment (GRACE) satellite, groundwater storage variations have been observed at different spatial and temporal scales [17] and the WaterGAP v2.2d Hydrological Model (WGHM) has emerged as a robust and qualitatively high-performing dataset of groundwater variations. The temporal changes in gravity observed by twin satellites reflect the changes in terrestrial water storage (TWS) [18]. TWS, obtained by summing the groundwater, snow water equivalent (SWE), soil moisture (SM), and surface water storage, represents the water stored above and underneath the Earth's surface [19]. Therefore, to isolate GWCs from TWS estimated by GRACE, water storage components must be subtracted from other auxiliary datasets of hydrological models. Many researchers have already combined GRACE data with land-hydro models to study the inter-annual changes in TWS and GWCs over the Indus Basin [20–22] and other aquifer systems around the globe [23–29]. However, most of these studies analyzed TWS and GWC variations on the catchment scale; a comprehensive evaluation of GWCs over the Indus Basin is lacking at different tier levels, i.e., the basin, river reach, and region. Normally, TWS and GWC variations are mainly caused by the combined effect of climate variability and human interventions [30]. Climatic parameters, i.e., precipitation (P), runoff (R), evapotranspiration (ET), and glacier recession, are influential factors of the changes in TWS and GWC [31]. Anthropogenic activities, i.e., groundwater abstraction, land use, and irrigation, also impact TWS and GWCs in populated areas [32]. Hence, it is important to investigate GWC and its controlling factors in IRB to ensure sustainable development of the basin as a whole. The water management in IRB is administered under the Indus Water Treaty 1960 at the international level [33], and administrative regulations are implemented province-wide under the Water Apportionment Accord 1991 at the national level [34]. Therefore, in addition to basin-scale studies, it is imperative to investigate the spatiotemporal dynamics of groundwater and its controlling factors at the regional (provincial) level to provide management options for each management unit. Due to the heterogeneity that exists among different reaches of IRB in terms of topography, land use, climate and agroecological conditions, and surface water distribution quotas, IRB was divided into three reaches (i.e., upper, middle, and lower). Due to these heterogeneities, there is considerable variability in the socio-economic conditions of the above-mentioned reaches. Therefore, hypothetically, human and natural controlling factors do vary among these reaches. In this regard, it is imperative to examine the groundwater changes both in space and time at the reach level in IRB.

Considering the limitations of previous studies in the Indus Basin [5,17,20,21,35–39], the novelty being captured in this study includes the following: (i) investigation of the groundwater dynamics at three tier levels (TTLs), i.e., the basin, river reach, and region, using GRACE and GLDAS for the period 2003–2016 to explore their implications on regional water resources management; (ii) assessment of the controlling factors (climatic and anthropogenic) at three tier levels to mitigate the vulnerability of groundwater resources; and (iii) identification of the hotspots of groundwater change patterns by decomposing and analyzing time series of the GRACE-extracted GWC, the Potential Groundwater Use (PGWUSE), and the Potential Groundwater Withdrawal (PGWWW) to explore groundwater management options.

## 2. Materials and Methods

### 2.1. Study Area

The area coverage of IRB is approximately 1.12 million km$^2$. It is shared among Pakistan (47%) and India (39%), followed by China (8%) and Afghanistan (6%) [2] and is situated between 66.3–82.47° E longitude and 24.63–37.05° N latitude (Figure 1). The Indus River (approximately 3000 km long) originates in the Tibetan Plateau near Mount Kailash (altitude 5500 m) and traverses towards the Arabian Sea [40]. The tributaries of IRB are the Indus, Jhelum, Chenab, Ravi, Sutlej, and Bias Rivers. The eastward-flowing tributary is the Chitral River, which originates in Pakistan, flows towards Afghanistan, and then drains into the Indus River in Pakistan [41]. The basin's climate ranges from subtropical arid and semiarid to temperate sub-humid on the plains of the Sindh and Punjab provinces to alpine in the northern mountainous highlands [2]. The average annual precipitation varies from 100 to 500 mm in the lowlands to approximately 2000 mm in the mountainous ranges. Moreover, the snowfall at higher altitudes (>2500 m) accounts for a significant part of the river flows in IRB. Although the higher peaks in the Upper Indus Basin (UIB) restrict the intrusion of monsoons, the low-lying areas in the northwest of the basin are subjected to monsoon rainfall, influencing GWCs over the basin.

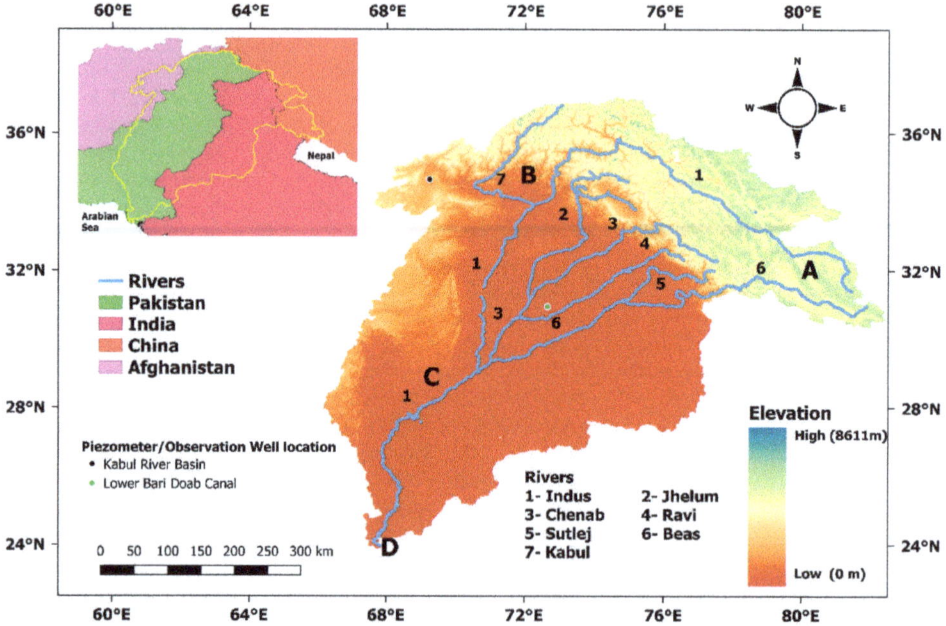


**Figure 1.** Transboundary Indus Basin with its tributaries and digital elevation model.

IRB is divided into upper, middle, and lower reaches based on its topography and climate. The upper reach includes Kashmir, Gilgit Baltistan, and Himachal Pradesh, which constitute the major watershed of IRB (points A to B). The middle reach (points B to C) comprises the major portions of Pakistani Punjab and Indian Punjab (the food basket of Pakistan and India), which produces most agricultural products and secures the region's food supply. The lowest part of the reach consists of Sindh (points C to Arabian sea), as shown in Figure 1. Agricultural production is influenced by waterlogged and saline agricultural lands in Sindh and some portions of Baluchistan.

Geologically, IRB is dominated by thick quaternary sediments. These sediments are mainly alluvial and deltaic deposits of fine- to medium-grained silt, sand, and clay. Coarser sands and gravels are present on the plain's upland margins while sand deposits can be found at the east of the Indus Plain (i.e., Thar and Cholistan). Cenozoic and Mesozoic sedimentary rocks can be found in a north–south region located at the west of the Indus Plain, which stretches from Peshawar to Karachi. Older sediments (Paleozoic) and crystalline base rocks (e.g., granites and metamorphic) are mainly found in the north, including Khyber Pakhtunkhwa, Jammu and Kashmir, and Gilgit regions. Groundwater yields from these sediments are typically in the range of 50–300 $m^3/h$ down to a depth of 150 m.

### 2.2. Data Sources

We extracted the monthly TWS data of GRACE RL-06 released by the National Aeronautics and Space Administration (NASA) from 2003 to 2016, covering the globe with a spatial resolution of $1^\circ \times 1^\circ$. The GRACE landmass datasets of three laboratories, namely, the Centre for Space Research (CSR), Jet Propulsion Laboratory (JPL), and Geo Forschungs Zentrum Potsdam (GFZ), were obtained from the NASA website (www.jpl.nasa.gov accessed on 17 December 2019) supported by the NASA MEASUREs Programme [42–45]. The gridded surface-mass anomalies were derived from GRACE mission measurements and estimated the Earth's mean gravity at the specified time. For the JPL datasets, scaling factors were applied to convert the 3-degree gaussian filter resolution to 1-degree spherical harmonics solutions. These grids represent the storage of the terrestrial water cycle and ice magnitudes over the entire land. Furthermore, the non-captured atmospheric and oceanic processes are included in the corresponding GRACE component. The SM, SWE, and canopy water storage (CWS) were collected from GLDAS_NOAH10_M_001 at a $1^\circ \times 1^\circ$ spatial resolution [46]. The data from WGHM by the Institut für Physische Geographie, University of Frankfurt, accurately simulates the changes in all forms of terrestrial water storage and GWC except glaciers [47] to observe the agreement with GRACE-based GWC. The outputs of this model have been widely used in recent years [31,48]. The present study used the recent GWC, PGWUSE, and PGWWW data collected from 2003 to 2016, with a monthly temporal resolution and a spatial resolution of $0.5^\circ \times 0.5^\circ$ [47]. To explore the controlling factors of TWS and GWC, we extracted the data of three hydrological fluxes (P, ET, and runoff (R)) from different sources. The precipitation data of the Tropical Rainfall Measuring Mission (TRMM, version TRMM-3B43 Level-3) with a temporal resolution of 1 month and a spatial resolution of $0.25^\circ \times 0.25^\circ$ were obtained from NASA [49]. The monthly ET, P, and R data, output of WGHM with a spatial resolution of $0.5^\circ \times 0.5^\circ$, were taken from data published on the Earth and environmental science website [47].

### 2.3. Methods

The TWS data of IRB was obtained from the three laboratories using a specialized program written in R and averaged to compute GWC. To calculate the monthly GWC, SM, SWE, and SWS were excluded from the average gridded TWS values of the JPL, CSR, and GFZ laboratories. The groundwater storage change/groundwater anomaly $\Delta GWC$ was calculated as given in Equation (1) [50,51]:

$$\Delta GWC = TWS - \Delta SM - \Delta SWE - \Delta CWS \quad (1)$$

where $\Delta SM$, $\Delta SWE$, and $\Delta CWS$ are the anomalies for soil moisture, snow water equivalent, and total canopy water, respectively. All variables in Equation (1) (downloaded on a rectangular grid) were imported into a geographic information system (GIS) to mask the irregular boundaries of the Indus River Basin. The centroids of the rectangular grids were obtained as the grid coordinates, and the variables of Equation (1) extracted by the R-code quantified the spatiotemporal groundwater dynamics at various levels. The TWS values for missing months were determined by a linear interpolation technique [52], and statistical analyses were performed by Mann–Kendall tests, regression, and Spearman's rho correlation.

### 2.3.1. Time-Series Decomposition

A filtering procedure that decomposes a time series into its seasonal (S), trend (T), and residual (R) components using locally weighted scatterplot smoothing is commonly known as the STL method [53]. Numerous hydrological studies [54,55] have used STL to clarify, refine, and obtain the detailed characteristics of variations in timing data. STL decomposition based on LOESS at the time $i$ in a time series $Yi$ is given as:

$$Yi = Si + Ti + Ri \qquad \text{where } i = (1, 2, 3 \ldots, N). \tag{2}$$

In this study, the time series of GRACE-based GWC, PGWUSE, and PGWWW were decomposed, and the characteristics of each decomposed component were analyzed to extract valuable information. Figure 2 summarizes the datasets and research methodology as a flowchart.

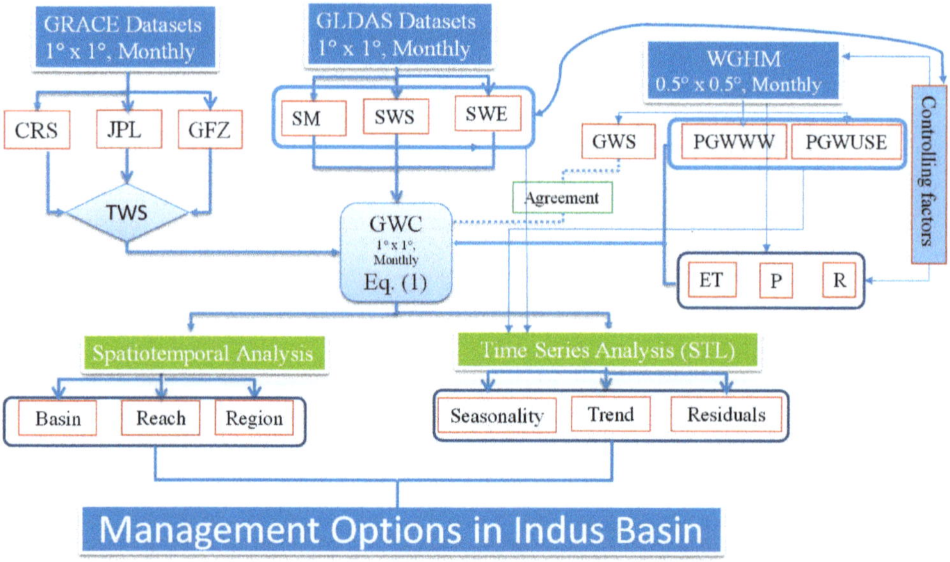


**Figure 2.** Flowchart of the datasets used and analyses performed.

### 2.3.2. Spatial and Temporal Analysis

Spatiotemporal analysis was performed to understand the groundwater dynamics of the basin using the available datasets. The 112 grids of the GRACE and GLDAS datasets covering the entire IRB were extracted using the GIS tool. The GRACE-based groundwater anomaly was averaged from January 2003 to December 2016 in each grid cell in the spatial analysis. The temporally averaged GWC values were then categorized into five classes based on GWC threshold values, providing more direct information for groundwater resource management options and planning [12]. The groundwater thresholds were identified by observing the range of GWC values over the IRB grids. The anomaly categorization is given in Table 1.

**Table 1.** Groundwater change categories in the Indus River Basin.

| Sr. No | Groundwater Changes (mm) | Indication of Groundwater Change |
|---|---|---|
| 1 | <−100 | Severe decline |
| 2 | −100 to −50 | Minor decline |
| 3 | −50 to 0 | Normal variation |
| 4 | 0 to 50 | Minor rise |
| 5 | >50 | Abnormal rise |

*2.4. Performance Assessment of the Groundwater Storage Changes*

GWCs estimated in this study across the wider IRB were evaluated at two random pixels both at the very extreme northwestern part (KRB) and at the downstream LBDC. The piezometric data of KRB was obtained from Taher et al. [56], which are monthly records and available for a period of 2005–2013 while for that of LBDC's, there are only two records available per year (i.e., pre and post monsoon) available from 2003 to 2015. The observation wells' data for LBDC was obtained from the Punjab Irrigation Department (Pakistan).

## 3. Results

The estimated spatial and temporal changes in GWC revealed a groundwater decline over the study period. The fluctuations in GWC were higher before 2010 compared to the later period (2010–2016), i.e., reflecting a substantial decrease in the GRACE-based groundwater storage anomaly in the earlier period (Figure 3a,c).

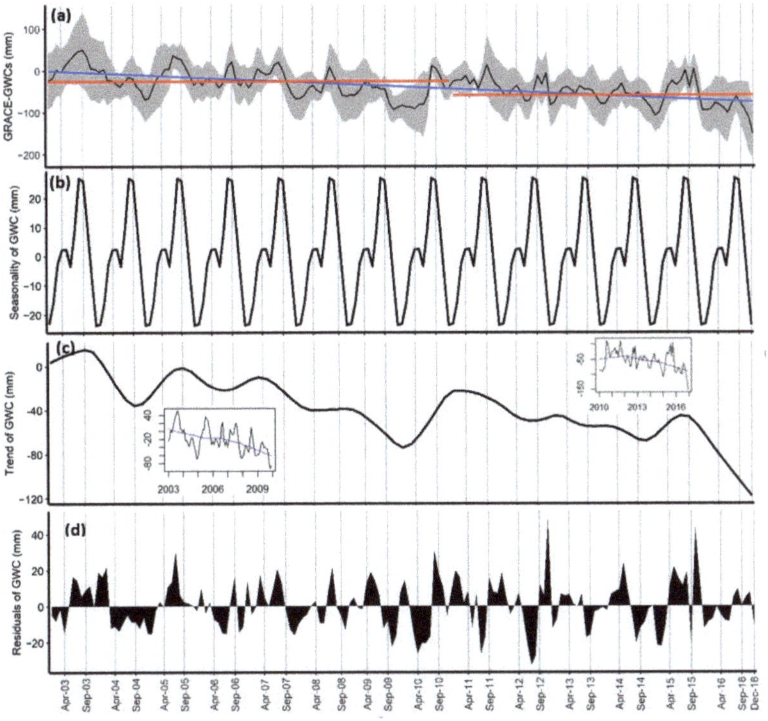


**Figure 3.** (**a**) GRACE-derived GWC time series, (**b**) seasonality, (**c**) trend, and (**d**) residuals of the decomposed GWC signals.

### 3.1. Spatiotemporal Groundwater Dynamics and Controlling Factors at the Basin Level

Figure 3a shows the temporal changes in the groundwater resources at the basin level. During the study period (January 2003 to December 2016), GWC decreased ($p = 2.2 \times 10^{-16}$; *t*-test) over the entire IRB. The mean GWC was −25.67 mm for the period of January 2003–December 2009 and −54.33 mm for January 2010–December 2016 (red lines in Figure 3a), confirming both a decline and shift in the mean GWC over the study period, and the fitted regression line (blue line in Figure 3a) showed an overall decreasing trend of 0.44 mm per month.

For the whole study period 2003–2016, the GWC trend declined at −0.44 mm/month in the GRACE model and −0.64 mm/month in WGHM. To identify the hotspots and understand the characteristics of GWC, the GRACE-based GWC time series was decomposed to observe its seasonality, trend, and abnormalities (see Figure 3). Across the entire IRB, the decomposed GWC signal showed a strong seasonality, with annual losses in the pre-monsoon period (around April) and annual gains in the post-monsoon period (around September) (Figure 3a,b). In each year, the seasonality curve dipped after April (Figure 3b) as excessive pumping for rice cultivation began. No significant changes were observed within the years in the seasonal component. The decomposed trend component (Figure 3c) of the GWC signal significantly declined until the end of the drought year (2009). Accordingly, groundwater changes (Figure 3c) were steeper in the first half of the study period, indicating an alarming decline in groundwater resources during this period as evident from the Mann–Kendall trend analysis, which indicated that the mean GWC was highly significant during the study period from January 2003 to December 2009 ($p = 1.65 \times 10^{-7}$) compared to January 2010–December 2016 ($p = 1.99 \times 10^{-4}$).

The groundwater level maximally declined in 2009, when the weak monsoon created drought conditions in the basin before the flood years [57]. Subsequently, it rose in the flood years, indicating the changes in groundwater mainly depend on the trend component. The residual component (a powerful indicator of abnormalities such as flood and drought events in the region) displayed abnormal signals in the winters of 2012 and 2015 (Figure 3d). In the Federal Flood Commission report [58], the years 2010–2016 were identified as flood years. Indeed, the groundwater levels showed a significant rise during these years, indicated by a wet period (Figure 3a) in the present analysis.

The spatial analysis revealed an overall decrease in groundwater resources throughout the basin. As shown in Figure 4, GWC declined from south to north and from west to east. A strip of severe GWCs was observed in Gilgit Baltistan, Kashmir, Himachal Pradesh, and Indian Punjab. In UIB, GWCs dominated in the range from −131 to −81 mm, indicating high groundwater storage variations with higher standard deviations. In the southwest part of IRB, the groundwater storages are minimum and with low variance. In general, the regions where the standard deviation of groundwater storage was high showed greater groundwater decline.

Groundwater storage changes were further explored using the SM, SWS, and SWE data from the GLDAS Noah model. Groundwater changes showed a declining trend in the study area (Figure 5a), and SWS contributed little to GWCs and their tendency (Figure 5b,c). The snow changes remained around zero because snow and ice concentrate in the mountainous part of the basin (upper IRB) and only cover parts of the Indus Basin. The SWE, which significantly affected the changes in groundwater storage, trended downward, particularly after 2005 (Figure 5c). The main contributor to GWC was SM, which is necessary for the growth and survival of flora. The occurrence of drought depends on when the current soil moisture cannot meet the plant's requirements. As shown in Figure 5c, SM and GWC exhibited the same declining trend patterns after the winter of 2007, indicating that GWC and SM are heavily dependent on each other in IRB.

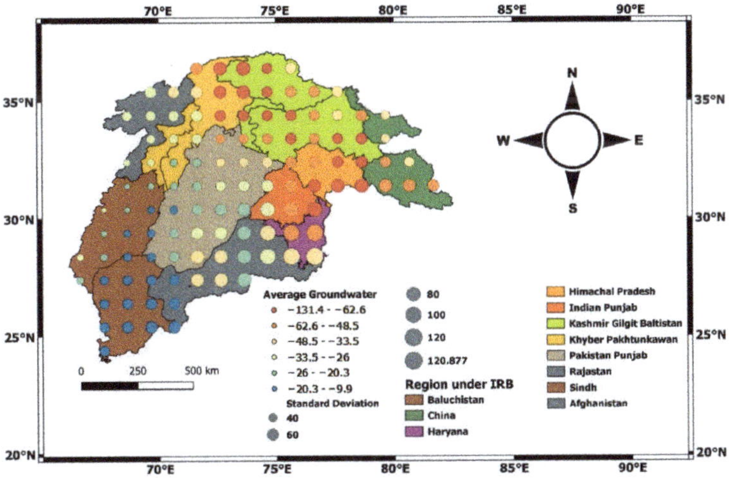


**Figure 4.** Spatial analysis of the time-averaged GWCs in IRB.

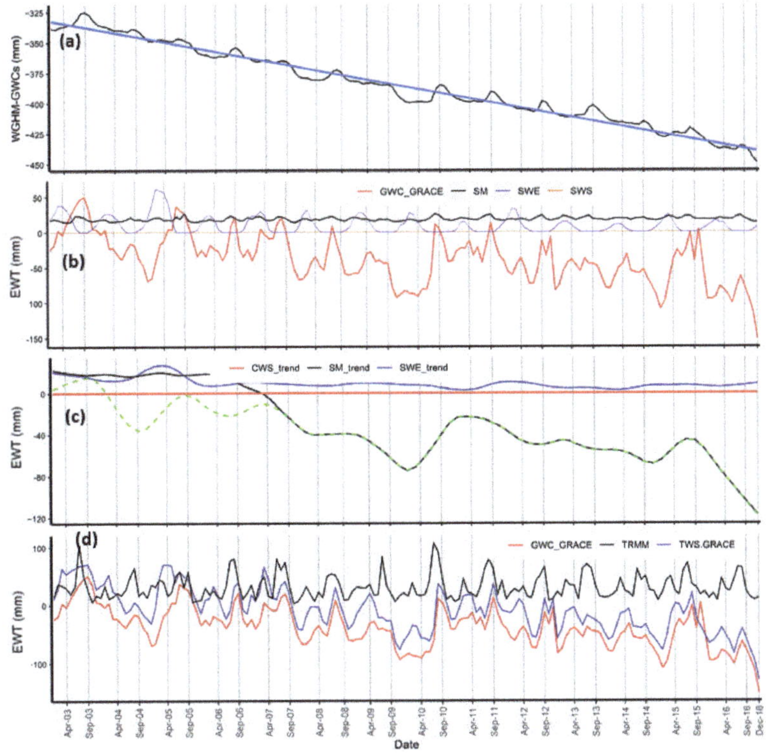


**Figure 5.** (**a**) Groundwater storage changes in WGHM (black line indicates the groundwater storage of WGHM and blue line indicates the fitted regression line); (**b**) changes and (**c**) trends of GWC, soil moisture (SM), canopy water storage (SWS), and snow water equivalent (SWE); (**d**) comparison of monthly time series of the GRACE-based GWC and TWS values and TRMM precipitation data in IRB.

The spatial differences in TWS and GWC were analyzed from the variabilities in ET, P, and R. We mainly analyzed the entire IRB, UIB, Middle Indus Basin (MIB), and Lower Indus Basin (LIB). In the Indus Basin, TWS and GWC declined at a rate of −0.545 and −0.640 mm/month, respectively; and ET, P, and R varied by a rate of 0.0417, 0.0659, and 0.034 mm/month, respectively (Figure 6a). The positive variability rates of ET, P, and R indicated that all these variables were sensitive to TWS and GWC. However, R is generated by precipitation and snowmelt in the study area, and contributed a minor amount to TWS and GWCs. Thus, we inferred that the ET increase was the main controlling factor of the groundwater and total water storage dynamics across IRB during the study period.

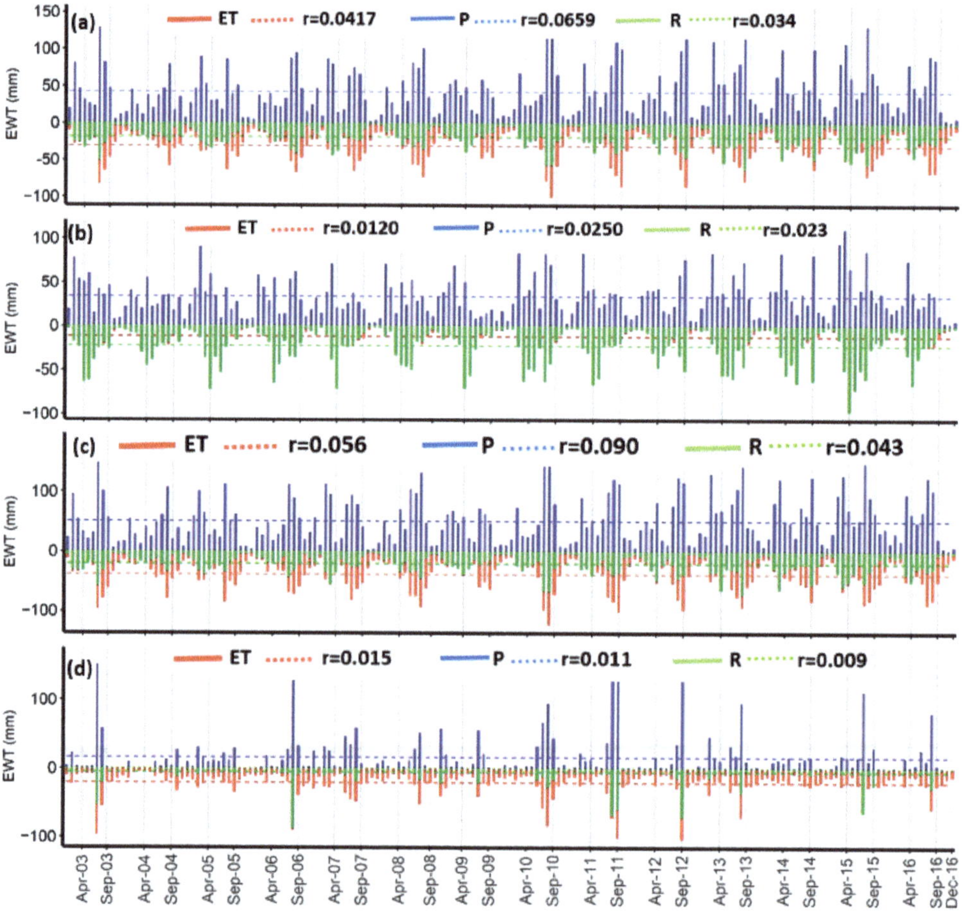


**Figure 6.** Time series analysis of P, ET, and R in (**a**) the entire Indus Basin, (**b**) Upper Indus Basin (UIB), (**c**) Middle Indus Basin (MIB), and (**d**) Lower Indus Basin (LIB). Here, 'r' (mm/month) denotes the rate of change from 2003 to 2016.

In UIB, TWS trended upward at an amount of 0.018 mm/month, whereas GWC declined at an amount of 0.22 mm/month. Both P and R showed a positive tendency with higher variability than ET (Figure 6b). P increased at an amount of 0.025 mm/month, whereas ET increased at an amount of 0.012 mm/month, indicating that the P increase mainly caused the TWS increase. The increasing trend of regional R was not consistent with the P variation (Figure 6b), suggesting that there is an additive water supply from snowmelt and glacier recession. Muhammad et al. [59] reported a negative mass balance in upper IRB, with the most negative values ranging from $-0.34 \pm 0.31$ to $0.44 \pm 0.27$ m w.e., with $-1$ in the Jhelum, Chenab, and Ravi sub-basins of the Himalaya. SWE in UIB changed by $-1.02$ mm/year (the glacier area was 115,675 $km^2$ in 2016). Therefore, our analysis suggests that glacier recession (not ET) is an important controlling factor of TWS and GWC variations in UIB.

In MIB, TWS and GWC significantly declined at an amount of $-0.89$ and $-0.9604$ mm/month, respectively. P and ET showed a positive tendency over time and higher variations in MIB than in the Indus Basin as a whole (Figure 6c). The rising ET trend accounts for most of the TWS loss. Meanwhile, the increasing trend in regional R (Figure 6c) is consistent with the P variation. Therefore, ET, groundwater usages and withdrawal were inferred to be the main factors that increase the TWS deficit and reduce the groundwater in MIB.

In LIB, TWS increased at a rate of 0.17 mm/month, whereas GWC declined at a rate of $-0.138$ mm/month. The variations in P, R, and ET were small and inconsistent with the TWS changes (Figure 6d), suggesting that precipitation and evapotranspiration are not the main factors leading to changes in the total water storage and GWC in this region. Instead, the surface water supply in terms of high water allowances to the canals provided by the Water Apportionment Accord (WAA) 1991 might explain the small groundwater and total water storage changes. The century-old irrigation system of IRB has allowed a variable allocation of surface water by its design [60]. The provincial bodies reviewed surface water allocations (WAs) at many stages to establish water rights and distribute the water of IRB among the four provinces of Pakistan, and these reviews led to a mutual consensus policy document: the Water Apportionment Accord (WAA) 1991 [60].

Precipitation markedly affects the spatial pattern of TWS and GWC and accelerates the spatial variation signal in the north–south direction. Figure 5d compares the mean monthly precipitation data from TRMM and the mean monthly TWS and GWC. The precipitation generated by TRMM varied from 2.28 (October 2007) to 107 mm/month (July 2010), whereas GWC ranged from $-152.43$ (December 2016) to 50.46 mm (September 2003).

As shown in Table 2, the TRMM, TWS, GWC, and SM data were positively correlated, indicating that the GWC values are affected not only by precipitation but also by parameters associated with anthropogenic activities, such as pumping. As shown in Figure 7a,b, the groundwater use in each year was lowest in February and highest in October. The PGWUSE started to increase after April as the rising temperature accelerated ET and rice cultivation activities in the basin. The decomposed trend components of PGWUSE and PGWWW showed highly reciprocal correlations with the trend of GWC (when PGWUSE was maximum, GWC was minimum; see Figure 7c). The monthly potential total water withdrawals from the groundwater resources were highly seasonal, minimal in April, and highest in November each year (Figure 7d,e). These results imply that much of the groundwater is withdrawn during cropping seasons and consumptive use. The trend component of the decomposed time series of PGWWW increased as the area became more dependent on groundwater (Figure 7f).

**Table 2.** Correlations between TRMM precipitation, TWS, and GWC.

| | | *Correlations* | | | |
|---|---|---|---|---|---|
| | | TRMM | TWS | GWC | SM |
| *Spearmen's Rho* | TRMM | 1.000 | 0.214 * | 0.188 * | 0.743 * |
| | TWS | 0.214 * | 1.000 | 0.907 * | 0.299 * |
| | GWC | 0.188 * | 0.907 * | 1.000 | 0.241 * |
| | SM | 0.743 * | 0.299 * | 0.241 * | 1.000 |

* Correlation is significant at the 0.01 level (two-tailed).

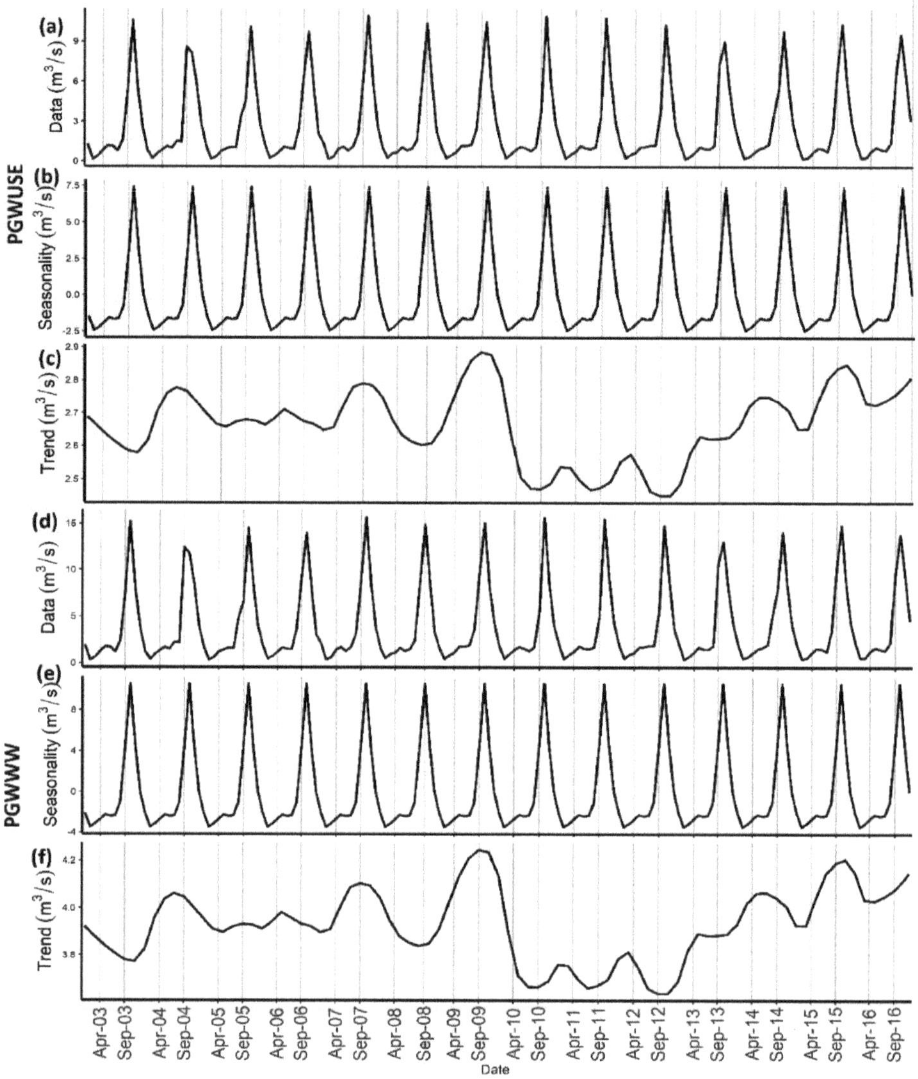


**Figure 7.** Time series: (**a**) data—PGWUSE; (**b**) seasonality—PGWUSE; (**c**) trend—PGWUSE; (**d**) data—PGWWW; (**e**) seasonality—PGWWW; and (**f**) trend—PGWWW over IRB during the 2003–2016 period.

### *3.2. Spatiotemporal Groundwater Dynamics in Different Reaches*

Reach-wise groundwater storage analysis revealed that changes in groundwater storage were diverse among the three reaches of IRB (Figure 8a) during the study period. The lower reach showed relatively higher groundwater storage than the middle reach while GWC in the upper reach was much smaller than in the middle reach (Figure 8a). These differences can be explained by variations in ET, R, P, and groundwater withdrawals as described above. GWCs in the upper and middle reaches declined more rapidly in later years than in previous years. Moreover, after 2009, GWC was higher in the middle reach than in the other reaches. Contrarily, GWC as determined for each of the reaches showed monotonous behavior during the study period. In the lower reach, the minor increases during 2007 and 2011 were contributed by groundwater recharge from flooding periods in previous years, explaining the smaller declining trend than the other reaches (Figure 8a).

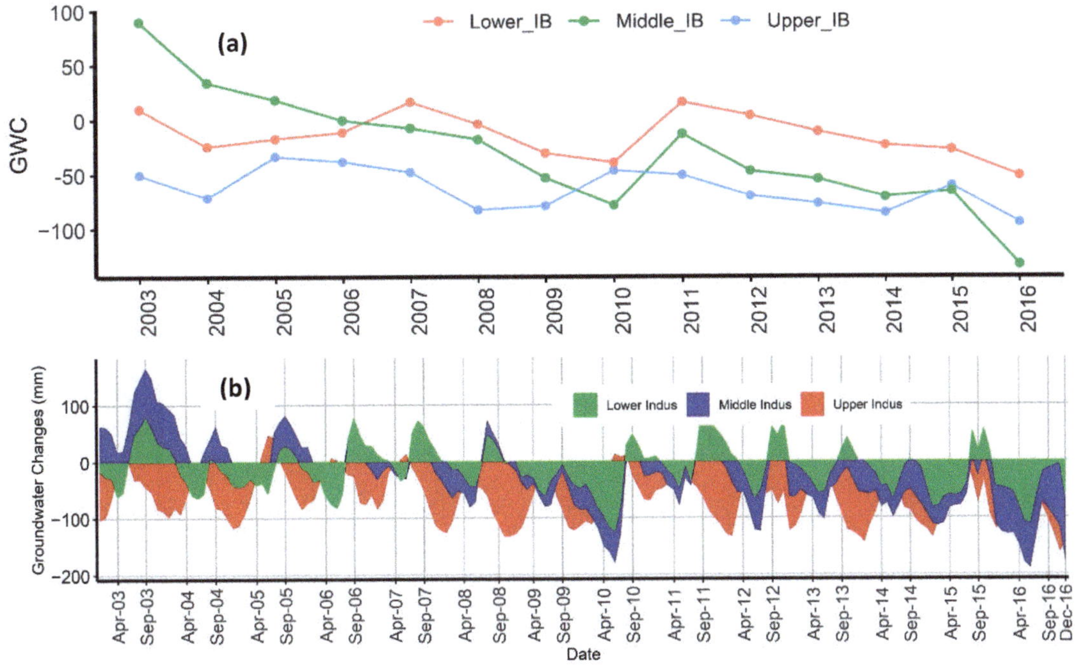


**Figure 8.** (**a**) Annual time series of the mean spatial groundwater storage changes in the three reaches of IRB. (**b**) Mean monthly GWC (mm/month) at different reaches in IRB.

Looking at the monthly scale groundwater changes (Figure 8b), we found no consistency in groundwater storage fluctuations among the three reaches during 2003–2009. However, the upper and lower reaches showed some consistency in their fluctuations during 2010–2016 (Figure 8b). In the upper reach, groundwater storage remained low in winter (~December) throughout the study period except for rising trends during the summer of 2005–2007 and 2010. The groundwater storage changes remained high during 2003–2008, and low in the rest of the study periods. The lower reach depicted high groundwater storage in the summer (~September). After the year 2009 (tipping year), groundwater storage values were continuously negative in the upper and middle reaches, indicating that the groundwater storage continuously remained lower when compared with the base period (2004–2009). Prior to 2010, all the reaches showed groundwater storage fluctuation in the range of −150–150 mm. After 2010, there was a less declining trend of the groundwater storage due to floods, and the annual minimum values were recorded during the study duration.

The most recent study year (2016) showed severe signs of depletion in groundwater storage, especially in the middle reach, as shown in Figure 9b. The upper reach showed surprisingly strong groundwater declines (<−100 mm) in all years except 2006, whereas the middle reach showed strong declines in groundwater storage from 2010 to 2016 onwards (Figure 9a,b). In general, the upper reach had the highest number of months that experienced severe groundwater declines compared with the other reaches during the study duration (Figure 9d). At the same time, there were no months with an abnormal rise (>50 mm) in the upper reach. The reach-specific analysis revealed that more rapid fluctuations in the upper reach make its ecosystem vulnerable, which may require further attention. Groundwater depletion in the upper reach should be mitigated with climate mitigation strategies.

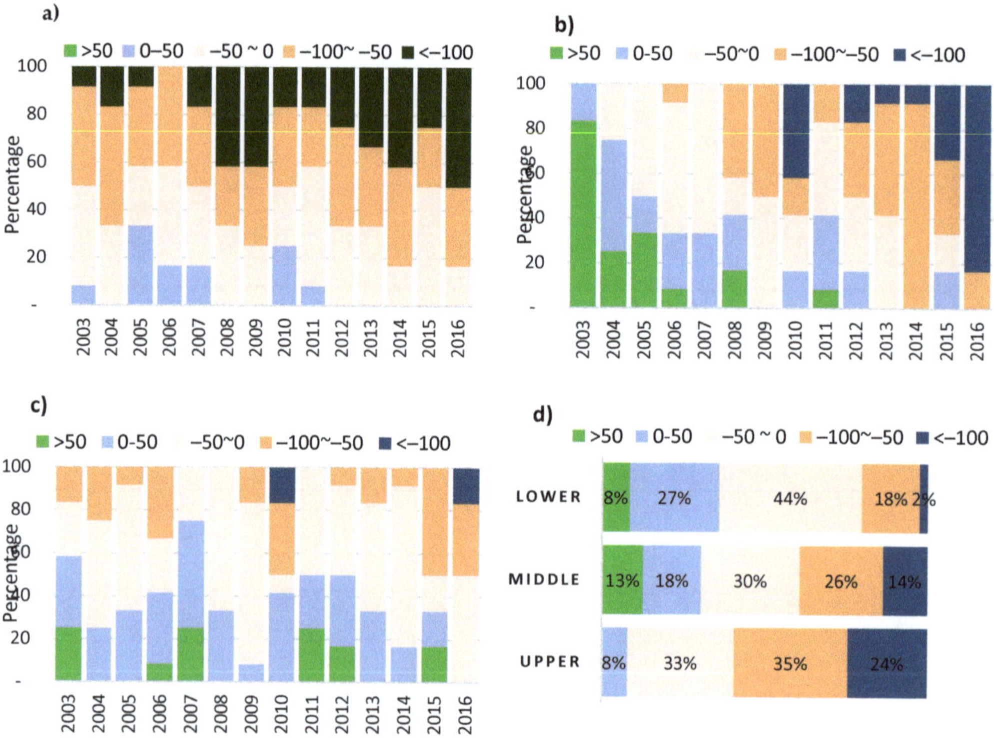


**Figure 9.** (**a**) Upper reach groundwater changes, (**b**) middle reach groundwater changes, (**c**) lower reach groundwater changes, and (**d**) annual severity of GWCs in the different reaches of IRB.

In the reach-specific analysis, 24%, 14%, and 2% of the time span of the groundwaters in the upper, middle, and lower reaches, respectively, were severely depleted over the entire period. These findings indicate the vulnerability of groundwater resources in the upper Indus reach (Figure 9d). During the same period, GWC rose abnormally in the middle reach. The reach-wise analysis revealed more rapid fluctuations in the upper reach, which may require further attention. In particular, groundwater depletion in the upper reach should be mitigated with climate mitigation strategies.

### *3.3. Spatiotemporal Groundwater Dynamics at the Regional Level*

To study GWCs in a spatially explicit way, the study area was subdivided into eight regions: Kashmir and Gilgit Baltistan, Tibet–China, Khyber Pakhtunkhwa, Afghanistan, Punjab, Sindh, Baluchistan, and Indian Punjab–Himachal Pradesh–Haryana (hereafter referred to as the Indian part), along the Indus River from upstream to downstream. All eight regions of IRB exhibited irregular annual patterns of continuous decline in GWC after 2003 (Figure 10). China, Afghanistan, KP, and Sindh showed consistent groundwater storage changes within the range from 0 to −50 mm except in 2016. Before 2009, groundwater changes were within −50–100 mm for all the regions while after 2010, all the regions showed a rise in groundwater storage in 2011 and then a continuous decline. In the Indian part, the groundwater storage declined until 2010, but in the years beyond 2010, GWCs were in the range of −150 mm. The low GWC values in the Tibet–China and Indian part of IRB in the second phase and particularly in the final year of the study period (2016) indicate a severe groundwater decline in these regions (see Figure 10). In the province of Sindh, the groundwater rose abnormally in 2011 and, after that, continuously declined at a lesser rate than in the other regions.

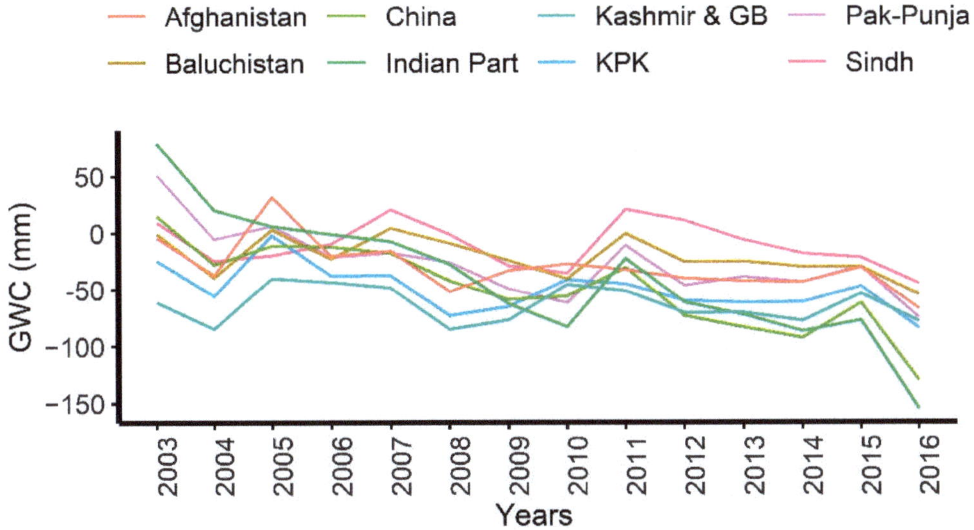


**Figure 10.** The annual time series of spatially averaged changes in groundwater storage in different regions.

The GRACE-based estimates of groundwater storage changes were compared to in situ measurements (1 piezometer data for KRB and 1 observation well for LBDC). In situ measurements for KRB and LBDC were available from 2003 to 2015 and 2005 to 2013, respectively. When the GWC and in situ measurements were compared, the correlation coefficients for LBDC and KRB were 0.78 and 0.35, respectively (Figure 11).

(**a**) Lower Bari Doab Canal (LBDC)

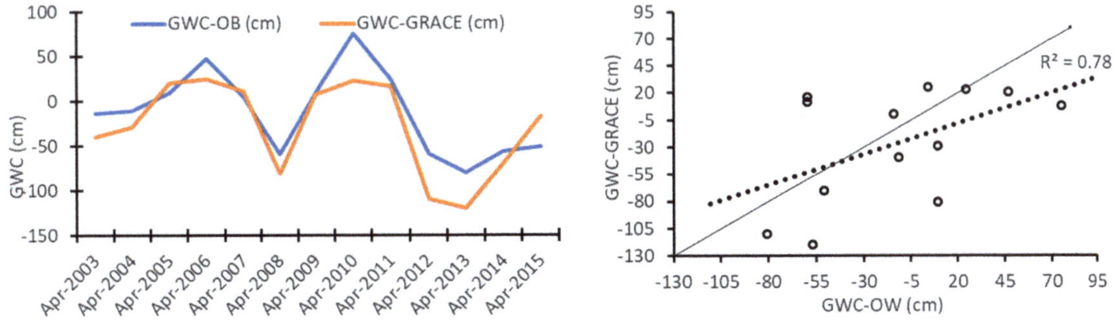


**Figure 11.** *Cont.*

(**b**) Kabul River Basin (KRB)

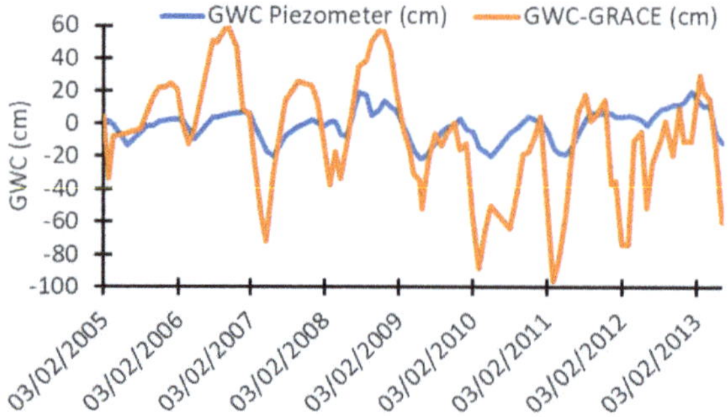


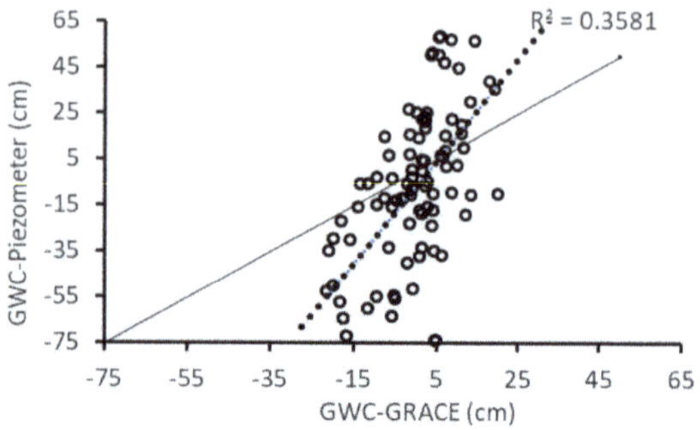


**Figure 11.** (**a**) Correlation of the observed (observation well data) and estimated (GRACE-based) groundwater storage anomalies in LBDC. (**b**) Correlation of the observed (piezometric data) and GRACE-based groundwater storage changes in the Kabul River Basin.

## 4. Discussion

Overall, our analysis revealed that regional GWCs in time and space could be derived from remote sensing data (GRACE) and data assimilation models (GLDAS). In the temporal analysis, a significant decline in GWC ($p = 2.2 \times 10^{-16}$) was observed over the entire study period (January 2003 to December 2016). The observed decreasing trend and fluctuations were larger before 2010 than in the second half of the study period, which is driven by the higher values in 2003 and very low numbers in 2010. The second phase of the time series is characterized by lower levels and extreme lower levels of GWC in 2016.

In the spatial analysis, a significantly declining trend in GWC was observed in the upper Indus reach, indicating a dominant snow receding effect [61]. On the regional scale, the aquifer storage showed a declining trend in upper Punjab (especially in Gujranwala, Sialkot, and Hafizabad) because the pumping rate was high in these regions. Similar results were reported by Qureshi et al. [62], Shoaib et al. [37], Arshad et al. [39], and Iqbal et al. [20], who mentioned that water is over-pumped at an alarming rate in areas where high delta crops (i.e., rice) are grown at a high tube well density. The groundwater depletion was moderate in the districts (Faisalabad, Vehari, Pakpattan, and Khanewal) of Pakistani Punjab because groundwater is overexploited to meet the crop water requirements (Iqbal et al. (2017)). Qureshi [63] reported that groundwater is declining to below 6 m in more than 50% of the irrigated area of Punjab, causing groundwater quality deterioration and increased pumping costs.

Groundwater storage changes were severe in the Indian part of IRB, reaching a maximum in 2016. Dominant changes in groundwater were observed in the south-eastern regions of the IRB, covering Himachal Pradesh, Indian Punjab, Haryana, and Rajasthan (the northern Indian states, which account for 95% of groundwater use in the region). High groundwater-based irrigation can explain these large changes, which has expanded to meet the wheat and rice demands of Indian government policy [64]. Other causes of groundwater decline are exponential population growth and rapid economic development [65], which have increased the pumping of groundwater [5,66], in addition to the provision of subsidies on electricity and diesel for groundwater abstraction, an increased number and density of tube wells (>30/km$^2$ in different districts), augmented groundwater draft, expansion of rice and wheat areas [39], and spikes in cropping intensity [67,68]. Similarly, Rodell et al. [17] detected a significant groundwater depletion in the north-western part of India based on the GRACE datasets. They calculated a mean depletion rate of $40 \pm 10$ mm/year equivalent height of water, whereas Asoka et al. [69] estimated a 20 mm/year groundwater decline rate in the same region. Singh and Bhakar [70] reflected upon the alarming groundwater situation in Rajasthan and concluded that there is an uncertain future for this commodity.

Steenbergen et al. [71] attributed groundwater exhaustion in the Baluchistan province of Pakistan to the intensive use of water for agricultural purposes. The Sindh province of Pakistan, the southwest part of IRB, showed lower GWC fluctuations than the other parts of the basin and similar results were revealed by Arshad et al. [39]. Basharat and Azhar [60] attributed the lower GWCs to the higher water allowance of canals under WAA 1991. Therefore, the authority responsible for the WAA 1991 should rationalize water allowances over IRB to counteract waterlogging and salinity in the lower reach (i.e., the Sindh province of Pakistan) and over-abstraction of groundwater in the upper part of the basin. The observed in situ groundwater anomalies were compared with the estimated GRACE-based groundwater storage anomalies (Figure 11). The results were consistent with the study conducted by Akhtar et al. [38] and Ali et al. [37].

Due to drought conditions, the groundwater declined maximally in 2009 while during flood periods, groundwater storages were improved. Chen et al. [18] reported in Songhua River Basin that flood and drought conditions characterize the fluctuation in groundwater storage. GWS recovery was attributed to an increase in precipitation [72] in the Tasmania region, Australia, after 2010. However, Indus Basin exhibited a decline in groundwater storage that was inconsistent with the precipitation variation till 2010 (Figure 5c). Afterwards, it showed a continuous decline due to the significant impact of anthropogenic activities, i.e., mainly pumping in agriculture (Figure 7c,f). In addition, Feng et al. [73] also reported that the pumping of groundwater for irrigation and coal mining activities was a major contributing factor in North China.

The spatiotemporal patterns of groundwater behavior extracted from the GLDAS and GRACE data might assist the implementation of sustainable groundwater management by policymakers and managers. Admittedly, the spatial resolution of the GRACE mission is too coarse to monitor the finer surface mass changes in large water bodies such as dams and lakes. Although the applied datasets are the best current data sources for determining GWCs in this area, they are insufficient for understanding groundwaters in canal command areas. Moreover, the mission ended in 2017, necessitating a new perspective of groundwater dynamics. To this end, statistical modeling of the GRACE time series and downscaling of GRACE to the local scale of groundwater dynamics are required. By connecting the data of GRACE, the latest launched GRACE-FO, and statistical modeling, we can ensure the continuity of the groundwater dynamics for prediction, especially when data are missing during the operation period of GRACE-FO. Issues such as sea water intrusion, how surface water bodies affect groundwater quality and quality, and land subsidence are not addressed by GRACE [74]. The uncertainties in hydrological models, i.e., GRACE and GLDAS, could be considered in the perspective studies.

## 5. Conclusions

The GRACE and GLDAS datasets were combined into a novel algorithm that estimates the spatial and temporal changes in groundwater, thereby providing water-resource management options at the basin, reach, and regional levels. Groundwater resources in the whole basin have declined over time but exhibited a spatially heterogeneous change pattern. The main conclusions of this study and outlooks for future research and policy intervention are summarized as follows:

- When analyzed over time, almost all GWC values were negative, with an overall decreasing trend from January 2003 to December 2016. The fluctuations in GWCs were stronger before 2010 than in the second half of the study period. An increase in evapotranspiration was found to be the main controlling factor for the groundwater and total water storage changes across IRB.
- In the spatial analysis, GWC trended downward over the whole Indus Basin. The decline was severe (−131 to −108 mm) in UIB, highlighting a need for sub-basin analysis. We recommend the development of quantitative groundwater aquifer-related maps to visualize water withdrawals, groundwater recharge, groundwater storage changes, groundwater levels, aquifer thickness and yields, and tube well density over the whole

IRB. These maps provide useful support to develop area-specific groundwater management strategies.
- The reach-specific analysis showed that the upper reach is more vulnerable to groundwater decline compared to the other reaches due to glacier recession. We recommend the adaption of climate mitigation strategies to cater for the climate change impact on glacier recession. This study concludes that groundwater abstraction and groundwater consumption are the most significant driving factors in the middle reach. These results highlight the need for a robust mechanism to monitor the quantity and quality of groundwater abstractions in the middle reach. The outcomes for the lower reach reveal that the high water allowances provided to the canals under WAA 1991 are the main driver of lower groundwater storage change. So, surface water allowances in Pakistan should be allocated based on a detailed understanding of groundwater withdrawals and depletion.
- On the regional scale, the groundwater decline was severe in Himachal Pradesh, Indian Punjab, Haryana, upper Punjab of Pakistan, and China. Due to these regional differences, it is evident that a region-specific groundwater strategic framework is necessary. Hence, a comprehensive mechanism for monitoring and establishing regulatory and permitting arrangements and pricing structures should be developed to envisage sustainable management of the land area and water resources. Because the number and density of tube wells and boreholes have largely increased in recent years, we propose the construction of a database containing georeferenced locations of all existing and new tube wells and boreholes in the regions. Such a database would not only support but also improve management decisions.
- The correlation between the observed and estimated GWC both at KRB (i.e., $R^2 = 0.35$) and LBDC (i.e., $R^2 = 0.78$) shows that there are spatial heterogeneities across IRB, which needs to be considered during planning initiatives.

In a nutshell, basin-wide groundwater monitoring infrastructures must be established using the latest technologies (GIS, remote sensing, and Heliborne technology). Such infrastructures are essential for the development of a sound database and maps for groundwater regulation and sustainable management. These will also enable robust monitoring of the quantity and quality of groundwater resources and their use.

**Author Contributions:** Conceptualization, K.M. and B.T.; methodology, K.M.; software, K.M.; validation, K.M; formal analysis, K.M.; investigation, K.M.; data curation, K.M.; writing—original draft preparation, K.M.; writing—review and editing, M.F. and M.U.; visualization, M.U.; supervision, M.F. All authors have read and agreed to the published version of the manuscript.

**Funding:** This research received no external funding.

**Data Availability Statement:** Not applicable.

**Acknowledgments:** The first author is thankful to the Higher Education Commission, Pakistan, and the German Academic Exchange Service (DAAD), Germany, for funding this study under the Overseas Scholarship Program. The first author is also thankful to the University of Agriculture-Faisalabad for granting study leave to pursue Ph.D. studies. The authors also thank Hannes Müller Schmied, affiliated with the University of Frankfurt, Germany, for providing access to the data. Thanks to the Center for Development Research (ZEF) and the University of Bonn for facilitation and support.

**Conflicts of Interest:** The authors declare no conflict of interest.

## References

1. Kinzelbach, W.; Bauer, P.; Siegfried, T.; Brunner, P. Sustainable Groundwater Management—Problems and Scientific Tools. *Episodes* **2003**, *26*, 279–284. [CrossRef] [PubMed]
2. Food and Agriculture Organization, Faoa. *Irrigation in Southern and Eastern Asia in Figures—AQUASTAT Survey—2011: Indus River Basin*; Food and Agriculture Organization: Rome, Italy, 2011; pp. 1–14.
3. Siebert, S.; Faures, J.M.; Frenken, K.; Hoogeveen, J. Groundwater Use for Irrigation—A Global Inventory. *Hydrol. Earth Syst. Sci.* **2010**, *14*, 1863–1880. [CrossRef]

4. Usman, M.; Qamar, M.U.; Becker, R.; Zaman, M.; Conrad, C.; Salim, S. Numerical Modelling and Remote Sensing Based Approaches for Investigating Groundwater Dynamics under Changing Land-Use and Climate in the Agricultural Region of Pakistan. *J. Hydrol.* **2020**, *581*, 124408. [CrossRef]
5. Cheema, M.J.M.; Immerzeel, W.W.; Bastiaanssen, W.G.M. Spatial Quantification of Groundwater Abstraction in the Irrigated Indus Basin. *Groundwater* **2014**, *52*, 25–36. [CrossRef] [PubMed]
6. Lytton, L.; Ali, A.; Garthwaite, B.; Punthakey, J.F.; Basharat, S. Groundwater in Pakistan's Indus Basin: Present and Future Prospects. 2021, p. 164. Available online: http://hdl.handle.net/10986/35065 (accessed on 16 April 2021).
7. Cheema, M.J.M.; Qamar, M.U. *Transboundary Indus River Basin: Potential Threats to Its Integrity*; Elsevier Inc.: Amsterdam, The Netherlands, 2019; ISBN 9780128127827.
8. Arfan, A.; Zhang, Z.; Zhang, W.; Gujree, I. Long-Term Perspective Changes in Crop Irrigation Requirement Caused by Climate and Agriculture Land Use Changes in Rechna Doab, Pakistan. *Water* **2019**, *11*, 1567. [CrossRef]
9. Sun, S.; Zhou, T.; Wu, P.; Wang, Y.; Zhao, X.; Yin, Y. Impacts of Future Climate and Agricultural Land-Use Changes on Regional Agricultural Water Use in a Large Irrigation District of Northwest China. *L. Degrad. Dev.* **2019**, *30*, 1158–1171. [CrossRef]
10. Watto, M.A. The Economics of Groundwater Irrigation in the Indus Basin, Pakistan: Tube-Well Adoption, Technical and Irrigation Water Efficiency and Optimal Allocation. Ph.D. Thesis, University of Western Australia, Crawley WA, Australia, 2015; pp. 1–218.
11. Qureshi, A.S.; Gill, M.A.; Sarwar, A. Sustainable Groundwater Management in Pakistan: Challenges and Pportunities. *Irrig. Drain.* **2010**, *59*, 107–116. [CrossRef]
12. Lin, M.; Biswas, A.; Bennett, E.M. Spatio-Temporal Dynamics of Groundwater Storage Changes in the Yellow River Basin. *J. Environ. Manage.* **2019**, *235*, 84–95. [CrossRef] [PubMed]
13. Mahmood, R.; JIA, S.; Lv, A.; Zhu, W. A Preliminary Assessment of Environmental Flow in the Three Rivers' Source Region, Qinghai Tibetan Plateau, China and Suggestions. *Ecol. Eng.* **2020**, *144*, 105709. [CrossRef]
14. Young, W.J.; Arif, A.; Bhatti, T.; Borgomeo, E.; Davies, S.; Garthwaite, W.R., III; Gilmont, E.M.; Leb, C.; Lytton, L.; Makin, I.; et al. *Pakistan: Getting More from Water*; World Bank: Washington, DC, USA, 2019; p. 163.
15. Yin, W.; Han, S.C.; Zheng, W.; Yeo, I.Y.; Hu, L.; Tangdamrongsub, N.; Ghobadi-Far, K. Improved Water Storage Estimates within the North China Plain by Assimilating GRACE Data into the CABLE Model. *J. Hydrol.* **2020**, *590*, 125348. [CrossRef]
16. Liu, F.; Kang, P.; Zhu, H.; Han, J.; Huang, Y. Analysis of Spatiotemporal Groundwater-Storage Variations in China from Grace. *Water* **2021**, *13*, 2378. [CrossRef]
17. Rodell, M.; Velicogna, I.; Famiglietti, J.S. Satellite-Based Estimates of Groundwater Depletion in India. *Nature* **2009**, *460*, 999–1002. [CrossRef] [PubMed]
18. Chen, H.; Zhang, W.; Nie, N.; Guo, Y. Long-Term Groundwater Storage Variations Estimated in the Songhua River Basin by Using GRACE Products, Land Surface Models, and in-Situ Observations. *Sci. Total Environ.* **2019**, *649*, 372–387. [CrossRef]
19. Tapley, B.D.; Bettadpur, S.; Watkins, M.; Reigber, C. The Gravity Recovery and Climate Experiment: Mission Overview and Early Results. *Geophys. Res. Lett.* **2004**, *31*, L09607. [CrossRef]
20. Iqbal, N.; Hossain, F.; Lee, H.; Akhter, G. Integrated Groundwater Resource Management in Indus Basin Using Satellite Gravimetry and Physical Modeling Tools. *Environ. Monit. Assess.* **2017**, *189*, 128. [CrossRef] [PubMed]
21. Iqbal, N.; Hossain, F.; Lee, H.; Akhter, G. Satellite Gravimetric Estimation of Groundwater Storage Variations over Indus Basin in Pakistan. *IEEE J. Sel. Top. Appl. Earth Obs. Remote Sens.* **2016**, *9*, 3524–3534. [CrossRef]
22. Tang, Y.; Hooshyar, M.; Zhu, T.; Ringler, C.; Sun, A.Y.; Long, D.; Wang, D. Reconstructing Annual Groundwater Storage Changes in a Large-Scale Irrigation Region Using GRACE Data and Budyko Model. *J. Hydrol.* **2017**, *551*, 397–406. [CrossRef]
23. Fallatah, O.A.; Ahmed, M.; Save, H.; Akanda, A.S. Quantifying Temporal Variations in Water Resources of a Vulnerable Middle Eastern Transboundary Aquifer System. *Hydrol. Process.* **2017**, *31*, 4081–4091. [CrossRef]
24. Ghebreyesus, D.; Temimi, M.; Fares, A.; Bayabil, H. A Multi-Satellite Approach for Water Storage Monitoring in an Arid Watershed. *Geosciences* **2016**, *6*, 33. [CrossRef]
25. Huang, X.; Gao, L.; Crosbie, R.S.; Zhang, N.; Fu, G.; Doble, R. Groundwater Recharge Prediction Using Linear Regression, Multi-Layer Perception Network, and Deep Learning. *Water* **2019**, *11*, 1879. [CrossRef]
26. Verma, K.; Katpatal, Y.B. Groundwater Monitoring Using GRACE and GLDAS Data after Downscaling Within Basaltic Aquifer System. *Ground Water* **2019**, *58*, 143–151. [CrossRef]
27. Srivastava, S.; Dikshit, O. Seasonal and Trend Analysis of TWS for the Indo-Gangetic Plain Using GRACE Data. *Geocarto Int.* **2020**, *35*, 1343–1359. [CrossRef]
28. Papa, F.; Frappart, F.; Malbeteau, Y.; Shamsudduha, M.; Vuruputur, V.; Sekhar, M.; Ramillien, G.; Prigent, C.; Aires, F.; Pandey, R.K.; et al. Satellite-Derived Surface and Sub-Surface Water Storage in the Ganges-Brahmaputra River Basin. *J. Hydrol. Reg. Stud.* **2015**, *4*, 15–35. [CrossRef]
29. Moghim, S. Assessment of Water Storage Changes Using GRACE and GLDAS. *Water Resour. Manag.* **2020**, *34*, 685–697. [CrossRef]
30. Li, H.; Lu, Y.; Zheng, C.; Zhang, X.; Zhou, B.; Wu, J. Seasonal and Inter-Annual Variability of Groundwater and Their Responses to Climate Change and Human Activities in Arid and Desert Areas: A Case Study in Yaoba Oasis, Northwest China. *Water* **2020**, *12*, 303. [CrossRef]
31. Zhu, Y.; Liu, S.; Yi, Y.; Qi, M.; Li, W.; Saifullah, M.; Zhang, S.; Wu, K. Spatio-Temporal Variations in Terrestrial Water Storage and Its Controlling Factors in the Eastern Qinghai-Tibet Plateau. *Hydrol. Res.* **2021**, *52*, 323–338. [CrossRef]

32. Huang, Y.; Salama, M.S.; Krol, M.S.; Su, Z.; Hoekstra, A.Y.; Zeng, Y.; Zhou, Y. Estimation of Human-Induced Changes in Terrestrial Water Storage through Integration of GRACE Satellite Detection and Hydrological Modeling: A Case Study of the Yangtze River Basin. *Water Resour. Res.* **2015**, *51*, 8494–8516. [CrossRef]
33. World Bank The Indus Waters Treaty 1960 (with Annexes). Signed at Karachi, on 19 September 1960. *United Nations—Treaty Ser.* **1962**, *6032*, 1–85.
34. IRSA Apportionment of Waters of Indus River System between the Provinces of Pakistan. *Indus River Syst. Auth. Gov. Pak.* **1992**, 1–47.
35. Ali, S.; Liu, D.; Fu, Q.; Cheema, M.J.M.; Pham, Q.B.; Rahaman, M.M.; Dang, T.D.; Anh, D.T. Improving the Resolution of Grace Data for Spatio-Temporal Groundwater Storage Assessment. *Remote Sens.* **2021**, *13*, 3513. [CrossRef]
36. Shekhar, S.; Mao, R.S.K.; Imchen, E.B. Groundwater Management Options in North District of Delhi, India: A Groundwater Surplus Region in over-Exploited Aquifers. *J. Hydrol. Reg. Stud.* **2015**, *4*, 212–226. [CrossRef]
37. Ali, S.; Wang, Q.; Liu, D.; Fu, Q.; Mafuzur Rahaman, M.; Abrar Faiz, M.; Jehanzeb Masud Cheema, M. Estimation of Spatio-Temporal Groundwater Storage Variations in the Lower Transboundary Indus Basin Using GRACE Satellite. *J. Hydrol.* **2022**, *605*, 127315. [CrossRef]
38. Akhtar, F.; Nawaz, R.A.; Hafeez, M.; Awan, U.K.; Borgemeister, C.; Tischbein, B. Evaluation of GRACE Derived Groundwater Storage Changes in Different Agro-Ecological Zones of the Indus Basin. *J. Hydrol.* **2022**, *605*, 127369. [CrossRef]
39. Arshad, A.; Mirchi, A.; Samimi, M.; Ahmad, B. Combining Downscaled-GRACE Data with SWAT to Improve the Estimation of Groundwater Storage and Depletion Variations in the Irrigated Indus Basin (IIB). *Sci. Total Environ.* **2022**, *838*, 156044. [CrossRef] [PubMed]
40. Archer, D.R.; Forsythe, N.; Fowler, H.J.; Shah, S.M. Sustainability of Water Resources Management in the Indus Basin under Changing Climatic and Socio Economic Conditions. *Hydrol. Earth Syst. Sci.* **2010**, *14*, 1669–1680. [CrossRef]
41. Iqbal Abdul Rauf Environmental Issues of Indus River Basin: An Analysis. *ISSRA Pap. Inst. Strateg. Stud. Res. Anal. (ISSRA) Natl. Def. Univ. Islam. Pak.* **2013**, *5*, 91–112.
42. Landerer, F.W.; Swenson, S.C. Accuracy of Scaled GRACE Terrestrial Water Storage Estimates. *Water Resour. Res.* **2012**, *48*, 1–11. [CrossRef]
43. Landerer, F. GFZ TELLUS GRACE Level-3 Monthly LAND Water-Equivalent-Thickness Surface-Mass Anomaly Release 6.0 in NetCDF/ASCII/GeoTIFF Formats. Ver. 6.0. PO.DAAC, CA, USA. Available online: Https://Doi.Org/10.5067/TELND-3AG06 (accessed on 17 December 2019).
44. Landerer, F. CSR TELLUS GRACE Level-3 Monthly LAND Water-Equivalent-Thickness Surface-Mass Anomaly Release 6.0 in NetCDF/ASCII/GeoTIFF Formats. Ver. 6.0. PO.DAAC, CA, USA. Available online: Https://Doi.Org/10.5067/TELND-3AC06 (accessed on 17 December 2019).
45. Landerer, F. JPL TELLUS GRACE Level-3 Monthly LAND Water-Equivalent-Thickness Surface-Mass Anomaly Release 6.0 in NetCDF/ASCII/GeoTIFF Formats. Ver. 6.0. PO.DAAC, CA, USA. Available online: Https://Doi.Org/10.5067/TELND-3AJ06 (accessed on 4 December 2019).
46. Rodell, M.; Beaudoing, H. *GLDAS CLM Land Surface Model L4 Monthly 1.0 × 1.0 Degree V001*; Goddard Earth Sciences Data and Information Services Center: Greenbelt, MD, USA, 2007.
47. Müller Schmied, H.; Cáceres, D.; Eisner, S.; Flörke, M.; Herbert, C.; Niemann, C.; Peiris, T.A.; Popat, E.; Portmann, F.T.; Reinecke, R.; et al. The Global Water Resources and Use Model WaterGAP v2.2d: Model Description and Evaluation. *Geosci. Model Dev. Discuss.* **2021**, *14*, 1037–1079. [CrossRef]
48. Feng, W.; Shum, C.K.; Zhong, M.; Pan, Y. Remote Sensing Groundwater Storage Changes in China from Satellite Gravity: An Overview. *Remote Sens.* **2018**, *10*, 674. [CrossRef]
49. Huffman, G.J.; Bolvin, D.T.; Nelkin, E.J. Integrated Multi-SatellitE Retrievals for GPM (IMERG), Version 4.4. NASA's Precipitation Processing Center. Available online: https://gpm.nasa.gov/data/policy (accessed on 6 February 2021).
50. Voss, K.A.; Famiglietti, J.S.; Lo, M.; De Linage, C.; Rodell, M.; Swenson, S.C. Groundwater Depletion in the Middle East from GRACE with Implications for Transboundary Water Management in the Tigris-Euphrates-Western Iran Region. *Water Resour. Res.* **2013**, *49*, 904–914. [CrossRef]
51. Frappart, F.; Ramillien, G. Monitoring Groundwater Storage Changes Using the Gravity Recovery and Climate Experiment (GRACE) Satellite Mission: A Review. *Remote Sens.* **2018**, *10*, 829. [CrossRef]
52. Yin, W.; Hu, L.; Jiao, J.J. Evaluation of Groundwater Storage Variations in Northern China Using GRACE Data. *Geofluids* **2017**, *2017*, 8254824. [CrossRef]
53. Cleveland, R.B.; William, S.; Cleveland, J.E.; McRAe, I.T. STL: A Seasonal-Trend Decomposition Procedure Based on Loess. *J. Off. Stat.* **1990**, *6*, 3–73.
54. Shamsudduha, M.; Chandler, R.E.; Taylor, R.G.; Ahmed, K.M. Recent Trends in Groundwater Levels in a Highly Seasonal Hydrological System: The Ganges-Brahmaputra-Meghna Delta. *Hydrol. Earth Syst. Sci.* **2009**, *13*, 2373–2385. [CrossRef]
55. Buma, W.G.; Lee, S.I.; Seo, J.Y. Hydrological Evaluation of Lake Chad Basin Using Space Borne and Hydrological Model Observations. *Water* **2016**, *8*, 205. [CrossRef]
56. Taher, M.R.; Chornack, M.P.; Mack, T.J. *Groundwater Levels in the Kabul Basin, Afghanistan, 2004–2013*; U.S. Department of the Interior, U.S. Geological Survey: Reston, VA, USA, 2014.

57. Yu, W.; Yang, Y.-C.; Savitsky, A.; Alford, D.; Brown, C.; Wescoat, J.; Debowicz, D.; Robinson, S. *The Indus Basin of Pakistan*; The World Bank: Washington, DC, USA, 2013.
58. Federal Flood Commission, Ministry of Water Resources. Annual Flood Report 2017. 2017. Available online: https://mowr.gov.pk/SiteImage/Misc/files/2017%20Annual%20Flood%20Report%20of%20FFC.pdf (accessed on 20 February 2020).
59. Muhammad, S.; Tian, L.; Khan, A. Early Twenty-First Century Glacier Mass Losses in the Indus Basin Constrained by Density Assumptions. *J. Hydrol.* **2019**, *574*, 467–475. [CrossRef]
60. Basharat, M.; Ali, S.U.; Azhar, A.H. Spatial Variation in Irrigation Demand and Supply across Canal Commands in Punjab: A Real Integrated Water Resources Management Challenge. *Water Policy* **2014**, *16*, 397–421. [CrossRef]
61. Ali, S.; Cheema, M.J.M.; Waqas, M.M.; Waseem, M.; Awan, U.K.; Khaliq, T. Changes in Snow Cover Dynamics over the Indus Basin: Evidences from 2008 to 2018 MODIS NDSI Trends Analysis. *Remote Sens.* **2020**, *12*, 2782. [CrossRef]
62. Qureshi, A.S.; Shah, T.; Akhtar, M. *The Groundwater Economy of Pakistan*; IWMI: Colombo, Sri Lanka, 2003; p. 31.
63. Qureshi, A.S. Groundwater Governance in Pakistan: From Colossal Development to Neglected Management. *Water* **2020**, *12*, 3017. [CrossRef]
64. Mishra, V.; Asoka, A.; Vatta, K.; Lall, U. Groundwater Depletion and Associated $CO_2$ Emissions in India. *Earth's Futur.* **2018**, *6*, 1672–1681. [CrossRef]
65. NASA NASA Satellites Unlock Secret to Northern India's Vanishing Water. Available online: https://www.nasa.gov/topics/earth/features/india_water.html (accessed on 3 February 2021).
66. Chen, J.; Famigliett, J.S.; Scanlon, B.R.; Rodell, M. Groundwater Storage Changes: Present Status from GRACE Observations. *Surv. Geophys.* **2016**, *37*, 397–417. [CrossRef]
67. Baweja, S.; Aggarwal, R.; Brar, M. Groundwater Depletion in Punjab, India. *Encycl. Soil Sci. Third Ed.* **2017**, 1–5. [CrossRef]
68. Shah, T.; Roy, A.D.; Qureshi, A.S.; Wang, J. Sustaining Asia's Groundwater Boom: An Overview of Issues and Evidence. *Nat. Resour. Forum* **2003**, *27*, 130–141. [CrossRef]
69. Asoka, A.; Gleeson, T.; Wada, Y.; Mishra, V. Relative Contribution of Monsoon Precipitation and Pumping to Changes in Groundwater Storage in India. *Nat. Geosci.* **2017**, *10*, 109–117. [CrossRef]
70. Singh, A.P.; Bhakar, P. Development of Groundwater Sustainability Index: A Case Study of Western Arid Region of Rajasthan, India. *Environ. Dev. Sustain.* **2020**, *23*, 1844–1868. [CrossRef]
71. van Steenbergen, F.; Kaisarani, A.B.; Khan, N.U.; Gohar, M.S. A Case of Groundwater Depletion in Balochistan, Pakistan: Enter into the Void. *J. Hydrol. Reg. Stud.* **2015**, *4*, 36–47. [CrossRef]
72. Yin, W.; Li, T.; Zheng, W.; Hu, L.; Han, S.C.; Tangdamrongsub, N.; Šprlák, M.; Huang, Z. Improving Regional Groundwater Storage Estimates from GRACE and Global Hydrological Models over Tasmania, Australia. *Hydrogeol. J.* **2020**, *28*, 1809–1825. [CrossRef]
73. Feng, T.; Shen, Y.; Chen, Q.; Wang, F.; Zhang, X. Groundwater Storage Change and Driving Factor Analysis in North China Using Independent Component Decomposition. *J. Hydrol.* **2022**, *609*, 127708. [CrossRef]
74. Alley, W.M.; Konikow, L.F. Bringing GRACE Down to Earth. *Groundwater* **2015**, *53*, 826–829. [CrossRef]

water

MDPI

*Article*

# Impact of Boundary Conditions Dynamics on Groundwater Budget in the Campania Region (Italy)

**Mattia Gaiolini [1], Nicolò Colombani [1,*], Gianluigi Busico [2], Fabrizio Rama [3] and Micòl Mastrocicco [2]**

1 Department of Material, Environmental Sciences and Urban Planning, Polytechnic University of Marche, Via Brecce Bianche 12, 60131 Ancona, Italy
2 Department of Environmental, Biological and Pharmaceutical Sciences and Technologies, Second University of Naples, Via Vivaldi 43, 81100 Caserta, Italy
3 Jealott's Hill International Research Centre, Syngenta Ltd., Warfield, Bracknell RG42 6EY, UK
* Correspondence: n.colombani@univpm.it; Tel.: +39-071-220-4710

**Abstract:** Groundwater budgets and fluxes are affected by human activities and climate change. Numerical models are cost-effective tools to investigate the different components of the hydrologic cycle. In this study, a groundwater flow model of the unconfined aquifers of the Campania region (Italy) has been developed and calibrated in Processing Modflow 11, resulting in an accurate assessment of groundwater fluxes and their trends over fifteen years (2000–2015). The model was implemented using a high-resolution grid to capture small hydrogeological features such as wells and rivers and informed by time variable datasets used as boundary conditions (i.e., river and sea levels, aquifer recharge, evapotranspiration, and discharge from adjacent systems). Good calibration and validation performances were achieved for piezometric heads ($R^2$ = 0.958). A set of scenarios was developed using constant boundary conditions (i.e., constant sea-level BC, uniform extinction depth BC), and the outputs were compared, quantitively assessing differences in groundwater fluxes. Simulations pointed out that using time series to inform boundary conditions in the model does not always result in a significant change in the computed fluxes. Overall, non-uniform extinction depth was the most influential condition, while both rivers and sea level conditions barely affected groundwater budgets. In addition, results highlighted the need for an accurate estimation of spatiotemporal variations of both recharge and evapotranspiration, due to their strong seasonal variability and their massive contribution to the hydrogeological cycle. Finally, a marked increase of evapotranspiration fluxes controlled by interannual variability of precipitation and atmospheric temperatures has been quantified over the modelled period.

**Keywords:** flow model; groundwater budget; remote sensing; recharge; evapotranspiration

## 1. Introduction

The Mediterranean region, where exceptional levels of biodiversity occur [1], has been identified as one of the most prominent "Hot-Spots" in actual and future climate change (CC) projections [2]. Climatic disturbance in terms of increasing temperatures [3,4], global hydrological cycle disturbance [5], and sea-level rise [6–8] could greatly affect both groundwater systems and groundwater-dependent ecosystems. Indeed, groundwater inputs support or compensate for surface water inputs and play a vital role in maintaining healthy ecosystems [9]. Since CC has a significant effect on the hydrological cycle [10], is mandatory for water managers to plan appropriate measures to avoid negative potential impacts on water resources [11]. Most of the studies, published so far focused on the above-ground components of the hydrological cycle, analysing both historical and projected changes, while the scientific knowledge on the subsurface components, such as recharge, groundwater levels, aquifer fluxes, and groundwater quality have been only recently emerging in literature [12–14]. The main tools that the research community has developed to face these issues are observation networks, which allow to record the hydraulic parameters, and mathematical/physical models which simulate their evolution in time and space. Indeed, the implementation of numerical groundwater flow models is crucial to investigating groundwater-related issues by evaluating the trends of groundwater fluxes over a simulation time and thus quantifying the ongoing CC effects on the subsurface hydrological cycle [15]. Since its release in 1988, MODFLOW has become the worldwide standard for groundwater modelling because of its flexible modular structure, complete coverage of hydrogeological processes, and free availability [16]. MODFLOW can also be integrated with a geographic information system (GIS) to provide a visual environment for groundwater resources evaluation and management [17]. Application of groundwater flow models to large domains started a long time ago, in 1978, with the Regional Aquifer System Analysis (RASA) program of the U.S. Geological Survey [18], but only a few studies, so far, aimed to numerically evaluate changes in model outputs derived from constant or variable boundary conditions (BCs).

Coelho et al. [19] developed a study focusing on the influence of different BCs on numerical simulations of an unconfined aquifer in Brazil to define the most suitable BC, without considering scenarios with constant or variable BCs. Bastani and Harter [20] developed a three-dimensional transient flow and nitrate-transport model, comparing reference results to temporally upscaled models and considering annual-averaged flow and steady-state stresses across different scenarios. However, that study only focused on the concentrations of the nitrate discharging points without quantitively addressing changes in the groundwater budget. Moreover, employing proper spatial and temporal scales for both flow and transport simulations remains one of the most important challenges in numerical modelling [21–23]. One of the main tasks in transient groundwater modelling is the need for large amounts of hydrogeological data for both model implementation and calibration/validation; those constraints have often limited the use of transient numerical flow models to well-characterized field sites, while transient large-scale models are still rare [24]. Despite this, transient groundwater flow modelling has been recognized as one of the hydrological research fields extremely relevant for scientific progress [25]. In recent years, open datasets on hydrology and hydrogeology have been released by many research groups and international organizations, but their widespread use in numerical models still faces problems such as the lack of universally accepted guidelines or insufficient resolution [26]. To explore the new opportunities (and limitations) offered by open datasets to represent variable BCs in large-scale groundwater modelling, the unconfined aquifers of the Campania region (Italy) were selected as representative of potentially vulnerable to CC groundwater bodies. In fact, newly released open datasets on hydrology and hydrogeology can theoretically increase the model's capability to quantify the variations of groundwater fluxes induced by spatially and temporally variable BCs. Over the last few years, several studies have focused on developing conceptual and numerical models of the Campania region (Italy) to tackle a series of groundwater-related issues [27–30], but without investigating the impact of BCs variability on the model outputs. This study aimed to fill this gap, assessing the influence of different implementation of BCs (e.g., constant versus variable sea-level BC, uniform versus variable extinction depth BC, etc.) over the computed groundwater flow budget from 2000 to 2015. This kind of numerical analysis would also help in reducing both human and computational efforts in similar settings, by inferring what datasets are worth being simplified (e.g., averaged over time and space) and which need to be treated as spatially variable and transient over time.

## 2. Materials and Methods

### 2.1. The Study Area

The Campania region, located in the southern part of the Italian Peninsula, covers an area of about km$^2$. The region embraces a wide part of the Apennines chain, including composite geological and geomorphological units. The western slope of the Apennines is characterized by graben-like coastal depressions [31] and the plains alongside the coastal strip result from the aggradation of such structural depressions, whose bed is several thousand meters lower than the Apennine massifs [32]. The lithological and structural complexity of the region controls the hydrogeological fluxes.

The domain extends from the Tyrrhenian Sea to approximately 50 km inland, delimited in the South by the Cilento mountain range (Figure 1). The study area includes different alluvial plains (e.g., Garigliano, Volturno, Sarno, and Sele plains), characterized by debris and alluvial deposits. An important part is also occupied by limestone (e.g., M. Massico and Lattari Mts.) and volcanic units caused by the aggregation of volcanic material released during historical eruptions in the area (i.e., Vesuvio, Phlegraean Fields, and Roccamonfina volcanos are located within the model domain). In addition, the Volturno plain is characterized by a vast lowland area, subjected to land reclamation since the seventeenth century, which promoted the development of agriculture and farming along with strong coastal urbanization [33]. For toponyms used in the text refer to Figure S1.

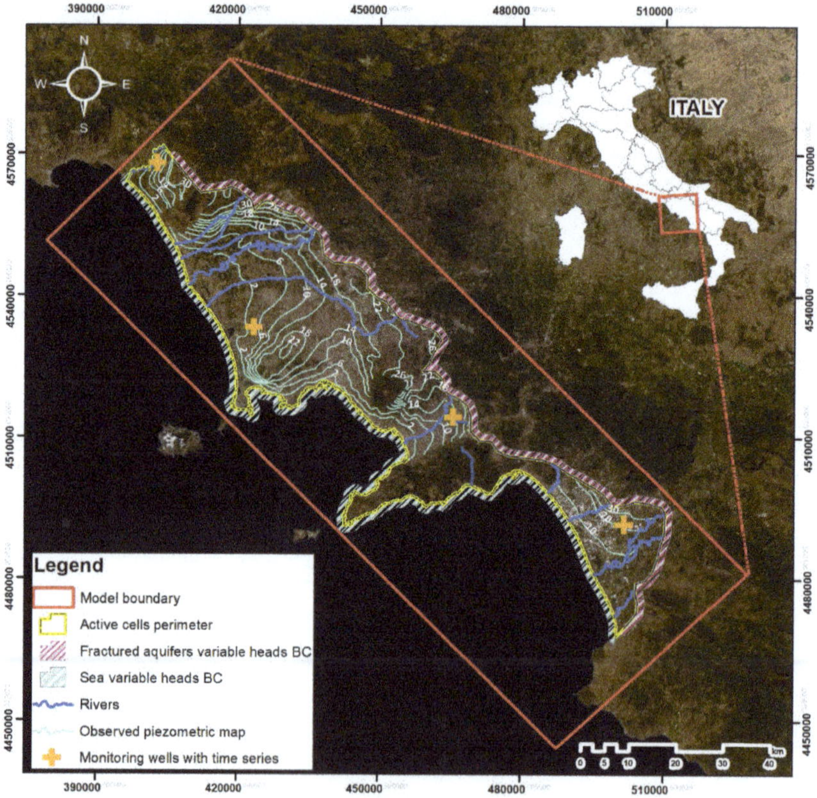


**Figure 1.** Model location with initial heads and boundary conditions (BCs).

## 2.2. Field Characterization and Boundary Conditions

Before the implementation of the groundwater flow model, all available datasets have been collected to define the conceptual model of the study area and the BCs (Table 1).

**Table 1.** National and worldwide open online datasets used to set up the model.

| Dataset | Parameter | Repository | URL |
|---|---|---|---|
| Soil | Grain size distribution | ISRIC | https://maps.isric.org/ accessed on 17 July 2022 |
| Geology | Aquifer thickness | ISPRA | http://sgi2.isprambiente.it/viewersgi2 accessed on 17 July 2022 |
| Hydrogeology | Hydraulic conductivity | ISPRA | http://sgi2.isprambiente.it/viewersgi2 accessed on 17 July 2022 |
| CORINE Land Cover | Land use | COPERNICUS | https://land.copernicus.eu/pan-european/corine-land-cover accessed on 17 July 2022 |
| SEALEVELEURP HYL4MY008068 | Sea levels | CMEMS | https://doi.org/10.48670/moi-00141 accessed on 13 June 2022 |
| Hydrometric station dependent | Rivers stages | CFM | http://centrofunzionale.regione.campania.it/#/pages/sensori/archivio-idrometrici accessed on 17 July 2022 |
| MYD16A3GF | Evapotranspiration | MODIS | https://doi.org/10.5067/MODIS/MOD16A2GF.006 accessed on 10 May 2022 |
| PCRGLOBWB 2 | Recharge | UU-HYDRO | https://github.com/UU-Hydro/PCR-GLOBWB_model accessed on 12 June 2022 |
| BIGBANG40 | Evapotranspiration | ISPRA | https://groupware.sinanet.isprambiente.it/bigbang-data/library/bigbang40/grids/it_evapotraspirazione_ae accessed on 17 July 2022 |
| BIGBANG40 | Recharge | ISPRA | https://groupware.sinanet.isprambiente.it/bigbang-data/library/bigbang40/grids/it_ricarica_degli_acquiferi_gr accessed on 17 July 2022 |

The regional SRTM Digital Terrain Model with a spatial resolution of 20 × 20 m cells was used to retrieve topographic information and represent morphology. The data files contain the elevation in a digital format, averaged on the model grid. Vegetation, buildings, and other cultural features were removed digitally, resulting in a proper terrain model.

The hydraulic conductivity of lithological units was obtained from permeability tests, by mining data from 91 borehole reports available on the online database of Instituto Superiore per la Protezione e Ricerca Ambientale [34]. All the data about pumping well rates, effective radius, screen length, and groundwater drawdown were retrieved from the same reports. The depth of the unconfined aquifer for each borehole was collected along with locations (latitude, longitude, and UTM coordinates). Other information on the aquifer's depth has been retrieved from previous studies carried out in the Campania region, such as logs on the Volturno river plain [35] or aquifer information in Cuma [36,37], Bagnoli, Giuliano, and San Valentino Torio.

Measurments on sea levels were downloaded from the open dataset Servizio Mareografico Nazionale (SMN). SMN exploits a network of 36 measuring stations uniformly distributed over the national territory and located mainly inside the port facilities. The hydrometric level is monitored since 2010 using a new microwave (radar) level meter with millimetric precision, coupled with a second level meter with a floater. Each level is referred to the absolute altimetric network of Istituto Geografico Militare (IGM). To gain a better understanding of the sea level behaviour across the entire region, Napoli, Salerno, Gaeta, and Palinuro stations were considered. However, since SMN open dataset does not cover the entire period of simulation, the datasets of the Copernicus Marine Environment Monitoring Service (CMEMS) were also exploited for sea-level trends at the same locations. CMEMS altimetric satellite data were elaborated together with the SMN ones allowing to cover the entire period of study.

The Moderate Resolution Imaging Spectroradiometer (MODIS) products have been deployed to obtain the trends of evapotranspiration in the study area for the period 2000–2015. Data were downloaded using the AppEEARS [38] interface and the MYD16A3GF model [39] to obtain spatially interpolated values. The algorithm used in MODIS is based on the Penman-Monteith equation, which includes daily meteorological data along with remotely sensed variables, such as vegetation dynamics, albedo, and land cover [40]. Land use and soil texture maps (Table 1) were overlapped via GIS to derive a two-dimensional matrix representing the evapotranspiration extinction depth [41].

Time series of groundwater recharge has been extracted from the outputs of PCR-GLOBWB v2.0 (Utrecht University, Utrecht, The Netherland) [42], a grid-based global hydrology and water resources model. Global hydrological models (GHMs) are tools able to quantify the global terrestrial water cycle, simulating the distributed hydrological response to weather and climate variations at high resolution.

Hydrometric levels were downloaded from the Centro Funzionale Multirischi (CFM), a Civil Protection Agency of the Campania region, to represent the variation of the river stage over the simulation period. CFM provides a list of all hydrometric stations of the region recording data with a daily frequency. The closest stations to the model's boundaries were considered representative of the baseflow entering the domain from the fractured aquifers. Hydraulic head fluctuation in the fractured aquifers was modified according to the average baseflow variations over the simulation time.

### 2.3. Numerical Modelling

To estimate the groundwater flow budget at a regional scale, assessing the influence of variable BCs, all the data were elaborated in a GIS environment and processed via Processing Modflow 11. The flow simulation was carried out using the MODFLOW-2005 code [43]. The equation that stands at the base of MODFLOW-2005 is the partial-differential equation that describes the three-dimensional movement of groundwater of constant density through a porous matrix:

$$\frac{\partial}{\partial x}\left(K_{xx}\frac{\partial h}{\partial x}\right)+\frac{\partial}{\partial y}\left(K_{yy}\frac{\partial h}{\partial y}\right)+\frac{\partial}{\partial z}\left(K_{zz}\frac{\partial h}{\partial z}\right)+W=S_s\frac{\partial h}{\partial t} \quad (1)$$

where $K_{xx}$, $K_{yy}$ and $K_{zz}$ are values of hydraulic conductivity along x, y and z coordinate axes ($LT^{-1}$); h is the potentiometric head (L); W is a volumetric flux per unit volume representing sources and/or sinks of water ($T^{-1}$); $S_s$ is the specific storage of the porous material ($L^{-1}$); and t is time (T). The standard version of MODFLOW-2005 treating the fractured units as equivalent porous media and not complex ones such as the Conduit Flow Package [44] was used given the lack of information on the location and dimensions of such zones. Thus, for the principle of parsimony in numerical modelling, a simpler representation was here adopted [45].

The model domain consisted of 656 rows, 335 columns, and 3 layers with variable width and thickness. The top of the grid represents the local ground surface, ranging from −3.49 m to 1642 m above sea level (a.s.l.). The bottom of the grid represents the bottom of the unconfined aquifers and ranges from −83 m below sea level (b.s.l.) to 144 m a.s.l.

Initially, a steady-state simulation has been run and the model has been calibrated via PEST [46] against an observed piezometric map of the Campania region for the year 2000 [47]. Given that the piezometric map was available only as a vector map, it was sampled along the contour lines to create 500 observation points that were compared versus calculated ones with an estimated error of ±1.0 m. Tyrrhenian Sea levels have been initially set to 0 m as constant head cells. Values of hydraulic conductivity, corresponding to boreholes in the same cluster, have been averaged to get an average value for each geological formation in the study area. In this way, the model domain was subdivided into polygons whose extension corresponds to the extension of the different geological formations reported in the geological map. Figure 2 highlights the heterogeneity of the Campania region, where the horizontal hydraulic conductivity values range from $1 \times 10^{-6}$ m/s in the coastal part of the Volturno plain, where clay layers are present, to $6.6 \times 10^{-4}$ m/s in the Vesuvio area, where highly fractured volcanic rocks are present. The value of the vertical hydraulic conductivity was set equal to the value of the horizontal hydraulic conductivity assigned to that cell in fractured units, while it was set as a tenth of the horizontal hydraulic conductivity in sedimentary units, to account for internal layering. It must be stressed, however, that local scale heterogeneities such as intensively faulted zones, small lenses, etc. have not been explicitly incorporated in this large-scale model.

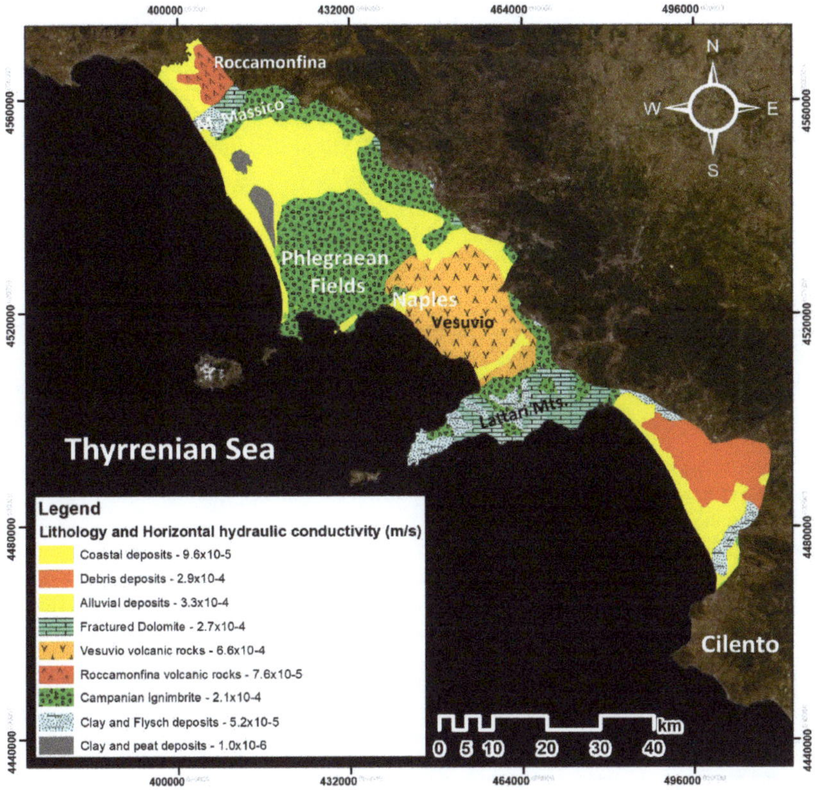


**Figure 2.** Distribution of the lithological units and their horizontal hydraulic conductivity values in layer 1 of the model domain.

For the steady-state simulation, evapotranspiration rates for the year 2000 were averaged from the MODIS dataset and interpolated on the numerical grid; while the recharge rates were averaged from the PCR-GLOBWB v2.0 for the same period and interpolated on the numerical grid. The Processing Modflow polyline input method in the River Package was employed to simulate, from North to South, Garigliano, Savone, Agnena, Volturno, Regi Lagni, Sarno, Picentino, Tusciano, Sele, and Calore rivers, drawing their actual pathways and defining river parameters. The riverbed conductance per unit length $C_{rivl}$ ($LT^{-1}$), the river stage $h_{riv}$ (L), and the elevation of the riverbed bottom $B_{riv}$ (L) have been assigned to the most upstream and downstream vertices of each river polyline and then applied to the model cells employing a linear interpolation. River stages were assumed close to the assigned starting head value and kept constant for the entire steady-state simulation. The flow rate $Q_{riv}$ ($L^3T^{-1}$) between each river reach and the model cell is calculated by the following equations:

$$Q_{riv} = \begin{cases} C_{riv} \cdot (h_{riv} - h), & \text{if } h > B_{riv} \\ C_{riv} \cdot (h_{riv} - B_{riv}), & \text{if } h < B_{riv} \end{cases} \tag{2}$$

where h (L) is the calculated hydraulic head in the cell containing the reach.

The wells analysed from the online ISPRA database were entered into the model through the Well Package. Industrial wells were set with continuous pumping rates, while the agricultural wells were set with discontinuous pumping rates, active only during the spring and summer seasons. Since the specified wells penetrate more than one model layer, the total pumping rate has not been distributed over the penetrated layers, but it has been assigned to the lowest cell of the well, belonging to the third model layer. The Drain Package was used to simulate the reclamation channels in the Marina di Licola area and also to simulate the effect of the underground metro lines in Naples, the mean height of each feature a.s.l. was derived from the topographical online database, while the cell conductance $C_d$ ($L^2T^{-1}$) was set to a very high value (0.01) to ensure optimal groundwater withdrawal from the aquifer. The location of the depth of the different metro lines was sourced from Russo et al. [48]. The discharge rate $Q_d$ ($L^3T^{-1}$) from the model cell to the drain is calculated by:

$$Q_d = \begin{cases} C_d \cdot (h - h_d), & \text{if } h > h_d \\ 0, & \text{if } h \leq h_d \end{cases} \quad (3)$$

The horizontal hydraulic conductivity was estimated by inverse modelling using the numerical code PEST, assigning a parameter number to it. PEST searches optimum parameter values for which the sum of squared deviations between model-calculated and observed values from the piezometric map of the region is reduced to a minimum. Such comparison has been performed using a scatter diagram of the simulated and observed data (Figure 3).

To investigate possible changes in the groundwater system associated with temporal variations of evapotranspiration and recharge patterns, a transient simulation was carried out. Sixty-four stress periods of seasonal length (3 months), subdivided into three time steps of 1 month each, were used. As recommended by Harbaugh [43], the starting hydraulic heads for the transient simulation have been set equal to the head values calculated at the end of the steady-state simulation. Table 2 shows the specific storage ($S_s$) and specific yield ($S_y$) values, depending on the geologic material, together with the riverbed conductance ($C_{riv}$) gained after the calibration procedure.

**Table 2.** $S_s$ and $S_y$ depending on the geologic material (upper panel), and $C_{riv}$ gained after the calibration procedure (lower panel).

| **Material** | **$S_s$ ($m^{-1}$)** | **$S_y$ (-)** |
|---|---|---|
| Clayey and calcareous clayey units | 0.001 | 0.2 |
| Debris, alluvial, fluviolacusters and fluvioglacial deposits | $1 \times 10^{-4}$ | 0.33 |
| Lavas, ignibrites, pyroclastics | $1 \times 10^{-5}$ | 0.21 |
| Organogenic, neritic, and platform limestone | $1 \times 10^{-5}$ | 0.14 |
| **River** | **$C_{riv}$ ($m^2/s$)** | |
| Garigliano | from 0.021 to 0.235 | |
| Savone | from 0.017 to 0.058 | |
| Agnena | from 0.001 to 0.033 | |
| Volturno | from 0.009 to 0.044 | |
| Regi Lagni | from 0.007 to 0.072 | |
| Sarno | from 0.007 to 0.040 | |
| Picentino | from 0.009 to 0.011 | |
| Tusciano | from 0.011 to 0.019 | |
| Sele | from 0.014 to 0.116 | |
| Calore | from 0.060 to 0.086 | |

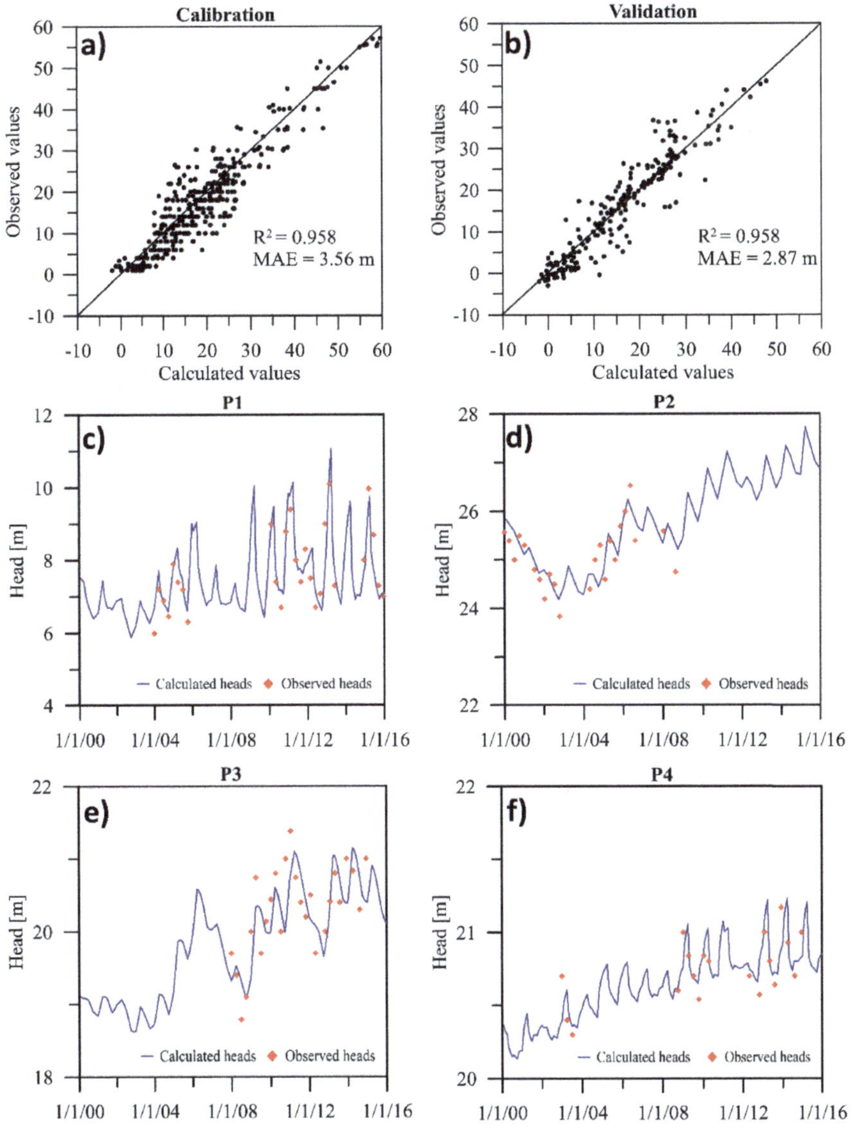


**Figure 3.** Head Scatter Diagrams of the calibration (**a panel**) and validation processes (**b panel**), and comparison between calculated and observed heads at piezometers P1, P2, P3 and P4 (**c–f panels**), represented in Figure 1.

To consider the variability of the evapotranspiration over time, the MODIS data have been spatialized using the kriging technique and inserted into the model with seasonal frequency. The values of recharge rate from the global numerical hydrologic model published by de Graaf et al. [49] have been processed in python to obtain spatially interpolated values for the period 2000–2015, which have been assigned to the model with seasonal frequency. The calibrated model was implemented with the before-mentioned transient datasets accounting for the seasonal changes in hydrometric levels, Tyrrhenian Sea levels, and fractured aquifer seasonal head fluctuations as variable BCs. The Evapotranspiration

Package was also integrated with spatially variable extinction depths dependent on the land use and soil texture maps.

The validation of the transient state model was performed by exploiting a piezometric map of the Campania region for the year 2013, which was derived from a secondary elaboration of the global groundwater depth dataset [50] to match modelled and measured hydraulic heads. Besides, selected monitoring wells (Figure 1) with piezometric trend data were compared versus model results.

Both the model calibration and validation were assessed by calculating and comparing the value of the coefficient of determination ($R^2$) and the mean absolute error (MAE).

The Processing Modflow ZoneBudget routine [43] can produce water budgets for user-defined zones within the model domain. Zones are defined by assigning zone numbers to model cells, allowing for to calculation of the volumetric water budget in a different part of the domain, accounting for evapotranspiration, fractured aquifers, rivers, and the Tyrrhenian Sea component separately. Each trend calculated by the model, considering all the transient input datasets as variable BCs (transient model), was compared with the respective trend calculated by an upscaled model considering a uniform extinction depth, a constant head at the fractured aquifers, and constant sea and river levels (basic model). Moreover, to investigate which datasets are worth being averaged, a set of scenarios have been developed applying to the transient model each of the averaged BCs individually:

- constant head at the fractured aquifers BC (scenario S1);
- uniform extinction depth BC (scenario S2);
- constant Tyrrhenian Sea level BC (scenario S3);
- constant river levels BC (scenario S4).

In addition, to numerically quantify the impact of the averaged conditions, the 15-year average flux of each groundwater budget component has been estimated and differences between basic, transient, S1, S2, S3, and S4 simulations were graphically compared.

Finally, since the pivotal contribution to the groundwater budget of both recharge and evapotranspiration, an additional recharge and evapotranspiration rates dataset has been investigated. Alternatively, to the PCR-GLOBWB v.2.0 published by Sutanudjaja et al. and the MODIS products, the BIGBANG (*Bilancio Idrogeologico GIS BAsed a scala Nazionale su Griglia regolare*) [51] distributed model was employed, and results compared.

## 3. Results and Discussion

### 3.1. Flow Model Performance

Given the extremely challenging regional scale of the domain (3672 $km^2$), reasonably good calibration and validation were achieved for hydraulic heads, resulting in an $R^2$ of 0.958 (Figure 3). MAE values of 3.56 m for calibration and 2.87 m for validation, were judged acceptable in the wide range of the simulated piezometric levels (>100 m). Observations of groundwater levels from selected monitoring wells were used to verify the ability of the model to describe water table fluctuations in the unconfined aquifers (Figure 3). Unfortunately, a widespread monitoring network of wells is not available in the region, and this is a limitation for the model performance. Indeed, monitoring groundwater levels at regional scale may enable a better assessment of groundwater depletion, as well as inferring changes in flow directions over time, as recently pointed out by Brakkee et al. [52]. Moreover, piezometric data are missing in some karst and fractured areas of the domain (Massico mountain, Vesuvio volcano, and Monti Lattari) consequently model reliability is lower on those areas than in alluvial zones. By means of the Processing Modflow ZoneBudget routine [43] a defined zone number was assigned to those areas and their importance in relation to the overall model domain was assessed in terms of water budget component percentages. As expected, groundwater recharge played a very important role in karst environments, representing 12.7% of the overall domain, while evapotranspiration was just 1.5% with respect to the total, given that in mountain areas the groundwater depth from the surface is higher than the extinction depth. The storage capacity turns out to be consistent, whereas all the other budget components resulted to be negligible. The general high permeability grade due to fracturing leads to a high groundwater recharge rate [53] that varies between 48% and 78% of the mean annual effective precipitation [54,55], depending on the relative abundance of limestone and dolomites lithologies.

The validated transient model allowed analysing and quantifying relevant stresses in different portions of the domain at different times. Figure 4 shows an example of both the three-dimensional and planar view of the calculated hydraulic heads map at the end of the simulation. It pointed out regions of the domain characterized by high hydraulic gradient, especially the northern part in correspondence of the Roccamonfina Volcano, which acted as a groundwater mounding area, as well as the Vesuvio, Phlegraean Fields, and Monti Lattari. Hydraulic heads resulted to be very low in the Volturno river plain, following the topographic gradient. Near the coast, the water level was very shallow, due to reclamation channels located in the Marina di Licola area (lower Volturno plain).

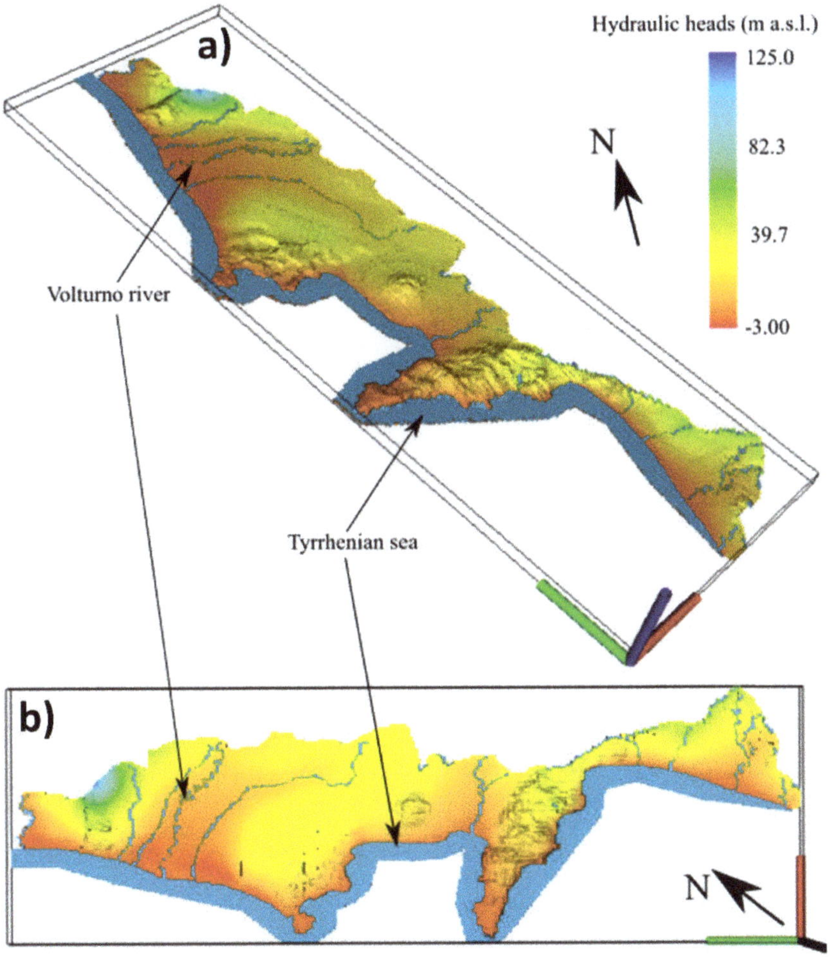


**Figure 4.** Three-dimensional sketch (**a panel**) and plan view (**b panel**) of the model calculated hydraulic heads map at the end of the simulation time.

In addition, the model produced accurate estimates of the groundwater budget with a mean error of less than 0.3% among in and out components. This allowed focusing on the single water budget components, whose fluxes are difficult to estimate without a physically based model even though they are crucial for better management of the hydrogeological basins.

*3.2. Water Fluxes Trends and Comparative Analysis*

The fluxes trends have been estimated accounting for the inflow from the fractured aquifers, the evapotranspiration, the river-groundwater interactions, and the Tyrrhenian Sea component, separately. To highlight the influence of the time variable datasets, the basic model calculated trends (red line), the transient model calculated trends (blue line), and the scenarios (S1, S2, etc.) calculated trends (dashed green line) associated with the budget component have been plotted together.

Figure 5a displays the inflow component from the fractured aquifers. The transient model calculated trend shows a quite higher flux with a wider seasonal variation throughout the simulation time, with respect to the basic and S1 model.

The evapotranspiration component represents the most important outflow contribution. Figure 5b clearly shows its seasonal variability with increasing fluxes over the warm seasons, reaching values up to 20 $m^3/s$. Moreover, since the extinction depth affects the evapotranspiration fluxes, the S2 trend perfectly reflects the basic model, underlying the importance of considering a non-uniform extinction depth over the domain to account for different land cover and different soil types that directly drive the extinction depth [41].

The seawater intrusion component (Figure 5c) is characterized by a very low flux and can be considered negligible with respect to submarine groundwater discharge. However, importing a transient sea-level dataset leads the seawater intrusion flux of the transient model to display positive peaks due to the seasonal oscillation of the Tyrrhenian Sea. This could be further investigated at the local scale by performing a variable-density flow and transport simulation. There are no evident changes between the basic, transient, and S3 scenarios in submarine groundwater discharge, which shows a mean flux over the simulation time between 5 and 6 $m^3/s$ (Figure 5d).

Both river inflow and outflow (Figure 5e,f), which represent the losing and gaining flow rate from and to the river, show no significant changes between basic, transient, and S4 simulations. Therefore, considering the effort to import all required data in the models, a constant river level dataset averaged over the simulation time should be considered for large-scale simulations.

To numerically quantify the impact of time variable input datasets on groundwater budget, the 15-year average flux of each component has been estimated and differences between basic, transient, S1, S2, S3, and S4 simulations were represented in Figure 6. A difference of ~1.67 $m^3/s$ was found in the mean evapotranspiration between the basic and the transient simulations. Almost the same difference (~1.70 $m^3/s$) was found between S2 and the transient model. This confirmed that extinction depth is the most influencing parameter, requiring a precise spatiotemporal estimation to obtain accurate predictions of the groundwater budget.

The difference in the mean inflow from the fractured aquifers component (~0.65 $m^3/s$) between the basic and transient models did not seem to be entirely due to the associated transient inflow parameters, but also accounted for the variable river level dataset.

Even if the difference in the averaged seawater intrusion component was relatively high (almost doubled from the basic to transient model), it should be considered negligible in absolute terms (0.02 $m^3/s$). The mean submarine groundwater discharge showed no appreciable changes between the scenarios, both in relative and absolute terms. Effectively, that component seems to be more affected by the non-uniform extinction depth dataset (0.16 $m^3/s$ difference) than by the transient Tyrrhenian Sea levels (0.06 $m^3/s$ difference), suggesting that sea level could be kept constant over the simulation time.

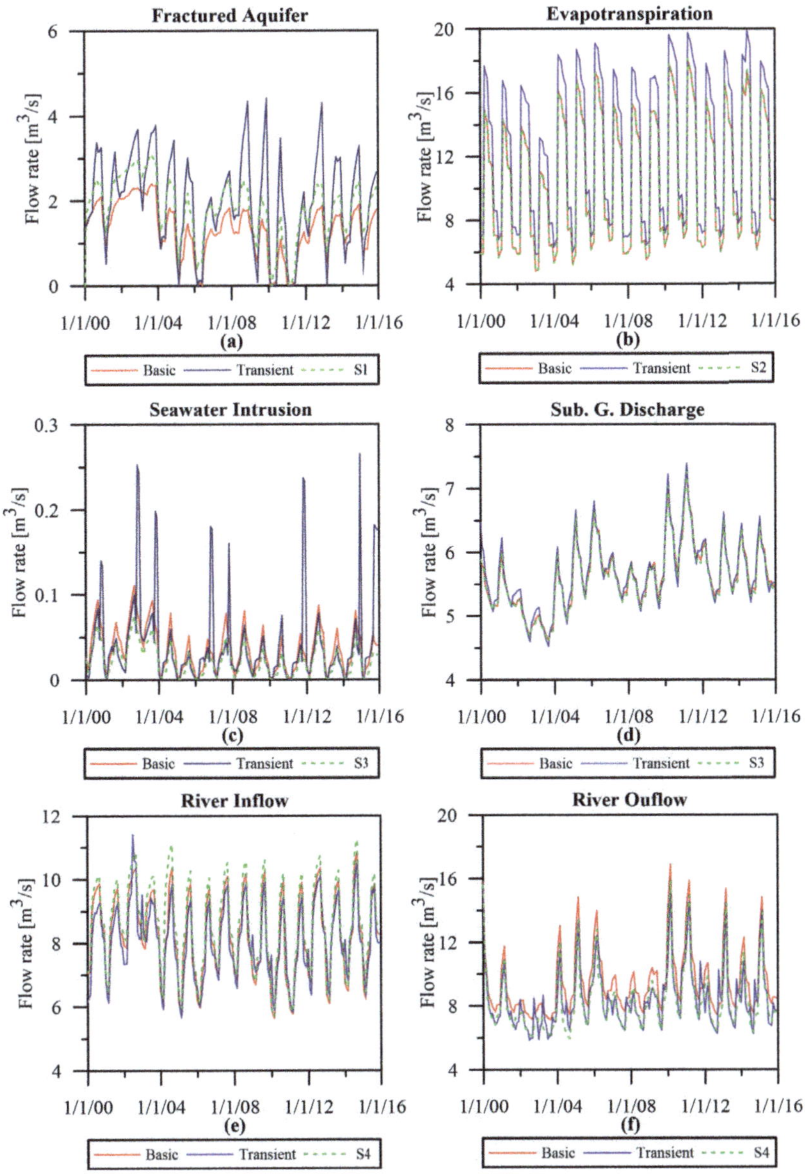


**Figure 5.** Comparison of the fluxes' trends over the simulation time among the basic (red line), the transient (blue line) and the S1, S2, S3, and S4 (green dashed lines) simulations: (**a**) Inflow from the fractured aquifer, (**b**) Evapotranspiration, (**c**,**d**) Seawater intrusion and Submarine groundwater discharge, (**e**,**f**) River inflow and outflow.

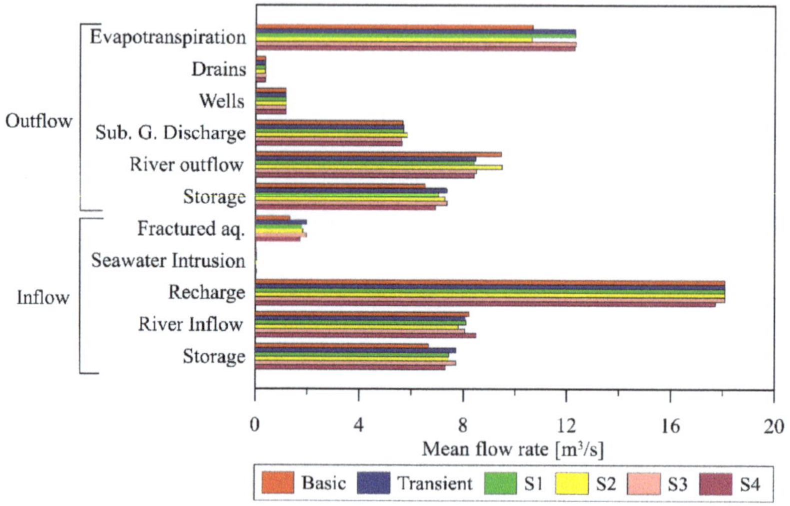


**Figure 6.** Comparison of the average flow rates and storage over the simulation time between: Basic, Transient, S1, S2, S3, and S4 simulations.

The same analysis can be conducted considering the mean groundwater flux towards the river (river outflow). The difference of 0.98 $m^3/s$ between the basic and the transient simulations was mostly due to the non-uniform extinction depth rather than to the variable river level dataset: the value rose to 1.02 $m^3/s$ between the S2 and the transient model, while it dropped to 0.05 $m^3/s$ between the S4 and the transient model. Considering also that the mean river flux towards the unconfined aquifers (river inflow) slightly differs among the various scenarios, accounting for a variable river level dataset does not seem to be crucial.

The comparative analysis pointed out the fundamental role of the extinction depth in shifting the allocation of groundwater fluxes among different compartments, altering the contribution of each component with respect to the total budget.

Moreover, since the evaluation of each groundwater contribution underlines the pivotal role of both the recharge (with a mean flux of 18.1 $m^3/s$) and evapotranspiration components, the additional BIGBANG dataset has been investigated. The derived recharge and evapotranspiration trends have been calculated and compared with the transient model results (Figure 7).

The recharge flow rates strongly differ among the two models, while the evapotranspiration fluxes were more similar. The linear temporal trends show an increase in the recharge flux over time for both the PCR-GLOBWB v2.0 and the BIGBANG products, reflecting the positive trends of the hydraulic heads observed in the monitoring wells P1, P2, P3, and P4 (Figure 3). The process of groundwater recharge is driven by a decadal variability of the North Atlantic Oscillation and its robust correlation with the rainfall patterns [56]. Both the calculated evapotranspiration fluxes, controlled by interannual variability of precipitation and air temperature, showed increasing trends with a small positive shift (~0.10 $m^3/s$ per year), highlighting that CC impact on groundwater resources is confirmed by the two independent databased here employed. Furthermore, because of the overestimation of the recharge flux due to topographic and geological heterogeneities of the Campania region, running the transient model using the BIGBANG products slightly decrease the performance parameters, with an $R^2 = 0.948$ and an MAE = 3.42 m, producing flooded areas during the peak recharge seasons that were not physically based.

Finally, it must be stressed that all other scenarios showed very little variation of the model performance indicators with an $R^2$ and an MAE that never exceeded 0.947 and 2.91 m, respectively. This is due to the low sensitivity of piezometric heads in capturing the changes induced by evapotranspiration at the basin scale [57], while to solve this complex issue, monitoring wells with continuous heads dataloggers should be employed coupled with baseflow measurements [58].

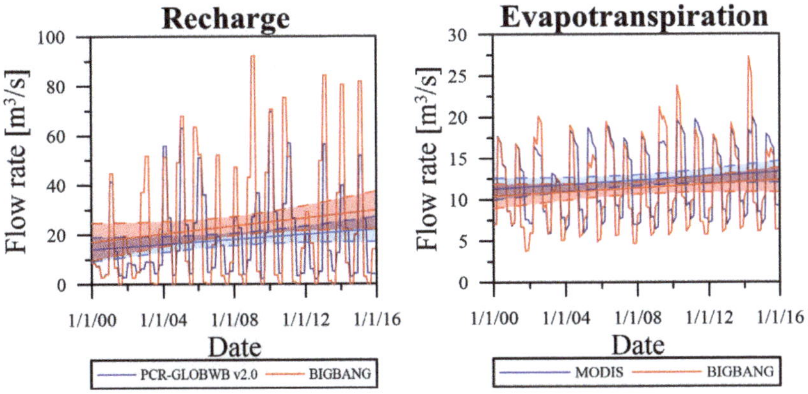


**Figure 7. Left panel**: comparison between the recharge rates obtained by PCR-GLOBWB v2.0 and BIGBANG, with their linear trends (solid lines); **Right panel**: comparison between the evapotranspiration rates obtain by MODIS products and BIGBANG, with their linear trends (solid lines). The coloured areas between the dashed lines represent the 95% confidence interval.

## 4. Conclusions

Open online datasets, along with historical observations and literature review were used to untangle the complexity of regional hydrogeological settings, supporting the implementation of a robust numerical model to describe groundwater fluxes for the unconfined aquifers of the Campania region. The quantification of the total flux of each component in the budget highlighted the pivotal role of recharge and evapotranspiration rates, with a mean flux of 18.1 and 12.3 $m^3/s$, respectively. Their variable patterns throughout the simulation period affected the whole hydrogeological budget, pointing out the need for their accurate estimation over time, to account for their high seasonal variability and climatic trends. Recharge can reach a value of above 60 $m^3/s$ during wet periods, while evapotranspiration variability ranges from 5 to maximum values up to 20 $m^3/s$ over the warm seasons. Time variable input datasets had very different effects on the components of the groundwater budget, controlling water fluxes between different compartments. Simulations pointed out that both fractured aquifers head fluctuations and non-uniform extinction depth datasets used as BCs should not be averaged. Considering transient data of extinction depth had a massive effect on the budget results, not only influencing evapotranspiration rates directly but also affecting all the other components of the groundwater budget, indirectly. Conversely, the inclusion of other datasets, namely variable river and sea levels, barely impacted the groundwater budget, not making it worth the extra effort. Therefore, the inclusion of transient BCs in a numerical model should be based on expert judgment, balancing accuracy and complexity and considering the human and computational efforts required. Finally, the results have underlined that continuous groundwater level and baseflow estimates are crucial to minimize model uncertainties and increase the reliability of groundwater budgets at the basin scale. Therefore, such measurements need to be included in an ideal groundwater monitoring network at a regional scale.

**Supplementary Materials:** The following supporting information can be downloaded at: https://www.mdpi.com/article/10.3390/w14162462/s1, Figure S1: The domain is limited by the Tyrrhenian Sea (light blue) and by the Apennines and the Cilento mountain range (light green). The area includes the Garigliano, Volturno, Sarno, and Sele plains (yellow), the volcanic units Vesuvio, Phlegraean Fields, and Roccamonfina (white) and the limestone units Mount Massico, Mount Maggiore, and Mounts Lattari (white). More in detail: the Garigliano plain hosts the Garigliano river, the Volturno plain hosts the Savone, Agnena, Volturno, and Regi Lagni rivers, the Sarno plain hosts the Sarno rivers, and the Sele plain hosts the Picentino, Tusciano, Sele and Calore rivers.

**Author Contributions:** Conceptualization, M.M. and N.C.; methodology, N.C., M.M. and M.G.; software, M.G. and N.C.; validation, M.M. and N.C.; formal analysis, M.G.; investigation, M.G.; data curation, M.G., F.R. and G.B.; writing—original draft preparation, M.G.; writing—review and editing, N.C., M.M. and F.R.; visualization, M.G. and G.B.; supervision, N.C. All authors have read and agreed to the published version of the manuscript.

**Funding:** This research received no external funding.

**Institutional Review Board Statement:** Not applicable.

**Informed Consent Statement:** Not applicable.

**Data Availability Statement:** Data will be made available by the authors upon request.

**Acknowledgments:** Simone Tavoni is acknowledged for helping to gather the raw data.

**Conflicts of Interest:** The authors declare no conflict of interest.

## References


1. Blondel, J.; Aronson, J. *Biology and Wildlife of the Mediterranean Region*, 1st ed.; Oxford University Press: Oxford, UK, 1999; p. 328.
2. Lionello, P.; Scarascia, L. The relation between climate change in the Mediterranean region and global warming. *Reg. Environ. Chang.* **2018**, *18*, 1481–1493. [CrossRef]
3. Cramer, W.; Guiot, J.; Fader, M.; Garrabou, J.; Gattuso, J.P.; Iglesias, A.; Xoplaki, E. Climate change and interconnected risks to sustainable development in the Mediterranean. *Nat. Clim. Chang.* **2018**, *8*, 972–980. [CrossRef]
4. Hallegatte, S. Strategies to adapt to an uncertain climate change. *Global. Environ. Chang.* **2009**, *19*, 240–247. [CrossRef]
5. IPCC. *Climate Change 2021: The Physical Science Basis. Contribution of Working Group I to the Sixth Assessment Report of the Intergovernmental Panel on Climate Change*; Masson-Delmotte, V.P., Zhai, A., Pirani, S.L., Connors, C., Péan, S., Berger, N., Caud, Y., Chen, L., Goldfarb, M.I., Gomis, M., et al., Eds.; Cambridge University Press: Cambridge, UK; New York, NY, USA, 2021; in press. [CrossRef]
6. FitzGerald, D.M.; Fenster, M.S.; Argow, B.A.; Buynevich, I.V. Coastal impacts due to sea-level rise. *Earth Planet. Sci.* **2008**, *36*, 601–647. [CrossRef]
7. Carrasco, A.R.; Ferreira, Ó.; Roelvink, D. Coastal lagoons and rising sea level: A review. *Earth Sci.* **2016**, *154*, 356–368. [CrossRef]
8. Benjamin, J.; Rovere, A.; Fontana, A.; Furlani, S.; Vacchi, M.; Inglis, R.H.; Gehrels, R. Late Quaternary sea-level changes and early human societies in the central and eastern Mediterranean Basin: An interdisciplinary review. *Quat. Int.* **2017**, *449*, 29–57. [CrossRef]
9. Erostate, M.; Huneau, F.; Garel, E.; Ghiotti, S.; Vystavna, Y.; Garrido, M.; Pasqualini, V. Groundwater dependent ecosystems in coastal Mediterranean regions: Characterization, challenges and management for their protection. *Water Res.* **2020**, *172*, 115461. [CrossRef]
10. Famiglietti, J.S. The global groundwater crisis. *Nat. Clim. Chang.* **2014**, *4*, 945–948. [CrossRef]
11. Erturk, A.; Ekdal, A.; Gurel, M.; Karakaya, N.; Guzel, C.; Gonenc, E. Evaluating the impact of climate change on groundwater resources in a small Mediterranean watershed. *Sci. Tot. Environ.* **2014**, *499*, 437–447. [CrossRef] [PubMed]
12. Butscher, C.; Huggenberger, P. Modeling the temporal variability of karst groundwater vulnerability, with implications for climate change. *Environ. Sci. Technol.* **2009**, *43*, 1665–1669. [CrossRef]
13. Oude Essink, G.H.P.; Van Baaren, E.S.; De Louw, P.G. Effects of climate change on coastal groundwater systems: A modeling study in the Netherlands. *Water Resour. Res.* **2010**, 46. [CrossRef]
14. Mastrocicco, M.; Gervasio, M.P.; Busico, G.; Colombani, N. Natural and anthropogenic factors driving groundwater resources salinization for agricultural use in the Campania plains (Southern Italy). *Sci. Total Environ.* **2021**, *758*, 144033. [CrossRef]
15. Allen, D.M.; Mackie, D.C.; Wei, M. Groundwater and climate change: A sensitivity analysis for the Grand Forks aquifer, southern British Columbia, Canada. *Hydrogeol. J.* **2004**, *12*, 270–290. [CrossRef]
16. Zhou, Y.; Li, W. A review of regional groundwater flow modeling. *Geosci. Front.* **2011**, *2*, 205–214. [CrossRef]
17. Aghlmand, R.; Abbasi, A. Application of MODFLOW with boundary conditions analyses based on limited available observations: A case study of Birjand plain in East Iran. *Water* **2019**, *11*, 1904. [CrossRef]

18. Sun, R.J.; Johnston, R.H. *Regional Aquifer-System Analysis Program of the US Geological Survey, 1978–1992*; US Government Printing Office: Denver, CO, USA, 1994; p. 126. [CrossRef]
19. Coelho, C.D.; Faria, A.C.S.; Marques, E.A.G. Comparative analysis of different boundary conditions and their influence on numerical hydrogeological modeling of Palmital watershed, southeast Brazil. *J. Hydrol.* **2017**, *12*, 210–219. [CrossRef]
20. Bastani, M.; Harter, T. Effects of upscaling temporal resolution of groundwater flow and transport boundary conditions on the performance of nitrate-transport models at the regional management scale. *Hydrogeol. J.* **2020**, *28*, 1299–1322. [CrossRef]
21. Cushman, J.H.; Bennethum, L.S.; Hu, B.X. A primer on upscaling tools for porous media. *Adv. Water Resour.* **2002**, *25*, 1043–1067. [CrossRef]
22. Dentz, M.; Le Borgne, T.; Englert, A.; Bijeljic, B. Mixing, spreading and reaction in heterogeneous media: A brief review. *J. Contam. Hydrol.* **2011**, *120*, 1–17. [CrossRef] [PubMed]
23. Scheibe, T.D.; Murphy, E.M.; Chen, X.; Rice, A.K.; Carroll, K.C.; Palmer, B.J.; Wood, B.D. An analysis platform for multiscale hydrogeologic modeling with emphasis on hybrid multiscale methods. *Groundwater* **2015**, *53*, 38–56. [CrossRef] [PubMed]
24. Hellwig, J.; de Graaf, I.E.M.; Weiler, M.; Stahl, K. Large-scale assessment of delayed groundwater responses to drought. *Water Resour. Res.* **2020**, *56*, e2019WR025441. [CrossRef]
25. Fatichi, S.; Vivoni, E.R.; Ogden, F.L.; Ivanov, V.Y.; Mirus, B.; Gochis, D.; Downer, C.W.; Camporese, M.; Davison, J.H.; Ebel, B.A.; et al. An overview of current applications, challenges, and future trends in distributed process-based models in hydrology. *J. Hydrol.* **2016**, *537*, 45–60. [CrossRef]
26. Rau, G.C.; Cuthbert, M.O.; Post, V.E.A.; Daniel Schweizer, D.; Acworth, R.I.; Andersen, M.S.; Blum, P.; Carrara, E.; Rasmussen, T.C.; Ge, S. Future-proofing hydrogeology by revising groundwater monitoring practice. *Hydrogeol. J.* **2020**, *28*, 2963–2969. [CrossRef]
27. De Vita, P.; Fusco, F.; Tufano, R.; Cusano, D. Seasonal and event-based hydrological and slope stability modeling of pyroclastic fall deposits covering slopes in Campania (Southern Italy). *Water* **2018**, *10*, 1140. [CrossRef]
28. Corniello, A.; Ducci, D.; Ruggieri, G.; Iorio, M. Complex groundwater flow circulation in a carbonate aquifer: Mount Massico (Campania Region, Southern Italy). Synergistic hydrogeological understanding. *J. Geochem. Explor.* **2018**, *190*, 253–264. [CrossRef]
29. Mastrocicco, M.; Busico, G.; Colombani, N.; Vigliotti, M.; Ruberti, D. Modelling actual and future seawater intrusion in the Variconi coastal wetland (Italy) due to climate and landscape changes. *Water* **2019**, *11*, 1502. [CrossRef]
30. Lancia, M.; Petitta, M.; Zheng, C.; Saroli, M. Hydrogeological insights and modelling for sustainable use of a stressed carbonate aquifer in the Mediterranean area: From passive withdrawals to active management. *J. Hydrol.* **2020**, *32*, 100749. [CrossRef]
31. Brancaccio, L.; Cinque, A.; Romano, P.; Rosskopf, C.; Russo, F.; Santangelo, N.; Santo, A. Geomorphology and neotectonic evolution of a sector of the Tyrrhenian flank of the southern Apennines (Region of Naples, Italy). *Z. Geomorphol.* **1991**, *82*, 47–58.
32. Pappone, G.; Alberico, I.; Amato, V.; Aucelli, P.P.C.; Di Paola, G. Recent evolution and present-day conditions of the Campanian Coastal plains (South Italy): The case history of Sele River Coastal plain. *WIT Trans. Ecol. Envir.* **2011**, *149*, 15–27. [CrossRef]
33. Ruberti, D.; Vigliotti, M. Land use and landscape pattern changes driven by land reclamation in a coastal area: The case of Volturno delta plain, Campania region, southern Italy. *Environ. Earth Sci.* **2017**, *76*, 694. [CrossRef]
34. ISPRA. National Geological Database. 2021. Available online: http://sgi2.isprambiente.it/viewersgi2/ (accessed on 10 October 2021).
35. Corniello, A.; Ducci, D. Hydrogeochemical characterization of the main aquifer of the "Litorale Domizio-Agro Aversano NIPS" (Campania—Southern Italy). *J. Geochem. Explor.* **2014**, *137*, 1–10. [CrossRef]
36. Allocca, V.; Coda, S.; De Vita, P.; Di Rienzo, B.; Ferrara, L.; Giarra, A.; Mangoni, O.; Stellato, L.; Trifuoggi, M.; Arienzo, M. Hydrogeological and hydrogeochemical study of a volcanic-sedimentary coastal aquifer in the archaeological site of Cumae (Phlegraean Fields, southern Italy). *J. Geochem. Explor.* **2018**, *185*, 105–115. [CrossRef]
37. Stellato, L.; Coda, S.; Arienzo, M.; De Vita, P.; Di Rienzo, B.; D'Onofrio, A.; Ferrara, L.; Marzaioli, F.; Trifuoggi, M.; Allocca, V. Natural and anthropogenic groundwater contamination in a coastal volcanic-sedimentary aquifer: The case of the archaeological site of Cumae (Phlegraean Fields, Southern Italy). *Water* **2020**, *12*, 3463. [CrossRef]
38. AppEEARS Team. *Application for Extracting and Exploring Analysis Ready Samples (AppEEARS); Ver. 2.54.1. NASA EOSDIS Land Processes Distributed Active Archive Center (LP DAAC)*; USGS/Earth Resources Observation and Science (EROS) Center: Sioux Falls, SD, USA, 2020. Available online: https://lpdaacsvc.cr.usgs.gov/appeears (accessed on 10 October 2021).
39. Running, S.; Mu, Q.; Zhao, M.; Moreno, A. MOD16A2GF MODIS/Terra Net Evapotranspiration Gap-Filled 8-Day L4 Global 500m SIN Grid V006. NASA EOSDIS Land Processes DAAC. 2019. Available online: https://doi.org/10.5067/MODIS/MOD16A2GF.006 (accessed on 10 October 2021).
40. Mu, Q.; Zhao, M.; Running, S.W. Improvements to a MODIS global terrestrial evapotranspiration algorithm. *Rem. Sens Environ.* **2011**, *115*, 1781–1800. [CrossRef]
41. Shah, N.; Nachabe, M.; Ross, M. Extinction depth and evapotranspiration from ground water under selected land covers. *Groundwater* **2007**, *45*, 329–338. [CrossRef]
42. Sutanudjaja, E.H.; van Beek, R.; Wanders, N.; Wada, Y.; Bosmans, J.H.C.; Drost, N.; van der Ent, R.J.; de Graaf, I.E.M.; Hoch, J.M.; de Jong, K.; et al. PCR-GLOBWB 2: A 5 arcmin global hydrological and water resources model. *Geosci. Model Dev.* **2018**, *11*, 2429–2453. [CrossRef]
43. Harbaugh, A.W. *MODFLOW-2005, the US Geological Survey Modular Ground-Water Model: The Ground-Water Flow Process; U.S. Geological Survey Techniques and Methods 6-A16*; US Department of the Interior; US Geological Survey: Reston, VA, USA, 2005; p. 253.

44. Reimann, T.; Hill, M.E. MODFLOW-CFP: A New Conduit Flow Process for MODFLOW-2005. *Groundwater* **2009**, *47*, 321–325. [CrossRef]
45. Hill, M.C. The practical use of simplicity in developing ground water models. *Groundwater* **2006**, *44*, 775–781. [CrossRef]
46. Doherty, J. PEST-Model-Independent Parameter Estimation. Version 12. Watermark Computing. Australia. 2010. Available online: http://www.pesthomepage.org/ (accessed on 10 October 2021).
47. De Vita, P.; Allocca, V.; Celico, F.; Fabbroncino, S.; Cesaria, M.; Monacelli, G.; Musilli, I.; Piscopo, V.; Scalise, A.R.; Summa, G.; et al. Hydrogeology of continental southern Italy. *J. Maps* **2018**, *14*, 230–241. [CrossRef]
48. Russo, G.; Viggiani, C.; Viggiani, G. Geotechnical design and construction issues for Lines 1 and 6 of the Naples Underground/Geotechnische Planung und Ausführung der U-Bahnlinien 1 und 6 in Neapel. *Geomech. Tunnelbau* **2012**, *5*, 300–311. [CrossRef]
49. de Graaf, I.E.; van Beek, R.L.; Gleeson, T.; Moosdorf, N.; Schmitz, O.; Sutanudjaja, E.H.; Bierkens, M.F. A global-scale two-layer transient groundwater model: Development and application to groundwater depletion. *Adv. Water Resour.* **2017**, *102*, 53–67. [CrossRef]
50. Fan, Y.; Li, H.; Miguez-Macho, G. Global patterns of groundwater table depth. *Science* **2013**, *339*, 940–943. [CrossRef] [PubMed]
51. Braca, G.; Bussettini, M.; Ducci, D.; Lastoria, B.; Mariani, S. Evaluation of national and regional groundwater resources under climate change scenarios using a GIS-based water budget procedure. *Rend. Fis. Acc. Lincei* **2019**, *30*, 109–123. [CrossRef]
52. Brakkee, E.; Van Huijgevoort, M.H.; Bartholomeus, R.P. Improved understanding of regional groundwater drought development through time series modelling: The 2018–2019 drought in the Netherlands. *Hydrol. Earth Syst. Sci.* **2022**, *26*, 551–569. [CrossRef]
53. Ruggieri, G.; Allocca, V.; Borfecchia, F.; Cusano, D.; Marsiglia, P.; De Vita, P. Testing Evapotranspiration Estimates Based on MODIS Satellite Data in the Assessment of the Groundwater Recharge of Karst Aquifers in Southern Italy. *Water* **2021**, *13*, 118. [CrossRef]
54. Allocca, V.; Manna, F.; De Vita, P. Estimating annual groundwater recharge coefficient for karst aquifers of the southern Apennines (Italy). *Hydrol. Earth Syst. Sci.* **2014**, *18*, 803–817. [CrossRef]
55. Allocca, V.; De Vita, P.; Manna, F.; Nimmo, J.R. Groundwater recharge assessment at local and episodic scale in a soil mantled perched karst aquifer in southern Italy. *J. Hydrol.* **2015**, *529*, 843–853. [CrossRef]
56. De Vita, P.; Allocca, V.; Manna, F.; Fabbrocino, S. Coupled decadal variability of the North Atlantic Oscillation, regional rainfall and karst spring discharges in the Campania region (southern Italy). *Hydrol. Earth Syst. Sci.* **2012**, *16*, 1389–1399. [CrossRef]
57. Colombani, N.; Gaiolini, M.; Busico, G.; Postacchini, M. Quantifying the impact of evapotranspiration at the aquifer scale via groundwater modelling and MODIS data. *Water* **2021**, *13*, 950. [CrossRef]
58. Lehr, C.; Lischeid, G. Efficient screening of groundwater head monitoring data for anthropogenic effects and measurement errors. *Hydrol. Earth Syst. Sci.* **2020**, *24*, 501–513. [CrossRef]

*water*

MDPI

*Article*

# Hybrid Fuzzy Multi-Criteria Analysis for Selecting Discrete Preferable Groundwater Recharge Sites

**Christopher Papadopoulos [1], Mike Spiliotis [1,*], Fotios Pliakas [1], Ioannis Gkiougkis [1], Nerantzis Kazakis [2] and Basil Papadopoulos [1]**

[1] Department of Civil Engineering, School of Engineering, Democritus University of Thrace, 67100 Xanthi, Greece; cpapadp@civil.duth.gr (C.P.); fpliakas@civil.duth.gr (F.P.); jgiougis@civil.duth.gr (I.G.); papadob@civil.duth.gr (B.P.)

[2] Department of Geology, School of Sciences, Aristotle University of Thessaloniki, 54124 Thessaloniki, Greece; kazakis@geo.auth.gr

* Correspondence: mspiliot@civil.duth.gr; Tel.: +30-25410-79613

**Abstract:** This study proposes a hybrid fuzzy multi-criteria methodology for the selection of the most preferable site for applying managed aquifer recharge (MAR) systems by utilizing floodwaters. The use of MAR can increase water resources for later water utilization in case of drought. In this multi-criteria problem, seven recharge sites are under consideration, based on nine criteria, aiming to make a final list of their relative ranking. A fuzzy analytic hierarchy process (FAHP) based on the logarithmic fuzzy preference programming (LFFP) method is used to determine the weights of criteria. LFFP is an optimization-based method that produces a priority vector from a fuzzy pairwise comparison matrix. Furthermore, fuzzy inference systems (FIS) based on the Mamdani approach are used to estimate the rating of each alternative with respect to the criterion examined, and then the final evaluation of the alternatives is obtained. A FIS is a fuzzy if–then rule-based system where the experts' qualitative knowledge is translated into numerical reasoning for each individual criterion. The proposed methodology is applied in the aquifer system of the agricultural plain located to the southeast of the city of Xanthi in the Prefecture of Xanthi, NE Greece.

**Keywords:** fuzzy analytic hierarchy process; fuzzy inference systems; managed aquifer recharge; Xanthi plain

## 1. Introduction

In recent decades, there has been growing concern about water availability, mainly due to the development of human societies and climate change [1]. Regarding the last one, trends show an increasing frequency and intensity of natural extreme events related to temperature and precipitation, such as droughts and floods [2–5].

A way of tackling low water availability challenges is the storage of water in artificial surface structures. However, this can meet only part of the increasing demand for water, mainly due to limited capacity, high evaporation losses, and competing objectives [6–9]. An alternative is groundwater recharge by utilizing excess surface water such as flood waters. Aquifers are more resistant to water supply variability related to drought than surface water supplies; however, they are under growing stress [6,10–13], especially in arid and semi-arid areas [14].

In recent years, managed aquifer recharge (MAR) systems have been increasingly used to increase groundwater availability for later use in dry seasons [15–19]. A variety of MAR techniques can be used that mainly depend on local conditions [20,21].

Scanlon et al. [12] state that two basic approaches for managing groundwater storage include: (1) conjunctive use (CU) of surface water and groundwater, and (2) managed aquifer recharge (MAR), which can be considered as an extension of CU. According to Foster and van Steenbergen [22], conjunctive groundwater use with surface water resources, in terms of practical water management, represents one of the most important responses to improving drought water supply security and for long-term climate change adaptation. Maliva [23] mentions that MAR can be viewed as a means for optimizing the use of aquifers. Dillon et al. [24] observed that the "objectives of groundwater management relate to maximizing economic utility of aquifers while sustaining the environment and providing security for meeting human needs". Maliva [23] also states that depending upon the system, recharge is performed by either applying water onto a land surface (surface spreading), subsurface discharge into the vadose zone using wells, galleries, and trenches, or by injection using wells into either confined or unconfined aquifers. Particularly, surface spreading by using floodwater (excess winter water from a river) can be an effective MAR method [25]. A common technique for this method is the use of infiltration basins, while the suitability of an area for establishing floodwater spreading depends on various factors, such as geomorphology, hydrogeology, land use, socioeconomic issues, environmental impacts, etc., [26]. Recharge may also be induced by pumping groundwater close to connected surface water bodies (induced recharge). Modifications of the land surface and stream channels, such as by the removal (or change of) vegetation and construction of dams and levees, are also used to intentionally increase aquifer recharge.

A common issue that needs to be dealt with in applying MAR systems is the selection of suitable sites. Obviously, this is a problem of multi-criteria nature, in which an alternative or several alternative potential site/sites for applying MAR are considered based on a series of criteria, which may be divided into various categories. A variety of studies address this decision problem by the use of remote sensing and/or GIS techniques to produce a map with suitable zones for applying MAR (e.g., [27–29]). Other research provides advanced integrated methodologies by combining the GIS capabilities with multi-criteria methods (e.g., [30–35]). In such surveys, a weighted thematic layer for each criterion is produced. The criteria have been initially divided into classes (ranges) and each point of the grid obtains a score based on its classification into the classes. Then, a final map can be generated with overlay techniques using crisp operators of Boolean logic. Although these approaches are very significant and have been widely used, they are not always applicable. The main reason is the low data availability. The generation of thematic layers is based on interpolation methods, and therefore a small sample with respect to each criterion may diminish the reliability of interpolation results.

Fuzzy logic and fuzzy set theory were introduced by Zadeh [36] as a generalization of Aristotle's logic and/or classical set theory. In Aristotle's logic, a statement can be only true or false and, in classical set theory, an element $x$ (or an individual) either belongs {1} or does not belong {0} to a given set and thus, it is described by two values [37]. In contrast to dual logic, fuzzy logic is many-valued and manipulates the concept of partial truth [38] and, in fuzzy set theory, an element $x$ can partially participate in a (fuzzy) set, which means that it can obtain all the values in a closed unit interval [0, 1]. The principles of fuzzy logic and sets are the mathematical representation of human thinking and reasoning where truth may appear as an inference from inaccurate or partial knowledge, dealing with uncertainty due to vagueness. A variety of methodologies based on the principles of fuzzy logic and sets have been developed which incorporate the uncertainty of complex issues by the use of either autonomous fuzzy tools or hybrid analytical methods.

In particular, fuzzy logic and the theory of fuzzy sets have had an extensive use in hydrologic analysis and the analysis of hydrological multi-criteria problems since the early application of Bogardi et al. [39], in which fuzziness was used for partial satisfaction of objectives in regional water resources management, until the present [40]. The recent studies of Kazakis et al. [41], Kourgialas et al. [42], and Spiliotis et al. [43] are given as some examples of hybrid fuzzy multi-criteria methodologies. In these studies, the analysis can be carried out in order to either spatially address the examined multi-criteria problem, producing suitable fuzzy maps, or categorize points (alternatives) that can belong to several categories to some degree. This wide use of fuzzy logic and sets is due to the ability to simulate human reasoning in its use of approximate information, and to mathematize the uncertainty and vagueness in decision-making [44,45]. Furthermore, the use of fuzzy logic and sets in multi-criteria decision-making (MCDM) can express the gray zone of the decision and manage imprecise information (e.g., the score of criteria, the available resources), etc.

Regarding the selection problem of suitable recharge sites, most research uses fuzzy operators and/or fuzzy if–then rule-based systems in a GIS environment in order to produce a map with fuzzy suitable zones (e.g., [46–52]). However, the use of such models in combination with hybrid fuzzy multi-criteria methods to address this problem is an open issue.

The main proposal of this study is the ranking of seven alternatives representing suitable sites for applying MAR systems using floodwaters via a hybrid fuzzy multi-criteria methodology, in the context of increasing the local groundwater resources for later water use in case of drought. As described in the following section, the local conditions of the case study favor the application of floodwater spreading using infiltration basins. The evaluation of the alternatives was based on nine criteria that were selected based on the local conditions and the type of MAR. The weights of criteria were obtained through a fuzzy analytic hierarchy process (FAHP)-based methodology; however, they are crisp numbers. Fuzzy inference systems (FIS) were used in the rating of each alternative with respect to each criterion. The use of FIS was preferred since the objective function regarding the criteria was unknown. The composition of criteria was achieved through a simple aggregated model where a final ranking of the alternatives was obtained. The case study refers to the aquifer system of the agricultural plain located to the southeast of the city of Xanthi in northern Greece.

## 2. Case Study

The study area is the aquifer system of the agricultural plain located to the southeast of the city of Xanthi in the Prefecture of Xanthi, NE Greece (Figure 1). The wider region of study area is located within the boundaries of the tertiary Vistonida basin [53], while as presented in Figure 2 (modified map based on geological map of [54]), the aquifer system is hosted in the alluvial plain consisting of loam, clays, sands, gravels, etc. Gravels and sands of Pleiocene–Pleistocene origin occur in the north of study area, while Pleiocene brackish sediments, mainly consisting of sand and clay, with no wide surface spread, occur near Vistonida Lagoon. Clastic sediments of the Eocene and Oligocene, consisting of conglomerate breccia, limestones, and mollasic formations, occur in both the south and north of study area, while the background consists of igneous rocks that underlie andesitic rocks of the same geological period.

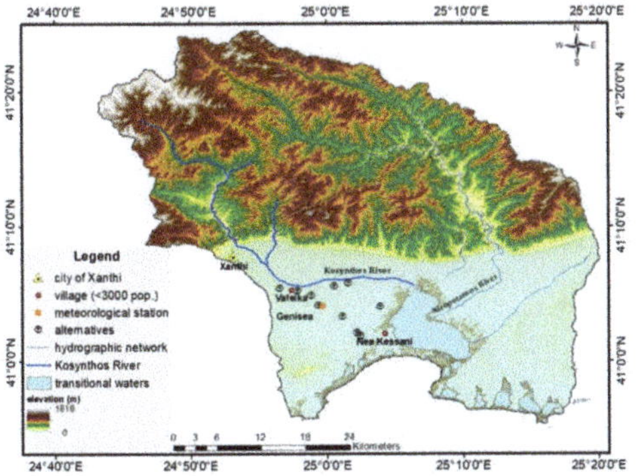


Figure 1. Geomorphology of the wider area and the alternative sites.

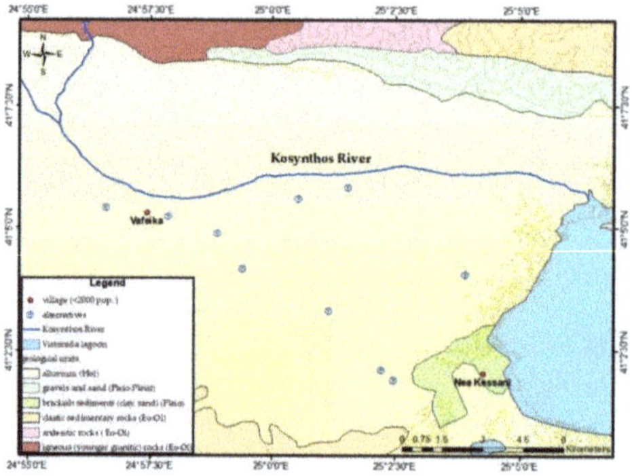


**Figure 2.** Geological formations of the area under consideration (modified map based on geological map of [54].

Geological formations of the alluvial plain range from fine to silty clay in the SE–E direction towards Vistonida Lagoon [53,55]. Regarding the geological structure and hydrogeological conditions, relevant assessments have shown that [56,57]:

(1) Three main geological formations are presented: (a) the upper formation (8–80 m in thickness), of low permeability, consisting of clayey sand which interchanges at certain locations with gravel sand of little thickness, (b) the intermediate aquifer formation (10–70 in thickness), consisting of permeable gravel sand, considered as a shallow confined aquifer, in some locations changing to semi-confined aquifer, and (c) the lower impermeable formation consisting of clayey silt in depth of 30–90 m;

(2) Transmissivity, T, ranges from 90 to 915 ($m^2$/day), while the storage coefficient, S, ranges from $2.4 \times 10^{-4}$ to $8.75 \times 10^{-2}$;

(3) The shallow aquifer system is naturally recharged, mainly from direct infiltration of precipitation and partially from percolation of the River Kosynthos (upstream zone of coarse-grained deposits);

(4) The groundwater flow has a SE direction, being almost identical to that of the old riverbed's direction in the area;

(5) The groundwater main recharge axes appear to be in a S–SE direction towards Vistonida Lagoon;

(6) The high quality of the water of the River Kosynthos gives rise to its use for artificial recharge purposes.

The climatic conditions of the wider area differ in the lowlands from those in the mountains [58]. Based on precipitation and temperature records from the meteorological station of Genisea (41°04′07″ N–24°59′44″ E) located in the study area, the mean annual cumulative precipitation is 605.6 mm in the period 1966–2018. The available temperature data concern the period of 1988–2018, in which the mean annual temperature was 14.4 °C, while mean temperature of the summer months was 23.9 °C. In the wider region, both wet periods (that caused flood events such as the flood of the Kosynthos River in 1996) and drought periods have occurred. In fact, a drought period might have had an effect on groundwater levels of the examined aquifer. Particularly, the groundwater level in January of the shallow aquifer is significantly related to drought in the previous hydrological year and drought in the first trimester of the current hydrological year [59].

The exploitation of the aquifer system mainly takes place in the south of the riverbed of the Kosynthos River where there is a significant number of irrigation wells. Most of these wells there are not deeper than 50 m, while there are a few shallow wells (up to 10 m) that are not in use any more [58,60]. According to Diamantis et al. [60], the annual pumping volume is estimated at $48 \times 10^6$ $m^3$ approximately. The spatial distribution of the irrigation wells and other detailed information regarding the study area can be found in the studies of Diamantis et al. [60] and Pisinaras [58].

## 3. Materials and Methods

The development of the proposed hybrid fuzzy methodology refers to the intermediate formation of the case study. It can only be applied in unconfined aquifers and where there is an underlying drainage axis. In addition, the application of the methodology requires the presence of an adjacent river whose excess water during the flood period can be utilized without burdening the ecological supply. Last, in order for each alternative to be evaluated, the lithological profile and hydraulic characteristics of both the vadose zone and saturated zone had to be available. Each alternative represents a potential site for MAR application where there is a well with available hydrogeological data. The alternatives were considered through the nine criteria described below in ascending order of importance. The selected criteria can be added or subtracted depending on the specific conditions of an area, while there is no limit in number of the examined alternatives.

### 3.1. Criteria Description

Ecological status ($C_1$): This refers to the status of the Kosynthos River. Based on River Basin Management Plans [61], the riverbed of Kosynthos is divided into sections characterized by different ecological status. Each alternative may be supplied water for applying MAR by a different section. The worst case is unknown ecological status, while the best case is high ecological status.

Slope (%) ($C_2$): Slope is a commonly-used criterion in identification of suitable sites for applying MAR. Obviously, the steeper the slope, the less desirableare conditions for applying MAR. The dividing of slope into classes is based on previous studies (e.g., [35,62]); however, the classes of slope were adjusted to the case study area, which is characterized by smooth slopes.

Distance from hydrographic network (m) ($C_3$): This refers to the distance between each alternative and a source with available water for MAR. Only the distance from the Kosynthos River was considered since the water of Vistonida Lagoon is unsuitable due to its connection with coastal water surfaces (Figure 3). In the research of Farhadian et al. [63], the distance from rivers is also used, while in other research (e.g., [35]) the distance from drainage networks is taken into account.

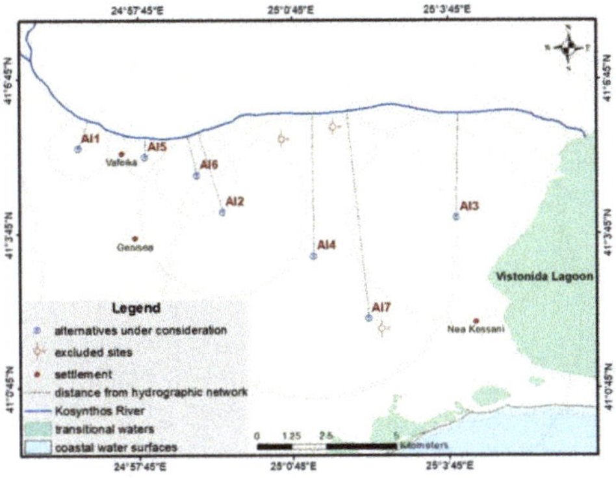


**Figure 3.** Distance from the hydrographic network (from the Kosynthos River) of each alternative.

Distance from road network (m) ($C_4$): This criterion may affect the suitability of an area for applying MAR either in a negative or a positive way. For instance, it is desirable a recharge site is a minimum distance from highways and freeways in order for potential groundwater pollution to be avoided [52]. However, this holds in the case of short distance (a few hundred meters) between the recharge site under consideration (alternative) and the potential pollution source. On the other hand, if the distance between the alternatives and road network is in the order of kilometers, where there is in fact no danger of groundwater pollution, then an easy and fast access to the recharge site is desirable [63]. In this study, the distance of the alternatives from the highway was in the order of kilometers; therefore, the shorter distance, the higher the rating of the alternative. Figure 4 presents a map with the distances in kilometers of each alternative from the highway.

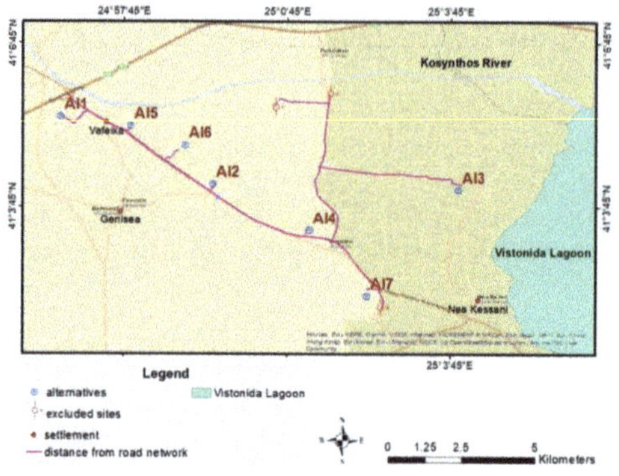


**Figure 4.** Distance from the road network of each alternative (based on World Street map).

Transmissivity ($m^2$/day) ($C_5$): This is a common hydrogeological parameter used in studies regarding the selection of suitable recharge sites [32,35,52,64], since it takes into account both the (horizontal) hydraulic conductivity ($K$) and the saturated thickness of the aquifer (b). High transmissivity values favor the recharge process since transmissivity ($T$) is defined as T = Kb ($m^2$/day). According to Pisinaras [58], in the study area, T ranges between 25 and 2950 ($m^2$/day).

Storativity ($C_6$): Another hydrogeological criterion used in site selection for applying MAR is storativity (S) [65]. The storativity of each alternative in the study area was obtained based on previous research [56–58], where storage coefficient values varied from $10^{-2}$ to $10^{-4}$. The lower the S-values, the more unfavorable conditions are for applying MAR.

Hydraulic resistance ($C_7$): This is a critical hydrological criterion because in its calculation, $c = \sum_{i=1}^{n} d_i/K_i$, the (vertical) hydraulic conductivity ($K$) (m/days) of each sedimentary layer and its thickness ($d$) (m) are introduced. Kazakis [35] suggests the use of (log values) hydraulic resistance to take into account the permeability of the entire thickness of the vadose zone. High $c$-values indicate unsuitable recharge sites. It is noted that in the case of confined aquifers, this criterion cannot be used. The estimation of c-values of the alternatives was based on the research of Pisinaras [58], in which stratigraphic columns regarding the examined alternatives are available.

Type of aquifer ($C_8$): As aforementioned, the hydraulic resistance does not refer to confined aquifers. Thus, the classes of this criterion are considered to be crisp. That is, an alternative has the highest rating (10) when its aquifer is unconfined (or/and leakage aquifer), otherwise it is zero. Based on the available stratigraphic columns [58], there were confined underlying aquifers in the cases of three alternatives and, thus, these alternatives were excluded from the analysis. The aquifers of the other seven alternatives were unconfined or/and leakage (semi-confined) aquifers.

Piezometry ($C_9$): Piezometry and the type of aquifer are essential (hydrogeological) criteria for applying MAR systems. When a recharge axis is underlying an alternative, then this recharge site is unsuitable and vice versa. The identification of the study area regarding the recharge axes and the drainage axes was based on previous research [56–58].

On that point, it should be noted that the proposed methodology has been developed in order to analyze both large and small datasets with various hydrogeological parameters. Hence, some parameters such as aquifer type (e.g., confined) have been included in order to help to this analysis. Initially, all available information (lithological profiles, hydraulic parameters, etc.) was selected, and after the first screening, the alternatives with no available profiles were eliminated and thus, ten alternatives remained. Based on the lithological profiles, three of ten alternatives overlie confined aquifers, and hence, these alternatives were excluded from the analysis (Figures 3 and 4). On balance, the following preparatory steps were carried out (Figure 5):

(1) Literature review and building of a detailed database including hydrogeological, morphological, etc., data;
(2) The points with lithological profiles and hydraulic data of the aquifer and vadose zone were chosen for the application and development of the methodology. The other sites were eliminated;
(3) Tensites were then chosen, of which 3 referred to confined aquifers, and for this reason they were excluded from the analysis;
(4) The method was applied to the final chosen sites (7 in total) and the most suitable for MAR application was determined.

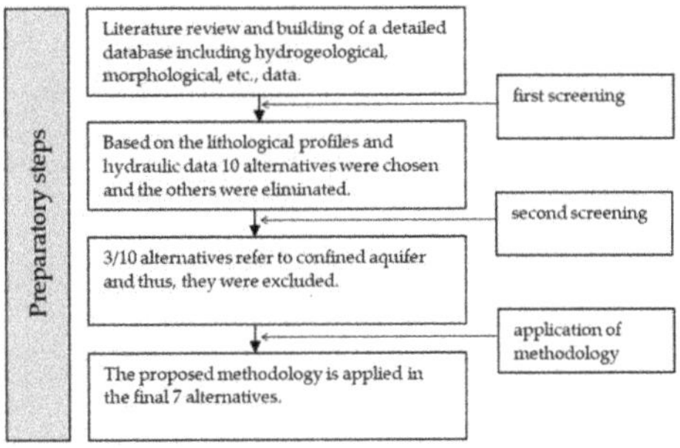


**Figure 5.** Preparatory steps for screening of the final alternatives.

*3.2. Fuzzy Analytic Hierarchy Process—FAHP*

As aforementioned, fuzzy logic has a wide use in multi-criteria analysis and has been applied to several hydrological issues. The basic concepts of fuzzy logic and sets, such as the membership function, $\alpha$-cuts, fuzzy number [66], extension principle [67,68] and max–min composition of fuzzy relations [69], are presented in Appendix A. The fuzzy methods used in this study are based on these concepts.

One of the most used methods for quantification of the significance of criteria is the analytic hierarchy process (AHP) suggested by Saaty [70]. As is known, the AHP is based on pairwise comparisons between criteria (or/and alternatives), where a positive reciprocal pairwise comparison matrix $\boldsymbol{A}_{nxn}$ is constructed. A fundamental scale is used [70] in order to numerically express the initial experts' judgments regarding the relative criteria importance (ratios of weights $(w_i/w_j)$); however, it is not the only one [71]. Based on the eigenvalue theory from linear algebra, Saaty [70] developed the eigenvector method (EVM) in order to obtain a unique solution to derive the weights of criteria (or alternatives) from a $\boldsymbol{A}_{nxn}$. This is the principal eigenvector $\boldsymbol{w}$ determined by solving the system of equations:

$$(\boldsymbol{A}_{nxn} - \lambda_{\max}\boldsymbol{I})\boldsymbol{w} = 0 \Leftrightarrow \boldsymbol{A}_{nxn}\boldsymbol{w} = \lambda_{\max}\boldsymbol{w} \tag{1}$$

where $\boldsymbol{I}$ is the unit matrix and $\lambda_{\max}$ is the principle eigenvalue of $\boldsymbol{A}_{nxn}$.

Given that the initial judgments cannot be identical to the ratios of weights, the $\boldsymbol{A}_{nxn}$ is not fully consistent. Therefore, the validation of AHP is based on the measure of the consistency of the $\boldsymbol{A}_{nxn}$. For this reason, the consistency index CI = $(\lambda_{\max} - n)/(n-1)$ is estimated. Then, the CI-index is divided with a predefined index (RI), the value of which depends on the number of criteria $n$.

Fuzzy set theory is the most widely used approach integrated with the AHP [72]. The use of fuzziness can incorporate the inherent uncertainty in the assessments of pairwise comparison between criteria, derived by the subjectivity in the quantification of the ratios expressing the relative importance of the decision elements (weights of criteria and ratings/scores of alternatives).

To cope with the uncertainty of imprecise judgments, Chang [73] used a fuzzy pairwise comparison matrix and then he used extent analysis to extract a crisp priority vector. An advantage of the Chang method is its computational simplicity.

However, even if extent analysis has been widely used (e.g., [74–77]), it has a major drawback. This is the fact that a weight of a criterion may obtain zero value, and hence it makes no sense for such a criterion to be under consideration [78]. In addition, extent analysis produces a priority vector, which may be quite different from the true one, and thus the weights obtained are not representative regarding the relative importance of decision criteria.

Mikhailov [79], extending his previous research [80], suggests a method for derivation of crisp weights from a matrix with fuzzy judgments, by applying a fuzzy preference programming (FPP)-based nonlinear method (optimization approach). In fact, the method of Mikhailov [79] is based on (fuzzy) flexible programming; that is, the uncertainty appears in the constant term of the fuzzy constraints [68].

Let us consider the following reciprocal matrix $\widetilde{A}_{nxn}$ with fuzzy pairwise comparisons between $n$ criteria:

$$\widetilde{A} = \left(\widetilde{a}_{ij}\right)_{nxn} \begin{array}{c} \begin{array}{ccccc} C_1 & C_2 & C_3 & \cdots & C_n \end{array} \\ \left[\begin{array}{ccccc} 1 & (l_{12}, m_{12}, u_{12}) & (l_{13}, m_{13}, u_{13}) & \cdots & (l_{1n}, m_{1n}, u_{1n}) \\ (l_{21}, m_{21}, u_{21}) & 1 & (l_{23}, m_{23}, u_{23}) & \cdots & (l_{2n}, m_{2n}, u_{2n}) \\ (l_{31}, m_{31}, u_{21}) & (l_{32}, m_{32}, u_{22}) & 1 & \cdots & (l_{3n}, m_{3n}, u_{3n}) \\ \cdots & \cdots & \cdots & \cdots & \cdots \\ (l_{n1}, m_{n1}, u_{n1}) & (l_{n2}, m_{n2}, u_{n2}) & (l_{n2}, m_{n2}, u_{n2}) & \cdots & 1 \end{array}\right] \end{array} \begin{array}{c} C_1 \\ C_2 \\ C_3 \\ \vdots \\ C_n \end{array} \tag{2}$$

where $C_i$ (for $i = 1, \ldots, n$) are the criteria, $\widetilde{a}_{ij} = 1/\widetilde{a}_{ji} = w_i/w_j$, that is $l_{ij} = 1/u_{ji}$, $m_{ij} = 1/m_{ji}$ and $u_{ij} = 1/l_{ji}$ for $l_{ij} \leq m_{ij} \leq u_{ji}$ and for all $i, j = 1, 2, \ldots, n,\ j \neq i$.

In order to obtain crisp weights whose ratios are approximately included in the initial fuzzy judgments ($l_{ij} \widetilde{\leq} w_i/w_j \widetilde{\leq} u_{ij}$), the following membership function is introduced by Mikhailov [79]:

$$\mu_{ij}(w_i/w_j) = \begin{cases} \frac{(w_i/w_j) - l_{ij}}{m_{ij} - l_{ij}}, & l_{ij} < (w_i/w_j) \leq m_{ij} \\ \frac{u_{ij} - (w_i/w_j)}{u_{ij} - m_{ij}}, & u_{ij} > (w_i/w_j) \geq m_{ij} \end{cases} \tag{3}$$

where $\mu_{ij}(w_i/w_j)$ denotes the degree of satisfaction (membership) to which the ratio $(w_i/w_j)$ belongs to fuzzy judgment $\widetilde{a}_{ij}$.

In order to determine a global evaluation from individual membership functions [81], a proper fuzzy operator, that is the min intersection, is used with respect to the above fuzzy inequalities. The min intersection is preferred because it secures a common satisfaction of all the selected membership functions [82,83]. Furthermore, in this study, the use of min intersection can linearize the decision problem

$$\lambda = \min\{\mu_{ij}(w_i/w_j) | i = 1, \ldots, n-1, j = i+1, \ldots, n\} \tag{4}$$

where $\lambda$ is the degree of the common satisfaction to which the crisp priority vector satisfies simultaneously each fuzzy pairwise comparison.

Assuming that the decision-maker looks forward to the highest membership degree of the ratio of weights in the fuzzy judgments $\widetilde{a}_{ij}$, Mikhailov [79] and Mikhailov and Tsvetinov [84] maximize $\lambda$ under the constraint that the total sum of weights must be equal to the unit (normalization constraint) and the positivity constraint in their values:

$$\max \lambda$$
$$s.t.$$
$$\begin{cases} \min_{ij}\left[\mu_{ij}(w_i/w_j)\right] \geq \lambda, \ i=1,\ldots,n-1, \ j=i+1,\ldots,n \\ \sum_{i=1}^{n} w_i = 1 \\ w_i \geq 0, \ i=1,\ldots,n \end{cases} \tag{5}$$

In the case of a positive $\lambda$-value, all solution ratios completely satisfy the fuzzy judgments, while a negative $\lambda$-value indicates strongly inconsistent initial judgments [79].

This paper applies the logarithmic fuzzy preference programming (LFPP)-based methodology for fuzzy AHP priority derivation proposed by Wang and Chin [85].The LFPP-based nonlinear method is preferred over the FFP-based nonlinear method because (a) it can ensure nonnegative values of membership function and hence, it produces a unique optimal solution and, (b) the same priority vectors are derived; either the upper triangular elements of a fuzzy pairwise comparison matrix are used or the lower ones.

Wang and Chin [85] addressed the drawbacks of Mikhailov's method [79]. Initially, they used logarithmed values in fuzzy judgments, which can be justified based on the extension principle, and the strict monotony of the (increasing) function $\ln \tilde{a}_{ij}$ [86]:

$$\ln \tilde{a}_{ij} \approx (\ln l_{ij}, \ln m_{ij}, \ln u_{ij}) \tag{6}$$

Equation (6) is an approximation of the fuzzy judgments $\ln \tilde{a}_{ij}$ considered as fuzzy triangular numbers, since due to the logarithmic transformation the linearity does not hold. The membership function $\mu_{ij}(\ln \tilde{a}_{ij})$ of the triangular fuzzy number can be expressed by the following Equation (7) [85].

$$\mu_{ij}(\ln(w_i/w_j)) = \begin{cases} \frac{\ln(w_i/w_j) - \ln l_{ij}}{\ln m_{ij} - \ln l_{ij}}, & \ln(w_i/w_j) \leq \ln m_{ij} \\ \frac{\ln u_{ij} - \ln(w_i/w_j)}{\ln u_{ij} - \ln m_{ij}}, & \ln(w_i/w_j) \geq \ln m_{ij} \end{cases} \tag{7}$$

Following Mikhailov's approach [79], it would be desirable to construct an optimization problem (Equation (8)) in which the parameter $\lambda = min\{\mu_{ij}(\ln(w_i/w_j))$ for $i = 1, \ldots,$ $n-1$ and $j = i+1, \ldots, n$ to be maximized.

$$\begin{cases} \max\lambda \\ s.t. \ \mu_{ij}(\ln(w_i/w_j)) \geq \lambda \\ w_i \geq 0, \ i=1,\ldots,n \end{cases} \Leftrightarrow \begin{cases} \max 1-\lambda \\ s.t. \ \ln w_i - \ln w_j - \lambda(\ln(m_{ij}/l_{ij})) \geq \ln l_{ij} \\ \ln w_j - \ln w_i - \lambda(\ln(u_{ij}/m_{ij})) \geq -\ln u_{ij} \\ w_i \geq 0, \ i=1,\ldots,n \end{cases} \tag{8}$$

For simplicity reasons, in the above optimization problem, the normalization constraint might be carried out after the solution. Nevertheless, a negative $\lambda$-value can still occur, which means that the inequalities of Equation (8) do not simultaneously hold. In other words, there is no an optimal solution (priority vector) to satisfy all the fuzzy judgments at the same time. Thus, according to Wang and Chin [85], two nonnegative deviation variables, $\delta_{ij}$ and $\eta_{ij}$, are introduced whose values must obviously be minimum. Therefore, the optimization problem applied in this research takes the following form:

$$\text{Minimize } (1-\lambda)^2 + M\sum_{i=1}^{n-1}\sum_{j=i+1}^{n}\left(\delta_{ij}^2 + n_{ij}^2\right)$$
$$s.t.$$
$$\begin{cases} x_i - x_j - \lambda\left(\ln\left(\frac{m_{ij}}{l_{ij}}\right)\right) + \delta_{ij} \geq lnl_{ij}, \ i=1,\ldots,n-1, \ j=i+1,\ldots,n \\ x_j - x_i - \lambda\left(\ln\left(\frac{u_{ij}}{m_{ij}}\right)\right) + n_{ij} \geq -lnu_{ij}, \ i=1,\ldots,n-1, \ j=i+1,\ldots,n \\ \lambda, x_i \geq 0, \ i=1,\ldots,n-1, \ j=i+1,\ldots,n \end{cases} \tag{9}$$

where $M$ is a specific constant term aiming for nonnegativity of $\lambda$, and to hold the values of deviation variables low, $x_i = lnw_i, lnw_j = x_j$.

The nonlinear objective function of Equation (9) focuses on minimizing the deviation variables, and after the priorities' derivation, the normalization of weights takes place (Equation (10)) in order to produce optimal normalized weight values.

$$w_i = \frac{\exp(x_i^*)}{\sum_{j=1}^{n} \exp\left(x_j^*\right)}, \; i = 1, \ldots, n \tag{10}$$

where $x_i^*(i = 1, \ldots, n)$ is the optimal solution of Equation (9).

The uniqueness of the obtained optimal priority vector holds due to the convexity of the $\lambda$ and the linearity of constraints. The other advantage of the LFFP based-nonlinear method (Equations (6)–(10)), which is the alignment of priorities regardless of the triangular part of the comparison matrix used, can be easily proved [85] after some simple algebraic transformation.

### 3.3. Fuzzy Inference Systems for the Evaluation of the Alternatives' Rating

This research uses the fuzzy inference system (FIS) based on Mamdani's approach [87] for the evaluation of the alternatives. In this study, the use of such a system was preferred because the function for the criteria synthesis was unknown. With the use of FIS, the ranges of criteria can be fuzzified as well. The input variable of the FIS was the examined criterion and the output variable was the rating of each alternative with respect to the examined criterion. Both the input and the output variables of the FIS were divided into classes (ranges), which were considered fuzzy sets (fuzzy trapezoidal numbers) and described by linguistic variables. Briefly, the linguistic variables (Figure 6) quantify the natural language characterizing the words by fuzzy sets defined in the universe of discourse in which the variable is defined [69]. Thus, each alternative, based on its numeric value regarding the examined criterion, takes a membership degree of belonging to each fuzzy class (range) of the criterion. The rating (R) of an alternative, concerning the examined criterion, depended on the activation degree of the fuzzy if–then rules designed by the experts. It is pointed out that a FIS was designed for each criterion that was examined.

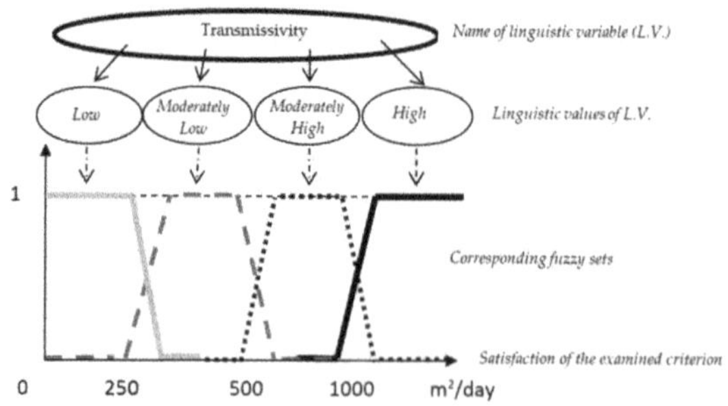


**Figure 6.** Transmissivity divided into four fuzzy classes described by linguistic variables.

The general structure of a FIS consists of the following components [88,89]:

(1) A fuzzifier. The degree to which the input data (the numeric value of the alternative with respect the examined criterion) satisfy the fuzzy rule is calculated. In this step, the classes of the input variables (criteria) are fuzzified.

(2) An inference engine module. A series of fuzzy if–then rules is modulated. Based on the satisfaction degree described in the previous step and the selected fuzzy implication, the fuzzy rules fire strength to infer knowledge. Thus, a fuzzy conclusion is calculated for each rule.

(3) Aggregation. All fuzzy conclusions inferred by all fuzzy rules are combined into a final fuzzy conclusion.
(4) A defuzzifier. The fuzzy output (inferred knowledge) is translated to crisp output (rating of each alternative with respect to the examined criterion).

From a mathematical point of view, a FIS, based on the Mamdani model, is composed of the premise (antecedent if–part) and conclusion (consequent then–part). It can be formulated as follows [82]:

$$Ru(k): If\ x_1\ is\ A_1^k\ and\ x_2\ is\ A_2^k \ldots and\ x_z\ is\ A_z^k\ then\ y\ is\ B^k \quad (11)$$

where $A_z^k$ is a fuzzy set in Universe $U \subset \mathbb{R}$ (input variables of the model), $B^k$ is a fuzzy set in Universe $Z \subset \mathbb{R}$ (output variables of the model), $(x_1, x_2, \ldots, x_z)^T \in U, y \in Z$ and $k$ is the fuzzy rule $k$th. The above linguistic rule describes a mapping from $U_1 \times U_2 \times \ldots \times U_z$ to $Z$ [88].

In this research, the antecedent of the fuzzy if–then rule includes only one input variable:

$$Ru(k):\ If\ x\ is\ A^k\ then\ y\ is\ B^k \quad (12)$$

The contribution of the rule $k$th, $Ru(k)$, to the output (inferred conclusion) of the model is also a fuzzy set which may be represented by a fuzzy implication. Here, the common engineering fuzzy implications of the algebraic product (Equation (13)) and the min implication (Equation (14)) are used, which are suited to inference on the basis of phenomenological information [90,91]. The min fuzzy implication is recommended in the case that there is no distinction between cause and effect [90].

$$\mu_{Ru^k}(x,y) = \mu_{A^k}(x) \cdot \mu_{B^k}(y) \quad (13)$$

$$\mu_{Ru^k}(x,y) = \min\{\mu_{A^k}(x),\ \mu_{B^k}(y)\} \quad (14)$$

Let a fuzzy rule consists of two fuzzy propositions as follows:

$$\left.\begin{array}{l} Fuzzy\ proposition\ 1:\ x\ is\ A'^k \\ Fuzzy\ proposition\ 2:\ Ru->If\ x\ is\ A^k\ then\ y\ is\ B^k \end{array}\right\} Conclusion:\ y\ is\ B'^k \quad (15)$$

Then, according to the generalized modus ponens (GMP) concept [66], a new fuzzy proposition should be inferred such that the closer $A'^k$ to $A^k$, the closer $B'^k$ to $B^k$. Therefore, based on the concept of GMP, a fuzzy conclusion can be defined for each fuzzy rule with respect to the given input:

$$\mu_{B'^k}(y) = \sup_{x \in X} t\ [\mu_{A'^k}(x),\ \mu_{Ru^k}(x,y)] \quad (16)$$

Other $t$-norms, instead of the algebraic product, could be used (Equation (17)). The following Equation (17) is in fact a max–min composition in case the input is a vector. Practically, it holds:

$$\mu_{B'^k}(y) = \max_{x \in X}\ [\mu_{A'^k}(x) \wedge \mu_{Ru^k}(x,y)] = \mu_{A'^k \circ Ru^k}(y) \quad (17)$$

In the case that the input data of the model are crisp numbers (fuzzy singleton), it holds:

$$\mu_{A'^k}(x) = \begin{cases} 1\ if\ x = x* \\ 0\ otherwise \end{cases} \quad (18)$$

where $x^*$ is a point in its universe.

Consequently, based on the fuzzy implication process, a fuzzy conclusion corresponds to each fuzzy rule. In order to obtain a final fuzzy conclusion regarding the examined criterion, all the fuzzy conclusions must be combined. This procedure is called the aggregation process and is carried out by using an operator which, in this research, is the max union.

In conclusion, in a fuzzy inference system based on the Mamdani model, the fuzzy implication of algebraic product, the algebraic product for all t-norms, and the fuzzy max union are selected. In the case of crisp input data, the Mamdani's product inference engine is described in Equation (19) [69]. In the case that the min fuzzy implication, min for all t-norm operators and the max union are selected, then the fuzzy output of the system is expressed by Equation (20).

$$\mu_{B'}(y) = \max_{k=1}^{L}[\mu_{A^k}(x_i*) \cdot \mu_{B^k}(y)] \tag{19}$$

$$\mu_{B'}(y) = \max_{k=1}^{L}\left[\min\left\{\mu_{A_i^k}(x_i*), \mu_{B^k}(y)\right\}\right] \tag{20}$$

As mentioned above, the output of the described fuzzy inference system is a fuzzy set $B'^k$. The last process leading to the decision (which, in this research, is the rating of an alternative with respect to the examined criterion) is the defuzzification of the fuzzy output. Here, the defuzzification method of centroid calculation is selected, which counts the gravity center (centroid) $y^*$ as the output of the fuzzy rule-based system.

$$y* = \frac{\sum_{i=1}^{n} y_i \mu(y_i)}{\sum_{i=1}^{n} \mu(y_i)} \tag{21}$$

**4. Implementation and Results**

A hybrid fuzzy multi-criteria methodology was implemented in this research to solve the problem of selection of the more suitable site for applying MAR, choosing between seven (7) alternatives (Al) under the consideration of nine (9) criteria (C). Thus, the research's final goal was a relative rank order list of the suitability of the examined alternatives. The methodology included the following steps:

(1) Based on the experts' knowledge, a fuzzy reciprocal pairwise comparison matrix $\widetilde{A}_{9x9}$ (Equation (2)) was generated regarding the examined (nine) criteria. Table 1 presents the elements of Saaty's scale [70] as triangular fuzzy numbers (TFN) (Equation (6)), while the fuzzy judgments of the matrix $\widetilde{A}_{9x9}$ are presented in Equation (24).

(2) The optimization problem described in Equation (9) was built and solved by the use of the optimization software LINGO and the weights of criteria (priority vector) were estimated as crisp numbers. Then, the priority vector was normalized based on Equation (10). For space saving reasons, only an example of a pair of inequalities is given regarding the criteria of transmissivity ($C_5$) and piezometry ($C_9$). Based on Equation (24) presented below, transmissivity was considered moderate less important in comparison to piezometry; that is $\ln(w_5/w_9) \approx \left(\frac{1}{4}, \frac{1}{3}, \frac{1}{2}\right)$. Therefore, it holds:

$$\begin{array}{l} \ln w_5 - \ln w_9 - \lambda(\ln((1/3)/(1/4))) + \delta_{59} \geq \ln(1/4) \\ \ln w_9 - \ln w_5 - \lambda(\ln((1/2)/(1/3)) + n_{59} \geq -\ln(1/2) \end{array}$$

As aforementioned, the parameter $M$ aims to achieve a nonnegative value $\lambda$ and a small deviation from the fuzzy inequalities constraints which define the fuzzy spread (support set) of each fuzzy judgment. The $M$-value should ensure a balance between the $\lambda$-value and the values of deviation variables [86]. The ideal solution is $\lambda$ = 1 and $\sum_{i=1}^{9-1}\sum_{j=i+1}^{9}\left(\delta_{ij}^2 + n_{ij}^2\right)$ = 0; however, in the real world this is a rare condition [86]. The optimization problem was solved for several $M$-values. For $M$ = 0.2 it holds: $\lambda$ = 0.6069 and $\sum_{i=1}^{9-1}\sum_{j=i+1}^{9}\left(\delta_{ij}^2 + n_{ij}^2\right)$ = 2.763, for $M$ = 0.5 and hence: $\lambda$ = 0.317 and $\sum_{i=1}^{9-1}\sum_{j=i+1}^{9}\left(\delta_{ij}^2 + n_{ij}^2\right)$ = 1.807.

For or $M = 1$ it holds: $\lambda = 0.0407$ and $\sum_{i=1}^{9-1}\sum_{j=i+1}^{9}\left(\delta_{ij}^2 + n_{ij}^2\right) = 1.172$, while for $M > 2$ the solutions of deviations converged on $\sum_{i=1}^{9-1}\sum_{j=i+1}^{9}\left(\delta_{ij}^2 + n_{ij}^2\right) = 1.096$ and, the parameter $\lambda = 0$.

Therefore, the $M$-value selected in this study was $M = 0.5$ since it led to a balanced solution. The solution of the priority vector for $M = 0.5$ before the normalization was $\mathbf{W}_{\tilde{A}}^{*} = (21.7949, 21.1237, 15.3198, 15.3198, 10.8694, 4.0643, 3.9769, 3.8704, 1.6641)$, while the normalized weights are presented in Table 2.

Let us verify the above fuzzy constraints regarding the criteria $C_5$ and $C_9$. According to the results (for $M = 0.5$), the deviation variables $\delta_{59}$ and $n_{59}$ are zero, therefore:

$$\ln(10.8694) - \ln(21.7949) - 0.3174(\ln((1/3)/(1/4))) + 0.0 \geq \ln(1/4) \Leftrightarrow -0.78703 \geq -1.38629$$
$$\ln(21.7949) - \ln(10.8694) - 0.3174(\ln((1/2)/(1/3))) + 0.0 \geq -\ln(1/2) \Leftrightarrow 1.35574 \geq 0.69315$$

which holds.

(3) A Mamdani fuzzy inference system, included in the Fuzzy Logic Toolbox of MATLAB, was designed for each criterion. Initially, based on the experts' judgments, the classes of the examined criteria were fuzzified. The number of classes was determined by using the empirical rule of Sturges [92], which is a common technique in hydrology (Equation (22)):

$$q = 1.33 \log N \tag{22}$$

where $q$ is the number of classes and $N$ is the number of data (the numeric value of each alternative regarding the examined criterion, i.e., $N = 7$).

Then, a system of fuzzy if–then rules (with one input variable, e.g., if slope is low then rating is high) was built (Equation (12)). Both the algebraic product and the min implication were selected to be used in the fuzzy inference process (Equations (19) and (20)). All fuzzy output variables of all rules were combined into a fuzzy conclusion with the max union. Finally, the aggregated fuzzy output was defuzzified based on Equation (21), which is the rating of the each alternative, with respect to each criterion. For the calculation of the rating ($R$) based on the designed FIS, a simple code was written in MATLAB, which fed the FIS with crisp input data.

(4) The rating ($R$) of each alternative was multiplied with the weight of the examined criterion. The final evaluation (final score) of each alternative with respect of all criteria was achieved based on an aggregative model described as follows:

$$Al_{u.c.} = \sum_{i=1}^{n} w_i \times R_i \tag{23}$$

where $Al_{u.c.}$ is the alternative under consideration, $w_i$ is the weight of the criterion $i$ (for $n = 9$) and $R_i$ is the estimated (by the corresponding FIS) rating of $Al_{u.c.}$ regarding the criterion $i$.

(5) Based on Equation (23), a rank list of the alternatives (from the most suitable site to the least suitable site for applying MAR) was obtained by using both the algebraic product (Equation (19)) and min implication (Equation (20)) (Table 3). It is noted that although the $M$-value adopted in this study is equal to 0.5, the results based on $M = 1$ are also presented in Table 3.

As shown in the above Table 1, the fuzzy spread increased as the domination between two criteria became stronger. Particularly, a lower fuzzy spread (0.5) corresponds to a judgment of equal importance, "about 1", while a higher fuzzy spread (2.0) corresponds to a judgment of extreme importance, "about 9". The authors used this increasing fuzziness on judgments because a wide fuzzy spread in an area of weak domination (i.e., from equal up to moderate importance) may lead to a large overlapping and therefore, two criteria may be considered of "about equal" importance even though they have a weak domination each other (e.g., $w_i/w_j$ = "about 2" or "about 3"). Furthermore, it seems more convenient to accurately *adjudicate* if two conditions are equal or about equal than to determine the degree of how important a condition is against the other in the case that the importance of the two conditions significantly differs. For instance, a decision maker (DM) might decide more comfortably that $\widetilde{a}_{1,2}$ = "about 2", which means that $w_1/w_2$ might be equal to 3 or equal to 1. On the other hand, in the case that a criterion is much more important than another, making such a judgment that $\widetilde{a}_{1,7}$ = "about 7", the real ratio of weights may be equal to 5 or 9, which also denotes significant importance.

**Table 1.** Triangular fuzzy numbers (TFN) for the pairwise comparison of the criteria regarding their relative importance.

| Relative Importance of Two Criteria | TFN |
|---|---|
| Equal | (0.5, 1, 1.5) |
| Moderate | (2, 3, 4) |
| Strong | (3.75, 5, 6.25) |
| Very strong | (5.5, 7, 8.5) |
| Extreme | (7, 9, 9) |
| Intermediate values | (1.25, 2, 2.75), (3, 4, 5), (4.75, 6, 7.25), (6.5, 8, 9) |

It is worth mentioning that several fuzzy spreads were tested. Particularly, constant spreads for the whole scale of Table 1 were used when either the fuzziness was equal to one (e.g., 2, 3, 4) or equal to two (e.g., (1, 3, 5). The tests showed that the weights of the less important criteria increased as the fuzziness increased.

$$\widetilde{A} = \begin{array}{c} \\ \end{array}\begin{bmatrix} C1 & C2 & C3 & C4 & C5 & C6 & C7 & C8 & C9 \\ 1 & \left(\frac{1}{4},\frac{1}{3},\frac{1}{2}\right) & \left(\frac{1}{5},\frac{1}{4},\frac{1}{3}\right) & \left(\frac{1}{6.25},\frac{1}{5},\frac{1}{3.75}\right) & \left(\frac{1}{8.5},\frac{1}{7},\frac{1}{5.5}\right) & \left(\frac{1}{9},\frac{1}{8},\frac{1}{6.5}\right) & \left(\frac{1}{9},\frac{1}{8},\frac{1}{6.5}\right) & \left(\frac{1}{9},\frac{1}{9},\frac{1}{7}\right) & \left(\frac{1}{9},\frac{1}{9},\frac{1}{7}\right) \\ (2,3,4) & 1 & \left(\frac{1}{2.75},\frac{1}{2},\frac{1}{1.25}\right) & (2,3,4) & \left(\frac{1}{5},\frac{1}{4},\frac{1}{3}\right) & \left(\frac{1}{6.25},\frac{1}{5},\frac{1}{3.75}\right) & \left(\frac{1}{6.25},\frac{1}{5},\frac{1}{3.75}\right) & \left(\frac{1}{8.5},\frac{1}{7},\frac{1}{5.5}\right) & \left(\frac{1}{9},\frac{1}{8},\frac{1}{6.5}\right) \\ (3,4,5) & (1.25,2,2.75) & 1 & \left(\frac{1}{2.75},\frac{1}{2},\frac{1}{1.25}\right) & \left(\frac{1}{5},\frac{1}{4},\frac{1}{3}\right) & \left(\frac{1}{6.25},\frac{1}{5},\frac{1}{3.75}\right) & \left(\frac{1}{6.25},\frac{1}{5},\frac{1}{3.75}\right) & \left(\frac{1}{7.25},\frac{1}{6},\frac{1}{4.75}\right) & \left(\frac{1}{7.25},\frac{1}{6},\frac{1}{4.75}\right) \\ (3.75,5,6.25) & \left(\frac{1}{4},\frac{1}{3},\frac{1}{2}\right) & (1.25,2,2.75) & 1 & \left(\frac{1}{4},\frac{1}{3},\frac{1}{2}\right) & \left(\frac{1}{5},\frac{1}{4},\frac{1}{3}\right) & \left(\frac{1}{5},\frac{1}{4},\frac{1}{3}\right) & \left(\frac{1}{6.25},\frac{1}{5},\frac{1}{3.75}\right) & \left(\frac{1}{6.25},\frac{1}{5},\frac{1}{3.75}\right) \\ (5.5,7,8.5) & (3,4,5) & (3,4,5) & (2,3,4) & 1 & \left(\frac{1}{2.75},\frac{1}{2},\frac{1}{1.25}\right) & \left(\frac{1}{2.75},\frac{1}{2},\frac{1}{1.25}\right) & \left(\frac{1}{4},\frac{1}{3},\frac{1}{2}\right) & \left(\frac{1}{4},\frac{1}{3},\frac{1}{2}\right) \\ (6.5,8,9) & (3.75,5,6.25) & (3.75,5,6.25) & (3,4,5) & (1.25,2,2.75) & 1 & \left(\frac{1}{1.5},1,\frac{1}{0.5}\right) & \left(\frac{1}{2.75},\frac{1}{2},\frac{1}{1.25}\right) & \left(\frac{1}{2.75},\frac{1}{2},\frac{1}{1.25}\right) \\ (6.5,8,9) & (3.75,5,6.25) & (3.75,5,6.25) & (3,4,5) & (1.25,2,2.75) & (0.5,1,1.5) & 1 & \left(\frac{1}{2.75},\frac{1}{2},\frac{1}{1.25}\right) & \left(\frac{1}{2.75},\frac{1}{2},\frac{1}{1.25}\right) \\ (7,9,9) & (5.5,7,8.5) & (4.75,6,7.25) & (3.75,5,6.25) & (2,3,4) & (1.25,2,2.75) & (1.25,2,2.75) & 1 & \left(\frac{1}{1.5},1,\frac{1}{0.5}\right) \\ (7,9,9) & (6.5,8,9) & (4.75,6,7.25) & (3.75,5,6.25) & (2,3,4) & (1.25,2,2.75) & (1.25,2,2.75) & (0.5,1,1.5) & 1 \end{bmatrix} \quad (24)$$

The following Figure 7 illustrates the scale of the fuzzy judgments of Table 1. For instance, based on Equation (24), the fuzzy pairwise comparison $\widetilde{a}_{5,9} = \widetilde{\frac{1}{3}} \Leftrightarrow w_5/w_9 = \widetilde{\frac{1}{3}} \Leftrightarrow \widetilde{a}_{9,5} = \widetilde{3} \Leftrightarrow w_9/w_5 = \widetilde{3}$ between the criteria $C_5$ (transmissivity) and $C_9$ (piezometry) means that piezometry was "about 3" times more significant than transmissivity.

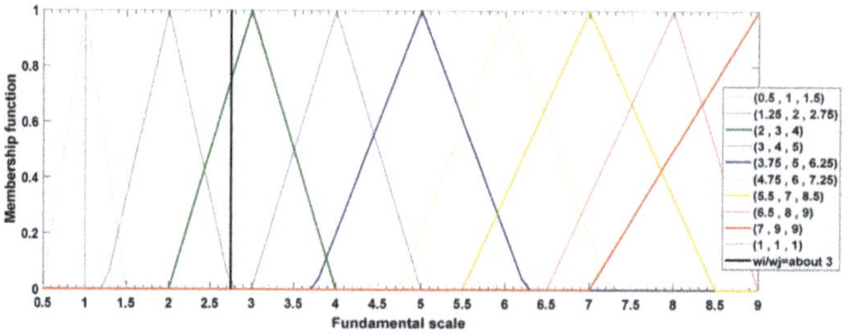


**Figure 7.** Graphical representation of the intensities of the relative importance of two criteria as triangular fuzzy numbers (TFN).

However, an ideal weight ratio ($w_5/w_9$) between these two criteria exists, which is a crisp number. As aforementioned, the ideal weight ratio of two criteria must be included in the corresponding fuzzy judgment (i.e., the support set of the TFN), or the deviations from the fuzzy thresholds should be small. In the investigation at hand, the black line in Figure 7 is assumed to be this ideal crisp weight ratio $w_1/w_5$.

The rating of alternatives, the weights before and after normalization, as well as the initial ranges (classes) with their corresponding linguistic variables of the criteria, are presented in Table 2 below. As aforementioned, in the FAHP of Wang and Chin [85], the final weights of criteria are crisp numbers. Based on the experts' knowledge, the criteria were divided into four crisp classes based on Equation (22). Then, the classes of each criterion were fuzzified and described by linguistic variables (low, moderately low, moderately high, high). These fuzzy classes are illustrated in Figures 8–10.

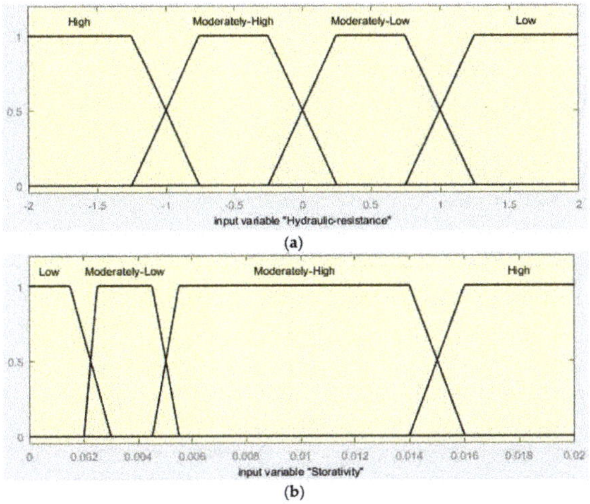


**Figure 8.** Fuzzy classes described by linguistic variables regarding the criteria (**a**) hydraulic resistance ($C_7$) and (**b**) storativity ($C_6$).

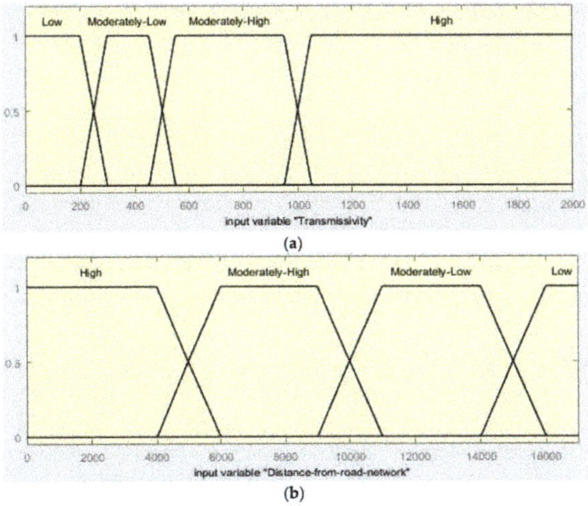


**Figure 9.** Fuzzy classes described by linguistic variables regarding the criteria (**a**) transmissivity ($C_5$) and (**b**) distance from road network ($C_4$).

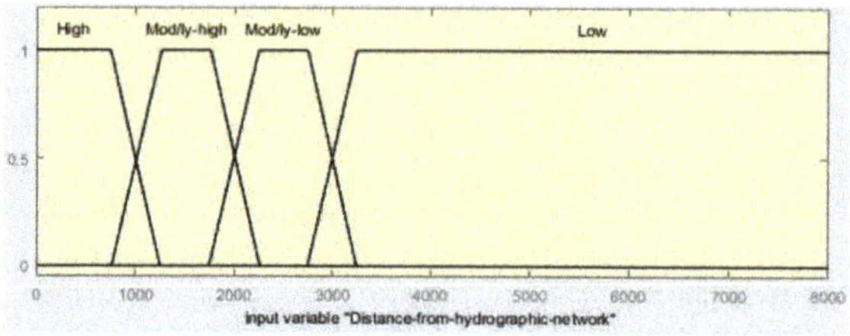


(a)

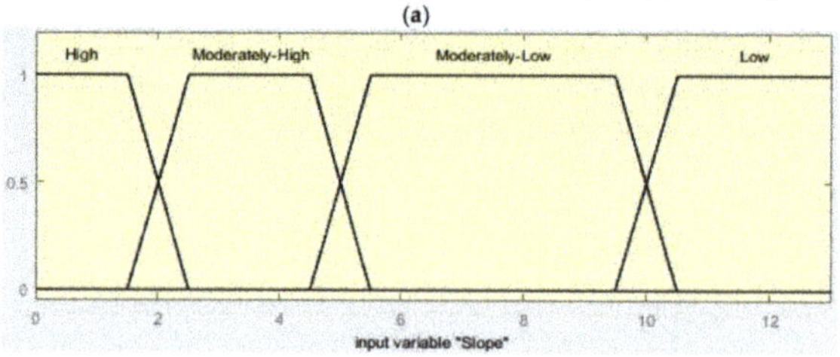


(b)

**Figure 10.** Fuzzy classes described by linguistic variables regarding the criteria (**a**) distance from hydrographic network ($C_3$) and (**b**) slope ($C_2$).

**Table 2.** Crisp weights of criteria and the rating of alternatives with respect to the classes of criteria. The crisp classes have been fuzzified in FIS.

| Criteria | Classes | Linguistic Variable | Rating | Weight | Normalized Weight |
|---|---|---|---|---|---|
| $C_1$ | | High<br>Good<br>Moderate<br>Deficient<br>Worse<br>Unknown | 10<br>8<br>6<br>4<br>2<br>0 | 1.6641 | 0.0169 |
| $C_2$ | 0–2<br>2–5<br>5–10<br>>10 | High<br>Moderately high<br>Moderately low<br>Low | 10<br>7.5<br>5<br>2.5 | 3.8704 | 0.0394 |
| $C_3$ | <1000<br>1000–2000<br>2000–3000<br>>3000 | High<br>Moderately high<br>Moderately low<br>Low | 10<br>7.5<br>5<br>2.5 | 3.9769 | 0.0405 |
| $C_4$ | 0–5000<br>5000–10,000<br>10,000–15,000<br>>15,000 | High<br>Moderately high<br>Moderately low<br>Low | 10<br>7.5<br>5<br>2.5 | 4.0643 | 0.0414 |
| $C_5$ | >1000<br>500–1000<br>250–500<br>0–250 | High<br>Moderately high<br>Moderately low<br>Low | 10<br>7.5<br>5<br>2.5 | 10.8694 | 0.1109 |
| $C_6$ | >0.015<br>0.005–0.015<br>0.0025–0.005<br>0–0.0025 | High<br>Moderately high<br>Moderately low<br>Low | 10<br>7.5<br>5<br>2.5 | 15.3198 | 0.1563 |
| $C_7$ | <−1<br>−1–0<br>0–1<br>>1 | High<br>Moderately high<br>Moderately low<br>Low | 10<br>7.5<br>5<br>2.5 | 15.3198 | 0.1563 |

**Table 2.** *Cont.*

| Criteria | Classes | Linguistic Variable | Rating | Weight | Normalized Weight |
|---|---|---|---|---|---|
| $C_8$ | 1<br>0 | High<br>Low | 10<br>0 | 21.1237 | 0.2155 |
| $C_9$ | 1<br>0 | High<br>Low | 10<br>0 | 21.7949 | 0.2223 |

As noted, the classes of piezometry, type of aquifer, and ecological status are crisp. Regarding the criteria of piezometry and type of aquifer, two classes were considered and are presented in Appendix B (Figure A2). Therefore, the rating of these alternatives can be either 10 or 0. The criterion of ecological status was divided into six classes based on the relevant national River Basin Management Plans [60]. Since these three criteria divided into crisp classes, the ratings of the alternatives were calculated without the use of FIS.

Table 3 and Figure 11 present the final rating of the alternatives resulting from using either the min implication or the fuzzy implication of the algebraic product, for $M = 1$ and $M = 2$. As shown in Table 3, the ranking order of the alternatives ($Al_5 > Al_2 > Al_1 > Al_4 > Al_7 > Al_6 > Al_3$) was identical in almost all cases. A slight variation between the latter options took place when the min implication was used, and the $M$-value was equal to $M = 1$ ($Al_5 > Al_2 > Al_1 > Al_4 > Al_7 > Al_3 > Al_6$).

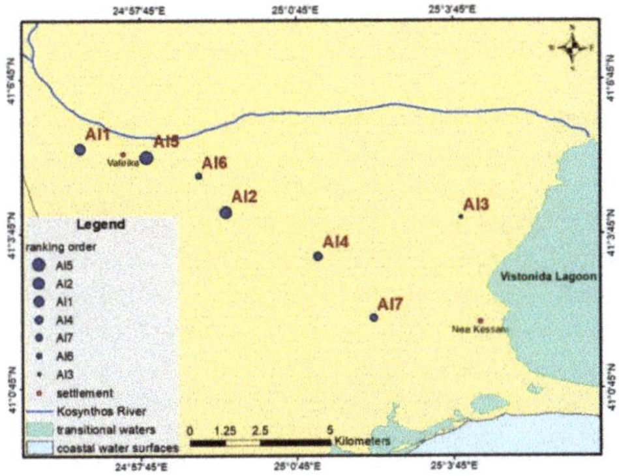


**Figure 11.** Final evaluation of alternatives.

**Table 3.** Final evaluation of the alternatives based on Mamdani product and min implication.

| Ranking Order/ Implication | Alg. Prod. ($M = 0.5$) | Min ($M = 0.5$) | Alg. Prod. ($M = 1$) | Min ($M = 1$) |
|---|---|---|---|---|
| $Al_5$ | 8.0317 | 8.0320 | 7.9846 | 7.9792 |
| $Al_2$ | 7.9859 | 7.9932 | 7.9355 | 7.9370 |
| $Al_1$ | 7.7936 | 7.7877 | 7.7329 | 7.7191 |
| $Al_4$ | 7.3382 | 7.3611 | 7.2765 | 7.2939 |
| $Al_7$ | 6.5145 | 6.5179 | 6.4232 | 6.4228 |
| $Al_6$ | 4.6199 | 4.6045 | 4.5888 | 4.5712 |
| $Al_3$ | 4.5822 | 4.5829 | 4.5508 | 4.7090 |

## 5. Discussion

According to the results (Table 3 and Figure 11), the more favorable sites ($Al_5 > Al_2 > Al_1$) for applying floodwater spreading were located, in general, in the northwest of the study area. Particularly, $Al_5$ was the most preferable alternative in all cases. In contrast, the final ratings of the alternatives located in the southeast part of the area under investigation were low (less favorable alternatives). This may be partly explained by the composition of geological formations which range from fine to silty clay in the SE–E direction towards Vistonida Lagoon [53,55]. However, this is not an absolute condition due to the significant variability of alluvial composition in the study area [58] on the one hand, and the contribution of the other criteria in the analysis on the other. For instance, the alternatives $A_1$ and $Al_6$ were less preferable than $Al_2$. Certainly, in the case of evaluating just a few alternatives, the best alternative might be identified on the basis of traditional hydrogeological analysis even though the criteria weights would not be taken into account. However, the more alternatives and criteria, the more complexity for the selection of the most preferable recharge site by an expert. Moreover, it is desirable that the MAR application is as effective as possible given its not negligible cost [93], and thus the solution of the selection problem should really be the best. As pointed out, the proposed methodology can be applied on both a large and small number of alternatives/criteria and this is a merit of the proposed methodology.

The local conditions of the case study (e.g., sparsely populated flat area) favor the application of flood spreading methods through infiltration basins, inasmuch as the excess winter water of the Kosynthos River might be used. Thus, the criteria were selected according to this type of MAR focusing on the hydrogeological local characteristics. Other criteria, such as socioeconomic and technical criteria, were not taken into account, as they are more in line with the design and implementation of MAR. It is worth mentioning that all alternatives were located south of the Kosynthos River. This is due to the fact that the Kosynthos River is the north hydraulic boundary of the aquifer system in the Xanthi plain [56–58].

The criteria weights were obtained through the FAHP–LFFP-based nonlinear method. Thus, in contrast to the conventional AHP, fuzziness was used to incorporate the uncertainty from the subjectivity of the initial experts' judgments. Using the FAHP–LFFP-based nonlinear method, a potential zero weight value was avoided and the non negativity of the $\lambda$-value was ensured, generating a unique optimal crisp priority vector. Besides, by using the FAHP–LFFP-based nonlinear method, the analysis could be performed either on the upper triangular elements of the fuzzy pairwise comparison matrix or on the lower ones. However, fuzzy weights can be obtained with the use of other methods (e.g., [94]).

An interesting point for discussion is the fuzziness of the initial judgments. In this study, an increasing fuzzy spread (width) was adopted regarding the scale of relative importance between two criteria. A reason for this was to avoid a large overlapping in the weak domination area that could lead to assigning almost equal importance to criteria even though there was a weak or moderately weak domination between the two criteria. Moreover, it seems more comfortable for a DM to assess that two conditions are equal or about equal than to assign the degree of their relative importance when their importance significantly differs. It is worth mentioning that after several tests on fuzzy spreads, which for space saving reasons are not presented, it was found that by increasing the fuzziness, the weights of the less important criteria increased (i.e., distance from road network, distance from hydrographic network, slope, and ecological status).

Regarding the parameter $M$ used in the FAHP–LFFP method, it aims to ensure the non-negativity of parameter $\lambda$ and to allow little deviation from fuzzy inequalities constraints, and hence an optimal solution that simultaneously satisfies all the fuzzy judgments can be achieved. An interesting point is that for $M > 2$, the solutions of deviations converged on a unit which had the lowest value in comparison with the solutions of deviation for $M < 2$. However, the $\lambda$-value was zero, which means that the inequalities of Equation (8) did not simultaneously hold, and thus these values were not used. The $M$-value selected was $M = 0.5$ because it produced a more balanced solution between the $\lambda$-value and the total deviations than those that were produced when $M = 1$ and $M = 0.2$ were used. In the calculation of the $\lambda$-value, the fuzzy min intersection was used because using the other widely used t-norm (algebraic product) resulted in a lower satisfaction degree of the fuzzy pairwise comparisons.

The ratings of the alternatives with respect to each criterion were based on using fuzzy inferences systems based on the Mamdani approach. The use of such systems was preferred in this study in order for the ranges (classes) of criteria to be fuzzified. The ranges of criteria were modulated on the basis of the available information (e.g., min–max values of each criterion regarding the case study) and of the experts' knowledge. Therefore, it seemed more reasonable to use fuzzy thresholds of the ranges than crisp ones for taking into account the uncertainty derived from the lack of information and imprecise knowledge.

Another reason for using FIS is that, there are not precise and well-known objective functions for each criterion, as in the usual multi-objective problems.

On the other hand, the selection problem examined in this study could be solved based on the exclusive use of FIS; however, this choice was not preferred because it would lead to a complex system of fuzzy if-then rules.

In this study, each FIS is based on a simple system of fuzzy if-then rules, where only an input variable, has been used (Equation (12)). Therefore, neither t-conorms nor t-norms were used in the antecedent if-part. As the fuzzy implication is concerned, both the two widely used fuzzy implications were used. That is, the Mamdani implication, in which the t-norm of algebraic product is used, and the min implication, in which the t-norm of min intersection is used. In general, in case of no distinction between cause and effect, the min fuzzy implication is recommended. However, several other fuzzy implications can be used [95], the appropriate selection of which is an open issue [90]. For the composition of the fuzzy rules, the widely used t-conorm of max union was preferred. However, other t-conorms can also be tested. The fuzzy output of each FIS method is defuzzified by selecting the centroid method, which is considered a balanced solution given that it counts the gravity center.

The synthesis of criteria with respect of each alternative was achieved based on a simple additive model, in which the rating (the defuzzified output of the FIS) of each alternative is multiplied with the normalized weight with respect of each criterion. The total sum is the final evaluation (final score) of each alternative. The alternative with the highest score is the solution to the decision-making problem. Other (fuzzy) multi attributes decision-making (MADM) methods might be used for the synthesis of criteria and the final ranking of alternatives using the weights derived by the FAHP–LFFP-based nonlinear method (e.g., fuzzy TOPSIS, fuzzy ELECTRE, fuzzy pattern recognition). However, the use of such methods would have increased the complexity of the solution process on one hand and more information would have been required on the other.

The FAHP–LFFP method and the FIS were applied to the examined problem for the first time. According to Table 3, the solution (ranking order of the alternatives) did not changed regardless of which $M$-value or fuzzy inference engine was used, except in the case of min implication, in which the less preferable alternatives exchanged order. Thus, based on the knowledge of previous studies [53,55–58,60] regarding the hydrogeological conditions of the case study, and by taking into account the stability of the solution, the results of the proposed methodology are considered reliable. It would be of great interest if these results were compared with those of other (hybrid) multi-criteria methods that could be applied in the composition of criteria (e.g., fuzzy ELECTRE, fuzzy pattern recognition).

## 6. Conclusions

This research applied a hybrid fuzzy multi-criteria methodology to select preferable site for applying MAR systems in the context of increasing the local groundwater availability for later water use in case of drought occurrence. Fuzzy set theory and fuzzy logic were used both in the determination of criteria weights and in the evaluation of alternatives, thus incorporating the uncertainties derived from the subjectivity and the lack of information.

The proposed methodology was applied in a flat agricultural area in the southeast of Xanthi city, northern Greece, where a significant number of irrigation wells draw from the shallow aquifer system. The local conditions favor the implementation of flood spreading through infiltration basins utilizing excess winter water of the Kosynthos River.

The results show that:

(1) The more preferable sites are located in the northwestern part of the study area near the Kosynthos River, while the most preferable is the $Al_5$ alternative. This mainly holds due to the hydrological conditions of the northwestern part, which are more favorable than the southeastern ones for groundwater recharge. However, this is not an absolute condition due to the intense diversity of geological formations. In addition, the contribution of the analysis of the other criteria also affected the final rank list.

(2) The proposed methodology can be used for distinguishing discrete preferable points (alternatives) for MAR application without limitation in number of the examined alternatives. In addition, the selected criteria can be added or subtracted depending on specific local conditions. In case of application of another type of MAR the selected criteria should be adjusted as well. Certainly, in the case of a small number (<10) of alternatives to study, a classical hydrogeological analysis may lead to the selection of the most preferable alternative. However, in that case, the criteria weights would not be taken into account and thus, there would be the possibility of not obtaining an efficient and comprehensive solution. On the other hand, in the case of a large number of examined alternatives and criteria, where the complexity increased, the proposed methodology might be also a useful tool for ranking discrete alternatives. Other than that, it could be used as an alternative way to identify suitable recharge sites in case of low data availability.

(3) The applicability of the methodology requires lithological profile and hydraulic characteristics of both vadose and saturated zones, while it can be applied only in unconfined aquifers and where there is an underlying drainage axis. Furthermore, the application of this type of MAR requires the presence of a river whose excess waters can be utilized. Thus, after a preparatory screening, seven alternatives satisfied all the assumptions and criteria were finally evaluated.

(4) In general, the use of fuzzy logic in AHP can incorporate the uncertainty from the subjectivity of the initial experts' judgments, while the fuzzy version of AHP implemented in this paper (FAHP–LFFP-based nonlinear method) can ensure a unique and optimal solution.

(5) Fuzzy inference systems (FIS) based on Mamdani's approach are used in order to determine the rating of each alternative, since the objective function (or value function) regarding each examined criterion is unknown. In addition, with use of FIS, the classes of criteria are fuzzified, and this fact is more reasonable to describe the real conditions.

**Author Contributions:** Conceptualization, C.P., M.S., F.P., I.G., N.K. and B.P.; methodology, C.P., M.S., F.P., I.G., N.K. and B.P.; software, C.P. and M.S.; investigation, C.P., M.S., F.P., I.G., N.K. and B.P.; resources, C.P., M.S., F.P., I.G., N.K. and B.P.; writing—original draft preparation, C.P., M.S., F.P., I.G., N.K. and B.P.; writing—review and editing, C.P., M.S., F.P., I.G., N.K. and B.P. All authors have read and agreed to the published version of the manuscript.

**Funding:** This research was co-financed by Greece and the European Union (European Social Fund—ESF) through the Operational Programme "Human Resources Development, Education and Lifelong Learning" in the context of the project "Strengthening Human Resources Research Potential via Doctorate Research" (MIS-5000432), implemented by the State Scholarships Foundation (IKY).

**Conflicts of Interest:** The authors declare no conflict of interest.

## Appendix A

Each mapping from the Universe X into the [0, 1] is a *fuzzy set*. An element $x$ of a fuzzy set $\widetilde{A}$ is expressed either by the membership function, $\mu_{\widetilde{A}}(x)$, or by $\alpha$-cuts, ${}^{\alpha}\widetilde{A}$. The $\alpha$-cuts are crisp sets defined as follows:

$$^{\alpha}\widetilde{A} = \{x \in X : \mu_{\widetilde{A}}(x) \geq \alpha\} \tag{A1}$$

If $\alpha = 0$, then it holds $\mu_{\widetilde{A}}(x) > 0$ for all the elements $x \in X$.

A special kind of fuzzy sets comprises *fuzzy numbers*. A *fuzzy number* $\widetilde{Z} \subset \mathbb{R}$ satisfies the following properties [65]:

(1) $\widetilde{Z}$ is a normal set, i.e., $\exists x \in X$ such that $\mu_{\widetilde{Z}}(x) = 1$;
(2) ${}^{\alpha}Z$ must be a closed interval $\forall \alpha \in (0, 1]$;
(3) the strong zero cut, ${}^{0+}Z$, which is called the support set of $\widetilde{Z}$, must be bounded.

The second property implies that a fuzzy number $\widetilde{Z}$ is a convex set as well.

There are many types of fuzzy numbers. Trapezoidal fuzzy numbers are based on four points (elements of the set of real numbers $\mathbb{R}$), while triangular fuzzy numbers are based on three points (elements of $\mathbb{R}$). Typical shapes of these fuzzy numbers are presented in the Appendix A in Figure A1a,b, correspondingly. The membership function of a trapezoidal fuzzy number is described as follows:

$$\mu_{\widetilde{Z}}(x) = \begin{cases} (x-l)/c_1 & if\ l \leq x < m \\ 1 & if\ m \leq x \leq n \\ (u-x)/c_2 & if\ n \leq x \leq u \\ 0 & otherwise \end{cases} \tag{A2}$$

where $(x - l)/c_1$ is a continuous from the left and increasing function from $(l, m) \to [0, 1]$ and $(u - x)/c_2$ is a continuous from the right and decreasing function from $(n, u) \to [0, 1]$. The parameters $c_1$ and $c_2$ denote the fuzzy spreads of the fuzzy numbers. In the case of trapezoidal fuzzy number (Figure A1a), it holds $c_1 = m - l$ and $c_2 = u - n$. In the case of triangular fuzzy numbers (Figure A1b), it holds $m = n$, and therefore, $c_1 = m - l$ and $c_2 = u - m$.

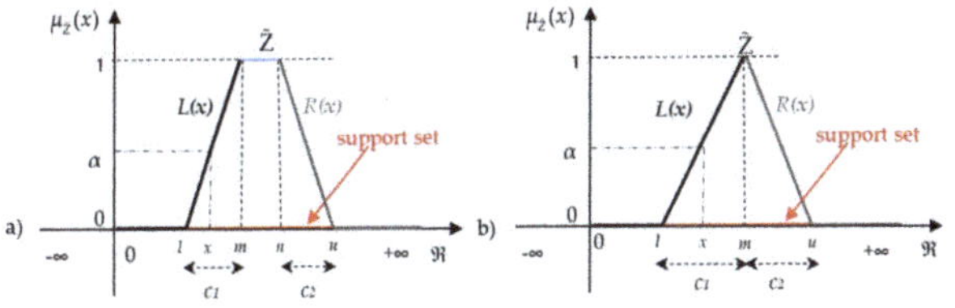


**Figure A1.** Typical shapes of (**a**) trapezoidal and (**b**) triangular fuzzy numbers.

In fuzzy set theory, the principle that extends all crisp mathematical concepts to fuzzy sets is named *extension principle*. With the use of the extension principle, all operations of the crisp functions between fuzzy sets (fuzzy numbers) are allowed [66].

Let $X$ be a Cartesian product of universes $X = X_1 \times \ldots \times X_n$ and $\widetilde{A}_1, \ldots, \widetilde{A}_n$ be fuzzy sets in $X_1 \times X_2 \times \ldots \times X_n$, correspondingly. If $f : X_1 \times \ldots \times X_n \to Y, y = f(x_1, \ldots, x_n)$, then a fuzzy set $\widetilde{B}$ in $Y$ can be defined as follows [67]:

$$\widetilde{B} = \{(y, \mu_B(y)) | y = f(x_1, \ldots x_n), (x_1, \ldots x_n) \in X\} \tag{A3}$$

where

$$\mu_{\widetilde{B}}(y) = \begin{cases} \sup\limits_{(x_1,\ldots,x_n)\in f^{-1}(y)} & \min\{\mu_{\widetilde{A}_1}(x_1),\ldots,\mu_{\widetilde{A}_n}(x_n)\ for\ f^{-1}(y) \neq \varnothing \\ 0 & otherwise \end{cases} \tag{A4}$$

where $f^{-1}$ is the inverse of $f$.

In the case of $n = 1$, then $f : X \to Y$, therefore it holds:

$$\widetilde{B} = f(\widetilde{A}) = \{(y, \mu_B(y))|y = f(x) \in X\} \tag{A5}$$

where

$$\mu_{\widetilde{B}}(y) = \begin{cases} \sup\limits_{x\in f^{-1}(y)} \mu_{\widetilde{A}_1}(x), & forf^{-1}(y) \neq \varnothing \\ 0 & otherwise \end{cases} \tag{A6}$$

*The extension principle* allows all algebraic operations between fuzzy sets to be performed. The relation between fuzzy sets is described by a crisp function.

In the general case, a fuzzy function is performed and a fuzzy relation (Equation (A7)), which is a fuzzy set, is produced even though the input is crisp number.

$$\widetilde{R} = \{((x,y), \mu_{\widetilde{R}}(x,y))|(x,y) = f(x) \in XxY\} \tag{A7}$$

The generalization of the extension principle, i.e., a fuzzy mapping from $X$ to universe $Y$, $\widetilde{f} : X \to Y$, is achieved with the use of *composition* of $R_1(X, Y)$ and $R_2(Y, Z)$:

$$\mu_{\widetilde{R}_1\circ\widetilde{R}_2}(z) = \underset{y\in Y}{\cup}\left[\min\{\mu_{\widetilde{R}_1}(x,y), \mu_{\widetilde{R}_2}(y,z)\}\right] x \in X, y \in Y, z \in Z \tag{A8}$$

where $\widetilde{R}_1(x,y)$, $(x,y) \in XxY$ and $\widetilde{R}_2(y,z)$, $(y,z) \in YxZ$ are two fuzzy relations [68]. By using the minimum intersection, the widely used *max–min composition* is produced. The use of algebraic product as fuzzy intersection in the composition is common too. The membership function $\mu_{\widetilde{R}_1\circ\widetilde{R}_2}(z)$ denotes the membership grades of the elements of the related fuzzy sets.

## Appendix B

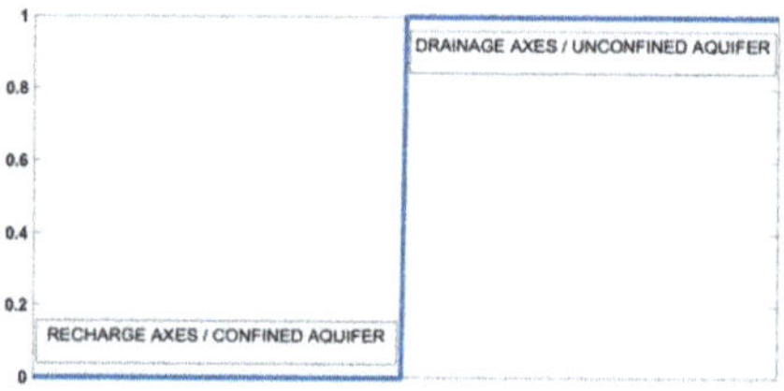


**Figure A2.** Typical shape of the crisp ranges regarding the criteria of piezometry ($C_9$) and type of aquifer ($C_8$).

## References


1. Vörösmarty, C.J.; Green, P.; Salisbury, J.; Lammers, R.B. Global Water Resources: Vulnerability from Climate Change and Population Growth. *Science* **2000**, *289*, 284–288. [CrossRef]
2. Tebaldi, C.; Hayhoe, K.; Arblaster, J.M.; Meehl, G.A. Going to the extremes. *Clim. Chang.* **2006**, *79*, 185–211. [CrossRef]
3. Taylor, R.G.; Scanlon, B.; Döll, P.; Rodell, M.; Van Beek, R.; Wada, Y.; Longuevergne, L.; Leblanc, M.; Famiglietti, J.S.; Edmunds, M.; et al. Ground water and climate change. *Nat. Clim. Chang.* **2013**, *3*, 322–329. [CrossRef]
4. Zwiers, F.W.; Alexander, L.V.; Hegerl, G.C.; Knutson, T.R.; Kossin, J.P.; Naveau, P.; Nicholls, N.; Schär, C.; Seneviratne, S.I.; Zhang, X. Climate extremes: Challenges in estimating and understanding recent changes in the frequency and intensity of

extreme climate and weather events. In *Climate Science for Serving Society*; Asrar, G.R., Hurrell, J.W., Eds.; Springer: Dordrecht, The Netherlands, 2013; pp. 339–389, ISBN 978-94-007-6691-4.
5. Wuebbles, D.; Meehl, G.; Hayhoe, K.; Karl, T.R.; Kunkel, K.; Santer, B.; Wehner, M.; Colle, B.; Fischer, E.M.; Fu, R.; et al. CMIP5Climate Model Analyses: Climate Extremes in the United States. *Bull. Am. Meteor. Soc.* **2013**, *95*, 571–583. [CrossRef]
6. Bouwer, H. Artificial recharge of groundwater:hydrogeology and engineering. *Hydrogeol. J.* **2000**, *10*, 121–142.
7. Maliva, R.G.; Guo, W.; Missimer, T.M. Aquifer storage and recovery: Recent hydrogeological advances and system performance. *Water Environ. Res.* **2006**, *78*, 2428–2435. [CrossRef]
8. Jager, H.I.; Smith, B.T. Sustainable Reservoir Operation: Can We Generate Hydropower and Preserve Ecosystem Values? *River Res. Appl.* **2008**, *24*, 340–352. [CrossRef]
9. Minsley, B.J.; Ajo-Franklin, J.; Mukhopadhyay, A.; Morgan, F.D. Hydrogeophysical Methods for Analyzing Aquifer Storage and Recovery Systems. *Groundwater* **2011**, *49*, 250–269. [CrossRef]
10. Giordano, M. Global Groundwater? Issues and Solutions. *Annu. Rev. Environ. Resour.* **2009**, *34*, 153–178. [CrossRef]
11. Kundzewicz, Z.W.; Doll, P. Will groundwater ease freshwater stress under climate change? *Hydrol. Sci. J.* **2009**, *54*, 665–675. [CrossRef]
12. Scanlon, B.R.; Reedy, R.C.; Faunt, C.C.; Pool, D.; Uhlman, K. Enhancing drought resilience with conjunctive use and managed aquifer recharge in California and Arizona. *Environ. Res. Lett.* **2016**, *11*, 035013.
13. de Graaf, I.E.M.; Gleeson, T.; Rens van Beek, L.P.H.; Sutanudjaja, E.H.; Bierkens, M.F.P. Environmental Flow Limits to Global Groundwater Pumping. *Nature* **2019**, *574*, 90–94. [CrossRef]
14. Dillon, P. Future management of aquifer recharge. *Hydrogeol. J.* **2005**, *13*, 313–316. [CrossRef]
15. Niswonger, R.G.; Morway, E.D.; Triana, E.; Huntington, J.L. Managed aquifer recharge through off-season irrigation in agricultural regions. *Water Resour. Res.* **2017**, *53*, 6970–6992. [CrossRef]
16. California Department of Water Resources. Using Flood Water for Managed Aquifer Recharge to Support Sustainable Water Resources. White Paper. 26 June 2018, p. 54. Available online: https://ui.adsabs.harvard.edu/abs/2018AGUFM.H51X1679M/abstract (accessed on 30 December 2021).
17. Dillon, P.; Stuyfzand, P.; Grischek, T.; Lluria, M.; Pyne, R.D.G.; Jain, R.C.; Bear, J.; Schwarz, J.; Wang, W.; Fernandez, E.; et al. Sixty years of global progress in managed aquifer recharge. *Hydrogeol. J.* **2019**, *27*, 1–30. [CrossRef]
18. Wendt, D.E.; Van Loon, A.F.; Scanlon, B.R.; Hannah, D.M. Managed Aquifer Recharge as a Drought Mitigation Strategy in Heavily-Stressed Aquifers. *Environ. Res. Lett.* **2021**, *16*, 014046. [CrossRef]
19. Zhao, M.; Boll, J.; Adam, J.C.; King, A.B. Can Managed Aquifer Recharge Overcome Multiple Droughts? *Water* **2021**, *13*, 2278. [CrossRef]
20. Gale, I.N.; Dillon, P. *Strategies for Managed Aquifer Recharge (MAR) in Semi-Arid Areas*; United Nations Educational, Scientific and Cultural Organisation (UNESCO): Paris, France, 2005; p. 30.
21. Ringleb, J.; Sallwey, J.; Stefan, C. Assessment of Managed Aquifer Recharge through Modeling—A Review. *Water* **2016**, *8*, 579. [CrossRef]
22. Foster, S.; van Steenbergen, F. Conjunctive groundwater use: A 'lost opportunity' for water management in the developingworld? *Hydrogeol. J.* **2011**, *19*, 959–962. [CrossRef]
23. Maliva, R.G. *Antropogenic Aquifer Recharge*; WSP Methods in Water Resources Evaluation Series No. 5; Springer: Cham, Switzerland, 2020; p. 861. [CrossRef]
24. Dillon, P.; Escalante, E.F.; Tuinhof, A. *Management of Aquifer Recharge and Discharge Processes and Aquifer Storage Equilibrium*; Groundwater Governance Thematic Paper 4; GEF-FAO: Rome, Italy, 2012.
25. Pakparvar, M.; Walraevens, K.; Cheraghi, S.A.M.; Cornelis, W.; Gabriels, D.; Kowsar, S.A. Assessment of groundwater recharge influenced by floodwater spreading: An intergrated approach with limited accessible data. *Hydrol. Sci. J.* **2017**, *62*, 147–164. [CrossRef]
26. Hashemi, H.; Berndtsson, R.; Persson, M. Artificial recharge by floodwater spreading estimated by water balancesand groundwater modeling in arid Iran. *Hydrol. Sci. J.* **2015**, *60*, 336–350. [CrossRef]
27. Saraf, A.K.; Choudhury, P.R. Integrated remote sensing and GIS for ground waterexploration and identification of artificial recharge sites. *Int. J. Remote Sens.* **1998**, *19*, 1825–1841. [CrossRef]
28. Al Saud, M. Mapping potential areas for groundwater storage in Wadi Aurnah Basin, western Arabian Peninsula, using remote sensing and geographic information system techniques. *Hydrogeol. J.* **2010**, *18*, 1481–1495. [CrossRef]
29. Senanayake, I.P.; Dissanayake, D.M.D.O.K.; Mayadunna, B.B.; Weerasekera, W.L. An approach to delineate groundwater recharge potential sites in Ambalantota, Sri Lanka using GIS techniques. *Geosci. Front.* **2016**, *7*, 115–124. [CrossRef]
30. Chenini, I.; Ben Mammou, A.; El May, M. Groundwater Recharge Zone Mapping Using GIS-Based Multi-criteria Analysis: A Case Study in Central Tunisia (Maknassy Basin). *Water Resour. Manag.* **2010**, *24*, 921–939. [CrossRef]
31. Chowdhury, A.; Jha, M.K.; Chowdary, V.M. Delineation of groundwater recharge zones and identification of artificial recharge sites in West Medinipur district, West Bengal, using RS, GIS and MCDM techniques. *Environ. Earth Sci.* **2010**, *59*, 1209–1222. [CrossRef]
32. Nasiri, H.; DarvishiBoloorani, A.; FarajiSabokbar, H.A.; Jafari, H.R.; Hamzeh, M.; Rafi, Y. Determining the most suitable areas for artificial groundwater recharge via an integrated PROMETHEE II-AHP method in GIS environment (case study: Garabaygan Basin, Iran). *Environ. Monit. Assess.* **2013**, *185*, 707–718. [CrossRef]

33. Valverde, J.P.B.; Blank, C.; Roidt, M.; Schneider, L.; Stefan, C. Application of a GIS Multi-Criteria Decision Analysis for the Identification of Intrinsic Suitable Sites in Costa Rica for the Application of Managed Aquifer Recharge (MAR) through Spreading Methods. *Water* **2016**, *8*, 391. [CrossRef]
34. Selvarani, A.G.; Maheswaran, G.; Elangovan, K. Identification of artificial recharge sites for Noyyal River Basin using GIS and remote sensing. *J. Indian Soc. Remote Sens.* **2017**, *45*, 67–77. [CrossRef]
35. Kazakis, N. Delineation of Suitable Zones for the Application of Managed Aquifer Recharge (MAR) in Coastal Aquifers Using Quantitative Parameters and the Analytical Hierarchy Process. *Water* **2018**, *10*, 804. [CrossRef]
36. Zadeh, L.A. Fuzzy sets. *Inf. Control* **1965**, *8*, 338–353.
37. Zimmermann, H.-J. Advanced Review. Fuzzy set theory. *WIREs Comput. Stat.* **2010**, *2*, 317–332. [CrossRef]
38. Garg, H. A Hybrid GA-GSA Algorithm for Optimizing the Performance of an Industrial System by Utilizing Uncertain Data. In *Handbook of Research on Artificial Intelligence Techniques and Algorithms*; Pandian, V., Ed.; IGI Global: Hershey, PE, USA, 2015; pp. 620–654. [CrossRef]
39. Bogardi, I.; Duckstein, L.; Bardossy, A. Regional management of an aquifer under fuzzy environmental objectives. *Water Resour. Res.* **1983**, *19*, 1394–1402. [CrossRef]
40. Kambalimath, S.; Deka, P.C. A basic review of fuzzy logic applications in hydrology and water resources. *Appl. Water Sci.* **2020**, *10*, 191. [CrossRef]
41. Kazakis, N.; Spiliotis, M.; Voudouris, K.; Pliakas, F.K.; Papadopoulos, B. A fuzzy multicriteria categorization of the GALDIT method to assess seawater intrusion vulnerability of coastal aquifers. *Sci. Total Environ.* **2018**, *621*, 524–534. [CrossRef]
42. Kourgialas, N.N.; Anyfanti, I.; Karatzas, G.P.; Dokou, Z. An integrated method for assessing drought prone areas-Water efficiency practices for a climate resilient Mediterranean agriculture. *Sci. Total Environ.* **2018**, *625*, 1290–1300. [CrossRef]
43. Spiliotis, M.; Iglesias, A.; Garrote, L. A multicriteria fuzzy pattern recognition approach for assessing the vulnerability to drought: Mediterranean region. *Evol. Syst.* **2021**, *12*, 109–122. [CrossRef]
44. Chan, F.T.S.; Kumar, N. Global supplier development considering risk factors using fuzzy extended AHP-based approach. *Omega* **2007**, *35*, 417–431. [CrossRef]
45. Chan, F.T.S.; Kumar, N.; Tiwari, M.K.; Lau, H.C.W.; Choy, K.L. Global supplier selection: A fuzzy-AHP approach. *Int. J. Prod. Res.* **2008**, *46*, 3825–3857. [CrossRef]
46. Saravi, M.M.; Malekian, A.; Nouri, B. Identification of Suitable Sites for Groundwater Recharge. In Proceedings of the 2nd International Conference on Water Resources & Arid Environment, Riyadh, Saudi Arabia, 26–29 November 2006; pp. 1–5.
47. Ghayoumian, J.; MohseniSaravi, M.; Feiznia, S.; Nouri, B.; Malekian, A. Application of GIS techniques to determine areas most suitable for artificial groundwater recharge in a coastal aquifer in southern Iran. *J. Asian Earth Sci.* **2007**, *30*, 364–374. [CrossRef]
48. Riad, P.H.; Billib, M.H.; Hasan, A.A.; Omar, M.A. Overlay Weighted Model and Fuzzy Logic to Determine the Best Locations for Artificial Recharge of Groundwater in a Semi-Arid Area in Egypt. *Nile Basin Water Sci. Eng. J.* **2011**, *4*, 24–35.
49. Malekmohammadi, B.; Mehrian, M.R.; Jafari, H.R. Site selection for managed aquifer recharge using fuzzy rules: Integrating geographical information system (GIS) tools and multi-criteria decision making. *Hydrogeol. J.* **2012**, *20*, 1393–1405. [CrossRef]
50. Dashtpagerdi, M.M.; Vagharfard, H.; Honarbakhsh, A.; Khoorani, A. GIS Based Fuzzy Logic Approach for Identification of Groundwater Artificial Recharge Site. *Open J. Geol.* **2013**, *3*, 379–383. [CrossRef]
51. Selvarani, A.G.; Maheswaran, G.; Elangovan, K. Identification of Artificial Recharge Sites For Noyyal Watershed Using GIS And Fuzzy Logic. *Int. J. Appl. Eng. Res.* **2015**, *10*, 11189–11208.
52. Ghasemi, A.; Saghafian, B.; Golian, S. Optimal location of artificial recharge of treated wastewater using fuzzy logic approach. *J. Water Supply Res. Technol.* **2017**, *66*, 141–156. [CrossRef]
53. Pliakas, F.; Kallioras, A.; Damianidis, P.; Kostakakis, P. Evaluation of precipitation effects on groundwater levels in a Mediterranean alluvial plain based on hydrogeological conceptualization. *Environ. Earth Sci.* **2015**, *74*, 3573–3588. [CrossRef]
54. Dimadis, E.; Zachos, S. *Geological Map of Rhodope Massif*; Institute of Geology and Mineral Exploration (IGME), Regional Subdivision of Eastern Macedonia—Thrace: Xanthi, Greece, 1986.
55. Sakkas, I.; Diamantis, I.; Pliakas, F. *Groundwater Artificial Recharge Study of Xanthi-Rhodope Aquifers (in Thrace, Greece)*; Greek Ministry of Agriculture Research Project, Final Report; Sections of Geotechnical Engineering and Hydraulics of the Civil Engineering Department of Democritus University of Thrace: Xanthi, Greece, 1998. (In Greek)
56. Pliakas, F.; Diamantis, I.; Petalas, C. Results from a Groundwater Artificial Recharge Application in Polysitos Aquifer at Xanthi Plain Region (Greece) by Reactivating Old Stream Beds. In Proceedings of the 5th Greek National Hydrogeological Conference, Nicosia, Cyprus, 12—14 November 1999; pp. 97–113. (In Greek).
57. Pliakas, F.; Petalas, C.; Diamantis, I.; Kallioras, A. Modeling of groundwater artificial recharge by reactivating an on old streambed. *Water Resour. Manag.* **2005**, *19*, 279–294.
58. Pisinaras, V. Ανάπτυξη ΕνόςΠλαισίου ΟλοκληρωμένηςΔιαχείρισης Πολύπλοκων Συστημάτων ΥπόγειωνΥδατικών πόρων [Development of a Methodology Framework for Integrated Management of Complex Groundwater Systems]. Ph.D. Thesis, Democritus University of Thrace, Xanthi, Greece, June 2008. (In Greek).
59. Papadopoulos, C.; Spiliotis, M.; Gkiougkis, I.; Pliakas, F.; Papadopoulos, B. Fuzzy linear regression analysis for groundwater response to meteorological drought in the aquifer system of Xanthi plain, NE Greece. *J. Hydroinform.* **2021**, *23*, 1112–1129. [CrossRef]

60. Diamantis, I.; Pliakas, F.; Petalas, C. The contribution of artificial recharge as a remedy procedure regarding the impacts of various interventions on the natural environment of plain regions. The case study in Thrace, Greece. In Proceedings of the International Symposium on Engineering Geology and the Environment, Organized by the Greek National Group of the International Association for Engineering Geology and the Environment (IAEG), Athens, Greece, 23–27 June 1997; pp. 2663–2667.
61. Special Secretariat for Water. *River Basin Management Plans of Thrace Water District (GR12)*; Ministry of Environmental and Energy: Athens, Greece, 2017.
62. Krishnamurthy, J.; Kkumar, N.V.; Jayaraman, V.; Manivel, M. An approach to demarcate ground water potential zones through remote sensing and geographic information system. *Int. J. Remote Sens.* **1996**, *17*, 1867–1884.
63. Farhadian, M.; Bozorg-Haddad, O.; Pazoki, M.; Loaiciga, H.A. Locating and Prioritizing Suitable Places forthe Implementation of Artificial Groundwater Recharge Plans. *J. Irrig. Drain. Eng.* **2017**, *143*, 04017018.
64. Ghayoumian, J.; Ghermezcheshmeh, B.; Feiznia, S.; Noroozi, A.A. Integrating GIS and DSS for identification of suitable areas for artificial recharge, case study Meimeh Basin, Isfahan. Iran. *Environ. Geol.* **2005**, *47*, 493–500.
65. Jasrotia, A.S.; Kumar, R.; Saraf, A.K. Delineation of groundwater recharge sites using integrated remote sensing and GIS in Jammu district, India. *Int. J. Remote Sens.* **2007**, *28*, 5019–5036.
66. Klir, G.J.; Yuan, B. *Fuzzy Sets and Fuzzy Logic. Theory and Applications*; Prentice Hall: Hoboken, NJ, USA, 1995.
67. Ross, T.J. *Fuzzy Logic with Engineering Applications*, 3rd ed.; John Wiley & Sons: West Sussex, UK, 2010; p. 585.
68. Zimmerman, H.-J. *Fuzzy Set Theory and Its Applications*, 4th ed.; Springer Science & Business Media: New York, NY, USA, 2001; p. 514.
69. Wang, L.-X. *A Course in Fuzzy Systems and Control*; Prentice Hall: Hoboken, NJ, USA, 1997; p. 424.
70. Saaty, T.L. *The Analytic Hierarchy Process*; McGraw-Hill: New York, NY, USA, 1980.
71. Pöyhönen, M.A.; Hämäläinen, R.P.; Salo, A.A. An Experiment on the Numerical Modeling of Verbal Ratio Statements. *J. Multi-Criteria Decis. Anal.* **1997**, *6*, 1–10.
72. Ho, W.; Ma, X. The state-of-the-art integrations and applications of the analytic hierarchy process. *Eur. J. Oper. Res.* **2018**, *267*, 399–414. [CrossRef]
73. Chang, D.-Y. Applications of the extent analysis method on fuzzy AHP. Theory and Methodology. *Eur. J. Oper. Res.* **1996**, *95*, 649–655.
74. Kwong, C.K.; Bai, H. Determining the importance weights for the customer requirements in QFD using a fuzzy AHP with anextent analysis approach. *IIE Trans.* **2003**, *35*, 619–626.
75. Kahraman, C.; Ruan, D.; Dogan, I. Fuzzy group decision making for facility location selection. *Inf. Sci.* **2003**, *157*, 135–153.
76. Kahraman, C.; Ertay, T.; Buyukozkan, G. A fuzzy optimization model for QFD planning process using analytic network approach. *Eur. J. Oper. Res.* **2006**, *171*, 390–411.
77. Srdjevic, B.; Medeiros, Y.D.P. Fuzzy AHP Assessment of Water Management Plan. *Water Resour. Manag.* **2008**, *22*, 877–894. [CrossRef]
78. Wang, Y.-M.; Luo, Y.; Hua, Z. On the extent analysis method for fuzzy AHP and its applications. *Eur. J. Oper. Res.* **2008**, *186*, 735–747. [CrossRef]
79. Mikhailov, L. Deriving priorities from fuzzy pairwise comparison judgements. *Fuzzy Sets Syst.* **2003**, *134*, 365–385. [CrossRef]
80. Mikhailov, L. A fuzzy programming method for deriving priorities in the analytic hierarchy process. *J. Oper. Res. Soc.* **2000**, *51*, 341–349. [CrossRef]
81. Li, L.; Lai, K.K. A fuzzy approach to the multiobjective transportation problem. *Comput. Oper. Res.* **2000**, *27*, 43–57. [CrossRef]
82. Tsakiris, G.; Spiliotis, M. Multicriteria ranking of water development scenario using a fuzzy rule based system. In Proceedings of the International Symposium on Water Resources Management: Risks and Challenges for the 21st Century, Izmir, Turkey, 2–4 September 2004.
83. Gurav, J.B.; Regulwar, D.G. Multi Objective Sustainable Irrigation Planning with Decision Parameters and Decision Variables Fuzzy in Nature. *Water Resour. Manag.* **2012**, *26*, 3005–3021. [CrossRef]
84. Mikhailov, L.; Tsvetinov, P. Evaluation of services using a fuzzy analytic hierarchy process. *Appl. Soft Comput.* **2004**, *5*, 23–33. [CrossRef]
85. Wang, Y.-M.; Chin, K.-S. Fuzzy analytic hierarchy process: A logarithmic fuzzy preference programming methodology. *Int. J. Approx. Reason.* **2011**, *52*, 541–553. [CrossRef]
86. Spiliotis, M.; Skoulikaris, C. A fuzzy AHP-outranking framework for selecting measures of river basin management plans. *Desalination Water Treat.* **2019**, *167*, 398–411. [CrossRef]
87. Mamdani, E.H.; Assilian, S. An Experiment in Linguistic Synthesis with a Fuzzy Logic Controller. *Int. J. Mach. Stud.* **1975**, *7*, 1–13. [CrossRef]
88. Yen, J.; Langari, R. *Fuzzy Logic. Intelligence, Control, and Information*; Prentice Hall: New York, NJ, USA, 1999; pp. 32–43.
89. Ojha, V.; Abraham, A.; Snasel, V. Heuristic Design of Fuzzy Inference Systems: A Review of Three Decades of Research. *Eng. Appl. Artif. Intell.* **2019**, *85*, 845–864. [CrossRef]
90. Papadopoulos, B.; Trasanides, G.; Hatzimichailidis, A.G. Optimization Method for the Selection of the Appropriate Fuzzy Implication. *J. Optim. Theory Appl.* **2007**, *134*, 135–141. [CrossRef]
91. Balopoulos, V.; Hatzimichailidis, A.; Papadopoulos, B. Distance and similarity measures for fuzzy operators. *Inf. Sci.* **2007**, *177*, 2336–2348. [CrossRef]

92. Sturges, H.A. The choice of a class interval. *J. Am. Stat. Assoc.* **1926**, *21*, 65–66. [CrossRef]
93. Amineh, Z.B.A.; Hashemian, S.J.A.D.; Magholi, A. Integrating Spatial Multi Criteria Decision Making (SMCDM) with Geographic Information Systems (GIS) for delineation of the most suitable areas for aquifer storage and recovery (ASR). *J. Hydrol.* **2017**, *551*, 577–595. [CrossRef]
94. Buckley, J.J. Fuzzy hierarchical analysis. *Fuzzy Sets Syst.* **1985**, *17*, 233–247.
95. Vemuri, N.R.; Jayaram, B. Thecomposition of fuzzy implications. *Fuzzy Sets Syst.* **2015**, *275*, 58–87. [CrossRef]

water

MDPI

Reply

# Reply to Louloudis et al. Comments on "Tzampoglou, P.; Loupasakis, C. Hydrogeological Hazards in Open Pit Coal Mines–Investigating Triggering Mechanisms by Validating the European Ground Motion Service Product with Ground Truth Data. *Water* 2023, *15*, 1474"

**Constantinos Loupasakis [1,*] and Ploutarchos Tzampoglou [1,2]**

1 Laboratory of Engineering Geology and Hydrogeology, Department of Geological Sciences, School of Mining and Metallurgical Engineering, National Technical University of Athens, 15780 Athens, Greece; tzampoglou.ploutarchos@ucy.ac.cy
2 Department of Civil & Environmental Engineering, University of Cyprus, Nicosia 1678, Cyprus
* Correspondence: cloupasakis@metal.ntua.gr

The comments of Louloudis, Roumpos, Mertiri, and Kostaridis [1] refer to data related to previous studies and not the one presented in the commented paper [2]. These specific studies have been published and evaluated in numerous articles and international conferences, certifying their validity. Also, a multitude of researchers and organizations align with our viewpoint that overexploitation of the aquifer leads to the activation of land subsidence phenomena [3–7]. The scope of this study is to combine the already verified and published ground truth datasets acquired for the study of mining geohazards with the European Ground Motion Service (EGMS) InSAR dataset, aiming to cross-verify them.

Before responding, we wish to point that the authors of the comments have disclosed a clear conflict of interest, as they were employed by the Public Power Corporation of Greece and were involved as hydrogeological, mining, or geological experts in the engineering study of the Amyntaion mine.

Focusing on the current study, its scope is to combine the ground truth datasets acquired for the study of mining geohazards, namely land subsidence and landslide, with the European Ground Motion Service (EGMS) InSAR dataset, aiming to cross-verify their outcomes. The main objectives of the study are to (a) estimate the spatial distribution of the vertical deformation due to land subsidence and landslide by evaluating the EGMS InSAR datasets; (b) correlate the above data with the data obtained through the geotechnical research highlighting opposite driving failures mechanisms and main causal factors and (c) estimate the interaction between these two catastrophic events.

Regarding the accuracy of the ground water level measurements, the extent of the depression cone, and the equal-drawdown contour maps, we would like to emphasize that the wider mining area has been thoroughly studied for the past 20 years by the authors, and several field work campaigns were conducted in which over 40 boreholes were measured at the wider perimeter of the mine. The equal-drawdown contour lines maps were designed by subtracting the piezometric curves before the start of the mining activities, as reported in the PhD thesis [8] of, at that time, employee of the hydrogeological department of PPC, with the piezometric surfaces that resulted from our campaigns. These maps, along with the methodology used to create them, have been presented in five peer-reviewed papers [9–13] and six international conferences [14–19].

 https://doi.org/10.3390/w17162436

Regarding the irrigation drills, a drills census was conducted during our field work campaigns. The census proved that almost all the drills at the perimeter of the mine, and at more than 2 km, were inactive due to the intensive ground water drawdown. Aiming to cover the irrigation water needs, the farmers at the perimeter of the mine began pumping water from the draining canal network using electricity from the old power supply network, installed initially for the operation of the deep wells' pumps. So, the electricity consumption, used by the authors of the comments [1] as proof of pumping activity, provides false information. This fact has been repeatedly reported to the PPC's scientific personnel and it was also introduced during common on-site field visits.

As far as the extent of the depression cone caused by the mining activities, besides our above reported accurately designed piezometric contour maps, it is clearly presented in maps and cross-sections included in a recent publication [20] cowritten by the, at that time, director of the hydrogeological department of PPC, claiming that it was caused by the practically inactive irrigation drills. So, we both agree that the depression cone exists, although we disagree on its cause.

Please note that it is not just the authors' opinion that the water table drawdown extends to a distance of 2 km, affecting the settlements located at the perimeter of the mine. It is commonly accepted and published in several papers and studies [3–7] that the open pit coal mine hydraulically resembles a huge well with a diameter of 3 to 4 km and appears to be the main cause of the groundwater table level drop, not only in its immediate vicinity but also in a wider area [6], triggering land subsidence phenomena.

Regarding the landslide event, in our publication [2], we identify several preparatory and triggering factors contributing to the triggering of the failure. Although each factor was individually studied, the synergistic interaction among them was not adequately assessed, ultimately leading to the failure's manifestation. As far as the geometry of the slopes, it is provided by simple calculations. Knowing, from high-resolution satellite images, the horizontal distance between the base of the slope and the crown of the benches, as well as the height of the benches and the total depth of the mine, the steepness was calculated. A clear description of the applied procedure can be found in a previous publication [13] and is not included in this manuscript as this is not in the scope of this paper.

The statement that "*the groundwater pore pressure at this depth* (at the depth of 200 m) *is not influenced by the mine's pumping activities in the upper aquifer. Consequently, the 2017 landslide is not related to the pumping operations at the perimeter of the mine.*" proves that, as abovementioned, the multiple causal and triggering factors were not properly evaluated. According to their statement, it is clear that they do not consider the interaction of the extensive faults intersecting the site, the high piezometric loads underneath the floor of the open pit, originating from a deeper karstic aquifer, or the occurrence of the pre-existing sheared zones within the Neogene formations that all together interact and co-act with the pore pressure of the shallow semi-confined aquifer. Also, if the stability of the slopes, as stated by the comments [1], "*is not related to the pumping operations at the perimeter of the mine*", they should not have applied all the draining and pumping measures from the beginning of its operation, avoiding at the same time the occurrence of the land subsidence.

Our findings—derived through a rigorous series of processes—are thoroughly corroborated by EGMS satellite data records, confirming their validity. This particular aspect, which also constitutes the central aim of the original article under discussion, was completely overlooked by the authors of the comment report.

Finaly, we underscore that the EGMS datasets constitute a significant and reliable resource for disentangling the various causal mechanisms of ground deformations that have impacted the study area. Their provision by an independent and internationally recognized authority ensures that the conclusions drawn are scientifically robust and impartial.

**Author Contributions:** Both authors analyzed the comments of Louloudis, Roumpos, Mertiri, and Kostaridis and wrote the reply. All authors have read and agreed to the published version of the manuscript.

**Funding:** This research received no external funding.

**Data Availability Statement:** Data will be made available by the authors upon request.

**Conflicts of Interest:** The authors declare no conflicts of interest.

## References

1. Louloudis, G.; Roumpos, C.; Mertiri, E.; Kostaridis, P. Comment on Tzampoglou, P.; Loupasakis, C. Hydrogeological Hazards in Open Pit Coal Mines–Investigating Triggering Mechanisms by Validating the European Ground Motion Service Product with Ground Truth Data. *Water* 2023, *15*, 1474. *Water* **2025**, *17*, 2343. [CrossRef]
2. Tzampoglou, P.; Loupasakis, C. Hydrogeological Hazards in Open Pit Coal Mines–Investigating Triggering Mechanisms by Validating the European Ground Motion Service Product with Ground Truth Data. *Water* **2023**, *15*, 1474. [CrossRef]
3. Soulios, G.; Tsapanos, T.; Voudouris, K.; Kaklis, T.; Mattas, C.; Sotiriadis, M. Ruptures on surface and buildings due to land subsidence in Anargyri village (Florina Prefecture, Macedonia). *Environ. Earth Sci.* **2011**, *5*, 505–512.
4. Tsourlos, P. *Research for the Appearance of Surface Cracks in the Valtonera Village, Amyntaio Municipality, Florina Region*; School of Geology of the Aristotle University of Thessaloniki: Thessaloniki, Greece, 2015; p. 17.
5. Gudulas, K. Investigation of the Mechanisms of Operation of Lakes Cheimaditida and Zazari, Located in the Amyntaio Basin, of Florina County, from a Hydrogeological and Environmental Point of View. Ph.D. Thesis, Aristotle University of Thessaloniki, Thessaloniki, Greece, 2012; p. 277.
6. Stamos, A.; Giannoulopoulos, P. *Hydrogeological Report, Geotechnical Work to the Anargiri Village, Aetos Municipality, Florina Prefecture, Greece*; Institution of Geology and Mineral Exploration: Athens, Greece, 2010; p. 34.
7. Tsourlos, P.; Papazaxos, C.; Bargmeziw, G.; Koutalou, B. *Investigation and Suggestions for the Problem Created at the Fault Which as Appeared in Municipalities of Fanos and Ksino Nero*; School of Geology of the Aristotelian University of Thessaloniki: Thessaloniki, Greece, 2007.
8. Dimitrakopoulos, D. Hydrogeological Conditioning at the Amyndeon Mine. Problems Occuring During the Exploitation and their Mitigation Actions. Ph.D. Thesis, National Technical University of Athens, Athens, Greece, 2001.
9. Loupasakis, C.; Angelitsa, V.; Rozos, D.; Spanou, N. Mining geohazards—Land subsidence caused by the dewatering of opencast coal mines: The case study of the Amyntaio coal mine, Florina, Greece. *Nat. Hazards* **2014**, *70*, 675–691. [CrossRef]
10. Tzampoglou, P.; Loupasakis, C. Mining geohazards susceptibility and risk mapping: The case of the Amyntaio open-pit coal mine, West Macedonia, Greece. *Environ. Earth Sci.* **2017**, *76*, 542. [CrossRef]
11. Tzampoglou, P.; Loupasakis, C. Evaluating geological and geotechnical data for the study of land subsidence phenomena at the perimeter of the Amyntaio coalmine, Greece. *Int. J. Min. Sci. Technol.* **2018**, *28*, 601–612. [CrossRef]
12. Tzampoglou, P.; Loupasakis, C. Numerical simulation of the factors causing land subsidence due to overexploitation of the aquifer in the Amyntaio open coal mine, Greece. *HydroResearch* **2019**, *1*, 8–24. [CrossRef]
13. Loupasakis, C. Contradictive mining–Induced geocatastrophic events at open pit coal mines: The case of Amintaio coal mine, West Macedonia, Greece. *Arab. J. Geosci.* **2020**, *13*, 582. [CrossRef]
14. Tzampoglou, P.; Loupasakis, C. New data regarding the ground water level changes at the Amyntaio basin-Florina Prefecture, Greece. *Bull. Geol. Soc. Greece* **2016**, *50*, 1006–1015. [CrossRef]
15. Tzampoglou, P.; Loupasakis, C. Updated ground water piezometry data of the Amyntaio sub-basin and their effect to the manifestation of the land subsidence phenomena. In Proceedings of the 11th International Hydorgeological Congress of Greece, Athens, Greece, 4–6 October 2017.
16. Tzampoglou, P.; Loupasakis, C. Land subsidence susceptibility and hazard mapping: The case of Amyntaio Basin, Greece. In Proceedings of the Fifth International Conference on Remote Sensing and Geoinformation of the Environment (RSCy2017), Paphos, Cyprus, 20–23 March 2017; p. 104441L.
17. Tzampoglou, P.; Loupasakis, C. Land subsidence due to the overexploitation of the aquifer at the Valtonera village. *Bull. Geol. Soc. Greece* **2017**, *50*, 1006. [CrossRef]
18. Tzampoglou, P.; Loupasakis, C. Investigation of land subsidence due to the overexploitation of the aquifer at the Amyntaio basin, Greece. In Proceedings of the 3rd European Regional Conference of IAEG, Euroengeo 2020, Athens, Greece, 8–12 April 2021.

19. Tzampoglou, P.; Loupasakis, C. Land subsidence due to the overexploitation of the aquifer at the Amyntaio basin. In Proceedings of the 12th International Hydrogeological Conference, Nicosia, Cyprus, 20–22 March 2022.
20. Dimitrakopoulos, D.; Koumantakis, I. Hydrodynamic regime of Amynteon basin. Influence of open lignite mines. In Proceedings of the 11th International Hydrogeological Congress of Greece, Athens, Greece, 4–6 October 2017; pp. 101–112.

*water*

MDPI

*Comment*

# Comment on Tzampoglou, P.; Loupasakis, C. Hydrogeological Hazards in Open Pit Coal Mines–Investigating Triggering Mechanisms by Validating the European Ground Motion Service Product with Ground Truth Data. *Water* 2023, *15*, 1474

**Georgios Louloudis *, Christos Roumpos *, Eleni Mertiri and Petros Kostaridis**

Department of Mining Engineering and Closure Planning, Public Power Corporation of Greece, 10432 Athen, Greece; e.mertiri@ppcgroup.com (E.M.); p.kostaridis@ppcgroup.com (P.K.)
* Correspondence: g.louloudis@ppcgroup.com (G.L.); c.roumpos@ppcgroup.com (C.R.); Tel.: +30-697-000-5090 (G.L.); +30-697-979-9291 (C.R.)

**Abstract**

The commented paper uses arbitrary and unsubstantiated hypotheses to attribute land subsidence phenomena in the Amyntaion basin to the operations of the Public Power Corporation (PPC) surface coal mine, disregarding, or at least grossly underestimating, the effect of about 600 pumped deep wells for irrigation purposes all over the basin. In addition to the huge difference in the pumped quantities of water from the aquifer, ground water table lowering due to the PPC mine has negligible influence at distances over 500 m from the edge of the mine, while the areas examined in the paper are at distances of several kilometers from the edge of the mine. Furthermore, the authors attribute the landslide that occurred in the mine in 2017 to the steep excavation slopes of the mine and the increased groundwater pore pressure due to reduced peripheral pumping, which is completely inaccurate. To build their case, the authors of the commented paper disregard multiple references in research publications on the above issues, as explained in the main text of this discussion.

**Keywords:** land subsidence; Amyntaion basin; Copernicus; irrigation wells; pit slopes

## 1. Introduction

The commented paper by Tzampoglou and Loupasakis purports [1] to provide "*an excellent case study to demonstrate the reliability and value of the EGMS product*", which is not the case, in our opinion. The article does not review the reliability of the Earth Observation data from the European Ground Motion Service (EGMS) of the Copernicus European Union's Earth observation program objectively, but selectively interprets the EGMS data to confirm the main argument of the paper, which is that the mining activities are responsible for the ongoing steady ground subsidence in an area extending at distances more than 2 km from the edge of the mine, and thus affecting the villages in this area. Our claim is that extensive deep pumping for irrigation purposes in the whole valley is responsible for the ground subsidence and damage in the villages.

The "*Introduction*" section of the commented paper does not adequately place the study in a broad context, and the current state of the research field is not thoroughly reviewed, with key publications not highlighted. Although the article investigates subsidence phenomena in the Amynteon basin and the hydrogeological hazards in Open-Pit Coal

Mines, reference [2] concerns environmental contamination, and reference [3] does not refer to mines or open pits but to tailings dam failures. There are also other references that do not correspond to the relevant text. For instance, it is mentioned that "*land subsidence slowly affects extensive areas around mines*" in references [4,5]. In contrast, reference [4] refers to mine collapse and affected buildings at a 200 m distance within the mining area, and reference [5] refers to a study investigating the possibility of land subsidence occurrence and mentions that, as of the publishing date, no damages had been recorded.

Also, the key publications used to investigate the failure mechanisms triggering natural hazards are those of the authors. Similarly, the authors' publications are also the key publications cited to relate the mining dewatering wells to the groundwater level drop extending up to 2 km around the mine, and their impact on the villages of Valtonera and Anargiroi.

Overall, the paper is based on unjustified statements and misconceptions, while the arguments are inconsistent throughout the text and have serious shortcomings, as clearly demonstrated in the following paragraphs.

## 2. The Authors' Attempt to Attribute Ongoing Steady Subsidence in an Area Extending Several Kilometers Around the Mine to the Mining Activities of PPC, Based on Insufficient Information, Leading to Arbitrary and Unsubstantiated Hypotheses

(a) The authors claim that farming activities cause minor settlements. However, the article lacks critical information [6–9] about the locations, density, and depth of the irrigation wells, and ignores the huge difference in pumped quantities for irrigation purposes compared to the pumping quantities for mining operations. According to undisputed data (electricity consumption for irrigation purposes), the approximately 600 irrigation wells, reaching depths up to 120 m all over the basin [10,11], pump about 40–45 million cubic meters of water per year over the last 25 years, while the PPC mine removed only 4–5 million cubic meters of water per year over the last 15 years (until 2020) via a small number of wells at the perimeter of the mine. In the Amynteon basin, the greatest pressure in the aquifer comes from the irrigation wells and is estimated to be in the order of 80–85% [6]. The authors fail to include and evaluate vital data, as key publications are ignored.

Note that the "control points" of vertical displacement measurements provided by Copernicus are located in areas surrounded by a large number of irrigation wells (Figure 1), presenting an average pumping rate of 25 $m^3/h$ (open data of irrigation wells' permits), where there is an absolute absence of mines' dewatering wells. Based on the pumping rate, the rate of irrigation consumption in Valtonera is estimated to be approximately 0.40 $m^3/m^2$, which corresponds to an average annual value of 8 million $m^3$ of groundwater abstraction for irrigation use.

Additionally, in Section 2.2 of the commented paper, "Hydrogeologic Setting", the authors mention that "It is clear that during the last three decades the operation of the mine and, secondarily, the increase of the agricultural activities radically changed the groundwater dynamics of the basin"; however, in Section 3.1, "The Land Subsidence Mechanism", it is mentioned that "the general hydrogeological conditions remain practically unchanged since 1992. The open pit keeps on operating as the main draining source while no extra irrigation wells have been drilled at the wider area". This is contradictory as it is clearly stated initially in the manuscript that there has been an increase in agricultural activities in recent decades, and thus, the groundwater dynamics are influenced by this increase.

In their reply to our comments, the authors state that "Aiming to cover the irrigation water needs the farmers at the perimeter of the mine pump water from the draining canals network using electricity from the old power supply network installed initially for the operation of the drills' pumps. So, the electricity consumption provides false information about the pumping activities from the aquifers". This is inaccurate, as documented above, and can be disputed by the open data of the irrigation wells' permits (see also Figure 1).

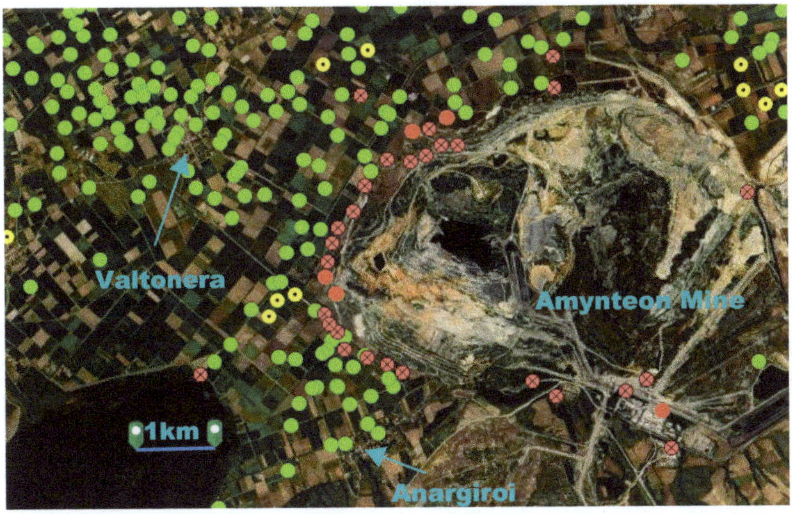


**Figure 1.** Irrigation pumping wells (green color) in the broader area of Amynteon (Greece) lignite mine, community pumping wells for other uses (yellow) and inactive PPC wells (have been decommissioned and sealed after the mine was decommissioned) in the perimeter (red/pink color) [7].

(b) In addition to the huge difference in the pumped quantities of water from the aquifer, ground water table lowering due to the PPC mine has negligible influence at distances over 500 m from the edge of the mine, while the areas examined in the paper are located at distances 2–8 km from the edge of the mine. The authors overlook the critical influencing factors of constant irrigation pumping over the years and the limited influence zone of the mine's dewatering measures around the mine's perimeter. Neither the mine's dewatering measures nor the mine itself can cause ground deformations at distances greater than an influence zone of a maximum of 500 m [8,12,13].

(c) Many piezometers located closer to the mine show significantly smaller groundwater table drawdown than piezometers located farther away in the valley [10]. It is thus evident that the larger values of groundwater table drawdown in the valley are caused by the operation of hundreds of irrigation wells rather than the peripheral wells of the mine [7]. This statement is also supported by the open data on groundwater table measurements in boreholes located north–northwest of the mine (Figure 2), at Valtonera village and the irrigation area between Valtonera and Pedino villages [14]. This confirms the importance of the hundreds of irrigation wells in lowering groundwater tables.

(d) The authors claim that "currently the open pit along with the surrounding draining wells operate as an oversized (4 km wide) well, continuously draining a big part of the Amyntaio basin". This is not correct as the dewatering wells were terminated in April 2020, and thus, the mine was decommissioned [8,9]. Also, according to the National Water Monitoring Network [14], it is evident that the groundwater level has been stabilized (as indicated by measurements conducted from 2018 to 2020, Figure 2).

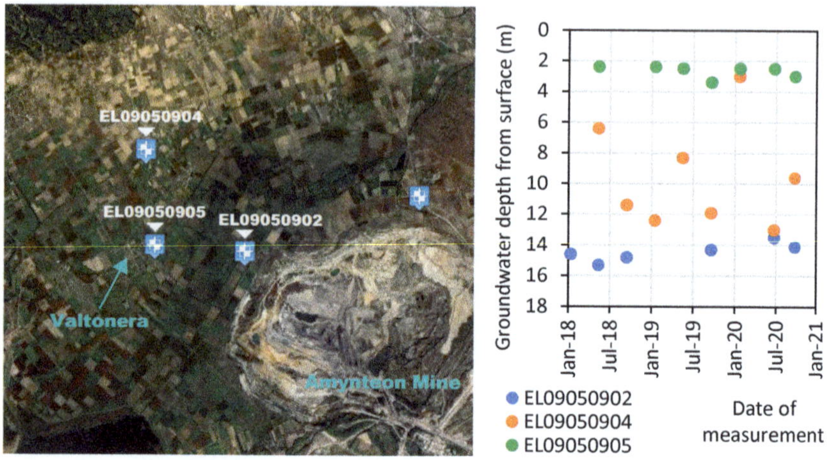


**Figure 2.** Monitoring of groundwater via the National Water Monitoring Network in the area near the mine. The open data concern period: 2018–2020 [14].

In their reply to our comments, the authors state that "*it is clearly presented in maps and cross sections included at a recent publication* [13] *written by the, at that time, director of the hydrogeological department of PPC, claiming that it was caused by the practically inactive irrigation drills*". This is also not correct, as in [13], it is clearly stated that the groundwater level dropdown in the valley (away from the mine's perimeter) is caused by the irrigation wells. Also, the same reference mentions that the negative water balance in the Amynteon aquifer was noticed before the beginning of the mine's pumping wells.

Also, in the authors' reply, it is stated that "It is commonly accepted that, the open pit coal mine resembles hydraulically to a huge well with a diameter of 3 to 4 km and appears to be the main cause of the groundwater table level drop not only in its immediate vicinity but also in a wider area, triggering land subsidence phenomena". This statement is supported only by a PhD thesis and a Technical Report, which, although mentioning this phenomenon, fail to justify it.

(e) Moreover, the authors mention that "*the data from May 1992 relate to the period before the operation of the open pit coal mine* [15]"; however, this is not correct as the mine was operating during that time. Based on the authors' reference [15], the groundwater drawdown during 1986 and 1992 (6-year time period) was 4–5 m and locally 6 m due to the overexploitation of the Amynteon aquifer by farmers for irrigation purposes. Note that dewatering measures were adopted to protect the surface mine during its operation after 1993. Therefore, based on the authors' reference [15], agricultural activities had a great influence on groundwater, which was evident in the area before the beginning of the mine's dewatering measures around the mine's perimeter. The authors fail to include and evaluate vital data from the publication [15] cited in their article, while other key publications, such as [6–9,14], are ignored.

(f) Figure 3 in the commented paper presents the groundwater level dropdown between certain periods, for a 23- and 24-year time period. This dropdown should be an outcome of the data processing and evaluation via piezometric maps (authors' own measurements in a three-year (3) period in 37 points). It is easy to verify that the dropdown contour lines are erroneous and do not correspond to the piezometric maps. For instance, according to the authors' piezometry, there is practically no groundwater level dropdown in the Valtonera and Anargiroi villages between the years 2014 and 2016. While comparing these measurements to those of 1992, a level dropdown of up to 5 m can be estimated. However, the calculated dropdown level presented by the authors is 10–20 m. It is evident that the outcome of the groundwater level dropdown is not verified from the "Ground truth datasets" used in the article. The same applies to Figure 4 in the commented paper for the incorrect groundwater level dropdown maps. Additionally, the statement in Section 3.1 of the commented paper, "*. . .at Valtonera village. The observed water level drop in this area of around 10 to 20 m. . .*" is inaccurate and inconsistent with the authors' own water level measurements.

Following the argument above, it is evident that there is practically no groundwater level dropdown in the Valtonera and Anargiroi villages. Therefore, it is derived that there is no correlation between the mentioned groundwater level dropdown and the vertical deformations provided by the Copernicus Land Monitoring Service. Thus, all the related statements in the article are incorrect.

(g) The borehole network where the groundwater measurements occur is important and should be considered in the evaluation process. For the year 1992, no network is presented. According to the authors' reference [15], in 1992, the network where the measurements occurred was scarce, while there was no network in a ~3 km radius around Valtonera village, nor between Valtonera and Anargiroi villages.
(h) The authors present the groundwater table contour lines in different time periods. However, the extent of the mine presented in the satellite images does not match the real extent during the respective time period. This is crucial for conducting proper piezometric maps.

## 3. The Authors Attribute the Landslide Which Occurred in the Mine in 2017 to the Steep Excavation Slopes of the Mine and the Increased Groundwater Pore Pressure Due to Reduced Peripheral Pumping, Which Is Completely Inaccurate

(a) The authors relate the land subsidence phenomenon and the landslide that occurred in 2017 to the dewatering of the mine. This is inaccurate and inconsistent with the authors' own water level measurements, which clearly show that there is practically no groundwater level dropdown in the Valtonera and Anargiroi villages, as documented above. Therefore, the correlation with the vertical deformations provided by the Copernicus Land Monitoring Service does not apply, and thus, all related statements in Tzampoglou and Loupasakis (2023) are not supported by the data. Note that the locations where the vertical deformations are presented (see authors' InSAR data in relation to Figure 1 above) are contiguous with irrigation wells and at a significant distance from the mine's dewatering wells. Also note that the dewatering measures stopped in April 2020, while the irrigation wells have not yet stopped.
(b) The authors mention that "*the pumping activities directly adjacent to Anargiroi village, according to PPC* [13]*, were essentially stopped in 2016, aiming to eliminate the deformation caused by overpumping at the nearby village*". However, according to the same reference, this is not accurate, as no such information is presented. Thus, the writers' outcome, where the mine's pumping activities are related to the landslide, is not supported.
(c) The authors' statement "*The drop in the groundwater level, combined with the area's general geotectonic setting, triggered extensive land subsidence in an area extending up to 2 km around the mine, causing damage to villages and infrastructure since 2002*" is not correct, because the effect of any large well is related to the magnitude of the drawdown (which in our case is 80–100 m close to the mine's perimeter) and is independent of the well diameter. Concerning the influence zone, this is also evident in hydrogeological cross sections that include water levels in boreholes [11,13]. Please note that any dropdown

in groundwater level can only enhance the stability of slopes and cannot trigger the landslide phenomenon.

(d) Based on the actual geometric data of the mining operations, all statements in the paper regarding the pit slopes and the geometry of the mine benches are inaccurate. Specifically, based on all available data and satellite images, there is no evidence to suggest that steep slopes were applied, or the excavation of the lower benches of the coal seams was prioritized, or that the excavation of the middle benches was systematically neglected. This is also verified in the Audit Findings report of the committee experts [16].

Moreover, no geometrical data are presented to support this statement. In their reply, the authors do not answer our comments and do not provide any data but only mention that they applied the Pythagorean theorem to calculate the steepness, knowing "*the horizontal distance between the base of the slope and the crown of the benches as well as the high of the benches and the total depth of the mine*". This statement is completely irrelevant to our arguments, and, of course, it is obvious that no application of the Pythagorean theorem is related to the overall slope calculation.

(e) The mine's pumping was limited at the upper aquifer (i.e., up to a depth of 100 m) that lies above the impermeable lignite strata [9], while the total depth of the mine pit exceeds 200 m, and the sliding occurred along a sub-horizontal lignite-to-clay/marl interface [17] at depth about 200 m. Thus, the groundwater pore pressure at this depth is not influenced by the mine's pumping activities in the upper aquifer. Consequently, the 2017 landslide is not related to the pumping operations at the perimeter of the mine.

In conclusion, the commented paper is based on inaccurate information and fails to present and evaluate the controversial and diverging aspects, leading to a misinterpretation of the data and, eventually, a misjudgment about the real causes of the described events. We point out that, due to the great importance and potential legal implications of these events for the operator of the mine (PPC), groundwater table levels around the mine have been monitored and studied for many years by geologists and engineers of PPC and by independent experts [10,11]; they all reach conclusions opposing the conclusions presented in the commented paper.

**Author Contributions:** All authors (C.R., G.L., E.M. and P.K.) analyzed the article of Tzampoglou and Loupasakis and designed and wrote the comment paper. The findings are based on the original work and analysis of all authors. All authors have read and agreed to the published version of the manuscript.

**Funding:** This research received no external funding.

**Data Availability Statement:** Data will be made available by the authors upon request.

**Conflicts of Interest:** All authors have been involved as hydrogeological, mining, or geological experts in the mining engineering study of the Amyntaion mine.

## References

1. Tzampoglou, P.; Loupasakis, C. Hydrogeological Hazards in Open Pit Coal Mines–Investigating Triggering Mechanisms by Validating the European Ground Motion Service Product with Ground Truth Data. *Water* **2023**, *15*, 1474. [CrossRef]
2. Passariello, B.; Giuliano, V.; Quaresima, S.; Barbaro, M.; Caroli, S.; Forte, G.; Carelli, G.; Iavicoli, I. Evaluation of the environmental contamination at an abandoned mining site. *Microchem. J.* **2002**, *73*, 245–250. [CrossRef]
3. WISE_Uranium_Project. Chronology of Major Tailings Dam Failures. Available online: https://www.wise-uranium.org/mdaf.html (accessed on 8 June 2023).
4. Sun, Y.; Zhang, X.; Mao, W.; Xu, L. Mechanism and stability evaluation of goaf ground subsidence in the third mining area in Gong Changling District, China. *Arab. J. Geosci.* **2015**, *8*, 639–646. [CrossRef]

5. Wolkersdorfer, C.H.; Thiem, G. Ground water withdrawal and land subsidence in northeastern Saxony (Germany). *Mine Water Environ.* **1999**, *18*, 81–92. [CrossRef]
6. Ministry for the Environment, Energy and Climate Change. Approved River Basin Management Plan 1st Update. Available online: http://wfdver.ypeka.gr/en/management-plans-en/# (accessed on 8 June 2023).
7. Ministry for the Environment, Energy and Climate Change. Irrigation Wells. Available online: http://lmt.ypeka.gr/public_view.html (accessed on 8 June 2023).
8. Kavvadas, M.; Roumpos, C.; Servou, A.; Paraskevis, N. Geotechnical Issues in Decommissioning Surface Lignite Mines—The Case of Amyntaion Mine in Greece. *Mining* **2022**, *2*, 278–296. [CrossRef]
9. Louloudis, G.; Louloudis, E.; Roumpos, C.; Mertiri, E.; Kasfikis, G.; Chatzopoulos, K. Forecasting Development of Mine Pit Lake Water Surface Levels Based on Time Series Analysis and Neural Networks. *Mine Water Environ.* **2021**, *41*, 458–474. [CrossRef]
10. Kavvadas, M.J.; Agioutantis, Z.; Skourtsos, E.; Steiakakis, C. *Evaluation of the Causes of Surface Ruptures and Differential Settlements that Have Been Observed in the Amynteon Basin, with Emphasis on the Area Around the Mine and at Valtonera Village, Municipality of Amyntaio;* Study of the Hydrogeological Regime and the Changes in Piezometry in the Area the Amynteon Basin; Public Power Corporation: Athens, Greece, 2021; Unpublished. (In Greek)
11. Konstantopoulou, G.; Nikolaou, N.; Sabatakakis, P.; Tzima, M.; Mattheopoulos, D.; Parcharidis, I.; Karavias, A.; Gatsios, T.; Babousis, K. *Monitoring the Ground Movements in the Wider Area of the Amynteon Mine*; EAGME: Athens, Greece, 2021; Unpublished. (In Greek)
12. Stefouli, M.; Panagiotopoulou, A.; Charou, E.; Spastra, Y.; Bratsolis, E.; Madamopoulos, N.; Perantonis, S. Assessment of the Added Value of Sentinel1&2 for Mapping and Monitoring Surface Mining. In Proceedings of the 14th International Symposium of Continuous Surface Mining, Thesaloniki, Greece, 23–26 September 2018; Sideri, D., Paraskevis, N., Eds.; pp. 73–87.
13. Dimitrakopoulos, D.; Koumantakis, I. Hydrodynamic regime of Amynteon basin. Influence of the open lignite mines. In Proceedings of the 11th International Hydrogeological Congress, The Geological Society of Greece, Athens, Greece, 4–6 October 2017; pp. 101–112. (In Greek)
14. Ministry for the Environment, Energy and Climate Change. National Monitoring Water Network. Available online: http://nmwn.ypeka.gr/?q=en/content/home (accessed on 8 June 2023).
15. Dimitrakopoulos, D. Hydrogeological Conditions of Amyndeon Mine. Problems During Exploitation and Overcoming Them. Ph.D. Dissertation, National Technical University of Athens School of Mining and Metallurgical Engineering, Athens, Greece, 2001. (In Greek)
16. Papageorgiou, C.; Kavvadas, M.; Marinos, P.; Gazetas, G.; Kolovos, C.; Roumpos, C.; Machtis, K.; Chaloulos, K.; Pavlidis, S. *Experts Evaluation Report of the Committee on the Landslide of Amynteon Mine on June 10th 2017*; National Technical University of Athens (NTUA), Aristotle University of Thessaloniki (AUTH), Public Power Corporation (PPC): Athens, Greece, 2017; Unpublished. (In Greek)
17. Kavvadas, M.; Roumpos, C.; Schilizzi, P. Stability of Deep Excavation Slopes in Continuous Surface Lignite Mining Systems. *Geotech. Geol. Eng.* **2020**, *38*, 791–812. [CrossRef]

MDPI AG
Grosspeteranlage 5
4052 Basel
Switzerland
Tel.: +41 61 683 77 34

*Water* Editorial Office
E-mail: water@mdpi.com
www.mdpi.com/journal/water

www.ingramcontent.com/pod-product-compliance
Lightning Source LLC
LaVergne TN
LVHW070540071126
851858LV00011BA/251
* 9 7 8 3 7 2 5 8 8 3 7 5 2 *